# Ecophysiology of Tropical Plants

Plants in tropical regions are coping with enormous challenges of physiological stresses owing to changing environmental and climatic conditions. Rapid growth of human population and rampant exploitation of fossil fuels and other developmental activities are actively contributing to such perturbations. The Intergovernmental Panel on Climate Change has projected a sustained increase in carbon dioxide ($CO_2$) emissions and thereby a rise in global temperature in the coming decades. The resultant changes in precipitation patterns are now evident across the globe due to intensification of hydrological cycle. Moreover, gaseous and particulate pollutants are also an immense challenge for tropical plants. Such vagaries in environmental conditions have significant impacts on the ecophysiological traits of plants, resulting from altered interactions of tropical plants with each other, as well as other biotic and abiotic components within the ecosystem. Books available in the market that particularly focus on ecophysiological responses of tropical plants to abiotic and biotic environmental factors under climate change are limited. This book intends to fill this knowledge gap and provides a detailed analysis on ecophysiological responses of tropical plants to these environmental challenges, as well as suggesting some approachable measures for plant adaptations to these challenges. The book is equally applicable to undergraduate and postgraduate students, researchers, teachers and forest managers, and policy makers.

Salient features of the book are:

1. A comprehensive discussion on adaptive mechanisms of plants through their ecophysiological responses to various biotic and abiotic stresses.
2. Elaboration on the recent techniques involved in ecophysiological research.
3. A detailed account of evolutionary responses of plants to changing climate.
4. Discussion of recent research results and some pointers to future advancements in ecophysiological research.
5. Presentation of information in a way that is accessible for students, researchers, and teachers practicing in plant physiology and ecology.

# Ecophysiology of Tropical Plants

## Recent Trends and Future Perspectives

Edited By

Sachchidanand Tripathi
*Department of Botany, Deen Dayal Upadhyaya College,
University of Delhi, New Delhi, India*

Rahul Bhadouria
*Department of Environmental Studies, Delhi College
of Arts and Commerce, University of Delhi, New Delhi, India*

Pratap Srivastava
*Department of Botany, University of Allahabad, Prayagraj, India*

Rishikesh Singh
*Department of Botany, Panjab University, Chandigarh, India*

Rajkumari Sanayaima Devi
*Department of Botany, Deen Dayal Upadhyaya College,
University of Delhi, New Delhi, India*

CRC Press is an imprint of the
Taylor & Francis Group, an **informa** business

First edition published 2024
by CRC Press
6000 Broken Sound Parkway NW, Suite 300, Boca Raton, FL 33487–2742

and by CRC Press
4 Park Square, Milton Park, Abingdon, Oxon, OX14 4RN

© 2024 selection and editorial matter, Sachchidanand Tripathi, Rahul Bhadouria, Pratap Srivastava, Rishikesh Singh, Rajkumari Sanayaima Devi; individual chapters, the contributors

*CRC Press is an imprint of Taylor & Francis Group, LLC*

Reasonable efforts have been made to publish reliable data and information, but the author and publisher cannot assume responsibility for the validity of all materials or the consequences of their use. The authors and publishers have attempted to trace the copyright holders of all material reproduced in this publication and apologize to copyright holders if permission to publish in this form has not been obtained. If any copyright material has not been acknowledged please write and let us know so we may rectify in any future reprint.

Except as permitted under U.S. Copyright Law, no part of this book may be reprinted, reproduced, transmitted, or utilized in any form by any electronic, mechanical, or other means, now known or hereafter invented, including photocopying, microfilming, and recording, or in any information storage or retrieval system, without written permission from the publishers.

For permission to photocopy or use material electronically from this work, access www.copyright.com or contact the Copyright Clearance Center, Inc. (CCC), 222 Rosewood Drive, Danvers, MA 01923, 978–750–8400. For works that are not available on CCC please contact mpkbookspermissions@tandf.co.uk

*Trademark notice*: Product or corporate names may be trademarks or registered trademarks and are used only for identification and explanation without intent to infringe.

*Library of Congress Cataloging-in-Publication Data*
Names: Tripathi, Sachchidanand, editor. | Bhadouria, Rahul, 1982– editor. | Srivastava, Pratap, editor. |
    Singh, Rishikesh, editor. | Devi, Rajkumari Sanayaima, editor.
Title: Ecophysiology of tropical plants : recent trends and future perspectives / edited by Sachchidanand Tripathi
    (Associate Professor, Department of Botany, Deen Dayal Upadhyaya College, University of Delhi, Delhi, India),
    Rahul Bhadouria (Assistant Professor at the Department of Environmental Studies, Delhi College of Arts and
    Commerce, University of Delhi, New Delhi, India), Pratap Srivastava (Assistant Professor,
    Department of Botany, University of Allahabad, India), Rishikesh Singh (National Post-doctoral Fellow (NPDF) at
    Panjab University, Chandigarh, India), Rajkumari Sanayaima Devi (Associate Professor, Department of
    Botany, Deen Dayal Upadhyaya College (University of Delhi), Delhi, India).
Description: First edition. | Boca Raton : CRC Press, 2024. | Includes bibliographical references and index.
Identifiers: LCCN 2023015339 (print) | LCCN 2023015340 (ebook) | ISBN 9781032370446 (hardback) |
    ISBN 9781032370477 (paperback) | ISBN 9781003335054 (ebook)
Subjects: LCSH: Tropical plants. | Tropical plants—Ecophysiology. | Tropical plants—Climatic factors.
Classification: LCC QK936 .E258 2024 (print) | LCC QK936 (ebook) | DDC 581.70913—dc23/eng/20230626
LC record available at https://lccn.loc.gov/2023015339
LC ebook record available at https://lccn.loc.gov/2023015340

ISBN: 978-1-032-3-7044-6 (hbk)
ISBN: 978-1-032-3-7047-7 (pbk)
ISBN: 978-1-003-3-3505-4 (ebk)

DOI: 10.1201/9781003335054

Typeset in Times
by Apex CoVantage, LLC

# Contents

Preface ..................................................................................................................................................................xvi
Sachchidanand Tripathi, Rahul Bhadouria, Pratap Srivastava, Rishikesh Singh, and Rajkumari S. Devi
Editor Biographies ............................................................................................................................................xviii
List of Contributors ............................................................................................................................................xix
Foreword ...........................................................................................................................................................xxiv
S.C. Bhatla

## SECTION 1  Tropical plants and changing climate scenarios

**Chapter 1**  Plant Adaptations in Dry Tropical Biomes: An Ecophysiological Perspective ............................... 3

S. Oyedeji

1.1  Introduction ................................................................................................................. 3
1.2  Ecology of Dry Tropical Biomes: Dry Forests and Savannas .................................... 3
    1.2.1  Seasonality in Tropical Dry Biomes ............................................................. 4
    1.2.2  Structure and Diversity of Tropical Dry Forests and Savannas .................... 5
1.3  Adaptation of Plants in Drought Conditions in the Dry Tropics: Mechanisms versus Traits ................ 5
    1.3.1  Adaptive Strategies in Grasses ..................................................................... 6
        1.3.1.1  $C_4$ Photosynthesis ....................................................................... 6
        1.3.1.2  Bunch-Forming Morphology ..................................................... 6
        1.3.1.3  Drought Escape as Seeds ............................................................ 7
    1.3.2  Adaptive Strategies in Woody Plants ........................................................... 7
        1.3.2.1  Strategies for Increased Water Capture ..................................... 7
        1.3.2.2  Strategies for Water Storage ...................................................... 7
        1.3.2.3  Water Management ..................................................................... 8
    1.3.3  Other Plant Adaptive Traits in Dry Tropical Biomes ................................... 9
        1.3.3.1  Extensive Root Systems .............................................................. 9
        1.3.3.2  Leaf Folding or Rolling .............................................................. 9
        1.3.3.3  Other Leaf Modifications ........................................................... 9
1.4  Ecophysiological Regulations of Germination, Dormancy and Flowering/Fruiting in Dry Tropical Biomes ................. 9
1.5  Climate Change and Plant Responses ....................................................................... 10
1.6  Conclusion ................................................................................................................. 11
References ............................................................................................................................ 11

**Chapter 2**  Evolutionary Responses of Tropical Plants to Changing Climate .................................................. 15

Zirwa Sarwar, Maria Hasnain, Maria Hanif, Huma Waqif, and Neelma Munir

2.1  Climate Change ......................................................................................................... 15
2.2  Plant Responses to the Environment ......................................................................... 15
2.3  Trophic Interaction .................................................................................................... 16
2.4  Thermal Interaction ................................................................................................... 16
2.5  Plant–Herbivore Interaction ...................................................................................... 17
    2.5.1  Rising Atmospheric $CO_2$ and Plant–Herbivore Interaction ....................... 18
2.6  Temperature ............................................................................................................... 18
2.7  Drought ...................................................................................................................... 18
2.8  Fertilization and Nitrogen Deposition ....................................................................... 19
2.9  Co-Evolutionary Responses ...................................................................................... 20

|  | 2.10 | Host–Pathogen Interactions | 20 |
|---|---|---|---|
|  | 2.11 | Deciphering the Role of Phytoalexins in Plant–Micro-organism Interactions and Human Health | 20 |
|  | 2.12 | Conclusion | 20 |
|  |  | References | 21 |

**Chapter 3** Ecophysiological Responses of Tropical Plants to Changing Climate ............................................................24

*Ranjan Pandey, Harminder Pal Singh, and Daizy R. Batish*

| | 3.1 | Introduction | 24 |
|---|---|---|---|
| | 3.2 | Tropics: The Most Diverse Ecosystem of the Earth | 24 |
| | 3.3 | Climate Change: A Real Challenge | 26 |
| | 3.4 | Ecophysiological Responses of Tropical Plants | 27 |
| | | 3.4.1 Responses to Increases in Temperature | 28 |
| | | 3.4.2 Responses to Increases in $CO_2$ | 30 |
| | | 3.4.3 Combined Effects of Temperature and $CO_2$ | 31 |
| | 3.5 | Conclusions | 32 |
| | 3.6 | Future Recommendations | 32 |
| | | References | 32 |

# SECTION 2  Tropical plants responses to atmospheric deposition and air pollutants

**Chapter 4** Impacts of Air Pollutants on the Ecophysiology of Tropical Plants ..............................................................37

*Pallavi Singh, Jigyasa Prakash, Harshita Singh, Shashi Bhushan Agrawal, and Madhoolika Agrawal*

| | 4.1 | Introduction | 37 |
|---|---|---|---|
| | 4.2 | Methodology | 37 |
| | 4.3 | Sources and Modes of Deposition of Pollutants | 37 |
| | 4.4 | Predominant Air Pollutants in the Tropics | 39 |
| | | 4.4.1 Gaseous Pollutants | 39 |
| | |     4.4.1.1 $SO_2$ and $NO_2$ | 39 |
| | |     4.4.1.2 Tropospheric Ozone | 40 |
| | | 4.4.2 Particulate Matter | 41 |
| | 4.5 | Impacts of Air Pollutants on Ecophysiology of Perennial Vegetation and Crop Plants | 41 |
| | | 4.5.1 Foliar Injury | 41 |
| | | 4.5.2 Physiology | 42 |
| | | 4.5.3 Biochemistry | 44 |
| | | 4.5.4 Growth, Biomass and Productivity | 46 |
| | 4.6 | Models Evaluating the Effects of Rising $O_3$ Concentrations on Tropical Crops and Trees | 48 |
| | | 4.6.1 Stomatal $O_3$ Flux Models and Dose–Response Relationship for Trees | 48 |
| | | 4.6.2 $O_3$ Dose–Response Relationships for Crops | 48 |
| | 4.7 | Conclusions and Future Concerns | 48 |
| | | Acknowledgements | 49 |
| | | References | 49 |

**Chapter 5** Impact of Nitrogen Oxides on Tropical Plants ..............................................................................................57

*Aisha Kamal, Nida Sultan, Shazia Siddiqui, and Farhan Ahmad*

| | 5.1 | Introduction | 57 |
|---|---|---|---|
| | 5.2 | Positive Impacts of $NO_x$ on Tropical Plants | 57 |
| | | 5.2.1 Effects of $NO_x$ on Plant Physiology | 58 |
| | | 5.2.2 $NO_x$ Effects on Post-Harvest Quality | 59 |
| | | 5.2.3 Protective Role of $NO_x$ under Plant Abiotic Stress Conditions | 59 |

Contents vii

    5.3   Negative Impact of $NO_x$ on Tropical Plants ................................................................................. 60
        5.3.1   Effect of $NO_x$ on Growth and Morphology of Tropical Plants ............................................ 60
        5.3.2   Effect of $NO_x$ on Plant Anatomy ........................................................................................ 61
        5.3.3   Effect of $NO_x$ on Physiology of Plants ............................................................................. 61
        5.3.4   Effect of $NO_x$ on Nitrogen Assimilation ........................................................................... 61
    5.4   Conclusions .................................................................................................................................. 62
    References .............................................................................................................................................. 62

**Chapter 6**   Impact of Particulate Matter on the Ecophysiology of Plants ........................................................ 66

    *Somdutta Sinha Roy, Saloni Bahri, Laishram Sundari Devi, and Sushma Moitra*

    6.1   Introduction ................................................................................................................................. 66
    6.2   Plants and Bioremediation of PM ............................................................................................... 66
    6.3   Physiological and Biochemical Effects of PM on Plants ........................................................... 67
    6.4   Toxic Effects of PM-Related Heavy Metals in Plants ................................................................ 68
    6.5   Effect of PM on Phyllosphere and Rhizosphere Processes ........................................................ 70
    6.6   Conclusions ................................................................................................................................. 71
    References ............................................................................................................................................. 71

**Chapter 7**   Ecophysiological Responses of Tropical Plants to Rising Air Pollution: A Perspective for Urban Areas .... 73

    *Sadhna, Pallavi B. Dhal, Sachchidanand Tripathi, Rajkumari Sanayaima Devi,*
    *Rishikesh Singh, and Rahul Bhadouria*

    7.1   Introduction ................................................................................................................................. 73
        7.1.1   Urbanization and Air Pollution ........................................................................................... 73
        7.1.2   Pollutants and Ecophysiology ............................................................................................ 73
    7.2   PFTs and Ecophysiological Responses ...................................................................................... 74
    7.3   Biochemical Analysis ................................................................................................................. 75
        7.3.1   Chlorophyll Concentration .................................................................................................. 75
        7.3.2   Ascorbic Acid Concentration .............................................................................................. 75
        7.3.3   Proline Content .................................................................................................................... 76
        7.3.4   Sugar Content ...................................................................................................................... 76
    7.4   Physiological Parameters ............................................................................................................ 76
        7.4.1   pH ......................................................................................................................................... 76
        7.4.2   Relative Water Content ....................................................................................................... 76
    7.5   Impact of Dust Load on Ecophysiological Attributes of Urban Plants ...................................... 77
    7.6   Role of Plants in Urban Areas .................................................................................................... 77
    7.7   Conclusion and Ways Forward ................................................................................................... 77
    References ............................................................................................................................................. 78

## SECTION 3   *Tropical plants responses to varrying resource availability*

**Chapter 8**   Ecophysiological Responses of Tropical Plants to Varying Resources Availability ..................... 85

    *Wajiha Sarfraz, Mujahid Farid, Noreen Khalid, Allah Ditta, Ujala Ejaz, Zarrin Fatima Rizvi,*
    *Nighat Raza, and Shafaqat Ali*

    8.1   Introduction ................................................................................................................................. 85
        8.1.1   Influence of Light Exposure and the Ecophysiological Constraints
                of Tropical Forest Plants ..................................................................................................... 86
        8.1.2   Effects of Light Saturation on Photosynthesis ................................................................... 87
        8.1.3   Effect of Plant Properties on Light Absorption in a Tropical Forest .................................. 87
    8.2   Impact of Temperature on Ecophysiological Responses of Plants in Tropical Environments ........... 87
        8.2.1   Temperature and Tree Mortality in Tropical Land Surfaces .............................................. 88
        8.2.2   Role of Temperature in Phytophysiology and Adaptation to Environmental Stress ............ 89
    8.3   Carbon Dynamics and Ecophysiological Impact in Tropical Forests ........................................ 89

        8.3.1    Microbe-Centric Approach to Soil Carbon Dynamics ........................................................ 89
        8.3.2    Impact of Atmospheric $CO_2$ on Photosynthesis ................................................................ 89
        8.3.3    Effect of $CO_2$ on Plants' Biochemical and Morphological Responses .............................. 90
  8.4    Changes in Nutrient Availability and the Effects on Species Survival and Phytochemistry in Tropical Rainforests ..................................................................................................................... 90
        8.4.1    Nutrient Availability and Translocation ............................................................................. 90
        8.4.2    Seasonal Variation in Nutrient Availability ....................................................................... 90
  8.5    Drought Effects on Tropical Rainforest Biota................................................................................. 91
        8.5.1    Impact of Rainforest Loss on Climate Change and Sustainability..................................... 91
        8.5.2    Drought-Induced Effects in Tropical Forests ..................................................................... 91
        8.5.3    Effects of Drought on Large Trees and Their Ecological and Biological Status ............... 91
        8.5.4    Tree Growth during Extreme Droughts and Their Implications ........................................ 92
  8.6    Hormonal Stress Responses of Plants ............................................................................................. 92
  8.7    Conclusion ...................................................................................................................................... 93
  References ................................................................................................................................................. 93

**Chapter 9**   Soil Nutrient Reservoir in a Changing Climate Scenario ................................................................. 98

*Janaki Subramanyan*

  9.1    Soil Health ....................................................................................................................................... 98
  9.2    Climate Change ............................................................................................................................... 98
  9.3    Effects of Climate Change on the Soil ........................................................................................... 98
        9.3.1    Mineralization and Nutrient Availability ........................................................................... 99
  9.4    Role of Carbon Sequestration .......................................................................................................... 99
        9.4.1    Permafrost ......................................................................................................................... 100
  9.5    Soil Biodiversity ............................................................................................................................ 100
        9.5.1    Microbiomes ..................................................................................................................... 100
        9.5.2    Rhizosphere Microflora ................................................................................................... 100
        9.5.3    Mycorrhizae ...................................................................................................................... 100
  9.6    Challenges and the Future ............................................................................................................. 101
  9.7    Impact on Food Security and Safety ............................................................................................. 101
  9.8    Mitigation ...................................................................................................................................... 101
  9.9    Conclusions................................................................................................................................... 102
  References ............................................................................................................................................... 102

**Chapter 10**   Abiotic Stress Responses of a Tropical Plant: Sugarcane (*Saccharum* species) ........................... 104

*R. Mishra, P. Agarwal, R. Soni, and G. Singh*

  10.1    Introduction .................................................................................................................................. 104
  10.2    Abiotic Stress Response Mechanisms in Sugarcane ..................................................................... 104
        10.2.1    Water Stress..................................................................................................................... 105
                10.2.1.1    Drought .......................................................................................................... 105
                10.2.1.2    Waterlogging ................................................................................................. 106
        10.2.2    Salt Stress ........................................................................................................................ 106
        10.2.3    Temperature ..................................................................................................................... 106
  10.3    Omics Approaches to Explicate Abiotic Stress Responses in Sugarcane ..................................... 106
  10.4    Abiotic Stress Resistance in Sugarcane......................................................................................... 107
        10.4.1    Salt Stress Tolerance in Sugarcane.................................................................................. 108
        10.4.2    Heat Stress Tolerance in Sugarcane ................................................................................ 108
        10.4.3    Drought Tolerance Mechanisms...................................................................................... 108
                10.4.3.1    Breeding as an Alternative to Screen Drought-Resistant Sugarcane Cultivars ........ 109
                10.4.3.2    Somaclonal Variation for Developing Drought-Tolerant Sugarcane Cultivars ........ 109
                10.4.3.3    Genetic Modifications in Sugarcane ............................................................ 109
                10.4.3.4    Differential Gene Expression in Sugarcane ................................................. 109
                10.4.3.5    Abscisic Acid Signaling as an Outcome of Drought Stress.......................... 110
  10.5    Engineering Sugarcane for Abiotic Stress Tolerance .................................................................... 110

## Contents

**Chapter 11** Effects of Rising Temperature on Flower Production and Pollen Viability in a Widespread Tropical Tree Species, *Muntingia calabura* .................................................................................................................... 119

*Martijn Slot, Natanja Schuttenhelm, Chinedu E. Eze, and Klaus Winter*

- 10.6 Regulatory Role of miRNAs in Abiotic Responses ........................................................................... 112
- 10.7 Conclusions and Future Prospects ..................................................................................................... 113
- References .................................................................................................................................................... 113

- 11.1 Introduction ........................................................................................................................................ 119
- 11.2 Methods and Materials ....................................................................................................................... 119
  - 11.2.1 Study Species ........................................................................................................................ 119
  - 11.2.2 Experimental Setup ............................................................................................................... 120
  - 11.2.3 Sample Collection and Processing ....................................................................................... 121
  - 11.2.4 Data Analysis ........................................................................................................................ 121
- 11.3 Results ................................................................................................................................................ 122
- 11.4 Discussion .......................................................................................................................................... 123
  - 11.4.1 Viability Decreases with Duration of Warming ................................................................... 124
  - 11.4.2 Which Step in Reproduction Is Most Sensitive to Warming? .............................................. 124
  - 11.4.3 Flower Production ................................................................................................................. 125
  - 11.4.4 Long-Term Effects of Atmospheric and Climate Change .................................................... 125
- 11.5 Conclusions ........................................................................................................................................ 125
- References .................................................................................................................................................... 126

**Chapter 12** Ecophysiological and Morphological Adaptations of Plants under Temperature Stress: Influence of Phytohormones .................................................................................................................... 128

*Ghalia S.H. Alnusairi, Abbu Zaid, Harvinder Kour, Khadiga Alharbi, and Mona H. Soliman*

- 12.1 Introduction ........................................................................................................................................ 128
- 12.2 Phytohormones and Temperature Stress: A Burning Issue ............................................................... 129
- 12.3 Adaptive Role of Phytohormones under Temperature Stress ........................................................... 130
- 12.4 Salicylic Acid ..................................................................................................................................... 130
- 12.5 Nitric Oxide ........................................................................................................................................ 132
- 12.6 Brassinosteroids .................................................................................................................................. 132
- 12.7 Conclusion .......................................................................................................................................... 134
- References .................................................................................................................................................... 134

**Chapter 13** Ecophysiological Response of Dipterocarp Seedlings to Ectomycorrhizal Colonisation: A Fungicide Addition Study ..................................................................................................................... 138

*Francis Q. Brearley*

- 13.1 Introduction ........................................................................................................................................ 138
- 13.2 Methods .............................................................................................................................................. 138
  - 13.2.1 Study Species ........................................................................................................................ 138
  - 13.2.2 Experimental Set-Up ............................................................................................................. 139
  - 13.2.3 Seedling Measurements ........................................................................................................ 139
  - 13.2.4 Ectomycorrhizal Colonisation .............................................................................................. 139
  - 13.2.5 Statistical Analyses ............................................................................................................... 139
- 13.3 Results ................................................................................................................................................ 140
  - 13.3.1 Ectomycorrhizal Colonisation .............................................................................................. 140
  - 13.3.2 Growth Rates, Biomass and Biomass Allocation ................................................................. 140
  - 13.3.3 Foliar Nutrients ..................................................................................................................... 140
- 13.4 Discussion .......................................................................................................................................... 142
- 13.5 Conclusions ........................................................................................................................................ 144
- Acknowledgements ...................................................................................................................................... 144
- References .................................................................................................................................................... 144

**Chapter 14** Ecophysiological Responses of Tropical Plants to Changing Concentrations of Carbon Dioxide .............. 146

*Rupali Jandrotia, Ipsa Gupta, Riya Raina, and Daizy R. Batish*

- 14.1 Introduction ................................................................................................................... 146
- 14.2 Why the Tropics Are Important from a Climate Change Perspective ........................... 146
- 14.3 Species Responses to Elevated $CO_2$ Concentrations ..................................................... 147
  - 14.3.1 Ecological Responses of Plant Species to Increasing $CO_2$ Concentrations ........ 147
  - 14.3.2 Whole Plant Responses of Plant Species to Elevated $CO_2$ ................................. 148
    - 14.3.2.1 Phenological Changes ........................................................................... 149
    - 14.3.2.2 Changes in Plant Growth ...................................................................... 149
    - 14.3.2.3 Effect of Elevated $CO_2$ on Crucial Biological Processes of Tropical Plants ........ 150
  - 14.3.3 Elevated $CO_2$ and Plant Biotic Factors ................................................................ 152
    - 14.3.3.1 Pests ....................................................................................................... 152
    - 14.3.3.2 Diseases and Soil Microbes .................................................................. 152
- 14.4 Phenotypic Plasticity vs. Genetic Differentiation ........................................................... 153
- 14.5 Microclimates as Buffers of Climate Change ................................................................. 153
- 14.6 Conclusion and Priorities for Future Research .............................................................. 153
- Acknowledgments ..................................................................................................................... 154
- References ................................................................................................................................. 154

# SECTION 4  Ecophysiological responses of tropical plants to disturbance events

**Chapter 15** Fire and Ecophysiological Responses of Tropical Plants ................................................ 161

*Shikha Singh and Tanu Kumari*

- 15.1 Introduction ................................................................................................................... 161
- 15.2 Fire in Tropical Ecosystems ........................................................................................... 162
  - 15.2.1 Causes of Forest Fires in the Tropics ................................................................... 162
    - 15.2.1.1 Climate Change ..................................................................................... 163
    - 15.2.1.2 Agriculture and Livestock Farming ...................................................... 163
    - 15.2.1.3 Rural and Urban Expansion .................................................................. 163
    - 15.2.1.4 Energy Production ................................................................................ 163
    - 15.2.1.5 Fire Exclusion and Suppression ............................................................ 163
    - 15.2.1.6 Exotic Species ....................................................................................... 163
    - 15.2.1.7 Arson ..................................................................................................... 163
  - 15.2.2 Fire Regime Types in the Tropics ........................................................................ 163
- 15.3 Environmental Changes and Plant Responses ............................................................... 165
- 15.4 Ecophysiological Responses of Plants to Fire ................................................................ 165
  - 15.4.1 Responses Based on Fire Regimes ....................................................................... 166
  - 15.4.2 Responses Based on Level of Organization ........................................................ 167
    - 15.4.2.1 Individual Level .................................................................................... 167
    - 15.4.2.2 Population Level ................................................................................... 168
    - 15.4.2.3 Community Level ................................................................................. 168
  - 15.4.3 Response Based on Resistance ............................................................................. 168
- 15.5 Plant Adaptations to Fire ................................................................................................ 169
  - 15.5.1 Fire-Adaptive Traits Associated with Resistance ................................................. 169
  - 15.5.2 Fire-Adaptive Traits Associated with Resprouters .............................................. 170
  - 15.5.3 Fire-Adaptive Traits Associated with Recruiters ................................................. 170
- 15.6 Conclusion ...................................................................................................................... 171
- References ................................................................................................................................. 171

**Chapter 16** Climate Change-Mediated Fire Effects on Community Structure and the Physio-Anatomical Adaptations of Plants in Tropical Savannas ...................................................................................................... 175

*S. Oyedeji, C.O. Ogunkunle, S.A. Adeniran, O.O. Agboola, and P.O. Fatoba*

16.1 Introduction ................................................................................................................. 175
16.2 Fire Disturbances in Tropical Savannas ....................................................................... 175
    16.2.1 Fire Regimes in Tropical Savannas................................................................. 176
    16.2.2 Determinants of Fire Frequency and Intensity in Tropical Savanna .............. 176
        16.2.2.1 Seasonal or Environmental Dryness ............................................... 177
        16.2.2.2 Ignition Timing................................................................................ 177
        16.2.2.3 Presence of 'Fire-Fighter' or 'Fire-Sensitive' Species .................... 177
        16.2.2.4 Litter or Fuel Load........................................................................... 177
        16.2.2.5 Determinants of Fire Intensity in Tropical Savannas...................... 177
16.3 Fire-Mediated Modifications in Savanna Ecosystems ................................................. 178
    16.3.1 Direct Fire Impact on Savanna Vegetation ..................................................... 178
    16.3.2 Indirect Fire Impact on Savanna Vegetation................................................... 179
16.4 Plant Diversity in Fire-Exposed Savannas ................................................................... 180
    16.4.1 Species Richness and Diversity....................................................................... 180
    16.4.2 Wildfire and Plant Traits ................................................................................. 181
16.5 Adaptation Strategies of Plant Species to Fire Effects................................................. 181
    16.5.1 Physiological Strategies................................................................................... 182
    16.5.2 Anatomical Strategies ..................................................................................... 183
16.6 Conclusion .................................................................................................................... 185
References ............................................................................................................................... 185

# SECTION 5  Ecophysiology and geoclimatic factors

**Chapter 17** Adaptation of Fruit Trees to Different Elevations in the Tropical Andes................................... 193

*Gerhard Fischer, Helber Enrique Balaguera-López, Alfonso Parra-Coronado, and Stanislav Magnitskiy*

17.1 Introduction ................................................................................................................. 193
17.2 Climatic Characteristics with Increasing Tropical Elevation....................................... 194
17.3 Tropical Elevation Ranges for Different Fruit Species ................................................ 195
17.4 Adaptation of Fruit Trees to Conditions in Tropical Highlands................................... 196
    17.4.1 Morphological Adaptations............................................................................. 196
    17.4.2 Physiological Adaptations ............................................................................... 198
    17.4.3 Adaptation of Fruits to the High Incidence of UV Light and Low Temperatures at High Elevations ..........................................................................................200
17.5 First Studies on the Andean Blueberry at High Elevation ........................................... 201
17.6 Deciduous Fruit Trees in the High Tropics: A Special Situation ................................ 203
17.7 Influence of Elevation on Feijoa, a Fruit Introduced into a Tropical Region ...................203
17.8 Conclusions...................................................................................................................204
References ...............................................................................................................................204

**Chapter 18** Impact of Altitudinal Shifts and Climatic Changes on Ecophysiological Responses of Tropical Plants ...................................................................................................................209

*Nagaraj Nallakaruppan, Kalaivani Thiagarajan, and Rajasekaran Chandrasekaran*

18.1 Introduction .................................................................................................................209
18.2 Plant Diversity of Tropical Regions in India................................................................ 210

18.3 Climate Characteristics in Tropical Highlands ................................................................... 211
18.4 Effect of Altitudinal Shift and Climate Change Factors on the Ecophysiology
of Tropical Plants ............................................................................................................... 211
   18.4.1 Temperature ........................................................................................................... 211
   18.4.2 Atmospheric Pressure ............................................................................................ 213
   18.4.3 Precipitation .......................................................................................................... 213
   18.4.4 Primary Productivity ............................................................................................. 213
   18.4.5 Biotic Interactions ................................................................................................. 213
18.5 Variation of Plant Metabolites under Altitudinal Shift and Climate Change ..................... 213
18.6 Conclusion ......................................................................................................................... 214
Acknowledgments ....................................................................................................................... 214
References ................................................................................................................................... 214

# SECTION 6  Emerging techniques in Ecophysiological Research

**Chapter 19** An Overview of Emerging Techniques in Ecophysiological Research ....................... 219

*Surbhi Sharma, Joat Singh, Neeru Bala, Priyanka Sharma, Shalini Bahel,
and Jatinder Kaur Katnoria*

19.1 Introduction ....................................................................................................................... 219
19.2 Ecophysiological Response of Tropical Plants against Various Stresses .......................... 219
   19.2.1 Drought .................................................................................................................. 220
   19.2.2 Atmospheric Carbondioxide ($CO_2$) ..................................................................... 221
   19.2.3 Temperature ........................................................................................................... 221
   19.2.4 Environmental Pollution ....................................................................................... 221
   19.2.5 Heavy Metals ......................................................................................................... 222
   19.2.6 Salinity ................................................................................................................... 222
   19.2.7 Agronomic Processes ............................................................................................ 222
   19.2.8 Radiations .............................................................................................................. 222
   19.2.9 Pathogens ............................................................................................................... 223
   19.2.10 Ozone ................................................................................................................... 223
19.3 Various Emerging Techniques Used in Ecophysiological Research ................................. 223
   19.3.1 Hybridization Method ........................................................................................... 223
   19.3.2 Thermal Imaging ................................................................................................... 223
      19.3.2.1 Water Loss, Transpiration, Stomatal Conductance and Stress Indices .... 224
      19.3.2.2 Temperature Variability ........................................................................... 224
      19.3.2.3 Irrigation Scheduling ................................................................................ 224
      19.3.2.4 Biophysical and Aerodynamic Properties ................................................ 225
      19.3.2.5 Metabolic Processes ................................................................................. 225
      19.3.2.6 Disease and Infection ............................................................................... 225
      19.3.2.7 Pollution and Agronomic Effects ............................................................. 225
   19.3.3 Metabolomics, Proteomics and QSARS/SARS .................................................... 225
      19.3.3.1 Metabolomics ........................................................................................... 225
      19.3.3.2 Proteomics ................................................................................................ 226
      19.3.3.3 QSARS/SARS .......................................................................................... 226
   19.3.4 Isotopes Techniques for Plant and Soil Studies .................................................... 226
      19.3.4.1 Isotope Technique Using Carbon ............................................................. 226
      19.3.4.2 Isotopes Techniques Using Hydrogen and Oxygen ................................. 227
      19.3.4.3 Isotopes Techniques Using Hydrogen and Nitrogen ............................... 227
   19.3.5 Flow Cytometry ..................................................................................................... 227
      19.3.5.1 Instrumentation in Flow Cytometers ........................................................ 228
      19.3.5.2 The Use of Flow Cytometry in Ecophysiological Research .................... 228
19.4 Conclusion ......................................................................................................................... 228
Conflict of Interest ...................................................................................................................... 228
References ................................................................................................................................... 229

Contents

**Chapter 20** A Critical Review of Different Methods of Estimation of the Above-Ground Biomass and Carbon Stocks in India .................................................................................................................233

*Dipti Karmakar, Srimanta Gupta, and Pratap Kumar Padhy*

- 20.1 Introduction .................................................................................................................................233
- 20.2 Various Methods for Estimation of Carbon Stocks and Above-Ground Tree Biomass of Forests .........................................................................................................234
  - 20.2.1 Destructive Method .......................................................................................................234
  - 20.2.2 Non-Destructive Method ...............................................................................................234
  - 20.2.3 Remote Sensing or Satellite Method .............................................................................234
- 20.3 Estimation of Biomass in Different Zonal Councils of India ....................................................235
  - 20.3.1 North Zonal Council .....................................................................................................235
    - 20.3.1.1 Haryana ..........................................................................................................235
    - 20.3.1.2 Himachal Pradesh ..........................................................................................235
    - 20.3.1.3 Jammu and Kashmir ......................................................................................235
  - 20.3.2 North-Eastern Council ..................................................................................................236
    - 20.3.2.1 Assam .............................................................................................................236
    - 20.3.2.2 Arunachal Pradesh .........................................................................................237
    - 20.3.2.3 Manipur ..........................................................................................................237
    - 20.3.2.4 Meghalaya ......................................................................................................237
  - 20.3.3 Central Zonal Council ...................................................................................................237
    - 20.3.3.1 Chhattisgarh ...................................................................................................237
    - 20.3.3.2 Madhya Pradesh .............................................................................................237
    - 20.3.3.3 Uttarakhand ....................................................................................................237
    - 20.3.3.4 Uttar Pradesh .................................................................................................237
  - 20.3.4 Eastern Zonal Council ..................................................................................................238
    - 20.3.4.1 Bihar ...............................................................................................................238
    - 20.3.4.2 Jharkhand .......................................................................................................238
    - 20.3.4.3 Odisha ............................................................................................................238
    - 20.3.4.4 West Bengal ...................................................................................................238
  - 20.3.5 Western Zonal Council .................................................................................................244
    - 20.3.5.1 Maharashtra ...................................................................................................244
    - 20.3.5.2 Gujarat ...........................................................................................................244
    - 20.3.5.3 Rajasthan .......................................................................................................244
  - 20.3.6 Southern Zonal Council ................................................................................................244
    - 20.3.6.1 Andhra Pradesh ..............................................................................................244
    - 20.3.6.2 Karnataka .......................................................................................................244
    - 20.3.6.3 Kerala .............................................................................................................245
    - 20.3.6.4 Tamil Nadu ....................................................................................................245
- 20.4 Comparison of Biomass among the Zonal Councils of India ....................................................245
- 20.5 Trends in Measurement of Tree Biomass in India .....................................................................245
- 20.6 Advantages and Limitations of Different Methods ....................................................................245
- 20.7 Conclusions .................................................................................................................................247
- Declaration on Conflicts of Interest .....................................................................................................247
- Acknowledgments ................................................................................................................................247
- References ...........................................................................................................................................247

**Chapter 21** Brassinosteroid Hormones: A Promising Strategy for Abiotic Stress Management in Plants under Changing Climate ..................................................................................................................253

*Sandeep Kumar*

- 21.1 Introduction .................................................................................................................................253
- 21.2 Discovery and Structure of Brassinosteroids ..............................................................................254
- 21.3 Role of Brassinosteroids in Growth and Development ..............................................................255
- 21.4 Brassinosteroids in Reproduction and Fruit Development ........................................................255
- 21.5 Photosynthesis, Water Relations, and Nutrient Uptake .............................................................256
- 21.6 Synergistic Role of Brassinosteroids Metabolism and Signaling ...............................................256

|  |  |  |
|---|---|---|
| | 21.7 Abiotic Stresses: Challenges in a Changing World | 257 |
| | 21.8 Brassinosteroids in Abiotic Stress Management | 258 |
| | 21.9 Brassinosteroids in Biotic Stress Management | 258 |
| | 21.10 Conclusions | 258 |
| | References | 260 |

**Chapter 22** Phytohormones: Role in Ecophysiological Responses of Tropical Plants to Varying Resource Availability ............................................................................................................................. 265

*Pallavi B. Dhal, Sadhna, Rajkumari Sanayaima Devi, Rahul Bhadouria, and Sachchidanand Tripathi*

22.1 Introduction ............................................................................................................................ 265
22.2 Ethylene .................................................................................................................................. 266
22.3 Cytokinin ................................................................................................................................ 271
    22.3.1 Cytokinin in Triggering the Development of Lateral Roots in Response to Changes in Nitrogen Levels ............................................................................................ 271
    22.3.2 Cytokinin in Response to Nitrogen Availability .................................................. 271
    22.3.3 Role of Cytokinin in Shade Avoidance Response ............................................... 272
    22.3.4 Cytokinin in Regulating the Cambial Activity .................................................... 272
    22.3.5 Cytokinin in Tuber Development ........................................................................ 272
22.4 Gibberellin .............................................................................................................................. 272
    22.4.1 Gibberellin in Regulating the Conducting Elements of the Vascular System ...... 272
    22.4.2 Gibberellin in Shade Avoidance Response .......................................................... 272
22.5 Auxin ...................................................................................................................................... 272
    22.5.1 Auxin in Vascular Development .......................................................................... 272
    22.5.2 Auxin in Shade Avoidance ................................................................................... 273
    22.5.3 Auxin in the Regulation of Root Architecture ..................................................... 273
    22.5.4 Auxin in Secondary Root Growth ........................................................................ 273
    22.5.5 Auxin in Response to Sulfur Deficit Conditions .................................................. 273
    22.5.6 Auxin in Response to Potassium Deficit Conditions ........................................... 274
    22.5.7 Auxin in Response to Phosphate Deficit .............................................................. 274
22.6 Abscisic Acid .......................................................................................................................... 274
    22.6.1 Abscisic Acid in Response to Shade Avoidance ................................................. 274
    22.6.2 Abscisic Acid in Response to Water Stress ......................................................... 274
    22.6.3 Abscisic Acid in Response to Nitrate Deficit Conditions ................................... 274
    22.6.4 Abscisic Acid in Response to Low Phosphorus Conditions ............................... 274
22.7 Jasmonic Acid in Plant Defense Mechanisms ....................................................................... 274
22.8 Strigolactone in Response to Various Nutrient Deficit Conditions ...................................... 275
22.9 Cross-Talk Related to Various Phytohormones in Response to the Ecophysiology of Tropical Plants ................................................................................................................... 275
    22.9.1 Cytokinin–Auxin Cross-Talk ............................................................................... 275
    22.9.2 Auxin–Strigolactone Cross-Talk Governing Stolon Architecture ....................... 275
    22.9.3 Brassinosteroids, Auxin, and Ethylene Cross-Talk .............................................. 275
    22.9.4 Gibberellin and Auxin Cross-Talk ....................................................................... 277
    22.9.5 Salicylic Acid–Auxin Cross-Talk ........................................................................ 277
22.10 Research Gaps and Future Perspectives ................................................................................ 277
22.11 Conclusion .............................................................................................................................. 277
References ......................................................................................................................................... 277

**Chapter 23** Next-Generation Techniques in Ecophysiology: Metabolomics, Proteomics, SAR/QSAR ........................ 284

*Priyanka Rathore and Rashmi Shakya*

23.1 Introduction ............................................................................................................................ 284
23.2 Proteomics in Plant Stress ...................................................................................................... 284
23.3 Proteomics Techniques ........................................................................................................... 285
    23.3.1 2-DE-MS and LC–MS ......................................................................................... 286

| | 23.4 | Application of Proteomics in Plant Stress Responses | 287 |
| | 23.5 | Metabolomics in Plant Stress Response | 287 |
| | 23.6 | Metabolomic Techniques | 289 |
| | | 23.6.1 Chromatography Coupled with MS | 289 |
| | | 23.6.2 Nuclear Magnetic Resonance (NMR) | 289 |
| | | 23.6.3 Application of Metabolomics Approaches for the Study of Plant Response to Stress | 290 |
| | 23.7 | Structure–Activity Relationship (SAR)/Quantitative Structure–Activity Relationship (QSAR) | 290 |
| | | 23.7.1 Introduction to SAR and QSAR | 290 |
| | | 23.7.2 History of QSAR | 291 |
| | 23.8 | QSAR Methodology | 291 |
| | | 23.8.1 Quality of Data | 291 |
| | | 23.8.2 Selection of Molecular Descriptors | 292 |
| | | 23.8.3 Statistical Methods | 292 |
| | 23.9 | Application of QSAR Approaches for the Study of Plant Response to Stress | 292 |
| | | 23.9.1 Designing and Development of Coumarin Derivatives as New Plant Protection Products (PPP) | 292 |
| | | 23.9.2 QSAR Analysis for Inhibition of Soybean Lipoxygenase (LOX) Activity by Coumarin Derivatives | 292 |
| | 23.10 | Summary | 293 |
| | References | | 293 |

**Chapter 24** Markers of Oxidative Stress in Plants .................................................................................298

*Rashmi Shakya and Deepali*

| | 24.1 | Introduction | 298 |
| | 24.2 | Markers of Oxidative Stress in Plants | 298 |
| | 24.3 | Enzymatic Markers of Oxidative Stress | 301 |
| | | 24.3.1 Superoxide Dismutase (SOD) | 301 |
| | | 24.3.2 Catalase (CAT) | 301 |
| | | 24.3.3 Glutathione Peroxidases (GPX) | 301 |
| | | 24.3.4 Ascorbate Peroxidase (APX) | 302 |
| | | 24.3.5 Monodehydroascorbate Reductase (MDAR) | 302 |
| | | 24.3.6 Dehydroascorbate Reductase (DHAR) | 302 |
| | | 24.3.7 Glutathione Reductase (GR) | 303 |
| | 24.4 | Non-Enzymatic Markers of Oxidative Stress | 303 |
| | | 24.4.1 Hydrogen Peroxide ($H_2O_2$) | 303 |
| | | 24.4.2 Ascorbic Acid (AsA) | 303 |
| | | 24.4.3 Glutathione (GSH) | 305 |
| | | 24.4.4 Malondialdehyde (MDA) | 305 |
| | | 24.4.5 Proline | 305 |
| | | 24.4.6 Carotenoids | 306 |
| | | 24.4.7 α-Tocopherols | 306 |
| | | 24.4.8 Flavanoids | 306 |
| | | 24.4.9 Ophthalmic Acid (OPH): A Novel Oxidative Stress Marker in Plants | 306 |
| | 24.5 | Summary | 306 |
| | References | | 307 |

**Index** .........................................................................................................................................................311

# Preface

Tropical plants are facing the enormous challenge of physiological stresses in recent times due to changing environmental and climatic conditions. Various factors—predominantly the rampant exploitation of fossil fuels and other resources by the growing human population—are responsible for such perturbations. According to IPCC (2013[1], 2018[2]), carbon dioxide ($CO_2$) is projected to increase up to 700 µL L$^{-1}$ from the current 401 µL L$^{-1}$, giving an approximate rise in global temperature by 4°C. Further, the precipitation patterns are expected to alter variably across spatial scales due to the intensification of the hydrological cycle, which is expected to create extremes of dry and wet weather conditions. On the other hand, gaseous and particulate pollutants are also posing an immense challenge for tropical plants. Such vagaries in environmental conditions are already exhibiting significant impact on the ecophysiological traits of plants, resulting from altered interactions of tropical plants with each other, as well as other biotic and abiotic components within the ecosystem. Thus, rapid change in climate (moisture-temperature interaction) and associated abiotic changes, including nutrient availability and fire, may influence the evolutionary process in the tropical region, particularly tropic and co-evolutionary interactions (such as plant–herbivore, host–pathogen, rhizospheric associations, pollen viability-temperature interaction, etc.), imposing selection pressure on the plants and leading to plant anatomical and physiological adaptation with the help of phytohormones such as brassinosteroids. Furthermore, studies indicate that tropical plants are experiencing unexpected environmental conditions under changing climates that are beyond their physiological limits and adaptation range.

The ecophysiology of plants significantly affects their growth, reproduction, survival, and geographical distribution, and it is closely associated with their physicochemical and biological interactions. The ecophysiological mechanisms at the levels of physiology, biochemistry, molecular biology, and biophysics need to be explored to understand the dynamics of tropical forest ecosystems in the era of climate change. Selective markers have been identified for specific abiotic stresses among these plants which may advance our understanding for monitoring and management of these forests in a sustainable manner. Genes conferring resistance and tolerance among these plants against abiotic stresses are being researched. It could be a powerful tool to understand the costs, benefits, and consequences of anthropocentric modification of the environment on tropical plants and ecosystems. Overall, the well-being of tropical plants can be better assessed through their ecophysiological responses to these environmental and climatic disturbances. Emerging new techniques and approaches, such as remote sensing and molecular biology, may help in the understanding ecological performance of tropical plants, particularly in the adaptation of plants to extreme habitats. Ecophysiological studies are now being seriously recognized worldwide, though in limited pockets. However, comprehensive exploration, analysis and interpretation in tropical regions is a research area limited to a larger extent for a clear prediction of specific responses of tropical plants to environmental change across the levels of study.

The rapid change in climate is impacting the ecology and evolutionary trajectories of every species present on the earth, and it is profoundly altering biological communities and ecosystems. In this respect, ecophysiological concepts and techniques can provide a powerful and meaningful approach to understand the effect of the dynamic interaction of various environmental drivers on tropical plant growth and development under climate change. For instance, a plant's response to growth-limiting factors under natural conditions may help to identify relevant physiological traits—such as drought and salinity tolerance and nutrient use efficiency—relevant in tree plantation, as well as forest management. Recent techniques have enabled scientists to design specific experiments to understand the response of tropical plants to changing environmental conditions and physiological stressors. Owing to distinct and diverse life histories and physiologies of relatively evolved tropical plants, these ecological communities have organized variously across latitudes. These plants may respond differently to climate and environmental change, contrary to plant species in the temperate and polar regions. Though the impacts of climate change have been observed from the poles to the equator, most studies have predominantly focused on biotic impacts at high-altitude regions due to immense visible impact of temperature change in this region. As the effects of climate change depend on species' sensitivity to the shift in moisture–temperature interaction and associated environmental (biotic and abiotic) variables, it is projected to be greater in the tropical region. Therefore, it would be an interesting and important area of scientific investigation to understand the ecophysiological response of tropical plants in the time to come. It would help us manage these pristine and ecologically important ecosystems for a sustainable and healthy biosphere. The book intends to provide a comprehensive discussion on the responses of tropical plants to various biotic and abiotic environmental concerns in the backdrop of changing climate.

The book contains a total of 24 chapters, summarized in the following.

S. Oyedeji explores the plant adaptability in tropical dry biomes through an ecophysiological perspective in Chapter 1.

In Chapter 2, Zirwa Sarwar et al. provide the current state of knowledge on trophic interaction, thermal interaction, plant–herbivore interaction, co-evolutionary responses, and host–pathogen interactions in tropical regions under a changing climate.

Ranjan Pandey et al. in Chapter 3 elaborate upon the physiological mechanisms displayed by tropical plants in coping up with climate change–associated environmental stresses.

In Chapter 4, Pallavi Singh et al. explore the impacts of common air pollutants on the ecophysiology of the dominant trees and major crops of tropical regions.

Aisha Kamal et al. investigate the status of oxides of nitrogen ($NO_X$) and their impact on the ecophysiology of tropical plants in Chapter 5.

In Chapter 6, Somdutta Sinha Roy et al. evaluate the impact of particulate matter on various ecophysiological parameters of tropical plants.

Sadhna et al.'s Chapter 7 presents the impact of air pollutants on ecophysiology of tropical plants in urban settlements.

Wajiha Sarfraz et al. discuss the determinants of ecophysiology of tropical forests, along with their conservation status and implications, in Chapter 8.

Janaki Subramanyan elaborates on the importance of soil nutrient reservoirs for tropical plants in changing climates in Chapter 9.

In Chapter 10, R. Mishra et al. emphasize the abiotic stress responses of different sugarcane species and discuss the important genes that confer enhanced abiotic stress tolerance in response to diverse abiotic stress conditions.

Martijn Slot et al. investigate in Chapter 11 the effects of high temperatures on flower production and pollen viability of a tropical tree species, *Muntingia calabura*, in lowland Panama.

In Chapter 12, Ghalia S.H. Alnusairi et al. provide a detailed review of the influence of phytohormones on ecophysiological and morphological adaptations of plants under varying temperature availability.

Francis Q. Brearley investigates in Chapter 13 the contrasting responses of two dipterocarp species to reductions in ectomycorrhizas colonization after adding fungicide, which may influence their regeneration under natural conditions.

In Chapter 14, Rupali Jandrotia et al. assess the ecophysiological responses of tropical plants to changing levels of $CO_2$, a potent greenhouse gas. Plant adjustment to elevated $CO_2$ concentration in terms of whole plant responses and biological, molecular, cellular, and hormonal responses are also elaborated.

Shikha Singh and Tanu Kumari discuss in Chapter 15 the ecophysiological mechanisms and adaptation strategies of tropical plants in response to fire.

In Chapter 16, S. Oyedeji highlights the impact of climate change–mediated fire events on community structure and the physio-anatomical adaptations of plants in tropical savannas.

Chapter 17, by Gerhard Fischer et al., emphasizes the physiological adaptations of fruit trees to different altitudes in the tropical Andes.

Nagaraj Nallakaruppan et al. review in Chapter 18 that altitude and climate change widely affect the ecophysiological response of tropical plants.

Chapter 19, by Surbhi Sharma et al., provides an overview on various emerging techniques available for ecophysiological research.

Dipti Karmakar et al. compiled in Chapter 20 various methods used for estimating the above-ground biomass in different zonal councils of India and the advantages and limitations of the commonly used methods like non-destructive, destructive, and remote sensing or satellite-based methods.

Chapter 21, by Sandeep Kumar, elaborates upon the role of brassinosteroids in ameliorating various abiotic stresses.

Pallavi B. Dhal et al. outline in Chapter 22 the recent updates on the multifaceted involvement of phytohormones in environmental adaptability and ecophysiological responses of tropical plants to availability of various resources.

Chapter 23, by Priyanka Rathore and Rashmi Shakya, provides an overview of next-generation techniques available for ecophysiology research, particularly in terms of metabolomics, proteomics, and SAR/QSAR.

Rashmi Shakya and Deepali provide in Chapter 24 an updated account of factors responsible for oxidative stress in plants and a detailed account of various enzymatic and non-enzymatic markers reported to date.

**Sachchidanand Tripathi**
**Rahul Bhadouria**
**Pratap Srivastava**
**Rishikesh Singh**
**Rajkumari Sanayaima Devi**

## NOTES

1 IPCC, 2013: Climate Change 2013: The Physical Science Basis. Contribution of Working Group I to the Fifth Assessment Report of the Intergovernmental Panel on Climate Change [T. F. Stocker, D. Qin, G.-K. Plattner, M. Tignor, S. K. Allen, J. Boschung, A. Nauels, Y. Xia, V. Bex and P. M. Midgley (eds.)]. Cambridge University Press, Cambridge, United Kingdom and New York, NY, USA, 1535 p, doi:10.1017/CBO9781107415324.

2 IPCC, 2018: Global warming of 1.5°C. An IPCC Special Report on the impacts of global warming of 1.5°C above pre-industrial levels and related global greenhouse gas emission pathways, in the context of strengthening the global response to the threat of climate change, sustainable development, and efforts to eradicate poverty [V. Masson-Delmotte, P. Zhai, H. O. Pörtner, D. Roberts, J. Skea, P. R. Shukla, A. Pirani, W. Moufouma-Okia, C. Péan, R. Pidcock, S. Connors, J. B. R. Matthews, Y. Chen, X. Zhou, M. I. Gomis, E. Lonnoy, T. Maycock, M. Tignor and T. Waterfield (eds.)]. In Press.

# Editor Biographies

**Sachchidanand Tripathi, Ph.D.,** is an Associate Professor at the Department of Botany, Deen Dayal Upadhyaya College, University of Delhi, New Delhi, India. Dr. Tripathi obtained his doctoral degree from the Department of Botany, Banaras Hindu University, India. The areas of his interest are plant ecology, soil ecology, ecophysiology, and urban ecology. He has published more than 50 publications (including research publications, books and book chapters, and conference proceedings) with reputed international journals and publishers.

**Rahul Bhadouria, Ph.D.,** is an Assistant Professor at the Department of Environmental Studies, Delhi College of Arts and Commerce, University of Delhi, New Delhi, India. Dr. Bhadouria obtained his doctoral degree from the Department of Botany, Banaras Hindu University, Varanasi, India. He has published more than 22 papers, 22 book chapters, and eight edited books in internationally reputed journals/publishers. His current research areas are, management of soil carbon dynamics to mitigate climate change, a perspective on tree seedling survival and growth attributes in tropical dry forests under the realm of climate change, plant community assembly, functional diversity and soil attributes along the forest–savanna–grassland continuum in India, recovery of degraded mountains in the central Himalayas, and urban ecology.

**Pratap Srivastava, Ph.D.,** is an Assistant Professor at the Department of Botany, University of Allahabad, Prayagraj, India. Dr. Srivastava completed his Ph.D. in botany from Banaras Hindu University, India. His area of research includes soil carbon dynamics, biochar technology, plant ecology, and waste management. He has published more than 50 research publications in international journals/books in the fields of soil carbon dynamics, carbon sequestration, environmental contaminant removal, and waste management. He is actively involved as an ad hoc reviewer of several international journals published by reputed publishers.

**Rishikesh Singh, Ph.D.,** is a National Post-doctoral Fellow (NPDF) at the Department of Botany, Panjab University, Chandigarh, India. Dr. Singh obtained his doctoral degree from Institute of Environment & Sustainable Development, Banaras Hindu University, Varanasi. He is an environmental scientist with his research interests in soil carbon dynamics, land-use change and management, waste management, environmental contaminants, biochar, and carbon sequestration. He has published several research and review articles, and is a reviewer of several international journals of Elsevier, Wiley, Taylor and Francis, Frontiers, PLoS, and Springer Nature groups. He has published six books in leading international publishers such as Elsevier, Springer Nature, and Wiley.

**Rajkumari Sanayaima Devi, Ph.D.,** is an Associate Professor at the Department of Botany, Deen Dayal Upadhyaya College, University of Delhi, New Delhi, India. Dr. Devi obtained her doctoral degree from the Department of Botany, University of Delhi, India. The area of her research is plant conservation and cryobiology. She has published more than 30 publications with international journals and publishers.

# Contributors

**S.A. Adeniran**
Plant Anatomy and Taxonomy Unit
Department of Plant Biology
University of Ilorin
Ilorin, Nigeria

**P. Agarwal**
Department of Botany
Gargi College
University of Delhi
New Delhi, India

**O.O. Agboola**
TETFUND Centre of Excellence in Biodiversity
 Conservation and Ecosystem Management
 (TCEBCEM)
University of Lagos
Akoka, Lagos State, Nigeria

**Madhoolika Agrawal**
Laboratory of Air Pollution and Global Climate Change
Department of Botany
Institute of Science
Banaras Hindu University
Varanasi, India

**Shashi Bhushan Agrawal**
Laboratory of Air Pollution and Global Climate Change
Department of Botany
Institute of Science
Banaras Hindu University
Varanasi, India

**Farhan Ahmad**
Department of Biotechnology
Ashoka Institute of Technology and Management
Varanasi, India

**Khadiga Alharbi**
Department of Biology
College of Science
Princess Nourah bint Abdulrahman University
Riyadh, Saudi Arabia

**Shafaqat Ali**
Department of Environmental Sciences and Engineering
Government College University
Faisalabad, Pakistan
and
Department of Biological Sciences and Technology
China Medical University
Taichung, Taiwan

**Ghalia S.H. Alnusairi**
Department of Biology
College of Science
Jouf University
Sakaka, Saudi Arabia

**Shalini Bahel**
Department of Electronics Technology
Guru Nanak Dev University
Amritsar, India

**Saloni Bahri**
Department of Botany
Miranda House
University of Delhi
New Delhi, India

**Neeru Bala**
Department of Botanical and Environmental
 Sciences
Guru Nanak Dev University
Amritsar, India

**Helber Enrique Balaguera-López**
Universidad Nacional de Colombia
Bogotá DC, Colombia

**Daizy R. Batish**
Department of Botany
Panjab University
Chandigarh, India

**Rahul Bhadouria**
Department of Environmental Studies
Delhi College of Arts and Commerce
University of Delhi
New Delhi, India

**Francis Q. Brearley**
Department of Animal and Plant Sciences
University of Sheffield
Western Bank, Sheffield, UK
and
Department of Natural Sciences
Manchester Metropolitan University
Manchester, UK

**Rajasekaran Chandrasekaran**
Department of Biotechnology
School of Bio Sciences and Technology
Vellore Institute of Technology
Vellore, India

**Deepali**
Department of Botany
Miranda House
University of Delhi
New Delhi, India

**Laishram Sundari Devi**
Department of Botany
Miranda House
University of Delhi
New Delhi, India

**Rajkumari Sanayaima Devi**
Deen Dayal Upadhyaya College
University of Delhi
New Delhi, India

**Pallavi B. Dhal**
Deen Dayal Upadhyaya College
University of Delhi
New Delhi, India
and
Department of Botany
University of Delhi
New Delhi, India

**Allah Ditta**
Department of Environmental Sciences
Shaheed Benazir Bhutto University
Sheringal, Khyber Pakhtunkhwa, Pakistan
and
School of Biological Sciences
University of Western Australia
Perth, Western Australia, Australia

**Ujala Ejaz**
Department of Plant Sciences
Quaid-i-Azam University
Islamabad, Pakistan

**Chinedu E. Eze**
Smithsonian Tropical Research Institute
Apartado, Balboa, Ancón, Republic of Panama
and
Department of Agronomy
Michael Okpara University of Agriculture
Umudike, Abia State, Nigeria

**P.O. Fatoba**
Plant Ecology and Environmental Botany Unit
Department of Plant Biology
University of Ilorin
Ilorin, Nigeria

**Mujahid Farid**
Department of Environmental Sciences
University of Gujrat
Hafiz Hayat Campus
Gujrat, Pakistan

**Gerhard Fischer**
Universidad Nacional de Colombia
Bogotá DC, Colombia

**Ipsa Gupta**
Department of Botany
Panjab University
Chandigarh, India

**Srimanta Gupta**
Department of Environmental Science
The University of Burdwan
Golapbag, Bardhaman, West Bengal, India

**Maria Hanif**
Department of Biotechnology
Lahore College for Women University
Lahore, Pakistan

**Maria Hasnain**
Department of Biotechnology
Lahore College for Women University
Lahore, Pakistan

**Rupali Jandrotia**
Department of Botany
Panjab University
Chandigarh, India

**Aisha Kamal**
Department of Bioengineering
Integral University
Lucknow, India

**Dipti Karmakar**
Department of Environmental Science
University of Burdwan
Golapbag, Bardhaman, West Bengal, India

**Jatinder Kaur Katnoria**
Department of Botanical and Environmental Sciences
Guru Nanak Dev University
Amritsar, India

**Noreen Khalid**
Department of Botany
Government College Women University
Sialkot, Pakistan

# Contributors

**Harvinder Kour**
Department of Botany
Govt. Gandhi Memorial Science College
Cluster University
Jammu, Jammu and Kashmir, India

**Sandeep Kumar**
Department of Botany
Deen Dayal Upadhyaya College
University of Delhi
New Delhi, India

**Tanu Kumari**
Integrative Ecology Laboratory
Institute of Environment & Sustainable Development
Banaras Hindu University
Varanasi, India

**Stanislav Magnitskiy**
Universidad Nacional de Colombia
Bogotá DC, Colombia

**R. Mishra**
Department of Botany
Gargi College
University of Delhi
New Delhi, India

**Sushma Moitra**
Department of Botany
Miranda House
University of Delhi
New Delhi, India

**Neelma Munir**
Department of Biotechnology
Lahore College for Women University
Lahore, Pakistan

**Nagaraj Nallakaruppan**
Department of Biotechnology
School of Bio Sciences and Technology
Vellore Institute of Technology
Vellore, India

**C.O. Ogunkunle**
Plant Ecology and Environmental
 Botany Unit
Department of Plant Biology
University of Ilorin
Ilorin, Nigeria

**S. Oyedeji**
Plant Ecology and Environmental Botany Unit
Department of Plant Biology
University of Ilorin
Ilorin, Nigeria

**Pratap Kumar Padhy**
Department of Environmental Studies
Institute of Science
Visva-Bharati Santiniketan
Birbhum, West Bengal, India

**Ranjan Pandey**
Department of Environment Studies
Panjab University
Chandigarh, India

**Alfonso Parra-Coronado**
Universidad Nacional de Colombia
Bogotá DC, Colombia

**Jigyasa Prakash**
Laboratory of Air Pollution and Global Climate Change
Department of Botany
Institute of Science
Banaras Hindu University
Varanasi, India

**Riya Raina**
Department of Environment Studies
Panjab University
Chandigarh, India

**Priyanka Rathore**
Department of Botany
Miranda House
University of Delhi
New Delhi, India

**Nighat Raza**
Department of Food Science and Technology
Muhammad Nawaz Sharif University of Agriculture
Multan, Pakistan

**Zarrin Fatima Rizvi**
Department of Botany
Government College Women University
Sialkot, Pakistan

**Somdutta Sinha Roy**
Department of Botany
Miranda House
University of Delhi
New Delhi, India

**Sadhna**
Deen Dayal Upadhyaya College
University of Delhi
New Delhi, India
and
Department of Botany
University of Delhi
New Delhi, India

**Wajiha Sarfraz**
Department of Botany
Government College Women University
Sialkot, Pakistan

**Zirwa Sarwar**
Department of Biotechnology
Lahore College for Women University
Lahore, Pakistan

**Natanja Schuttenhelm**
Smithsonian Tropical Research Institute
Apartado, Balboa, Ancón, Republic
 of Panama
and
Ecology and Biodiversity Group
Department of Biology
Utrecht University
Padualaan 8, Utrecht, The Netherlands

**Rashmi Shakya**
Department of Botany
Miranda House
University of Delhi
New Delhi, India

**Priyanka Sharma**
Department of Botanical and Environmental
 Sciences
Guru Nanak Dev University
Amritsar, India

**Surbhi Sharma**
Department of Botanical and Environmental
 Sciences
Guru Nanak Dev University
Amritsar, India

**Shazia Siddiqui**
Department of Bioengineering
Integral University
Lucknow, India

**G. Singh**
Department of Botany
Gargi College
University of Delhi
New Delhi, India

**Harminder Pal Singh**
Department of Environment Studies
Panjab University
Chandigarh, India

**Harshita Singh**
Laboratory of Air Pollution and Global
 Climate Change
Department of Botany
Institute of Science
Banaras Hindu University
Varanasi, India

**Joat Singh**
Department of Botanical and Environmental Sciences
Guru Nanak Dev University
Amritsar, India

**Pallavi Singh**
Laboratory of Air Pollution and Global Climate Change
Department of Botany
Institute of Science
Banaras Hindu University
Varanasi, India

**Rishikesh Singh**
Department of Botany
Panjab University
Chandigarh, India

**Shikha Singh**
Integrative Ecology Laboratory (IEL)
Institute of Environment & Sustainable Development
 (IESD)
Banaras Hindu University
Varanasi, India

**Martijn Slot**
Smithsonian Tropical Research Institute
Apartado, Balboa, Ancón, Republic of Panama

**Mona H. Soliman**
Botany and Microbiology Department
Faculty of Science
Cairo University
Giza, Egypt

**R. Soni**
Department of Botany
Gargi College
University of Delhi
New Delhi, India

**Janaki Subramanyan**
Department of Botany
Miranda House
University of Delhi
New Delhi, India

**Nida Sultan**
Department of Bioengineering
Integral University
Lucknow, India

**Kalaivani Thiagarajan**
Department of Biotechnology
School of Bio Sciences
  and Technology
Vellore Institute of Technology
Vellore, India

**Sachchidanand Tripathi**
Department of Botany
Deen Dayal Upadhyaya College
University of Delhi
New Delhi, India

**Huma Waqif**
Department of Biotechnology
Lahore College for Women University
Lahore, Pakistan

**Klaus Winter**
Smithsonian Tropical Research Institute
Apartado, Balboa, Ancón, Republic of Panama

**Abbu Zaid**
Department of Botany
Government Degree College Doda
Jammu and Kashmir, India
and
Department of Botany
Govt. Gandhi Memorial Science College
Cluster University
Jammu, Jammu and Kashmir, India

# Foreword

Due to extremes of climatic conditions, tropical vegetation is biologically more diverse than vegetation occurring in other latitudes. Seasonal droughts can impact the ecophysiology of tropical vegetation significantly. Consequently, tropical vegetation is prone to exhibit noteworthy morphological and physiological adaptations. Editors have aptly chosen to consolidate the available information on the ecophysiology of tropical plants into this volume, entitled *Ecophysiology of Tropical Plants: Recent Trends and Future Perspectives*, edited by Drs. Sachchidanand Tripathi, Rahul Bhadouria, Pratap Srivastava, Rishikesh Singh, and Rajkumari Sanayaima Devi. Broadly, the contributions to this book can be considered under four categories: (1) adaptations of tropical plants and associated evolutionary trends; (2) ecophysiological responses of tropical plants as affected by soil nutrients and ectomycorrhizal colonization; (3) impact of air pollutants, particulate matter, temperature, and fire; and (4) emerging techniques in ecophysiological research, including metabolomics and proteomics. Abiotic stress conditions lead to oxidative stress in plants, which has also been adequately discussed.

The editors and author contributors of this CRC publication deserve appreciation for providing significant technical inputs and their consolidation in a very logical manner. All of them possess the required expertise to meaningfully discuss the specific subjects.

I am sure that the international scientific community will benefit a lot from this very informative compilation on the subject.

Best wishes.

**Professor Satish C Bhatla**
*F.N.A.Sc.*
*Fellow, Alexander von Humboldt Foundation (Germany)*
**Department of Botany**
*University of Delhi*
*New Delhi, India*
*April 1, 2023*

# Section 1

*Tropical plants and changing climate scenarios*

# 1 Plant Adaptations in Dry Tropical Biomes
## *An Ecophysiological Perspective*

S. Oyedeji

## 1.1 INTRODUCTION

*Biome* is a synonym for the *biotic community*, according to Frederick Clements, who first used the term in 1916 (Mucina, 2019). Despite the fact that flora and fauna constitute the biotic community, biomes are typically described in terms of the dominant plant community, since plants are fixed to the soil and must necessarily adapt to their environment, and to the influences of biotic and abiotic factors, especially that of climate. Climatic factors, average annual rainfall, mean temperatures and relative humidity are important determinants of the biome. Edaphic factors such as soil structure, nutrient dynamics and topography, as well as disturbance arising from humans, animals (typically herbivores) and natural histories of fires, drought and flooding, also play critical roles in determining the structure of plant community (Dexter et al., 2018).

Biomes constitute large collection of plants that are characterized in terms of the decipherable structure/arrangement of the dominant species (Woodward et al., 2004) and can be defined as *major formations of the plant community (vegetation) with distinctive physiognomies (physical structure/forms) and ecological processes that can be characterized on a global scale* (Pennington et al., 2018). Succinctly put, biomes are areas characterized by similar life forms (Woodward et al., 2004). Of all environmental variables, the climate is a major driver for biotic formations, associations and classifications of terrestrial biomes (Mucina, 2019).

Robert H. Whittaker identified major terrestrial biomes to include tropical rainforests, tropical dry forests/savanna, subtropical desert, temperate rainforest, temperate dry forest, temperate grassland/desert, boreal forests, and tundra (Whittaker, 1975). Among the biomes in the tropics, attention is well focused on the rainforest, probably because of its rich biodiversity, while less attention has been given to the study of other tropical biomes such as the dry tropical biomes (Siyum, 2020).

Dry tropical biomes, including tropical dry forests and savannas, are unique ecosystems that experience seasonal climates—typically, a short rainy season with marked dry conditions (Gottsberger & Silberbauer-Gottsberger, 2008; Bhadouria et al., 2018b; Pennington et al., 2018) which expose plants in these vegetations to drought stress.

According to the Holdridge system of life zone system categorization, tropical dry forests and savannas are found in areas where the mean annual precipitation range from 250–2000 mm with average annual temperatures exceeding 17°C. The biomes are widespread across the globe and constitute the transition zone between the moist tropical rainforest and dry desert (Murphy & Lugo, 2008). Dry tropical biomes also support a wide variety of flora and fauna; however, these biomes have received less attention when compared to tropical rainforests, which are considered home to biodiversity (Pennington et al., 2018).

## 1.2 ECOLOGY OF DRY TROPICAL BIOMES: DRY FORESTS AND SAVANNAS

Tropical dry forests (TDFs) occur in partly dry to dry ecosystems, with notable long periods of dry months (Siyum, 2020). TDFs occur between 10° and 25° of latitude and are distributed towards the north and south of the tropical rainforest biome. They constitute over 40% of the forests in the tropics and is well distributed in Africa, Asia (particularly in India), North America (particularly in Mexico), eastern South America, and northern Australia (Van Bloem et al., 2004). Forests (both open and closed) occupy approximately 40% of the earth's tropical and subtropical landscape, with dry forest constituting 42%, moist forest 33% and rainforest 25% (Murphy & Lugo, 2008). The major fraction of the dry forest biome occurs in Africa, where they constitute 70–80% of the forest cover (including forested areas of tropical islands across the world). TDFs vary greatly in structure from other tropical biomes. They often consist of a closed forest canopy of tall trees in areas bordering moist forests to short scrubs with open canopy in drier areas (Pennington et al., 2000). Unlike in savannas, where annual fire disruption occurs, TDFs are often not exposed to recurrent fires due to the low abundance of grasses in the biome. This, however, does not mean that the biome is completely insulated from fire events. In vegetation where grasses are the dominant flora, such as in savannas, fire events are often triggered by accumulated biomasses (leaves, straw and roots) that act as viable fuel for burning (Kauffman et al., 1994). The absence of frequent fires in TDFs allows for the occurrence of large populations of fire-intolerant species, such as cacti

(Mooney et al., 1995; Dexter et al., 2018). The vegetation also consists of tall trees that form a closed canopy during the wet season (which open up in the dry season after their leaves have been shed) and other drought-tolerant plants including aleos, bromeliads, orchids and other succulents. The seasonality of rainfall in this biome allows for active growth to occur only during the wet (rainy) season. In the dry season, plants exhibit various mechanisms to conserve water. For example, the vast majority of the trees drop their leaves to reduce water loss via transpiration. This strategy of the dominant flora called *deciduous* habitat has been used to name the forest as *tropical deciduous forest*.

Similar to TDFs, tropical savannas also occupy areas with a warm climate that experiences seasonality in precipitation. The word *savanna* was derived from the Spanish word *sabana*—a term used to describe tropical grassland areas with more or less scattered dense trees (Lulla, 1987). Savannas are defined by the dominance of grasses and sedges (the herbaceous cover) that are interspersed by woody vegetation (trees and shrubs). The grasses reach above 30 cm in height during the rainy season. The woody vegetation in savannas usually forms a small component of its biomass (Sarmiento, 2014), while the growth of the herbaceous component is often seasonal with a decrease necessitated by water shortage during the dry season (Sternberg, 2001). Grasses, the dominant component in savannas, support fires in the vegetation; hence, the tree species are well adapted to recurring fires that help to maintain the structure of the vegetation (Durigan & Ratter, 2016; Abreu et al., 2017). People often confuse savannas with dry forests (Dexter et al., 2018), but savannas distinguishably have predominant grass cover (among the herbaceous components including sedges) with trees and shrubs at varying densities, depending on the mean annual precipitation of the region (Sankaran & Ratnam, 2013). Ecologists have also argued that savannas are composed mainly of $C_4$ grasses and not just any member of the Poaceae (Scholes & Hall, 1996; Lehmann et al., 2011). The tropical savanna is a relatively young biome compared to other tropical biomes. The savanna biome was birthed around the same period that $C_4$ grasses expanded in their distribution (Lehmann et al., 2011; Sankaran & Ratnam, 2013). Savannas represent the world's second-largest biome, occupying approximately 33 million square kilometers, which makes up about 20% of the lithosphere (Beerling & Osborne, 2006). The biome is distributed across Africa, Asia, Australia and South America (Sankaran & Ratnam, 2013). Savannas cover a great deal (over 50%) of all southern continents, occupying 45% of South America, 60% of Australia and 65% of Africa (Lulla, 1987).

### 1.2.1 Seasonality in Tropical Dry Biomes

Tropical regions are known to experience high temperatures which normally remain above 18°C throughout the year (Spaargaren & Deckers, 2005). Unlike in rainforest ecosystems which experience high rainfall, rainfall in dry tropical biomes is highly seasonal and poses a great influence on biological activities and diversity (Davies et al., 2015).

Tropical dry forests receive an annual rainfall of between 500 mm and 1500 mm (Center for International Forestry Research [CIFOR], 2014) but sometimes could reach 2000 mm distributed around 5–8 months during the summer. The mean annual temperature is well over 17°C almost always (Woodward & Foy, 2022). In Africa, the dry forests in the west of the continent experiences a longer dry season. The dry season which falls within the winter period is pronounced and lasts for 4–6 months. In Asia, the forest is influenced by the monsoon and rainfall could reach 2500 mm. The dry season is long and extends from November to April, and sometimes with less than 100 mm of rain (Woodward & Foy, 2022). In Central America, the rainfall rarely exceeds 800 mm annually with over 80% falling between June and October (with a peak in September). The temperature range in the region is 17–33°C, with an annual mean of about 25–27°C (Maass et al., 2017). The climate of savannas likewise varies considerably across continents. In Australia, the savanna vegetation falls within latitude 10°–20° South, and the rainy season lasts 5–6 months (December to March), while the dry season is usually from May to October (Alex, 2020). In Africa, the rainy season lasts for 5–7 months with a mean annual rainfall of 600–1500 mm. The period of rainfall is from April to October in West Africa, and October to March in southern Africa. Drier regions in both West Africa and southern Africa typically experience less than 1000 mm of rainfall in a year. The average annual temperatures in savannas range from 20–30°C (Holzman, 2008). In southern Africa, the average temperature is between 15°C and 25°C but may reach 37°C or more when it peaks (Rees et al., 2006). The dry season in dry forests, which is often prolonged and may last for half a year or more, is the period for leaf fall and inactivity in most plants. Leaf shedding in TDFs is in response to drought and a means to conserve water by the plants, unlike in temperate deciduous forests where plants respond to winter chilling. This is followed by a season of heavy rainfall. In Asia, monsoon influences the dry spell or rainfall causing a shift in the weather—sometimes causing a prolonged dry spell or a period of heavy rainfall. Hence, the forest is sometimes called a *tropical monsoon forest*. During the start of the rainy (monsoon) season, the forest undergrowth such as lianas, herbaceous plants and tree saplings take advantage of the increasing lighting to the forest floor (caused by leaf fall during the dry spell) and the abundant moisture from the rain to grow quickly before the canopy closes up again. Leaf flushing in trees and regrowth of epiphytes aid the build-up of the canopy, during which the growth of the understory plants is slowed down.

The dry season in savannas marks the start of an extensive period when plant growth is highly suppressed. During the period, grasses, sedges and other herbaceous plants limit or stop the production of new leaves, seed heads of grasses fall, the above-ground portion of a number of forbs drop and leaves of trees and shrubs turn out to be limited (in evergreen plants) or completely lost in deciduous species (Alberts et al., 2005). However, plant productivity is not completely shut down in tropical savannas as in those of temperate grasslands. Tropical savannas differ from tropical forests in having less amount of precipitation. The rainfall is also seasonal and erratic (Bourliere & Hadley, 1983; Solbrig et al., 1996).

### 1.2.2 Structure and Diversity of Tropical Dry Forests and Savannas

Tropical dry forests exhibit a typical structure (Hasnat & Hossain, 2020), which is determined by the crown layers and the presence or absence of epiphytes (especially orchids), climbers and lianas (de la Rosa-Manzano et al., 2014). The structure of the canopy formed by the crown of trees/shrubs is directly correlated with annual precipitation received by the vegetation (Gillespie et al., 2000). In regions with higher rainfall, the trees are taller with a relatively closed canopy, while the canopy is less dense with lower rainfall (Stan & Sanchez-Azofeifa, 2019). Gentry (1995) posited that once the critical precipitation threshold required to maintain a closed canopy in TDFs is reached, increases in the amount of precipitation have a negligible effect on species richness until precipitation is high enough to maintain a moist forest. The author observed that species diversity increases in Bolivian and Mexican forests the drier they are.

In the dry forests of Costa Rica and Nicaragua in Central America, members of the families Burseraceae, Fabaceae and Bignoniaceae are well represented. Some species in the families Sapindaceae, Rubiaceae and Euphorbiaceae can also be found there, although in smaller proportions. In these forests, *Bursera simaruba*, *Spondias mombin*, *Cochlospermum vitifolium* and *Simarouba amara* have been reported (Gillespie et al., 2000). Common tree species found in the Mesoamerican dry forest include *Bursera* spp. (Hasnat & Hossain, 2020), *Chloroleucon manguense*, *Eugenia axillaris*, *Gliricidia sepium*, *Guettarda elliptica*, *Gymnopodium floribundum*, *Jatropha* spp., *Malpighia lundellii*, *Sebastiana adenophora*, *Jatropha gaumeri* and *Thouinia paucidentata* (de la Rosa-Manzano et al., 2014). Notable trees in TDFs of Africa include species of *Adansonia*, *Albizia*, *Alchornea*, *Brachylaena*, *Dalbergia*, *Dichrostachys*, *Euphorbia*, *Grewia* and *Stereospermum* (Hasnat & Hossain, 2020). Generally, epiphytes are fewer in TDFs when compared with moist forests. Among the epiphytic plants, orchids and bromeliads are common but their abundance and diversity are biased toward a few host species of trees and shrubs (Vergara-Torres et al., 2010), as the survival of these epiphytes is linked to the traits of the host. For example, deciduous habit creates the opening of the canopy to permit high photon flux during the dry seasons which promotes the growth and survival of epiphytes. Orchids including *Encyclia nematocaulon*, *Cohniella* spp., *Laelia rubescens* and *Lophiaris oerstedii* were reported in the dry forests of Yucatan, Mexico (de la Rosa-Manzano et al., 2014). Lianas such as *Cissus fuliginea*, *Combretum farinosum*, *Gouania polygama*, *Mimosa tenuiflora* and *Tetracera volubilis* have been documented in dry forests (Werden et al., 2017). Dry forests have a wide variety of growth forms of plant species; however, their frequency depends on climate and disturbance. Generally, cacti and euphorbs are common in drier locations, while the lianas and other vines are frequent in areas with higher rainfall and disturbances that open canopy gaps to allow for access of these plants to insolation (Van Bloem et al., 2004).

Savannas vary greatly in their structure and diversity. They encompass grassy vegetation with variable trees composition (scanty trees to near woodlands), fluctuating characteristics of the dominant trees (fine-leaved trees to broad-leaved trees), life history of the dominant vegetation (deciduous to evergreen trees; annual to perennial grasses) and herbaceous cover (short grasses to tall grasses). Savanna is very different from steppe based on herbaceous cover—it is closed in the former is closed and open in the steppe (including desert scrub). The vegetation is frequently burned by wildfires, typically induced by humans. This effect is a key factor accountable for the modification of the structure and spatial diversity of savannas (Graz, 2004). The International Union of Biological Sciences (IUBS) Working Group as published by Frost et al. (1986) described savannas to include all tropical and subtropical ecosystems with a continuous cover of herbaceous plants mostly heliophylous $C_4$ grasses and sedges that exhibit seasonal growth in relation to water stress. Woody species such as trees and shrubs, as well as palms, exist in the vegetation but rarely form an unbroken cover parallel to that of the grass cover. Grasses, especially perennials, dominate most savanna landscapes. The available tree species vary in size and abundance in biome—and this bears correlation with climate, except for neotropical savannas, e.g., Brazilian cerrado, which is significantly influenced by edaphic factors (Lehmann et al., 2011). Asian savannas are known to support dense tall trees—the physiognomy of bear semblance with forests rather than open grassy landscapes with scanty trees found across Africa (Ratnam et al., 2016). In African savannas, tall grasses such as *Andropogon tectorum*, *A. gayanus* and *Hyparrhenia* spp. are common in zones such as the Guinea savanna, that borders with the rainforest. Predominant trees in the region include *Adansonia Adansonia digitata*, *Daniellia oliveri*, *Parkia biglobosa* and *Vitellaria paradoxa* (Oyedeji et al., 2016). In drier regions such as the Sahelian and Sudanian-type savannas, the grasses become shorter and the trees are fewer when compared to the Guinean-type savanna (Keay, 1949).

## 1.3 ADAPTATION OF PLANTS IN DROUGHT CONDITIONS IN THE DRY TROPICS: MECHANISMS VERSUS TRAITS

Adaptation in biology refers to the phenotypic features of organisms in relation to environmental selection requirements (Bock, 1980). Plants and other organisms acquire new features at the morphological, physiological, behavioral and molecular levels to better survive in a changing environment. Adaptive features are properties of forms/functions which allow the organism to maintain synergy between their biological role and the role imposed upon the organism by a selection force (Luo & Zhang, 2014).

Plants in dry tropical biomes possess unique characteristics that help them to adapt to contrasting extremes of wetness and dryness experienced in the region. The availability of water—timing and distribution—is the major factor influencing the dynamics of tropical dry biomes. The dry period, which may last for almost half of the year or more, is the period when growth activities in plants are paused or proceed at a slow rate. Studies have revealed that resistance to drought is more vital to plant species in dry tropical biomes than resistance to fire in vegetation prone to burning such as savannas. The plants in these biomes devise many strategies to utilize the available water during dry conditions in order to survive the period of drought or escape from it. The survival mechanisms of plants endow dry tropical biomes with their characteristic appearance. Plants in dry biomes employ varying strategies to survive drought conditions. These mechanisms may be drought escaping (when the species complete its growth cycle to avoid drought conditions), drought evading (when the species develop strategies to reduce water loss) or drought enduring (when the species can survive under conditions of low water availability).

### 1.3.1 Adaptive Strategies in Grasses

#### 1.3.1.1 C$_4$ Photosynthesis

Plants using C$_4$ photosynthesis have been reported to be more productive (up to 100%) than C$_3$ plants (Brown, 1978). This they achieve by altering their structural anatomy/morphology. Recent studies have shown that C$_4$ grasses alter the shape, size and structure of their organs. C$_4$ plants have important advantages in dry and infertile soils which characterize savannas and grasslands across the globe (Osborne & Sack, 2012). Grasses in dry tropical landscapes employ an efficient concentration mechanism, the C$_4$ pathway, to mitigate transpiration while improving photosynthesis. According to Christin et al. (2008), the C$_4$ pathway evolved separately and at different periods within the grass family. C$_4$ photosynthesis is more effective than C$_3$ photosynthetic pathway in terms of water use, light capture and nitrogen utilization (Schmidt et al., 2011). Plants that utilize C$_4$ photosynthesis are able to cope in environments with high light intensity and high temperatures which limit growth in C$_3$ plants. Among C$_4$ plants, three subtypes (NAD-ME, NADP-ME and PCK) have been identified based on their anatomy and physiological pathways. Studies have shown that the C$_4$ photosynthesis subtypes have different ecological preferences (Hattersley, 1983; Vogel et al., 1986; Schulze et al., 1996; Taub, 2000) due to variances in their water-use efficiency (Ghannoum et al., 2002, Schmidt et al., 2011). Drought tolerance and water-use efficiency of grasses using different subtypes of the C$_4$-photosynthetic pathway are outlined in Figure 1.1. Increased water-use efficiency here translates to adaptation to a drier environment.

#### 1.3.1.2 Bunch-Forming Morphology

Caespitose or Tussock-forming plants are typically perennials that grow as singular plants in bunches, clumps, tufts or hummocks. Tussock grasses have shown great success in a wide range of habitats and this could be attributed to their extreme plasticity in response to disturbances (Fetcher

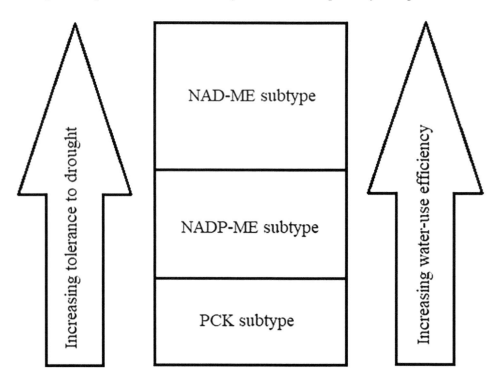

**FIGURE 1.1** Drought tolerance and water-use efficiency of subtypes of C$_4$ photosynthesis.

*Source*: Modified from Taub (2000)

& Shaver, 1982). During the dry season in tropical biomes, a bulk of the above-ground portions of perennial grasses dry up; however, some green leaf areas are left behind for photosynthesis and other activities to continue during this period of reduced growth. Tussocks tolerate drought using their tillers, such that their bunch basal growth protects the underground growing points (near the roots) from desiccation even when the aerial (above-ground) parts dry up. The tillers also help grasses to quickly recover and regenerate when the growing condition becomes favorable again. Tillering in tussocks in seasonally dry environments usually starts at the beginning of the rainy season to allow for a well-developed tussock before the dry spell. *Hyparrhenia* and *Andropogon* have shown great adaptation in savannas and other dry biomes and are typical examples of tussock grasses (Baruch, 2016; Haddad et al., 2021).

### 1.3.1.3 Drought Escape as Seeds

In seasonally dry biomes, especially in areas with short rainy seasons, e.g., Sahelian type savannas, most grasses are ephemerals—having a short life cycle—and quickly produce seeds during the short growing season so as to pass the unfavorable dry spell as seeds (Medina, 1982). Plants that adopt drought avoidance or escape strategy as seeds are called *Therophytes* (Sarmiento, 1992). Many savanna grasses accomplish their reproductive cycle during the rainy season. In these plants, new leaves—which serve for carbon assimilation and floral initiation—for seed production are produced in quick succession. In this way, the grass energy demands increase exponentially. These energy-demanding processes must be regulated at the end of the growing season and reduced to a semi-dormant state in readiness for the dry spell. This life history pattern has exposed some partitioning among grasses in dry tropical biomes, with *precocious species* blossoming even before the onset of the rainy season, *early and intermediate bloomers* blossoming early into the rainy season and *late bloomers* blossoming toward the ending of the rainy season or just at the commencement of the dry season, from which they germinate and persist in the vegetation. Seeds of the grasses are buried in the soil as a seed bank. Some grasses use hygroscopically active awns and pointed calluses to drive their seeds into the soil in order to protect them from excessive heat and fire. The germination of the seed bank is stimulated by early-season rainfall so that the grass can complete its life cycle during the wet season (Ramirez, 2002). The seeds grow inside bulky protective coverings/casings that help them to survive the dry spell until the first rainfall before germinating.

### 1.3.2 Adaptive Strategies in Woody Plants

Water is a key resource for plant growth and survival. Plants, especially in seasonally dry environments, adapt to changes in water availability. Most woody plants in tropical ecosystems, especially savannas and dry forests, exhibit specialized strategies for water capture, storage and management to cope with water deficit during the dry spell.

### 1.3.2.1 Strategies for Increased Water Capture

Trees and shrubs in seasonally dry biomes develop deep root systems to uptake water from profound soil depth where its availability is regular throughout the year. Trees in savannas and dry forests often rely on deep water sourcing to compensate for the water deficit in their tissues during the dry season as their weak stomatal control allows for increased transpiration during the period (Goldstein et al., 2008). Oliveira et al. (2005) reported that water stored in a deep soil layer (1–7.5 m depth) in *Cerrado denso*—woodland savanna vegetation in Brazil—accounted for about 82% of total water in the soil profile. This means that only about 18% of water is available in the upper soil layer (< 1 m depth). Savannas are among the most deeply rooted biomes (Jackson et al., n.d.). Deep rooting also allows for tree/grass coexistence since groundwater is partitioned by depth such that grasses and other herbaceous plant roots absorb water from the upper soil layer, while the roots of trees and shrubs dominate the deeper soil layers (Case et al., 2020). Apart from facilitating water uptake, deep rooting also performs other phenological functions such as pre-rain leaf flushing among some tree species (Zhou et al., 2020). In TDFs, the higher photosynthetic efficiency of lianas over trees species has been linked to their effectiveness in maintaining leaf water status through greater rooting depth (Werden et al., 2017).

During dry conditions, mycorrihizal fungi facilitate water uptake by plants, thus improving their resistance to water shortage. The association of ectomycorhizae with the roots of woody plants increases the absorptive area of roots by forming a dense sheath (mantle) around the roots from which a mass of hypha grows and reaches the soil for water and minerals. Endomycorrhizae (Arbuscular mycorrhizae) fungi also aid water capture by growing embedded within the root tissue, making the roots more effective for water sourcing in soil micropores (Boutasknit et al., 2020). Studies have shown that arbuscular mycorrhizal fungi (AMF) greatly improves drought resistance in some tree species in dry regions of Africa (Birhane et al., 2012; Bouten & Sevink, 2000).

### 1.3.2.2 Strategies for Water Storage

Trees in drought-prone environments have evolved different strategies to store water against dry spells. Of particular importance are stem-succulent tree species which are distributed in several plant taxa. Stem water storage (succulence) is common in tropical drylands. Most species with the feature, besides having improved water storage, also possess photosynthetic stems (with green bark) that allow for carbon assimilation with reduced transpiration (i.e. improved water-use efficiency), a process called *stem net photosynthesis* (SNP), or no water loss through re-assimilation of internally respired $CO_2$ (carbon dioxide) via a process called *stem recycling photosynthesis* (SRP). In most plant taxa, the stored water in the stem is channeled toward developing fresh leaves prior to the start of the rainy season. This confers an advantage to the plant such that they develop full canopy cover while other are still leafless. Typical examples of stem-succulent trees found in African dry forests and savannas include *Adansonia* (Figure 1.2),

**FIGURE 1.2** African baobab (*Adansonia digitata* L.) possess succulent stem for water storage.

*Source*: Courtesy of Bukola H. Oyedeji

*Bombax*, *Hildegardia*, *Moringa* and *Plumeria* species (Eleinis & Ezcurra, 2016). Some plants, such as *Adansonia digitata*, exhibit root succulence (swollen roots) together with swollen stems, which allow them to draw from saved water to endure the dry spell. Van den Bilcke et al. (2013) reported that root succulence in place of stem succulence accounted for drought tolerance of *A. digitata* as the tap root alone can store up to 68% of total plant water.

### 1.3.2.3 Water Management

Water management is one of the most important adaptations among plants in dry forests and savannas during extended droughts. Most woody plants in dry forests and savannas conserve water by shedding leaves annually, typically at the end of the rainy season as depicted in Figure 1.3. This mechanism, referred to as *deciduousness*, halts the processes of leaf photosynthesis and transpiration through which a large volume of water is lost from the plant. Deciduous species evade the cost of maintaining their leaves during the dry spell but have to bear the cost of producing new leaves annually (February & Higgins, 2016). Deciduousness is a physiological or phenotypic adaptation to survive the tough seasonal periods of water scarcity that is characteristics of tropical dry biomes (Arndt et al., 2015). Deciduousness is both high carbon- and nitrogen-demanding because these resources are deployed to build new leaves during seasonal flushing. The leaves of deciduous plants are nitrogen-rich to allow for the high rate of carbon sequestration such that the leaf flushing is completed within a short period. Deciduous plants often conserve nitrogen in their leaves by resorption (withdrawal) of the nutrient from the senescing leaves, prior to leaf fall, and its storage in woody parts of the plant (Ratnam et al., 2008). Species such as *Dichrostachys cinerea* and *Sclerocarya birrea* found in South African miombo woodland survive by deciduous strategy. *Acacia*

**FIGURE 1.3** Deciduous plants lose their leaves during the dry season and regrow them during the rainy season.
*Source*: Modified from Alberts et al. (2005)

species, especially *A. senegal* (gum Arabic) are deciduous and predominant in dry regions of Africa.

### 1.3.3 Other Plant Adaptive Traits in Dry Tropical Biomes

#### 1.3.3.1 Extensive Root Systems

Some plants can mitigate drought by developing an extensive root system as a crucial tool for increased access to water and nutrients from their surroundings (Shoaib et al., 2022). These plants develop long tap roots with increased lateral roots to scavenge water over a large area and increased depth. Studies have shown that plants growing in water-limited environments increase the flow of biomass toward their roots, thus increasing the root-to-shoot ratio. This allows for root growth and extension, thereby increasing the root access to water, as well as elevating the water harvest index of the roots (Polania et al., 2017).

#### 1.3.3.2 Leaf Folding or Rolling

Most plants, especially native grasses from semi-arid ecosystems, roll their leaves in response to water stress during dry periods. Rolling occurs due to the loss of turgor in the bulliform cells located on the upper leaf (adaxial) surface, leading to contraction of the subsurface sclerenchyma and mesophyll cells of the plant and contributing to inward folding of the leaves (involution) (Redmann, 1985). It is important to note that this adaptive mechanism is not limited to plants with bulliform cells, as other groups lacking the cells are also known to share this feature. Leaf rolling as an adaptive trait reduces transpiration and promotes heat dissipation by minimizing the specific leaf area of the plant (Yan et al., 2019).

#### 1.3.3.3 Other Leaf Modifications

Leaf modification is an important adaptation to dry conditions. In some species, a reduction in the size of the leaf blade or the complete absence of the leaf blade is another means of adapting to drought. In *Acacia senegal* and some species of *Euphorbia* such *E. hirta*, *E. prostrata* and *E. granulate*, the leaf blade is reduced. In these plants, other modifications such as the development of thick cuticles, dense hair coverings (trichophylly) and low stomata counts per unit of leaf surface or sunken stomata may also act to complement leaf size reduction (Kumar & Sen, 1985). In other plants, such as *Acacia auriculiformis* and *A. magnum*, the leaf blades are completely lost and the leaf petioles become flattened and widened in a typical leaf-like appearance called *phyllodes*. These phyllodes perform photosynthetic functions like leaves but with reduced potential for water loss by transpiration. Some savanna species such as *Acacia tortilis* also possess leaves with reflective surfaces (Munyati et al., 2018), while others tilt their leaves at an angle to evade direct surface heating of the leaves to minimize the damage of photosynthetic pigments.

## 1.4 ECOPHYSIOLOGICAL REGULATIONS OF GERMINATION, DORMANCY AND FLOWERING/FRUITING IN DRY TROPICAL BIOMES

Seed germination is the first and vital stage in the life cycle of plants. It begins with the imbibition of water by the inactive dry seed, which culminates in the protrusion of the radicle from the seed coat (Makhaye et al., 2021). The physiological process of germination is dependent on environmental variables including moisture, temperature, light, pH and nutrients (Humphries et al., 2018). Among these conditions, water is a basic requirement for germination as it is essential for the activation of enzymes that facilitate the breakdown, translocation and utilization of the reserve storage material in the seed. Seeds are naturally dormant until the critical moisture level sufficient to activate germination is attained in the seed. The critical seed moisture at which germination is initiated in seeds differs across species and even among species. For instance, 32% moisture level will initiate germination in rice, while 34% must be attained in maize for germination to be initiated (Kim & Jeon, 2009). When germination has been initiated, the process cannot be reversed even if soil moisture level and water potential drop. Under such conditions, the sprouted seeds are deprived of adequate moisture to meet the necessary physiological conditions, and they die. During dry conditions, the soil moisture content and water potential is low, and for most plant species, the critical water level for the seeds is hardly attained and the seeds remain dormant. Seeds of plants in dry tropical environments are adversely affected by drought that is associated with the dry season. Therefore, most plant seeds germinate and grow during the rainy season. Soil moisture limitation does not only reduce seed germination but also increases seedling mortality (Bhadouria et al., 2016; 2020). Some plant seeds lose viability under extreme drought conditions but persistent seeds remain dormant sometimes for over a year in the soil seed bank. Seed dormancy is therefore an adaptive approach by which seeds are preserved under seasonal and/or unpredictable climatic conditions that are unfavorable to growth or too short for seedlings to be established (Martins et al., 2019).

In most tropical plants, the initiation of floral buds typically comes after a check in the vegetative phase of growth which is commonly experienced in periods with low water supply (Chaikiattiyos et al., 1994). It has been reported in *Arabidiopsis thaliana* that drought stress induces flowering under long day-length conditions but delays the process under short day-length conditions (Takeno, 2016). Floral induction under conditions of water stresses has been reported for mandarin (*Citrus unshiu*) (Koshita & Takahara, 2004), star fruit (*Averrhoa carambola*) (Wu et al., 2017), lemon (*Citrus limon*), avocado (*Persea americana*) and mango (*Mangifera indica*) (Chaikiattiyos et al., 1994). Under drought stress, the stressed plants do not wait for the season when photoperiodic conditions are proper for blossoming, instead exhibit precocious blossoming which may serve to preserve the species in the environment.

Water stress has reducing effects on physiological processes such as nitrogen uptake, chlorophyll fluorescence, stomata conductance and ultimately on carbon assimilation. The decreases in the physiological processes have been linked with the upregulation of some floral induction genes in some plants such as the citrus flowering locus T (CiFT) gene in sweet orange (*Citrus sinensis*) (Thammatha et al., 2021).

In some tree species in savanna ecosystems such as baobab (*Adansonia digitata*) and shea butter tree (*Vitellaria paradoxa*), the flowering typically starts at the closing stages of the dry season or just before the first rain when new leaves start to emerge (Okullo et al., 2004). In tropical dry forests, many plants also concentrate their blossoming toward the start of the rainy season, during which they fruit (McLaren & McDonald, 2005). In this way, the fruit produced is well nourished, as photosynthesis and other physiological processes are optimum during the period of active growth.

## 1.5 CLIMATE CHANGE AND PLANT RESPONSES

Climate change refers to modifications in the earth's climate leading to changes in the average temperature and/or inconsistency in weather patterns over an extended period (Hahn, 2019). These changes are brought by the increase in greenhouse gases (GHGs)—including carbon dioxide, methane and nitrous oxide—that cause the average surface temperature to rise and lead to increased incidences of extreme events, including heavy precipitation, heat waves and prolonged drought (Seneviratne et al., 2012). The assessment of global warming by the Intergovernmental Panel on Climate Change (IPCC) has shown that the earth's average surface temperature has increased by 0.6°C (±0.2°C) in 1900 and is expected to increase even further to as much as 5.8°C in 2100. A rise in temperature by at least 2.5°C and a concomitant decrease in rainfall by a minimum of 50 mm on yearly basis have also been predicted to occur by 2055 (Hasnat, 2020).

Climate change is a critical issue, especially for dry tropical environments (Bhadouria et al., 2017). As the climate gets warmer and the environment becomes drier in dry tropics due to greenhouse effects and reduced annual precipitation, aridity becomes inevitable as a result of the increased rate of evapotranspiration; decreased soil fertility due to increased nutrient losses, especially carbon and nitrogen; increased rates of soil respiration and soil salinization; decreased nutrient cycling; and loss of soil structure. These problems affect vegetation dynamics as seed germination, seedling establishment, and net photosynthetic rates in plants are adversely affected (Bhadouria et al., 2018a; Tripathi et al., 2020).

The ability of plants to cope with these challenges associated with climate modification will involve the evolution of various complex resistance and adaptive mechanisms, which often differ from species to species. Many species in dry tropical environments possess adaptive mechanisms that improve their tolerance to water stress and increased temperature conditions. These features that allow for the reduction in transpiration loss while maximizing photosynthesis are achieved through morphological, physiological and biochemical modifications. These include an altered mechanism for carbon fixation (with a switch/change from $C_3$ to $C_4$ or CAM); change in stomata distribution, anatomy and conductance; modification in the structure of the leaf (rolling, folding and reduction to spines); change in mechanism for biomass allocation (root-to-shoot ratio); improved transpiration efficiency; increased osmotic and hormonal regulation; and delayed senescence.

Plants also respond to elevated $CO_2$ in the atmosphere through improved water-use efficiency. Studies (Prior et al., 2011; Wang et al., 2022) have shown elevated $CO_2$ to be beneficial to plants as it alleviates the adverse drought effects by improving plant water relation, net carbon assimilation, allocation to below-ground biomass and ultimately the growth of the plant. Elevated $CO_2$ also slows down the transpiration rate through the induction of partial closure in the stomata. These effects have been reported to facilitate the development of robust root systems in cotton, as well as increased root biomass in sorghum and soybean by up to 38% and 44%, respectively (Prior et al., 2003). Elevated $CO_2$ levels have also been shown to improve the carbon-to-nitrogen (C:N) ratio in plants due to reduced nitrogen under elevated temperatures. Leaf thickening and enlarged leaf area per unit plant as a result of increased $CO_2$ levels have also been reported (Prior et al., 2004).

Extreme heat and prolonged dry seasons associated with climate change may also increase the senescence of plant tissues, abscission, litter build-up and large-scale death among plants. Low rate of decomposition coupled with the high rate of litter deposition also increased litter accumulation and the risk of wildfires, especially in fire-exposed vegetations like savannas. Most plants in fire-prone ecosystems have developed anatomical modifications to suppress fire impacts. These include bark thickness to protect the internal tissues (Pausas, 2015), bud protection through deep embedment of buds in the stem, and underground storage organs to facilitate resprouting after fire events (Diaz-Toribio & Putz, 2021). Fire disturbances can lead to changes in vegetation structure and dynamics through loss of susceptible species and the dominance of fire-tolerant ones (Oyedeji et al., 2021).

## 1.6 CONCLUSION

Dry forests and savannas constitute widely distributed tropical ecosystems with a characteristic seasonal climate that exposes plants to alternating wet and dry conditions. An overview of this chapter has shown the diversity in structure and composition of flora, as well as the capacity of tropical plants to acclimatize to prolonged dry conditions associated with these biomes. With the increasing influence of climate change in these ecosystems and the associated impacts such as elevated temperatures, prolonged dry situation and increased $CO_2$ levels, plants in tropical dry biomes are exposed to intense stress conditions that dictate modifications in their adaptive traits including morpho-anatomical features, physiological and biochemical pathways that allow for improved water capture, utilization and storage. Plants are required to adapt to the prevailing environmental conditions of their surroundings—dry biomes included. To ensure continued survival of the plant in these environments, the adaptive strategies employed by a species must be passed on from one generation to the next. The knowledge of plants adaptation is therefore important in understanding and predicting the dynamics of vegetations.

## REFERENCES

Abreu, R. C. R., Hoffmann, W. A., Vasconcelos, H. L., Pilon, N. A., Rossatto, D. R., & Durigan, G. (2017). The biodiversity cost of carbon sequestration in tropical savanna. *Science Advances*, *3*, e1701284. https://doi.org/10.1126/sciadv.1701284

Alberts, S. C., Hollister-Smith, J. A., Mututua, R. S., Sayialel, S. N., Muruthi, P. M., Warurere, J. K., & Altmann, J. (2005). Seasonality and long-term change in a savanna environment. In D. K. Brockman & C. P. van Schaik (Eds.), *Seasonality in Primates: Studies of Living and Extinct Human and Non-Human Primates* (pp. 157–195). Cambridge University Press.

Alex, P. (2020). *Wet-dry Tropical Biomes*. www.blueplanetbiomes.org/savanna_climate.php

Arndt, S. K., Sanders, G. J., Bristow, M., Hutley, L. B., Beringer, J., & Livesley, S. J. (2015). Vulnerability of native savanna trees and exotic Khaya senegalensis to seasonal drought. *Tree Physiology*, *35*(7), 783–791. https://doi.org/10.1093/treephys/tpv037

Baruch, Z. (2016). Responses to drought and flooding in tropical forage grasses II: Leaf water potential, photosynthesis rate and alcohol dehydrogenase activity. *Plant and Soil*, *164*(1), 97–105.

Beerling, D. J., & Osborne, C. P. (2006). The origin of the savanna biome. *Global Change Biology*, *12*, 2023–2031.

Bhadouria, R., Singh, R., Srivastava, P., & Raghubanshi, A. S. (2016). Understanding the ecology of tree-seedling growth in dry tropical environment: A management perspective. *Energy, Ecology and Environment*, *1*(5), 296–309. https://doi.org/10.1007/s40974-016-0038-3

Bhadouria, R., Srivastava, P., Singh, R., Tripathi, S., Siingh, H., & Raghubanshi, A. S. (2017). Tree seedling establishment in dry tropics: An urgent need for interaction studies. *Environment, Systems and Decisions*, *37*(1), 88–100. https://doi.org/10.100710669-017-9625-x

Bhadouria, R., Srivastava, P., Singh, R., Tripathi, S., Verma, P., & Raghubanshi, A. S. (2020). Effects of grass competition on tree seedlings growth under different light and nutrient availability conditions in tropical dry forests in India. *Ecological Research*, *35*(5), 807–818.

Bhadouria, R., Srivastava, P., Singh, S., Singh, R., Raghubanshi, A. S., & Singh, J. S. (2018a). Effects of light, nutrient and grass competition on growth of seedlings of four tropical tree species. *Indian Forester*, *144*(1), 54–65.

Bhadouria, R., Tripathi, S. D., & Rao, K. S. (2018b). Understanding plant community assemblage, functional diversity and soil attributes of Indian savannas through a continuum approach. *Tropical Ecology*, *59*(4), 545–554.

Birhane, E., Sterck, F. J., Fetene, M., Bongers, F., & Kuyper, T. W. (2012). Arbuscular mycorrhizal fungi enhance photosynthesis, water use efficiency, and growth of frankincense seedlings under pulsed water availability conditions. *Oecologia*, *169*, 895–904. https://doi.org/10.1007/s00442-012-2258-3

Bock, W. J. (1980). The definition and recognition of biological adaptation. *American Zoology*, *20*, 217–227.

Bourliere, F., & Hadley, M. (1983). Present-day savannas: An overview. In F. Bourliere (Ed.), *Ecosystems of the World 13: Tropical Savannas* (pp. 1–17). Elsevier Scientific.

Boutasknit, A., Baslam, M., Ait-El-Mokhtar, M., Anli, M., Ben-Laouane, R., Douira, A., Modafar, C., El Mitsui, T., Wahbi, S., & Meddich, A. (2020). Arbuscular mycorrhizal fungi mediate drought tolerance and recovery in two contrasting carob (Ceratonia siliqua L.) ecotypes by regulating stomatal, water relations, and (in)organic adjustments. *Plants*, *9*(1), 80. https://doi.org/10.3390/plants9010080

Bouten, W., & Sevink, J. (2000). Gross rainfall and its partitioning into throughfall, stemflow and evaporation of intercepted water in four forest ecosystems in western Amazonia. *Journal of Hydrology*, *237*, 40–57.

Brown, R. H. (1978). Difference in N use efficiency in C3 and C4 plants and its implications in adaptation and evolution. *Crop Science, 18*, 93–98.

Case, M. F., Nippert, J. B., Holdo, R. M., & Staver, A. C. (2020). Root-niche seperation between savanna trees and grasses is greater on sandier soils. *Journal of Ecology, 108*, 2298–2308. https://doi.org/10.1111/1365-2745.13475

Center for International Forestry Research (CIFOR). (2014). Tropical dry forests: Under threat & under-researched. *Research Program on Forests, Trees and Agroforestry*, 1–4. www.cifor.org/publications/pdf_files/WPapers/DPBlackie 1401.pdf

Chaikiattiyos, S., Menzel, C. M., & Rasmussen, T. S. (1994). Floral induction in tropical fruit trees: Effects of temperature and water supply. *Journal of Horticultural Science, 69*, 3397–3415.

Christin, P. A., Besnard, G., Samaritani, E., Duvall, M. R., Hodkinson, T. R., Savolainen, V., & Salamin, N. (2008). Oligocene CO2 decline promoted C-4 photosynthesis in grasses. *Current Biology, 18*(1), 37–43.

Davies, A. B., Eggleton, P., Rensburg, B. J. Van, & Parr, C. L. (2015). Seasonal activity patterns of African savanna termites vary across a rainfall gradient. *Insectes Sociaux, 62*, 157–165. https://doi.org/10.1007/s00040-014-0386-y

de la Rosa-Manzano, E., Andrade, J. L., Zotz, G., & Reyes-García, C. (2014). Epiphytic orchids in tropical dry forests of Yucatan, Mexico—Species occurrence, abundance and correlations with host tree characteristics and environmental conditions. *Flora, 209*(2), 100–109. https://doi.org/10.1016/j.flora.2013.12.002

Dexter, K. G., Pennington, R. T., Oliveira-Filho, A. T., Bueno, M. L., Silva de Miranda, P. L., & Neves, D. M. (2018). Inserting tropical dry forests into the discussion on biome transitions in the tropics. *Frontiers in Ecology and Evolution, 6*. https://doi.org/10.3389/fevo.2018.00104

Diaz-Toribio, M. H., & Putz, F. E. (2021). Underground carbohydrate stores and storage organs in fire-maintained longleaf pine savannas in Florida, USA. *American Journal of Botany, 108*(3), 432–442. https://doi.org/10.1002/ajb2.1620

Durigan, G., & Ratter, J. A. (2016). The need for a consistent fire policy for Cerrado conservation. *Journal of Applied Ecology, 53*, 11–15. https://doi.org/10.1111/1365-2664.12559

Eleinis, Á., & Ezcurra, E. (2016). Stem-succulent trees from the old and new world tropics. In G. Goldstein & L. S. Santiago (Eds.), *Tropical Tree Ecology* (pp. 45–65). Springer International Publishing. https://doi.org/10.1007/978-3-319-27422-5

February, E. C., & Higgins, S. I. (2016). Rapid leaf deployment strategies in a deciduous savanna. *PLoS ONE, 11*(6). https://doi.org/10.1371/journal.pone.0157833

Fetcher, N., & Shaver, G. R. (1982). Growth and tillering patterns within tussocks of eriophorum vaginatum published by : Wiley on behalf of Nordic Society Oikos Stable (URL: www.jstor.org/stable/3682455 Growth and tillering patterns within tussocks of Eriophorum vaginatum). *Holarctic Ecology, 5*(2), 180–186.

Frost, P., Medina, E., Menaut, J., Solbrig, O., Swift, M., & Walker, B. (1986). Responses of savannas to stress and disturbance: Proposal for a collaborative programme of research. *Biology International, 10*.

Gentry, A. H. (1995). Diversity and floristic composition of neotropical dry forests. In S. H. Bullock, H. A. Mooney & E. Medina (Eds.), *Seasonally Dry Tropical Forests* (pp. 146–194). Cambridge University Press.

Ghannoum, O., von Caemmerer, S., & Conroy, J. P. (2002). The effect of drought on plant water use efficiency of nine NAD-ME and nine NADP-ME Australian C-4 grasses. *Functional Plant Biology, 29*(11), 1337–1348.

Gillespie, T. W., Grijalva, A., & Farris, C. N. (2000). Diversity, composition, and structure of tropical dry forests in Central America. *Plant Ecology, 147*(1), 37–47. https://doi.org/10.1023/A:1009848525399

Goldstein, G., Meinzer, F. C., Bucci, S. J., Scholz, F. G., Franco, A. C., & Hoffmann, W. A. (2008). Water economy of neotropical savanna trees: Six paradigms revisited. *Tree Physiology, 28*, 395–404.

Gottsberger, G., & Silberbauer-Gottsberger, I. (2008). Gottsberger 2008 tropical savannas introduction.pdf. In *Encyclopedia of Life Support Systems (EOLSS): Vol. X* (Tropical B, pp. 1–30). UNESCO.

Graz, F. P. (2004). *Structure and Diversity of the Dry Woodland Savanna of Northern Namibia*. Universitat Gottingen.

Haddad, C. R., Foord, S. H., & Whitehead, L. (2021). Tussock circumference, land use type and drought variably influence spider assemblages associated with Hyparrhenia hirta grass tussocks. *African Entomology, 29*(1), 150–164. https://doi.org/10.4001/003.029.0150

Hahn, C. (2019). *Seasonal Effects of Drought on the Productivity and Fodder Quality of Temperate Grassland Species*. University of Basel.

Hasnat, G. N. T. (2020). Climate change effects, adaptation, and mitigation techniques in tropical dry forests. In R. Bhadouria, T. Sachchidanand, P. Srivastava & P. Singh (Eds.), *Handbook of Research on the Conservation and Restoration of Tropical Dry Forests* (pp. 42–64). IGI Global. https://doi.org/10.4018/978-1-7998-0014-9.ch0003

Hasnat, G. N. T., & Hossain, M. K. (2020). Global overview of tropical dry forests. In R. Bhadouria, T. Sachchidanand, P. Srivastava & P. Singh (Eds.), *Handbook of Research on the Conservation and Restoration of Tropical Dry Forests* (pp. 1–23). IGI Global. https://doi.org/10.4018/978-1-7998-0014-9.ch001

Hattersley, P. W. (1983). The distribution of C3 and C4 grasses in Australia in relation to climate. *Oecologia, 57*, 113–128.

Holzman, B. A. (2008). Tropical forest and woodland biomes. In S. L. Woodward (Ed.), *Greenwood Guide to Biomes of the World*. Greenwood Press.

Humphries, T., Chauhan, B. S., & Florentine, S. K. (2018). Environmental factors effecting the germination and seedling emergence of two populations of an aggressive agricultural weed; Nassella trichotoma. *PLoS ONE, 5*, 1–25. https://doi.org/10.1371/journal.pone.0199491 J

Jackson, B. B., Canadell, J., Ehleringer, J. R., Mooney, H. A., Sala, O. F., & Schulze, F. D. (n.d.). A global analysis of root distributions for terrestrial biomes. *Oecologia, 108*, 389–411.

Kauffman, J. B., Cummings, D. L., & Ward, D. E. (1994). Relationships of fire, biomass and nutrient dynamics along a vegetation gradient in the Brazilian cerrado. *Journal of Ecology, 82*(3), 519–531.

Keay, R. W. J. (1949). An example of sudan zone vegetation in Nigeria. *Journal of Ecology, 37*(2), 335–364.

Kim, S. H., & Jeon, Y. S. (2009). Critical seed moisture content for germination in crop species. *The Journal of the Korean Society of International Agriculture, 25*(3), 159–164.

Koshita, Y., & Takahara, T. (2004). Effect of water stress on flower-bud formation and plant hormone content of satsuma mandarin (Citrus unshiu Marc.). *Scientia Horticulturae, 99*, 301–307.

Kumar, S., & Sen, D. N. (1985). Survival adaptations of three Euphorbia spp. in Arid ecosystem. *Folia Geobotanica and Phytotaxonomica*, *20*(1), 57–66.

Lehmann, C. E. R., Archibald, S. A., Hoffmann, W. A., & Bond, W. J. (2011). Deciphering the distribution of the savanna biome. *New Phytologist*, *191*(1), 197–209. https://doi.org/10.1111/j.1469-8137.2011.03689.x

Lulla, K. (1987). Savanna climate. In *Climatology. Encyclopedia of Earth Science*. Springer. https://doi.org/10.1007/0-387-30749-4_152

Luo, L., & Zhang, W. (2014). A review on biological adaptation : With applications in engineering science. *Selforganizology*, *1*(1), 23–30.

Maass, M., Ahedo-Hernández, R., Araiza, S., Verduzco, A., Martínez-Yrízar, A., Jaramillo, V. J., Pascual, F., García-Méndez, G., & Sarukhán, J. (2017). Forest ecology and management long-term (33 years) rainfall and runoff dynamics in a tropical dry forest ecosystem in Western Mexico: Management implications under extreme hydrometeorological events ☆. *Forest Ecology and Management*, 1–11. https://doi.org/10.1016/j.foreco.2017.09.040

Makhaye, G., Mofokeng, M. M., Tesfay, S., Aremu, A. O., Van Staden, J., & Amoo, S. O. (2021). Influence of plant biostimulant application on seed germination. In *Biostimulants for Crops from Seed Germination to Plant Development: A Practical Approach* (pp. 109–135). Academic Press. https://doi.org/10.1016/B978-0-12-823048-0.00014-9

Martins, A. A., Opedal, O. H., Armbruster, W. S., & Pelabon, C. (2019). Rainfall seasonality predicts the germination behavior of a tropical dry-forest vine. *Ecology and Evolution*, *9*, 5196–5205. https://doi.org/10.1002/ece3.5108

McLaren, K. P., & McDonald, M. A. (2005). Seasonal patterns of flowering and fruiting in a dry tropical forest in Jamaica. *Biotropica*, *37*(4), 584–590. https://doi.org/10.1111/j.1744-7429.2005.00075.x

Medina, E. (1982). Physiological ecology of neotropical savanna plants. In B. J. Huntley (Ed.), *Ecology of Tropical Savannas* (pp. 309–335). Springer-Verlag.

Mooney, H. A., Bullock, S. H., & Medina, E. (1995). Introduction. In S. H. Bullock, H. A. Mooney & E. Medina (Eds.), *Seasonally Dry Tropical Forests* (pp. 1–8). Cambridge University Press.

Mucina, L. (2019). Biome: Evolution of a crucial ecological and biogeographical concept. *New Phytologist*, *222*(1), 97–114. https://doi.org/10.1111/nph.15609

Munyati, C., Malomane, L. E., & Malahlele, O. E. (2018). Evaluating the effects of leaf characteristics on spectral signatures of savannah woody species on remotely sensed imagery. *South African Journal of Geomatics*, *7*(3), 319–330. https://doi.org/http://dx.doi.org/10.4314/sajg.v7i3.9

Murphy, P. G., & Lugo, A. E. (2008). Ecology of tropical dry forest. *Annual Review of Ecology and Systematics*, *17*(1986), 67–88.

Okullo, J. B. L., Hall, J. B., & Obua, J. (2004). Leafing, flowering and fruiting of Vitellaria paradoxa subsp. Nilotica in savanna parklands in Uganda. *Agroforestry Systems*, *60*, 71–91.

Oliveira, R. S., Bezerra, L., Davidson, E. A., Pinto, F., Klink, C. A., Nepstad, D. C., & Moreira, A. (2005). Deep root function in soil water dynamics in cerrado savannas of central Brazil. *Functional Ecology*, *19*, 574–581. https://doi.org/10.1111/j.1365-2435.2005.01003.x

Osborne, C. P., & Sack, L. (2012). Evolution of C 4 plants : A new hypothesis for an interaction of $CO_2$ and water relations mediated by plant hydraulics. *Philosophical Transactions of the Royal Society*, *367*, 583–600. https://doi.org/10.1098/rstb.2011.0261

Oyedeji, S., Agboola, O. O., Oriolowo, T. S., Animasaun, D. A., Fatoba, P. O., & Isichei, A. O. (2021). Early-season effects of wildfire on soil nutrients and weed diversity in two plantations. *Scientia Agriculturae Bohemica*, *52*(1), 1–10. https://doi.org/10.2478/sab-2021-0001

Oyedeji, S., Onuche, F. J., Animasaun, D. A., Ogunkunle, C. O., Agboola, O. O., & Isichei, A. O. (2016). Short-term effects of early-season fire on herbaceous composition, dry matter production and soil fertility in Guinea savanna, Nigeria. *Archives of Biological Sciences*, *68*(1). https://doi.org/10.2298/ABS150526002O

Pausas, J. G. (2015). Bark thickness and fire regime. *Functional Ecology*, *29*, 315–327. https://doi.org/10.1111/1365-2435.12372

Pennington, R. T., Lehmann, C. E. R., & Rowland, L. M. (2018). Tropical savannas and dry forests. *Current Biology*, *28*(9), R541–R545. https://doi.org/10.1016/j.cub.2018.03.014

Pennington, R. T., Prado, D. E., & Pendry, C. A. (2000). Neotropical seasonally dry forests and Quaternary vegetation changes. *Journal of Biogeography*, *27*, 261–273. https://doi.org/10.1046/j.1365-2699.2000.00397.x

Polania, J., Rao, I. M., Cajiao, C., Grajales, M., Rivera, M., Velasquez, F., Raatz, B., & Beebe, S. E. (2017). Shoot and root traits contribute to drought resistance in recombinant inbred lines of MD 23–24 × SEA 5 of common bean. *Frontiers in Plant Science*. https://doi.org/10.3389/fpls.2017.00296

Prior, S. A., Pritchard, S. G., & Runion, G. B. (2004). Leaves and the effects of elevated carbon dioxide levels. In R. M. Goodman (Ed.), *Encyclopedia of Plant and Crop Science* (pp. 648–650). M. Dekker. https://doi.org/10.1081/E-EPCS 12001061

Prior, S. A., Runion, G. B., Marble, S. C., Rogers, H. H., Gilliam, C. H., & Torbert, H. A. (2011). A review of elevated atmospheric $CO_2$ effects on plant growth and water relations: Implications for horticulture. *Hortscience*, *46*(2), 158–162.

Prior, S. A., Torbert, H. A., Runion, G. B., & Rogers, H. H. (2003). Implications of elevated $CO_2$ induced changes in agroecosystem productivity. *Journal of Crop Productivity*, *8*, 217–244.

Ramirez, N. (2002). Reproductive phenology, life-forms, and habitats of the Venezuelan central plain. *American Journal of Botany*, *89*(5), 836–842.

Ratnam, J., Sankaran, M., Hanan, N. P., Grant, R. C., & Zambatis, N. (2008). Nutrient resorption patterns of plant functional groups in a Tropical savanna: Variation and functional significance. *Oecologia*, *157*(1), 141–151.

Ratnam, J., Tomlinson, K. W., Rasquinha, D. N., & Sankaran, M. (2016). Savannahs of Asia : Antiquity, biogeography, and an uncertain future. *Philosophical Transactions of the Royal Society B*, *371*, 20150305. https://doi.org/http://dx.doi.org/10.1098/rstb.2015.0305

Redmann, R. E. (1985). Adaptation of grasses to water stress-leaf rolling and stomate distribution. *Annals of the Missouri Botanical Garden*, *72*(4), 833–842. www.jstor.org/stable/2399225

Rees, R. M., Wuta, M., Furley, P. A., & Li, C. (2006). Nitrous oxide fluxes from savanna (miombo) woodlands in Zimbabwe. *Journal of Biogeography*, *3*(3), 424–437.

Sankaran, M., & Ratnam, J. (2013). African and Asian savannas. *Encyclopedia of Biodiversity*, *1*, 58–74. https://doi.org/10.1016/B978-0-12-384719-5.00355-5

Sarmiento, G. (1992). Adaptive strategies of perennial grasses in South American savannas. *Journal of Vegetation Science*, *3*(3), 325–336.

Sarmiento, G. (2014). *The Ecology of Neotropical Savannas*. Harvard University Press.

Schmidt, M., König, K., Müller, J. V, Brunken, U., & Zizka, G. (2011). Modelling the distribution of photosynthetic types of grasses in Sahelian Burkina Faso with high-resolution satellite data. *Ecotropica*, *17*, 53–63.

Scholes, R. J., & Hall, D. O. (1996). The carbon budget of tropical savannas, woodlands and grasslands. In A. I. Breymeyer, D. O. Hall, J. M. Melillo & G. I. Agren (Eds.), *Global Change: Effects on Coniferous Forests and Grasslands* (pp. 69–100). Wiley.

Schulze, E. D., Ellis, R., Schulze, W., & Trimborn, P. (1996). Diversity, metabolic types and delta C-13 carbon isotope ratios in the grass flora of Namibia in relation to growth form, precipitation and habitat conditions. *Oecologia*, *106*(3), 352–369.

Seneviratne, S. I., Nicholls, N., Easterling, D., Goodess, C. M., Kanae, S., Kossin, J., Luo, Y., Marengo, J., McInnes, K., Rahimi, M., Reichstein, M., Sorteberg, A., Vera, C., & Zhang, X. (2012). Changes in climate extremes and their impacts on the natural physical environment. In C. B. Field, V. Barros, T. F. Stocker, D. Qin, D. J. Dokken, K. L. Ebi, M. D. Mastrandrea, K. J. Mach, G.-K. Plattner, S. K. Allen, M. Tignor & P. M. Midgley (Eds.), *Managing the Risks of Extreme Events and Disasters to Advance Climate Change Adaptation* (Climate, A, pp. 109–230). Cambridge University Press.

Shoaib, M., Banerjee, B. P., Hayden, M., & Kant, S. (2022). Roots' drought adaptive traits in crop improvement. *Plants*, *11*(17), 2256. https://doi.org/10.3390/plants11172256

Siyum, Z. G. (2020). Tropical dry forest dynamics in the context of climate change: Syntheses of drivers, gaps, and management perspectives. *Ecological Processes*, *9*, 25. https://doi.org/10.1186/s13717-020-00229-6

Solbrig, O. T., Medina, E., & Silva, J. F. (1996). Determinants of tropical savannas. In O. T. Solbrig, E. Medina & J. F. Silva (Eds.), *Biodiversity and Savanna Ecosystem Processes. Ecological Studies* (Vol. 121). Springer. https://doi.org/10.1007/978-3-642-78969-4_2

Spaargaren, O., & Deckers, J. (2005). Factors of soil formation: Climate. In *Encyclopedia of Soils in the Environment* (pp. 512–520). Academic Press. https://doi.org/10.1016/B0-12-348530-4/00012-6

Stan, K., & Sanchez-Azofeifa, A. (2019). Tropical dry forest diversity, climatic response, and resilience in a changing climate. *Forests*, *10*(5). https://doi.org/10.3390/f10050443

Sternberg, L. D. S. L. (2001). Savanna-forest hysteresis in the Tropics. *Global Ecology and Biogeography*, *10*(4), 369–378.

Takeno, K. (2016). Stress-induced flowering: The third category of flowering response. *Journal of Experimental Botany*, *67*(17), 4925–4934. https://doi.org/10.1093/jxb/erw272

Taub, D. R. (2000). Climate and the US distribution of C-4 grass subfamilies and decarboxylation variants of C-4 photosynthesis. *American Journal of Botany*, *87*(8), 1211–1215.

Thammatha, P., Lapjit, C., Tarinta, T., Techawongstien, S., & Techawongstien, S. (2021). The responses of physiological characteristics and flowering related gene to the different water stress levels of red-flesh pummelo cultivars (Citrus grandis (L.) Osbeck) own-rooted by air layering propagation under two growing conditions. *Horticulturae*, *7*(579), 1–12. https://doi.org/10.3390/horticulturae7120579

Tripathi, S., Bhadouria, R., Srivastava, P., Devi, R. S., Chaturvedi, R., & Raghubanshi, A. S. (2020). Effects of light availability on leaf attributes and seedling growth of four tree species in tropical dry forest. *Ecological Processes*, *9*(1), 1–16.

Van Bloem, S. J., Lugo, A. E., & Murphy, P. G. (2004). Regional forest types: Tropical dry forests. In *EFORS* (Vol. 00176). Forestry and Environmental Conservation, Clemson University Tigerprints. https://doi.org/10.1890/0012-9658(1997)078[0323:TDF]2.0.CO;2

Van den Bilcke, N., De Smedt, S., Simbo, D. J., & Samson, R. (2013). Sap flow and water use in African baobab (Adansonia digitata L.) seedlings in response to drought stress. *South African Journal of Botany*, *88*(2013), 438–446. https://doi.org/10.1016/j.sajb.2013.09.006

Vergara-Torres, C., Pacheco-Álvarez, M. C., & Flores-Palacios, A. (2010). Host preference and host limitation of vascular epiphytes in a tropical dry forest of central Mexico. *Journal of Tropical Ecology*, *26*, 563–570.

Vogel, J. C., Fuls, A., & Danin, A. (1986). Geographical and environmental distribution of C3 und C4 grasses in the Sinai, Negev, and Judean deserts. *Oecologia*, *70*, 258–265.

Wang, Z., Wang, C., & Liu, S. (2022). Elevated $CO_2$ alleviates adverse effects of drought on plant water relations and photosynthesis: A global meta-analysis. *Journal of Ecology*. https://doi.org/10.1111/1365-27445.13988

Werden, L. K., Waring, B. G., Smith-Martin, C. M., & Powers, J. S. (2017). Tropical dry forest trees and lianas differ in leaf economic spectrum traits but have overlapping water-use strategies. *Tree*, *38*, 517–530. https://doi.org/10.1093/treephys/tpx135

Whittaker, R. H. (1975). *Communities and Ecosystems* (2nd Edition). Macmillan.

Woodward, F. I., Lomas, M. R., & Kelly, C. K. (2004). Global climate and the distribution of plant biomes. *Philosophical Transactions of the Royal Society B: Biological Sciences*, *359*(1450), 1465–1476. https://doi.org/10.1098/rstb.2004.1525

Woodward, S. L., & Foy, A. (2022). Seasonally dry tropical forests. *Biomes of the World*. Department of Geoscience, Radford University. https://php.radford.edu/~swoodwar/biomes/?page_id=102

Wu, P., Wu, C., & Zhou, B. (2017). Drought stress induces flowering and enhances carbohydrate accumulation in Averrhoa carambola. *Horticultural Plant Journal*, *3*(2), 60–66.

Yan, Y., Liu, Q., Zhang, Q., Ding, Y., & Li, Y. (2019). Adaptation of dominant species to drought in the inner Mongolia grassland—Species level and functional type level analysis. *Frontiers in Plant Science*, *10*, 1–10. https://doi.org/10.3389/fpls.2019.00231

Zhou, Y., Wigley, B. J., Case, M. F., Coetsee, C., & Staver, A. C. (2020). Rooting depth as a key woody functional traits in savannas. *New Phytologist*, *227*, 1350–1361. https://doi.org/10.1111/nph.16613

# 2 Evolutionary Responses of Tropical Plants to Changing Climate

*Zirwa Sarwar, Maria Hasnain, Maria Hanif, Huma Waqif, and Neelma Munir*

## 2.1 CLIMATE CHANGE

Climate change is pushing the world's tropical plants to their limits (Sentinella et al., 2020). Because plants are unable to move, they must adapt to abiotic stresses such as drought, salt stress, and extreme temperatures, all of which have a direct influence on plant growth, fertility, and production. This chapter focuses on recent developments trophic interaction, thermal interaction, plant–herbivore interaction, co-evolutionary responses, and host–pathogen interactions impacts of climate change in tropical areas (Fahad et al., 2017; Tripathi et al., 2020).

Climate change has an impact on species ecology and evolution, as well as biological groupings and ecosystems. From the poles to the equator, climate change has been detected, although studies and research evaluations have concentrated on the biotic consequences in mid- and high-latitude locations where temperatures are predicted to rise the most (Feeley et al., 2017; Tripathi et al., 2020). Different plant species show different responses to climatic changes. Furthermore, species will encounter a variety of abiotic changes in the coming years, not only temperature increases (Deutsch et al., 2008; Bhadouria et al., 2016).

Tropical species may be more sensitive to some climatic changes, such as global warming, because they emerged in areas with less temporal temperature fluctuation (Janzen, 1967; Sheldon et al., 2018). Even minor temperature increases at low latitudes could have a substantial influence on tropical species' fitness (Bhadouria et al., 2018).

Tropical species may adapt to climate change differently than temperate and polar species due to differences in physiology and life history (Weiskopf et al., 2020). Visible indications of climate change, or fingerprints, have been identified at higher latitudes. Significant, directional changes in $CO_2$ (carbon dioxide) and temperature, shifts in moisture regimes, and rises in the frequency and severity of weather conditions are all evidence of climate change in the tropics.

## 2.2 PLANT RESPONSES TO THE ENVIRONMENT

Tropical forest environments are rapidly altering due to spurt in growth of human populations and the indiscriminate developmental activities for the growth of economies. Tropical forests also have an outsized role in the global carbon and energy cycles, and they are home to half of all known species, as well as unknown species (Taveira & Moreno, 2003; Wan et al., 2020). Recognizing global climate change and environmental protection requires evaluation of anthropogenic human change in tropical forests (Bhadouria et al., 2017; Martinet et al., 2021). There are two basic kinds of anthropogenic influences on tropical forests. The first is deforestation and fragmentation, and the second is over-exploitation of natural resources. Local implications include changes in land cover, invasive species, and the harvest of wood and bush meat. Climate and temperature changes induced mostly by the use of fossil fuels, as well as remote land cover change, are examples of global effects. The issue is evaluating the total impact of this confluence of anthropogenic causes, especially because anthropogenic forcing is expected to increase. Tropical nations' populations increased from 1.81 billion in 1950 to 4.90 billion in 2000, with another two billion anticipated by 2030, and both tropical and global economies are likely to grow much quicker by focusing on the tropical plants with high economic value (Wang et al., 2020); hence, biotic reactions to global climate change are clear at high latitudes, while they are sparse and contentious at low latitudes. Many temperate and boreal species are reproducing sooner, having longer growing seasons, moving to higher latitudes or elevations, or increasing their ranges. Similar shifts have occurred on tropical slopes, where persistent cloud formation has increased in response to global warming (Wright et al., 2006; Bhadouria et al., 2017).

Most plants require a similar combination of resources to ensure maximum development: energy, water, and mineral nutrients. An interdisciplinary approach is required to understand plant responses to resource imbalances (Bhadouria et al., 2018). The fields of community or ecosystem science, ecology, micrometeorology, and soil science are all concerned with resource availability. Plant responses have metabolic and structural basis and are investigated in physiology, biochemistry, and functional anatomy. In order to focus on the connection between organisms and the environment, plant physiological ecology relies extensively on all of these topics. The majority of past research has concentrated on plant reactions and responses to certain environmental variables, despite the fact that plants in nature usually confront many stressors (Tripathi et al., 2018; Heyduk, 2022). It is improbable that the impacts of interacting environmental factors will be cumulative. High-light or photo-inhibition damage, for example, is greatly exacerbated when combined with water deficiency/stress or extremely lower or higher temperatures (Pearce-Higgins et al., 2022).

Because it integrates approaches for monitoring the environment, measuring multiple components of the response of plants, and integrating plant responses to the degree of ecological success or agricultural productivity, physiological ecology is particularly suited to the study of plant responses to many elements. We are interested in the interplay between carbon (as a measure of the energy stored in molecules) and nitrogen resources at various levels of an organization, from the cell to the ecosystem. These resources are highlighted because they typically limit plant development and exhibit the concepts necessary for expanding the technique (Chapin et al., 1987; Tripathi et al., 2018).

## 2.3 TROPHIC INTERACTION

The effect of climate change helps to modify the three-dimensional vegetation structure at higher elevations and favors the coexistence of plant species. The reorganized trophic interactions may play an important role in the future climate challenges in driving changes in the plant community (Guo et al., 2022). The ecological equilibrium could be disrupted under climate change, supplementing a direct effect of temperature and biotic interactions which represent a neglected but major driver of modifications of the ecosystem (Lukeneder & Lukeneder, 2022).

The response of individual tropical species and the entire population can be predicted by the most commonly employed methods, i.e., by estimating their continuous and qualitative traits (Pakeman & Quested, 2007). These methods incorporate all fundamental ecological and evolutionary processes that may acts as a major contributing factor to response of the species to climate change (Pinsky et al., 2022). These responses include the evolutionary adaptation of the species when interacting with climate change. While the process of evolutionary adaptation is not rapid for some tropical taxa which then exhibits the risk of extinction, tropical interactions play an essential role in understanding the responses to a temperature shift, especially the evolutionary responses (O'Brien et al., 2022; Tripathi et al., 2020).

The additive effect of herbivores on a producer resource system exhibits the impacts of change in the climate by directly altering the environment and solidity of the system. The herbivore addition have the eco-evolutionary impressions by following the direction of the collection of tropical plants, as well as their adaptation rate, which is facilitated through the ecological aspects (Gaitán-Espitia & Hobday, 2021). The niches of the related species establish their comebacks to environmental changes. Large niche leads to maintain a large tropical population and extends the persistence time for the species in the fluctuating ecosystem (Hamann et al., 2021).

The availability of intermediate niche indicates the extended persistence of particular tropical varieties. The trophic interaction of species shows greater extinction rates in the case of the intermediate niche. The qualitative interaction of the tropical species and the herbivore presents incorporating evolution (Rauschkolb et al., 2022). The distribution and occurrence of rain events also alters the size and perseverance of the tropical plants. Such alternations may affect the trophic interactions of local communities in terms of composition. Many researchers have used the bromeliads as a tropical aquatic ecosystem in order to predict rainfall changes in tropical communities. The structural modeling investigates the combined effects of rainfall changes and litter diversity on tropical interactions (Angulo et al., 2022).

The changes in the rain events disrupted the tropical relationships despite minor species abundance, richness, and composition of the community. This change also alters the way in which species interact with each other and decreases their linkages among the tropic groups (Abrego et al., 2021). Such reduction of the biotic interactions under climate change reveals the critical outcomes for the proper functioning of tropical ecosystems (Shinzato et al., 2021).

## 2.4 THERMAL INTERACTION

The exchange of energy through the process of heat in a thermodynamic system is referred as thermal interaction (Kumari et al., 2021). The combined effect of urbanization and the rise of global temperature reveals an efficient temperature change. The urban vegetation influences the climate conditions and the outdoor thermal relief (Meili et al., 2021). The climatic variables such as relative humidity, precipitation, and the temperature of the air plays an important role in varying open-air thermal relief. These interactions improve plant resilience to the environmental stresses (Trivedi et al., 2022).

Thermal contact with tropical vegetation can exchange energy in heat transfer and thermodynamics, and it minimizes the energy destruction. These thermal interactions are also known as abiotic stress for plants (Yadav et al., 2022). While the tropical vegetation decreases the air temperature, increases the environmental humidity, and reduces the mean radiant temperature, in tropical cities, the perfect thermal isolation is always in thermal contact with environment to some extent. The tropical vegetation structure, interception, and physiological traits are motivated by thermal interaction (Leger, 2021).

The traits of the plants are characterized by their morphological, biochemical, physiological, and phenological features, and they are inclined by the selection of the tropical vegetation species (Tarvainen et al., 2022). In the land surface model, the analysis of systematic sensitivity can identify the essential standard factors. The supply of food is highly challenged by rising temperatures, as tropical crops are severely impacted by heat stress. The global and local temperature rise can drastically affect crop production (Mutamiswa et al., 2022).

The prevalence of high thermal interaction is the cause of stress in tropical vegetation and an imbalance in the production of plant yield. The comparative effect of high temperature and dearth stresses reduces the productivity of crop and affects food yield (Alés et al., 2022). An increase in the severity of abiotic factors like climate change affect the production potential of crops. Even a single-degree change in the temperature threshold can cause many vital changes in the growth and development of tropical plants. These vital changes trigger the physiological, morphological, and biochemical changes in the growth and development of tropical plants (Xiong & Yao, 2021). Table 2.1 illustrates the yield losses caused by drought and heat stresses in some major tropical legumes.

Although thermal stresses can have negative impact from the germination stage to the reproductive stage, they are regarded as quite severe due to the incidence of root diseases at the germination stage (Kumar & Sharma, 2022). The genetic potential of tropical plants allows them to acclimate to adverse natural environments, but this genetic potential modifies according to the variety and management system they are exposed to (Sivageerthi et al., 2022). A slight variation from the normal condition can cause deviation in the genetic pattern of the germplasm and disturbs global plant yield. Figure 2.1 shows the molecular response of tropical plants towards thermal stress and the physiological changes in the plant which occur due to thermal interaction.

## 2.5 PLANT–HERBIVORE INTERACTION

Plant–herbivore interaction contours the ecosystem as they consist of nearly 50% of the living organisms on this earth. Herbivores consume the land plants then plants evolve different defenses mechanisms against them and herbivores also developed traits to help evade these defenses (Wirth et al., 2008). Evolutionary diminuendos generate interactions leading to species diversity and functional qualities. The following Figure 2.2 illustrates the plant–herbivore interaction in the abiotic environment. The abiotic atmosphere intensely stimulates the trait expression of plants and herbivores and modifies this environmentally significant interaction (Hay & Fenical, 1988).

**TABLE 2.1**
**Yield Losses Caused by Drought and Heat Stresses in Some Major Tropical Legumes**

| Legumes | Stress | Yield losses (%) | References |
| --- | --- | --- | --- |
| Chickpea (*Cicer arietinum*) | Heat and drought | 40–45 | Rani et al. (2020) |
| Lentil (*Lens culinaris*) | Heat and drought | 57 | Choukri et al. (2020) |
| Soyabean (*Glycine max*) | Heat | 72–82 | Wei et al. (2018) |
| Cowpea (*Vigna unguiculata*) | Heat | 29 | Kyei-Boahen et al. (2017) |
| Field pea (*Pisum sativum*) | Heat | 20–54 | Kumar et al. (2016) |

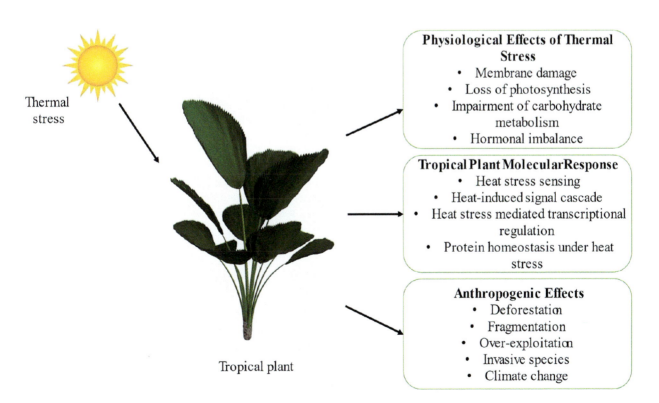

**FIGURE 2.1** Molecular response of a tropical plant due to the physiological effects of thermal stress.

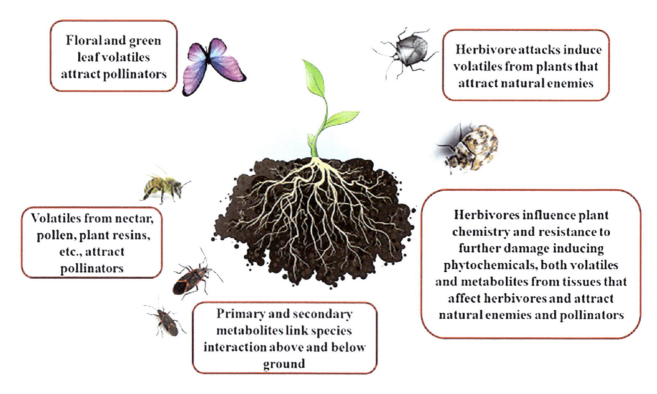

**FIGURE 2.2** Plant–herbivore interaction.

### 2.5.1 RISING ATMOSPHERIC $CO_2$ AND PLANT–HERBIVORE INTERACTION

Atmospheric $CO_2$ increased from 280 ppm to 410 ppm in the duration of 80–100 years and could be above 600 ppm by the 21st century (Roberts et al., 2022). Higher $CO_2$ has less direct effects on insects than herbivores but it can affect herbivores indirectly by changing plant chemistry, enhancing the carbon-to-nitrogen (C:N) ratio, shrinkage of leaves and reduction of nitrogen concentration as nitrogen is a limiting nutrient for insects and reducing proteins content in leaves, so that nutritive quality of plants is reduced (Johnson et al., 2020). To recompense, herbivore enhance their food consumption. In a study, it was reported that C:N ratio in flowers was increased by up to 19% while protein was reduced to 10%; insects increased their ingestion rate up to 14% with a 15% reduction in conversion efficacy (Hall et al., 2020). Plants have the ability to store more biomass and improve growth to compensate for the loss of leaves (Figure 2.2 and Figure 2.3), though numerous climate parameters are fluctuating concurrently and improving the growth of the herbivore population. An enlarged herbivory executes a sturdy assortment on plant populations for constitutive and inducible defenses. Moreover, imminent research can appraise more evolutionary responses via assessing genetic variation in defenses in plants by assessing the rate of gene flow and resurgence methods to examine unswervingly for adaptive responses against climate variation and enlarged herbivory (Johnson et al., 2022).

## 2.6 TEMPERATURE

Temperature is one of the most important abiotic factors affecting plant–herbivore interactions (Ho & Pennings, 2013). Since 1880, the temperature has increased by 1°C globally, which affects the physiology and metabolism of herbivores directly (Dusenge et al., 2019). Faster herbivore metabolic rates under raised temperatures lead to greater ingestion and growth, which enhance their population, overwinter persistence, thermal plasticity, and insect epidemics. High temperatures have a progressive impact on insects, as their performance is enhanced with increasing temperature up until getting a maximum (thermal tolerant insects) and then quickly declining (less thermal tolerance) (Jactel et al., 2019). Climate change affects herbivore life cycle as insects reproduce faster in hotter temperatures. These fluctuations increase plant damage but still do not prompt voltinism in all species. As demonstrated in Figure 2.2, the modest increase in temperature-enhanced plant producibility and productivity of secondary metabolites by increasing adenosine monophosphate in the biomass, which increased root length, hair, and biomass to improve the uptake of phosphorus and micronutrients (Guyer et al., 2021).

## 2.7 DROUGHT

After climate change, drizzling increased in many regions of the world, but still remaining regions are suffering from drought. High temperatures and heat waves affect the

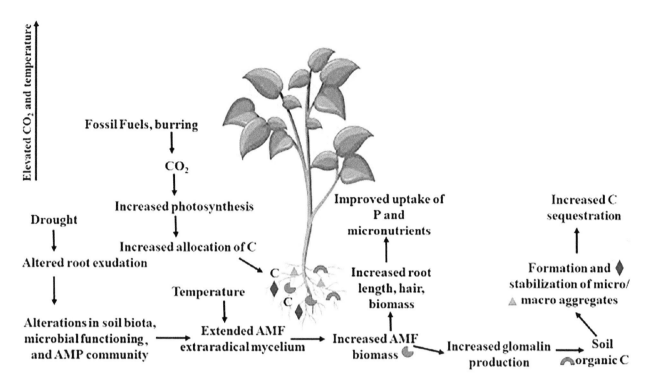

**FIGURE 2.3** Plant adaptation in elevated $CO_2$ and temperature.

**TABLE 2.2**
**Plant–Herbivore Interaction in Drought**

| Interacting Species | Interaction at Low Moisture | Interaction at High Moisture | Effect | References |
|---|---|---|---|---|
| *A. karroo*—goats | Herbivory | Mutualism | Seedling growth | Scogings & Gowda (2019) |
| *S. cacti*—wood rats | | | | McCluney et al. (2012) |
| *S. cacti*—moths | Mutualism | | Production of fruit and seed | McCluney et al. (2012) |
| Vegetation—sheep | Commensal | | Vegetation biomass | |
| Cottonwood leaves—consumer | Herbivory | Neutral | Net consumption | Bronstein et al. (2007) |
| Cottonwood leaves—crickets | | | | |
| Cottonwood saplings—voles | | | Sapling persistence | Zhong et al. (2021) |
| Seeds—rodents | | Herbivory | Seeds caches taken | Marazzi et al. (2015) |
| Seeds—rodents | | | Giving-up density | Hamann et al. (2021) |
| Nubian ibex—alfalfa pellets | | | Giving-up density | Ulappa et al. (2020) |

ecosystems. Drought affects plants seriously by changing their nutritive value and digestibility, as well as their defense systems (Peng et al., 2022). Insects perform well under drought conditions and cause outbursts of forest defoliation and crop damage due to nitrogen-containing compounds in plants which increased their growth (Karthik et al., 2021). Moreover, drought badly affects phloem and sap feeders. Modest drought increases plant defense mechanisms, while severe drought reduces plant defense; in both cases, limitation of nutritional compounds occur which reduce herbivore enactment (Hamann et al., 2021). Plant–herbivore interactions in drought are listed in Table 2.2.

## 2.8 FERTILIZATION AND NITROGEN DEPOSITION

Fertilization is very important aspect in agriculture to increase nitrogen deposition, which triggers growth of plants as well their nutritional value. The triggering effect of nitrogen deposition is less in magnitude than the adversative effect of high $CO_2$ such as climate change which still spoils herbivore growth. Moreover, nitrogen deposition permits plants to allot additional resources to herbivore defenses and impair herbivory due to the stresses. Hence, it is vital to fertilize with the amalgamation. By combining different types of fertilizers, such as nitrogen-based, phosphorus-based, and potassium-based fertilizers, a balanced nutrient supply can be provided to meet the specific needs of the plants (Anderson et al., 2020).

## 2.9 CO-EVOLUTIONARY RESPONSES

Plants show intra-specific variations among the different species in their morphological and chemical responses (Ullah et al., 2021). Co-evolution result in significant contributions to biodiversity (Wan et al., 2020). There is a 45% rise in the concentration of atmospheric $CO_2$ due to the industrial evolution (Karthik et al., 2021). The increase in $CO_2$ concentration may cause the extreme weather alteration and rapid increase in temperature (Knapp et al., 2015). Global climatic change somehow affects the physiological functions of plants such as the rate of assimilation, plant growth, fruiting time, germination, and leaf senescence (Fu & Waltman, 2018). Plants can overcome these climatic changes by adapting to these harsh conditions, or by the alteration in their phenotypic characteristics (Aitken & Whitlock, 2013).

Climatic changes can affect plant germination rates (Ullah et al., 2021). The impacts of harsh environmental condition on seedlings may last longer and can have a negative effect on growth and development of plants (Nonogaki et al., 2010). The high climatic temperature and shortage of water may affect the timings of seed germination (Hoyle et al., 2015). The seed viability and dormancy process may be slowed down due to the increase in environmental temperature (Jaganathan, 2020). Various other abiotic factors may alter the germination rate. Still, only a little is known about how change in climate affects genetic and physiological parameters of seed viability and germination (Cochrane et al., 2020).

The growth and photosynthesis rate of plants is also affected by drastic climatic change. It has been observed by Anderson et al. (2020) that an increase in temperature and $CO_2$ concentration leads to improved plant growth in herbaceous species and deciduous trees. The natural biomes present at higher elevations are also positively affected by the change in temperature (Jactel et al., 2019). The availability of less nitrogen to plants may affect the fertilizing effect of $CO_2$ (Reich et al., 2018). Usually, plant models did not consider some species from tropical regions. Climate has decreased the photosynthetic potential and growth rate of tropical plants in specific regions (Marazzi et al., 2015). It is needed to moderate plants according to environmental conditions like extreme temperature and water and nutrient limitations. The drought stress in plants is increased due to the deficit of vapor pressure under warm temperatures (Hatfield & Prueger, 2015).

Another evolutionary response due to climate change is earlier reproduction in plants. This usually affects flowering angiosperms (Hamann et al., 2021). It has been revealed by one study that change in boreal, alpine, and subalpine climate may affect the reproductive phenology of plants. The summer temperature does not affect the flowering phenology as compared to the spring climate (Hall et al., 2020).

## 2.10 HOST–PATHOGEN INTERACTIONS

Another parameter that can affect host–pathogen interactions is the changing growth pattern and rate that can reduce the pathogenicity of the host (Madhusudhan et al., 2019). The rate of adaptation is dependent on the conditions of competition. The competitional responses like predator versus prey and host versus pathogen equally effect the evolutionary reactions to the climate change by either increasing or decreasing rate of adaptation (Munguia, 2022). By exploring relations of predators and their prey, it was found that predators help prey to extend the extinction time by allowing the stronger selection pressure on the adapted individuals (Portalier et al., 2022). As a result of host plant interactions, a number of metabolic plants are produced which play a significant role in host plant interaction (Angulo et al., 2022).

Plants are sensitive to a wide range of micro-organisms, yet they are become resistant to many of them as a result of evolutionary response that may affect the immune response of plants to pathogens. When plants are attached by pathogenic and non-pathogenic organisms like bacteria, fungi, or viruses, they accumulate secondary metabolites that boost the immune response of plants. Compounds like Phytoalexin production appears to be a common way for plants to fight themselves against bacteria and maybe other pests (Hall et al., 2020).

## 2.11 DECIPHERING THE ROLE OF PHYTOALEXINS IN PLANT–MICRO-ORGANISM INTERACTIONS AND HUMAN HEALTH

The last decade has seen significant developments in plant pathology, some of which have been exceptionally interesting. Virods were found, and their discovery expanded the range of plant infections to the smallest imaginable limit: pathogens that are single-chemical molecules capable of being replicated by the host plant (Angulo et al., 2022). The crown gall bacteria was genetically examined to expose its deception. It transfers some of its DNA to the host genome, causing the host plant to create unusual amino acids that the bacteria can absorb but the host plant cannot. Exciting discoveries such as these have been followed by continuous accumulation of knowledge across a broad front, and plant pathology in practically all of its numerous divisions, from bacterial genetics to curative fungicides, is now a lot better science than it was ten years ago. This is particularly true in case of host–pathogen interactions (Ullah et al., 2021).

## 2.12 CONCLUSION

The change in climate leads to various alterations in plants from an evolutionary viewpoint. In order to survive in the changing climatic conditions, plants adapt themselves with various morphological and metabolic changes. Plants also shift their geographical distribution. However, it is still unknown if these changes that occur due to increasing temperatures have a long-term effect on plants or not. In order to know the facts, we have to study the plant life cycle thoroughly from seed germination to mature plant, which includes multiple climatic factors. Few studies have been carried out to reveal which traits have rapid adaptation to climate change. The genetic alteration was also studied using various techniques. Genomic study and transcriptomic tools can be used to determine the genetic alteration in plants due to environmental changes. These genetic-based studies will also be a clue about how to enhance the efficiency of conservation strategies.

## REFERENCES

Abrego, N., Roslin, T., Huotari, T., Ji, Y., Schmidt, N. M., Wang, J., . . . & Ovaskainen, O. (2021). Accounting for species interactions is necessary for predicting how arctic arthropod communities respond to climate change. *Ecography*, *44*(6), 885–896.

Aitken, S. N., & Whitlock, M. C. (2013). Assisted gene flow to facilitate local adaptation to climate change. *Annual Review of Ecology, Evolution, and Systematics*, *44*(10), 367–388.

Alés, R. G., Acosta, J. C., Astudillo, V., & Córdoba, M. (2022). Season-sex interaction induces changes in the ecophysiological traits of a lizard in a high altitude cold desert, Puna region. *Journal of Thermal Biology*, *103*, 103152.

Anderson, R., Bayer, P. E., & Edwards, D. (2020). Climate change and the need for agricultural adaptation. *Current Opinion in Plant Biology*, *56*, 197–202.

Angulo, V., Beriot, N., Garcia-Hernandez, E., Li, E., Masteling, R., & Lau, J. A. (2022). Plant–microbe eco-evolutionary dynamics in a changing world. *New Phytologist*, *234*(6), 1919–1928.

Bhadouria, R., Singh, R., Srivastava, P., & Raghubanshi, A. S. (2016). Understanding the ecology of tree-seedling growth in dry tropical environment: A management perspective. *Energy, Ecology and Environment*, *1*(5), 296–309.

Bhadouria, R., Singh, R., Srivastava, P., Tripathi, S., & Raghubanshi, A. (2017). Interactive effect of water and nutrient on survival and growth of tree seedlings of four dry tropical tree species under grass competition. *Tropical Ecology*, *58*(3), 611–621.

Bhadouria, R., Srivastava, P., Singh, S., Singh, R., Raghubanshi, A., & Singh, J. (2018). Effects of Effects of light, nutrient and grass competition on growth of seedlings of four tropical tree species. *Indian Forester*, *144*(1), 54–65.

Bronstein, J. L., Huxman, T. E., & Davidowitz, G. (2007). Plant-mediated effects linking herbivory and pollination. *Ecological Communities: Plant Mediation in Indirect Interaction Webs*, 75–103.

Choukri, H., Hejjaoui, K., El-Baouchi, A., El Haddad, N., Smouni, A., Maalouf, F., . . . & Kumar, S. (2020). Heat and drought stress impact on phenology, grain yield, and nutritional quality of lentil (Lens culinaris Medikus). *Frontiers in Nutrition*, *7*, 596307.

Cochrane, G., Mitchell, A. L., Almeida, A., . . . & Richardson, L. J. (2020). Mgnify: The microbiome analysis resource in 2020. *Nucleic Acids Research*, *48*(1), 570–578.

Deutsch, C. A., Tewksbury, J. J., Huey, R. B., Sheldon, K. S., Ghalambor, C. K., & Haak, D. C. (2008). Impacts of climate warming on terrestrial ectotherms across latitude. *Proceedings of the National Academy of Sciences*, *105*(18), 6668–6672.

Dusenge, M. E., Duarte, A. G., & Way, D. A. (2019). Plant carbon metabolism and climate change: Elevated CO 2 and temperature impacts on photosynthesis, photorespiration and respiration. *New Phytologist*, *221*(1), 32–49.

Fahad, S., Nguyen, T. L. H., & Shun, B. Y. (2017). Farmers' perception, awareness and adaptation to climate change: Evidence from northwest Vietnam. *International Journal of Climate Change Strategies and Management*, *9*(4), 555–576.

Feeley, K. J., Stroud, J. T., & Perez, T. M. (2017). Most 'global' reviews of species' responses to climate change are not truly global. *Diversity and Distributions*, *23*(3), 231–234.

Fu, H. Z., & Waltman, L. (2018). A large-scale bibliometric analysis of global climate change research between 2001 and 2018. *Climatic Change*, *170*(4), 36.

Gaitán-Espitia, J. D., & Hobday, A. J. (2021). Evolutionary principles and genetic considerations for guiding conservation interventions under climate change. *Global Change Biology*, *27*(3), 475–488.

Guo, S., Tao, C., Jousset, A., Xiong, W., Wang, Z., Shen, Z., . . . & Liu, S. (2022). Trophic interactions between predatory protists and pathogen-suppressive bacteria impact plant health. *The ISME Journal*, 1–12.

Guyer, A., van Doan, C., Maurer, C., Machado, R. A., Mateo, P., Steinauer, K., . . . & Erb, M. (2021). Climate change modulates multitrophic interactions between maize, a root herbivore, and its enemies. *Journal of Chemical Ecology*, *47*(10), 889–906.

Hall, C. R., Mikhael, M., Hartley, S. E., & Johnson, S. N. (2020). Elevated atmospheric CO2 suppresses jasmonate and silicon-based defences without affecting herbivores. *Functional Ecology*, *34*(5), 993–1002.

Hamann, E., Blevins, C., Franks, S. J., Jameel, M. I., & Anderson, J. T. (2021). Climate change alters plant–herbivore interactions. *New Phytologist*, *229*(4), 1894–1910.

Hamann, E., Denney, D., Day, S., Lombardi, E., Jameel, M. I., MacTavish, R., & Anderson, J. T. (2021). Plant eco-evolutionary responses to climate change: Emerging directions. *Plant Science*, *304*, 110737.

Hatfield, J. L., & Prueger, J. H. (2015). Temperature extremes: Effect on plant growth and development. Weather and climate extremes. *Weather and Climate Extremes*, *10*, 4–10.

Hay, M. E., & Fenical, W. (1988). Marine plant-herbivore interactions: The ecology of chemical defense. *Annual Review of Ecology and Systematics*, 111–145.

Heyduk, K. (2022). Evolution of Crassulacean acid metabolism in response to the environment: Past, present, and future. *Plant Physiology*, *190*(1), 19–30.

Ho, C.-K., & Pennings, S. C. (2013). Preference and performance in plant–herbivore interactions across latitude–A study in US Atlantic salt marshes. *PloS One*, *8*(3), e59829.

Hoyle, G. L., Steadman, K. J., Good, R. B., McIntosh, E. J., Galea, L. M., & Nicotra, A. B. (2015). Seed germination strategies: An evolutionary trajectory independent of vegetative functional traits. *Frontiers in Plant Science*, *6*(7), 731.

Jactel, H., Koricheva, J., & Castagneyrol, B. (2019). Responses of forest insect pests to climate change: Not so simple. *Current Opinion in Insect Science*, *35*, 103–108.

Jaganathan, G. K. (2020). Defining correct dormancy class matters: Morphological and morphophysiological dormancy in Arecaceae. *Annals of Forest Science*, *77*(3), 1–6.

Janzen, D. H. (1967). Why mountain passes are higher in the tropics. *The American Naturalist*, *101*(919), 233–249.

Johnson, S. N., Cibils-Stewart, X., Waterman, J. M., Biru, F. N., Rowe, R. C., & Hartley, S. E. (2022). Elevated atmospheric CO2 changes defence allocation in wheat but herbivore resistance persists. *Proceedings of the Royal Society B*, *289*(1969), 20212536.

Johnson, S. N., Waterman, J. M., & Hall, C. R. (2020). Increased insect herbivore performance under elevated CO2 is associated with lower plant defence signalling and minimal declines in nutritional quality. *Scientific Reports*, *10*(1), 1–8.

Karthik, S., Reddy, M. S., & Yashaswini, G. (2021). Climate change and its potential impacts on insect-plant interactions. In S. Harris (Ed.), *The Nature, Causes, Effects and Mitigation of Climate Change on the Environment*. IntechOpen.

Knapp, A. K., David, L. H., Kevin, R. W., Meghan, L. A., Sally, E. K., Kimberly, J., La, P., Michael, E. L., Yiqi, L., Osvaldo, E. S., & Melinda, D. S. (2015). Characterizing differences in precipitation regimes of extreme wet and dry years: Implications for climate change experiments. *Global Change Biology*, *21*(7), 2624–2633.

Kumar, K., Solanki, S., Singh, S., & Khan, M. (2016). Abiotic constraints of pulse production in India. *Disease of Pulse Crops and their Sustainable Management*, 23–39.

Kumar, P., & Sharma, A. (2022). Assessing the monthly heat stress risk to society using thermal comfort indices in the hot semi-arid climate of India. *Materials Today: Proceedings*, *61*, 132–137.

Kumari, V. V., Roy, A., Vijayan, R., Banerjee, P., Verma, V. C., Nalia, A., ... & Reja, M. (2021). Drought and heat stress in cool-season food legumes in sub-tropical regions: Consequences, adaptation, and mitigation strategies. *Plants*, *10*(6), 1038.

Kyei-Boahen, S., Savala, C. E., Chikoye, D., & Abaidoo, R. (2017). Growth and yield responses of cowpea to inoculation and phosphorus fertilization in different environments. *Frontiers in Plant Science*, *8*, 646.

Leger, R. J. S. (2021). Insects and their pathogens in a changing climate. *Journal of Invertebrate Pathology*, *184*, 107644.

Lukeneder, A., & Lukeneder, P. (2022). Taphonomic history and trophic interactions of an ammonoid fauna from the Upper Triassic Polzberg palaeobiota. *Scientific Reports*, *12*(1), 1–14.

Madhusudhan, P., Sinha, P., Rajput, L., Bhattacharya, M., Sharma, T., Bhuvaneshwari, V., ... & Singh, A. (2019). Effect of temperature on Pi54-mediated leaf blast resistance in rice. *World Journal of Microbiology and Biotechnology*, *35*(10), 1–11.

Marazzi, B., Bronstein, J. L., Sommers, P. N., Lopez, B. R., Ortega, E. B., Burquez, A., ... Franklin, K. (2015). Plant biotic interactions in the Sonoran Desert: Conservation challenges and future directions. *Journal of the Southwest*, 457–501.

Martinet, B., Zambra, E., Przybyla, K., Lecocq, T., Anselmo, A., Nonclercq, D., ... Hennebert, E. (2021). Mating under climate change: impact of simulated heatwaves on the reproduction of model pollinators. *Functional Ecology*, *35*(3), 739–752.

McCluney, K. E., Belnap, J., Collins, S. L., González, A. L., Hagen, E. M., Nathaniel Holland, J., & ... Wolf, B. O. (2012). Shifting species interactions in terrestrial dryland ecosystems under altered water availability and climate change. *Biological Reviews*, *87*(3), 563–582.

Meili, N., Acero, J. A., Peleg, N., Manoli, G., Burlando, P., & Fatichi, S. (2021). Vegetation cover and plant-trait effects on outdoor thermal comfort in a tropical city. *Building and Environment*, *195*, 107733.

Munguia, S. B. (2022). *Demographic Approaches to Understand Plant Community Responses to Climate Change* (pp. 1–24). Princeton University.

Mutamiswa, R., Chikowore, G., Nyamukondiwa, C., Mudereri, B. T., Khan, Z. R., & Chidawanyika, F. (2022). Biogeography of cereal stemborers and their natural enemies: Forecasting pest management efficacy under changing climate. *Pest Management Science*, *78*(11), 4446–4457.

Nonogaki, H., Bassel, G. W., & Bewley, J. D. (2010). Germination—still a mystery. *Plant Science*, *179*(6), 574–581.

O'Brien, E. K., Walter, G. M., & Bridle, J. (2022). Environmental variation and biotic interactions limit adaptation at ecological margins: Lessons from rainforest Drosophila and European butterflies. *Philosophical Transactions of the Royal Society B*, *377*(1848), 20210017.

Pakeman, R. J., & Quested, H. M. (2007). Sampling plant functional traits: What proportion of the species need to be measured? *Applied Vegetation Science*, *10*(1), 91–96.

Pearce-Higgins, J. W., Antão, L. H., Bates, R. E., Bowgen, K. M., ... & Gregory, R. D. (2022). A framework for climate change adaptation indicators for the natural environment. *Ecological Indicators*, *136*(87), 108690.

Peng, D., Zhang, Y., Wang, J., & Pennings, S. C. (2022). The opposite of biotic resistance: Herbivory and competition suppress regeneration of native but not introduced mangroves in Southern China. *Forests*, *13*(2), 192.

Pinsky, M. L., Comte, L., & Sax, D. F. (2022). Unifying climate change biology across realms and taxa. *Trends in Ecology & Evolution*, *37*(8), 672–682.

Portalier, S. M., Candau, J. N., & Lutscher, F. (2022). A temperature-driven model of phenological mismatch provides insights into the potential impacts of climate change on consumer–resource interactions. *Ecography*, e06259.

Rani, A., Devi, P., Jha, U. C., Sharma, K. D., Siddique, K. H., & Nayyar, H. (2020). Developing climate-resilient chickpea involving physiological and molecular approaches with a focus on temperature and drought stresses. *Frontiers in Plant Science*, *10*, 1759.

Rauschkolb, R., Li, Z., Godefroid, S., Dixon, L., Durka, W., Májeková, M., ... & Scheepens, J. (2022). Evolution of plant drought strategies and herbivore tolerance after two decades of climate change. *New Phytologist*, *235*(2), 773–785.

Reich, P. B., Hobbie, S. E., Lee, T. D., & Pastore, M. A. (2018). Unexpected reversal of $C_3$ versus $C_4$ grass response to elevated $CO_2$ during a 20-year field experiment. *Science*, *360*(6386), 317–320.

Roberts, A. J., Crowley, L. M., Sadler, J. P., Nguyen, T. T., Hayward, S. A., & Metcalfe, D. B. (2022). Effects of elevated atmospheric CO2 concentration on insect herbivory and nutrient fluxes in a mature temperate forest. *Forests*, *13*(7), 998.

Scogings, P. F., & Gowda, J. H. (2019). Browsing herbivore–woody plant interactions in savannas. *Savanna Woody Plants and Large Herbivores*, 489–549.

Sentinella, A. T., Warton, D. I., Sherwin, W. B., Offord, C. A., & Moles, A. T. (2020). Tropical plants do not have narrower temperature tolerances but are more at risk from warming because they are close to their upper thermal limits. *Global Ecology and Biogeography*, *29*(8), 1387–1398.

Sheldon, K., Huey, R. B., Kaspari, M., Sanders, N. J. (2018). Fifty years of mountain passes: A perspective on Dan Janzen's classic article. *The American Naturalist*, *191*(5), 553–565.

Shinzato, C., Khalturin, K., Inoue, J., Zayasu, Y., Kanda, M., Kawamitsu, M., ... & Satoh, N. (2021). Eighteen coral genomes reveal the evolutionary origin of Acropora strategies to accommodate environmental changes. *Molecular Biology and Evolution*, *38*(1), 16–30.

Sivageerthi, T., Sankaranarayanan, B., Ali, S. M., & Karuppiah, K. (2022). Modelling the relationships among the key factors affecting the performance of coal-fired thermal power plants: Implications for achieving clean energy. *Sustainability*, *14*(6), 3588.

Tarvainen, L., Wittemann, M., Mujawamariya, M., Manishimwe, A., Zibera, E., Ntirugulirwa, B., . . . & Spetea, C. (2022). Handling the heat–photosynthetic thermal stress in tropical trees. *New Phytologist*, *233*(1), 236–250.

Taveira, M. D. C., & Moreno, M. L. R. (2003). Guidance theory and practice: The status of career exploration. *British Journal of Guidance and Counselling*, *31*(2), 189–208.

Tripathi, S., Bhadouria, R., Srivastava, P., Devi, R. S., Chaturvedi, R., & Raghubanshi, A. (2020). Effects of light availability on leaf attributes and seedling growth of four tree species in tropical dry forest. *Ecological Processes*, *9*(1), 1–16.

Tripathi, S., Bhadouria, R., Srivastava, P., Singh, R., & Raghubanshi, A. S. (2018). The effects of interacting gradient of irradiance and water on seedlings of five tropical dry forest tree species. *Tropical Ecology*, *59*(3), 489–504.

Trivedi, P., Batista, B. D., Bazany, K. E., & Singh, B. K. (2022). Plant–microbiome interactions under a changing world: Responses, consequences and perspectives. *New Phytologist*, *234*(6), 1951–1959.

Ulappa, A. C., Shipley, L. A., Cook, R. C., Cook, J. G., & Swanson, M. E. (2020). Silvicultural herbicides and forest succession influence understory vegetation and nutritional ecology of black-tailed deer in managed forests. *Forest Ecology and Management*, *470*, 118216.

Ullah, A., Bano, A., & Khan, N. (2021). Climate change and salinity effects on crops and chemical communication between plants and plant growth-promoting microorganisms under stress. *Frontiers in Sustainable Food Systems*, 161.

Wan, N.-F., Zheng, X.-R., Fu, L.-W., Kiær, L. P., Zhang, Z., Chaplin-Kramer, R., . . . & Hu, Y.-Q. (2020). Global synthesis of effects of plant species diversity on trophic groups and interactions. *Nature Plants*, *6*(5), 503–510.

Wang, W., Chen, S., Guo, W., Li, Y., & Zhang, X. (2020). Tropical plants evolve faster than their temperate relatives: A case from the bamboos (Poaceae: Bambusoideae) based on chloroplast genome data. *Biotechnology & Biotechnological Equipment*, *34*(1), 482–493.

Wei, Y., Jin, J., Jiang, S., Ning, S., & Liu, L. (2018). Quantitative response of soybean development and yield to drought stress during different growth stages in the Huaibei Plain, China. *Agronomy*, *8*(7), 97.

Weiskopf, S. R., Rubenstein, M. A., Crozier, L. G., Gaichas, S., Griffis, R., Halofsky, J. E., . . . & Muñoz, R. C. (2020). Climate change effects on biodiversity, ecosystems, ecosystem services, and natural resource management in the United States. *Science of the Total Environment*, *733*, 137782.

Wirth, R., Meyer, S. T., Leal, I. R., & Tabarelli, M. (2008). *Plant Herbivore Interactions at the Forest Edge Progress in Botany* (pp. 423–448). Springer.

Wright, S., Keeling, J., & Gillman, L. (2006). The road from Santa Rosalia: A faster tempo of evolution in tropical climates. *Proceedings of the National Academy of Sciences*, *103*(20), 7718–7722.

Xiong, L., & Yao, Y. (2021). Study on an adaptive thermal comfort model with K-nearest-neighbors (KNN) algorithm. *Building and Environment*, *202*, 108026.

Yadav, M. R., Choudhary, M., Singh, J., Lal, M. K., Jha, P. K., Udawat, P., . . . & Maheshwari, C. (2022). Impacts, tolerance, adaptation, and mitigation of heat stress on wheat under changing climates. *International Journal of Molecular Sciences*, *23*(5), 2838.

Zhong, Z.-W., Li, X.-F., & Wang, D.-L. (2021). Research progresses of plant-herbivore interactions. *Chinese Journal of Plant Ecology*, *45*(10), 1036.

# 3 Ecophysiological Responses of Tropical Plants to Changing Climate

*Ranjan Pandey, Harminder Pal Singh, and Daizy R. Batish*

## 3.1 INTRODUCTION

Climate change is one of the most important challenges being faced by the earth today. It has drawn the attention of the entire world, so much so that it has emerged as a 'hot topic' in most of the global conferences pertaining to the environment and its conservation (www.un.org/en/conferences/environment; https://research.un.org/en/docs/environment/conferences, last accessed on 25 December 2022). The continuously worsening climatic conditions are posing a survival threat to all kinds of living beings, from marine aquatic animals to the terrestrial plants, birds, and animals, and of course, to humans (Goshua et al., 2021; Maulu et al., 2021). As a result, each country has set goals and targets to cut down on emissions of greenhouse gases (GHGs), and emphasize the rational utilization of non-renewable resources. in an attempt to slow down the rate of accelerating temperature of the earth and to mitigate the effects of climate change. Nevertheless, countering the ever-worsening climate has posed a significant challenge, and continues to torment almost all the life forms with respect to food, reproduction, and eventually their survival (Cavicchioli et al., 2019). Animals and humans, to some extent, have an option of migrating to an optimal, more suitable climatic zone if they are unable to adapt in the prevailing one (Goshua et al., 2021), but most of the plants have not been blessed with the characteristic of motility. Therefore, the only option for them is to adapt to whatever threats they face in their existing ecosystem. One of such hurdles of survival is also being faced by the tropical plants (Garrett et al., 2011). Due to the peculiar conditions of the tropics which support a big portion of the biodiversity on the planet, it becomes even more heightened to study and understand the numerous adaptations which tropical plant species are acclimatizing to as a coping mechanism to survive in the changing climatic conditions.

Thus, this chapter begins by outlining the ecosystem in the tropics, the diversity encountered therein and the characteristics that differentiate the tropics from the temperate and frigid regions. Afterwards, a brief overview of the climate change and the threats it poses to the biotic community is provided. Finally, the chapter identifies the various challenges the tropical plants encounter, primarily the elevations in temperature and carbon dioxide, and discusses the various adaptations these plants have undergone to overcome these challenges.

## 3.2 TROPICS: THE MOST DIVERSE ECOSYSTEM OF THE EARTH

The tropical region is geographically defined as the region between the Tropic of Cancer (23°26′22″ N) and the Tropic of Capricorn (23°26′22″ S). This includes the equator of the earth and the surrounding regions. The main classical features of the tropical region are high temperatures, due to which it is sometimes also referred to as the 'torrid zone', and the heavy rainfall (Sylla et al., 2016). The tropical region is a unique ecosystem in itself. Although it comprises only 40% of the earth's surface area, it serves as a home to almost two-thirds of the total biodiversity of the world (Table 3.1) (Bhardwaj, 2019). In the year AD 1807, Alexander von Humboldt, the Father of Ecology, stated that "The nearer we approach the tropics, the greater the increase in the variety of structure, grace of form, and mixture of colors, as also in perpetual youth and vigor of organic life", which supports the theory of latitudinal gradient in species diversity (Stevens & Tello, 2018). The tropics are blessed with heavy rainfall and ample sunlight, which provide favorable conditions to the variety of biotic systems that predominate in this region. This variety includes high mountains, deserts, rainforests, savannas, mangroves, wetlands, aquatics, and many more such features (Bhardwaj, 2019). Out of the 17 countries in the world that have been identified as mega-biodiverse countries, 15 are—completely or partially—located in the tropical region, i.e., northern Australia, Brazil, China, Colombia, Democratic Republic of the Congo, Ecuador, India, Indonesia, Madagascar, Malaysia, Mexico, Papua New Guinea, Peru, Philippines, and Venezuela (Table 3.1) (Ali et al., 2013). Additionally, more than half of the ecological biodiversity hotspots of the world—such as the Atlantic forests, the Caribbean islands, the Andes mountainous region, and the Western Ghats—are situated in the tropical region (Table 3.2) (Myers et al., 2000).

The rainforests in the tropical region are one of the oldest habitats on the earth. Although rainforest covers only 6% of the world's land area, still it encompasses more than half of the species on earth by the virtue of its low seasonal

## TABLE 3.1
### List of Megadiverse Nations which Are in Tropical Climate Zones of the Earth

**Megadiverse Countries in Tropical Biome**

1. Brazil
2. China
3. Colombia
4. Democratic Republic of the Congo
5. Ecuador
6. India
7. Indonesia
8. Madagascar
9. Malaysia
10. Mexico
11. Northern Australia
12. Papua New Guinea
13. Peru
14. Philippines
15. Venezuela

*Source:* Based on Ali et al. (2013)

## TABLE 3.2
### List of Ecological Hotspots which Are in Tropical Climate Zones of the Earth

**Ecological Hotspots in Tropical Biome**

1. Atlantic Forest, South America
2. Caribbean Islands, North and Central America
3. Cerrado, South America
4. Coastal forests of Eastern Africa, Africa
5. East Melanesian Islands, Southeast Asia and Asia-Pacific
6. Eastern Afromontane, Africa
7. Guinean Forests of West Africa, Africa
8. Horn of Africa, Africa
9. Indo-Burma, South Asia
10. Madagascar and the Indian Ocean Islands, Africa
11. New Caledonia, Southeast Asia and Asia-Pacific
12. Philippines, Southeast Asia and Asia-Pacific
13. Polynesia–Micronesia, Southeast Asia and Asia-Pacific
14. Succulent Karoo, Africa
15. Sundaland, Southeast Asia and Asia-Pacific
16. Tropical Andes, South America
17. Tumbes-Choco-Magdalena, South America
18. Wallacea, Southeast Asia and Asia-Pacific
19. Western Ghats and Sri Lanka, South Asia

*Source:* Based on Myers et al. (2000)

temperature variation (Poker & MacDicken, 2016). These forests are largely comprised of various vegetation layers, and each of these layers serves as a habitat for a wide variety of species. The trees which outgrow the forest and tower above the rest are exposed to intensive sunlight and strong winds. This layer is called the 'emergent layer' where bats, eagles, monkeys, insects, insect-eating birds, and snakes can be spotted (Leal & Resener, 2022). The second layer is the 'canopy cover', which consists of tall trees with large and leathery leaves allowing only 20% of sunlight to reach the ground. It is full of fruits, nuts, and other edible jewels of the forest. Species such as monkeys, sloths, tree frogs, margay cats, ants, beetles, bats, toucans, parrots, hummingbirds, snakes, and lizards live here (Leal & Resener, 2022). The next layer of the rainforest is the 'undercover' (named so as it is covered by the emergent and canopy layers), where mostly sciophytes are present as they thrive in low-lit areas. This 'undercover' is home to great number of butterflies, termites, toads, frogs, snakes, lizards, beetles, and parakeets (Andresen et al., 2018). Finally, the 'forest cover' which receives only scattered rays of light, provides shelter to the flowering plants, edible roots and tubers, and animal species such as tapirs, armadillos, peccaries, slugs, centipedes, cockroaches, termites, beetles, and many decomposers also live in this layer (Andresen et al., 2018).

All flora and fauna found in this tropical region are interlinked together, keeping the ecosystem sustainable. They not only coexist, but have developed biotic interactions to function and evolve together (Andresen et al., 2018; Leal & Resener, 2022). More specifically, they assist in the pollination of each other. The tallest trees and lianas rely on winds for their dispersal of their pollens and seeds; however, many flowering plants and tress—such as Bombacaceae, Passifloraceae, and Fabaceae—are dependent on bats for their pollination (Bawa, 1990). Plants attract pollinators with flashy color displays, alluring scents, and nutritious pollen as rewards. Plants with small flowers—Acanthaceae, Bromeliaceae, Gesneriaceae, Marantaceae, Musaceae, Rubiaceae, and Zingiberaceae—are often pollinated by hummingbirds (Bawa, 1990). Dispersal of seeds of the plants and trees helps in expansion and distribution of the population. Sometimes, these seeds flow with wind currents; other times, they ges stuck on fur and feathers; seeds can thus be dispersed by birds to other geographical areas while traveling or eating the fruit. *Bertholletia excelsa*, commonly known as the Brazil nuts in the Amazon Basin, gets pollinated by Euglossines and seed dispersal is done by agoutis (Mertens et al., 2021). Fig trees of Borneo Island are pollinated by fig wasps, one of the smallest creatures, but the seeds are dispersed by orangutans, one of the large great ape primates (Reyes et al., 2021). In Madagascar, the traveler's palm tree is pollinated by lemurs, which are small primates endemic to the island, and the seeds are dispersed by parrots (Mertens et al., 2021). There are numerous aforesaid examples hidden in the topical regions which suggest all these species coexist perform important ecological roles for each other, benefit from each other, and sustain life on earth. Some species of animals and reptiles which are found in tropics of the earth are sloths, tamarins, tapirs, jaguars, ocelots, kinkajous, lemurs, agouti, frogs, salamanders, snakes, and lizards (Andresen et al., 2018). Hummingbirds, harpy eagles, spectacled owls, toucans, macaws, quetzals, hornbills, and finches are types of birds which thrive in the tropical regions. Flora such as orchids, rubber trees, poinsettias, cacao trees, Venus flytraps, passion flowers, peace lilies, Brazil nut trees, mahogany trees, lianas, strangler fig and quinine are known to be tropical species (www.arborday.org/programs/rainforest/animals-and-plants-of-the-rain-forest.cfm; last accessed on 25 December 2022).

Due to this diversity in their biotic environments, the tropical rainforests are considered to be the most productive habitats on the terrestrial ecosystem with the highest net primary productivity (NPP) rating, i.e., rainforests store more carbon per hectare than any other vegetation type. Some studies suggest that the tropics have maintained high diversification and speciation because they have experienced fewer disturbances as compared to other biomes which have undergone massive calamities in the past (Ewel, 1980; Deharveng & Bedos, 2019).

## 3.3 CLIMATE CHANGE: A REAL CHALLENGE

In present times, the earth is facing several challenges such as climate change, deforestation, logging, urbanization, and change in land use patterns. Among the total nine planetary boundaries, the crucial one which human beings have overstepped the most is change in climatic pattern of the earth (Rockström et al., 2021). Climate change is defined as long-term alterations in the temperature and weather conditions of the earth (Olabi & Abdelkareem, 2022). These changes can be driven by natural or anthropogenic factors. Variations in solar intensity and volcanic eruptions can be such natural factors, but the influence of these factors is either negligible or too slow to explain the rapid warming seen in recent times (Fawzy et al., 2020). Since the beginning of industrialization, it is the human activities which have become the main reason for crossing the planetary boundaries. Activities such as fossil fuel burning, urbanization, industrialization, and consumptive behavior have collectively led to variation in the climate at unprecedented rate (Fawzy et al., 2020; Olabi & Abdelkareem, 2022). The GHGs such as $CO_2$ (carbon dioxide), $CH_4$ (methane), $N_2O$ (nitrous dioxide), and chlorofluorocarbons (CFCs) are released massively in human actions which in turn trap the infrared radiation of sunrays and rise the temperature of the earth (Cronin et al., 2018). This ultimately causes drastic variations in the climate and roots of global warming. According to statistics from the World Research Institute (WRI) and the Intergovernmental Panel on Climate Change (IPCC), $CO_2$ in atmosphere has already exceeded 418 parts per million which is surge of 46% since pre-industrial times 250 times faster than it did from natural sources after the last Ice Age (IPCC, 2021; Boehm et al., 2021). There has been rise in temperature of the earth by 1.1°C since 1808 and it is expected to rise by 1.5°C in the next two decades (IPCC, 2021). The last decade has been recorded to be the warmest decade since industrialization. Further, the U.S. National Aeronautics and Space Administration (NASA) reports that polar ice sheets are decreasing at a rate of 427 billion metric tons per year, which has caused sea level to rise by 4 inches since 1993 (NASA, 2018). Burning of fossil fuels accounts for 75% of total greenhouse emissions and 90% of $CO_2$ concentration in the environment (Boehm et al., 2021). Other than the primary cause for global warming, deforestation and land system change are decreasing the natural sinks of carbon. Some other reasons for building up of GHGs in the atmosphere are increased livestock, landfills for garbage, fertilizer production, unsustainable production and consumption practices, construction, and agriculture (Post et al., 2019).

Global warming and climate change are not only restricted to rise in temperature but are actually just the tip of the iceberg. As everything on the earth is interconnected, a slight variation in one system brings out the change in other systems in multiple ways (Figure 3.1). With rise in the average temperature, the hydrological cycle on earth is affected, resulting in incidences of intense rainfall and flooding and simultaneously severe drought conditions in water-stressed regions (Kahn et al., 2019). Natural disasters such as storms, cloud-bursts, cyclones, hurricanes, and typhoons have also become more frequent due to changes in the water cycle (Cronin et al., 2018). Such events destroy homes and communities, causing deaths and huge economic losses. Rising temperatures also fuel wildfires which start more easily and spread faster in warmer conditions, gutting hundreds to thousands hectares of land and wildlife and converts them into ashes. In 2019, more than 5.8 million hectares burned in Australia and showed devastating after-effects (Boer et al., 2020). The increased heat is melting the snow cover of the planet and causing increase in the sea levels, and a soaring threat to coastal countries and island communities (Figure 3.1). Oceans are natural sinks of carbon and absorb almost one-third of the emissions from fossil fuels. Increased absorption has turned oceans 30% more acidic than it was before industrialization (IPCC, 2021). The acidification of the oceans has started showing consequences in the form of bleaching of the coral reefs, an ecosystem that support more than 25% of all marine life, and decreasing the population of oysters, clams, and fish (Kompas et al., 2018; Gissi et al., 2021). The long-term effects of such events will upset the livelihood of people who are dependent on seafood for their survival and also can call for economic ruins.

Surging of extreme climate conditions is shrinking crop production and hitting the global supply of food. Altered rainfall patterns, temperature range, concentration of GHGs, and frequent disasters are wiping out entire crops over wide areas and leading to poor nutrition and increased hunger (Krishnan et al., 2020). Eruption of pests and infectious diseases is an additional branch of effects that has been linked with climate change. Mounting temperature and pollution together are resulting in rising asthma, heatstroke, heart and kidney ailments, allergies of airborne pollen and mold, and tormenting hay fever (Cavicchioli et al., 2019). The World Health Organization (WHO) states "Climate change

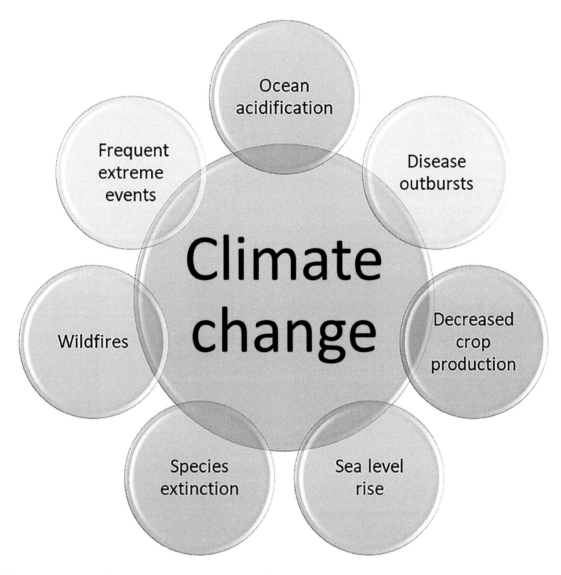

FIGURE 3.1 Summary of effects of climate change on the earth.

is expected to cause approximately 250,000 additional deaths per year between 2030 and 2050" (https://www.who.int/news-room/fact-sheets/detail/climate-change-and-health; Last accessed on 6th July 2023). Another effect the world is witnessing with climbing temperatures is species extinction. With an outburst of diseases, pests, drought conditions, floods, and habitat destruction, along with intensifying temperatures, wildlife is finding it difficult to cope up with quickly changing environment (Kompas et al., 2018). Many species are migrating toward poles seasonally or do not even come back as the temperature is suitable there. At the present rate of temperature increasing, a quarter of land animals, birdlife and plants may soon become extinct. According to the Intergovernmental Panel on Climate Change, "Taken as a whole, the range of published evidence indicates that the net damage costs of climate change are likely to be significant and to increase over time" (IPCC, 2021).

## 3.4 ECOPHYSIOLOGICAL RESPONSES OF TROPICAL PLANTS

As mentioned in the previous sections, various effects of changing climate are affecting biodiversity. These effects are further exacerbated in the tropical settings, on account of the extremes of temperature, coupled usually with excessive humidity. Tropical regions are characterized by high temperatures during the day, along with direct exposure of the plants to sunlight; and even temperatures toward the higher side of the normal range during the night (Sylla et al., 2016). Unlike animals and birds, which have the option of migrating to a different habitat if the existing one seems too stressful, plants cannot move—and therefore, they have to develop adaptive mechanisms to tolerate these climatic conditions and survive in them (Marques et al., 2022). Moreover, climate is not a static entity. It keeps on changing with time and even may vary across the different seasons in the area. Therefore, it becomes imperative for the plants to develop adaptive mechanisms if they intend to survive in these conditions.

Various studies have been conducted in order to elucidate the adaptive mechanisms in plants that tend to enhance their survival in the hostile climatic conditions in the tropical regions. Such studies have formed the basis of whatever little understanding we possess today on the adaptability of plants (Hasanuzzaman et al., 2013; Sakai et al., 2019; Li et al., 2021). Much of the knowledge on this topic still remains unexplored and unexplained; and therefore, the need for further studies on this topic is necessary. The available data points toward temperature and carbon dioxide as two of the most common stressors to plants in the tropical areas. The further sections explore the role played by these stressors independently, as well as in combination; and the mechanisms that plants have evolved to adapt to these unfavorable conditions.

### 3.4.1 Responses to Increases in Temperature

Temperature, as a stressor, is considered to be extremely important in the survival of plants (Zandalinas et al., 2018). This is more likely to be crucial in tropical plants, which bear the brunt of the increased temperature by virtue of their location (Hinojosa et al., 2019). The ever-increasing trend in global temperatures further complicates the problem, as plants not only need to sustain their protective efforts, but add up on it so as to protect themselves from the continual increase in temperatures (Figure 3.2). The precise mechanisms as to how these plant processes and activities are adversely affected by high temperatures are not completely understood. However, one of the possible mechanisms attributes these detrimental outcomes on the effects of high temperatures on proteins. Proteins act as structural components of the plant cell wall and cell membrane, as also the intracellular cytoskeleton. They also act as enzymes that catalyze vital reactions in the plant cells. On exposure to high temperatures, the proteins may coagulate, resulting in their loss of tertiary structures, and hence, alteration in their conformations. As a result, the structure of the cell is disrupted, which may lead to cell death. Additionally, the vital metabolic and enzymatic processes may slow down or come to a halt, resulting in deficiencies in various plant activities (Hasanuzzaman et al., 2013).

In particular, plant reproduction and pollination are one of the processes that has been suggested to be extremely sensitive to the effect of increased temperatures (Lippmann et al., 2019). An increase in temperature can reduce the reproductive capacity of plants, thus decreasing their yield. As per an IPCC report on climate change in 2021, even modest increases in temperatures of 1–2°C can reduce the yield of tropical plants (particularly cereals), which would further worsen by the end of the 21st century (IPCC, 2021). This has been demonstrated in maize, where the grain yield decreased by as much as 80–90% when subjected to ranges of higher temperatures, and this yield decreased proportionally to the increase in temperature (Hatfield & Prueger, 2015). This reduction in yield can be further exacerbated by water deficits, which can occur directly as a result of higher temperatures.

Plant pollination has also been shown to be dependent on temperature, and an increase in temperature may reduce the ability of the plant to pollinate. When exposed to daytime high temperatures, the pollination in rice (*Orzya sativa* L.) was found to significantly decrease, which may be due to poor pollination due to insufficient pollen grain deposition or unsatisfactory fertilization on the stigma (Wu et al., 2020). As pollination is an important step in plant reproduction, reduced pollination may impact the ability of the plant to procreate, thus reducing its yield. Apart from plant pollination, the natural development of plant seeds is also susceptible to changes in temperature (Suriyasak et al., 2020). An optimal temperature is required for the various biological activities that take place within a seed, ultimately leading to its germination. The nutritional content needs to be properly regulated so that the developing seed is not starved and so that the growth is not stunted. However, reduced transcriptions of genes such as auxin (ADR12), and β-glucosidase, which are involved in normal seed development in soybean (*Glycine max* L. Merr. cv. 'Bragg') have been reported at day/night temperatures exceeding 28°C/18°C (Thomas et al., 2003). These modifications are associated with reductions in the concentrations of carbohydrates, lipids, nitrogen, and phosphorus within the seed at elevated temperatures. This indicates that the development of seed is negatively affected when the plant is exposed to higher temperatures.

Another interesting finding—and a possible mechanism that might explain the changes in plant yield on exposure to adverse temperatures—is the effect of temperature on symbiotic interactions between plants and other organisms. As an example, colonization by a symbiotic fungus *Piriformospora indica* has been reported to enhance the resistance of *Musa acuminata* cv. Tianbaojiao (banana) to cold (Li et al., 2021). This is made possible by an up-regulation of the enzymes involved in free radical scavenging and plant cell protection. Hence, the banana plant is able to better withstand the oxidative stress induced by the reduced temperatures. Therefore, a concern with an increase in global temperature is that the protective effect afforded by this symbiotic interaction may be lost. Either a reduced fungal colonization of the banana trees or a modification in the oxidative stress—or both in combination—may render the banana plants more susceptible to damage by intracellular free radicals, thus impacting their survival and hence their yield (Li et al., 2021). As a consequence of these damaging effects of high temperatures, plants tend to undergo genetic mutations and modifications as protective responses to enable themselves to overcome the stressing factor, and hence survive the increased global temperatures. One of the examples can be stated as the development of antioxidant capacity to overcome intracellular oxidative

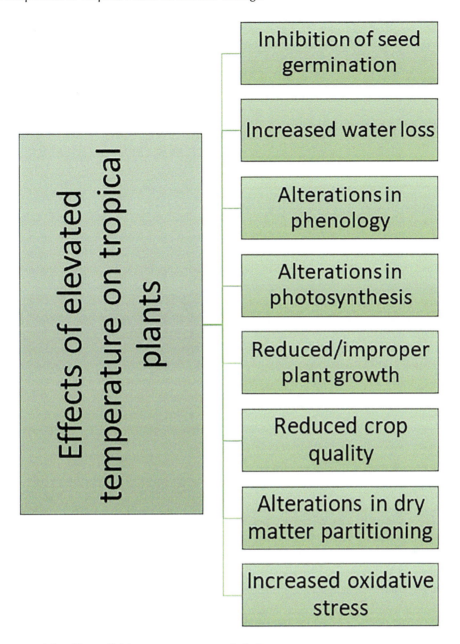

FIGURE 3.2 Summary of the effects of high temperature on tropical plants.

stress (Hasanuzzaman et al., 2013). As a consequence of cell damage in account of increases in temperatures, various pro-oxidant molecules are released that contribute to destruction and death of the cells. These pro-oxidants can be countered by the intracellular antioxidants, which resist their potential for oxidative damage (Cai et al., 2021).

The expression of antioxidant genes was found to be up-regulated in the invasive plant *Sphagneticola trilobata* L., with corresponding increased activity of antioxidant enzymes and reduced accumulation of the reactive oxygen species, on exposure to high temperatures (Cai et al., 2021). This was in comparison with *Sphagneticola calendulacea*, a non-invasive species of the same genus. Thus, on account of this genetic modification, increased global temperatures may actually benefit the survival of *S. trilobata*, with an enhanced invasion into the surroundings (Cai et al., 2021). However, such an invasion by the plant can further lead to significant changes in the surrounding ecosystem, which are yet to be studied. The increases in global temperatures also seem to affect the plant palatability. An experimental increase in temperature from 15–25°C was found to reduce the palatability of the aquatic plant shining pondweed (*Potamogeton lucens*) in the pond snails (*Lymnaea stagnalis*) (Zhang et al., 2019). This can possibly be attributed to the underlying chemical changes in plant stoichiometry and mineral contents. Although results from a single study cannot be completely generalized, this finding points toward a significant effect on the food chain, which can potentially be disturbed by global warming. However, such effects have not been replicated in other marine plants such as macroalgae (Poore et al., 2016) or terrestrial plants such as *Quercus pubescens* (Backhaus et al., 2014). Hence, the precise nature of an interaction between the elevated temperatures and plant palatability cannot be commented upon or discussed with the existing body of research.

### 3.4.2 Responses to Increases in $CO_2$

Carbon dioxide is released into the atmosphere during respiration by all living organisms, including plants. The level of $CO_2$ in the earth's atmosphere is currently increasing as a result of global warming (IPCC, 2021). The increased temperature around the immediate surface of the earth tends to retain gases like $CO_2$, increasing the exposure of various plants to $CO_2$. As can be expected, the higher temperatures in the tropical regions can lead to a greater accumulation of $CO_2$ in the environment in these regions. Thus, similar to an increase in temperature, an increased exposure to $CO_2$ in the surrounding environment also threatens the survival of the plants (Figure 3.3). As a consequence, plants undergo genetic manipulation by themselves as a protective response against these environmental stresses. These adaptations are expected to favor the growth and yield of the plants by enhancing their survival in these conditions. Studies have reported an enhanced survival among plants, when chronically exposed to high $CO_2$ concentrations in controlled conditions (Tubiello et al., 2007; Wassmann et al., 2010; Ghini et al., 2015; Magrach & Ghazoul, 2015).

Conducting free-air $CO_2$ enhancement (FACE) experiments is a valid study design to determine the effects of elevated $CO_2$ concentrations on plants without enclosure (hence, the term 'free-air', in which $CO_2$ levels are increased using an array of vent pipes that release jets of $CO_2$). Although there are limitations and criticisms of the FACE technique—primarily that the conditions tend to be controlled and therefore may not mimic the natural conditions in the wild—the technique has been very widely studied and used for experimental purposes (Allen et al., 2020). One of the major modifications that tend to occur when exposure to $CO_2$ on a long-term basis results in increases in plant photosynthesis, water-use efficiency, and productivity, along with increases in their biomass and yield (Kimball et al., 2002; Nowak et al., 2004; Norby et al., 2016). These benefits have been consistently demonstrated in the FACE experiments. These are considered to be the protective mechanisms that enable the plants to tide over the adverse situation of increased exposure to $CO_2$. Moreover, studies suggest that these changes are more pronounced and better characterized, and are hence more likely to benefit the $C_4$ plants than the $C_3$ plants (Kimball et al., 2002; Ainsworth & Long, 2005). Similar increases in photosynthesis have also been reported in other tropical plants. Eddos (*Colocasia antiquorum*), a tuber, is native to the tropical regions. When it was cultivated under elevated $CO_2$ conditions in a temperature gradient chamber, at both low and high temperatures, the yield of the plant photosynthesis was found to be significantly increased (Bin Zaher et al., 2021). This increase in photosynthesis is generally accompanied by increases in the dry weights of the roots, petioles, leaf blades, and aboveground parts of the plant, as well as the number of leaves and the total leaf area, indicating an overall anabolic stimulation by the excessive $CO_2$ levels. Similar to eddos, the photosynthetic yield in other plants such as rice is also generally increased upon exposure to elevated $CO_2$ levels. Japanese rice has been classified into five different varieties, termed as 'cultivars', namely 'Aikoku', 'Norin8', 'Koshihikari', Akihikari', and 'Akidawara'. These cultivars differ in various properties, such as the optimal growth conditions, the nutritional profile, and other characteristics. Therefore, their responses to adverse conditions such as increased $CO_2$ also tend to differ. However, all five cultivars have demonstrated an increase in photosynthesis under elevated $CO_2$ conditions in FACE experiments over two years, albeit ranging from as low as 16% to as high as 31% (Sakai et al., 2019). The strongest response was seen with Norin8, which also demonstrated increases in the percentage of ripe grains and spikelet density, reflecting an increase in nitrogen uptake. Thus, the rice undergoes adaptations in order to increase its productivity when exposed to difficult survival conditions.

Similar to rice, an increase in the yield of vegetables such as tomatoes, potatoes, peppers, eggplant, etc., can also occur under elevated $CO_2$ conditions. Exposure of tomatoes to high $CO_2$ concentrations of 1000 ppm has shown an early increase in the net assimilation rate by 100% (Bencze et al., 2011). An increase in the average plant height and stem diameter of the eggplant were maximally increased by 28.62% and 5.20%, respectively, when treated with elevated $CO_2$ concentrations (Xu et al., 2020). These increases are also estimated to be due to similar mechanisms discussed previously under eddos, wherein concomitant increases in the photosynthetic activity, and the dry weight of the plants confer a survival benefit to the plant by increasing its yield under adverse conditions. In general, an increased yield of vegetables has been summarized in reviews, when exposed to high $CO_2$ concentration of 700–1000 ppm (Gruda & Tanny, 2014). A similar effect is also observed on the cash crops, both food and non-food. An increase in photosynthesis was reported in the Latin American region among the commercial coffee cultivars (Catuaí and Obatã) in FACE experiments, accompanied by increases in water-use efficiency, plant growth, and overall yield under elevated $CO_2$ concentrations over two years (Ghini et al., 2015). Thus, similar to the cereals and vegetables, the coffee crops in the tropical climates do respond positively when $CO_2$ concentrations are elevated.

Besides the rate of photosynthesis, the $CO_2$ levels can also affect the nutritional quality of the edible parts of the plants. The total content of sugars and antioxidants is increased by exposure to elevated $CO_2$ concentrations (Dong et al., 2018). These increases in sugars and antioxidants are supposed to be beneficial, as they enhance the tolerability of the plants to the $CO_2$ stress by regulating the metabolism and oxidative stress (Drake et al., 2011; Huang & Xu, 2015). Antioxidants directly combat the

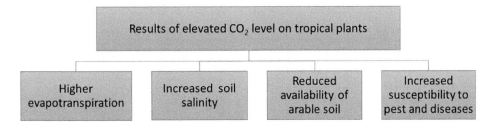

**FIGURE 3.3** Summary of adverse effects of elevated $CO_2$ on tropical plants.

oxidative stress, while the metabolites of sugars provide the necessary substrates for these reactions, as the antioxidant molecules are replenished intracellularly when carbohydrates are broken down to simple sugars. The actual effect on the increase in nutritional quality needs to be questioned, however, as the proteins and essential elements in the plants are decreased following exposure to elevated $CO_2$ levels (Dong et al., 2018). Therefore, it is possible that the increase in nutritional quality may not be complete or wholesome.

Additionally, making general assumptions or drawing general conclusions based on the effect of temperature and $CO_2$ on plant growth, reproduction, or photosynthesis on the basis of these experimental conditions can be fraught with error, as there can be multiple other factors that can independently impact the growth and yield of plants such as air quality, soil, water, the presence of pests, density of other plants that compete for resources, etc. (Garrett et al., 2011; Zandalinas et al., 2021). An experimental setting tends to modulate the factor of interest while keeping the others constant, so that the exact effect of a particular factor can be studied. For example, when $CO_2$ is increased in the FACE experiments, the other factors such as temperature and moisture content are maintained (unless they are changed by the $CO_2$ increase itself). As expected, these scenarios are not exact replicas of what occurs naturally, as when the plants grow in the wild, all these factors influence together simultaneously and the net effect therefore tends to be somewhat different—or sometimes even totally the opposite of what was observed in the experimental studies. For instance, the increase in photosynthesis in sweetgum and loblolly pine trees under experimental conditions was observed to be as high as 30–67% over a period of three years (Herrick & Thomas, 2001; Ellsworth et al., 2012). However, the increase was much lower at 19% when Eucalyptus trees were exposed to high $CO_2$ concentrations over three years (Ellsworth et al., 2017). This lower level of increase was attributed to the deficiency of phosphorus in the soil where the Eucalyptus trees were grown, indicating that $CO_2$ is not the only parameter that can affect the rate of photosynthesis; rather, it is a combined effect of multiple factors that brings about the ultimate outcome. Hence, the other factors also need to be taken into consideration and caution should be exercised while making direct inferences from studies on experimental scenarios—and these effects may have to be further determined and validated in real-world settings (Ellsworth et al., 2017).

### 3.4.3 Combined Effects of Temperature and $CO_2$

Unfortunately, most of these studies conducted in the research have considered the effects of temperature and $CO_2$ on plant growth and functions independently. This can serve as another source of error, as never do plants face one of these in the absence of the other. The ultimate effect is a combination of exposure to both the stressors (and other stressors, also, in reality). Therefore, studies that evaluate the simultaneous impact of both temperature and $CO_2$ tend to be more informative on how exactly plants might behave in the wild in response to adverse climatic conditions. For example, Marques et al. (2020) reported that the expression of genes that regulate photosynthesis and antioxidant pathways is increased in coffee plants native to Brazil (*Coffea arabica* and *Coffea canephora*) on exposure to elevated $CO_2$ concentrations (Marques et al., 2020). The increased potential for photosynthesis is expected to enable these tropical coffee plants in surviving the excessive stress induced by $CO_2$ elevation. However, on the other hand, the same plants demonstrated a down-regulation of genes involved in photosynthesis, in response to elevated temperatures, up to 42°C, regardless of the concentrations of $CO_2$ (Marques et al., 2022). Therefore, the benefit of increase in photosynthesis achieved with increased $CO_2$ concentrations may be offset by higher temperatures, which are usually the norm in the tropical regions.

Similarly, the elevation of temperature from 18°C to 23°C and 28°C reduced the total isoflavone content in a dwarf soyabean species (*Glycine max* L. Merrill) by approximately 65% and 90%, respectively, in a controlled environmental chamber (Caldwell et al., 2005). This reduction was partially reversed when the plant was simultaneously exposed to elevated $CO_2$ (up to 700 ppm) and drought conditions. These changes in isoflavone contents can affect the nutritional value of the plant foods, as isoflavones constitute a major component of the plant fruits and other edible structures. Thus, another example supports the possibility that increase in the yield of plants by one stressor may be counterbalanced by increases in another stressor.

Temperature and $CO_2$ also appear to work together to regulate the flowering time in plants. Elevated temperatures can influence the flowering time and anthesis, as well as regulate the expression of numerous transcription factors, flowering suppressors, and autonomous pathway genes in the *Arabidopsis* plants (Jagadish et al., 2016). *Arabidopsis* represents the mustard (Brassicaceae family), which includes commonly cultivated vegetables such as radishes and cabbage. These effects are further modestly augmented by elevated $CO_2$ levels. Thus, the yields of these plants can be simultaneously regulated by both temperature and $CO_2$, wherein an increase in both can affect the plants positively. The opposite effect is also possible, i.e., the increase in temperature may offset the positive effects of increase in $CO_2$. For example, the yield of wheat exposed to high $CO_2$ concentrations (at 450 ppm) was found to be increased at a temperature level of 0.8°C (Guoju et al., 2005). However, when the temperature was increased beyond 1.5°C, the yield actually decreased. This was postulated to be due to an increase in the water demand offset by the increase in temperature, as the same study reported that additional irrigation at temperatures beyond 1.5°C managed to restore the reduced yield toward normal (Guoju et al., 2005). Thus, the availability of water for growth is also an important consideration that can work in tandem with temperature and $CO_2$ to regulate plant growth and survival. Therefore, the combined effect of elevated temperature and $CO_2$ on tropical plants is extremely complicated and needs to be further studied, preferably in real-life scenarios that reflect the changes and their effects better than in experimental conditions—although experimental conditions can provide more controlled and homogeneous situations to draw inferences from, as previously discussed, they may not accurately reflect the real-world scenario.

## 3.5 CONCLUSIONS

To summarize, the tropical region boasts of an ecosystem of its own, due to the unique characteristics it possesses. Climate change tends to possess a challenge to these organisms—plants in particular—for survival. In order to overcome these challenges, the plants undergo genetic modifications that enable them to tolerate the changing climatic conditions. These changes have been extensively studied for elevations in temperature and $CO_2$, which have been shown to affect various plant processes such as photosynthesis, pollination, reproduction, and seed germination, and therefore affect the plant yield and the nutritive values of the food plants. These genetic modifications, however, have been studied primarily under controlled conditions in experimental settings, which have evaluated the effect of change in a single selected parameter. Nevertheless, these studies do shed a light and provide a basic understanding of the various adaptive alterations that occur at the genetic, molecular, and cellular levels in the tropical plants on exposure to increased climatic stresses.

## 3.6 FUTURE RECOMMENDATIONS

Based on the existing research, certain lacunae in the knowledge regarding the adaptation of tropical plants are identified which need to be further explored. As mentioned previously, studies conducted so far have evaluated the effect of changes in a single parameter under controlled experimental conditions. Real-world studies, where the effect to changes in multiple stresses that occur simultaneously and therefore might affect the plants differently, need to be further encouraged to give a broader view of what actually goes on in the wilderness. Moreover, studies have so far elucidated only a few of the genetic changes that occur in tropical plants when exposed to stresses. There can be a multitude of other adaptive mechanisms that have so far been hidden from human knowledge. Further exploration can help us understand these mechanisms and how they might work in tandem to enhance the plant survival in the face of increasing stresses. These might include changes in the consumption of water and nutrients to conserve them in the face of shortages, or changes that enable them to grow in altered soil compositions, or changes in the reproductive capacity, and so on. Finally, the pace of climate change also needs to be considered while experimenting. The adaptations in the plants need to keep pace with that of the climate change—if they are to survive.

## REFERENCES

Ainsworth, E. A., & Long, S. P. (2005). What have we learned from 15 years of free-air $CO_2$ enrichment (FACE)? A meta-analytic review of the responses of photosynthesis, canopy properties and plant production to rising $CO_2$. *New Phytologist*, 165(2), 351–372.

Ali, M. K., Badman, T., Bertzky, B., Engels, B., Hughes, A., & Shi, Y. (2013). *Terrestrial Biodiversity and the World Heritage List: Identifying Broad Gaps and Potential Candidate Sites for Inclusion in the Natural World Heritage Network*. https://policycommons.net/artifacts/1374541/terrestrial-biodiversity-and-the-world-heritage-list/1988784/

Allen, L. H., Kimball, B. A., Bunce, J. A., Yoshimoto, M., Harazono, Y., Baker, J. T., . . . & White, J. W. (2020). Fluctuations of $CO_2$ in Free-Air CO2 Enrichment (FACE) depress plant photosynthesis, growth, and yield. *Agricultural and Forest Meteorology*, 284, 107899.

Andresen, E., Arroyo-Rodríguez, V., & Escobar, F. (2018). Tropical biodiversity: The importance of biotic interactions for its origin, maintenance, function, and conservation. In *Ecological Networks in the Tropics* (pp. 1–13). Springer, Cham.

Backhaus, S., Wiehl, D., Beierkuhnlein, C., Jentsch, A., & Wellstein, C. (2014). Warming and drought do not influence the palatability of Quercus pubescens Willd. Leaves of four European provenances. *Arthropod-Plant Interactions*, 8(4), 329–337.

Bawa, K. S. (1990). Plant-pollinator interactions in tropical rain forests. *Annual Review of Ecology and Systematics*, 399–422.

Bencze, S., Keresztényi, I., Varga, B., Kőszegi, B., Balla, K., Gémesné-Juhász, A., & Veisz, O. (2011). Effect of $CO_2$ enrichment on canopy photosynthesis, water use efficiency and early development of tomato and pepper hybrids. *Acta Agronomica Hungarica*, 59(3), 275–284.

Bhardwaj, Y. (2019). Tropics. In Vonk, J., & Shackelford, T. (eds), *Encyclopedia of Animal Cognition and Behavior*. Springer, Cham. https://doi.org/10.1007/978-3-319-47829-6_358-1

Bin Zaher, M. A., Kumagai, E., Yabiku, T., Nakajima, M., Matsunami, T., Matsuyama, N., et al. (2021). Effects of elevated atmospheric CO2 concentration on growth and photosynthesis in eddo at two different air temperatures. *Plant Production Science*, 24(3), 363–373.

Boehm, S., Lebling, K., Levin, K., Fekete, H., Jaeger, J., Waite, R., . . . & Galvin, M. (2021). *State of Climate Action 2021: Systems Transformations Required to Limit Global Warming to 1.5°C*. https://www.wri.org/research/state-climate-action-2021

Boer, M. M., Resco de Dios, V., & Bradstock, R. A. (2020). Unprecedented burn area of Australian mega forest fires. *Nature Climate Change*, 10(3), 171–172.

Cai, M., Lin, X., Peng, J., Zhang, J., Chen, M., Huang, J., . . . & Peng, C. (2021). Why is the invasive plant *Sphagneticola trilobata* more resistant to high temperature than its native congener? *International Journal of Molecular Sciences*, 22(2), 748.

Caldwell, C. R., Britz, S. J., & Mirecki, R. M. (2005). Effect of temperature, elevated carbon dioxide, and drought during seed development on the isoflavone content of dwarf soybean [Glycine max (L.) Merrill] grown in controlled environments. *Journal of Agricultural and Food Chemistry*, 53(4), 1125–1129.

Cavicchioli, R., Ripple, W. J., Timmis, K. N., Azam, F., Bakken, L. R., Baylis, M., . . . & Webster, N. S. (2019). Scientists' warning to humanity: Microorganisms and climate change. *Nature Reviews Microbiology*, 17(9), 569–586.

Cronin, J., Anandarajah, G., & Dessens, O. (2018). Climate change impacts on the energy system: A review of trends and gaps. *Climatic Change*, 151(2), 79–93.

Deharveng, L., & Bedos, A. (2019). Biodiversity in the tropics. In *Encyclopedia of Caves* (pp. 146–162). Academic Press, Cambridge.

Dong, J., Gruda, N., Lam, S. K., Li, X., & Duan, Z. (2018). Effects of elevated CO2 on nutritional quality of vegetables: A review. *Frontiers in Plant Science*, 9, 924.

Drake, J. E., Gallet-Budynek, A., Hofmockel, K. S., Bernhardt, E. S., Billings, S. A., Jackson, R. B., . . . & Finzi, A. C. (2011). Increases in the flux of carbon belowground stimulate nitrogen uptake and sustain the long-term enhancement of forest productivity under elevated CO2. *Ecology Letters*, 14(4), 349–357.

Ellsworth, D. S., Anderson, I. C., Crous, K. Y., Cooke, J., Drake, J. E., Gherlenda, A. N., . . . & Reich, P. B. (2017). Elevated $CO_2$ does not increase eucalypt forest productivity on a low-phosphorus soil. *Nature Climate Change*, 7(4), 279–282.

Ellsworth, D. S., Thomas, R., Crous, K. Y., Palmroth, S., Ward, E., Maier, C., . . . & Oren, R. (2012). Elevated $CO_2$ affects photosynthetic responses in canopy pine and subcanopy deciduous trees over 10 years: A synthesis from Duke face. *Global Change Biology*, 18(1), 223–242.

Ewel, J. (1980). Tropical succession: Manifold routes to maturity. *Biotropica*, 2–7.

Fawzy, S., Osman, A. I., Doran, J., & Rooney, D. W. (2020). Strategies for mitigation of climate change: A review. *Environmental Chemistry Letters*, 18(6), 2069–2094.

Garrett, K. A., Forbes, G. A., Savary, S., Skeley, P., Sparks, A. H., Valdivia, C., . . . & Yuen, J. (2011). Complexity in climate-change impacts: An analytical framework for effects mediated by plant disease. *Plant Pathology*, 60(1), 15–30.

Ghini, R., Torre-Neto, A., Dentzien, A. F., Guerreiro-Filho, O., Iost, R., Patrício, F. R., . . . & DaMatta, F. M. (2015). Coffee growth, pest and yield responses to free-air CO2 enrichment. *Climatic Change*, 132(2), 307–320.

Gissi, E., Manea, E., Mazaris, A. D., Fraschetti, S., Almpanidou, V., Bevilacqua, S., . . . & Katsanevakis, S. (2021). A review of the combined effects of climate change and other local human stressors on the marine environment. *Science of the Total Environment*, 755, 142564.

Goshua, A., Gomez, J., Erny, B., Burke, M., Luby, S., Sokolow, S., . . . & Nadeau, K. (2021). Addressing climate change and its effects on human health: A call to action for medical schools. *Academic Medicine*, 96(3), 324–328.

Gruda, N., & Tanny, J. (2014, August). Protected crops–recent advances, innovative technologies and future challenges. In *XXIX International Horticultural Congress on Horticulture: Sustaining Lives, Livelihoods and Landscapes (IHC2014): 1107* (pp. 271–278). Acta Horticulturae, Belgium. https://doi.org/10.17660/ActaHortic.2015.1107.37

Guoju, X., Weixiang, L., Qiang, X., Zhaojun, S., & Jing, W. (2005). Effects of temperature increase and elevated CO2 concentration, with supplemental irrigation, on the yield of rain-fed spring wheat in a semiarid region of China. *Agricultural Water Management*, 74(3), 243–255.

Hasanuzzaman, M., Nahar, K., Alam, M. M., Roychowdhury, R., & Fujita, M. (2013). Physiological, biochemical, and molecular mechanisms of heat stress tolerance in plants. *International Journal of Molecular Sciences*, 14(5), 9643–9684.

Hatfield, J. L., & Prueger, J. H. (2015). Temperature extremes: Effect on plant growth and development. *Weather and Climate Extremes*, 10, 4–10.

Herrick, J. D., & Thomas, R. B. (2001). No photosynthetic down-regulation in sweetgum trees (Liquidambar styraciflua L.) after three years of CO2 enrichment at the Duke Forest FACE experiment. *Plant, Cell & Environment*, 24(1), 53–64.

Hinojosa, L., Matanguihan, J. B., & Murphy, K. M. (2019). Effect of high temperature on pollen morphology, plant growth and seed yield in quinoa (*Chenopodium quinoa* Willd.). *Journal of Agronomy and Crop Science*, 205(1), 33–45.

Huang, B., & Xu, Y. (2015). Cellular and molecular mechanisms for elevated CO2–regulation of plant growth and stress adaptation. *Crop Science*, 55(4), 1405–1424.

Intergovernmental Panel on Climate Change. (2021). https://climate.selectra.com/en/news/ipcc-report-2022

Jagadish, S. K., Bahuguna, R. N., Djanaguiraman, M., Gamuyao, R., Prasad, P. V., & Craufurd, P. Q. (2016). Implications of high temperature and elevated CO2 on flowering time in plants. *Frontiers in Plant Science*, 7, 913.

Kahn, M. E., Mohaddes, K., Ng, R. N., Pesaran, M. H., Raissi, M., & Yang, J. C. (2019). *Long-Term Macroeconomic Effects of Climate Change: A Cross-Country Analysis* (No. w26167). National Bureau of Economic Research, Cambridge.

Kimball, B. A., Kobayashi, K., & Bindi, M. (2002). Responses of agricultural crops to free-air CO2 enrichment. *Advances in Agronomy*, 77, 293–368.

Kompas, T., Pham, V. H., & Che, T. N. (2018). The effects of climate change on GDP by country and the global economic gains from complying with the Paris climate accord. *Earth's Future*, 6(8), 1153–1173.

Krishnan, R., Sanjay, J., Gnanaseelan, C., Mujumdar, M., Kulkarni, A., & Chakraborty, S. (2020). *Assessment of Climate Change Over the Indian Region: A Report of the Ministry of Earth Sciences (MOES), Government of India* (p. 226). Springer Nature, London.

Leal, J. M., & Resener, J. (2022). *Vertical Stratification of Mammals in the Amazon*. https://ecommons.luc.edu/ures/2022/2022/111/

Li, D., Bodjrenou, D. M., Zhang, S., Wang, B., Pan, H., Yeh, K. W., . . . & Cheng, C. (2021). The endophytic fungus *Piriformospora indica* reprograms banana to cold resistance. *International Journal of Molecular Sciences*, 22(9), 4973.

Lippmann, R., Babben, S., Menger, A., Delker, C., & Quint, M. (2019). Development of wild and cultivated plants under global warming conditions. *Current Biology*, *29*(24), R1326–R1338.

Magrach, A., & Ghazoul, J. (2015). Climate and pest-driven geographic shifts in global coffee production: Implications for forest cover, biodiversity and carbon storage. *PloS ONE*, *10*(7), e0133071.

Marques, I., Fernandes, I., David, P. H., Paulo, O. S., Goulao, L. F., Fortunato, A. S., . . . & Ribeiro-Barros, A. I. (2020). Transcriptomic leaf profiling reveals differential responses of the two most traded coffee species to elevated [CO2]. *International Journal of Molecular Sciences*, *21*(23), 9211.

Marques, I., Ribeiro-Barros, A., & Ramalho, J. C. (2022). Editorial: Tropical plant responses to climate change. *International Journal of Molecular Sciences*, *23*, 7236.

Maulu, S., Hasimuna, O. J., Haambiya, L. H., Monde, C., Musuka, C. G., Makorwa, T. H., . . . & Nsekanabo, J. D. (2021). Climate change effects on aquaculture production: Sustainability implications, mitigation, and adaptations. *Frontiers in Sustainable Food Systems*, *5*, 609097.

Mertens, J. E., Brisson, L., Janeček, Š., Klomberg, Y., Maicher, V., Sáfián, S., . . . & Tropek, R. (2021). Elevational and seasonal patterns of butterflies and hawkmoths in plant-pollinator networks in tropical rainforests of Mount Cameroon. *Scientific Reports*, *11*(1), 1–12.

Myers, N., Mittermeier, R. A., Mittermeier, C. G., Da Fonseca, G. A., & Kent, J. (2000). Biodiversity hotspots for conservation priorities. *Nature*, *403*(6772), 853–858.

NASA. (2018). *Goddard Institute for Space Studies*. https://science.gsfc.nasa.gov/earth/giss/

Norby, R. J., De Kauwe, M. G., Domingues, T. F., Duursma, R. A., Ellsworth, D. S., Goll, D. S., . . . & Zaehle, S. (2016). Model–data synthesis for the next generation of forest free-air CO 2 enrichment (FACE) experiments. *New Phytologist*, *209*(1), 17–28.

Nowak, R. S., Ellsworth, D. S., & Smith, S. D. (2004). Functional responses of plants to elevated atmospheric CO2–do photosynthetic and productivity data from FACE experiments support early predictions? *New Phytologist*, *162*(2), 253–280.

Olabi, A. G., & Abdelkareem, M. A. (2022). Renewable energy and climate change. *Renewable and Sustainable Energy Reviews*, *158*, 112111.

Poker, J., & MacDicken, K. (2016). Tropical forest resources: Facts and tables. *Tropical Forestry Handbook*, 3–45.

Poore, A. G., Graham, S. E., Byrne, M., & Dworjanyn, S. A. (2016). Effects of ocean warming and lowered pH on algal growth and palatability to a grazing gastropod. *Marine Biology*, *163*(5), 1–11.

Post, E., Alley, R. B., Christensen, T. R., Macias-Fauria, M., Forbes, B. C., Gooseff, M. N., . . . & Wang, M. (2019). The polar regions in a 2C warmer world. *Science Advances*, *5*(12), eaaw9883.

Rockström, J., Gupta, J., Lenton, T. M., Qin, D., Lade, S. J., Abrams, J. F., . . . & Winkelmann, R. (2021). Identifying a safe and just corridor for people and the planet. *Earth's Future*, *9*(4), e2020EF001866.

Reyes, H. C., Draper, D., & Marques, I. (2021). Pollination in the rainforest: Scarce visitors and low effective pollinators limit the fruiting success of tropical orchids. *Insects*, *12*(10), 856.

Sakai, H., Tokida, T., Usui, Y., Nakamura, H., & Hasegawa, T. (2019). Yield responses to elevated CO2 concentration among Japanese rice cultivars released since 1882. *Plant Production Science*, *22*(3), 352–366.

Stevens, R. D., & Tello, J. S. (2018). A latitudinal gradient in dimensionality of biodiversity. *Ecography*, *41*(12), 2016–2026.

Suriyasak, C., Oyama, Y., Ishida, T., Mashiguchi, K., Yamaguchi, S., Hamaoka, N., . . . & Ishibashi, Y. (2020). Mechanism of delayed seed germination caused by high temperature during grain filling in rice (Oryza sativa L.). *Scientific Reports*, *10*(1), 1–11.

Sylla, M. B., Elguindi, N., Giorgi, F., & Wisser, D. (2016). Projected robust shift of climate zones over West Africa in response to anthropogenic climate change for the late 21st century. *Climatic Change*, *134*(1), 241–253.

Thomas, J. M. G., Boote, K. J., Allen, L. H., Gallo-Meagher, M., & Davis, J. M. (2003). Elevated temperature and carbon dioxide effects on soybean seed composition and transcript abundance. *Crop Science*, *43*(4), 1548–1557.

Tubiello, F. N., Soussana, J. F., & Howden, S. M. (2007). Crop and pasture response to climate change. *Proceedings of the National Academy of Sciences*, *104*(50), 19686–19690.

Wassmann, R., Nelson, G. C., Peng, S. B., Sumfleth, K., Jagadish, S. V. K., Hosen, Y., & Rosegrant, M. W. (2010). Rice and global climate change. *Rice in the Global Economy: Strategic Research and Policy Issues for Food Security*, 411–432.

Wu, C., Cui, K., Tang, S., Li, G., Wang, S., Fahad, S., . . . & Ding, Y. (2020). Intensified pollination and fertilization ameliorate heat injury in rice (Oryza sativa L.) during the flowering stage. *Field Crops Research*, *252*, 107795.

Xu, X., Wu, P., Song, H., Zhang, J., Zheng, S., Xing, G., . . . & Li, M. (2020). Identification of candidate genes associated with photosynthesis in eggplant under elevated CO2. *Biotechnology & Biotechnological Equipment*, *34*(1), 1166–1175.

Zandalinas, S. I., Fritschi, F. B., & Mittler, R. (2021). Global warming, climate change, and environmental pollution: Recipe for a multifactorial stress combination disaster. *Trends in Plant Science*, *26*(6), 588–599.

Zandalinas, S. I., Mittler, R., Balfagón, D., Arbona, V., & Gómez-Cadenas, A. (2018). Plant adaptations to the combination of drought and high temperatures. *Physiologia Plantarum*, *162*(1), 2–12.

Zhang, P., Grutters, B. M., Van Leeuwen, C. H., Xu, J., Petruzzella, A., Van den Berg, R. F., & Bakker, E. S. (2019). Effects of rising temperature on the growth, stoichiometry, and palatability of aquatic plants. *Frontiers in Plant Science*, *9*, 1947.

## WEBSITES

1) www.un.org/en/climatechange/science/causes-effects-climate-change
2) www.un.org/en/conferences/environment
3) https://research.un.org/en/docs/environment/conferences
4) www.arborday.org/programs/rainforest/animals-and-plants-of-the-rain-forest.cfm
5) https://www.who.int/news-room/fact-sheets/detail/climate-change-and-health

# Section 2

*Tropical plants responses to atmospheric deposition and air pollutants*

# 4 Impacts of Air Pollutants on the Ecophysiology of Tropical Plants

*Pallavi Singh, Jigyasa Prakash, Harshita Singh, Shashi Bhushan Agrawal, and Madhoolika Agrawal*

## 4.1 INTRODUCTION

One of the most grievous challenges of the present time which human civilization is facing is air pollution. The concentration, toxicity and complexity of pollution in the atmosphere has increased alarmingly in due course of time. Air pollution can be described as an intricate mixture of gases and compounds present in the air at an unusual concentration which is potentially harmful to plants, ecosystems and humans. The predominant pollutants include oxides of sulfur ($SO_X$) and nitrogen ($NO_X$), carbon monoxide (CO), volatile organic carbons (VOCs), ozone ($O_3$) and particulate matter ($PM_{2.5}$ and $PM_{10}$) (Kumar et al., 2019; Opio et al., 2021). The World Health Organization (WHO) recently published data indicating an increase in pollutant levels in most parts of the world. Almost 99% of the world's population lives in areas where airborne pollutants exceed WHO permissible levels (WHO, 2022). In 2021, the WHO published air quality guidelines (AQGs) as a new global update after 2005. AQGs act as an indicative and informative monitoring tool for policy-makers to amend the legislation and policies accordingly, thereby reducing the levels of air pollutants and further decreasing the burden of diseases that results from exposure to air pollutants worldwide (WHO, 2021) (Table 4.1).

Plants actively participate in various ecosystem services including the recycling of nutrients and gases, specifically oxygen and carbon dioxide and provide impingement area on their leaves, allowing accumulation and absorption of pollutants to reduce pollution load in the atmosphere, and thus play an important role in sustaining and maintaining the ecological balance (Escobedo et al., 2008). As the richest biological ecosystems, the tropics are already suffering from various natural and anthropogenic risks owing to modifications in climatic conditions, thus making the assessment of the responses of tropical plants to air pollution important. The current chapter is concerned with the source apportionment for various air pollutants and describes their consequent effects on growth, physiology and biochemistry, along with biomass allocation and yield attributes of dominant crops and trees of the tropical region.

## 4.2 METHODOLOGY

An organized overview was performed by utilizing the Internet (World Wide Web) with keywords such as air pollution and its sources, sulfur dioxide and nitrogen oxides, ozone, particulate matter and effects on tropical plants and vegetation. The application of specific filters such as tropical forest, tropics and subtropics in Google Scholar, Springer Link, PubMed and ScienceDirect was done to screen the studies that were carried out in a tropical or subtropical climate, specifically in parts of North America (Mexico) and South America (Brazil); West, Central and East Africa; along with Southeast Asia. Articles were further screened for different response types such as visible symptoms, leaf functional traits, foliar bioaccumulation, growth and morphology, physiology and biochemical characteristics for trees, grasses, shrubs, ornamental and crop plants for assessment of tropical and subtropical plant responses to air pollution.

## 4.3 SOURCES AND MODES OF DEPOSITION OF POLLUTANTS

Pollutants in the air are emitted from a variety of sources. To regulate the emission of these pollutants, it is important to determine the nature of sources, along with their contributing factors (Figure 4.1). The process of combustion is the major contributor to primary as well as secondary pollutants. In order to clear land and destroy crop residue, extensive biomass is burned, which includes controlled and uncontrolled forest and savannah fires. Outdoor combustion sources include transportation on air, water and land, as well as industrial and power generation (Agrawal & Singh, 2001). Volcanic eruptions and incomplete combustion of fossil fuel are the other sources specifically for $SO_2$. Local sources include manufacturing, electricity generation and transportation. Domestic activities such as cooking with coal or wood, briquettes or firewood, and garbage burning also emit gases and particulate matter locally (Bateebe, 2011; Wamoto, 2013).

The process of $O_3$ formation entails photochemical reactions of $NO_X$, CO and VOCs, which basically have the anthropogenic origin and primarily result from fossil fuel combustion, biomass burning and industrial emissions.

**TABLE 4.1**

**WHO-Recommended Long- and Short-Term Air Quality Guidelines for 2005 and 2021**

| Pollutant | Averaging time | 2005 AQGs | 2021 AQGs |
|---|---|---|---|
| CO, mg m$^{-3}$ | 24-hour | – | 4 |
| NO$_2$, µg m$^{-3}$ | Annual | 40 | 10 |
|  | 24-hour | – | 25 |
| O$_3$, µg m$^{-3}$ | Annual | – | 60 |
|  | Peak season | 100 | 100 |
| SO$_2$, µg m$^{-3}$ | 24-hour | 20 | 40 |
| PM$_{2.5}$, µg m$^{-3}$ | Annual | 10 | 5 |
|  | 24-hour | 25 | 15 |
| PM$_{10}$, µg m$^{-3}$ | Annual | 20 | 15 |
|  | 24-hour | 50 | 45 |

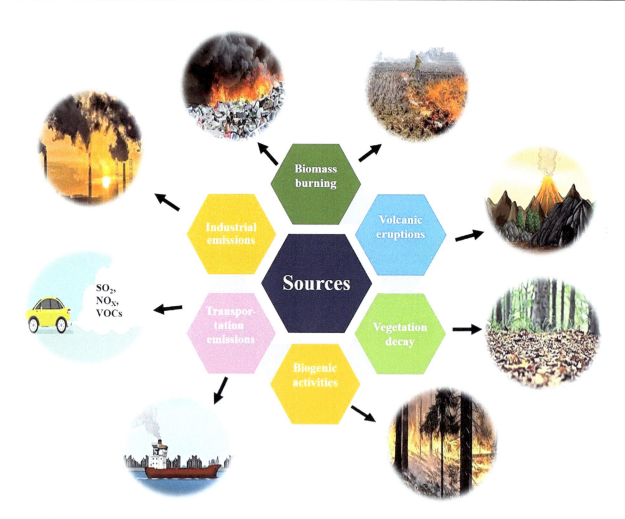

**FIGURE 4.1** Schematic diagram showing various sources of gaseous and particulate pollutants.

Further, the identified source profiles for PM are wood burning, soil dust, brick kilns, road dust, motor vehicles, sea salt, biomass burning and metal smelters (Begum & Hopke, 2019). Gurjar et al. (2016) identified industries and power plants as the main anthropogenic sources of SPM and PM$_{10}$ emissions (20–80%) in three Indian megacities. Small-scale industries, traffic, construction activities, resuspended soil and road dust, residential coal burning and biomass burning are additional sources. Light duty vehicles were identified as the main source of PAH emissions in PM, followed by industrial plants, according to Cavalcante et al. (2017). In tropical rainforest in Manaus, Brazil, biomass or

# Impacts of Air Pollutants on Ecophysiology of Tropical Plants

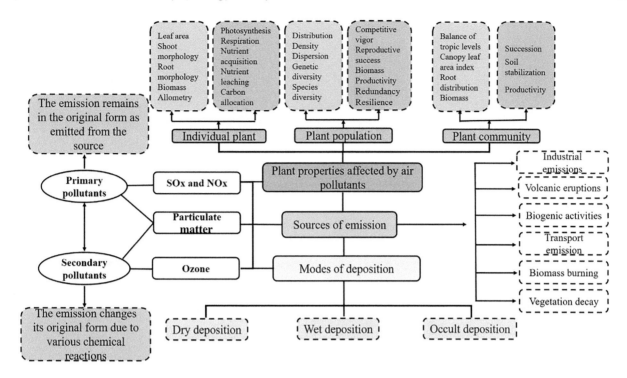

**FIGURE 4.2** Detailed descriptions of sources and modes of deposition of predominant air pollutants and their probable effects.

fossil fuel combustion was identified as the primary source of atmospheric PAHs (Krauss et al., 2005).

The deposition of air pollutants from the atmosphere follows three different modes: namely wet deposition, dry deposition and occult deposition (Figure 4.2), which refers to the latent form which remains unmeasured (by fog, cloud water, dew and mist interception) (Grantz et al., 2003). In dry deposition, the pollutants are deposited due to wind or resuspension from anthropogenic disturbances and are highly regulated by the surface features of plants, as well as meteorological factors. Dry deposition depends upon the distance and strength of emission from the source. Wet deposition, mainly facilitated by rain and snow, contributes to more deposition of pollutants than dry. The epidermal tissue is coated by cuticular wax, which checks the pollutant uptake through acid rain. Atmospheric conditions equally contribute to the strength of dry deposition and the major factors are wind speed, stability, mixing height, temperature, humidity and dew formation. Along with these, leaf traits like leaf pubescence, leaf shape, plant density and branch spacing, as well as the vegetation conditions, i.e., surface wetness, salt and organic exudates, also influence the pollutant accumulation potential of vegetation.

## 4.4 PREDOMINANT AIR POLLUTANTS IN THE TROPICS

### 4.4.1 Gaseous Pollutants

#### 4.4.1.1 $SO_2$ and $NO_2$

The emission of atmospheric nitrogen dioxide ($NO_2$) and sulfur dioxide ($SO_2$) has been reported to show an increasing trend in tropical and subtropical regions during the last few decades (Krotkov et al., 2016). $SO_2$ and $NO_X$ are principal air pollutants formed during the combustion of fossil fuels. The corresponding proportions of nitrogen oxides (NO, $N_2O$, $NO_2$, $N_2O_3$, $N_2O_5$) are highly variable and extremely unstable, both temporarily and spatially in the atmosphere. $NO_X$ plays a notable role in global warming and the thinning of the protective ozone layer. Among different nitrogen oxides, $NO_2$ is considered to be the most potent form affecting plants, while other forms can get converted to $NO_2$ in the presence of sunlight and conducive atmospheric conditions (Wang et al., 2019). Being a colorless and water-soluble gas, $SO_2$ regarded as the most injurious gases to plants and human health. $SO_2$ and $NO_2$ also act as precursors in the genesis of sulphate ($SO_4^{2-}$) and nitrate ($NO_3^-$) aerosols, respectively, leading to acid rain formation (Opio et al., 2021; Prakash et al., 2023) (Figure 4.3). The atmospheric lifetime of $SO_2$ and $NO_2$ is dynamic and ranges from days to a few months, depending on the relative height of species in the atmosphere (Myhre et al., 2013; Seinfeld & Pandis, 2016).

Among major Indian cities, Delhi had the highest $SO_2$ emissions, followed by Mumbai (Sharma & Mauzerall, 2022); however, after 2018, both cities evidenced lowered levels of $SO_2$. Delhi, Hyderabad and Kolkata also showed comparatively high $NO_2$ with annual average concentrations exceeding the National Ambient Air Quality Standard (NAAQS) of 40 μg m$^{-3}$ for residential areas (Sharma & Mauzerall, 2022; Mandal et al., 2022). Areas around Varanasi recorded $NO_2$ concentrations (49–157 μg m$^{-3}$) higher than NAAQS, whereas $SO_2$ (3–21 μg m$^{-3}$) level was below the NAAQS (50 μg m$^{-3}$) (Mukherjee & Agrawal, 2016). An analysis by Kuttippurath et al. (2022) showed that $SO_2$ emission in India during 2010–2020 decreased as a result of the adaptation of new clean technologies, strict

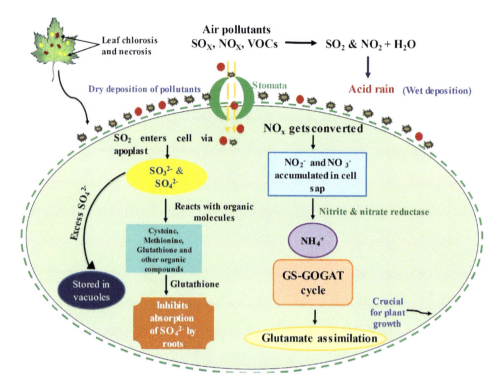

**FIGURE 4.3** Simplified scheme for entry, accumulation and metabolism of $SO_2$ and $NO_2$ in plants.

**TABLE 4.2**
**Variations in $SO_2$ and $NO_2$ Concentrations at Different Locations/Cities**

| Location | Period | $SO_2$ (µg m$^{-3}$) | $NO_2$ (µg m$^{-3}$) | References |
|---|---|---|---|---|
| Ahmedabad, India | 2004–2018 | 14.23 | 23.63 | Gupta and Dhir (2022) |
| Rajkot, India | 2004–2018 | 13.4 | 19.6 | Gupta and Dhir (2022) |
| Dhaka, Bangladesh | 2013–2017 | 28.5 | 31.7 | Rahman et al. (2019) |
| Northern India | 2015–2019 | 12 ± 7 | 35 ± 21 | Sharma and Mauzerall (2022) |
| Southern India | 2015–2019 | 12 ± 10 | 27 ± 16 | Sharma and Mauzerall (2022) |
| Delhi, India | 2017 | 9 ± 5 | 59 ± 22 | Mandal et al. (2022) |
| Kolkata, India | 2020 | <10 | 40–50 | Biswas et al. (2020) |
| Chennai, India | 2015–2017 | 3–78 | 5.6–99.6 | Prabakaran and Manikandan (2018) |
| Varanasi, India | 2014–2018 | 2–51 | 6–195 | Singh et al. (2021a) |
| Ho Chi Minh, Vietnam | 2015–2019 | 50 | 60 | Ho et al. (2022) |
| Visakhapatnam, India | 2018–2019 | 14.43 | 49.33 | Chandu and Dasari (2020) |
| South Africa | 1995–2015 | 2.9 | 2.4 | Swartz et al. (2020) |
| Petaling Jaya, Malaysia | 2008–2012 | 4.21 | 32.75 | Zizi et al. (2018) |
| Northern Thailand | 2006–2016 | 2.9 ± 3.6 | 13.5 ± 15.9 | Janta et al. (2020) |

control measures and a shift toward renewable energy sources. Table 4.2 shows the variations in $SO_2$ and $NO_2$ concentrations in different cities of India, as well as tropical countries of the world.

### 4.4.1.2 Tropospheric Ozone

Tropospheric ozone ($O_3$), a strong greenhouse gas and a secondary air pollutant, though having a shorter life time than $CO_2$ (carbon dioxide), is competent enough to persuade global warming and climate change (Alexander et al., 2013). A handful of studies have suggested the detrimental effects of ozone on the growth, yield and productivity of annual and perennial species (Mills et al., 2018; Emberson et al., 2018; Agathokleous et al., 2020).

Research reporting the effect of $O_3$ on plants including agricultural crops and trees has helped us to figure out the modifications and adaptive strategies that the plants follow. Tropics are considered as the center of origin for the majority of food crops including *Zea mays, Solanum tuberosum, Ipomoea batatas, Sorghum bicolor, Lycopersicon esculentum, Capsicum annum, Cucumis sativa, Arachis hypogea, Hevea brasiliensis, Nicotiana tabacum, Gossypium hirsutum, Phaseolus vulgaris, Saccharum officinarum*, etc., that have been domesticated throughout the region from the ages because of their economic and ecological importance (Morales, 2009). For these reasons, tropical crops—especially cereals—have been much in the limelight for their responses to the increasing $O_3$ concentrations, but their studies are considerably fewer in number as compared with temperate crops. Variations in ozone tolerance/sensitivity have been reported for major crop species, i.e., *Triticum aestivum* (Yadav et al., 2021a; Fatima et al., 2019), *Oryza sativa* (Rai et al., 2008), *Z. mays* (Singh et al., 2014) and *Glycine max* (Jiang et al., 2018).

According to the current Indian scenario, recent $O_3$ monitoring data (2018–2019) showed comparatively higher AOT40 values in crop-growing seasons of *O. sativa* and *T. aestivum* (Yadav et al., 2020; Mukherjee et al., 2018). The existing $O_3$ levels, recorded with a multimodel approach, varied from 37.3–56.1 ppb over India (Hakim et al., 2019). In three important cities in Malaysia, $O_3$ levels exceeded the values of the National Ambient Air Quality Standard (Noor et al., 2018). Similarly, higher concentrations than the Brazilian and WHO standards were reported in São Paulo, Brazil (Moura et al., 2018) where the concentrations varied from 8.7–96.1 ppb. Rural sites are always the major hub for the highest $O_3$ concentrations than the semi-urban sites. The maximum hourly $O_3$ concentration in Thailand reached 123 ppb, which is above the standard concentration of 100 ppb (Kittipornkul et al., 2021). Further, the data from 24 monitoring stations revealed an increment in the average $O_3$ concentration in Thailand, particularly in the growth duration of off-season rice. At present, the monitoring data of $O_3$ in most regions of Africa is insufficient, but areas across Southern Africa are predicted to be at higher risk under a "policy fail" SRES A2-type emission scenario (Laban et al., 2018).

### 4.4.2 Particulate Matter

Particulate matter (PM) has gained considerable attention owing to its negative effects on the environment and human well-being, along with its positive impact on global climate change (Polichetti et al., 2009; Perron et al., 2010). Fine PM are of serious concern due to its increasing number of sources (Rajput & Agrawal, 2005). Fine PM can damage cells, obstruct stomata and alter the mechanisms of plant evapotranspiration and respiration (Rahul & Jain, 2014), while coarse PM are more likely to affect leaf surfaces by deposition, thereby impeding with gaseous diffusion and photosynthetic spectral quality. It is a heterogeneous combination of liquids and suspended solids with various chemical and physical properties that come from both natural and synthetic sources. The primary criterion used to characterize PM in air quality monitoring and assessment is its aerodynamic diameter. They have been characterized as $PM_{10}$, $PM_{2.5}$ and $PM_1$, and generally termed as coarse, fine and ultra-fine, respectively. Variability in combustion sources and atmospheric conditions have an immense effect on the chemical characteristics of PM, which vary widely (Daresta et al., 2015; Amodio et al., 2012; Solomon & Sioutas, 2008).

Plants are negatively impacted by PM, either advertently (via adherence on above-ground biomass or surface penetration of leaf) or inadvertently (via soil-root interaction) (Žalud et al., 2012). The adversity of PM deposition on plant leaves follows two major ways: (1) by the obnoxious effects of the trace elements adhered to the PM; and (2) by the shading effect of PM on the leaf surface (Pavlík et al., 2012). Dust alters the spectral quality and quantity of light available for photosynthesis, notably the surface reflection of short-wave infrared and visible light (Hope et al., 1991; Keller & Lamprecht, 1995). The presence of trace metals is one of the prime characteristics of particulate matter that is related to its toxicity (Rodriguez et al., 2010). Additionally, PM undermines food security by inflicting direct or indirect harm, such as crop stunting; damage to the cells, tissue and stomata; and causing chlorosis/necrosis of foliage.

## 4.5 IMPACTS OF AIR POLLUTANTS ON ECOPHYSIOLOGY OF PERENNIAL VEGETATION AND CROP PLANTS

### 4.5.1 Foliar Injury

Leaves are the most sensitive and effective pollutant-trapping systems by the process of absorption of particulates and gaseous pollutants ($SO_2$, $NO_X$ and other trace elements) (Balasubramanian et al., 2018) (Figure 4.2). Foliar injuries in form of chlorosis, necrosis, flecking and intercostal and discrete stipples have been identified and widely reported to be caused by gaseous air pollutants (Shaw et al., 1993; Mukherjee et al., 2019). Fumigation of plants with low concentrations of $SO_2$ led to the development of chronic injury due to sufficient $SO_4^{2-}$ accumulation in the cells, which causes yellowing of leaves or mild chlorosis (Agrawal & Agrawal, 1999) (Figure 4.3). Fumigation with 7 ppm and 14 ppm $SO_2$ leads to the development of chlorosis on the leaf of tree species (Han et al., 2022). Similarly, the development of necrosis was also observed on the needles of *Pinus sylvestris* when subjected to $SO_2$ concentration of 0.013–0.022 ppm for four hours up to one month (Shaw et al., 1993). Exposure to 0.3 ppm $SO_2$ resulted in the browning and shedding of leaves of *Aleurites montana* after 5–8 days (He et al., 1996). Visible injuries on leaves of deciduous trees (*Azadirachta indica, Cassia siamea, Sizygium cumini, Dalbergia sissoo, Tectona grandis* and *Butea monosperma*) in the form of chlorosis, necrosis and tip burning were found by Nayak et al. (2013).

Kateivas et al. (2022) observed leaf injury and alterations in anatomical features of tree species under pollutant plumes

(SO$_2$ and NO$_X$). The results further reported increases in the frequency of stomata and mesophyll thickness. Pollutants induced irregularity in the shape of stomata, deformities and thickening of the epidermis in *Mangifera indica*; thickening of the upper epidermis, mesophyll and palisade parenchyma in *Psidium cattleianum* were also reported (Kateivas et al., 2022).

Moura et al. (2018) found small brownish intercostal stipples in *Astronium graveolens* and discrete stipples on adaxial surface of *Piptadenia gonoacantha*, while *Croton floribundus* remained unaffected when exposed to 70 ppb of elevated O$_3$. However, when these seedlings were exposed to 200 ppb of O$_3$, all the species developed injuries like necrosis and stippling on the adaxial surface of the leaves (Moura et al., 2014). Further, symptoms like flecking and chlorosis were developed in *C. floribundus* under exposure of 100 ppb of O$_3$ (Dias et al., 2019). According to de Rezende and Furlan (2009) and Furlan et al. (2008), the foliar symptoms in *P. guajava* and *Tibouchina pulchra* were typically characterized by dark or dark-reddish colored interveinal stippling on the older leaves which became more intense with increasing exposure. Visible symptoms were also found on road side tree species like *Cinnamomum camphora*, *M. indica* and *Ficus altissima* (Long et al., 2018). *C. camphora* showed small gray spots, while brown spots appeared on leaves of *M. indica* and black punctate spots on *F. altissima*.

Cerón-Bretón et al. (2010a) found foliar symptoms in form of necrosis and chlorosis under higher concentrations of O$_3$ (110 ppb and 250 ppb) in four tropical tree species: *Haematoxylum campechianum*, *Tabebuia rosea*, *Cedrela odorata* and *Swietenia macrophylla*. However, *Phoenix dactylifera* fumigated with 200 ppb of O$_3$ did not show any typical foliar symptoms (Du et al., 2018).

Norby and Luxmoore (1983) reported no visible injury on leaves of soybean when fumigated to a combination of SO$_2$ and O$_3$. High concentrations of NO$_2$ also caused leaf injury and plasmolysis of cells leading to membrane destruction, metabolic inhibition and death (Alscher et al., 1987; Prakash et al., 2023). Short-term NO$_2$ exposure resulted in a significant increase in the thickness of the palisade tissue, which caused swelling of the thylakoids within the chloroplasts (Sheng & Zhu, 2018). A study of Agrawal et al. (1983) observed brownish necrotic bands on the upper and middle laminar regions of *Panicum miliaceum* leaves after eight days of exposure with 500 µg m$^{-3}$ SO$_2$.

Field studies conducted by Ahmad et al. (2013) found visible injury in *S. tuberosum*, *Allium cepa* and *G. hirsutum* at two sites in Peshawar, Pakistan. However, no ozone injury was found on *C. sativus*, *L. esculentum* or *Beta vulgaris*. Yellowing and chlorosis of leaves were found in *T. aestivum*, *Pisum sativum* and *Vigna radiate* under ambient O$_3$ levels at three different sites, characterized as high (291 ppb) and low (110 ppb) polluted as compared to the control site (Adrees et al., 2016a, 2016b).

Most of the studies conducted on African food crops such as *T. aestivum*, *Eleusine coracana*, *Pennisetum glaucum* and *P. vulgaris* showed visible injuries under varying levels of O$_3$ characterized as high (84–110 ppb), medium (67.3–80 ppb) and low (30–32.5 ppb) pollutant sites (Sharps et al., 2021; Hayes et al., 2020b). Interveinal chlorosis and reddening of leaves under high O$_3$ concentrations were also detected in two *Z. mays* cultivars (cvs. HQPM1 and DHM117) as a result of over-accumulation of anthocyanin (Singh et al., 2014). Preliminary screening of six cultivars of *Trifolium alexandrinum* revealed that among the six cultivars, Wardan, Bundel and JHB-146 showed the appearance of small pale yellow and brown flecks at the initial stages of growth which later turned into bifacial necrosis under ambient and elevated O$_3$ (Chaudhary & Agrawal, 2013), whereas Fahli, Saidi and Mescavi reflected less foliar injury that developed in their later growth stages. The damaging effect of PM is widely known, but no evidence of typical plant injuries is reported in the available research.

### 4.5.2 Physiology

It is important for plants to regulate their function with the changing environmental conditions to maintain their vital processes. Plants try to minimize their exposure to air pollutants either by stomatal closure or by membrane impermeability, which may lead to a series of changes in the physiological behavior of the plants (Figure 4.2). Plants responded negatively to air pollutants, and most of the physiological activities such as photosynthetic rate ($P_n$), stomatal conductance ($g_s$), $F_v/F_m$ (maximal photochemical efficiency), photosystem II (PSII) quantum yield and electron transport rate (ETR) were found to be reduced with the increasing pollutant load (Yadav et al., 2020).

Photosynthesis is one of the first processes to be harmed by high SO$_2$ levels (Ziegler, 1972). SO$_X$ and NO$_X$ caused a negative impact on the thylakoid membrane system (Le & Oanh, 2010). Han et al. (2022) found that the leaf temperature, green-peak reflectance and $F_v/F_m$ decreased significantly as the concentration of SO$_2$ was increased. Moreover, there was a consistent reduction in $F_v/F_m$ of *Ulmus pumila* compared to *Syringa oblata* and *Prunus cerasifera*. The order of the leaf resistance performance of tree species under different SO$_2$ concentrations was *P. cerasifera* > *S. oblata* > *U. pumila*, and the degree of resistance decreased with increasing concentration (Han et al., 2022). Effect of gaseous NO$_2$ was assessed (4 ppm) on *Populus alba* × *P. berolinensis* by Hu et al. (2015), which showed that $P_n$ was reduced significantly as compared to the control (ambient NO$_2$ <0.1 ppm). Similar effects have been also reported for increased fumigation frequency or increased NO$_X$ concentrations (<0.005, 0.025, 0.05, 0.091 and 0.187 ppm) at a given single application of four *Eucalyptus* species, viz., *E. globulus*, *E. microcorys*, *E. marginata* and *E. pilularis* (Murray et al., 1994).

Both acute (Moraes et al., 2004; Du et al., 2018) and chronic exposures of O$_3$ (Pina & Moraes, 2010; Pina et al., 2017) were reported to reduce the physiological activities of plants. A reduction of 50% in $P_n$ and 42% in $g_s$ were reported for seedlings of *Caesalpinia echinata* under elevated $O_3$ concentration of 68 ppb (Moraes et al., 2006). Similarly, $P_n$ and $g_s$ in *A. graveolens* were reduced by 40% and 36%, respectively, while respiration ($R_D$) was found to be increased under 70 ppb of $O_3$ (Moura et al., 2018). The damaged photosynthetic machinery under $O_3$ stress led to the decrease in $P_n$ while the increase in intercellular $CO_2$ ($C_i$) might be related to the decrease in $g_s$ (Hoshika et al., 2015). The increase in $R_D$ was probably due to the reprograming of mitochondrial metabolism, higher detoxification rates and increased repair of cellular damage (Kitao et al., 2009; Hoshika et al., 2013).

Da Silva Engela et al. (2021) found that the decline in $P_n$ (31–41%) of *Eugenia uniflora* was compensated by an increase in leaf number. Calatayud et al. (2006) reported an increase in $C_i$ concentration, along with decreases in $P_n$ (19%), $g_s$ (41%) and non-cyclic electron flow, in *Citrus unshiu* saplings under elevated $O_3$ concentration for a year. Percent reductions in $P_n$ varied between 16–39% in *F. infectoria*, followed by 7–31% in *Pongamia pinnata* and 6–26% in *Bauhinia variegata*, while RuBP carboxylase activity was reduced maximally in *B. variegata* (10–32%) followed by *F. infectoria* (10–23%) and *P. pinnata* (9–15%) under elevated $O_3$ (40–120 ppb) (Chapla & Kamalakar, 2004).

A study by Singh et al. (2021d) at Jharia coalfield evaluated the effect of air pollution on five native tree species (*F. benghalensis*, *B. monosperma*, *Adina cordifolia*, *Aegle marmelos* and *F. religiosa*). The study concluded that the decline in $g_s$ was maximum in *F. benghalensis*, which consequently checked the flux of other gaseous pollutants. A mangrove plant, *Avicennia marina*, showed that leaves coated with dust displayed decreased PSII quantum yield, ETR through PS II and Fv/Fm, along with gaseous exchange capacity of upper and lower leaf surfaces of the plant (Naidoo & Chirkoot, 2004). Pereira et al. (2009) reported reductions in $P_n$, $g_s$ and transpiration (E), and quantum yield of PS II, in a CAM plant, *Clusia hilariana*, under PM stress. A comparative study between polluted and unpolluted forests of Birbhum district, India by Saha and Padhy (2012) evaluated the impact of stone dust deposition on E in *Shorea robusta* plants, which showed leaf damage and dust accumulation at the stomatal surface leading to a decrease in E of the leaves at the polluted forest site. The E index decreased with higher PM concentrations and dust fall.

One of the pioneer studies by Majernik and Mansfield (1970) found that $SO_2$ of 1ppm promotes the opening of stomata and thereby increases the diffusion of $SO_2$ in the leaves of *Vicia faba*. In the same study, $SO_2$-fumigated plants showed a faster rate of stomatal opening and a larger stomatal aperture as compared to untreated plants. The stimulatory effect of $SO_2$ on stomatal opening was reversible in a short term but became irreversible after six hours. It has also been observed that the night opening of stomata occurs 1–2 hours earlier under $SO_2$ treatment as compared to the control. Similar observations have been also reported in *Z. mays* and *Hordeum vulgare* (Majernik & Mansfield, 1970).

According to Han et al. (2019), $SO_2$ acted as a beneficial regulator for plants against drought tolerance in *Setaria italica* seedlings. The leaf relative water content of drought-stressed seedlings was significantly improved by pre-exposure to 10.9 ppm $SO_2$, which was ascribed to reduced stomatal apertures and transpiration induced by $SO_2$ under drought conditions. Exposure to seasonal (winter/summer) fluctuations of air pollutants $SO_2$ (0.03–0.04 ppm), $NO_2$ (29–80 ppb), and $O_3$ (9–58 ppb) to different crop species, including *T. aestivum*, *Brassica campestris*, *B. vulgaris* and *V. radiata*, showed reduced rates of $P_n$ and $g_s$ in the suburban areas of Varanasi, India (Agrawal et al., 2003).

Physiological responses of plants to $NO_X$ are complex due to the diverse metabolic responses of plants and interactions with other concurrent air pollutants (Sutton & Bleeker, 2013). Inhibitory effects of $NO_2$ on $P_n$ and $R_D$ have been reported in *P. vulgaris* (Srivastava, 1974), *G. max* (Carlson, 1983), *Vulpia microstachys* (Vallano et al., 2012) and *Raphanus sativus* (Mazarura, 2012), whereas stimulatory effects on $R_D$ have been found for *G. max* (Sabaratnam et al., 1988) and *Lolium multiflorum* (Vallano et al., 2012). Moreover, micro-morphological analysis revealed reductions in stomatal number, size and index in all four tree species, while increases in the number and length of epidermal cells and trichome length were recorded in *F. platyphylla*, *M. indica*, *Polyalthia longifolia* and *Terminalia cattapa* at the polluted sites compared to the control (Uka & Belford, 2020).

Sheng and Zhu (2018) evaluated the physiological responses, along with stomatal and chloroplast features, in the leaves of *Carpinus putoensis* under exposure to $NO_2$ (6 ppm). The findings revealed that short-duration exposure with a high concentration of $NO_2$ led to an adverse impact on the leaf chlorophyll content and photosynthesis.

Several studies have documented the negative effects of $O_3$ on photosynthesis (Rai & Agrawal, 2008; Singh et al., 2009; Sarkar & Agrawal, 2010). The reduced assimilation in *T. aestivum* was found to be related to leaf longevity as well as increasing senescence under ambient $O_3$ concentration (Sarkar & Agrawal, 2010). Reductions in $P_n$, $g_s$ and E were also observed in Malaysian cultivars of *O. sativa* (MR 219) under ambient $O_3$ (Shahadan, 2005). No effect on the initial fluorescence of the cultivars indicated that the light-harvesting system and the antenna pigments remained unaffected by $O_3$ in their initial stages. However, $P_n$ was reduced under high $O_3$ treatment by 62%, and this decline was more related to the stomatal closure than light-harvesting system of the cultivars. Studies from Pakistan also concluded that reductions in $P_n$ and $g_s$ were found under ambient $O_3$ in all the studied crops (Adrees et al., 2016a, 2016b). Stomatal conductance of *T. aestivum* and *P. vulgaris* was affected by the increasing $O_3$ levels but $g_s$ of *E. coracana* and *P. glaucum* remained unaffected (Hayes et al., 2020a).

Physiological studies on varieties of *Brassica* (*B. campestris* and *B. juncea*) under both ambient and elevated $O_3$ reflected reductions in $P_n$, $g_s$ and $F_v/F_m$ ratio (Singh et al., 2009, 2013). Thwe et al. (2014) found that the cultivars of *L. esculentum* (Look Tor) showed sensitivity toward the acute $O_3$ doses (100, 175, and 250 ppb) despite the plant age. The reduction of $g_s$ damaged photosynthetic apparatus and impaired PSII activity, which negatively affected the photosynthesis. Ariyaphanphitak et al. (2005) observed that lower $P_n$ was due to the negative regulation of stomata and production of oxygen radicals by $O_3$ induced ETR system of the plant. Several studies have analyzed the physiological machinery of *T. aestivum* cultivars in India (Rai et al., 2007; Ghosh et al., 2020; Yadav et al., 2020). Results showed reductions in $P_n$, $g_s$, water-use efficiency (WUE), $F_v/F_m$, along with photosynthetic nitrogen use efficiency. Mishra and Agrawal (2015) reported that lower WUE in *V. radiata* cultivars was related to the severe reduction of $CO_2$ assimilation than the E under an elevated $O_3$ concentration of 68 ppb. Prasad and Inamdar (1990) studied the effect of PM on *V. munga* and observed a reduction in $P_n$ and an increase in $R_D$.

Duggar and Cooley (1914) conducted preliminary studies on the effects of PM on plants using crops grown for commercial purposes. The effects of different dust led to reduction in E of *L. esculentum*. A reduction in $g_s$ of *C. sativus* and *P. vulgaris* was observed by Hirano et al. (1995). By modifying the amount of sunlight that plants get or by inflicting phytotoxic harm to the plants, PM has both direct and indirect effects on agricultural output (Mina et al., 2018; Sonkar et al., 2019). As a result of PM deposition, the foliage surface is modified which further impairs gas exchange and overall plant development is hampered (Rai, 2016). Reductions in the rate of $P_n$ and $g_s$ were reported in *Ruellia tuberosa* plants when exposed to thermal power emissions in Aligarh, India (Nighat et al., 2000).

WUE decreased concomitantly with the decrease in PM in *Lactuca serriola*, along with the significant increase in $C_i$ after PM deposition on the leaves (Pavlík et al., 2012). Application of PM to *L. sativa* leaves resulted in 88% enhancement in E (Chartzoulakis et al., 2002). A study by Sharifi et al. (1997) revealed that a dust load of 40 gm$^{-2}$ increased the vegetation temperature by 2–3°C in a desert environment. PM deposition enhanced leaf temperature from 1–6% in *O. sativa* varieties Pusa Basmati-1509 (PB-1509) and Pusa Sugandh-5 (PS-5) (Mina et al., 2018). The assimilation rate was reduced by 15–44% in PB-1509 and 9–52% in PS-5, along with a reduction in the $g_s$ of 36–80% and 12–40% in PB-1509 and PS-5, respectively, while E declined by 8–46% in PB-1509 and 6–16% in PS-5. Chaurasia (2013) established a significant decline in the chlorophyll content of *A. hypogea*, *Sesmum indicum* and *Triticum* species upon exposure to cement dust. An interrelationship between the differential stomatal distribution and physiological impairment was established by Singh et al., 2022 on the tree species exposed to different PM doses. The tree includes dominant species of Indo-Gangetic plains (*M. indica*, *P. guajava*, *F. religiosa*, *D. sissoo* and *A indica*), and showed significant reduction in $P_n$, $g_s$, $C_i$, E and WUE with increasing PM load, but the reduction was low in case of *P. guajava* and *D. sissoo* owing to their hypostomatic stomatal distribution.

### 4.5.3 Biochemistry

Biochemical parameters of a plant are prime determinants of stress conditions. The antioxidants (both enzymatic and non-enzymatic) and defense-related metabolites are widely used to assess the impact of air pollutants on plants (Singh et al., 2023a) and alterations in these are the primary response of plants against air pollution (Figure 4.2). A consistent reduction in the chlorophyll content of *U. pumila* (36.68%), *S. oblate* (15.35%) and *P. cerasifera* (16.42%) was observed by Han et al. (2022) under varying $SO_2$ (7 ppm and 14 ppm) stress. A study dealing with the emissions around thermal power plants which predominantly contain $SO_2$ by Nayak et al. (2013) evaluated the biochemical responses of ten dominant trees in that area. It revealed a significant decrement in the chlorophyll content in descending order as *A. indica* (72%) > *C. siamea* (71.7%) > *S. cumini* (70.6%) > *T. grandis* (69.5%) > *B. monosperma* (69.3%), while others showed insignificant decline. Further, a decline in the pH of the leaf extract (pH 3) was observed in *A. indica* and *M. indica* during winter season. Total phenolics and proline contents in the leaf showed significant increments compared to control site which was positively correlated with concentrations of $SO_2$ and $NO_X$ at polluted sites (Nayak et al., 2013). A greenhouse-controlled experiment carried out (Storch-Böhm et al., 2022) on two indigenous Brazilian tree species (*Schinus terebinthifolius* and *Cupania vernalis*) showed increments in peroxidase (POD) activity in both species when exposed to $SO_2$ and $NO_2$. Zhao et al. (2021) observed increased levels of starch content in leaves of *Ligustrum lucidum* tree when subjected to high concentrations of $NO_2$ and $SO_2$. A study carried out by Qayoom Mir et al. (2009) on few medicinally dominant plants such as *Ocimum sanctum* and *Catharanthus roseus* grown at five different polluted sites having concentrations ranging from 0.0015–0.015 ppm of $SO_2$ and 0.0005–0.029 ppm of $NO_2$ showed significant increases in total flavonoids and phenolic content.

Under $O_3$ exposure, mesophyll cells of *P. guajava* accumulated a higher quantity of phenolic compounds, which acted as a marker for accelerated cell senescence (Alves et al., 2016). *P. guajava* has been documented as a bioindicator species for $O_3$ in most of the tropical studies (Furlan et al., 2007; Pina & Moraes, 2007; de Rezende & Furlan, 2009). *P. guajava* plant subjected to three $O_3$ concentrations—CF (AOT40 = 17 ppb h), NF (AOT40 = 542 ppb h) and NF + 40 (AOT40 = 7802 ppb h)—showed maximum increments in anthocyanins and tannins under NF + 40 ppb compared to CF treatment. A study with three different doses (50, 110 and 250 ppb) of $O_3$ on three mangrove species (*Rhizophora mangle*, *Laguncularia racemosa* and *Conocarpus erectus*) was conducted by Cerón-Bretón et al. (2010b). All the species were found sensitive to $O_3$, and the most sensitive was *R. mangle*. The photosynthetic pigments and soluble proteins decreased under elevated $O_3$. A study with four tropical trees (*H. campechianum*, *T. rosea*, *C. odorata* and *S. macrophylla*) showed that all the species were sensitive toward $O_3$, while *H. campechianum* and

*S. macrophylla* were most sensitive (Cerón-Bretón et al., 2010a). *Clementina mandarin* exposed to ambient and elevated concentrations of $O_3$ showed reductions in total chlorophyll, carotenoids and carbohydrate contents, along with increases in 1-aminocyclopropane-1-carboxylic acid (ACC) content and ethylene production (Iglesias et al., 2006). Moreover, there was a decrease in the ascorbate leaf pool but significantly higher solute leakage and lipid peroxidation were observed in plants at ambient levels.

Long et al. (2018) reported an enhancement in MDA content as a result of an increase in cell membrane permeability, but there was a decline in chlorophyll and superoxide dismutase (SOD) activity under elevated $O_3$ in *M. indica*, *C. camphora* and *F. altissima*. An increment in proline and soluble sugar content was correlated with the regulation of osmotic potential. These indices were used for screening the resistant trees of the tropical region and the tolerance intensity was in order *M. indica* > *C. camphora* > *F. altissima*. The effect of ambient $O_3$ on *F. insipida* was evident as decrease in phenolics and terpenoids, along with lower composition of membrane lipids (Schneider et al., 2017). Biochemical analysis of three tree species (*A. graveolens*, *P. gonoacantha* and *C. floribundus*) during wet and dry seasons revealed different patterns of $O_3$ responses (Aguiar-Silva et al., 2016). Catalase acted as leaf tolerance indicator mainly during the wet season. During the dry season, oxidative/antioxidative imbalances such as increased lipid peroxidation and decreased glutathione, as well as chlorophyll content, were the most effective oxidative stress indicators. In terms of oxidative stress tolerance, *C. floribundus* was found to be the most efficient species.

Most of the plants showed a decrease in both chlorophyll and carotenoid levels due to foliar dust deposition. Chlorophyll a and total chlorophyll in *F. benghalensis* were reduced by 46.5% and 45.4%, respectively, as a result of dust deposition (Prusty et al., 2005). Many organic and inorganic compounds found in the dust enhanced reactive oxygen species (ROS) formation predominantly harming membranes and pigments after entering the plant cells. In a study by Chaudhary and Rathore (2018), the decrease in photosynthetic pigments and increase in membrane permeability and stomatal index in trees (*A. indica*, *Albizia lebbeck*, *D. sissoo*, *C. fistula*, *F. religiosa*, *F. virens* and *F. benghalensis*) were reported under PM stress. A study on CAM plant (*C. hilariana*) observed a change in CAM photosynthetic pathway, increased relative membrane permeability and reduced CAT and SOD activities (Pereira et al., 2009). An increase in heme oxygenase activity in nine perennial species (*A. indica*, *Bougainvillea spectabilis*, *Colophospermum mopane*, *E. globulus*, *F. religiosa*, *Monoon longifolium*, *Salvadora persica*, *Tinospora cordifolia* and *Vachellia nilotica*) was reported by Popek et al. (2022) upon PM exposure.

A field experiment to study the effect of $SO_2$ (1 ppm) on the biochemical parameters of *Cicer arietinum* showed a significant reduction in total chlorophyll, along with a slight shift in leaf extract pH toward the acidic side (Sharma, 2018). A long-term investigation by Agrawal and Deepak (2003) on two cultivars of *T. aestivum* (cv. Malviya 234 and HP1209) fumigated with $SO_2$ (0.06 ± 0.005 ppm) showed significant reductions in total chlorophyll, protein and starch contents. However, the total soluble sugars and phenolics increased in response to $SO_2$. The exposure of $SO_2$ led to a significant increase of 34.7% and 19% in POD activity in Malviya 234 and HP1209, respectively, while the CAT activity showed a significant reduction of 17.5% in HP1209 only (Agrawal & Deepak, 2003). Extensive screening of garden plants to assess the physiological and biochemical responses under fumigation of $NO_2$ (6 ppm) showed decrease in chlorophyll content and increases in MDA content, POD activity and soluble protein content (Sheng & Zhu, 2019). It was also found that high concentration of $NO_2$ affected mineral ion concentrations of five macro-elements (Ca, K, N, Mg and P), indicating activation of defense and antioxidative protection (Sheng & Zhu, 2019).

Poonia (2019) observed a significant reduction in the alpha-amylase activity in germinating seeds of *V. mungo*, *P. sativum*, *Cajanus cajan* and *C. arietinum* when exposed to different concentrations of $SO_2$, viz., 0.25, 0.50, 0.99 and 1.5 ppm. The maximum reduction of 24% in amylase activity was observed in *V. mungo* under 1.5 ppm $SO_2$, while a minimal increase has been observed in *P. sativum* and *C. cajan* at 0.25 ppm. Furthermore, the $SO_2$ pretreatment significantly increased proline accumulation along with enhanced enzymatic activity of CAT and POD in the leaves of drought-stressed seedlings that led to the scavenging of hydrogen peroxide ($H_2O_2$) and alleviated drought-induced oxidative damage, as evidenced by a lower MDA level in $SO_2$-pretreated plants (Han et al., 2019).

Reduction in chlorophyll and carotenoid contents were reported in *O. sativa* (Agrawal et al., 1982), *T. aestivum*, *P. sativum* and *V. radiata* (Adrees et al., 2016a, 2016b) and in *L. esculentum* cultivars upon elevated $O_3$ exposure (Mina et al., 2010). Further, the cultivars also showed an increase in isoprene emission. A study on two cultivars of *O. sativa* (Saurabh 950 and NDR 97) showed that several enzymatic and non-enzymatic antioxidants were affected under ambient (Rai & Agrawal, 2008) and elevated $O_3$ concentrations (Ismail et al., 2015). Both the sensitive and tolerant cultivars of *V. radiata* showed an increment in antioxidants when exposed with an elevated $O_3$ dose (Mishra & Agrawal, 2015; Chaudhary & Agrawal, 2015). The level of lipid peroxidation, solute leakage, superoxide production rate and enzymatic antioxidants (APX, CAT, GR and SOD) were increased under elevated $O_3$. However, reduction in protein content was observed, whereas ascorbic acid and phenol contents increased in all the treated cultivars (Mishra & Agrawal, 2015). Similarly, a comparative study was done on four cultivars of *V. radiata* to assess the $O_3$ sensitivity of tolerant (Chainat 84-1-1 and Kampangsan 2) and sensitive cultivars (Chainat 3 and 4) at 70–100 ppb of $O_3$ (Kittipornkul et al., 2021). Chainat 4 was found to be the most sensitive one due to lower ascorbic acid, defensive proteins and antioxidants along with the higher accumulations of H$_2$O$_2$ and salicylic acid. In contrast, the most tolerant cultivar (Chainat 84–1–1) showed a higher level of ascorbic acid, an upregulation of defense-related proteins—especially biosynthesis and regeneration of ascorbic acid—and lower accumulation of salicylic acid, resulting in little visible injury (Kittipornkul et al., 2021). Singh et al. (2014) reported significant increase in superoxide radical (O$_2^-$) and hydrogen peroxide (H$_2$O$_2$) which led to peroxidation of lipids and higher accumulation of secondary metabolites such as phenol, anthocyanin and flavonoids in maize cultivars under 50–80 ppb of O$_3$.

A study conducted by Daresta et al. (2015) on *S. lycopersicum* showed that ROS generation in roots increased from 32–93% with increasing PM load. Total chlorophyll content was reported to decrease along with a marked decline in chl a/chl b and chl (a+b)/carotenoid with increasing PM load. The amino acids metabolized during the photorespiration—namely, the pools of free glutamate and asparagine—are found to be depleted upon PM contamination (Pavlík et al., 2012). It was further confirmed that tryptophan biosynthesis was diminished in the presence of PM stress.

### 4.5.4 Growth, Biomass and Productivity

Plants develop tolerance against various stresses and shift their assimilates between growth and defense. Pollutants cause leaf impairment, premature senescence, stomatal damage and reduced photosynthesis, and thus lead to diminished growth and yield of plants (Ashraf & Harris, 2013; Prakash et al., 2023) (Figure 4.2). A study by Gupta et al. (2022) on *Calendula officinalis* exposed to SO$_2$ at concentrations of 320, 667 and 1334 ppm at the pre-flowering stage showed significant reductions in leaf area and number, length of roots and shoots with increasing SO$_2$ concentration as compared to the control. Similarly, Agrawal and Singh (2000) established SO$_2$ and PM as the major emissions from coal power plants imposing adverse environmental impacts on the plants. The study further revealed marked changes in the foliar elemental concentrations of tropical trees.

In *Morus alba*, positive impacts of NO$_2$ (4 ppm) have been reported under low exposures that manifested the higher growth rate, nutrient uptake, more leaf area, above-ground biomass production and higher production of flowers and fruit (Wang et al., 2019). The NO$_2$ exposure enhanced photosynthetic efficiency and nitrogen metabolism by improving the absorption of light energy and protection of the photosynthetic apparatus (Wang et al., 2019) (Figure 4.3).

Storch-Böhm et al. (2022) studied the sensitivity of two indigenous Brazilian plant species, *C. vernalis* and *S. terebinthifolius*, against the phytotoxicity of air pollutants including SO$_2$ and NO$_2$. As the experiment progressed, the dry biomass for *S. terebinthifolius* and *C. vernalis* generally increased under exposure to SO$_2$ (0.5 ppm) and NO$_2$ (0.84 ppm) compared to control (SO$_2$: 0.0083 ppm; NO$_2$: 0.0085 ppm). For *C. vernalis*, no significant change in biomass was recorded between 0–30 days, which corresponded to the adaptation to the experimental conditions. On the other hand, at 90 days and 120 days, a significant decrease in dry biomass of *C. vernalis* was observed compared to the control.

A field study by Kateivas et al. (2022) evaluated the morpho-physiological responses of pot-grown tree species (namely *L. multiflorum*, *M. indica* and *P. cattleianum*) against the atmospheric pollution containing mainly NO$_X$ and SO$_2$ emitted from the mining industry located in the Brazilian city of Brumado. The growth characteristics such as leaf numbers, total leaf area and biomass of leaf, stem and root were lowered in *L. multiflorum* and *P. cattleianum* as compared to the reference site. Suppressed growth in terms of a decrease in plant height was more in *P. cattleianum* (62.7%) than *L. multiflorum* (18%) compared to the reference site (Kateivas et al., 2022).

Uka and Belford (2020) investigated the variations in macro and micro-morphological features that may improve the tolerance and survival of tropical tree leaves—specifically, *F. platyphylla*, *P. longifolia*, *T. cattapa* and *M. indica*—under vehicular emissions of CO, SO$_2$, NO$_2$ and VOCs. It was found that the leaf area of tree species at polluted sites was significantly lower as compared to the control site except in *F. platyphylla*. Significant increases in shoot height and biomass of *E. microcorys* were reported with increasing exposure of NO$_X$, while *E. pilularis* and *E. globulus* showed increased height at low levels of NO$_X$; however, effects were reversible under higher concentrations (Murray et al., 1994). NO$_X$ fumigation had no significant effect on biomass and shoot height of *E. marginata*.

Studies on O$_3$ exposure commonly reported a decrease in dry matter of plants (da Silva Engela et al., 2021; Moura et al., 2021). There was an initial decrease in height and dry mass in *A. graveolens* under ambient air for five months compared to filtered air (Cassimiro & Moraes, 2016). Similarly, reductions in height and root-to-shoot ratio were detected in *E. uniflora* when fumigated with twice the ambient O$_3$ concentration for 75 days in an O$_3$-FACE system (da Silva Engela et al., 2021). Significant effects of O$_3$ stress were also recorded in *Moringa oleifera* subjected to ambient (36 ppb), middle (53 ppb, i.e., 1.5 times of ambient) and high (64 ppb, i.e., twice of ambient) doses of O$_3$ (Moura et al., 2021). Leaf biomass was declined under middle and high O$_3$ concentrations compared to ambient. In contrast, *C. echinata* (Moraes et al., 2006) and *F. insipida* (Schneider et al., 2017) displayed insignificant changes in leaf, stem and root biomass, as well as in root-to-shoot ratio between the treatments. A significant reduction in the leaf biomass was also reported in seven urban tree species by Chaudhary and Rathore (2018). A specific feature of non-glandular emergences was observed on the leaf surface of *C. floribundus* on the entire surface of leaves, which acted as a barrier for O$_3$ entry into the leaf tissue. When the leaves with these emergences were exposed to 100 ppb of O$_3$, no alterations were identified, while leaves with removed emergences showed O$_3$-specific morphological symptoms (Dias et al.,

2019). According to the study of Singh et al. (2021b), long-term $O_3$ (59–79 ppb) exposure for two years reduced the leaf size of *Leucaena leucocephala*. The plant initially showed sensitiveness, but with time, it developed tolerance against $O_3$ stress. Productivity losses in the two hybrid varieties of *Mangifera indica* have been observed under 20 ppb of $EO_3$ above the ambient $O_3$ concentration (Singh et al., 2023c). The study further established that both the varieties invested more in growth therefore, compromised their reproductive potential under $EO_3$ exposure.

Prusty et al. (2005) examined the dust accumulation potential of the vegetation near a highway in Sambalpur, Orissa and concluded that *I. carnea*, *Tabernaemontana divaricata* and *P. pinnata* had higher dust accumulation rates than *F. religiosa* and *Quisqualis indica*. These results were correlated with the roughness of the leaf surface, the diminutive size of the petiole and the shorter height of the plants, which encouraged greater dust adherence. According to Prusty et al. (2005), plants with shorter heights accumulated more particulate compared to those with greater heights. Ajuru and Friday (2014) investigated the effects of PM on the anatomy of three tropical plants (*Alchornea cordifolia*, *Musa paradisiaca* and *Manihot esculenta*) in Port Harcourt, Nigeria, and found an increase in the number of vascular bundles, whereas the tissues of the cortex, epidermis and pith remained unaffected. Another recent study has reported that deposition of PM subsided the effect of low temperature in *P. guajava* by enhancing the deposition of anthocyanin in epidermal layers (Singh et al., 2023b).

Agrawal et al. (2003) reported negative impacts of air pollutants on plant biomass and seed yield of different crop species (*T. aestivum*, *B. campestris*, *B. vulgaris* and *V. radiata*) growing at sites with varying concentrations of $SO_2$ (0.03–0.04 ppm), $NO_2$ (0.03–0.08 ppm) and $O_3$ (9–58 ppb). The most sensitive plant was found to be *V. radiata*, showing decline in yield by 73%, followed by *T. aestivum* (25%) and *B. campestris* (20%) (Agrawal et al., 2003). Several studies, however, also reported an increase in the yield of various tropical plant species under varying levels of $NO_X$, including *Arabidopsis thaliana* (Takahashi et al., 2014), *N. plumbaginifolia* (Morikawa et al., 2003), *L. sativa*, *Helianthus annuus*, *C. sativus* and *Cucurbita moschata* (Adam et al., 2008).

Norby and Luxmoore (1983) investigated the impact of interactive effect of $SO_2 + O_3$ on the growth of greenhouse-grown *G. max*. It was found that root-to-shoot ratio, WUE and the number of root nodules reduced by 14%, 34% and 39%, respectively, when fumigated with $SO_2 + O_3$ (0.2 ppm + 100 ppb). Higher $SO_2$ concentrations (0.029 ppm and 0.034 ppm) caused detrimental effects on *Medicago sativa*, leading to decrease in leaf demography and growth when exposed to $SO_2$ for 45 consecutive days (Cotrozzi, 2020). The observed outcomes were correlated with reduced total dry weight and root-to-shoot ratio of plants due to disturbances in carbon allocation.

A field experiment was performed by Deepak and Agrawal (1999) using open-top chambers *T. aestivum* (cv. Malviya 234) from germination to maturity to analyze the impact of $SO_2$ (0.06 ppm) on the growth and yield parameters. It was observed that fumigation of $SO_2$ caused significant reductions in shoot length, leaf area and total plant height by 10%, 14.8% and 9%, respectively, at 60 days of plant age. The overall yield in terms of the number of ears planted (27%), number of grains/m$^2$ (19%) and grain yield (15%) also reduced in plants exposed to $SO_2$ (Deepak & Agrawal, 1999).

Reductions in root and shoot length, number of tillers, number of leaves and ears of *P. miliaceum* (Agrawal et al., 1983) and two cultivars of *O. sativa* (cultivars Malviya dhan 36 and Shivani) were reported under elevated $O_3$ level (Sarkar & Agrawal, 2012). Studies on *V. radiata* showed that growth parameters like plant height, the number of leaves and leaf area reduced both under ambient and elevated $O_3$ (Chaudhary & Agrawal, 2015). The total biomass was reduced by 50%, 24.2% and 42%, while the yield reduced by 47.5%, 56% and 48% in *T. aestivum*, *P. sativum* and *V. radiate*, respectively, under ambient $O_3$ (Adrees et al., 2016a, 2016b). No difference in plant height of two varieties of *V. unguiculata* was found between the treatments; however, significant dsecreases in leaf, stem, root biomass and root-to-shoot ratio were detected at the elevated level of $O_3$ (Tetteh et al., 2015).

Several studies reported the negative effect of $O_3$ on the growth attributes of different rice cultivars. The growth pattern of Thai rice cultivars found tillering stage more sensitive to $O_3$ than the booting and maturing stages. Elongation of root and shoot lengths was inhibited by an elevated dose of 50 ppb and 150 ppb $O_3$. The dry weight of the shoot was more affected at tillering stage and was reduced by 29% compared to the maturing stage (18%), whereas yield parameters reflected the deterioration of grain quality and quantity due to the imbalance of the nutrient accumulation in seeds under high $O_3$ dose (Ariyaphanphitak et al., 2003). Similarly, decreases in plant height and number of tillers and root and shoot dry mass were found in Malaysian rice cultivars under varying $O_3$ levels (Ishii et al., 2004; Shahadan, 2005). Yield reduction in rice cultivars was reported by Ismail et al. (2015) under high $O_3$ concentrations (60 ppb and 120 ppb) with more reduction under 120 ppb of $O_3$. Singh et al. (2021c) found that the duration of growth period was shortened by 14 days in *C. arietinum* under an elevated level of 60 ppb $O_3$. The decline in chlorophyll and nutrients led to the early maturity of leaves, thereby accelerating the process of senescence.

Effects of both acute (>100 ppb) and chronic (40–60 ppb) $O_3$ exposures were found in form of reduced root nodulation of legumes which in turn affected the activity of biological nitrogen fixation (Hewitt et al., 2016). Acute exposure of 137 ppb $O_3$ reduced the below-ground biomass of *Glycine max* by 25%, nodule biomass by 30% and biological nitrogen fixation by 21% (Biancari et al., 2021). Mina et al. (2010) reported reductions in shoot length (2–10.5%), shoot biomass (44%) and root biomass (8–65%) of *L. esculentum* cultivars under $O_3$ concentrations of 75 ppb and 150 ppb. Agrawal et al. (2003) also reported 22–26% of reductions in seed yield of *B. campesteris* when exposed to varying levels of $O_3$ and $SO_2$. A field experiment on *T. aestivum* grown in the most polluted areas around Varanasi showed that $O_3$ as the prominent pollutant led to 32% yield loss with poor seed quality (Rajput & Agrawal, 2005). Open-top chamber studies found 9–25% yield losses in *T. aestivum* (Rai et al., 2007; Sarkar & Agrawal, 2010) under elevated $O_3$. Hayes et al. (2020a) found yield reduction in *T. aestivum* and *P. vulgaris* with increasing $O_3$ load, while no significant effect was noticed on *E. coracana* and *P. glaucum*.

According to Garg and Singla (2011) and Luo et al. (2019), the accumulation of atmospheric PMs on plants can change their physical or biological functions, which inhibit their growth and lower the production of crops on quantitative or qualitative levels. Prasad and Inamdar (1990) reported significant decreases in biomass and net primary productivity of *V. mungo* under PM stress. The yield parameters such as spikelet count, spike length, weight per spike, the average number of grains and harvest index reflected a marked decline under PM stress (Sharma et al., 2020). In a study by Mina et al. (2018), the grain yield of *O. sativa* (PB-1509) was reduced by 7.5–14% under PM load of 183–3551 µg $cm^{-2}$. A significant decline in the yield, oil content and biomass under the impact of PM was reported in *B. campestris* by Shukla et al. (1990).

A study conducted by Daresta et al. (2015) to analyze the impact of $PM_{10}$ mainly on the roots of *S. lycopersicum* showed maximum reduction of 58% in the primary root elongation with increasing PM load, whereas fresh weight of root and shoot declined by 52.2–70% and 29–35%, respectively. The treated plants showed highly branched root system with the formation of thinner and longer lateral roots with increasing PM concentrations.

## 4.6 MODELS EVALUATING THE EFFECTS OF RISING $O_3$ CONCENTRATIONS ON TROPICAL CROPS AND TREES

Plant responses to phytotoxic potential of $O_3$ is assessed both through flux-based and concentration-based approaches. In European countries, concentration-based ozone metrics are mostly practiced e.g., SUM00 (sum of all hourly average concentrations) and AOT40 (accumulated ozone over a threshold of 40 ppb hours). However, the flux-based model is considered more appropriate for characterizing the threat to trees, crops and natural ecosystems (Mills et al., 2011; Büker et al., 2015) because AOT40 is based on the atmospheric $O_3$ concentration rather than considering stomatal uptake of the pollutant. Several models have been worked out and parameterized to predict the $O_3$ risk assessment.

### 4.6.1 Stomatal $O_3$ Flux Models and Dose–Response Relationship for Trees

Most of the regions may differ in their $O_3$ flux values, despite showing similar AOT40 values (Simpson et al., 2007). In tropical countries, the ozone flux-based model ($DO_3SE$ or deposition of ozone for stomatal exchange) was first time applied for *P. guajava* 'Paluma' in Sao Paulo, Brazil (Assis et al., 2015). Parameterization of $g_s$ was done and used for calculating different rate thresholds. The model performed well, showing a $r^2$ value of 0.56. Cassimiro et al. (2016) parameterized the model for *A. graveolens* under ambient $O_3$. The measured $g_s$ reflected significant relationship with the modeled conductance showing a $r^2$ value of 0.58. The model was found to be comprehensible and within the verified range that was identified for the other forest tree species. Leaf abscission was corelated to POD0 (phytotoxic ozone dose) while SUM00 (sum of all hourly average concentrations) was better correlated to severity. Fernandes et al. (2021) also validated that POD not only corresponded to the emergence of visible injury but the increased POD values, in turn increasing the intensity of $O_3$-induced visible injury. Other studies also parameterized the $O_3$ effect on $g_s$ of tropical tree species like *C. floribundus*, *P. gonoacantha* and *M. oleifera* (Moura et al., 2014; Moura et al., 2021).

### 4.6.2 $O_3$ Dose–Response Relationships for Crops

Leguminous crops such as *P. vulgaris* and *V. unguiculata* displayed similar flux-effect relationship with increasing $O_3$ flux while the yield of millets (*E. coracana* and *P. glaucum*) was not sensitive (Hayes et al., 2020a). Total yield was increased with the increasing $O_3$ flux in *E. coracana*. However, no significant relationship between yield and $O_3$ flux was obtained for *P. glaucum*. The different trend of response between leguminous crops and millets showed that the yield of leguminous crops was more sensitive, while the yield was not affected in case of millets.

In India, the first study based on $O_3$ flux model was conducted to assess the species-specific $O_3$ sensitivity in early and late sown cultivars of *T. aestivum* (Yadav et al., 2021b). $POD_6SPEC$ had a negative linear relationship with aboveground biomass and grain yield. The relationship revealed that 0.284 mmol $O_3$ $m^{-2}$ accumulated POD6SPEC is responsible for a 5% reduction of grain yield in early sown cultivars, while 0.393 mmol $O_3$ $m^{-2}$ accumulation is required for a 5% reduction in yield of late sown cultivars. An analysis of $O_3$ exposure dose and the relative yield for *O. sativa*, *T. aestivum* and *Z. mays* from experiments conducted in East Asian countries revealed that China had the highest relative yield loss at 33%, 23% and 9% for *T. aestivum*, *O. sativa*, and *Z. mays*, respectively. Furthermore, yield loss in hybrid is much greater than inbred rice (Feng et al., 2022).

## 4.7 CONCLUSIONS AND FUTURE CONCERNS

Anthropogenic activities have led to overutilization of resources in the tropics (which foster vast biodiversity of the world), which in turn has deteriorated the air quality and severely affected the structure and function of tropical ecosystems. PM, $O_3$, $NO_2$ and $SO_2$ have been recognized as the predominant air pollutants in the tropics which adversely affect the regional flora. Plants adapt themselves with the changing air quality by either increasing their tolerance capacity or restricting the pollutants' entry inside the cells.

Plants which showed more than 50% reduction in the biomass or yield were recognized to be most sensitive. In some plants, positive impacts were observed at low concentration of exposure to $SO_2$ and $NO_2$, while the impacts were negatively alternated on increasing the concentration. Among the studied plants, *P. cattleianum* showed a reduction of 62.7% in biomass on $SO_2$ and $NO_2$ exposure. *V. radiata* showed 73% decline in yield on combined exposure of $SO_2$, $NO_2$ and $O_3$, followed by *S. lycopersicum* (58% reduction in the yield) and *T. aestivum* (50% reduction in the yield) under the influence of PM and $O_3$, respectively. Conversely, *E. marginata*, *C. echinate*, *F. insipida*, *C. floribundus* and *V. unguiculata* are comparatively tolerant under air pollutant stress.

Lack of fundamental research work on air pollution response studies in the tropical regions with diverse vegetation limited the scope of this chapter to identify a broad range of air pollution effects in the tropics, although current evidence clearly highlights the negative influence of air pollution in different tropical regions of the world.

Tropics are the major contributors to the global carbon budget and centers of productivity; therefore, assessing the response of natural vegetation to prevailing air pollutants is of utmost important to trace their susceptibility. The detailed response and phenological studies of tropical plants under air pollution stress are comparatively much less than those in temperate regions. Studies focusing on visible injuries are restricted to gaseous pollutants, which need further extension in relation to PM. More studies on dose–response relationships of crops and trees to predict the future risk of air pollutants on yield and productivity will be helpful to increase our understanding of response relationships, as many of the species have significant commercial use and help in upgrading the economy of the tropical region.

## ACKNOWLEDGMENTS

Pallavi Singh and Jigyasa Prakash are thankful to the Council of Scientific and Industrial Research, New Delhi, for JRF and SRF (File No.09/013(0908)/2019-EMR-I and File No. 09/013(0934)/2020-EMR-I) fellowships. Harshita Singh is thankful to the Department of Science and Technology- INSPIRE (IF190187) for JRF and SRF fellowships. Madhoolika Agrawal is grateful to DST-SERB, New Delhi, for J.C. Bose Fellowship (JCB/2021/000040).

## REFERENCES

Adam, Suaad E. H., Jun Shigeto, Atsushi Sakamoto, Misa Takahashi, and Hiromichi Morikawa. "Atmospheric nitrogen dioxide at ambient levels stimulates growth and development of horticultural plants." *Botany* 86, no. 2 (2008): 213–217.

Adrees, Muhammad, Muhammad Ibrahim, Aamir Mehmood Shah, Farhat Abbas, Farhan Saleem, Muhammad Rizwan, Saadia Hina, Fariha Jabeen, and Shafaqat Ali. "Gaseous pollutants from brick kiln industry decreased the growth, photosynthesis, and yield of wheat (*Triticum aestivum* L.)." *Environmental Monitoring and Assessment* 188, no. 5 (2016a): 1–11.

Adrees, Muhammad, Farhan Saleem, Fariha Jabeen, Muhammad Rizwan, Shafaqat Ali, Sofia Khalid, Muhammad Ibrahim, Nazish Iqbal, and Farhat Abbas. "Effects of ambient gaseous pollutants on photosynthesis, growth, yield and grain quality of selected crops grown at different sites varying in pollution levels." *Archives of Agronomy and Soil Science* 62, no. 9 (2016b): 1195–1207.

Agathokleous, Evgenios, Zhaozhong Feng, Elina Oksanen, Pierre Sicard, Qi Wang, Costas J. Saitanis, Valda Araminiene et al. "Ozone affects plant, insect, and soil microbial communities: A threat to terrestrial ecosystems and biodiversity." *Science Advances* 6, no. 33 (2020): eabc1176.

Agrawal, Madhoolika, and Jyoti Singh. "Impact of coal power plant emission on the foliar elemental concentrations in plants in a low rainfall tropical region." *Environmental Monitoring and Assessment* 60 (2000): 261–282.

Agrawal, Madhoolika, and S. Singh Deepak. "Physiological and biochemical responses of two cultivars of wheat to elevated levels of $CO_2$ and $SO_2$, singly and in combination." *Environmental Pollution* 121, no. 2 (2003): 189–197.

Agrawal, Madhoolika, P. K. Nandi, and D. N. Rao. "Effect of ozone and sulphur dioxide pollutants separately and in mixture on chlorophyll and carotenoid pigments of Oryza sativa." *Water, Air, and Soil Pollution* 18, no. 4 (1982): 449–454.

Agrawal, Madhoolika, P. K. Nandi, and D. N. Rao. "Ozone and sulphur dioxide effects on Panicum miliaceum plants." *Bulletin of the Torrey Botanical Club* (1983): 435–441.

Agrawal, Madhoolika, B. Singh, M. Rajput, F. Marshall, and J. N. B. Bell. "Effect of air pollution on peri-urban agriculture: A case study." *Environmental Pollution* 126, no. 3 (2003): 323–329.

Agrawal, Madhoolika, and Raj Kumar Singh. "Effect of industrial emission on atmospheric wet deposition." *Water, Air, and Soil Pollution* 130, no. 1 (2001): 481–486.

Agrawal, Shashi Bhushan, and Madhoolika Agrawal, eds. *Environmental Pollution and Plant Responses*. CRC Press, 1999.

Aguiar-Silva, Cristiane, Solange E. Brandão, Marisa Domingos, and Patricia Bulbovas. "Antioxidant responses of Atlantic Forest native tree species as indicators of increasing tolerance to oxidative stress when they are exposed to air pollutants and seasonal tropical climate." *Ecological Indicators* 63 (2016): 154–164.

Ahmad, Muhammad Nauman, Patrick Büker, Sofia Khalid, Leon Van Den Berg, Hamid Ullah Shah, Abdul Wahid, Lisa Emberson, Sally A. Power, and Mike Ashmore. "Effects of ozone on crops in north-west Pakistan." *Environmental Pollution* 174 (2013): 244–249.

Ajuru, Mercy Gospel, and Upadhi Friday. "Effects of particulate matter on the anatomy of some tropical plants (Alchonea cordifolia, Musa paradisiaca, and Manihot esculenta)." *International Journal of Scientific & Technology Research* 3, no. 12 (2014): 304–308.

Alexander, Lisa, Simon Allen, and Nathaniel L. Bindoff. *Working Group I Contribution to the IPCC Fifth Assessment Report Climate Change 2013: The Physical Science Basis Summary for Policymakers* (No. Bajados de Internet/2013). OPCC, 2013.

Alscher, Ruth, Michael Franz, and C. W. Jeske. "Sulfur dioxide and chloroplast metabolism." *Phytochemical Effects of Environmental Compounds* (1987): 1–28.

Alves, Edenise Segala, Bárbara Baêsso Moura, Andrea Nunes Vaz Pedroso, Fernanda Tresmondi, and Silvia Rodrigues Machado. "Cellular markers indicative of ozone stress on bioindicator plants growing in a tropical environment." *Ecological Indicators* 67 (2016): 417–424.

Amodio, M., E. Andriani, G. De Gennaro, A. Demarinis Loiotile, A. Di Gilio, and M. C. Placentino. "An integrated approach to identify the origin of PM10 exceedances." *Environmental Science and Pollution Research* 19, no. 8 (2012): 3132–3141.

Ariyaphanphitak, W., A. Chidthaisong, E. Sarobol, V. N. Bashkin, and S. Towprayoon. "Effects of elevated ozone concentrations on Thai jasmine rice cultivars (Oryza sativa L.)." *Water, Air, and Soil Pollution* 167, no. 1 (2005): 179–200.

Ariyaphanphitak, W., A. Chidthaisong, E. Sarobol, and S. Towprayoon. "Responses of Thai rice cultivar Chainat I (*Oryza sativa* L.) to enhanced tropospheric ozone concentrations." 2003.

Ashraf, M. H. P. J. C., and Phil J. C. Harris. "Photosynthesis under stressful environments: An overview." *Photosynthetica* 51, no. 2 (2013): 163–190.

Assis, Pedro I. L. S., Rocío Alonso, Sérgio T. Meirelles, and Regina M. Moraes. "DO3SE model applicability and O3 flux performance compared to AOT40 for an O3-sensitive tropical tree species (Psidium guajava L.'Paluma')." *Environmental Science and Pollution Research* 22, no. 14 (2015): 10873–10881.

Balasubramanian, A., Prasath, C. H., Gobalakrishnan, K., and Radhakrishnan, S. (2018). "Air pollution tolerance index (APTI) assessment in tree species of Coimbatore urban city, Tamil Nadu, India." *International Journal of Environment and Climate Change*, 8(1), 27–38.

Bateebe, Irene Pauline. *Investigation of Probable Pollution from Automobile Exhaust Gases in Kampala City, Uganda*. PhD diss., School of Industrial Engineering and Management Energy, 2011.

Begum, Bilkis A., and Philip K. Hopke. "Identification of sources from chemical characterization of fine particulate matter and assessment of ambient air quality in Dhaka, Bangladesh." *Aerosol and Air Quality Research* 19, no. 1 (2019): 118–128.

Biancari, Lucio, Clara Cerrotta, Analía I. Menéndez, Pedro E. Gundel, and M. Alejandra Martínez-Ghersa. "Episodes of high tropospheric ozone reduce nodulation, seed production and quality in soybean (*Glycine max* (L.) merr.) on low fertility soils." *Environmental Pollution* 269 (2021): 116117.

Biswas, Kuntal, Arpita Chatterjee, and Jyotibrata Chakraborty. "Comparison of air pollutants between Kolkata and Siliguri, India, and its relationship to temperature change." *Journal of Geovisualization and Spatial Analysis* 4, no. 2 (2020): 1–15.

Büker, Patrick, Zaozhong Feng, Johann Uddling, Alain Briolat, Rocio Alonso, Sabine Braun, Susana Elvira et al. "New flux based dose–response relationships for ozone for European forest tree species." *Environmental Pollution* 206 (2015): 163–174.

Calatayud, A., Domingo J. Iglesias, Manuel Talón, and Eva Barreno. "Effects of long-term ozone exposure on citrus: Chlorophyll a fluorescence and gas exchange." *Photosynthetica* 44, no. 4 (2006): 548–554.

Carlson, Roger W. "Interaction between $SO_2$ and $NO_2$ and their effects on photosynthetic properties of soybean Glycine max." *Environmental Pollution Series A, Ecological and Biological* 32, no. 1 (1983): 11–38.

Cassimiro, Jéssica C., and Regina M. Moraes. "Responses of a tropical tree species to ozone: Visible leaf injury, growth, and lipid peroxidation." *Environmental Science and Pollution Research* 23, no. 8 (2016): 8085–8090.

Cassimiro, Jéssica C., Bárbara B. Moura, Rocio Alonso, Sérgio T. Meirelles, and Regina M. Moraes. "Ozone stomatal flux and $O_3$ concentration-based metrics for Astronium graveolens Jacq., a Brazilian native forest tree species." *Environmental Pollution* 213 (2016): 1007–1015.

Cavalcante, Rivelino M., Camille A. Rocha, Íthala S. De Santiago, Tamiris F. A. Da Silva, Carlos M. Cattony, Marcus V. C. Silva, Icaro B. Silva, and Paulo R. L. Thiers. "Influence of urbanization on air quality based on the occurrence of particle-associated polycyclic aromatic hydrocarbons in a tropical semiarid area (Fortaleza-CE, Brazil)." *Air Quality, Atmosphere & Health* 10 (2017): 437–445.

Cerón-Bretón, J. G., R. M. Cerón-Bretón, J. J. Guerra-Santos, C. Aguilar-Ucán, C. Montalvo-Romero, C. Vargas-Cáliz, V. Córdova-Quiroz, and R. Jiménez-Corzo. "Effects of Simulated Tropospheric Ozone on nutrients levels and photosynthetic pigments concentrations of three Mangrove species." *WSEAS Transactions on Environment and Development* 6, no. 2 (2010b): 133–143.

Cerón-Bretón, J. G., R. M. Cerón-Bretón, J. J. Guerra-Santos, A. V. Córdova-Quiroz, C. Vargas-Cáliz, L. G. Aguilar-Bencomo, K. Rodriguez-Heredia, E. Bedolla-Zavala, and J. Pérez-Alonso. "Effects of simulated tropospheric ozone on soluble proteins and photosynthetic pigments levels of four woody species typical from the Mexican Humid Tropic." *WSEAS Transactions on Environment and Development* 6, no. 5 (2010a): 335–344.

Chandu, Kavitha, and Madhavaprasad Dasari. "Variation in concentrations of PM2.5 and PM10 during the four seasons at the port city of Visakhapatnam, Andhra Pradesh, India." *Nature Environment and Pollution Technology* 19, no. 3 (2020): 1187–1193.

Chapla, J., and J. A. Kamalakar. "Metabolic responses of tropical trees to ozone pollution." *Journal of Environmental Biology* 25, no. 3 (2004): 287–290.

Chartzoulakis, K., A. Patakas, G. Kofidis, A. Bosabalidis, and A. Nastou. "Water stress affects leaf anatomy, gas exchange, water relations and growth of two avocado cultivars." *Scientia Horticulturae* 95, no. 1–2 (2002): 39–50.

Chaudhary, Indra Jeet, and Dheeraj Rathore. "Suspended particulate matter deposition and its impact on urban trees." *Atmospheric Pollution Research* 9, no. 6 (2018): 1072–1082.

Chaudhary, Nivedita, and S. B. Agrawal. "Intraspecific responses of six Indian clover cultivars under ambient and elevated levels of ozone." *Environmental Science and Pollution Research* 20, no. 8 (2013): 5318–5329.

Chaudhary, Nivedita, and S. B. Agrawal. "The role of elevated ozone on growth, yield and seed quality amongst six cultivars of mung bean." *Ecotoxicology and Environmental Safety* 111 (2015): 286–294.

Chaurasia, Sadhana. "Effect of cement industry pollution on chlorophyll content of some crops at Kodinar, Gujarat, India." *Proceedings of the International Academy of Ecology and Environmental Sciences* 3, no. 4 (2013): 288.

Cotrozzi, Lorenzo. "Leaf demography and growth analysis to assess the impact of air pollution on plants: A case study on alfalfa exposed to a gradient of sulphur dioxide concentrations." *Atmospheric Pollution Research* 11, no. 1 (2020): 186–192.

da Silva Engela, Marcela Regina Gonçalves, Claudia Maria Furlan, Marisia Pannia Esposito, Francine Faia Fernandes, Elisa Carrari, Marisa Domingos, Elena Paoletti, and Yasutomo Hoshika. "Metabolic and physiological alterations indicate that the tropical broadleaf tree *Eugenia uniflora* L. is sensitive to ozone." *Science of The Total Environment* 769 (2021): 145080.

Daresta, Barbara Elisabetta, Francesca Italiano, Gianluigi de Gennaro, Massimo Trotta, Maria Tutino, and Pasqua Veronico. "Atmospheric particulate matter (PM) effect on the growth of Solanum lycopersicum cv. Roma plants." *Chemosphere* 119 (2015): 37–42.

de Rezende, Fernanda Mendes, and Cláudia Maria, Furlan. "Anthocyanins and tannins in ozone-fumigated guava trees." *Chemosphere* 76, no. 10 (2009): 1445–1450.

Deepak, S. S., and Madhoolika Agrawal. "Growth and yield responses of wheat plants to elevated levels of $CO_2$ and $SO_2$, singly and in combination." *Environmental Pollution* 104, no. 3 (1999): 411–419.

Dias, Márcia Gonçalves, Bárbara Baêsso Moura, Giselle da Silva Pedrosa, Silvia Ribeiro de Souza, and Poliana Cardoso-Gustavson. "The role of non-glandular emergences in Croton floribundus (Euphorbiaceae) upon elevated ozone exposures." *Nordic Journal of Botany* 37, no. 11 (2019).

Du, Baoguo, Jürgen Kreuzwieser, Jana Barbro Winkler, Andrea Ghirardo, Jörg-Peter Schnitzler, Peter Ache, Saleh Alfarraj, Rainer Hedrich, Philip White, and Heinz Rennenberg. "Physiological responses of date palm (Phoenix dactylifera) seedlings to acute ozone exposure at high temperature." *Environmental Pollution* 242 (2018): 905–913.

Duggar, Benjamin M., and Jacquelin Smith Cooley. "The effect of surface films and dusts on the rate of transpiration." *Annals of the Missouri Botanical Garden* 1, no. 1 (1914): 1–22.

Emberson, Lisa D., Håkan Pleijel, Elizabeth A. Ainsworth, Maurits Van den Berg, Wei Ren, Stephanie Osborne, Gina Mills et al. "Ozone effects on crops and consideration in crop models." *European Journal of Agronomy* 100 (2018): 19–34.

Escobedo, Francisco J., John E. Wagner, David J. Nowak, Carmen Luz De la Maza, Manuel Rodriguez, and Daniel E. Crane. "Analyzing the cost effectiveness of Santiago, Chile's policy of using urban forests to improve air quality." *Journal of Environmental Management* 86, no. 1 (2008): 148–157.

Fatima, Adeeb, Aditya Abha Singh, Arideep Mukherjee, Tsetan Dolker, Madhoolika Agrawal, and Shashi Bhushan Agrawal. "Assessment of ozone sensitivity in three wheat cultivars using ethylenediurea." *Plants* 8, no. 4 (2019): 80.

Feng, Zhaozhong, Yansen Xu, Kazuhiko Kobayashi, Lulu Dai, Tianyi Zhang, Evgenios Agathokleous, Vicent Calatayud et al. "Ozone pollution threatens the production of major staple crops in East Asia." *Nature Food* 3, no. 1 (2022): 47–56.

Fernandes, Francine Faia, and Bárbara Baesso Moura. "Foliage visible injury in the tropical tree species, Astronium graveolens is strictly related to phytotoxic ozone dose (PODy)." *Environmental Science and Pollution Research* 28, no. 31 (2021): 41726–41735.

Furlan, Cláudia Maria, Regina M. Moraes, Patricia Bulbovas, Marisa Domingos, Antonio Salatino, and M. J. Sanz. "Psidium guajava 'Paluma'(the guava plant) as a new bio-indicator of ozone in the tropics." *Environmental Pollution* 147, no. 3 (2007): 691–695.

Furlan, Cláudia Maria, Regina M. Moraes, Patricia Bulbovas, Maria J. Sanz, Marisa Domingos, and Antonio Salatino. "Tibouchina pulchra (Cham.) Cogn., a native Atlantic Forest species, as a bio-indicator of ozone: Visible injury." *Environmental Pollution* 152, no. 2 (2008): 361–365.

Garg, Neera, and Priyanka Singla. "Arsenic toxicity in crop plants: Physiological effects and tolerance mechanisms." *Environmental Chemistry Letters* 9, no. 3 (2011): 303–321.

Ghosh, Annesha, Ashutosh Kumar Pandey, Madhoolika Agrawal, and Shashi Bhushan Agrawal. "Assessment of growth, physiological, and yield attributes of wheat cultivar HD 2967 under elevated ozone exposure adopting timely and delayed sowing conditions." *Environmental Science and Pollution Research* 27, no. 14 (2020): 17205–17220.

Grantz, D. A., J. H. B. Garner, and D. W. Johnson. "Ecological effects of particulate matter." *Environment International* 29, no. 2–3 (2003): 213–239.

Gupta, Abhishek, and Amit Dhir. "Spatial and temporal variations of air pollutants in urban agglomeration areas in Gujarat, India during 2004–2018." *Mapan* 37, no. 1 (2022): 215–226.

Gupta, Richa, Priyanka Pal, Chanchal Malhotra, and Vikas Sarsar. "Impact of sulphur dioxide ($SO_2$) stress on growth and physiological responses of Calendula officinalis L." *Bionature* (2022): 43–57.

Gurjar, Bhola Ram, Khaiwal Ravindra, and Ajay Singh Nagpure. "Air pollution trends over Indian megacities and their local-to-global implications." *Atmospheric Environment* 142 (2016): 475–495.

Hakim, Zainab Q., Scott Archer-Nicholls, Gufran Beig, Gerd A. Folberth, Kengo Sudo, Nathan Luke Abraham, Sachin Ghude, Daven K. Henze, and Alexander T. Archibald. "Evaluation of tropospheric ozone and ozone precursors in simulations from the HTAPII and CCMI model intercomparisons–a focus on the Indian subcontinent." *Atmospheric Chemistry and Physics* 19, no. 9 (2019): 6437–6458.

Han, Aru, Yongbin Bao, Xingpeng Liu, Zhijun Tong, Song Qing, Yuhai Bao, and Jiquan Zhang. "Plant Ontogeny Strongly Influences SO2 Stress Resistance in Landscape Tree Species Leaf Functional Traits." *Remote Sensing* 14, no. 8 (2022): 1857.

Han, Yansha, Hao Yang, Mengyang Wu, and Huilan Yi. "Enhanced drought tolerance of foxtail millet seedlings by sulfur dioxide fumigation." *Ecotoxicology and Environmental Safety* 178 (2019): 9–16.

Hayes, Felicity, Harry Harmens, Katrina Sharps, and Alan Radbourne. "Ozone dose-response relationships for tropical crops reveal potential threat to legume and wheat production, but not to millets." *Scientific African* 9 (2020a): e00482.

Hayes, Felicity, Katrina Sharps, Harry Harmens, Ieuan Roberts, and Gina Mills. "Tropospheric ozone pollution reduces the yield of African crops." *Journal of Agronomy and Crop Science* 206, no. 2 (2020b): 214–228.

He, P., A. Radunz, K. P. Bader, and G. H. Schmid. "Influence of $CO_2$ and $SO_2$ on growth and structure of photosystem II of the Chinese tung-oil tree Aleurites montana." *Zeitschrift für Naturforschung C* 51, no. 7–8 (1996): 441–453.

Hewitt, D. K. L., G. Mills, Felicity Hayes, D. Norris, M. Coyle, S. Wilkinson, and W. Davies. "N-fixation in legumes–An assessment of the potential threat posed by ozone pollution." *Environmental Pollution* 208 (2016): 909–918.

Hirano, Takashi, Makoto Kiyota, and Ichiro Aiga. "Physical effects of dust on leaf physiology of cucumber and kidney bean plants." *Environmental Pollution* 89, no. 3 (1995): 255–261.

Ho, Bang Quoc, Hoang Ngoc Khue Vu, Thoai Tam Nguyen, Thi-Thu Giang Nguyen, Thi-Quynh Nhu Diep, Phu Le Vo, and Thi-Kim Nhung Pham. "Modeling impacts of industrial park activity on air quality of surrounding area for identifying isolation distance: A case of Tan Tao Industrial

Park, Ho Chi Minh City, Viet Nam." In *IOP Conference Series: Earth and Environmental Science* (vol. 964, no. 1, p. 012023). IOP Publishing, 2022.

Hope, Allen S., Jeffrey B. Fleming, Douglas A. Stow, and Edward Aguado. "Tussock tundra albedos on the north slope of Alaska: Effects of illumination, vegetation composition, and dust deposition." *Journal of Applied Meteorology and Climatology* 30, no. 8 (1991): 1200–1206.

Hoshika, Yasutomo, Makoto Watanabe, Naoki Inada, and Takayoshi Koike. "Model-based analysis of avoidance of ozone stress by stomatal closure in Siebold's beech (Fagus crenata)." *Annals of Botany* 112, no. 6 (2013): 1149–1158.

Hoshika, Yasutomo, Makoto Watanabe, Mitsutoshi Kitao, Karl-Heinz Häberle, Thorsten EE Grams, Takayoshi Koike, and Rainer Matyssek. "Ozone induces stomatal narrowing in European and Siebold's beeches: A comparison between two experiments of free-air ozone exposure." *Environmental Pollution* 196 (2015): 527–533.

Hu, Yanbo, Nacer Bellaloui, Mulualem Tigabu, Jinghong Wang, Jian Diao, Ke Wang, Rui Yang, and Guangyu Sun. "Gaseous NO2 effects on stomatal behavior, photosynthesis and respiration of hybrid poplar leaves." *Acta Physiologiae Plantarum* 37, no. 2 (2015): 1–8.

Iglesias, Domingo J., Ángeles Calatayud, Eva Barreno, Eduardo Primo-Millo, and Manuel Talon. "Responses of citrus plants to ozone: Leaf biochemistry, antioxidant mechanisms and lipid peroxidation." *Plant Physiology and Biochemistry* 44, no. 2–3 (2006): 125–131.

Ishii, S., F. M. Marshall, and J. N. B. Bell. "Physiological and morphological responses of locally grown Malaysian rice cultivars (Oryza sativa L.) to different ozone concentrations." *Water, Air, and Soil Pollution* 155, no. 1 (2004): 205–221.

Ismail, Marzuki, Azrin Suroto, and Samsuri Abdullah. "Response of Malaysian Local Rice Cultivars Induced by Elevated Ozone Stress." *Environment Asia* 8, no. 1 (2015).

Janta, Radshadaporn, Kazuhiko Sekiguchi, Ryosuke Yamaguchi, Khajornsak Sopajaree, Bandhita Plubin, and Thaneeya Chetiyanukornkul. "Spatial and temporal variations of atmospheric PM10 and air pollutants concentration in upper Northern Thailand during 2006–2016." *Applied Science and Engineering Progress* 13, no. 3 (2020): 256–267.

Jiang, Lijun, Zhaozhong Feng, Lulu Dai, Bo Shang, and Elena Paoletti. "Large variability in ambient ozone sensitivity across 19 ethylenediurea-treated Chinese cultivars of soybean is driven by total ascorbate." *Journal of Environmental Sciences* 64 (2018): 10–22.

Kateivas, Katielle Silva Brito, Paulo Araquém Ramos Cairo, Pedro Henrique Santos Neves, Roger Sebastian Silva Ribeiro, Leohana Martins Machado, and Carlos André Espolador Leitão. "The impact of NOx and SO2 emissions from a magnesite processing industry on morphophysiological and anatomical features of plant bioindicators." *Acta Physiologiae Plantarum* 44, no. 8 (2022): 1–12.

Keller, J., and R. Lamprecht. "Road dust as an indicator for air pollution transport and deposition: An application of SPOT imagery." *Remote Sensing of the Environment* 54 (1995): 1–12.

Kitao, Mitsutoshi, Markus Löw, Christian Heerdt, Thorsten EE Grams, Karl-Heinz Häberle, and Rainer Matyssek. "Effects of chronic elevated ozone exposure on gas exchange responses of adult beech trees (Fagus sylvatica) as related to the within-canopy light gradient." *Environmental Pollution* 157, no. 2 (2009): 537–544.

Kittipornkul, Piyatida, Sucheewin Krobthong, Yodying Yingchutrakul, and Paitip Thiravetyan. "Mechanisms of ozone responses in sensitive and tolerant mungbean cultivars." *Science of The Total Environment* 800 (2021): 149550.

Krauss, Martin, Wolfgang Wilcke, Christopher Martius, Adelmar G. Bandeira, Marcos V. B. Garcia, and Wulf Amelung. "Atmospheric versus biological sources of polycyclic aromatic hydrocarbons (PAHs) in a tropical rain forest environment." *Environmental Pollution* 135, no. 1 (2005): 143–154.

Krotkov, Nickolay A., Chris A. McLinden, Can Li, Lok N. Lamsal, Edward A. Celarier, Sergey V. Marchenko, William H. Swartz et al. "Aura OMI observations of regional $SO_2$ and $NO_2$ pollution changes from 2005 to 2015." *Atmospheric Chemistry and Physics* 16, no. 7 (2016): 4605–4629.

Kumar, Prashant, Angela Druckman, John Gallagher, Birgitta Gatersleben, Sarah Allison, Theodore S. Eisenman, Uy Hoang et al. "The nexus between air pollution, green infrastructure and human health." *Environment International* 133 (2019): 105181.

Kuttippurath, Jayanarayanan, Vikas Kumar Patel, Mansi Pathak, and Ajay Singh. "Improvements in SO2 pollution in India: Role of technology and environmental regulations." *Environmental Science and Pollution Research* (2022): 1–13.

Laban, Tracey Leah, Pieter Gideon Van Zyl, Johan Paul Beukes, Ville Vakkari, Kerneels Jaars, Nadine Borduas-Dedekind, Miroslav Josipovic, Anne Mee Thompson, Markku Kulmala, and Lauri Laakso. "Seasonal influences on surface ozone variability in continental South Africa and implications for air quality." *Atmospheric Chemistry and Physics* 18, no. 20 (2018): 15491–15514.

Le, Hoang Anh, and Nguyen Thi Kim Oanh. "Integrated assessment of brick kiln emission impacts on air quality." *Environmental Monitoring and Assessment* 171, no. 1 (2010): 381–394.

Long, J. X., J. F. Liu, and H. Y. Cheng. "Physiological responses of three kinds of street trees to acute stress of Ozone." In *IOP Conference Series: Earth and Environmental Science* (vol. 146, no. 1, p. 012018). IOP Publishing, 2018.

Luo, Xiaosan, Haijian Bing, Zhuanxi Luo, Yujun Wang, and Ling Jin. "Impacts of atmospheric particulate matter pollution on environmental biogeochemistry of trace metals in soil-plant system: A review." *Environmental Pollution* 255 (2019): 113138.

Majernik, Ondrej, and T. A. Mansfield. "Direct effect of SO2 pollution on the degree of opening of stomata." *Nature* 227, no. 5256 (1970): 377–378.

Mandal, Papiya, Raju Sarkar, Neel Kamal, Manob Das, and Anubha Mandal. "Diurnal and seasonal variation of atmospheric particulate matter and trace gases in industrial area of Delhi: A study." *Bulletin of Environmental Contamination and Toxicology* (2022): 1–7.

Mazarura, U. "Effect of sequences of ozone and nitrogen dioxide on plant dry matter and stomatal diffusive resistance in radish." *African Crop Science Journal* 20 (2012): 371–384.

Mills, Gina, Håkan Pleijel, Sabine Braun, Patrick Büker, Victoria Bermejo, Esperanza Calvo, Helena Danielsson et al. "New stomatal flux-based critical levels for ozone effects on vegetation." *Atmospheric Environment* 45, no. 28 (2011): 5064–5068.

Mills, Gina, Katrina Sharps, David Simpson, Håkan Pleijel, Malin Broberg, Johan Uddling, Fernando Jaramillo et al. "Ozone pollution will compromise efforts to increase global wheat production." *Global Change Biology* 24, no. 8 (2018): 3560–3574.

Mina, Usha, T. K. Chandrashekara, S. Naresh Kumar, M. C. Meena, S. Yadav, S. Tiwari, Deepak Singh, Pranav Kumar, and Ram Kumar. "Impact of particulate matter on basmati rice varieties grown in indo-gangetic plains of India: Growth, biochemical, physiological and yield attributes." *Atmospheric Environment* 188 (2018): 174–184.

Mina, Usha, Pranav Kumar, and C. Varshney. "Effect of ozone exposure on growth, yield and isoprene emission from tomato (Lycopersicon esculentum L.) plants." *Vegetable Crops Research Bulletin* 72 (2010): 35.

Mishra, Amit Kumar, and S. B. Agrawal. "Biochemical and physiological characteristics of tropical mung bean (*Vigna radiata* L.) cultivars against chronic ozone stress: An insight to cultivar-specific response." *Protoplasma* 252, no. 3 (2015): 797–811.

Moraes, Regina M., Patricia Bulbovas, Cláudia M. Furlan, Marisa Domingos, Sérgio T. Meirelles, Welington B. C. Delitti, and Maria J. Sanz. "Physiological responses of saplings of Caesalpinia echinata Lam., a Brazilian tree species, under ozone fumigation." *Ecotoxicology and Environmental Safety* 63, no. 2 (2006): 306–312.

Moraes, Regina M., C. M. Furlan, P. Bulbovas, M. Domingos, S. T. Meirelles, A. Salatino, W. B. C. Delitti, and M. J. Sanz. "Photosynthetic responses of tropical trees to short-term exposure to ozone." *Photosynthetica* 42, no. 2 (2004): 291–293.

Morales, F. J. "Introduction to tropical agriculture and outlook for tropical crops in a globalized economy." *Tropical Biology and Conservation Management* 3 (2009): 1–27.

Morikawa, H., M. Takahashi, and Y. Kawamura. "Metabolism and genetics of atmospheric nitrogen dioxide control using pollutant-philic plants." *Phytoremediation: Transformation and Control of Contaminants* (2003): 765–786.

Moura, Bárbara Baêsso, Cecilia Brunetti, Marcela Regina Gonçalves da Silva Engela, Yasutomo Hoshika, Elena Paoletti, and Francesco Ferrini. "Experimental assessment of ozone risk on ecotypes of the tropical tree Moringa oleifera." *Environmental Research* 201 (2021): 111475.

Moura, Bárbara Baêsso, Edenise Segala Alves, Mauro Alexandre Marabesi, Silvia Ribeiro de Souza, Marcus Schaub, and Pierre Vollenweider. "Ozone affects leaf physiology and causes injury to foliage of native tree species from the tropical Atlantic Forest of southern Brazil." *Science of the Total Environment* 610 (2018): 912–925.

Moura, Bárbara Baêsso, Sílvia Ribeiro de Souza, and Edenise Segala Alves. "Response of Brazilian native trees to acute ozone dose." *Environmental Science and Pollution Research* 21, no. 6 (2014): 4220–4227.

Mukherjee, Arideep, and Madhoolika Agrawal. "Pollution response score of tree species in relation to ambient air quality in an urban area." *Bulletin of Environmental Contamination and Toxicology* 96, no. 2 (2016): 197–202.

Mukherjee, Arideep, Shashi Bhushan Agrawal, and Madhoolika Agrawal. "Intra-urban variability of ozone in a tropical city—characterization of local and regional sources and major influencing factors." *Air Quality, Atmosphere & Health* 11, no. 8 (2018): 965–977.

Mukherjee, Arideep, Bhanu Pandey, S. B. Agrawal, and Madhoolika Agrawal. "Responses of tropical and subtropical plants to air pollution." In *Tropical Ecosystems: Structure, Functions and Challenges in the Face of Global Change* (pp. 129–162). Springer, 2019.

Murray, F., R. Monk, and C. D. Walker. "The responses of shoot growth of Eucalyptus species to concentration and frequency of exposure to nitrogen oxides." *Forest Ecology and Management* 64, no. 1 (1994): 83–95.

Myhre, G., D. Shindell, F.-M. Bréon, W. Collins, J. Fuglestvedt, J. Huang, D. Koch, J.-F. Lamarque, D. Lee, B. Mendoza et al." Anthropogenic and natural radiative forcing." In Stocker, T.F., Qin, D., Plattner, G.-K., Tignor, M., Allen, S.K., Boschung, J., Nauels, A., Xia, Y., Bex, V., Midgley, P.M., Eds., *Climate Change 2013: The Physical Science Basis. Contribution of Working Group I to the Fifth Assessment Report of the Intergovernmental Panel on Climate Change.* Cambridge University Press, 2013.

Naidoo, G., and D. Chirkoot. "The effects of coal dust on photosynthetic performance of the mangrove, Avicennia marina in Richards Bay, South Africa." *Environmental Pollution* 127, no. 3 (2004): 359–366.

Nayak, Rekha, Debasis Biswal, and Rupnarayan Sett. "Biochemical changes in some deciduous tree species around Talcher thermal power station, Odisha, India." *Journal of Environmental Biology* 34, no. 3 (2013): 521.

Nighat, F. Mahmooduzzafar, and M. Iqbal." Stomatal conductance, photosynthetic rate, and pigment content in Ruellia tuberosa leaves as affected by coal-smoke pollution." *Biol Plant* 43 (2000): 263–267.

Noor, Norazian Mohamed, Nor Naimah Binti Mohamad, and Nur Izzah Mohamad Hashim. "Variation of ground-level ozone in the West Coast of Peninsular Malaysia." *EnvironmentAsia* 11, no. 3 (2018).

Norby, R. J., and R. J. Luxmoore. "Growth analysis of soybean exposed to simulated acid rain and gaseous air pollutants." *New Phytologist* 95, no. 2 (1983): 277–287.

Opio, Ronald, Isaac Mugume, and Joyce Nakatumba-Nabende. "Understanding the Trend of NO2, SO2 and CO over East Africa from 2005 to 2020." *Atmosphere* 12, no. 10 (2021): 1283.

Pavlík, Milan, Daniela Pavlíková, Veronika Zemanová, František Hnilička, Veronika Urbanová, and Jiřina Száková. "Trace elements present in airborne particulate matter—Stressors of plant metabolism." *Ecotoxicology and Environmental Safety* 79 (2012): 101–107.

Pereira, Eduardo Gusmão, Marco Antonio Oliva, Kacilda Naomi Kuki, and José Cambraia. "Photosynthetic changes and oxidative stress caused by iron ore dust deposition in the tropical CAM tree Clusia hilariana." *Trees* 23, no. 2 (2009): 277–285.

Perron, N., J. Sandradewi, M. R. Alfarra, P. Lienemann, R. Gehrig, A. Kasper-Giebl, V. A. Lanz et al. "Composition and sources of particulate matter in an industrialised Alpine valley." *Atmospheric Chemistry and Physics Discussions* 10, no. 4 (2010): 9391–9430.

Pina, Juliana M., and Regina M. Moraes. "Gas exchange, antioxidants and foliar injuries in saplings of a tropical woody species exposed to ozone." *Ecotoxicology and Environmental Safety* 73, no. 4 (2010): 685–691.

Pina, Juliana M., and Regina M. Moraes. "Ozone-induced foliar injury in saplings of Psidium guajava 'Paluma'in São Paulo, Brazil." *Chemosphere* 66, no. 7 (2007): 1310–1314.

Pina, Juliana M., S. R. Souza, Sérgio Tadeu Meirelles, and R. M. Moraes. "Psidium guajava Paluma responses to environmental conditions and ozone concentrations in the urban forest of São Paulo, SE-Brazil." *Ecological Indicators* 77 (2017): 1–7.

Polichetti, Giuliano, Stefania Cocco, Alessandra Spinali, Valentina Trimarco, and Alfredo Nunziata. "Effects of particulate matter (PM10, PM2.5 and PM1) on the cardiovascular system." *Toxicology* 261, no. 1–2 (2009): 1–8.

Poonia, Shefali. "Studies on Alpha–Amylase Activity in Germinating Seeds of Four Leguminous Crops in Response to Sulphur Dioxide." *Biotech Today* 9, no. 2 (2019): 51–53.

Popek, Robert, Lovely Mahawar, Gyan Singh Shekhawat, and Arkadiusz Przybysz. "Phyto-cleaning of particulate matter from polluted air by woody plant species in the near-desert city of Jodhpur (India) and the role of heme oxygenase in their response to PM stress conditions." *Environmental Science and Pollution Research* (2022): 1–14.

Prabakaran, P., and A. Manikandan. "Annual augmentation of respirable particulates (PM10 & PM2.5) and gaseous pollutants ($SO_2$, $NO_2$ and $O_3$) in Chennai City." *International Journal of Civil Engineering and Technology* 9, no. 7 (2018): 363–372.

Prakash, Jigyasa, Shashi Bhushan Agrawal, and Madhoolika Agrawal. "Global trends of acidity in rainfall and its impact on plants and soil." *Journal of Soil Science and Plant Nutrition* 23, no. 1 (2023): 398–419.

Prasad, M. S. V., and J. A. Inamdar. "Effect of cement kiln dust pollution on black gram (Vigna mungo (L.) Hepper)." *Proceedings: Plant Sciences* 100, no. 6 (1990): 435–443.

Prusty, B. A. K., P. C. Mishra, and P. A. Azeez. "Dust accumulation and leaf pigment content in vegetation near the national highway at Sambalpur, Orissa, India." *Ecotoxicology and Environmental Safety* 60, no. 2 (2005): 228–235.

Qayoom Mir, A., T. Yazdani, S. Ahmad, and M. Yunus. "Total flavonoids and phenolics in *Catharanthus roseus* L. and *Ocimum sanctum* L. as biomarkers of urban auto pollution." *Caspian Journal of Environmental Sciences* 7, no. 1 (2009): 9–16.

Rahman, Md Mostafijur, Shakil Mahamud, and George D. Thurston. "Recent spatial gradients and time trends in Dhaka, Bangladesh, air pollution and their human health implications." *Journal of the Air & Waste Management Association* 69, no. 4 (2019): 478–501.

Rahul, Jitin, and Manish Kumar Jain. "An investigation in to the impact of particulate matter on vegetation along the national highway: A review." *Research Journal of Environmental Sciences* 8, no. 356 (2014): e372.

Rai, Prabhat Kumar. "Impacts of particulate matter pollution on plants: Implications for environmental biomonitoring." *Ecotoxicology and Environmental Safety* 129 (2016): 120–136.

Rai, Richa, and Madhoolika Agrawal. "Evaluation of physiological and biochemical responses of two rice (Oryza sativa L.) cultivars to ambient air pollution using open top chambers at a rural site in India." *Science of the Total Environment* 407, no. 1 (2008): 679–691.

Rai, Richa, Madhoolika Agrawal, and S. B. Agrawal. "Assessment of yield losses in tropical wheat using open top chambers." *Atmospheric Environment* 41, no. 40 (2007): 9543–9554.

Rajput, Madhu, and Madhoolika Agrawal. "Biomonitoring of air pollution in a seasonally dry tropical suburban area using wheat transplants." *Environmental Monitoring and Assessment* 101, no. 1 (2005): 39–53.

Rodriguez, Judith Hebelen, María Luisa Pignata, A. Fangmeier, and A. Klumpp. "Accumulation of polycyclic aromatic hydrocarbons and trace elements in the bioindicator plants *Tillandsia capillaris* and *Lolium multiflorum* exposed at PM10 monitoring stations in Stuttgart (Germany)." *Chemosphere* 80, no. 3 (2010): 208–215.

Sabaratnam, Sashikala, Gian Gupta, and Charles Mulchi. "Effects of nitrogen dioxide on leaf chlorophyll and nitrogen content of soybean." *Environmental Pollution* 51, no. 2 (1988): 113–120.

Saha, Dulal C., and Pratap K. Padhy. "Effect of particulate pollution on rate of transpiration in Shorea robusta at Lalpahari forest." *Trees* 26, no. 4 (2012): 1215–1223.

Sarkar, Abhijit, and S. B. Agrawal. "Elevated ozone and two modern wheat cultivars: An assessment of dose dependent sensitivity with respect to growth, reproductive and yield parameters." *Environmental and Experimental Botany* 69, no. 3 (2010): 328–337.

Sarkar, Abhijit, and S. B. Agrawal. "Evaluating the response of two high yielding Indian rice cultivars against ambient and elevated levels of ozone by using open top chambers." *Journal of Environmental Management* 95 (2012): S19–S24.

Schneider, Gerald F., Alexander W. Cheesman, Klaus Winter, Benjamin L. Turner, Stephen Sitch, and Thomas A. Kursar. "Current ambient concentrations of ozone in Panama modulate the leaf chemistry of the tropical tree Ficus insipida." *Chemosphere* 172 (2017): 363–372.

Seinfeld, John H., and Spyros N. Pandis. "From air pollution to climate change." In *Atmospheric Chemistry and Physics*, 3rd ed. JohnWiley & Sons, Inc, 2016.

Shahadan, Nurul Azzura. *Measurement of Growth, Photosynthesis and Transpiration Rates of a Malaysian Local Rice Cultivar (MR 219) for Assessing Ambient Ozone Stress*. PhD diss., Universiti Putra Malaysia, 2005.

Sharifi, M. Rasoul, Arthur C. Gibson, and Philip W. Rundel. "Surface dust impacts on gas exchange in Mojave Desert shrubs." *Journal of applied Ecology* (1997): 837–846.

Sharma, Disha, and Denise Mauzerall. "Analysis of air quality data in India between 2015 and 2019." *Aerosol and Air Quality Research* 22 (2022): 210204.

Sharma, M. "Effect of Sulphur-dioxide fumigation on *Cicer arietinum* L." *Environment Conservation Journal* 19, no. 1–2 (2018): 89–91.

Sharma, Priyanka, Poonam Yadav, Chirashree Ghosh, and Bhupinder Singh. "Heavy metal capture from the suspended particulate matter by Morus alba and evidence of foliar uptake and translocation of PM associated zinc using radiotracer (65Zn)." *Chemosphere* 254 (2020): 126863.

Sharps, Katrina, Massimo Vieno, Rachel Beck, Felicity Hayes, and Harry Harmens. "Quantifying the impact of ozone on crops in Sub-Saharan Africa demonstrates regional and local hotspots of production loss." *Environmental Science and Pollution Research* 28, no. 44 (2021): 62338–62352.

Shaw, P. J. A., M. R. HOLLAND, N. M. Darrall, and A. R. McLeod. "The occurrence of SO2-related foliar symptoms on Scots pine (Pinus sylvestris L.) in an open-air forest fumigation experiment." *New Phytologist* 123, no. 1 (1993): 143–152.

Sheng, Qianqian, and Zunling Zhu. "Effects of nitrogen dioxide on biochemical responses in 41 garden plants." *Plants* 8, no. 2 (2019): 45.

Sheng, Qianqian, and Zunling Zhu. "Photosynthetic capacity, stomatal behavior and chloroplast ultrastructure in leaves of the endangered plant Carpinus putoensis WC Cheng during gaseous NO2 exposure and after recovery." *Forests* 9, no. 9 (2018): 561.

Shukla, J., V. Pandey, S. N. Singh, M. Yunus, N. Singh, and K. J. Ahmad. "Effect of cement dust on the growth and yield of Brassica campestris L." *Environmental Pollution* 66, no. 1 (1990): 81–88.

Simpson, David, M. R. Ashmore, L. Emberson, and J.-P. Tuovinen. "A comparison of two different approaches for mapping potential ozone damage to vegetation. A model study." *Environmental Pollution* 146, no. 3 (2007): 715–725.

Singh, Aditya Abha, S. B. Agrawal, J. P. Shahi, and Madhoolika Agrawal. "Investigating the response of tropical maize (Zea mays L.) cultivars against elevated levels of O3 at two developmental stages." *Ecotoxicology* 23, no. 8 (2014): 1447–1463.

Singh, Amarendra, Atul Kumar Srivastava, V. Varaprasad, Sunil Kumar, Virendra Pathak, and Arvind Kumar Shukla. "Assessment of near-surface air pollutants at an urban station over the central Indo-Gangetic Basin: Role of pollution transport pathways." *Meteorology and Atmospheric Physics* 133, no. 4 (2021a): 1127–1142.

Singh, E., S. Tiwari, and M. Agrawal. "Effects of elevated ozone on photosynthesis and stomatal conductance of two soybean varieties: A case study to assess impacts of one component of predicted global climate change." *Plant Biology* 11 (2009): 101–108.

Singh, Harshita, Pallavi Singh, Shashi Bhushan Agrawal, and Madhoolika Agrawal. "Assessment of the reverberations caused by predominant air pollutants on urban vegetation: A multi-site study in Varanasi located in Indo-Gangetic Plains." *Gases* 3, no. 2 (2023a): 57–76.

Singh, Harshita, Pallavi Singh, Shashi Bhushan Agrawal, and Madhoolika Agrawal. "Implications of foliar particulate matter deposition on the physiology and nutrient allocation of dominant perennial species of the Indo-Gangetic Plains." *Frontiers in Plant Science* 13 (2022).

Singh, Harshita, Pallavi Singh, Shashi Bhushan Agrawal, and Madhoolika Agrawal. "Particulate matter deposition on the foliage aided in low temperature acclimatization of *Psidium guajava* L. in winter." *Proceedings of the National Academy of Sciences, India Section B: Biological Sciences* (2023b): 1–11.

Singh, Pallavi, Harshita Singh, Shashi Bhushan Agrawal, and Madhoolika Agrawal. "Assessment of the differential trade-off between growth, subsistence, and productivity of two popular Indian hybrid mango varieties under elevated ozone exposure." *Science of the Total Environment* (2023c): 164275.

Singh, Pratiksha, Rekha Kannaujia, Shiv Narayan, Ashish Tewari, Pramod A. Shirke, and Vivek Pandey. "Impact of chronic elevated ozone exposure on photosynthetic traits and anti-oxidative defense responses of *Leucaena leucocephala* (Lam.) de wit tree under field conditions." *Science of The Total Environment* 782 (2021b): 146907.

Singh, R. N., Joydeep Mukherjee, V. K. Sehgal, P. Krishnan, Deb Kumar Das, Raj Kumar Dhakar, and Arti Bhatia. "Interactive effect of elevated tropospheric ozone and carbon dioxide on radiation utilisation, growth and yield of chickpea (*Cicer arietinum* L.)." *International Journal of Biometeorology* 65, no. 11 (2021c): 1939–1952.

Singh, Satyavan, Arti Bhatia, Ritu Tomer, Vinod Kumar, B. Singh, and S. D. Singh. "Synergistic action of tropospheric ozone and carbon dioxide on yield and nutritional quality of Indian mustard (*Brassica juncea* (L.) Czern.)." *Environmental Monitoring and Assessment* 185, no. 8 (2013): 6517–6529.

Singh, Siddharth, Bhanu Pandey, Lal Babu Roy, Sameer Shekhar, and Ranjeet Kumar Singh. "Tree responses to foliar dust deposition and gradient of air pollution around opencast coal mines of Jharia coalfield, India: Gas exchange, antioxidative potential and tolerance level." *Environmental Science and Pollution Research* 28, no. 7 (2021d): 8637–8651.

Solomon, Paul A., and Constantinos Sioutas. "Continuous and semicontinuous monitoring techniques for particulate matter mass and chemical components: A synthesis of findings from EPA's particulate matter supersites program and related studies." *Journal of the Air & Waste Management Association* 58, no. 2 (2008): 164–195.

Sonkar, Geetika, R. K. Mall, Tirthankar Banerjee, Nidhi Singh, TV Lakshmi Kumar, and Ramesh Chand. "Vulnerability of Indian wheat against rising temperature and aerosols." *Environmental Pollution* 254 (2019): 112946.

Srivastava, Hari Shanker. *Effects of Nitrogen Dioxide on Gas Exchange in Phaseolus Vulgaris Leaves*. PhD diss., University of British Columbia, 1974.

Storch-Böhm, Renata F., Cleder A. Somensi, Renan C. Testolin, Überson B. Rossa, Rogério Corrêa, Rafael Ariente-Neto, Gizelle I. Almerindo, Jean-François Férard, Sylvie Cotelle, and Claudemir M. Radetski. "Urban afforestation: Using phytotoxicity endpoints to compare air pollution tolerance of two native Brazilian plants Aroeira (Schinus terebinthifolius) and Cuvatã (Cupania vernalis)." *Environmental Science and Pollution Research* (2022): 1–13.

Sutton, Mark A., and Albert Bleeker. "The shape of nitrogen to come." *Nature* 494, no. 7438 (2013): 435–437.

Swartz, J-S., P. G. Van Zyl, J. P. Beukes, C. Labuschagne, E-G. Brunke, Thierry Portafaix, Corinne Galy-Lacaux, and J. J. Pienaar. "Twenty-one years of passive sampling monitoring of SO2, NO2 and O3 at the Cape Point GAW station, South Africa." *Atmospheric Environment* 222 (2020): 117128.

Takahashi, Misa, Takamasa Furuhashi, Naoko Ishikawa, Gorou Horiguchi, Atsushi Sakamoto, Hirokazu Tsukaya, and Hiromichi Morikawa. "Nitrogen dioxide regulates organ growth by controlling cell proliferation and enlargement in Arabidopsis." *New Phytologist* 201, no. 4 (2014): 1304–1315.

Tetteh, Rashied, Masahiro Yamaguchi, Yoshiharu Wada, Ryo Funada, and Takeshi Izuta. "Effects of ozone on growth, net photosynthesis and yield of two African varieties of Vigna unguiculata." *Environmental Pollution* 196 (2015): 230–238.

Thwe, Aye Aye, Gilles Vercambre, Hélène Gautier, Frédéric Gay, Jessada Phattaralerphong, and Poonpipope Kasemsap. "Response of photosynthesis and chlorophyll fluorescence to acute ozone stress in tomato (Solanum lycopersicum Mill.)." *Photosynthetica* 52, no. 1 (2014): 105–116.

Uka, N. U., and E. J. Belford. "Response of roadside tree leaves in a tropical city to automobile pollution." *Notulae Scientia Biologicae* 12, no. 3 (2020): 752–768.

Vallano, Dena M., Paul C. Selmants, and Erika S. Zavaleta. "Simulated nitrogen deposition enhances the performance of an exotic grass relative to native serpentine grassland competitors." *Plant Ecology* 213, no. 6 (2012): 1015–1026.

Wamoto, John Milikzitiek. *Determination of NOX and SO2 Concentration Levels in Nairobi City, Kenya by Use of Passive Samplers*. PhD diss., 2013.

Wang, Yue, Weiwei Jin, Yanhui Che, Dan Huang, Jiechen Wang, Meichun Zhao, and Guangyu Sun. "Atmospheric nitrogen dioxide improves photosynthesis in mulberry leaves via effective utilization of excess absorbed light energy." *Forests* 10, no. 4 (2019): 312.

World Health Organization. *WHO Global Air Quality Guidelines: Particulate Matter (PM2.5 and PM10), Ozone, Nitrogen Dioxide, Sulfur Dioxide and Carbon Monoxide*. World Health Organization, 2021.

World Health Organization (WHO). https://www.who.int/news/item/04-04-2022-billions-of-people-still-breathe-unhealthy-air-new-who-data, 2022.

Yadav, Durgesh Singh, S. B. Agrawal, and Madhoolika Agrawal. "Ozone flux-effect relationship for early and late sown Indian wheat cultivars: Growth, biomass, and yield." *Field Crops Research* 263 (2021b): 108076.

Yadav, Durgesh Singh, Bhavna Jaiswal, Shashi Bhushan Agrawal, and Madhoolika Agrawal. "Diurnal variations in physiological characteristics, photoassimilates, and total ascorbate in early and late sown Indian wheat cultivars under exposure to elevated ozone." *Atmosphere* 12, no. 12 (2021a): 1568.

Yadav, Durgesh Singh, Amit Kumar Mishra, Richa Rai, Nivedita Chaudhary, Arideep Mukherjee, S. B. Agrawal, and Madhoolika Agrawal. "Responses of an old and a modern Indian wheat cultivar to future O3 level: Physiological, yield and grain quality parameters." *Environmental Pollution* 259 (2020): 113939.

Žalud, Petr, Jiřina Száková, Jiřina Sysalová, and Pavel Tlustoš. "Factors influencing uptake of contaminated particulate matter in leafy vegetables." *Central European Journal of Biology* 7 (2012): 519–530.

Zhao, Xiping, Pingping Guo, Yongqiang Yang, and Haixin Peng. "Effects of air pollution on physiological traits of Ligustrum lucidum Ait. Leaves in Luoyang, China." *Environmental Monitoring and Assessment* 193, no. 8 (2021): 1–14.

Ziegler, I. "Effect of air-polluting gases on plant metabolism." *Environmental Quality and Safety (Germany, Federal Republic of)* 2 (1972).

Zizi, N. A. Mohd, N. Mohamed Noor, N. Izzah Mohamad Hashim, and S. Y. Yusuf. "Spatial and temporal characteristics of air pollutants concentrations in industrial area in Malaysia." In *IOP Conference Series: Materials Science and Engineering* (vol. 374, no. 1, p. 012094). IOP Publishing, 2018.

# 5 Impact of Nitrogen Oxides on Tropical Plants

*Aisha Kamal, Nida Sultan, Shazia Siddiqui, and Farhan Ahmad*

## 5.1 INTRODUCTION

Between the latitudes of the Tropic of Cancer (23.5°N) and the Tropic of Capricorn (23.5°S), the tropics encircle the planet. These regions are biologically the richest ecosystem, characterized by the prevalence of rain in the humid areas close to the equator which becomes more seasonal as distance moves further from the equator (Laurance et al., 2012). In tropical and subtropical climates, high solar irradiance and heat can enhance photochemical reactions that result in the production and accumulation of many secondary contaminants such as peroxyacetyl nitrate (PAN), smog, and nitrogen oxides ($NO_X$) within the troposphere (Liu et al., 2022). In urban and industrial areas, $NO_X$ are significant air pollutants. In the sunlight and the presence of other atmospheric gases, particularly $O_3$, $NO_X$ stands for the total concentrations of NO and $NO_2$ in the atmosphere, which are inter-convertible. Automobiles, power plants, factories, incinerators, and high-temperature home activities include the principal sources of $NO_X$ emissions. Significant increase in the use of automobiles and fossil fuels since the 20th century is principally responsible for an increase in nitrogenous air pollution (Brimblecombe & Stedman, 1982; Yunus et al., 1996, Perera, 2017). Agricultural soils are also considered as a major source for the emission of $NO_X$, the third-most-important greenhouse gas, caused by the microbial interactions with nitrogenous fertilizers (Bhadouria et al., 2019), and nearly 40% of the world's applications of nitrogen (N) fertilizers occur in tropical and subtropical areas (Xiankai et al., 2008). Over the past few decades, current nitrogen fluxes ($NO_X$, $NO_3^-$, $NH_3$, $NH_4^+$) are far higher than the expected natural flows to the atmosphere (Seinfeld & Pandis, 2016). The environment has suffered significantly as a result of these large-scale perturbations. The negative implications of this increased nitrogen flux include the acidification of soils, accelerated building and monument deterioration, and the development regarding secondary air pollutants (Tian & Niu, 2015; Mukherjee et al., 2019). In the major tropical West African city of Lagos, Nigeria, $CO:NO_X$ ratio was found to be 150–200 times greater than the earlier reports (Odu-Onikosi et al., 2022). In various areas of Varanasi City, India, a remarkable increase (25.6–72.2%) in $NO_2$ was also reported over the past 20 years (Pandey et al., 1992; Mukherjee & Agrawal, 2016).

$NO_X$ in the atmosphere has been recognized as a toxic pollutant and a crucial molecule in biological systems at the same time because of its roles as a regulator and messenger in plants and animals (Hayat et al., 2009). At higher concentrations, both NO and $NO_2$ act as phytotoxins, meaning that they can directly inhibit plant growth and decrease overall yield (WHO, 2000). It was reported that there are at least two indirect routes via which $NO_X$ can function. Primarily, it is a crucial precursor for the production of ozone ($O_3$) and a phytotoxin that is known to lower crop yields (Mills et al., 2018). Second, $NO_X$ is also a precursor to particulate matter aerosols which in presence of ammonia can result in increased concentrations of ammonium nitrate aerosols ($NH_4NO_3$), which is frequently the situation in agricultural areas due to the application of nitrogenous fertilizers like urea (Kharol et al., 2013; Lobell et al., 2022). These particles alter the radiation environment that crops experience and restrict access to radiation that is photosynthetically active, resulting in significant declines in crop quality and yield (Proctor et al., 2018: Proctor, 2021). It was also found in many investigations that $NO_2$ is more persistent and phytotoxic than NO, even at very less concentration. It causes visible damage and physiological aberrations (Mackenzie & El-Ashry, 1988). It was also found that nitrogen accumulation severely affects species dominance, richness, evenness, and abundance, and it has been considered as a major contributor to the loss of biodiversity (Bobbink et al., 2010; Vellend et al., 2017; Midolo et al., 2019). The damage caused by $NO_X$ is typically invisible but manifests as slower growth and diminished output in terms of value and amenity. Very few studies have examined the impact of atmospheric $NO_X$. This chapter emphasizes on the beneficial as well as the adverse impacts of $NO_X$ on the physiological processes of tropical plants, and is an effort to fill this gap and advanced knowledge of the issue.

## 5.2 POSITIVE IMPACTS OF $NO_X$ ON TROPICAL PLANTS

Nitrogen is one of the essential macronutrients for the development of plants. Atmospheric nitrogen gets into the system through the stomata or roots (in the form of nitrates) and is reduced to nitrite and then ammonia by ferredoxin or nicotinamide adenine dinucleotide phosphate (NADPH) before its utilization (Jiechen et al., 2021; Prassad, 1996; Paneque, 1964). It has been hypothesized that NO$_X$ may compete with carbon absorption for NADPH produced by the photoreaction and prevent carbon dioxide fixation in the dark reaction resulting in decrease in photosynthetic rate (Sabaratnam & Mulchi, 1988). NO$_X$ in the environment affects growth and development of plants, but these impacts are not always negative and are difficult to categorize. Numerous studies have shown that NO$_X$ plays the role of phytohormone that affects a variety of physiological processes; through nitrate assimilation route, nitrate metabolization and absorption are in charge of production of various organic nitrogenous compounds at lower concentrations (Pols et al., 2022). However, higher NO$_X$ concentrations resulted in detrimental physiological responses like the production of reactive oxygen species (ROS) and a reduction in the absorption of nitrogen, alteration in chlorophyll concentration of soluble proteins, lipid peroxidation, and antioxidant enzyme activity, leading to acute leaf damage, cell acidification, and excessive nitrite (NO$^{2-}$) accumulation which can cause whole-plant chlorosis or even mortality (Sukumaran, 2014; Sheng & Zhu, 2019). Changes in the morphological and anatomical properties of leaves and stems, in particular, were reported in plants growing beside roads in tropical areas with heavy traffic; this might be because of the strain caused by automobile exhaust emissions (Azmat et al., 2009; Rai & Mishra, 2013; Muthu et al., 2021). Since NO$_2$ levels were often highest in the winter, wheat and other winter crops were exposed to higher NO$_2$ levels than summer crops. Exposure was generally higher for plants of tropical and sub-tropical regions (Table 5.1).

## 5.2.1 Effects of NO$_X$ on Plant Physiology

Plant seed germination is aided by NO, which has been proven. In order to control the terminal transport step and pace of electron transport, it is also known to play a major role in signaling and energy metabolism. Various physiological and developmental processes in a variety of plants are influenced NO$_X$. Numerous authors have noted that NO$_X$ is necessary for the development and growth of plants, including fruit ripening, organ senescence, and respiratory metabolism. This comprises a plant's growth and development from germination through flowering (Liu et al., 2011). A lower concentration of NO demonstrated a protective effect against herbicide increased plant oxidative stress (Beligni et al., 2002). It was also observed that NO$_2$ may function as an airborne fertilizer. Its dissolution in water can result in the conversion of nitrate and nitrite, which are used in the nitrate metabolism by plants (Zeevaart, 1976). Additionally, NO is discovered to stimulate lateral root development probably by taking part in the auxin signaling transduction pathway (Correa-Aragunde et al., 2004). It was also found that a pertinent amount of NO$_2$ can favor the growth of tomato seedlings (Jiechen et al., 2021). With some exceptions, plants exposed to gaseous NO$_2$ experience an increase in nutrient uptake, nutrient metabolism,

### TABLE 5.1
### Collection of Various Studies Representing Effect of NO$_X$ on Some Tropical Plants

| S. No. | Plants | Impact of NO$_X$ | References |
| --- | --- | --- | --- |
| 1. | *Tilia cordata* (Mill.) and *Betula pendula* (Roth.) | Reduction in number and size of stomata. | Janjić et al. (2017); Neighbour et al. (1988) |
| 2. | *Fragaria annasa* (Strawberry) | Drastic reduction in photosynthesis protein, carbohydrate and sucrose content, and closure of stomata. | Muneer et al. (2014) |
| 3. | *Lycopersicum esculentum* | Reduction of net photosynthesis, Growth suppression, inhibits pigment content, and distortion of leaves. | Hayat et al. (2010); Taylor & Eaton (1966) |
| 4. | *Pseudotsuga mensiezii* | enhanced stomatal conductance and decline water use efficiency | Ripullone et al. (2004); Van Hove et al. (1992). |
| 5. | *Nicotiana tabacum* | Leaf necrosis. | Sifola and Postiglione (2003); Bush et al. (1962) |
| 6. | *Lolium perenne* | Significant impact acid/base regulation and nutrients status. | Wollenheber and Raven (1993). |
| 7. | *Pinus sylvestris* | Increased GS activity in pine, alterations in amino acid profile with no significant physiological changes. | Pérez-Soba et al. (1994); Näsholm (1991) |
| 8. | *Phaseolus vulgaris* | NO$_2$ stimulated the overall plant growth and crop yield; increase in transpiration, stunted growth distortion of leaves. | Sandhu and Gupta (1989); Ashenden (1979). |
| 9. | *Pisum sativum* | Closing of stomata and emission of stress ethylene. | Mehlhorn and Wellburn (1987); Rao et al. (1983) |
| 10. | *Medicago sativa, Avena sativa* | Decreased photosynthesis rate. | Hill and Bennet (1970) |

and photosynthesis that causes their total leaf area, shoot biomass, and contents of C (carbon), N (nitrogen), P (phosphorus), K (potassium), Ca (calcium), Mg (magnesium), S (sulfur), and Fe (iron), free amino acids, and crude proteins to roughly double compared to control plants (Adam et al., 2008; Vallano & Sparks, 2008; Takahashi et al., 2005, 2011, 2014). Similarly, NO has also been linked to a variety of events, including germination, senescence, photosynthesis, and cellular redox equilibrium (Fatma et al., 2016; Sehar et al., 2019; Jahan et al., 2021; Iqbal et al., 2021).

### 5.2.2 NO$_X$ Effects on Post-Harvest Quality

It was observed that NO$_X$ improves the post-harvest quality of some fruits and vegetables. Bananas, tomatoes, and kiwifruit, as well as several climacteric and non-climacteric fruits, all benefitted from low NO levels in terms of shelf life and nutritive qualities (Pristijono et al., 2006). Triacylglycerol and diacylglycerol levels were also lowered by NO (Zhu & Zhou, 2006). NO$_X$ at a lower level postponed the pericarp browning of apples and longan fruit (*Dimocarpus longa*) by reducing the amount of soluble solids and ascorbic acid, thereby increasing the shelf life (Xuewu et al., 2007). It was exhibited that NO and N$_2$O treatments extend the storage life of fruits such as bananas and starfruit without compromising on physico-chemical quality attributes. It therefore may be able to prevent the ripening and also effectively reduce the browning incidence of fruits after harvest (Palomer et al., 2005; Sembok & Yen, 2017).

### 5.2.3 Protective Role of NO$_X$ under Plant Abiotic Stress Conditions

NO$_X$'s function as a crucial mediator in a number of physiological activities in plants is well known (Hao & Zhang, 2010). The way that plants react to various abiotic stresses, including ozone, UV-B radiation, toxic metals, salt, drought, high or low temperatures, and toxic metals has indeed been linked to NO$_X$ in recent years (Figure 5.1) (Palavan-Unsal & Arisan, 2009; Ahmad & Kamal, 2022). However, it was postulated that NO$_X$, particularly NO, have antioxidant effects on its own and may act as a cue to trigger the action of ROS-scavenging enzymes to diverse abiotic stresses (Arora et al., 2016). It is postulated that in stress circumstances, NO induces the expression of the alternative oxidase (AOX), and there is evidence that AOX can control mitochondrial NO synthesis (Gupta et al., 2018). Moreover, NO helps to improve the abiotic stress resistance and mitigates stress-induced toxicity more effectively by improving photosynthetic-NUE (nitrogen use efficiency) and -SUE (sulfur-use efficiency), and by fortifying the antioxidant defense system (Gautam et al., 2021). Even at a very low

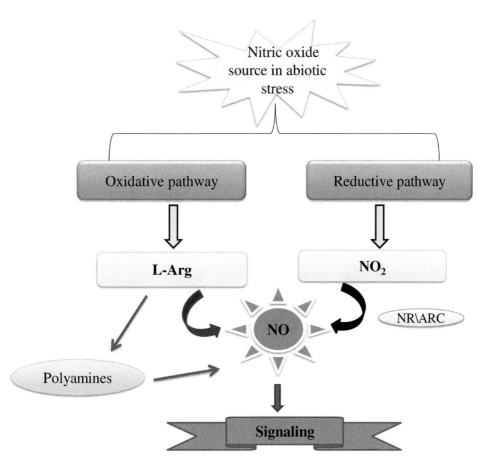

**FIGURE 5.1** NO-mediated signaling mechanisms in abiotic stress conditions.

concentration, NO can enhanced tolerance to a variety of conditions including metal toxicity, salt, drought, and high temperatures by primarily influencing the key elements of the antioxidant defense system and so lowering enhanced ROS-acquired oxidative stress (Siddiqui et al., 2011; Ahmad et al., 2018; Jahan et al., 2020; Sehar et al., 2021).

## 5.3 NEGATIVE IMPACT OF NO$_X$ ON TROPICAL PLANTS

While NO$_X$ promotes growth and development in lower quantities, it can inhibit growth and metabolism significantly at high concentrations (Figure 5.2) (Beligni & Lamattina, 2001). Due to the fact that NO$_2$, NH$_3$, and NH$_4^+$ are all highly phytotoxic, they might be responsible for the harmful consequences of nitrogen-containing air pollution. Wellburn (1990) proposed that the unpaired electron •N=O contributes to NO$_X$'s phytotoxicity. Since NO$_2$ is more persistent than NO and has been shown to be more phytotoxic in numerous studies, its phytotoxicity has been studied more thoroughly than that of NO. Plants and plant cells suffer severe internal and external injuries caused by NO$_X$ pollution (Gheorghe & Ion, 2011). Exposure to high concentration of NO$_X$ in general and NO$_2$ in particular is potentially toxic to plants and can cause chlorosis, tip burn in leaves, and reduced growth. Older needles on pine trees exhibit bleaching, followed by distinct red or brown bands separating necrotic and healthy tissue (Gheorghe & Ion, 2011). Spruce experience immediate defoliation of older needles. Ivory necrosis, herringbone necrosis, black necrosis, and red/brown necrosis have been observed in the older leaves of apple, beech and hazel (Plant Database, 2021). Rapid induction of $O_2^{-1}$ and $H_2O_2$ was reported in strawberry leaves by high dose of NO which enhanced the ROS production and can cause cell damage (Mittler et al., 2004; Muneer et al., 2014). Sheng and Zhu (2019) studied the impact of NO$_2$ on 41 garden species and observed that NO$_2$ exposure adversely affected leaf chlorophyll contents in most of the plants. Inhibition of photosynthesis was also recorded by higher concentration of both NO and NO$_2$ in normal condition. High levels of NO$_2$ can cause cell acidification (Schmutz et al., 1995) and excessive accumulation of nitrite (Okano & Totsuka, 1986), which adversely affect N-assimilation and plant growth brought on by an increase in the production of reactive oxygen species (ROS). Additional consequences of these impacts include acute leaf damage, whole plant chlorosis, and even plant death are all possible outcomes (Sheng & Zhu, 2019). NO$_X$ have a variety of hazardous and harmful effects on plant parameters as discussed in the following subsections.

### 5.3.1 Effect of NO$_X$ on Growth and Morphology of Tropical Plants

Stunted growth and morphological traits are common in the stressed plant, which might be due the fact that more energy may be allocated to defense responses. *Tibouchina pulchra*, a common species of tropical Atlantic forests, showed a considerable decline in growth metrics including plant height, leaf area, basal diameter, biomass, and plant biomass at the locations in Cubato, Brazil with higher NO$_X$ loads (Moraes et al., 2003). Similar observations were also noted by Pandey and Agrawal (1994) in widely planted tropical trees, including *Delonix regia*, *Casia fistula*, and *Carissa carandas*.

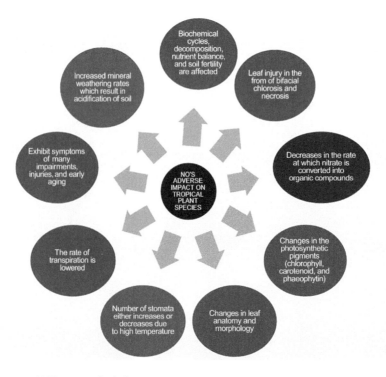

**FIGURE 5.2** Negative impact of NO$_X$ on tropical plants.

## 5.3.2 Effect of NO$_X$ on Plant Anatomy

Alterations in anatomy and morphology were observed—along with considerable modifications in the chlorophyll, carotenoid, and phaeophytin photosynthetic pigments, as well as the proportion of water and powerful antioxidant enzymes (catalase and peroxidase)—in plants exposed to a higher level of NO$_X$ (Verma & Chandra, 2014). According to Yunus and Ahmed (1979), greater levels of NO$_X$ in the environment caused an increase in the number of epidermal cells and stomata, necrotic lesions, and death of epidermal cells, as well as a considerable reduction in the size of the xylem and shoot axes on some tropical plants (Usha Shri & Haritha, 2019). Previously, Dubey et al. (1985) reported the significant decrease in leaf area of some plants such as *Azadirachta indica*, *Mangifera indica*, and *Clerodendrum indicum* when, due to a higher load of nitrogenous pollution, the stem cortex and pith's development greatly slowed down. The phloem parenchymatous cells were also severely impacted by NO$_X$ (Turgeon, 2010).

Wang and Weiger (1988) reported that increase in nitrogenous pollution causes a considerable increase in fiber length and vessel width, and a significant declination in the shoot axis and xylem area's circumference. Additionally, lower epidermis, higher epidermis, spongy parenchyma, and vascular bundles showed anatomical changes in leaves and stems. (Kumar & Nandini, 2013).

Studies on the xylem area, the pith, the cortical area, the shoot axis diameter, and the length of the tree's fibers in the vicinity of a clay company and automobile exhaust which contains a higher amount of NO$_X$ found a considerable decrease in the diameter of the xylem region and the shoot axis of the tropical plants *Abutilon indicum*, *Croton sparsiflorus*, and *Cassia occidentalis*. This decline may be due to the poor rate of xylem formation brought on by air pollution (Sukumaran, 2014). Similar results have already been observed for *Dalbergia sisso* and a few other woody trees (Ghouse et al., 1984; Holopanien et al., 2000). In the region rich in nitrogenous pollutants, conifer assimilatory organs and wood have suffered severe losses (Konopka et al., 1997). It was also noticed that compared to other elements, vessels are more vulnerable to pollution (Sukumaran, 2014).

Due to leaf damage and dust deposition at the stomatal surface, the rate of transpiration was lowered in the polluted tropical forest. The number of stomata has also either risen significantly or decreased due to high temperatures. The dilution of nitrogen from higher macronutrient composition and the reduced soil mineral uptake and water intake are two possible causes of the decrease in tissue nitrogen (Taub & Wang, 2008; Bloom & Burger, 2010). A study conducted on three tropical plants (*Carissa carandas* L., *Delonix regia* Rafin and *Cassia fistula* L.) exposed to the high NO$_X$ from the air were noted to have necrosis and bifacial chlorosis, especially at the leaf tips and edges. Additionally, it has been discovered that NO$_2$ in combination with SO$_2$ intensified the damage (Pandey & Agrawal, 1994).

## 5.3.3 Effect of NO$_X$ on Physiology of Plants

Plants must regulate their function in accordance with changes in environmental conditions in order to grow and develop properly. Plants possess the capability to efficiently utilize their resources to maintain growth by controlling physiological processes in response to environmental conditions. Under air pollution stress conditions, there are well-known variances in physiological processes such stomatal conductance, respiration, photosynthesis, transpiration, and photosynthetic efficiency (Moraes et al., 2002; Baek & Woo, 2010).

Moraes et al. (2002) discovered significant reduction in the rate of photosynthesis in *T. pulchra* in extremely contaminated areas of Cubato, Brazil. Similar decreases in stomatal conductance and photosynthetic efficiency were seen in *R. tuberosa* plants in Aligarh, India, which were exposed to thermal power station emissions (rich in NO$_2$) (Nighat et al., 2000). On the other hand, at locations with higher nitrogenous pollution loads, *Pterocarpus indicus* and *Erythrina orientalis* showed an increase in photosynthetic rate, as well as stomatal conductance (Baek & Woo, 2010). This might be due to foliar intake of NH$_3$, which in turn increases the chlorophyll content, RuBisCo levels, dark respiration, and photosynthesis. Stomatal conductance is also said to increase at night, boosting transpiration and drought sensitivity (Van Der Eerden & Pérez-Soba, 1992; Adrizal et al., 2006). Relatively high concentrations of NO and NO$_2$ under optimal temperature and high light conditions exhibited inhibition of photosynthesis and stomatal conductance (Srivastava et al., 1975; Caporn & Mansfield, 1976). It could be concluded that both NO$_X$ and NHy (consists of all oxides of nitrogen in which the oxidation state of the N atom is +2 or greater) cause growth stimulation at low concentrations and growth suppression at higher concentrations.

## 5.3.4 Effect of NO$_X$ on Nitrogen Assimilation

The rise in leaf N content, which reflects the use of ambient NO$_X$ and NH$^{4+}$ in N assimilation, is ascribed to the increase in pigment concentration (Mukherjee et al., 2019). After foliar uptake, gaseous nitrogen assimilated in plant's nitrogen metabolism, which is an essential process for plant's growth and metabolism (Bataung et al., 2012). The process involves the enzymatic conversion of nitrate to nitrite by nitrate reductase (NR), the conversion of nitrite to ammonium by nitrite reductase (NiR), and the incorporation of ammonium into amino acids. Changes in nitrogen metabolism brought on by NO$_X$ may also be a reflection of alterations in plant metabolism as a whole, particularly with regard to photosynthetic activity, which provides the reducing power needed to reduce nitrogen (Naik et al., 1982). In response to NO$_X$ exposure, the enzyme activity known as NR, which is involved in the conversion of nitrate to nitrite, invariably increases. These alterations were noticed in both wooded and herbaceous plants. Additionally, increased NO$_2$ modifies the composition of plant tissues. There seem to be three major ways that this greenhouse gas influences the environment: indication, mitigation, and ecoconservation. Because leaves are the first receptor for pollutants, they can directly cause toxicity by permeating them, damaging individual cells, and lowering their ability to do photosynthesis (Randhi & Reddy, 2012). Red Spruce needles exhibited the enhanced NR activity on exposure of $NO_2$ (Norby & Sigal, 1989). Similarly, tomato cultivars fumigated with NO exhibited increase in NiR. It was believed that this increase might be due to the accumulation of nitrite in cells (Wellburn, 1990).

## 5.4 CONCLUSIONS

The world's most diverse ecosystems can be found in tropical and subtropical areas. Nitrogen oxides ($NO_X$) enhance the soil's quality where there is a scarcity of nitrogen in the soil and the plants are thus less dependent on fertilizers. Nitrogen dioxide may be poisonous to plants, damage leaves, and decrease growth and production at high concentrations, whereas it is beneficial for plants at very low concentration. It improves the quality of plant and food production, whereas the dynamics of tropical ecosystems have been altered as a result of overexploitation and increasing demand for natural resources. In the tropics, $NO_X$ is one of the main causes of these alterations. The quality of the air in the tropical region has declined due to nitrogen deposition, particle matter, and other hydrocarbons that cause air pollution. The two main gaseous air pollutants in the tropics are $NO_X$ and $O_3$, which may lead to global warming, and acid rain, which in turn affects the xylem vessels, stomata, and transpiration of plants, and alters the growth of plant tissues. Thus, we should go for an alternate method in order to reduce nitrogenous pollution so that we can protect our tropical forest and its environment. This chapter might aid researchers in expanding and updating their understanding of the various ecological difficulties posed by nitrogenous pollutants in the tropical ecosystems which have been faced in recent years.

## REFERENCES

Adam, S. E. H., J. Shigeto, A. Sakamoto, M. Takahashi, H. Morikawa. (2008). Atmospheric nitrogen dioxide at ambient levels stimulates growth and development of horticultural plants. *Botany*, 86, 213–217.

Adrizal, P. H., Patterson, R. M. Hulet, R. M. Bates. (2006). Growth and foliar nitrogen status of four plant species exposed to atmospheric ammonia. *Journal of Environmentalscience and Health* (Part. B, Pesticides, food contaminants, and agricultural wastes), 41(6), 1001–1018.

Ahmad, F., A. Kamal. (2022). Crosstalk-interaction of nitric oxide in plant growth and development under abiotic stress. *Journal of Innovation in Applied Research*, 5(3), 13–25.

Ahmad, P., A. M. Abass, A. M. Nasser, L. Wijaya, P. Alam, M. Ashraf. (2018). Mitigation of sodium chloride toxicity in Solanum lycopersicum L. by supplementation of jasmonic acid and nitric oxide. *Journal of Plant Interactions*, 13, 64–72.

Arora, D., P. Jain, N. Singh, H. Kaur, S. C. Bhatla. (2016). Mechanisms of nitric oxide crosstalk with reactive oxygen species scavenging enzymes during abiotic stress tolerance in plants. *Free Radical Research*, 50(3), 291–303.

Ashenden, T. W. (1979). Effects of $SO_2$ and $NO_2$ pollution on transpiration in Phaseolus vulgaris L. *Environmental pollution*, 18, 45–50.

Azmat, R., S. Haider, H. Nasreen, F. Aziz, M. Riaz. (2009). A viable alternative mechanism in adapting the plants to heavy metal environment. *Pakistan Journal of Botany*, 41, 2729.

Baek, S. G., S. Y. Woo. (2010). Physiological and biochemical responses of two tree species in urban areas to different air pollution levels. *Photosynthetica*, 48, 23–29.

Bataung, M., Z. Xianjin, Y. G. Xianlong. (2012). Review: Nitrogen assimilation in crop plants and its affecting factors. *Canadian Journal of Plant Science*, 92(3), 399–405.

Beligni, M. V., A. Fath, P. C. Bethke, L. Lamattina, R. L. Jones. (2002). Nitric oxide acts as an antioxidant and delays programmed cell death in barley aleurone layers. *Plant Physiology*, 129, 1642–1650. doi: 10.1104/pp.002337.

Beligni, M. V., L. Lamattina. (2001). Nitric oxide: A non-traditional regulator of plant growth. *Trends in Plant Science*, 6, 508–509.

Bhadouria, R., R. Singh, V. K. Singh, A. Borthakur, A. Ahamad, G. Kumar, P. Singh. (2019). Agriculture in the era of climate change: Consequences and effects. In *Climate Change and Agricultural Ecosystems* (pp. 1–23). Woodhead Publishing.

Bloom, A. J., M. Burger. (2010). Carbon dioxide inhibits nitrate assimilation in wheat and Arabidopsis. *Science*, 328, 899–903.

Bobbink, R., K. Hicks, J. Galloway, T. Spranger, R. Alkemade, M. Ashmore, W. de Vries. (2010). Global assessment of nitrogen deposition effects on terrestrial plant diversity: A synthesis. *Ecological Applications*, 20(1), 30–59.

Brimblecombe, P., D. H. Stedman. (1982). Historical evidence for a dramatic increase in the nitrate component of acid rain. *Nature*, 298, 460–462.

Bush, A. F. et al. (1962). *The Effects of Engine Exhaust on the Atmosphere When Automobiles are Equipped with Afterburners*. University of California (Report 62–63).

Caporn, T. M., T. A. Mansfield. (1976). Inhibition of net photosynthesis in tomato in air polluted with NO and $NO_2$. *Journal of Experimental Botany*, 27, 1181–1186.

Correa-Aragunde, N., M. Graziano, L. Lamattina. (2004). Nitric oxide plays a central role in determining lateral root development in tomato. *Planta*, 218(6), 900–905.

Dubey, P. S., K. Pawar. (1985). Air pollution and plant responses Review of work done at Vikram University Centre. In D. N. Rao, K. J. Ahmed, M. Yunis, S. N. Singh, Eds., *Perspectives In Environmental Botany* (Vol. 1, pp. 101–117). Print House.

Fatma, M., A. Masood, T. S. Per, N. A. Khan. (2016). Nitric oxide alleviates salt stress inhibited photosynthetic performance by interacting with sulfur assimilation in mustard. *Frontiers in Plant Science*, 7, 521.

Gautam, H., Z. Sehar, M. T. Rehman, A. Hussain, M. F. AlAjmi, N. A. Khan. (2021). Nitric oxide enhances photosynthetic nitrogen and sulfur-use efficiency and activity of ascorbate-glutathione cycle to reduce high temperature stress-induced oxidative stress in rice (Oryza sativa L.) plants. *Biomolecules*, 11(2), 305.

Gheorghe, I. F., B. Ion. (2011). The effects of air pollutants on vegetation and the role of vegetation in reducing atmospheric pollution. In *The Impact of Air Pollution on Health, Economy, Environment and Agricultural Sources*. IntechOpen.

Ghouse, A. K. M., S. Siddique, F. A. Khan. (1984). Effect of air pollution on wood formation in Dalbergia sissco, a timber tree of gangetic plain. *Journal Tree Science*, 1, 43–44.

Gupta, K. J., A. Kumari, I. Florez-Sarasa, A. R. Fernie, U. Abir, A. U. Igamberdiev. (2018). Interaction of nitric oxide with the components of the plant mitochondrial electron transport chain. *Journal of Experimental Botany*, 69(14), 3413–3424.

Hao, G. P., J. H. Zhang. (2010). The role of nitric oxide as a bioactive signaling molecule in plants under abiotic stress. In S. Hayat, M. Mori, J. Pichtel, A. Ahmad, Eds., *Nitric Oxide in Plant Physiology* (pp 115–138). Wiley-VCH Verlag.

Hayat, S., S. A. Hasan, M. Mori, Q. Fariduddin, A. Ahmad. (2009). *Nitric Oxide in Plant Physiology* (pp. 1–16). Wiley-VCH Verlag GmbH & Co. KGaA.

Hayat, S., S. Yadav, B. Ali et al. (2010). Interactive effect of nitric oxide and brassinosteroids on photosynthesis and the antioxidant system of Lycopersicon esculentum. *Russian Journal of Plant Physiology*, 57, 212–221.

Hill, A. C., J. H. Bennet. (1970). Inhibition of apparent photosynthesis by nitrogen oxides. *Atmospheric Environment*, 4, 341–348.

Holopanien, T., S. Anttonen, A. Wuff, V. Palomaki, L. Karenlampi. (2000). Comparative evaluation of the effects of gaseous pollutants, acidic deposits and mineral deficiencies: Structural changes in the tissue of forest plants. *Agriculture, Ecosystems & Environment*, 42, 3–4.

Iqbal, N., S. Umar, N. A. Khan, F. J. Corpas. (2021). Nitric oxide and hydrogen sulphide coordinately reduce glucose sensitivity and decrease oxidative stress via ascorbate- luta- thione cycle in heat-stressed wheat (*Triticum aestivum L.*) plants. *Antioxidants*, 10, 108.

Jahan, B., M. F. Al Ajmi, M. T. Rehman, N. A. Khan. (2020). Treatment of nitric oxide supplemented with nitrogen and sulfur regulates photosynthetic performance and stomatal behavior in mustard under salt stress. *Physiologia Plantarum*, 168, 490–510.

Jahan, B., N. Iqbal, M. Fatma, Z. Sehar, A. Masood, A. Sofo, I. D'Ippolito, N. A. Khan. (2021). Ethylene supplementation combined with split application of nitrogen and sulfur protects salt-inhibited photosynthesis through optimization of proline metabolism and antioxidant system in mustard (*Brassica juncea L.*). *Plants*, 10, 1303.

Janjić N., D. Hasanagić, T. Maksimović. (2017). Stomatal apparatus response of *Tilia cordata* (Mill.) and *Betula pendula* (Roth.) to air quality conditions in Banjaluka (Bosnia and Herzegovina). *Biologia Serbica*, 39(2), 9–16.

Jiechen, W., W. Yue, Z. Huihui, G. Dandan, S. Guangyu. (2021). Atmospheric nitrogen dioxide at different concentrations levels regulates growth and photosynthesis of tobacco plants. *Journal of Plant Interactions*, 16(1), 422–431.

Kharol, S. K., R. V. Martin, S. Philip, S. Vogel, D. K. Henze, D. Chen, C. L. Heald. (2013). Persistent sensitivity of Asian aerosol to emissions of nitrogen oxides. *Geophysical Research Letters*, 40(5), 1021–1026.

Konopka, B., J. Paulenka, J. Konopka. (1997). Damages to coniferous stands in the spins region. *Lesnictri*, 43, 381–388.

Kumar, M., N. Nandini. (2013). Identification and evaluation of air pollution tolerance index of selected avenue tree species of urban Bangalore, India. *International Journal of Emerging Technologies in Computational and Applied Sciences*, 4, 388–390.

Laurance, W. F., D. Carolina Useche, J. Rendeiro. (2012). Averting biodiversity collapse in tropical forest protected areas. *Nature*, 489, 290–294.

Liu, X., L. Wang, L. Liu, Y. Guo, H. Ren. (2011). Alleviating effect of exogenous nitric oxide in cucumber seedling against chilling stress. *African Journal of Biotechnology*, 10, 4380–4386.

Liu, Y., H. Bing, Y. Wu, H. Zhu, X. Tian, Z. Wang, R. Chang. (2022). Nitrogen addition promotes soil phosphorus availability in the subalpine forest of eastern Tibetan Plateau. *Journal of Soils and Sediments*, 22, 1–11.

Lobell, D. B., Tommaso, S. D., Burney, J. A. (2022). Globally ubiquitous negative effects of nitrogen dioxide on crop growth. *Environmental Studies*, 8(22), eabm9909.

Mackenzie, J. J., T. El-Ashry. (1988). *Winds: Air Borne Pollution's Toil on Trees and Crops* (pp. 53–330). World Resource Institute.

Mehlhorn, H., A. R. Wellburn. (1987). Stress ethylene formation determines plant sensitivity to ozone. *Nature*, 327, 417–418.

Midolo, G., R. Alkemade, A. M. Schipper, A. Benítez-López, M. P. Perring, W. D. Vries. (2019). Impacts of nitrogen addition on plant species richness and abundance: A global meta-analysis. *Global Ecology Biogeography*, 28, 398–341.

Mills, G., H. Pleijel, C. S. Malley, B. Sinha, O. R. Cooper, M. G. Schultz, H. S. Neufeld, D. Simpson, K. Sharps, Z. Feng, G. Gerosa, H. Harmens, K. Kobayashi, P. Saxena, E. Paoletti, V. Sinha, X. Xu. (2018). Tropospheric Ozone Assessment Report: Present-day tropospheric ozone distribution and trends relevant to vegetation. *Elementa: Science of the Anthropocene*, 6, 47.

Mittler R., S. Vanderauwera, M. Gollery, F. Van-Breusegem. (2004). Reactive oxygen gene network of plants. *Trends in Plant Science*, 9, 490–498.

Moraes, R. M., W. B. C. Delitti, J. A. P. V. Moraes. (2003). Gas exchange, growth, and chemical parameters in a native Atlantic forest tree species in polluted areas of Cubatão, Brazil. *Ecotoxicology and Environmental Safety*, 54, 339–345.

Moraes, R. M., A. Klumpp, C. M. Furlan. (2002). Tropical fruit trees as bioindicators of industrial air pollution in southeast Brazil. *Environment International*, 28, 367–374.

Mukherjee, A., M. Agrawal. (2016). Pollution response score of tree species in relation to ambient air quality in an urban area. *Bulletin of Environmental Contamination and Toxicology*, 96, 1–6.

Mukherjee, A., B. Pandey, S. B. Agrawal, M. Agrawal. (2019). Responses of Tropical and Subtropical Plants to Air Pollution. In S. Garkoti, S. Van Bloem, P. Fulé, R. Semwal, Eds., *Tropical Ecosystems: Structure, Functions and Challenges in the Face of Global Change*. Springer.

Muneer, S., T. H. Kim, B. C. Choi, B. S. Lee, J. H. Lee. (2014). Effect of CO, NOx and $SO_2$ on ROS production, photosynthesis and ascorbate-glutathione pathway to induce *Fragaria×annasa* as a hyperaccumulator. *Redox Biology*, 2, 91–98.

Muthu, M., J. Gopal, D. H. Kim, I. Sivanesan. (2021). Reviewing the impact of vehicular pollution on road-side plants—Future perspectives. *Sustainability*, 13, 5114.

Naik, M. S., Y. P. Abrol, T. V. R. Nair, C. S. RamAbo. (1982). Nitrate assimilation—its regulation and relationship to reduced nitrogen in higher plants, *Phytochemistry*, 21, 495–504.

Näsholm, T. et al. (1991). Uptake of NOx by mycorrhizal and non-mycorrhizal Scots pine seedlings: Quantities and effects on amino acid and protein concentrations. *New Phytologist*, 119, 83–92.

Neighbour, E. A. et al. (1988). Effects of sulphur dioxide and nitrogen dioxide on the control of water loss by birch (*Betula* spp.). *New Phytologist*, 108, 149–157.

Nighat, F. M., M. Iqbal. (2000). Stomatal conductance, photosynthetic rate, and pigment content in *Ruellia tuberosa* leaves as affected by coal-smoke pollution. *Biol Plant*, 43, 263–267.

Norby, R. J., L. L. Sigal. (1989). Nitrogen fixation in the lichen *Lobaria pulmonaria* in elevated atmospheric carbon dioxide. *Oecologia*, 79, 566–568.

Odu-Onikosi, A., P. Herckes, M. Fraser, P. Hopke, J. Ondov, P. A. Solomon, O. Popoola, G. M. Hidy. (2022). Tropical air chemistry in Lagos, Nigeria. *Atmosphere*, 13, 1059.

Okano, K., T. Totsuka. (1986). Absorption of nitrogen dioxide by sunflower plants grown at various levels of nitrate. *New Phytologist*, 102, 551–562.

Palavan-Unsal, N., D. Arisan. (2009). Nitric oxide signaling in plants. *The Botanical Review*, 75, 203–229.

Palomer, X., I. Roig-Villanova, D. Grima-Calvo, M. Vendrell. (2005). Effects of nitrous oxide ($N_2O$) treatment on the postharvest ripening of banana fruit. *Postharvest Biology and Technology*, 36, 167–175.

Pandey, J., M. Agrawal. (1994). Evaluation of air pollution phytotoxicity in a seasonally dry tropical urban environment using three woody perennials. *New Phytologist*, 126, 53–61.

Pandey, J., M. Agrawal, N. Khanam et al. (1992). Air pollutant concentrations in Varanasi, India. *Atmospheric Environment. Part B, Urban Atmosphere*, 26, 91–98.

Paneque, A., J. M. Ramírez, F. F. Del Campo, M. Losada. (1964). Light and dark reduction of nitrite in a reconstituted enzymic system. *The Journal of Biological Chemistry*, 239, 1737.

Perera, F. (2017). Pollution from fossil-fuel combustion is the leading environmental threat to global pediatric health and equity: Solutions exist. *International Journal of Environmental Research and Public Health*, 15(1), 16.

Pérez-Soba, M. et al. (1994). Effect of atmospheric ammonia on the nitrogen metabolism of Scots pine. *Physiologia Plantarum*, 90, 629–636.

Plant Database. (2021). *ENVIS Centre on Plants and Pollution*. CSIR-National Botanical Research Institute.

Pols, S., B. Van de Poel, M. L. A. T. M. Hertog, B. M. Nicolaï. (2022). The regulatory role of nitric oxide and its significance for future postharvest applications. *Postharvest Biology and Technology*, 188, 111869.

Prassad, T. K. (1996). Mechanisms of chilling-induced oxidative stress injury and tolerance in developing maize seedlings: Changes in antioxidant system, oxidation of proteins and lipids, and protease activities. *The Plant Journal*, 10, 1017–1026.

Pristijono, P., R. Will, J. Golding. (2006). Inhibition of browning on the surface of apple slices by short term exposure to nitric oxide (NO) gas. *Postharvest Biology and Technology*, 42, 256–259.

Proctor, J. (2021). Atmospheric opacity has a nonlinear effect on global crop yields. *Nature Food*, 2, 166–173.

Proctor, J., S. Hsiang, J. Burney, M. Burke, W. Schlenker. (2018). Estimating global agricultural effects of geoengineering using volcanic eruptions. *Nature*, 560, 480–483.

Rai, P., R. M. Mishra. (2013). Effect of urban air pollution on epidermal traits of road side tree species, *Pongamia pinnata* (L.) Merr. *Journal of Environmental Science, Toxicology and Food Technology*, 2, 2319–2402.

Randhi, U. D., M. A. Reddy. (2012). Evaluation of tolerant plant species in urban environment: A case study from Hyderabad, India. *Universal Journal of Environmental Research & Technology*, 2, 300–304.

Rao, M. I., R. B. Amundson, R. Herman-Alscher, L. E. Anderson. (1983). Effect of $SO_2$ on stomatal metabolism in *Pisum sativum* L. *Plant Physiology*, 72, 573–577.

Ripullone, F., M. Lauteri, G. Grassi, M. Amato, M. Borghetti. (2004). Variation in nitrogen supply changes water-use efficiency of Pseudotsuga menziesii and Populus x euroamericana; A comparison of three approaches to determine water-use efficiency. *Tree Physiology*, 24(6), 671–679.

Sabaratnam, S. G., C. Mulchi. (1988). Effects of nitrogen dioxide on biochemical and physiological characteristics of soybean. *Environmental Pollution*, 55, 149–158.

Sandhu, R., G. Gupta. (1989). Effects of nitrogen dioxide on growth and yield of black turtle bean (*Phaseolus vulgaris* L.) cv. 'Domino'. *Environmental Pollution*, 59(4), 337–344.

Schmutz, P., D. Tarjan, M. S. Günthardt-Goerg, R. Matyssek, J. B. Bucher. (1995). Nitrogen dioxide—A gaseous fertilizer of Poplar trees. *Python*, 35, 219–232.

Sehar, Z., N. Iqbal, M. I. R. Khan, A. Masood, M. T. Rehman, A. Hussain, N. A. Khan. (2021). Ethylene reduces glucose sensitivity and reverses photosynthetic repression through optimization of glutathione production in salt-stressed wheat (*Triticum aestivum*). *Scientific Reports*, 11(1), 1–12.

Sehar, Z., A. Masood, N. A. Khan. (2019). Nitric oxide reverses glucose-mediated photosynthetic repression in wheat (*Triticum aestivum* L.) under salt stress. *Environmental and Experimental Botany*, 161, 277–289.

Seinfeld, J. H., S. N. Pandis. (2016). *Atmospheric Chemistry and Physics: From Air Pollution to Climate Change*. Wiley.

Sembok, W. Z. W., K. S. Yen. (2017). Effects of nitrogen gas fumigation on postharvest quality of minimally processed starfruit (*Averrhoa carambola* L.) stored at low temperature. *Malaysian Applied Biology Journal*, 46, 189–198.

Sheng, Q., Z. Zhu. (2019). Effects of Nitrogen Dioxide on Biochemical Responses in 41 Garden Plants. *Plants (Basel, Switzerland)*, 8(2), 45.

Siddiqui, M. H., M. H. Al-Whaibi, M. O. Basalah. (2011). Role of nitric oxide in tolerance of plants to abiotic stress. *Protoplasma*, 248(3), 447–455.

Sifola, M., L. Postiglione. (2003). The effect of nitrogen fertilization on nitrogen use efficiency of irrigated and non-irrigated tobacco (*Nicotiana tabacum* L.). *Plant and Soil*, 252, 313–323.

Srivastava, H. S. et al. (1975). The effects of environmental conditions on the inhibition of leaf gas exchange by $NO_2$. *Canadian Journal of Botany*, 53, 475–482.

Sukumaran, D. (2014). Effect of air pollution on the anatomy some tropical plants. *Applied Ecology and Environmental Sciences*, 2, 32–36.

Takahashi, M., T. Furuhashi, N. Ishikawa, G. Horiguchi, A. Sakamoto, H. Tsukaya, H. Morikawa. (2014). Nitrogen dioxide regulates organ growth by controlling cell proliferation and enlargement in Arabidopsis. *New Phytologist*, 201, 1304–1315.

Takahashi, M., M. Nakagawa, A. Sakamoto, C. Ohsumi, T. Matsubara, H. Morikawa. (2005). Atmospheric nitrogen dioxide gas is a plant vitalization signal to increase plant size and the contents of cell constituents. *New Phytologist*, 168, 149–154.

Takahashi, M., A. Sakamoto, H. Ezura, H. Morikawa. (2011). Prolonged exposure to atmospheric nitrogen dioxide increases fruit yield of tomato plants. *Plant Biotechnology Journal*, 8, 485–487.

Taub, D. R., X. Z. Wang. (2008). Why are nitrogen concentrations in plant tissues lower under elevated $CO_2$? A critical examination of the hypotheses. *Journal of Integrative Plant Biology*, 50, 1365–1374.

Taylor, O. C., F. M. Eaton. (1966). Suppression of plant growth by $NO_2$. *Plant Physiology*, 41, 132–135.

Tian, D., S. Niu. (2015). A global analysis of soil acidification caused by nitrogen addition. *Environmental Research Letters*, 10, 24019.

Turgeon, R. (2010). The role of phloem loading reconsidered. *Plant Physiology*, 152(4), 1817–1823.

Usha, S., P., Haritha. (2019). Structural changes in stomata in plants exposed to air pollution. *IOSR Journal of Environmental Science, Toxicology and Food Technology (IOSRJESTFT)*, 13(8), 66–70.

Vallano, D. M., J. P. Sparks. (2008). Quantifying foliar uptake of gaseous nitrogen dioxide using enriched foliar delta15N values. *New Phytologist*, 177, 946–955.

Van Der Eerden, L. J. M., M. Pérez-Soba. (1992). Physiological responses of Pinus sylvestris to atmospheric ammonia. *Trees*, 6, 48–53.

Van Hove, L. W. A. et al. (1992). Physiological effects of a long term exposure to low concentrations of $NH_3$, $NO_2$ and $SO_2$ on Douglas fir (*Pseudotsuga menziesii*). *Physiologia Plantarum*, 86, 559–567.

Vellend, M., L. Baeten, A. Becker-Scarpitta, V. Boucher-Lalonde, J. L. McCune, J. Messier, D. F. Sax. (2017). Plant biodiversity change across scales during the Anthropocene. *Annual Review of Plant Biology*, 68, 563–586.

Verma, V., N. Chandra. (2014). Biochemical and ultrastructural changes in Sida cordifolia L. and *Catharanthus roseus* L. to auto pollution. *International Scholarly Research Notices*, 1–11.

Wang, D., S. Weiger. (1988). Air pollution impact on Plants. *Atlas Science Animal Plant Science*, 1, 33–39.

Wellburn, A. R. (1990). Why are atmospheric oxides of nitrogen usually phytotoxic and not alternative fertilizers? Tansley Review 24. *New Phytologist*, 115, 395–429.

Wollenheber, B., J. A. Raven. (1993). Implications of N acquisition form atmospheric $NH_3$ for acid–base and cation–anion balance in *Lolium perenne*. *Physiologia Plantarum*, 89, 519–523.

World Health Organization. (2000). *Effects of Nitrogen Containing Air Pollutants: Critical Levels, in Air Quality Guidelines Europe* (2nd Edition, p. 288). World Health Organization.

Xiankai, L., M. Jiangming, D. Shaofeng. (2008). Effects of nitrogen deposition on forest biodiversity. *Acta Ecologica Sinica*, 28, 5532–5548.

Xuewu, D., S. Xinguo, Y. Yanli, Q. Hongxia, Y. Li, J. Yueming. (2007). Effect of nitric oxide on pericarp browning of harvested longan fruit in relation to phenolic metabolism. *Food Chemistry*, 104(2), 571–576.

Yunus, M., K. Ahmed. (1979). Use of epidermal traits of plants in pollution monitoring. In *Proceedings of the National Seminar on Environmental Pollution and its Control: A Status Review*. National Productivity Council, Bombay, India.

Yunus, M., N. Singh, M. Iqbal. (1996). *Global Status of Air Pollution: An Overview. In Plant Response to Air Pollution* (pp. 1–34). John Wiley & Sons.

Zeevaart, A. J. (1976). Some effects of fumigating plants for short periods with $NO_2$. *Environmental Pollution*, 11, 97–108.

Zhu, S., J. Zhou. (2006). Effects of nitric oxide on fatty acid composition in peach fruits during storage. *The Journal of Agricultural and Food Chemistry*, 54, 9447–9452. doi: 10.1021/jf062451u.

# 6 Impact of Particulate Matter on the Ecophysiology of Plants

*Somdutta Sinha Roy, Saloni Bahri, Laishram Sundari Devi, and Sushma Moitra*

## 6.1 INTRODUCTION

Air pollution is a bane of modern times, and urban spaces around the world are affected by the harmful effects of rising air pollution (Janhäll, 2015; Yin et al., 2020). Airborne particulate matter (PM) is one of the major air pollutants and is known to have adverse effects on human health. PM arises from various human activities like burning of fossil fuels like coal and petroleum and other organic wastes, vehicular pollution, and activities like mining and building construction. PMs have sizes which are in the order of diameter ranging from 0.01 μm to 100 μm (Przybysz et al., 2014). PM of size 2.5 ($PM_{2.5}$) and lower are called ultrafine particles (UFPs) and are especially dangerous as they are able to evade the barriers in nose and throat and enter the deep alveolar spaces and affect the functioning of the lung in humans, giving rise to various lung diseases. PM is thought to be one of the primary causes of the rise in lung diseases including chronic obstructive pulmonary disease (COPD) and various lung infections. Plants are effective in improving the air quality by removing the suspended PM from air (Beckett et al., 1998, 2000; Rahul & Jain, 2014). But similar to effects of PM on human health, suspended PM also affects plants growing along the roadsides which are exposed to PM pollution. Recently, there have been studies that show that sub-micron particles can coagulate, resulting in the formation of larger particles on the surface of the leaves (Yin et al., 2020) of the plants. These larger particles, which also contain other air pollutants like organic chemicals and heavy metals, then interact with the leaf tissue at a cellular level.

The interaction of vegetation with PM pollutants depends on both the architecture of the vegetation present in the area and the proximity to the source of pollution (Janhäll, 2015). Wind patterns and speed will also determine the amount of PM that is captured by the plants. Yang et al. (2005) demonstrated that trees were able to remove 1,261 tons of air pollutants, out of which 770 tons were $PM_{10}$. Similar studies were undertaken in various cities around the world (Cavanagh & Clemons, 2006; Nowak et al., 2014; Samek et al., 2018).

## 6.2 PLANTS AND BIOREMEDIATION OF PM

Plants are known to be effective biofilters for PM. Therefore, plants in general and trees in particular have been used for bioremediation of PM in the atmosphere (i.e. phytoremediation). Most of the phytoremediation is known to occur because of trapping of PM particles by the leaves of the plants. Various foliar factors like size, shape, number of trichome present on the surface, and amount of waxy cuticles on the surface affects PM absorption efficiency of the plants (Leonard et al., 2016) (Figure 6.1). A positive correlation between the PM load and the leaf surface area has been demonstrated by Liu et al. (2015). However, Leonard et al. (2016) conclusively proved that characters like the shape of the leaf and the length of petiole, which increases the movement of the leaf due to air movement, decreasing PM deposition on the leaves. Therefore, leaves with broader bases and those with shorter petioles tend to have more deposition of PM on them. Trichomes present on leaves also increase the PM load on the leaves. Trichomes not only increase the surface area available for the deposition, but prevent trapped PM from dislodging. Cuticles present on the leaf epidermal surface and wax depositions also interact with different pollutants in the atmosphere. The organic, lipophilic pollutants are known to get absorbed by the leaf by dissolving through the waxy layer. This includes polycyclic aromatic hydrocarbons (PAHs) found in burning coal and vehicular exhaust, causing oxidative stress in the plants (Volta et al., 2020). UFPs are thought to be more harmful than the medium and larger PM. It is due to their surface-associated environmentally persistent free radicals (EPFRs) which are generated as a result of reactive oxygen species (ROS) action.

It was shown that in plants like *Broussonetia papyrifera*, *Ulmus pumila*, and *Catalpa speciosa*, the number of trichomes in the leaf correlated positively with the amount of trapped $PM_{2.5}$ (Chen et al., 2017). PM with diameter 2 μm could be collected from the stomatal cavities in a study from Beijing, China (Song et al., 2015). Rai (2016) showed that PM is found associated with the reduction in size of stomata in 12 roadside plant species without affecting their growth, suggesting the potential of plants in absorbing PM pollution.

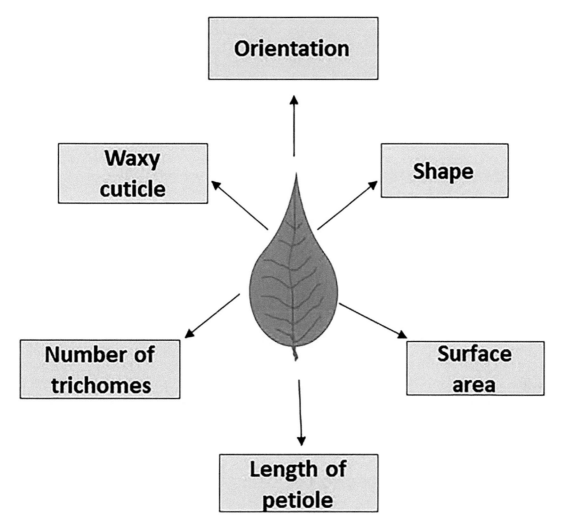

**FIGURE 6.1** Foliar factors affecting the absorption of PM by plants.

The quantity of PM deposited on leaves is given by the following formula (Janhäll, 2015).

Deposited amount of pollutant = LAI × $v_d$ × C × t

Where,
LAI = Leaf area index (i.e., amount of leaf surface area per m² of ground area)
$v_d$ = Deposition velocity
C = Concentration of the pollutant in the air
t = Time

Thus, in addition to the leaf characteristics, total leaf area available also determines the total amount of PM captured. Also, as evident from this formula, factors like temperature and wind velocity will also affect the amount of PM deposited. Thus, local weather patterns—as well as the density and height of the vegetation present near the pollution source—will determine the amount of PM filtered out of the air by plants.

## 6.3 PHYSIOLOGICAL AND BIOCHEMICAL EFFECTS OF PM ON PLANTS

Accumulation of PM on the leaves of a plant forms a physical layer on the leaves, called the shading effect, that might reduce the chlorophyll content in the leaf (Rai, 2016). It also builds up inside the open stomata which are the primary organ for gaseous exchange for the vital processes of respiration and photosynthesis; thus, these processes are affected by the deposition of PM. Chlorophyll content reduction may eventually lead to less primary productivity and less growth, and therefore can be used as an indicator of air pollution, but the effect of PM deposition is not limited to the physical barrier—the chemical composition of the particles also adversely affects other biochemical processes of the leaf. It has been shown that PM adversely affects flowering and hormonal balance (Farooqui et al., 1995) and photosynthesis rate (Armbrust, 1986). Shukla et al.

(1990) demonstrated in *Brassica campestris* that exposure to cement dust reduced photosynthetic pigments, oil yield, and overall growth as compare to control plants. Hubai et al. (2021) demonstrated the effect of $PM_{2.5}$ on chlorophyll content and total protein content of the leaf using tomato (*Lycopersicon esculentum*) as a model. PM pollution causes the oxidative stress leading to an increase in the ROS levels. Oxidative stress negatively impacts any biological system. Photosynthetic pigments and electron transport chains in the chloroplast are affected, and this may result in lowering of primary productivity of the plant (Karmakar & Padhy, 2019). The result may manifest by lowering of chlorophyll pigment and yellowing of leaves. It is known to have structural effect on PS II (photosystem II) of the chloroplast (Jin et al., 2017). Increase in free radicals also affects ascorbic acid content (Varshney & Varshney, 1984), interaction between ascorbate and glutathione (Foyer & Noctor, 2011), elevated activity of antioxidant enzymes like SOD (superoxide dismutase) and glutathione S-transferase, and sugar and protein content (Rai, 2016) (Figure 6.2). Observations suggest that certain plants with characters like stomata only on the lower surface, sunken stomata, thick cuticles, or numerous trichomes are known to capture higher amounts of PM. Once the physical barriers of the leaf are breached, biochemical pathways which are able to negate the negative effects of PM give an adaptive advantage to the plants growing on the roadsides, with heavy PM load. Some recent studies (Chen et al., 2022; Guo et al., 2023) suggest the same. However, there are reports of an increase in chlorophyll concentrations as a result of exposure to pollutants in certain trees like mango (*Mangifera*) (Tripathi & Gautam, 2007), *Callistemon*, and *Albizzia* (Seyyednejad et al., 2009).

## 6.4 TOXIC EFFECTS OF PM-RELATED HEAVY METALS IN PLANTS

Rapid development in the urban areas and industrialization in many developing countries has accelerated the anthropogenic emission of atmospheric PM and trace metals, especially heavy metals from combustion of biomass and fossil fuels (Popoola et al., 2018). Heavy metals include Ni (nickel), Cd (cadmium), Hg (mercury), Pb (lead), Fe (iron), Zn (zinc), Co (cobalt), Cu (copper), etc., and such metals are highly lethal and have high specific gravities, densities, or atomic weights (Duffus, 2002). Heavy metals are largely accumulated in soil and aquatic ecosystems; however, a minor proportion is present in the atmosphere as PM or aerosols (Nagajyoti et al., 2010). The atmospheric heavy metal particulates are deposited either from natural processes such as volcanic activity, re-suspension of metal dust, etc., or anthropogenic processes such as industrial and automobile activities (Shahid et al., 2019; Yaashikaa et al., 2022) (Figure 6.3). There are three major pathways of heavy metals entering into the environment: dumping of metal-rich sewage sludge, deposition of atmospheric heavy metal particulates, and byproducts of mining (Shrivastav, 2001; Popoola et al., 2018). Heavy metal particles are known to travel up to long distances and contaminate different types of ecosystems. Heavy metals drastically reduce the ecophysiological performances of plants like growth,

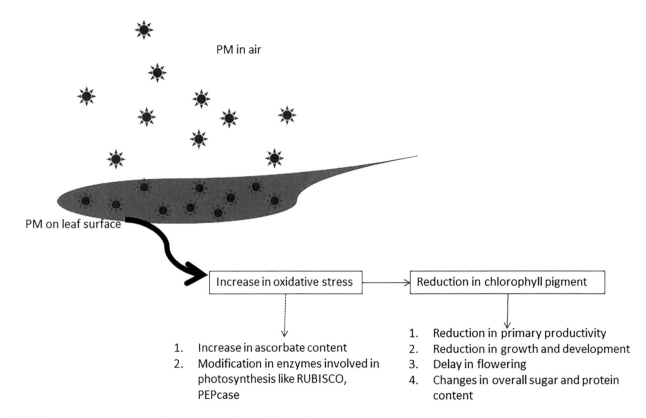

**FIGURE 6.2** Physiological and biochemical effects of PM on plants.

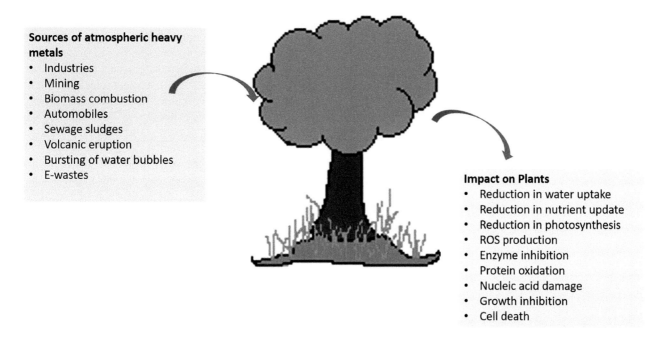

FIGURE 6.3  Different sources of atmospheric heavy metals and their impacts on plants.

development, productivity, competition, resource utilization, etc., of plants by imposing continuous stress conditions (Diaconu et al., 2020; Yaashikaa et al., 2022). The toxicity of heavy metals in plants varies depending on the species, type and concentration of metal, etc., as some of these heavy metals such as iron, copper and zinc are also the activators and cofactors of many enzymes. Such trace elements also take part in many physiological and metabolic reactions like electron transport chains, redox reactions, and nucleic acid metabolism (Nagajyoti et al., 2010). Nevertheless, some heavy metals like Cd (cadmium), As (arsenic), Pd (palladium), and Hg (mercury) are very poisonous to some enzymes, causing growth inhibition and even death of the plants (Figure 6.3). Non-essential heavy metals are accumulated at various trophic levels in the food chain and trigger biomagnification. Such pollutants cause morpho-physiological and biochemical impacts on the plants, and serious human health issues, as well (Rai, 2016; Shahid et al., 2019). Moreover, lichens and mosses are also very sensitive to such pollution.

Leaves of plants are the most exposed to atmospheric pollutants. Therefore, the foliar parts are the main depositing part of the plant for the atmospheric heavy metal particulates (Shahid et al., 2019). A study of ten trees and shrubs species in Jodhpur, India, near the Thar Desert, observed that the accumulation of PM and heavy metals are strongly correlated; however, a differential level of accumulation among different species was observed, depending on their adaptability (Popek et al., 2022). Continuous exposure to exceedingly high levels of heavy metals results in the reduction of photosynthesis, water and nutrient uptake, chlorosis, inhibition of growth, and ultimately the death of the plants (Jin et al., 2015; Khan et al., 2021) by metal-induced molecular and structural alterations in plant cells. A study on *Sesbania grandiflora* reported that under high concentration of mercury, DNA polymorphisms were detected by RAPD (random amplified polymorphic DNA) fingerprinting analysis (Malar et al., 2015). Though copper is one of the essential micronutrients of plants, prolonged exposure to higher concentrations induces oxidative stress and ROS (Wang et al., 2017). Oxidative stress and ROS are very destructive in nature for the metabolic pathways and macromolecules (Yaashikaa et al., 2022).

Mercury, another heavy metal, is exceptionally dangerous to many biological systems. It can be available in many forms to the plants (Ling et al., 2020). A substantial amount of mercury interferes with the function of mitochondria, triggering oxidative stress. Generation of ROS due to this oxidative stress results in disruption of membrane phospholipids, as well as other metabolic pathways in plants (Mondal et al., 2022). High concentrations of chromium (Cr) also cause much physiological destruction such as leaf chlorosis, under-development, root tip damage, and improper nutrient uptake (Ertani et al., 2017). It is also involved in the modification of the propagation cycle and regeneration of leaves, stem development, and root growth, ultimately leading to loss of biomass and yield (Singh et al., 2013). Plant metabolism is affected by chromium by generation elevated amount of ROS or inhibition of enzyme activities (Wakeel et al., 2020). Lead can be deposited from sewage sludge, industries, mining, explosives, and e-wastes. It exhibits an aggressive effect on the morphological, developmental and photosynthetic processes of plants (Cenkci et al., 2010). Higher concentrations of lead (Pb) also cause inhibition of enzymes, changes in plasma membrane porosity, and changes in uptake of water and minerals. Lead, like other heavy metals, stimulates oxidative stress by intensifying the ROS production in plants (Cenkci et al., 2010; Afaj et al., 2016). Arsenate also pursues the equivalent transporters in the plasmalemma of plant roots similar to phosphate.

Degree of arsenate tolerance varies in different plant species (Finnegan & Chen, 2012; Farooq et al., 2016).

## 6.5 EFFECT OF PM ON PHYLLOSPHERE AND RHIZOSPHERE PROCESSES

The most common effect of PM on the phyllosphere is the physical barriers that PM forms on the surface of the leaf. This may affect the gaseous exchange process via stomata, affecting biochemical and physiological functions (Hubai et al., 2021). PM is also known to bind to many phytotoxic chemicals from the atmosphere that are deposited on the plants by dry or wet depositions. These pollutants then reach the rhizosphere of the plant when the PM is washed off by water/rain. These depositions are known to modify the microbial communities found in the leaf surface and around the root. The phyllospheric/rhizospheric microbial communities and the endophytic bacteria found on the leaves and plant roots metabolize, degrade, or detoxify these pollutants, thus helping in the phytoremediation process (Weyens et al., 2015). Thus, the microbes associated with the plants contribute to the tolerance of the plants from these pollutants. Wei et al. (2017) isolated many bacterial species from the phyllosphere like *Sphingomonas*, *Klebsiella*, *Beijerinckia*, *Methylobacterium*, *Azotobacter*, and *Stigonema*, and fungi like *Taphrina*, *Cladosporium*, and *Aureobasidium*; however, the composition of the microbial communities differed from plant to plant, as well as with pollutants found in a particular area under study. A study of ten ornamental plants in Bangkok (Yutthammo et al., 2010) demonstrated that out of $2.31 \times 10^3$–$1.25 \times 10^5$ MPN $g^{-1}$ phyllospheric bacteria, PAH-degrading bacteria was observed to range between 1–10%. Most of the available studies suggested that a consortium of bacteria may be responsible for the degradation of pollutants in phyllosphere and rhizosphere. However, the interaction of the microorganisms with the plant and the pollutants may not be only restricted to degradation but also result in stress tolerance and an increase in leaf area index (LAI), thus increasing the area of leaves available for phytoremediation.

There is a lot of scope and stress on green technologies and practices in the urban environment. Living green walls, green roofs, and urban farming using hydroponics and aeroponics all are known to contribute toward the reduction in the PM load of the atmosphere (Wróblewska & Jeong, 2021). Identifying microbial consortiums which enhance the effectiveness of plants in filtering out PM pollutants provides another aspect for future research (Figure 6.4).

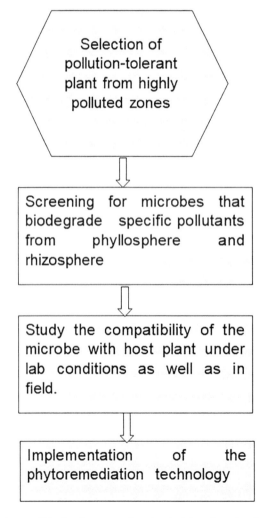

**FIGURE 6.4** Flow chart showing the possible development of a microbe/plant-based technology for phytoremediation.

## 6.6 CONCLUSIONS

Plants, especially trees, have been on the forefront of combating air pollution all over the world. Tree-planting drives are undertaken as a measure of mitigation of air pollution. Plants act as biofilters for the harmful particulate pollutants. Recently, there has been much advocating for living, green walls around traffic zones with high vehicular pollution. Even for the indoor particulate matter (PM) pollution which is on the rise, it has been suggested that having indoor plants can aid with mitigation. However, air pollutants like PMs, polycyclic aromatic hydrocarbons (PAHs) and heavy metals have profound effects on plants. Accumulation of large amounts of PM on leaves decreases the rate of photosynthesis, as there is reduction of absorption of light and blockage of stomatal pores. Exchange of gases gets affected, leading to alterations in the biochemical functions. Heavy metals negatively affect the ecophysiological processes of plants such as productivity, growth, and development. Effective bioremediation of air pollutants, especially PM, using plants is possible only if we understand the physico-chemical makeup of the plants and their associated microbes. Further studies should be carried out to discover newer strategies to combat pollution. These studies can also help in the identification of tolerant plant species which can be the potential biomonitors.

## REFERENCES

Afaj AH, Jassim AJ, Noori MM, Schuth C. 2016. Effects of lead toxicity on the total chlorophyll content and growth changes of the aquatic plant Ceratophyllum demersum L. *Int J Environ Stud.* 74: 119–128. https://doi.org/10.1080/00207233.2016.1220723

Armbrust DV. 1986. Effect of particulates (dust) on cotton growth, photosynthesis, and respiration. *Agron J.* 78(6): 1078–1081. https://doi.org/10.2134/agronj1986.00021962007800060027x

Beckett KP, Freer-Smith PH, Taylor G. 1998. Urban woodlands: Their role in reducing the effects of particulate pollution. *Environmental Pollution* 99: 347–360. https://doi.org/10.1016/s0269-7491(98)00016-5

Beckett KP, Freer-Smith PH, Taylor G. 2000. Effective tree species for local air quality management. *Journal of Arboriculture* 26: 12–18. https://doi.org/10.48044/jauf.2000.002

Cavanagh J, Clemons J. 2006. Do urban forests enhance air quality. *Aust. J. Environ. Manag.* 13: 120. https://doi.org/10.1080/14486563.2006.10648678

Cenkci S, Cigerci IH, Yildiz M, Ozay C, Bozdag A, et al. 2010. Lead contamination reduces chlorophyll biosynthesis and genomic template stability in *Brassica rapa* L. *Environ Exp Bot.* 67: 467–473. https://doi.org/10.1016/j.envexpbot.2009.10.001

Chen L, Liu C, Zhang L, Zou R, Zhang Z. 2017. Variation in tree species ability to capture and retain airborne fine particulate matter (PM2.5). *Sci. Rep.* 7: 3206. https://doi.org/10.1038/s41598-017-03360-1

Chen S, Yu H, Teng X, Dong M, Li W. 2022. Composition and size of retained aerosol particles on urban plants: Insights into related factors and potential impacts. *Sci of The Total Environ.* 853: 158656. https://doi.org/10.1016/j.scitotenv.2022.158656.

Diaconu M, Pavel LV, Hlihor RM, Rosca M, Fertu DI, et al. 2020. Characterization of heavy metal toxicity in some plants and microorganisms—A preliminary approach for environmental bioremediation. *Nat Biotechnol.* 56: 130–139. https://doi.org/10.1016/j.nbt.2020.01.003

Duffus JH. 2002. Heavy metals: A meaningless term? *Pure Appl Chem.* 74(5): 793–807. https://doi.org/10.1351/pac200274050793

Ertani A, Mietto A, Borin M, et al. 2017. Chromium in agricultural soils and crops: A Review. *Water Air Soil Pollut.* 228: 190. https://doi.org/10.1007/s11270-017-3356-y

Farooq MA, Islam F, Ali B, Najeeb U, Mao B, Gill RA, et al. 2016. Arsenic toxicity in plants: Cellular and molecular mechanisms of its transport and metabolism. *Environ Exp Bot.* 132: 42–52. https://doi.org/10.1016/j.envexpbot.2016.08.004

Farooqui A, Kulshreshtha K, Srivastava K, Singh SN, Farooqui SA, Pandey V, Ahmad PJ. 1995. Photosynthesis, stomatal response and metal accumulation in Cineraria maritima L. and Centauria moschata L. grown in metal-rich soil. *Sci of Total Environ.* 164(3): 203–207. https://doi.org/10.1016/0048-9697(95)04471-C.

Finnegan PM, Chen W. 2012. Arsenic toxicity: The effects on plant metabolism. *Front Physio.* 3: 182. https://doi.org/10.3389/fphys.2012.00182

Foyer CH, Noctor G. 2011. Ascorbate and glutathione: The heart of the redox hub. *Plant Physiology.* 155(1): 2–18. https://doi.org/10.1104/pp.110.167569

Guo K, Yan L, He Y, Li H, Lam SS, Peng W, Sonne C. 2023. Phytoremediation as a potential technique for vehicle hazardous pollutants around highways. *Environmental Pollution* 121130. https://doi.org/10.1016/j.envpol.2023.121130.

Hubai K, Kovats N, Teke G. 2021. Effects of urban atmospheric particulate matter on higher plants using *Lycopersicon esculentum* as model species. *SN Appl Sci* 3: 770. https://doi.org/10.1007/s42452-021-04745-8

Janhäll S. 2015. Review on urban vegetation and particle air pollution—deposition and dispersion. *Atmos Environ.* 105: 130–137. https://doi.org/10.1016/j.atmosenv.2015.01.052

Jin L, Che X, Zhang Z, Li Y, Gao H, Zhao S. 2017. The mechanisms by which phenanthrene affects the photosynthetic apparatus of cucumber leaves. *Chemosphere* 168: 1498–1505. https://doi.org/10.1016/j.chemosphere.2016.12.002

Jin M, Liu X, Wu L, Liu M. 2015. An improved assimilation method with stress factors incorporated in the WOFOST model for the efficient assessment of heavy metal stress levels in rice. *Int J Appl Earth Obs Geoinf.* 41: 118–129. https://doi.org/10.1016/j.jag.2015.04.023

Karmakar D, Padhy PK. 2019. Metals uptake from particulate matter through foliar transfer and their impact on antioxidant enzymes activity of S. robusta in a tropical forest, West Bengal. *India Arch Environ Con Tox.* 76: 605–616. https://doi.org/10.1007/s00244-019-00599-9

Khan I, et al. 2021. Effects of silicon on heavy metal uptake at the soil-plant interphase: A review. *Ecotoxicol Environ Saf.* 222: 112510. https://doi.org/10.1016/j.ecoenv.2021.112510

Leonard RJ, McArthur C, Hochuli, DF. 2016. Particulate matter deposition on roadside plants and the importance of leaf trait combinations. *Urban For Urban Green* 20: 249–253. https://doi.org/10.1016/j.ufug.2016.09.008

Ling Y, Wu J, Man X, Xu Y, Qi Y, et al. 2020. BiOIO3/graphene interfacial heterojunction for enhancing gaseous heavy metal removal. *Mater Res Bull.* 122: 110620. https://doi.org/10.1016/j.materresbull.2019.110620

Liu Z, Hu B, Wang L, et al. 2015. Seasonal and diurnal variation in particulate matter (PM10 and PM2.5) at an urban site of Beijing: Analyses from a 9-year study. *Environ Sci Pollut Res.* 22: 627–642. https://doi.org/10.1007/s11356-014-3347-0

Malar S, Sahi SV, Favas PJC, et al. 2015. Assessment of mercury heavy metal toxicity-induced physiochemical and molecular changes in Sesbania grandiflora L. *Int J Environ Sci Technol.* 12: 3273–3282. https://doi.org/10.1007/s13762-014-0699-4

Mondal S, Pramanik K, Ghosh SK, et al. 2022. Molecular insight into arsenic uptake, transport, phytotoxicity, and defense responses in plants: A critical review. *Planta.* 255: 87. https://doi.org/10.1007/s00425-022-03869-4

Nagajyoti PC, Lee KD, Sreekanth TVM. 2010. Heavy metals, occurrence and toxicity for plants: A review. *Environ Chem Lett.* 8: 199–216. https://doi.org/10.1007/s10311-010-0297-8

Nowak DJ, Hirabayashi S, Bodine A, Greenfield E. 2014. Tree and forest effects on air quality and human health in the United States. *Environ Pollut.* 193: 119. https://doi.org/10.1016/j.envpol.2014.05.028

Popek R, Mahawar L, Shekhawat GS, et al. 2022. Phyto-cleaning of particulate matter from polluted air by woody plant species in the near-desert city of Jodhpur (India) and the role of heme oxygenase in their response to PM stress conditions. *Environ Sci Pollut Res.* 29: 70228–70241. https://doi.org/10.1007/s11356-022-20769-y

Popoola LT, Adebanjo SA, Adeoye BK. 2018. Assessment of atmospheric particulate matter and heavy metals: A critical review. *Int J Environ Sci Technol.* 15: 935–948. https://doi.org/10.1007/s13762-017-1454-4

Przybysz A, Sæbø A, Hanslin H M, Gawronski S. 2014. Accumulation of particulate matter and trace elements on vegetation as affected by pollution level, rainfall and the passage of time. *Sci Tot Environ.* 481: 360–369. https://doi.org/10.1016/j.scitotenv.2014.02.072

Rahul J, Jain MK. 2014. An investigation in to the impact of particulate matter on vegetation along the national highway: A review. *Res J Environ Sci.* 8: 356–372. https://doi.org/10.3923/rjes.2014.356.372

Rai PK. 2016. Impacts of particulate matter pollution on plants: Implications for environmental biomonitoring. *Ecotoxicol Environ Saf.* 129: 120–136. https://doi.org/10.1016/j.ecoenv.2016.03.012

Samek L, et al. 2018. Seasonal contribution of assessed sources to submicron and fine particulate matter in a Central European urban area. *Environ Pollut.* 41: 406–411. https://doi.org/10.1016/j.envpol.2018.05.082

Seyyednejad SM, Niknejad M, Yusefi M. 2009. Study of air pollution effects on some physiology and morphology factors of Albizia lebbeck in high temperature conditions in Khuzestan. *J Plant Sci.* 41: 122–126. https://scialert.net/abstract/?doi=jps.2009.122.126

Shahid M, et al. 2019. Ecotoxicology of heavy metal(loid)-enriched particulate matter: Foliar accumulation by plants and health impacts. *Rev Environ Contam Toxicol.* 253. https://doi.org/10.1007/398_2019_38

Shrivastav R. 2001. Atmospheric heavy metal pollution: Development of chronological records and geochemical monitoring. *Resonance* 2: 62–68. https://doi.org/10.1007/BF02994594

Shukla J, Pandey V, Singh SN, Yunus M, Singh N, Ahmad KJ. 1990. Effect of cement dust on the growth and yield of Brassica campestris L. *Environ Pollut* 66(1): 81–88. https://doi.org/10.1016/0269-7491(90)90200-V

Singh HP, Mahajan P, Kaur S, et al. 2013. Chromium toxicity and tolerance in plants. *Environ Chem Lett.* 11: 229–254. https://doi.org/10.1007/s10311-013-0407-5

Song Y, Maher BA, Li F, Wang X, Sun X, Zhang, H. 2015. Particulate matter deposited on the leaves of five evergreen species in Beijing, China: Source identification and size distribution. *Atm Envion.* 105: 53. https://doi.org/10.1016/j.atmosenv.2015.01.032

Tripathi AK, Gautam M. 2007. Biochemical parameters of plants as indicators of air pollution. *J Environ Biol.* 28: 127–132. www.researchgate.net/publication/6122082

Varshney SR, Varshney CK. 1984. Effects of SO2 on ascorbic acid in crop plants. *Environ Pollut Ser A, Ecol Biol.* 35(4): 285–290. https://doi.org/10.1016/0143-1471(84)90074-6

Volta AS, Camurati C, Teoldi F, Maiorana S, et al. 2020. Ecotoxicological effects of atmospheric particulate produced by braking systems on aquatic and edaphic organisms. *Environ Int.* 137: 105564. https://doi.org/10.1016/j.envint.2020.105564

Wakeel A, Xu M, Gan Y. 2020. Chromium-induced reactive oxygen species accumulation by altering the enzymatic antioxidant system and associated cytotoxic, genotoxic, ultrastructural, and photosynthetic changes in plants. *Int J Mol Sci.* 21(3): 728. https://doi.org/10.3390/ijms21030728

Wang X, Ma R, Cui D, et al. 2017. Physio-biochemical and molecular mechanism underlying the enhanced heavy metal tolerance in highland barley seedlings pre-treated with low-dose gamma irradiation. *Sci Rep.* 7: 14233. https://doi.org/10.1038/s41598-017-14601-8

Wei X, Lyu S, Yu Y, Wang Z, Liu H, Pan D, Chen J, 2017. Phylloremediation of air pollutants: Exploiting the potential of plant leaves and leaf-associated microbes. *Front Plant Sci.* 8: 1318. https://doi.org/10.3389/fpls.2017.01318

Weyens N, Thijs S, Popek R, et al. 2015. The role of plant–microbe interactions and their exploitation for phytoremediation of air pollutants. *Int J Mol Sci.* 16: 25576. https://doi.org/10.3390/ijms161025576

Wróblewska, Jeong. 2021. Effectiveness of plants and green infrastructure utilization in ambient particulate matter removal. *Environ Sci Eur.* 33: 110. https://doi.org/10.1186/s12302-021-00547-2

Yaashikaa PR, Kumar PS, Jeevanantham S, Saravanan R. 2022. A review on bioremediation approaches for heavy metal detoxification and accumulation in plants. *Environ Pollution.* 301: 119035. https://doi.org/10.1016/j.envpol.2022.119035

Yang J, McBride J, Zhou J, Sun Z. 2005. The urban forest in Beijing and its role in air pollution reduction. *Urban For Urban Green.* 3: 65. https://doi.org/10.1016/j.ufug.2004.09.001

Yin S, et al. 2020. Coagulation effect of aero submicron particles on plant leaves: Measuring methods and potential mechanisms. *Environ Pollut.* 257: 113611. https://doi.org/10.1016/j.envpol.2019.113611

Yutthammo C, Thongthammachat N, Pinphanichakarn P, Luepromchai E. 2010. Diversity and activity of pah-degrading bacteria in the phyllosphere of ornamental plants. *Microb Ecol.* 59: 357. https://doi.org/10.1007/s00248-009-9631-8

# 7 Ecophysiological Responses of Tropical Plants to Rising Air Pollution
## A Perspective for Urban Areas

*Sadhna, Pallavi B. Dhal, Sachchidanand Tripathi, Rajkumari Sanayaima Devi, Rishikesh Singh, and Rahul Bhadouria*

## 7.1 INTRODUCTION

### 7.1.1 Urbanization and Air Pollution

Air pollution is a major concern in most of the tropical countries and is increasing rapidly due to increasing population, industrialization, urbanization, and numbers of automobiles. To meet the demands of this burgeoning population, the industrial sector is growing faster and resulting in a rise in air pollution (Pandey et al., 2015a; Sen et al., 2017). Particulate matter (PM) is the main component of polluted air (Pandey et al., 2015b). PM is generally emitted by anthropogenic activities (Ferreira-Baptista & De Miguel, 2005). Both fine and coarse PM may inflict a negative impact on humans, as well as plants; however, the effects may be more severe in the case of fine PM. Various detrimental effects of PM have been reported on the human health and physiology of plants. To reduce the air pollution emission, PM must be eliminated, but it is not feasible as no such device has been found yet which can remove the pollution completely (Gupta et al., 2016). Plants stabilize ecosystem by participating in nutrient cycling and gaseous exchange from leaves to the environment, and leaves provide a large surface area for the accumulation and assimilation of pollutants (Khalid & Khanoranga, 2019). Pollutants released in the atmosphere may be absorbed and metabolized by the plants, though differentially (Rai & Panda, 2014). Urban dust, along with vehicular pollutants, further aggravate the environmental conditions and may negatively impact the ecophysiology of the tropical plants. Leaves are the most efficient device to absorb atmospheric pollutants, PM, and secondary pollutants (Prajapati & Tripathi, 2008; Pathak et al., 2011). There is growing evidence that increasing levels of PM and atmospheric pollutants affect the morphology, biochemical composition, and ecophysiology of the plants (Kaur & Nagpal, 2017; Karmakar & Padhy, 2019). Due to excessive traffic, urbanization, and industrialization, primary and secondary air pollutants are growing rapidly, and the industrial sector may expand due to its favorable correlation with human population growth and rising consumer demand (Kanakidou et al., 2011). The ambient air quality of the surrounding areas is significantly influenced by the gaseous pollutants, such as sulphur oxide ($SO_X$), nitrogen oxide ($NO_X$), and suspended particulate matter (SPM), which are released by various industrial sectors (Sengupta, 2003). Urban plants remove ambient air by eliminating heavy metals and absorbing particulate matter (McDonald et al., 2007) by reducing the stomatal size and minimizing gaseous exchange and particle entry into the stomata (Wuytack et al., 2010). The particulate matter enters into the plant through the stomata and adversely affects the physiological functioning (photosynthesis) of the plant (Younis et al., 2013). In urban air, a variety of contaminants are present which vary in concentration and composition and have a significant impact on human health. Through dispersion and interception, urban vegetation can significantly contribute to lowering ambient dust pollution (Chaurasia et al., 2022). Classification of plant functional traits has been listed in Table 7.1.

### 7.1.2 Pollutants and Ecophysiology

The particulate matter and phytotoxic gases individually and in combination affect the ecophysiology of plants by altering the rate of photosynthesis, respiration, and transpiration, which further alter other functions of the plant (Sisodia & Dutta, 2016). Studies conclude that variation in seasons influences the concentration of pollutants; PM of all sizes were generally higher in the dry season as compared to the rainy season (Balogun & Orimoogunje, 2015). Particulate matter is mainly divided on the basis of size; studies show that PM >10 mm in size remained close to the hairs of the leaf and forms dense clusters (Kończak et al., 2021). PM with size <10 mm remained on the leaf blade and on the cuticle layer of the leaf, and roughness and coarseness also increases the deposition of fine particles on the leaf (Bell & Treshow, 2002). These particles ($PM_{2.5}$) bind to the leaf surface very easily and clog the pores and stomata and stop photosynthesis and transpiration (Kończak et al., 2021). The velocity of the movement of PM affects their deposition rate on the leaf. APTI is the most suitable indicator to ascertain air pollution tolerance of the plants (to different air pollutants) which may vary by species

(Nadgórska-Socha et al., 2017; Ogunkunle et al., 2015). Trees absorb air pollutants through stomatal pores and PM by leaf surface (Nowak et al., 2006). Trees are particularly good at removing particulate matter directly through wet and dry deposition processes due to their huge surface area, dense leaves, twigs, and branches (Junior et al., 2009; Yang et al., 2005). Some of the pollutants inhibit photosynthesis by causing damage to the chlorophyll pigments (Chauhan, 2015). In stress conditions, chlorophyll pigments undergo many chemical reactions which lead to changes in the ecophysiology of the plant (Giri et al., 2013). Air pollution can be reduced by plants in two ways: (1) direct reduction from the air; and (2) by indirect reduction by avoiding emission from air pollutants. In direct reduction, gaseous pollutants ($NO_2$, $SO_2$, $O_3$) are absorbed by plants through stomata and they dissolve water-soluble pollutants on the moist surface of the leaf (Nowak, 1994). Additionally, tree canopy has the ability to capture PM in the air (Beckett et al., 1998). Indirectly, through direct shading and evapotranspiration, trees can lower the air temperature in the summer season, which also lowers the amount of air pollutants released during the process of generating energy for cooling activities (Taha, 1996; Nowak et al., 2000). Some leaf functional traits and their definitions have been listed in the Table 7.2.

## 7.2 PFTs AND ECOPHYSIOLOGICAL RESPONSES

The plant species act as bioindicators because they are very sensitive to gaseous contaminants and continuously exchange various pollutants in and out of their foliar systems (Giri et al., 2013). The foliar surface may exhibit structural and functional variation in the presence of particulate matter which affects the morphological, physiological, and biochemical properties of the plants, thereby affecting the overall ecophysiology of the plant (Giri et al., 2013). PM causes stress condition in the plant, which is studied by the change in ascorbic acid content, proline concentration, pH, water holding capacity, photosynthetic rate, and rate of respiration of the plant (Sen et al., 2017). Further, studies have indicated significant changes in plant functional traits at different urban locations (Zhu et al., 2021). Functional traits do not work independently, however, but rather are interconnected in functioning (Zhu et al., 2021). Leaf and stem traits are the most important part of the plant for analyzing the impact of air pollutants present in the environment. Leaves are the main portion of the plant that receives most of the environmental stress, and it is the main organ that helps in purifying air pollution (Liu et al., 2012). It is one of the structures that have maximum contact with the environment and has great sensitivity to the surrounding environment. Therefore, leaves are frequently used to determine how the plant grows and how the external environment affects the plant (Zhu et al., 2021). The important functional traits that can influence a species' effectiveness in air phytoremediation are the cuticular properties of the leaf surface, the presence and density of glandular and non-glandular trichomes, the size and number of stomata, and physiological activity (Niinemets et al., 2014; Li et al., 2018; Zhang et al., 2018; Esposito et al., 2020). Dust particulate (air pollutants) pose a major hazard to urban people's respiratory systems, as well as to the regular development of urban vegetation (Yan et al., 2018). According to studies, stomatal closure and dust deposition on leaves are the main ways that dust affects trees (Yan et al., 2018; Hu et al., 2018). Plants present near the roadside are adversely affected by automobile pollution (Mitu et al. (2019). PM directly affects the epidermal cells of stomata, which

### TABLE 7.1
### Classification of Plant Functional Traits

| Types of Traits | Measurements of Traits | Measuring Units |
| --- | --- | --- |
| Morphological traits | Life form | categorical |
| | Growth form | categorical |
| | Plant height | m |
| | Canopy cover | categorical |
| Physiological traits | Leaf dry matter content (LDMC) | mg g$^{-1}$ |
| | Specific leaf area (SLA) | cm$^2$ kg$^{-1}$ |
| | Leaf carbon concentration (LCC) | cg |
| | Leaf nitrogen concentration (LNC) | mg g$^{-1}$ |
| | Leaf phosphorus concentration (LPC) | mg g$^{-1}$ |
| | Stomatal conductance (Gsmax) | Mmolm$^{-2}$ s$^{-1}$ |
| | Photosynthetic rate (Amax) | μmol m$^{-2}$ s$^{-1}$ |
| | Intrinsic water use efficiency (WUEi) | μmol mol$^{-1}$ |
| Biochemical traits | Chlorophyll concentration (Chl) | mg g$^{-1}$ |
| | Proline content | μmol gm$^{-1}$ FW |
| | Soluble sugar content | mg g$^{-1}$ |
| | Ascorbic acid concentration | mg g$^{-1}$ |

## TABLE 7.2
### Some Important Leaf Functional Traits, Their Characteristics, Measuring Units and Roles in the Ecosystem

| Leaf Traits | Definition/Formula | Measurement units | Role | References |
|---|---|---|---|---|
| Specific leaf area (SLA) | One side area of fresh leaf, divided by its oven-dry mass | $cm^2\ g^{-1}$ | To assess the reproductive strategy | Chen and Black (1992); Garnier et al. (2001) |
| Leaf thickness (LT) | Thickness of leaves | mm or μm | To predict plant health and carbon assimilation | Garnier and Laurent (1994); Wright et al. (2002); Vile et al. (2005); Poorter et al. (2009) |
| Leaf dry matter content (LDMC) | Oven-dry mass (mg) of leaf, divided by its water-saturated fresh mass | $mg\ g^{-1}$ | Indicator trait for resource use strategy | Garnier et al. (2001); Vaieretti et al. (2007) |
| Photosynthetic rate (A) | Oxygen production either as the amount of chlorophyll content or green plant tissue weight per unit mass or area | μmol oxygen evolved or $CO_2$ consumed per $m^2\ s^{-1}$ | Food production using light and $CO_2$ | Ellsworth and Reich (1992) |
| Stomatal conductance (Gs) | The rate of the flow of water and $CO_2$ through stomatal opening on the leaf | $mmol\ m^{-2}\ s^{-1}$ | Fundamental to leaf level calculations of transpiration | Yadav et al. (2022) |
| Leaf tissue density (LTD) | Leaf dry mass divided by leaf volume | $g\ ml^{-1}$ | | Zhu et al. (2021) |
| Leaf nutrient concentration (N, P) | Total amount of N and P, respectively, per unit of dry leaf mass | $mg\ g^{-1}$ | Related with plant growth and reproduction | Horneck and Miller (2019); Temminghoff and Houba (2004) |
| Water-use efficiency (WUE) | The ratio between net A and Gs | $\mu mol\ mol^{-1}$ | Percentage of water supplied to the plant that is effectively taken by the plant | Niinemets et al. (2009) |

affects the opening and closing of the stomata (Chaurasia, 2013), further affecting the gaseous exchange or respiration and reducing the biochemical activity of the leaf (Sarkar & Gupta, 2016).

## 7.3 BIOCHEMICAL ANALYSIS

### 7.3.1 Chlorophyll Concentration

Chlorophyll content is an important aspect for understanding the ecophysiology of the plant and it is affected by the $SO_X$ present, along with the PM, as $SO_2$ is involved in the formation of pheophytin which further reduces the chlorophyll content (Rao & LeBlanc, 1966; Mandloi & Dubey, 1988; Sen et al., 2017). To evaluate the effect of air pollutants on plants, the measurement of chlorophyll is a very important tool. Plant growth is directly proportional to the chlorophyll concentration of plants (Rai, 2016). The photosynthetic activity of the plant is governed by chlorophyll content, which ultimately leads to the growth and development of biomass (Khalid & Khanoranga, 2019). The chlorophyll content is affected by vehicular pollution, as trees present near the roadside are adversely affected (Iqbal et al., 2015). The plant chlorophyll content is dependent on the intensity of pollution, leaf age, and other biotic and abiotic conditions (temperature, drought, light intensity, salt stress). The chlorophyll content of plants also varies from one species to another (Katiyar & Dubey, 2001; Prajapati & Tripathi, 2008; Kaur & Nagpal, 2017; Banerjee et al., 2019; Karmakar & Padhy, 2019). Photosynthetic pigments have been significantly reduced in the trees present near the roadside and affected by vehicular pollution, as seen by (Iqbal et al., 2015).

### 7.3.2 Ascorbic Acid Concentration

Ascorbic acid is an important pollution tolerance index, as stress and pollution decrease the ascorbic acid content in the plant (Joshi & Swami, 2009; Sen et al., 2017). Ascorbic acid in plants protects them from oxidative damage and is an essential aspect of understanding the physiology of the plant by observing the concentration of ascorbic acid. Self-detoxicant properties of ascorbic acid protect plants from damage caused by pollution (Chauhan, 2015). As an antioxidant, ascorbic acid is found in large amounts in all the growing parts of plants and it also influences the resistance to adverse environmental conditions, including air pollution (Keller & Schwager, 1977; Pathak et al., 2011; Rai & Panda, 2014; Pandey et al., 2015b). Ascorbic acid plays an important role in the antioxidant system when it is associated with other components and protects the plant against various oxidative damages which result from aerobic metabolism, photosynthesis, and a series of pollutants (Smirnoff, 1996). Ascorbic content was reduced in polluted areas in the pre-monsoon season as compared to the post-monsoon season (Sen et al., 2017).

### 7.3.3 Proline Content

Proline content in the plant is found higher in the pollution-stressed site (Seyyednejad & Koochak, 2011). Due to stress conditions like $SO_2$ fumigation, salinity stress, and heavy metal stress, proline content increases in the leaf (Tankha & Gupta, 1992; Wang et al., 2009; Woodward & Bennett, 2005). The accumulation of proline concentration in the leaf acts as an indicator because the higher the accumulation of the proline concentration, the higher the stress tolerance of the species (Agbaire, 2016). Proline content was higher in the polluted pre-monsoon area as compared to post-monsoon season (Sen et al., 2017). Additionally, some researchers found that proline is also accumulated in the regenerative organs of various plant species, which also tells about the possibility of accumulation of proline content in non-stress conditions (Mattioli et al., 2009). Various studies observed that proline content is higher in polluted areas as compared to clean areas (Tantrey & Agnihotri, 2010; Kandziora-Ciupa et al., 2016; Kandziora-Ciupa et al., 2017).

### 7.3.4 Sugar Content

Soluble sugars act as nutrients, as well as signaling molecules, and are involved in various environmental stressors (Afzal et al., 2021). The concentration of soluble sugars in the leaf indicates the sensitivity of the plant against pollution stress (Tripathi & Gautam, 2007). Synthesis and breakdown of sugar molecules take place when photosynthesis and respiration occur, respectively. On the polluted site, the rate of photosynthesis reduces and high energy is required to reduce soluble sugar content (Rai, 2016). High concentrations of $SO_X$ and $NO_X$ are present in the polluted site, which is the reason behind the decrease in the soluble sugar content in the leaf (Rai & Panda, 2015). A decrease in soluble sugar content was observed due to the increase in respiration and decrease in $CO_2$ (carbon dioxide) fixation (Agbaire, 2016). The soluble sugar content can be accumulated or reduced in the polluted environment it depends on the sensitivity of the plant species toward air pollution (Agbaire, 2016).

## 7.4 PHYSIOLOGICAL PARAMETERS

### 7.4.1 pH

One of the aspects of determining the tolerance and sensitivity of the plant is pH, as the change in the pH of leaves affects the stomatal activity of the plant. Studies show that the alkaline pH of leaf extracts exhibits tolerance to pollution consisting of sulfur oxide ($SO_X$), nitrogen oxide ($NO_X$), and suspended particulate matter (SPM) (Sisodia & Dutta, 2016). Photosynthetic efficiency is strongly dependent on the pH content of leaves. When the leaf pH is low, the photosynthesis of the plant is also reduced (Türk & Wirth, 1975; Thakar & Mishra, 2010). At low pH, the leaf shows sensitivity toward air pollution and reduces the overall photosynthetic process (Enete et al., 2013; Thakar & Mishra, 2010). Generally, the enzymes obtain the highest efficiency in various biological activities of organisms at different ranges of pH. Some are efficient at higher pH and others even function under neutral pH (Achakzai et al., 2017). The pH content of the leaf extract is lowered in the presence of acidic pollutants containing $SO_2$ and $NO_X$ in the air, and the decline becomes larger in sensitive plant species as compared to tolerant plant species (Scholz & Reck, 1977; Kaur & Nagpal, 2017). Therefore, those plants which have high pH content of leaf extract in a polluted environment are to be considered tolerant species (Prajapati & Tripathi, 2008). Acidic pH increases the stomatal sensitivity to pollution thus, altering important physiological processes (i.e., rate of respiration and transpiration) of plants (Sen et al., 2017; Khalid & Khanoranga, 2019). So, it was observed that when the gaseous $SO_2$ entered the leaf, the pH changes toward the acidic range in most of the plant species (Kousar et al., 2014; Khalid & Khanoranga, 2019).

### 7.4.2 Relative Water Content

Relative water content (RWC) is the amount of water with respect to a fully swollen leaf. RWC is related to the protoplast permeability in the cell, which causes the loss of solvent (water and nutrients) and the impact of pollution on it (Sisodia & Dutta, 2016). Pollution reduces the transpiration rate by choking the stomatal pores and destroying the water suction mechanism from root to leaves created by leaves, which affects the transfer of other materials like minerals and affects the cooling of the leaf by affecting the rate of transpiration (Lohe et al., 2015). PM in the air increases the cell permeability (Keller, 1986; Sen et al., 2017; Khalid & Khanoranga, 2019), which causes loss of water and nutrients, and leads to early aging of the plant (Masuch et al., 1988; Khalid & Khanoranga, 2019). Studies show that high RWC during pollution stress may confer plant's tolerance against environmental pollutants (Kuddus et al., 2011). RWC is associated with protoplasmic permeability, and plants with higher values of RWC are thus perhaps more tolerant to pollution (Pandey et al., 2015b). Further, large amount of water (RWC) present in the plant body maintains physiological balances under pollution stress (González & González-Vilar, 2001). The RWC of foliar tissue may be decreased due to air pollutants indicating the disturbed physiological status of the plant (Sen et al., 2017; Khalid & Khanoranga, 2019). Increased levels of pollution increased the permeability of cells and dissolved nutrients, increasing the risk of early senescence (Khalid & Khanoranga, 2019). RWC is associated with protoplasmic permeability in a cell, which leads to the loss of water and nutrients and thus indicates the impact of pollution in the early stage (Sen et al., 2017). Furthermore, higher RWC in plants can dilute acidity inside the leaf cell sap and resist the drought condition in plants (Kaur & Nagpal, 2017). Higher percentage of RWC in plants increases the resistance in plants and makes the plant more tolerant under stress conditions due to pollution (Karmakar & Padhy, 2019). The significant reduction of RWC in plant leaves was noted due to pollution stress, the stomatal closure, and the resulting loss of leaf transpiration rate (Ghafari et al., 2021). Due to the reduction of leaf transpiration rate in plants, the water uptake from soil for photosynthesis is also reduced (Rai, 2016).

## 7.5 IMPACT OF DUST LOAD ON ECOPHYSIOLOGICAL ATTRIBUTES OF URBAN PLANTS

Dust particles in the air affect photosynthesis, respiration, and transpiration, as well as the gaseous pollutant lodging in the pores (Gupta et al., 2016, Chaurasia et al., 2022). Plants continuously exchange gaseous pollutants in and out of their leaves, which ultimately affects the overall physiology of the plants (Sen et al., 2017). Dust pollutants not only affect the trees but also affect the herbs grown on the roadside, which are continuously exposed to environmental stresses like temperature, dust, light, and limited space. All these factors may simultaneously affect the overall physiological well-being or ecophysiology of the plant (Chaurasia et al., 2022). Based on the concentration of pollutants, their exposure to the leaves, soil conditions, climatic conditions, and the type of plant species present, a drastic loss in the yield has been observed in urban as well as in sub-urban and rural areas (Mina et al., 2013). Plants which are grown in more dust-polluted sites show decline in photosynthetic efficiency transpiration rate, water-use-efficiency (WUE), and stomatal conductance (Sen et al., 2017). Reduction in stomatal activity may cause permanent blocking of the stomatal aperture, which further affects transpiration and photosynthetic rate—and thereby overall plant productivity (Naidoo & Chirkoot, 2004; Nanos & Ilias, 2007). PM is absorbed by the leaves, which act as a filter to protect the environment (Enete et al., 2013) as the leaves have the unique ability to absorb the PM present in the air which purifies the air, but the absorption of PM adversely affects the ecophysiology of the plant (Rao, 2007). Dust present in the air reduces the rate of photosynthesis by causing damage to the photosynthetic system and chlorophyll pigments (Prajapati & Tripathi, 2008; Joshi & Swami, 2009). PM present in air destructively affects plant life. Accumulation of particulate matter on the leaves of plants affects photosynthetic apparatus; the amount of particulate matter accumulation depends on the morphology of plant leaves like size, quality, etc. Also, the accumulation of particulates increases if the amount of epicuticle wax is higher on the leaves (Łukowski et al., 2020).

## 7.6 ROLE OF PLANTS IN URBAN AREAS

Urban vegetation is an important part of the urban ecosystem, and it plays a vital role in reducing air pollution (Singh et al., 2020). Different tree species have different levels of adaptations and mitigation potential for air pollution (Singh et al., 2020). The air pollution tolerance index (APTI) assessment provides whether the plant species is tolerant or sensitive and tells which tolerant species can act as mitigating species for environmental pollution and which sensitive species act as bioindicators of environmental stress (Rai, 2016). Those plants which give higher APTI value can be used for greenbelt development (Rai, 2016). Plants may act as filters by different mechanisms like absorption and accumulation of PM, and detoxification, and therefore are greenbelt components (Mondal et al., 2011). Urban trees in the greenbelt region act to remove PM and gaseous pollutants, making them the so-called "lungs of the urban areas" (Singh et al., 2020, p. 2). Some plant species which are used for the greenbelt plantation in urban areas are listed in Table 7.3.

## 7.7 CONCLUSION AND WAYS FORWARD

Particulate pollutants are effectively reduced by plants in urban green spaces. Air pollution is mainly reduced by roadside planting, which absorbs dust particles and alters the biochemical parameters so in this manner tree leaves are the bioindicators of air pollution. Large trees should not be planted close to each other because a dense canopy interferes with the diffusion of PM and intensifies air pollution. Large older trees are more helpful in mitigating the pollutants from the air as compared to young trees. The formation of shelterbelt (or windbreak) helps in improving the quality of air by absorbing harmful gaseous pollutants and PM. Absorption of heat islands in the urban system is also reduced by the shelterbelt. The introduction of green infrastructure (GI) is a valuable solution for improving quality of air in urban areas; this method reduces the concentration of PM at road level without putting any type of restrictions on road traffic. Trees absorb more pollutants and dust particles as compared to grasses. Use of plants phytoremediators has the additional benefit of removing all the pollutants—including gaseous pollutants, as well as heavy metals—simultaneously.

Roadsides could improve from the planting of tree species that are more resilient and efficient at reducing air pollution, but there is little data available to help choose which are the most suitable trees. However, it is unclear how plants react to various environments within cities because of urbanization, as some researchers focus on the adaptation of plant functional traits to the environment within urban–rural gradients. According to studies, the features of high resolution, a lot of information, strong data continuity, and ease of access are also underlined by the quick growth of spectral technology. Currently there has not been enough research on how urban atmospheric pollutants affect plant function traits and their proper functioning. Investigating

## TABLE 7.3
### Some Plant Species Preferred for Greenbelt Plantings in Urban Areas

| Site of Study | Trees Species Preferred | Reference |
| --- | --- | --- |
| Dehradun (India) | *Grevillea robusta* <br> *Mangifera indica* | Singh et al. (2020) |
| Beijing (China) | *Euonymus japonicus* | Zhu et al. (2021) |
| Santiniketan (India) | *Mangifera indica* <br> *Peltophorum pterocarpum* | Karmakar et al. (2021) |
| Norway, Poland (Europe) | *Betula pendula* <br> *Pinus sylvestris* | Sæbø et al. (2012) |
| Beijing (China) | *Betula pendula* | Łukowski et al. (2020) |
| Katowice, Poland (Europe) | *Betula pendula* | Kończak et al. (2021) |
| Jinju, Republic of Korea | *Quercus glauca* <br> *Quercus salicina* | Jin et al. (2021) |
| NCR of Delhi (India) | *Terminalia arjuna* <br> *Morus alba* | Gupta et al. (2016) |
| Chungcheongbuk-do (South Korea) | *Pinus densiflora* | Bui et al. (2021) |
| Warsaw, Poland (Europe) | *Spiraea japonica* | Dzierżanowski et al. (2011) |
| Rome, Italy (Europe) | *Quercus ilex* | Fusaro et al. (2021) |
| Singapore | *Grewia laevigata* <br> *Ficus aurata* <br> *Muntingia calabura* | Chiam et al. (2019) |

the PM capture capacity of various plant species is critical to maximizing their usage in varied urban situations, since the ability to successfully catch PM is a key factor in selecting the best plant species to be used for urban greening.

## REFERENCES

Achakzai, Khanoranga, Sofia Khalid, Muhammad Adrees, Aasma Bibi, Shafaqat Ali, Rab Nawaz, and Mohammad Rizwan. "Air pollution tolerance index of plants around brick kilns in Rawalpindi, Pakistan." *Journal of Environmental Management* 190 (2017): 252–258.

Afzal, Shadma, Nidhi Chaudhary, and Nand K. Singh. "Role of soluble sugars in metabolism and sensing under abiotic stress." In *Plant Growth Regulators*, pp. 305–334. Springer, 2021.

Agbaire, O. P. "Impact of air pollution on proline and soluble sugar content of selected plant species." *Chemistry and Materials Research* 8, no. 5 (2016): 72–76.

Balogun, Verere Sido, and Oluwagbenga Oluwapamilerin Isaac Orimoogunje. "An assessment of seasonal variation of air pollution in Benin City, Southern Nigeria." *Atmospheric and Climate Sciences* 5, no. 03 (2015): 209.

Banerjee, S., A. Banerjee, D. Palt, and P. Roy. "Assessment of vegetation under air pollution stress in urban industrial area for greenbelt development." *International Journal of Environmental Science and Technology* 16, no. 10 (2019): 5857–5870.

Beckett, K. Paul, P. H. Freer-Smith, and Gail Taylor. "Urban woodlands: Their role in reducing the effects of particulate pollution." *Environment Pollution* 99, no. 3 (1998): 347–360.

Bell, J. Nigel B., and Michael Treshow, eds. *Air Pollution and Plant Life*. John Wiley & Sons, 2002.

Bui, Huong-Thi, Uuriintuya Odsuren, Kei-Jung Kwon, Sang-Yong Kim, Jong-Cheol Yang, Na-Ra Jeong, and Bong-Ju Park. "Assessment of air pollution tolerance and particulate matter accumulation of 11 woody plant species." *Atmosphere* 12, no. 8 (2021): 1067.

Chauhan, Avnish. "Effect of SO2 on ascorbic acid content in crop plants-first line of defense against oxidative stress." *International Journal of Innovative Research & Development* 4, no. 11 (2015): 8–13.

Chaurasia, Meenakshi, Kajal Patel, Indu Tripathi, and Kottapalli Sreenivasa Rao. "Impact of dust accumulation on the physiological functioning of selected herbaceous plants of Delhi, India." *Environmental Science and Pollution Research* (2022): 1–16.

Chaurasia, Sadhana. "Effect of cement industry pollution on chlorophyll content of some crops at Kodinar, Gujarat, India." *Proceedings of the International Academy of Ecology and Environmental Sciences* 3, no. 4 (2013): 288.

Chen, J. M., and T. A. Black. "Defining leaf area index for non-flat leaves plant." *Cell & Environment* 15 (1992): 421–429.

Chiam, Zhongyu, Xiao Ping Song, Hao Ran Lai, and Hugh Tiang Wah Tan. "Particulate matter mitigation via plants: Understanding complex relationships with leaf traits." *Science of the Total Environment* 688 (2019): 398–408.

Dzierżanowski, Kajetan, Robert Popek, Helena Gawrońska, Arne Sæbø, and Stanislaw W. Gawroński. "Deposition of particulate matter of different size fractions on leaf surfaces and in waxes of urban forest species." *International Journal of Phytoremediation* 13, no. 10 (2011): 1037–1046.

Ellsworth, D. S., and P. B. Reich. "Leaf mass per area, nitrogen content and photosynthetic carbon gain in Acer saccharum seedlings in contrasting forest light environments." *Functional Ecology* (1992): 423–435.

Enete, I. C., V. U. Chukwudeluzu, and A. Okol. "Evaluation of air pollution tolerance index of plants and ornamental shrubs in Enugu City: Implications for urban heat island effect." (2013).

Esposito, F., V. Memoli, S. C. Panico, G. Di Natale, M. Trifuoggi, A. Giarra, and G. Maisto. "Leaf traits of Quercus ilex L. Affect particulate matter accumulation." *Urban Forestry & Urban Greening* 54 (2020): 126780.

Ferreira-Baptista, Lopes, and E. De Miguel. "Geochemistry and risk assessment of street dust in Luanda, Angola: A tropical urban environment." *Atmospheric Environment* 39, no. 25 (2005): 4501–4512.

Fusaro, Lina, Elisabetta Salvatori, Aldo Winkler, Maria Agostina Frezzini, Elena De Santis, Leonardo Sagnotti, Silvia Canepari, and Fausto Manes. "Urban trees for biomonitoring atmospheric particulate matter: An integrated approach combining plant functional traits, magnetic and chemical properties." *Ecological Indicators* 126 (2021): 107707.

Garnier, E., and G. Laurent. "Leaf anatomy, specific mass and water content in congeneric annual and perennial grass species." *New Phytologist* 128, no. 4 (1994): 725–736.

Garnier, E., B. Shipley, C. Roumet, and G. Laurent. "A standardized protocol for the determination of specific leaf area and leaf dry matter content." *Functional Ecology* (2001): 688–695.

Ghafari, Sara, Behzad Kaviani, Shahram Sedaghathoor, and Mohammad Sa"egh Allahyari. "Assessment of air pollution tolerance index (APTI) for some ornamental woody species in green space of humid temperate region (Rasht, Iran)." *Environment, Development and Sustainability* 23, no. 2 (2021): 1579–1600.

Giri, Sumitra, Deepali Shrivastava, Ketki Deshmukh, and Pallavi Dubey. "Effect of air pollution on chlorophyll content of leaves." *Current Agriculture Research Journal* 1, no. 2 (2013): 93–98.

González, L., and M. González-Vilar. "Determination of relative water content." In M. J. Reigosa Roger, ed., *Handbook of Plant Ecophysiology Techniques* (pp. 207–212), Springer, Dordrecht, 2001.

Gupta, Gyan Prakash, Bablu Kumar, and U. C. Kulshrestha. "Impact and pollution indices of urban dust on selected plant species for green belt development: Mitigation of the air pollution in NCR Delhi, India." *Arabian Journal of Geosciences* 9, no. 2 (2016): 1–15.

Horneck, D. A., and Miller, R. O. (2019). "Determination of total nitrogen in plant tissue." In *Handbook of Reference Methods for Plant Analysis* (pp. 75–83). CRC Press.

Hu, Z., X. Tang, C. Zheng, M. Guan, and J. Shen. "Spatial and temporal analyses of air pollutants and meteorological driving forces in Beijing–Tianjin–Hebei region, China." *Environmental Earth Sciences* 77, no. 14 (2018): 1–19.

Iqbal, M., M. Shafiq, S. Zaidi, and M. Athar. "Effect of automobile pollution on chlorophyll content of roadside urban trees." *Global Journal of Environmental Science and Management* 1, no. 4 (2015): 283–296.

Jin, Eon Ju, Jun Hyuck Yoon, Eun Ji Bae, Byoung Ryong Jeong, Seong Hyeon Yong, and Myung Suk Choi. "Particulate matter removal ability of ten evergreen trees planted in Korea urban greening." *Forests* 12, no. 4 (2021): 438.

Joshi, P. C., and Abhishek Swami. "Air pollution induced changes in the photosynthetic pigments of selected plant species." *Journal of Environmental Biology* 30, no. 2 (2009): 295–298.

Junior, Armando Molina Divan, Paulo Luiz de Oliveira, Carolina Trindade Perry, Vera Lúcia Atz, Letícia Nonnenmacher Azzarini-Rostirola, and Maria Teresa Raya-Rodriguez. "Using wild plant species as indicators for the accumulation of emissions from a thermal power plant, Candiota, South Brazil." *Ecological Indicators* 9, no. 6 (2009): 1156–1162.

Kanakidou, Maria, Nikolaos Mihalopoulos, Tayfun Kindap, Ulas Im, Mihalis Vrekoussis, Evangelos Gerasopoulos, Eirini Dermitzaki et al. "Megacities as hot spots of air pollution in the East Mediterranean." *Atmospheric Environment* 45, no. 6 (2011): 1223–1235.

Kandziora-Ciupa, Marta, Ryszard Ciepał, Aleksandra Nadgórska-Socha, and Gabriela Barczyk. "Accumulation of heavy metals and antioxidant responses in Pinus sylvestris L. needles in polluted and non-polluted sites." *Ecotoxicology* 25, no. 5 (2016): 970–981.

Kandziora-Ciupa, Marta, Aleksandra Nadgórska-Socha, Gabriela Barczyk, and Ryszard Ciepał. "Bioaccumulation of heavy metals and ecophysiological responses to heavy metal stress in selected populations of Vaccinium myrtillus L. and Vaccinium vitis-idaea L." *Ecotoxicology* 26, no. 7 (2017): 966–980.

Karmakar, Dipti, Kuheli Deb, and Pratap Kumar Padhy. "Ecophysiological responses of tree species due to air pollution for biomonitoring of environmental health in urban area." *Urban Climate* 35 (2021): 100741.

Karmakar, Dipti, and Pratap Kumar Padhy. "Air pollution tolerance, anticipated performance, and metal accumulation indices of plant species for greenbelt development in urban industrial area." *Chemosphere* 237 (2019): 124522.

Katiyar, V., and P. S. Dubey. "Sulphur dioxide sensitivity on two stages of leaf development in a few tropical tree species." *Indian Journal of Environment and Toxicology* 11, no. 2 (2001): 78–81.

Kaur, Mandeep, and Avinash Kaur Nagpal. "Evaluation of air pollution tolerance index and anticipated performance index of plants and their application in development of green space along the urban areas." *Environmental Science and Pollution Research* 24, no. 23 (2017): 18881–18895.

Keller, Theodore. "The electrical conductivity of Norway spruce needle diffusate as affected by certain air pollutants." *Tree Physiology* 1, no. 1 (1986): 85–94.

Keller, Theodore, and H. Schwager. "Air pollution and ascorbic acid." *European Journal of Forest Pathology* 7, no. 6 (1977): 338–350.

Khalid, Sofia, and Khanoranga. "Spatio-temporal variations in the PAH concentrations in the soil samples collected from functional brick kilns locations in Balochistan, Pakistan." *Polycyclic Aromatic Compounds* 41, no. 1 (2021): 184–198.

Kończak, B., M. Cempa, and M. Deska. "Assessment of the ability of roadside vegetation to remove particulate matter from the urban air." *Environmental Pollution* 268 (2021): 115465.

Kousar, Hina, K. D. Nuthan, K. Pavithra, and M. P. Adamsab. "Analysis of biochemical parameters as tolerance index of some chosen plant species of Bhadravathi town." *The International Journal of Environmental Science and Technology* 3, no. 11 (2014).

Kuddus, Mohammed, Rashmi Kumari, and Pramod W. Ramteke. "Studies on air pollution tolerance of selected plants in Allahabad city, India." *Journal of Environmental Research and Management* 2, no. 3 (2011): 042–046.

Li, S., T. Tosens, P. C. Harley, Y. Jiang, A. Kanagendran, I. M. Grosberg, . . . and Ü. Niinemets. "Glandular trichomes as a barrier against atmospheric oxidative stress: Relationships with ozone uptake, leaf damage, and emission of LOX products across a diverse set of species." *Plant, Cell & Environment* 41, no. 6 (2018): 1263–1277.

Liu, L., D. Guan, and M. R. Peart. "The morphological structure of leaves and the dust-retaining capability of afforested plants in urban Guangzhou, South China." *Environmental Science and Pollution Research* 19, no. 8 (2012): 3440–3449.

Lohe, R. N., B. Tyagi, V. Singh, Tyagi P. Kumar, D. R. Khanna, and R. Bhutiani. "A comparative study for air pollution tolerance index of some terrestrial plant species." (2015): 315–324.

Łukowski, Adrian, Robert Popek, and Piotr Karolewski. "Particulate matter on foliage of Betula pendula, Quercus robur, and Tilia cordata: Deposition and ecophysiology." *Environmental Science and Pollution Research* 27, no. 10 (2020): 10296–10307.

Mandloi, B. L., and P. S. Dubey. "Industrial emission and plant response at pithampur (MP)." *International Journal of Ecology and Environmental Sciences* (1988).

Masuch, Georg, H. G. Kicinski, A. Kettrup, and K. S. Boos. "Single and combined effects of continuous and discontinuous O3 and SO2 immission on norway spruce needles: I. Histological and cytological changes." *International Journal of Environmental Analytical Chemistry* 32, no. 3–4 (1988): 187–212.

Mattioli, Roberto, Paolo Costantino, and Maurizio Trovato. "Proline accumulation in plants: Not only stress." *Plant Signaling & Behavior* 4, no. 11 (2009): 1016–1018.

McDonald, A. G., W. J. Bealey, D. Fowler, U. Dragosits, U. Skiba, R. I. Smith, R. G. Donovan, H. E. Brett, C. N. Hewitt, and E. Nemitz. "Quantifying the effect of urban tree planting on concentrations and depositions of PM10 in two UK conurbations." *Atmospheric Environment* 41, no. 38 (2007): 8455–8467.

Mina, U., R. Sigh, and B. Chakrabarti. "Agricultural production and air quality: An emerging challenge." *International Journal of Environmental Science: Development and Monitoring* 4, no. 2 (2013): 80–85.

Mitu, K. J., M. A. Islam, P. Biswas, S. Marzia, and M. A. Ali. "Effects of different environmental pollutants on the anatomical features of roadside plants." *Progressive Agriculture* 30, no. 4 (2019): 344–351.

Mondal, Dali, Srimanta Gupta, and Jayanta Kumar Datta. "Anticipated performance index of some tree species considered for green belt development in an urban area." *International Research Journal of Plant Science* 2, no. 4 (2011): 99–106.

Nadgórska–Socha, Aleksandra, Marta Kandziora-Ciupa, Michał Trzęsicki, and Gabriela Barczyk. "Air pollution tolerance index and heavy metal bioaccumulation in selected plant species from urban biotopes." *Chemosphere* 183 (2017): 471–482.

Naidoo, G., and D. Chirkoot. "The effects of coal dust on photosynthetic performance of the mangrove, Avicennia marina in Richards Bay, South Africa." *Environmental Pollution* 127, no. 3 (2004): 359–366.

Nanos, George D., and Ilias F. Ilias. "Effects of inert dust on olive (Olea europaea L.) leaf physiological parameters." *Environmental Science and Pollution Research-International* 14, no. 3 (2007): 212–214.

Niinemets, Ü., A. Díaz-Espejo, J. Flexas, J. Galmés, and C. R. Warren. (2009). "Role of mesophyll diffusion conductance in constraining potential photosynthetic productivity in the field." *Journal of Experimental Botany*, 60(8), 2249–2270.

Niinemets, Ü., S. Fares, P. Harley, and K. J. Jardine. "Bidirectional exchange of biogenic volatiles with vegetation: Emission sources, reactions, breakdown and deposition." *Plant, Cell & Environment* 37, no. 8 (2014): 1790–1809.

Nowak, David J. "Air pollution removal by Chicago's urban forest." *Chicago's Urban Forest Ecosystem: Results of the Chicago Urban Forest Climate Project* (1994): 63–81.

Nowak, David J., Kevin L. Civerolo, S. Trivikrama Rao, Gopal Sistla, Christopher J. Luley, and Daniel E. Crane. "A modelling study of the impact of urban trees on ozone." *Atmospheric Environment* 34, no. 10 (2000): 1601–1613.

Nowak, David J., Daniel E. Crane, and Jack C. Stevens. "Air pollution removal by urban trees and shrubs in the United States." *Urban Forestry & Urban Greening* 4, no. 3–4 (2006): 115–123.

Ogunkunle, C. O., L. B. Suleiman, S. Oyedeji, O. O. Awotoye, and P. O. Fatoba. "Assessing the air pollution tolerance index and anticipated performance index of some tree species for biomonitoring environmental health." *Agroforestry Systems* 89 (2015): 447–454.

Pandey, Ashutosh Kumar, Mayank Pandey, Ashutosh Mishra, Ssiddhant Mohan Tiwary, and B. D. Tripathi. "Air pollution tolerance index and anticipated performance index of some plant species for development of urban forest." *Urban Forestry & Urban Greening* 14, no. 4 (2015a): 866–871.

Pandey, Ashutosh Kumar, Mayank Pandey, and B. D. Tripathi. "Air pollution tolerance index of climber plant species to develop vertical greenery systems in a polluted tropical city." *Landscape and Urban Planning* 144 (2015b): 119–127.

Pathak, Vinita, B. D. Tripathi, and V. K. Mishra. "Evaluation of anticipated performance index of some tree species for green belt development to mitigate traffic generated noise." *Urban Forestry & Urban Greening* 10, no. 1 (2011): 61–66.

Poorter, Hendrik, Ülo Niinemets, Lourens Poorter, Ian J. Wright, and Rafael Villar. "Causes and consequences of variation in leaf mass per area (LMA): A meta-analysis." *New Phytologist* 182, no. 3 (2009): 565–588.

Prajapati, Santosh Kumar, and B. D. Tripathi. "Anticipated Performance Index of some tree species considered for green belt development in and around an urban area: A case study of Varanasi city, India." *Journal of Environmental Management* 88, no. 4 (2008): 1343–1349.

Prajapati, Santosh Kumar, and B. D. Tripathi. "Seasonal variation of leaf dust accumulation and pigment content in plant species exposed to urban particulates pollution." *Journal of Environmental Quality* 37, no. 3 (2008): 865–870.

Rai, Prabhat Kumar. "Impacts of particulate matter pollution on plants: Implications for environmental biomonitoring." *Ecotoxicology and Environmental Safety* 129 (2016): 120–136.

Rai, Prabhat Kumar, and L. S. Panda. "Leaf dust deposition and its impact on biochemical aspect of some roadside plants of Aizawl, Mizoram, North East India." *International Research Journal of Environment Sciences* 3, no. 11 (2014): 14–19.

Rai, Prabhat Kumar, and L. S. Panda. "Roadside plants as bio indicators of air pollution in an industrial region, Rourkela, India." *International Journal of Advanced Research and Technology* 4, no. 1 (2015): 14–36.

Rao, C. S. *Environmental Pollution Control Engineering*. New Age International, 2007.

Rao, D. N., and Fabius LeBlanc. "Effects of sulfur dioxide on the lichen alga, with special reference to chlorophyll." *Bryologist* (1966): 69–75.

Sæbø, Arne, R. Popek, B. Nawrot, H. M. Hanslin, H. Gawronska, and S. W. Gawronski. "Plant species differences in particulate matter accumulation on leaf surfaces." *Science of the Total Environment* 427 (2012): 347–354.

Sarkar, S., and A. Gupta. "Biological monitoring of cement factory emissions in Badarpur, Assam, India, using Mangifera indica L." *Indian Journal of Applied Research* 6, no. 7 (2016): 391–393.

Scholz, F., and S. Reck. "Effects of acids on forest trees as measured by titration in vitro, inheritance of buffering capacity in Picea abies." *Water, Air, and Soil Pollution* 8, no. 1 (1977): 41–45.

Sen, Abhishek, Indrani Khan, Debajyoti Kundu, Kousik Das, and Jayanta Kumar Datta. "Ecophysiological evaluation of tree species for biomonitoring of air quality and identification of air pollution-tolerant species." *Environmental Monitoring and Assessment* 189, no. 6 (2017): 1–15.

Sengupta, B. "Guidelines for ambient air quality monitoring." In *National Ambient Air Quality Monitoring Series (NAAQMS)/2003–04*. Central Pollution Control Board, Ministry of Environment & Forests, 2003.

Seyyednejad, M. S., and Haniyeh Koochak. "A study on air pollution effects on Eucalyptus camaldulensis." In *International Conference on Environmental, Biomedical and Biotechnology*, vol. 16. IPCBEE, 2011.

Singh, Hukum, Mukta Yadav, Narendra Kumar, Amit Kumar, and Manoj Kumar. "Assessing adaptation and mitigation potential of roadside trees under the influence of vehicular emissions: A case study of *Grevillea robusta* and *Mangifera indica* planted in an urban city of India." *Plos One* 15, no. 1 (2020): e0227380.

Sisodia, A., and S. Dutta. "Air pollution tolerance index of certain plant species: A study of national highway no-8, India." *Journal of Environmental Research and Development* 10, no. 4 (2016): 723.

Smirnoff, Nicholas. "Botanical briefing: The function and metabolism of ascorbic acid in plants." *Annals of Botany* 78, no. 6 (1996): 661–669.

Taha, "Haider. "Modeling impacts of increased urban vegetation on ozone air quality in the South Coast Air Basin." *Atmospheric Environment* 30, no. 20 (1996): 3423–3430.

Tankha, K., and R. K. Gupta. "Effect of water deficit and sulphur dioxide on total soluble proteins, nitrate reductase activity and free proline content in sunflower leaves." *Biologia Plantarum* 34, no. 3 (1992): 305–310.

Tantrey, M. Shafi, and R. K. Agnihotri. "Chlorophyll and proline content of gram (Cicer arietinum L.) under cadmium and mercury treatments." *Agricultural Science Research Journal* 1, no. 2 (2010): 119–122.

Temminghoff, Erwin E. J. M., and Victor J. G. Houba, eds. *Plant Analysis Procedures*. Springer Netherlands, 2004.

Thakar, B. K., and P. C. Mishra. "Dust collection potential and air pollution tolerance index of tree vegetation around Vedanta Aluminium Limited, Jharsuguda." *The Bioscan* 3 (2010): 603–612.

Tripathi, A. K., and Mukesh Gautam. "Biochemical parameters of plants as indicators of air pollution." *Journal of Environmental Biology* 28, no. 1 (2007): 127.

Türk, R., and V. Wirth. "The pH dependence of SO2 damage to lichens." *Oecologia* 19, no. 4 (1975): 285–291.

Vaieretti, Maria Victoria, Sandra Diaz, Denis Vile, and Eric Garnier. "Two measurement methods of leaf dry matter content produce similar results in a broad range of species." *Annals of Botany* 99, no. 5 (2007): 955–958.

Vile, Denis, Eric Garnier, Bill Shipley, Gérard Laurent, Marie-Laure Navas, Catherine Roumet, Sandra Lavorel et al. "Specific leaf area and dry matter content estimate thickness in laminar leaves." *Annals of Botany* 96, no. 6 (2005): 1129–1136.

Wang, Feijuan, Bin Zeng, Zongxiu Sun, and Cheng Zhu. "Relationship between proline and Hg2+-induced oxidative stress in a tolerant rice mutant." *Archives of Environmental Contamination and Toxicology* 56, no. 4 (2009): 723–731.

Woodward, Andrew J., and Ian J. Bennett. "The effect of salt stress and abscisic acid on proline production, chlorophyll content and growth of in vitro propagated shoots of Eucalyptus camaldulensis." *Plant Cell, Tissue and Organ Culture* 82 (2005): 189–200.

Wright, Ian J., Mark Westoby, and Peter B. Reich. "Convergence towards higher leaf mass per area in dry and nutrient-poor habitats has different consequences for leaf life span." *Journal of Ecology* 90, no. 3 (2002): 534–543.

Wuytack, Tatiana, Kris Verheyen, Karen Wuyts, Fatemeh Kardel, Sandy Adriaenssens, and Roeland Samson. "The potential of biomonitoring of air quality using leaf characteristics of white willow (Salix alba L.)." *Environmental Monitoring and Assessment* 171 (2010): 197–204.

Yadav, Abhinav, Pramit Verma, and Akhilesh Singh Raghubanshi. "An overview of the role of plant functional traits in tropical dry forests." *Research Anthology on Ecosystem Conservation and Preserving Biodiversity* (2022): 92–117.

Yan, Dan, Yalin Lei, Yukun Shi, Qing Zhu, Li Li, and Zhien Zhang. "Evolution of the spatiotemporal pattern of PM2.5 concentrations in China–A case study from the Beijing-Tianjin-Hebei region." *Atmospheric Environment* 183 (2018): 225–233.

Yang, Jun, Joe McBride, Jinxing Zhou, and Zhenyuan Sun. "The urban forest in Beijing and its role in air pollution reduction." *Urban Forestry & Urban Greening* 3, no. 2 (2005): 65–78.

Younis, Uzma, Tasveer Zahra Bokhari, Muhammad Hasnain Raza Shah, Seema Mahmood, and Saeed Ahmad Malik. "Dust interception capacity and alteration of various biometric and biochemical attributes in cultivated population of Ficus carica L." *Journal of Pharmacy and Biological Sciences* 6, no. 4 (2013): 35–42.

Zhang, Weikang, Zhi Zhang, Huan Meng, and Tong Zhang. "How does leaf surface micromorphology of different trees impact their ability to capture particulate matter?" *Forests* 9, no. 11 (2018): 681.

Zhu, Jiyou, and Chengyang Xu. "Intraspecific differences in plant functional traits are related to urban atmospheric particulate matter." *BMC Plant Biology* 21, no. 1 (2021): 1–12.

Zhu, Jiyou, Jingliang Xu, Yujuan Cao, Jing Fu, Benling Li, Guangpeng Sun, Xinna Zhang, and Chengyang Xu. "Leaf reflectance and functional traits as environmental indicators of urban dust deposition." *BMC Plant Biology* 21, no. 1 (2021): 1–13.

# Section 3

*Tropical plants responses to varrying resource availability*

# 8 Ecophysiological Responses of Tropical Plants to Varying Resources Availability

*Wajiha Sarfraz, Mujahid Farid, Noreen Khalid, Allah Ditta, Ujala Ejaz, Zarrin Fatima Rizvi, Nighat Raza, and Shafaqat Ali*

## 8.1 INTRODUCTION

The study of physiological responses to the surroundings in plants is known as plant ecophysiology. It is concerned with explaining the physiological functions that underlie ecological observations (Battaglia et al., 1996). Ecophysiology is concerned with social and environmental issues concerning plant distribution, abundance, and interactions with other organisms, and it is based on causal and deterministic theories (Janzen, 1967; Read, 1990; Cunningham & Read, 2002; Mercado et al., 2018).

Tropical tree resource interaction research has traditionally focused on cultivating plants that are less susceptible to ecological stress, allowing them to resist periods of unfavorable weather or grow in less desirable ecosystems. Tropical plants develop organically in exceptionally infertile, dry, or saline environments with nutrient availability, light, temperature, hormonal interactions, and so on; ecophysiologists were interested in understanding the mechanisms that allow this to happen (Vårhammar et al., 2015). Some research findings show a relationship between growth and nutrients, while others recommend that various factors such as light, elevated $CO_2$ (carbon dioxide) levels, and lower moisture content cause drought and hormonal stress (Bucci et al., 2006). An increasing volume of the latest research explicitly examines the physiological changes of tropical plants to varying factors to answer these and other questions (Lovelock et al., 2006). As a result, agronomists and physiologists have investigated how plants respond to or resist environmental stresses. Plant microenvironment, water relations, and carbon exchange patterns are all measured in phytophysiology (Pasquini & Santiago, 2012; Bhadouria et al., 2016).

Among all forest ecosystems, tropical plants have the highest species diversity, productivity, and carbon sequestration, making them extremely important to global climate change. Climate change shows the impact on tropical trees as they interact with environmental change and feedback to the global environment (Beer et al., 2010; Saatchi et al., 2011). Forests under lower thermal ranges were considered less capable of acclimating to temperature changes than forests in higher thermal regions (Williams et al., 2007).

Furthermore, tropical forests expect to outcompete their climatological margins within the forthcoming decades (Diffenbaugh & Scherer, 2011). There is a significant level of uncertainty regarding the accuracy of large-scale carbon flux estimates for tropical forests (Booth et al., 2012; Mercado et al., 2018). As tropical trees have high biomass and store a substantial amount of carbon, addressing this research gap is essential to reducing global carbon emissions. Forests respond to atmospheric change at a variety of phenomenological scales. Tropical forests have a relatively low thermodynamic threshold for photosynthetic function (Pan et al., 2013; Cavaleri et al., 2015). It is possible for physicochemical changes in leaves, such as changes in $CO_2$ concentration or light intensity, to occur almost simultaneously, while changes in tree community structure caused by climate change over centuries can persist over the long term (Bonan & Levis, 2010; Mora et al., 2013). Currently, we are unsure to what extent nutrients limit the carbon-concentration feedback or whether tropical forests can enhance carbon uptake as $CO_2$ concentrations rise (Anderegg et al., 2015).

Different ecotypes reside in different habitat types, and individuals' morphological and physiological responses to ecological variation have been associated with the ecosystem (Saldaña et al., 2005). The availability of light and nutrients has a significant impact on tropical trees (Futuyma & Moreno, 1988). Several taxa have studied the link respectively phenotypic patterns to light accessibility and the ecological diversity of plant species (Bhadouria et al., 2018; Szabó et al., 2020). As a result, species classified as a limited context exhibit consistent light use, whereas ecologically diverse varieties use light effectively in shade and sunshine (Sultan et al., 1998; González & Gianoli, 2004). Considering nutrient limitations restrict our capability to anticipate tropical plants' responses to globalization, knowing tree nutrition limitations is more meaningful than ever (Phillips et al., 2010). In addition to rainforests as the hub for tropical plant diversity, open vegetation on mountain slopes serves as a biodiversity hotspot (Feeley et al., 2011). Despite significant advances in systemic, phytogeographic, and genetic structures, the ecology of these plants remains unclear (Lousada et al., 2011; Bitencourt & Rapini, 2014).

A drought-induced death of trees has brought tropical forests to the forefront of climate change discussion (Condit et al., 1995; Slik, 2004). Due to very high annual rainfall, the seasonal presence of drought is a determining factor in available resources in tropical areas. Seasonal variations in rainfall often improve plant reproduction and expansion in these forests (Medina & Fernandes, 2007). Tropical vegetation endures significant seasonal variations, with harsh dry winters and drizzly tropical summers (Richards, 1952; Lombardozzi et al., 2015). Water scarcity triggers different plant stress responses during the dry season, which is exacerbated by numerous stresses, including temperature, relative humidity, light, and mineral availability (Carvalho et al., 2012). Moreover, biotic components like pathogens and predators can pose a threat to and possibly harm plants, threatening species diversity (Carvalho et al., 2014; Bhadouria et al., 2017a).

This chapter focuses on the connection between water deficits, carbon balance, and plant survival under the influence of seasonal droughts. The abscisic acid (ABA) hormone proved to be beneficial during seasonal drought. In natural environments, stomatal closure, abiotic and biotic defense activation, and ABA-related regulation of nutrient accumulation in leaves all correspond to plant stressors. The purpose of this chapter is to summarize the known research about tropical tree responses to the resource availability listed in Figure 8.1 and to demonstrate the importance of using such metadata to overview how tropical trees can adapt to a changing environment.

### 8.1.1 Influence of Light Exposure and the Ecophysiological Constraints of Tropical Forest Plants

Tropical forests have experienced tremendous growth in ecophysiological studies owing to the emergence of various agricultural meteorology theories and techniques in previous decades (Lemeur & Blad, 1974; Myneni et al., 1989). Even though many environmental exposure gradients are visible in tropical forested areas, the frequency of irradiation level appears to have the most significant influence on the ecophysiological behavior of forest plants (Saldaña et al., 2005). Tropical plants respond differently to light variation depending on their availability to occupy different light environments (Givnish, 1988; Ellsworth & Reich, 1995; Tripathi et al., 2020). Humid forest understory plants rely heavily on light for survival and growth. Due to this, tropical trees are likely to vary widely in how they capture and use this resource in forest ecosystems (Saldaña & Lusk, 2003). Ultraviolet irradiance affects biological mechanisms based on the bandwidth used (Norman & Welles, 1983; Pearcy, 1990). Solar radiation (300–3000 nm) consists of approximately half the photosynthetically active waveband. UV-A (320–380 nm) and UV-B (280–320 nm) rays could affect or

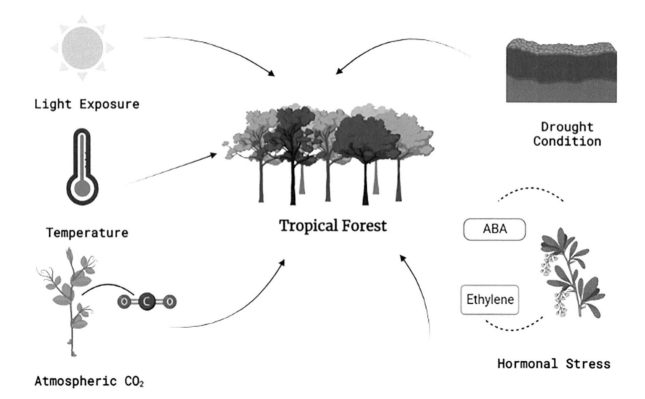

FIGURE 8.1  Resources availability in tropical rainforest

inhibit photosynthesis. The wavelengths of blue (400–500 nm), red (600–700 nm), and far-red (700–800 nm) may have an indirect effect on photosynthesis (Lee & Downum, 1991; Deng et al., 2012). Photosynthetically active irradiance (PAR) has a wavelength range of 400–700 nm. The intensity of PAR at 400–700 nm wavelengths varies considerably at the canopy top in full sunlight (Bhadouria et al., 2017b).

### 8.1.2 Effects of Light Saturation on Photosynthesis

Photo inhibition or photo death of photosynthetic apparatus may result from excessive or insufficient light. There is net $CO_2$ removal attributable to respiration in darkness (zero PAR). Net-$CO_2$ decreases gradually as light frequency rises until it reaches the light compensation point, at which point photosynthesizing $CO_2$ absorption equals respiratory $CO_2$ release. $CO_2$ uptake peaks around the time of light saturation (Vogelmann et al., 1996). Because light saturation approaches progressively, determining the light-saturation point can be difficult. Thus, half saturation of photosynthesis is frequently mentioned in alternative or additional ways. The near-linear component of the curve below exposure can be analyzed to calculate the apparent quantum yield of photosynthesis. When the chlorophyll content per unit area increases, PAR is more easily absorbed. Despite chlorophyll localization in chloroplasts, high internal scattering reduces the transmissivity of the chlorophyll solution by increasing the effective path length (Brown & Bouton, 1993). Even though shade gets higher priority over the sun for light absorption, chlorophyll concentrations per unit area do not differ significantly. Shade plant leaves contain much more chlorophyll per chloroplast than sun plant leaves (Bjorkman, 1981; Bhadouria et al., 2020). Nevertheless, shade leaves have slightly fewer chloroplasts per unit surface area. Chlorophyll b content in shady leaves is relatively high relative to chlorophyll a content in understory plants (Chow et al., 1988). The chloroplasts in these leaves contain extensive grana stacking that contribute to this enrichment (Bjorkman, 1981; Chazdon, 1992).

### 8.1.3 Effect of Plant Properties on Light Absorption in a Tropical Forest

Light absorption is affected by both leaf and plant properties. Light focusing through leaf intercellular spaces may increase irradiance frequencies in the mesophyll of tropical forest understory plants (Bjorkman, 1981). The morphology and physiology of leaves should allow plant species established across a variety of light conditions to produce leaves that will adapt to a broad range of lighting scenarios. The implications of photosynthesis capacity on plant development could be linked to ecological breadth. Plants optimized to shaded areas ought to have lower respirational carbon losses, and a reduction in dark respiration in low-light situations should compensate for the low assimilation rate (Walters & Field, 1987). Plants keep producing thin leaves with more specific leaf areas in low-light environments to maximize light absorption (Sultan et al., 1998). Leaves reflect and transmit very little visible light (400–700 nm), but most far-red light (>700 nm). However, in general, photosynthetic activity elicits the most responses. Solar radiation reaches the tree canopy or a vast clearing depending on atmospheric conditions, season, and time (Anderson et al., 1988). A narrow opening can be caused by a fall of a branch or tree, landslides, hurricane damage, selective logging, or clear-cutting. There can be a significant impact of atmospheric conditions on the infrared spectrum of radiation in open areas. Photosynthesis under the canopy is affected by the spectral composition of solar irradiance. The size of the canopy opening increases with higher irradiance at ground level, both overall and at peak. Openings in the forest canopy have a strong influence on the light surroundings (Fernandez & Fetcher, 1991).

## 8.2 IMPACT OF TEMPERATURE ON ECOPHYSIOLOGICAL RESPONSES OF PLANTS IN TROPICAL ENVIRONMENTS

Tropical rainforests only cover 7% of the earth's land area, but contain roughly half of all species of plants and animals and half of all forest ecosystem resources. Tropical forests produce twice as quickly as their counterparts in cooler regions. Although living longer than trees in other climates, they have an average lifespan of only 186 years. Climates in tropical rainforests range from 21–30°C on average (Diffenbaugh & Scherer, 2011; Corlett, 2011). Climate change may make tropical ecosystems more susceptible than previously thought. In the tropics, life expectancy is expected to decrease even further. Many tropics regions are rapidly warming, with large areas becoming significantly warmer than 25°C on average. Tropical trees cannot currently photosynthesize at temperatures above 38°C, which is the current average. However, the dome maintains high moisture levels, whereas the natural warming of forests would result in the surrounding air and excessive water loss from plants, hastening tree destruction (Locosselli et al., 2020). The various pollutants greenhouse gases and $CO_2$ can be significantly affected by even small changes in tropical forests. Deforestation and global warming together are putting additional strain on tropical forests, which raises concerns about forests' ability to offset $CO_2$ emissions if tropical trees die earlier due to the combustion of fossil fuels. Despite drastic emissions reduction measures, temperatures will continue to rise soon. The tropics are rapidly warming, and large areas will become significantly warmer than 25°C on average (PNAS, 2020). Conditions of drought will exacerbate the effect of temperature on tree longevity. Trees in tropical regions reside for a lesser duration than trees in temperate regions. The areas of tropical forests indicate that trees live shorter lives because of the higher temperatures. Tree longevity decreases in tropical lowlands when forests become drier (Figure 8.2). Furthermore, the average temperature exceeds 25.4°C (Sharkey & Monson, 2014).

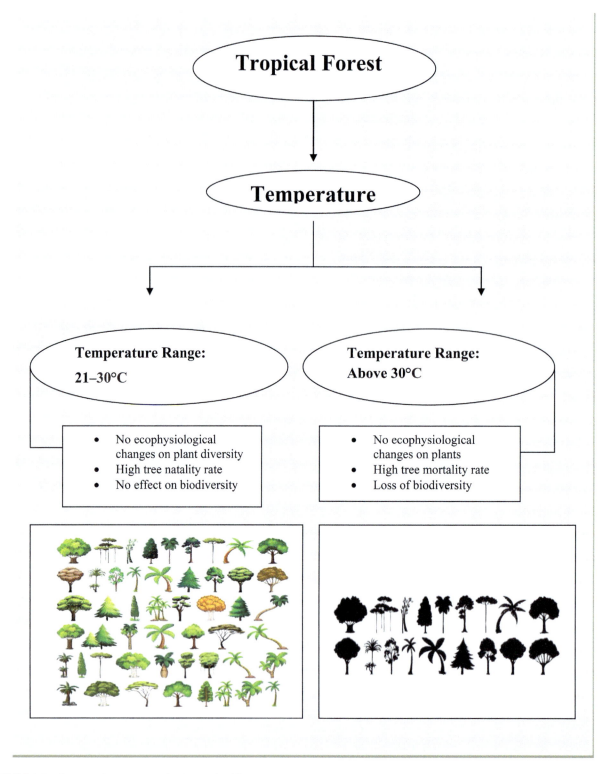

**FIGURE 8.2** Impact of temperature in the tropical forest.

### 8.2.1 Temperature and Tree Mortality in Tropical Land Surfaces

Temperatures on tropical land surfaces are expected to rise by 3–5°C during the 21st century. Tropical climates may experience less warming than higher latitudes, but their relatively limited seasonal temperature fluctuation may cause severe impacts. Furthermore, with higher temperature variability and more frequent extremes, tropical climates are likely to surpass their historical climatic limits much faster than other biomes (Mora et al., 2013). Temperature does have a significant influence on thermal niche occupation along latitudinal and elevation gradients. Increasing temperatures along all vast regions of the tropics are expected to expedite tree mortality, including in the Amazon, Pantanal, and Atlantic forests, affecting wildlife habitat, quality of air, and carbon stocks. Climate change will cause tropical trees to live shorter lives, posing a threat to global biodiversity and carbon stocks (Vogel et al., 2017). The climatological change will have an impact on the Congo Forest in West Africa, the world's largest tropical forest after the Amazon, as well as tropical rainforests in South America. Congo's temperatures are lower than those of the Amazon rainforest, whose temperatures have already reached this critical point. As temperatures rise, trees may begin to die more frequently (IPCC, 2021). Global temperatures are predicted to reach 2.5°C on average by 2050 (Locosselli et al., 2020).

### 8.2.2 Role of Temperature in Phytophysiology and Adaptation to Environmental Stress

Temperature is a dominant attraction in organism distribution due to its direct involvement in DNA, proteins, and/or supramolecular structures (chromosomes) (Counts et al., 2017). The rapid changes in ambient temperature can activate specialized routes of molecular and biological interactions in different cell sections, forming an inclusive reaction by the cell (Ruelland & Zachowski, 2010). Changing temperatures or other environmental elements cause stress levels that affect themetabolism and functioning of a species. An irreversible change can result from a severe and prolonged scenario. Sessile organisms like plants cannot move to avoid environmental stress. Adaptation mechanisms allow plants to survive under adverse conditions (Levitt, 1980; Larcher, 2006). The number of species may also decline, and biodiversity may change. Two types of environmental stress resistance mechanisms exist in plants. Avoidance techniques involve preventing extraneous stressors from eliciting adaptation mechanisms that alter plant function. Tolerance is also developed by activating or changing physiological systems that permit the plant to repair damage or resist stress (Luo, 2011). The ability of a plant to adapt to stress is determined by its tolerance mechanisms and its avoidance adaptations. Plants can be thermo- or cold-tolerant, based on their niche and adaptations. Temperature-sensitive species live in colder environments, while heat-tolerant species live in warmer regions (Raven et al., 2005). Forest trees are sensitive to extreme temperature conditions during the reproductive season. Increasing productivity of different strains is probably resulting in a low genetic variation caused by artificial selection (Chaudhary, 2013; De Pinto et al., 2015).

## 8.3 CARBON DYNAMICS AND ECOPHYSIOLOGICAL IMPACT IN TROPICAL FORESTS

Tropical forests absorb and store more carbon than any other ecosystem, accounting for over a quarter of all global soil carbon and two-thirds of all plant biomass carbon. Variations in temperature, precipitation, and disruption regimes may alter vegetation structure, as well as carbon distribution and storage. Changes in $CO_2$ concentrations or light intensities can have immediate physiological effects on leaves, but modifications in tree communities induced by a slowly changing climate can take years or even centuries (Walther et al., 2002). Climate change did not affect tropical soil C (carbon) before historical warming, but its environment may have influenced its presence. Trees can observe the response to climate change at different species levels. Furthermore, several reactions may be transitory and not reflect long-term ecological adaptations. Many researchers have investigated how climate change affects community composition, species diversity, and ecosystem function in these areas. Forest residues and sizable canopy gaps can significantly alter species composition and carbon cycle dynamics (Dale et al., 2001; Enquist, 2002).

### 8.3.1 Microbe-Centric Approach to Soil Carbon Dynamics

Tropical ecosystems have received very little attention compared to temperate, boreal, and arctic ecosystems. In stressful times, many tropical trees might have evolved to conserve water and store carbon. The soil of tropical forests' carbon contributed significantly to historical warming events (Pan et al., 2011). We are unlikely to be able to predict climate change in the future unless we understand how tropical forests respond to warming. However, empirical data are scarce, so we cannot accurately predict future weather patterns (Jackson et al., 2017). Biodiversity is primarily rooted in biological systems that store and cycle carbon. A microbe-centric approach to soil C dynamics can help us better understand how climate change will affect tropical forest soils (Chambers et al., 2000). Therefore, tropical forest soil microbe communities thrive in environments with high C inputs and metabolic rates, implying that fractional changes caused by warming and have a substantial influence on atmospheric $CO_2$ (Ruwaimana et al., 2020). The residence time of carbon in tissue types differs tremendously in tropical forests. While petioles and fine rhizomes decompose quickly, slow-growing trees in the central Amazon can live for over 1,000 years, and the trunks and branches of dead trees can take years to break down (Pan et al., 2011). Carbon storage will not benefit from productivity increases that raise only incredibly quickly tissues, whereas additional photosynthesizing will affect woody tissue production (Laurance et al., 1997, 1998).

### 8.3.2 Impact of Atmospheric $CO_2$ on Photosynthesis

Photosynthesis provides nutrients to the roots of plants. Plants are sensitive to atmospheric $CO_2$ levels because they produce oxygen through photosynthesis (Amthor, 2000). Photosynthesis directly produces carbohydrates. Changing the soluble sugar content of a plant is necessary to understand how it responds to different $CO_2$ concentrations. Several experiments have noted that $CO_2$ enrichment does not affect leaves' soluble sugar concentration (Tjoelker et al., 2001). $CO_2$ exposure has varying effects on plants, regardless of how long it lasts. In $C_3$ plants exposed to large $CO_2$ concentrations for an extended period, the net $CO_2$ assimilation rate does not always increase. Once contrasted to plants grown at ambient $CO_2$ concentrations, it may remain unchanged or even decrease in some cases (Jahnke & Krewitt, 2002; Cavaleri et al., 2015). Tropical plants often experience declines in photosynthetic carbon assimilation rates at midday and afterward due to moisture stress. Adding more photosynthate to woody tissues produces relatively large carbon storage, whereas maximizing productivity favors fast-cycling tissues that do not (Williams et al., 1998; Ter Steege et al., 2013).

### 8.3.3 Effect of $CO_2$ on Plants' Biochemical and Morphological Responses

$CO_2$ seems to have a minimal immediate impact on plant respiration, and any previous effects were merely experimental artifacts due to leaks. Root dry weight (DW), root-to- shoot ratio, root length, and root number were all primary root recognition variables to rising $CO_2$ levels. A root's relative DW increases the most when $CO_2$ is high. Understanding the effects of root systems on the entire plant is crucial (Winter et al., 2001). Besides storing carbon, roots also function as carbon sinks. Other aspects of plant roots were also discussed, including structure, function, and rhizosphere. An understanding of $CO_2$-enriched environments requires integrating above-ground and below-ground processes (Amthor et al., 2001). Leaf size and number change as $CO_2$ levels rise, potentially altering ecosystem leaf area index (LAI), and biochemical and morphological responses frequently interact (Norby et al., 1992). When LAI increases as stomatal conductance decreases, the total forest transpiration rate may not be affected, but tropical forests may have limited space for additional leaf area. Stomatal conductance was lower in leaves cultivated in $CO_2$-enriched air in most measurements (Bruhn et al., 2002; Hobbie & Hobbie, 2013; Hashimoto et al., 2015).

## 8.4 CHANGES IN NUTRIENT AVAILABILITY AND THE EFFECTS ON SPECIES SURVIVAL AND PHYTOCHEMISTRY IN TROPICAL RAINFORESTS

Tropical rainforests have the maximum plant species diversity of any habitat. The nutrients available across many tropical forests have changed over time (Winter et al., 2001). Elevated $CO_2$ levels influence plant responses to nutrient availability. A $CO_2$ soil management effect will not be detectable over the geological timescale meaningful to our ability to recognize it (Tinker & Nye, 2000). Phosphate and nitrogen accessibility pulses are frequently linked with rapid fluctuations in moisture suitability, resulting in soil and litter lysis. Microbial death reduces microbes' competition and relieves nutrients. The kind of modification in moisture abundance that causes microbiota death varies between forests (Vitousek & Denslow, 1986). A dramatic increase in moisture content at the start of the wet period kills microbes in monsoon and other seasonally dry forests. The availability of nutrients may also change over the year without being related to rapid changes in moisture (Lodge et al., 1994). Secondary minerals can form during phosphorus dilation with iron (Fe) and aluminum (Al) oxides and hydroxides, limiting plant growth. Because of the abundance of Fe and Al minerals in highly weathered tropical soils, these mechanisms consider being dominant ways of removing phosphorus (P) from a rapidly cycling labile pool (Vitousek, 1982, 1984).

### 8.4.1 Nutrient Availability and Translocation

These processes are more likely to influence nutritional communication in tropical forests than immediate variations in nutrient uptake. There is less N and P availability in tropical rainforests than in lowlands, and nutrient translocation is an effective mechanism for nutrient conservation. The translocation of P is generally inversely related to foliar concentration, with translocation increasing with elevation (Veneklaas, 1991). Low N availability limits net primary production (NPP) and inhibits $CO_2$ fertilizer application response in temperate ecosystems (Finzi et al., 2002). As evidenced by their high NPP and processes designed to improve P acquisition, forests may alter their native vegetation to take advantage of the abundance of resources available to them. Previous research does not support the idea that tropical forests can capture and uptake P in a relatively short period and sustain this growth. Lateritic soils in tropical forests have a high concentration of N, and P is usually the primary limiting element for NPP. The principal source of P is parent material weathering, whereas for N it is biological processes (McGill & Cole, 1981). Considering the circulation of N and P within the rainforest tree stand, and the relationship between it and sclerophylly, provides insight into N and P's limitation in rainforest tree leaves. Nitrogen is scarce in mountainous rainforests and nutrients in litterfall are limited, causing a high organic matter content per unit of nutrients absorbed (Uehara & Gillman, 1981).

### 8.4.2 Seasonal Variation in Nutrient Availability

Plant phenologies can be determined based on consistent alterations in nutrient levels. Plants in the wild can store and upcycle nutrients effectively, but this diminishes the impact of variations in nutrient availability over time (Tanner et al., 1990, 1992). The continual leakage of mineral and organic forms may not compensate for the loss of soil P due to rock weathering, atmospheric inputs, or other deposition processes over time (Chapin, 1980). Based on this seasonal variation in abundance, nutrients whose concentrations remain high until they are released from the decayed litter can be classified. In La Selva, Costa Rica, the availability of phosphate and ammonium was not seasonal. The seasonal patterns of decomposable nutrients were different on Barro Colorado Island. While potassium and sulfate levels did not change seasonally, phosphorus and ammonium levels did (Cornejo et al., 1994). N reabsorption should increase as elevation increases if soil N accessibility decreases with elevation in all mountain ranges (Marrs et al., 1988).

## 8.5 DROUGHT EFFECTS ON TROPICAL RAINFOREST BIOTA

Tropical regions have high annual rainfall and evapotranspiration rates. The precipitation and solar radiation in tropical areas vary significantly by season. Intertropical conversion zone latitudinal changes may result in atmospheric variations that cause soil drought conditions. A prolonged period of scant rainfall occurs during an environmental drought (Li et al., 2006). Every year, the majority of tropical forests experience dry periods. Interannual variation makes it difficult to detect long-term trends when tropical droughts are involved with inter-climatic cycles. Natural and artificial droughts have both led to tropical tree mortality.

This chapter defines drought as a long period of abnormally dry weather that causes a serious hydrological imbalance. Tropical rainforest research currently emphasizes the following three types of drought.

1. **Seasonal drought:** Rainfall variations cause seasonal droughts at a given location. This study examines long-term precipitation regimes. Climate change may have facilitated the evolution of seasonal droughts in tropical rainforest environments, culminating in species adapting to them. However, these seasonal changes affect the functioning of the ecosystem (Marengo et al., 2011).
2. **Extreme drought:** Rainfall is usually low during extreme droughts. Several extreme drought circumstances have struck tropical areas of the world in recent decades. The intensity, duration, location, and number of epicenters vary considerably. They continue to influence the governing of tropical rainforests and could even adversely impact them. Then, using current information about their implications to highlight the constraints that these ecosystems can adapt (Lewis et al., 2011).
3. **Experimental drought:** Experiments with drought comply with experiments with throughfall marginalization. Excluding a zone from rainfall allows researchers to observe the effects of rain on soil, trees, and ecosystems. A drawback of this approach is the relatively small size of the area under consideration (usually less than 1 hectare).

### 8.5.1 The Impact of Rainforest Loss on Climate Change and Sustainability

Rainforests influence local and global climates in a variety of ways. Winds near the ocean transport moisture from the tropical Atlantic Ocean into the Amazon. When it rains, the tropical forest quickly evaporates the moisture content and releases it into the atmosphere (Marengo et al., 2018). During the rainy season, water vapor will return to the rainforest, while the rest will move to nearby areas. Water evaporated from the rainforest accounts for 30–70% of the rainfall in the Amazon basin. Temperatures and rainfall in the rainforest can impact the climate of the entire continent (Marengo et al., 2011). The probability of a significant global climate emergency rises with the loss of rainforests. Researchers are concerned that the continued loss of the Amazon rainforest will remove a valuable $CO_2$ absorber from the atmosphere. It would also be dangerous for the Amazon region's natural rainfall cycle to be disrupted by rainforest loss, jeopardizing the forest's sustainability (Aragão et al., 2018). The decrease in rainwater may cause a drier, warmer climate, increasing the risk of fires and erosion in that zone.

### 8.5.2 Drought-Induced Effects in Tropical Forests

Tropical tree leaves have evolved to withstand seasonal water stress, and leaf removal is a predominant drought response. During abnormally prolonged dry periods, stems, roots, and leaves can also store water that can buffer daily fluctuations in the water supply. Deep-rooted trees can help to prevent or delay drought by providing access to groundwater. By concentrating their roots in the upper soil layers, rainforest trees keep their leaf water potential high during times of drought (Bonal et al., 2015). Species that are shade-tolerant and use more conservative resources in forests are less vulnerable to forest fires than those that come from wetter environments. Droughts could decrease tree efficiency by minimizing canopy cover, nutrient cycling, and the amount of sunlight that reaches land floors. Drought effects will vary by forest type, with perennially wet forests bearing the brunt of the damage. The level of nutrients in the soil and the environment can affect the biota and the services provided by forests. In a changing climate, droughts could cause tree mortality, causing drier-adapted species to replace wetter-adapted species (Lawrence & Vandecar, 2015).

### 8.5.3 Effects of Drought on Large Trees and Their Ecological and Biological Status

Large trees are particularly vulnerable to droughts since they have exposed canopies and tall trunks, which may lead to reduced growth and mortality. Understory trees are less susceptible to drought. The effects of drought and fire are additive for small, thin-barked trees (Grogan & Schulze, 2012). Large trees may be more vulnerable because they must move water from exfoliating soil to unprotected leaves at distances of up to 50 meters. A localized drought may cause small branches and leaves to drop from the upper canopy, with satellite microwave observations showing this damage can last for several years (Bennett et al., 2015). Understory and sub-canopy trees are less likely to be drought-deciduous than emergent tree species. Watering canopy trees is a challenging process. It can lead to dangerously high temperatures during hot, dry weather by reducing evaporation and mass transfer. Fruiting and flowering can be impacted during droughts, affecting wildlife (Rowland et al., 2015). Wildfires and droughts can destroy forests and spread fire-adapted nonnative plants. Furthermore, when stomata close, even drought-adapted species lose water (Brando et al., 2014).

### 8.5.4 Tree Growth during Extreme Droughts and Their Implications

Severe droughts have the same consequence on tropical plants' carbon storage as years with relatively low moisture in the soil deficiencies. Phillips et al. (2009) report halting long-term biomass increases for an increasing number of plots during extreme drought circumstances in Amazonia in 2005. Indirect observations do not support the claim that water stress causes topographic variation in tree growth response (Silva et al., 2013). Reduced transpiration indicates a reduction in transpiration regulation (reduced soil-root conductivity and water transport). Tropical rainforests experience significant seasonal changes in tree growth. Tree growth in inventory plots typically decreases during dry spells but increases during rainy seasons (Wagner et al., 2012, 2014). Climate change relies on the exchange of water and energy between rainforests and the atmosphere. Droughts of various severity disrupt the rainfall patterns in these ecosystems. During seasonal droughts, tropical rainforest ecosystems and trees lose evapotranspiration (Gloor et al., 2013).

Drought has both short-term and long-term effects on tropical forests, including the following.

1. **Drought tolerant:** The majority of tree species and animals are drought tolerant. The smaller, thicker leaves of tree species are an adaptive response that reduces water loss during dry periods (Burrowes et al., 2004).
2. **Changing species communities:** Drought can affect the reproduction rates of many plants, and other disturbances, like wildfires or insect outbreaks, can hasten death. It could cause long-term variations in species diversity (Barker & Rios-Franceschi, 2014).
3. **Species extinction:** Long-term droughts could lead to the loss of bird species dependent on seasonal rains for insects and fruits, like those found in the Guánica State Forest in southwest Puerto Rico. Several species are endemic and endangered in Puerto Rico's rainy cloud forests, which are home to the most vulnerable communities (Faaborg et al., 1984).

## 8.6 HORMONAL STRESS RESPONSES OF PLANTS

Besides rainforests, mountain tops are a significant hotspot of neotropical plant diversity. The ecophysiology of endemic plants is little documented despite extensive research on their systemic nature, geography, and genetic structure (Noctor et al., 2014; Bitencourt & Rapini, 2014). Plant hormones play an essential role in response to stress in the face of climate change by regulating various signaling mechanisms. Plants experience stress due to the following two main hormones.

1. **Abscisic acid (ABA):** Plants require ABA for growth, development, and defense responses. Plants depend on abscisic acid-mediated adaptive responses to survive drought, salinity, or cold stress. The reduction in water stress may be attributed to ABA-induced stoma closure during the dry season (Zhu, 2002). Stomatal closure by ABA is an efficient way for plant species to constrain water content and sustain in the absence of soil moisture. Additionally to closing pores and reducing water loss through transpiration, ABA is believed to activate antioxidant defenses (Zeevaart & Creelman, 1988).
2. **Ethylene:** Plants communicate through this gaseous substance. Ethylene is considered essential in the production of phytohormones. Under different abiotic and biotic climatic stresses, ethylene regulates germination, maturation, leaf growth, and seed senescence (Dubois et al., 2018). A signaling pathway exists between ethylene and plant growth. Abiotic stresses such as salinity, waterlogging, high temperatures, frost, contact with heavy metals, nutrient deficiencies, and drought affect ethylene synthesis (Klay et al., 2018). Numerous physiological and environmental stresses activate plant ethylene reaction factors (ERFs). Despite extensive research into abiotic stress, no specific signaling pathways have been identified (Deligios et al., 2019).

When confronted with challenging situations like drought, high light intensity, high temperatures, salinity, or micronutrient deficits, the plant produces more reactive oxygen species (ROS). Excess excitation energy (the difference between light absorption and light utilization) caused by soil moisture deficiency can boost the production of singlet oxygen and superoxide anion radical (Mittler, 2002; Lousada et al., 2011). Antioxidant defense mechanisms are enzyme-dependent and non-enzymatic in protecting plants from environmental stress and oxidative damage (Peñuelas et al., 2004; Bartels & Sunkar, 2005). Several research projects examined water status, stomatal closure, and antioxidant protection in temperate climate plants. Studies of mineral deficiencies and antioxidant mechanisms have revealed little about their significance (Munné-Bosch & Alegre, 2003; Zang et al., 2014). Macronutrients serve both structural and functional roles in plants, forming complex integrated systems that control their function (Tewari et al., 2006). Due to low nutrient availability, elements of the photosynthesis electron transport chain are reduced excessively. Ultimately, it causes slower photosynthesis and increased

ROS levels (Suzuki et al., 2012; Munné-Bosch et al., 2013). Therefore, macronutrient deficiencies can cause oxidative stress. Such disorders can cause a wide range of antioxidative responses and their severity (Mooney et al., 1991). A variety of ROS—depending on exposure type, timing, and severity—are most common. Singlet oxygen, hydrogen peroxide, superoxide anion radicals, and hydroxyl radicals are among them (Apel & Hirt, 2004).

## 8.7 CONCLUSION

We distinguish the crucial factors that contribute to a precise picture of tropical plant ecophysiological properties, such as the following.

1. Shade and sun plants respond differently to light exposure. Both leaf and plant properties affect light absorption.
2. Climate change may make tropical plants more vulnerable than previously assumed. There is a high probability that tropical regions will see their life expectancy decrease even further. Increasing temperatures can lead to trees dying more frequently. Tropical forests, however, are markedly influenced by latitudinal/elevation gradients of temperature between 20°C and 30°C.
3. The tropical forest absorbs and saves the most carbon, accounting for more than a quarter of all soil carbon and two-thirds of all plant biomass carbon. A slowly shifting climate can alter tree communities for centuries, though adjustments in $CO_2$ concentrations and light intensities can impact leaves. A microbe-focused approach to soil C dynamics can help to understand how climate change affects the soil of tropical forests. Furthermore, increasing $CO_2$ levels often interact with biochemical and morphological responses.
4. Tropical rainforests have the maximum diversity of plant species in any habitat. Several tropical forests' nutrient availability has changed over time. Rather than directly influencing nutrient intake, different processes regulate nutrient interactions in tropical forests. Biota and forest performance are affected by soil nutrient levels.
5. Annual precipitation and evapotranspiration are high in tropical regions. Most tropical forests experience drought every year. The decrease in rainfall may result in a drier, warmer climate, increasing the risk of fires and erosion in that area. Droughts can reduce canopy cover, the nutrient cycle, and sunlight from reaching the forest floor, reducing tree productivity. In perennially wet forests, the effects of the drought will be greater than in other types of forests.
6. Plant hormones can also cause stress responses in response to the changing climate. Drought, high light intensity, elevated temperatures, salinity, and mineral deficits cause plants to produce more reactive oxygen species (ROS). ROS shows a drastic impact on plant ecophysiological characteristics, as well.

The wide variety of resources available on the ecophysiology of tropical trees responds to distinguishing between their unique natural histories and underlying mechanisms. A diverse range of resources provided several important insights into the ecophysiological traits of tropical trees, which drive ecosystem responses to environmental change.

## REFERENCES

Amthor JS (2000) Direct effect of elevated $CO_2$ on nocturnal in situ leaf respiration in nine temperate deciduous tree species is small. *Tree Physiol.* 20: 139–144.

Amthor JS, Koch GW, Willms JR, Layzell DB (2001) Leaf $O_2$ uptake in the dark is independent of coincident CO2 partial pressure. *J. Exp. Bot.* 52: 2235–2238.

Anderegg WRL, Schwalm C, Biondi F, Camarero JJ, Koch G, Litvak M, Ogle K, Shaw JD, Shevliakova E, Williams AP, Wolf A, Ziaco E, Pacala S (2015) Pervasive drought legacies in forest ecosystems and their implications for carbon cycle models. *Science* 349: 528–532.

Anderson JM, Chow WS, Goodchild DJ (1988) Thylakoid membrane organization in sun/shade acclimation. *Aust J Plant Physiol* 15: 11–26.

Apel K, Hirt H (2004) Reactive oxygen species: Metabolism, oxidative stress, and signal transduction. *Annu Rev Plant Biol.* 55: 373–399.

Aragão LEOC, Anderson LO, Fonseca MG, Rosan TM, Vedovato LB, Wagner FH, Silva CVJ, Junior CHLS, Arai E, Aguiar AP, Barlow J, Berenguer E, Deeter MN, Domingues LG, Gatti L, Gloor M, Malhi Y, Marengo JA, Miller JB, Phillips OL, Saatchi S (2018) 21st Century drought-related fires counteract the decline of Amazon deforestation carbon emissions. *Nat Commun.* 9: 536. doi: 10.1038/s41467-017-02771-y

Barker BS, Rios-Franceschi A (2014) Population declines of mountain coqui (Eleutherodactylus portoricensis) in the Cordillera of central Puerto Rico. *Herpetol Conserv Biol* 9(3): 578–589.

Bartels D, Sunkar R (2005) Drought and salt tolerance in plants. *Crit Rev Plant Sci.* 24: 23–58.

Battaglia M, Beadle C, Loughhead S (1996) Photosynthetic temperature responses of Eucalyptus globus and Eucalyptus nitens. *Tree Physiol* 16: 81–89.

Beer C, Reichstein M, Tomelleri E, Ciais P, Jung M, Carvalhais N, Rodenbeck C, Arain MA, Baldocchi D, Bonan GB, Bondeau A, Cescatti A, Lasslop G, Lindroth A, Lomas M, Luyssaert S, Margolis H, Oleson KW, Roupsard O, Veenendaal E, Viovy N, Williams C, Woodward FI, Papale D (2010) Terrestrial gross carbon dioxide uptake: Global distribution and covariation with climate. *Science* 329: 834–838.

Bennett AC, McDowell NG, Allen CD, Anderson-Teixeria KJ (2015) Larger trees suffer most during drought in forests worldwide. *Na. Plants.* 1: 15139.

Bhadouria R, Singh R, Srivastava P, Raghubanshi AS (2016) Understanding the ecology of tree-seedling growth in dry tropical environment: A management perspective. *Energy Ecol Environ* 1(5): 296–309.

Bhadouria R, Singh R, Srivastava P, Tripathi S, Raghubanshi AS (2017b) Interactive effect of water and nutrient on survival and growth of tree seedlings of four dry tropical tree species under grass competition. *Trop Ecol* 58(3): 611–621.

Bhadouria R, Srivastava P, Singh R, Tripathi S, Singh H, Raghubanshi AS (2017a) Tree seedling establishment in dry tropics: An urgent need of interaction studies. *Environ Syst Decis* 37(11): 88–100.

Bhadouria R, Srivastava P, Singh R, Tripathi S, Verma P, Raghubanshi AS (2020) Effects of grass competition on tree seedlings growth under different light and nutrient availability conditions in tropical dry forests in India. *Ecol Res* 35(5): 807–818.

Bhadouria R, Srivastava P, Singh S, Singh R, Raghubanshi AS, Singh JS (2018) Effects of Effects of light, nutrient and grass competition on growth of seedlings of four tropical tree species. *Ind Forest* 144(1): 54–65.

Bitencourt C, Rapini A (2014) Centres of endemism in the Espinhaço Range: Identifying cradles and museums of Asclepiadoideae (Apocynaceae). *Syst Biodivers*. 11: 525–536.

Bjorkman O (1981) Responses to different quantum flux densities. In OL. Lange, PS. Nobel, CB. Osmond, H. Ziegler (eds.), *Encyclopedia of plant physiology*, N.S., Vol. 12A. Springer-Verlag, Berlin, pp. 57–107.

Bonal D, Burban B, Stahl C, Wagner F, Hérault B (2015) The response of tropical rainforests to drought-lessons from recent research and future prospects. *Ann For Sci*. 73: 27–44.

Bonan GB, Levis S (2010) Quantifying carbon-nitrogen feedbacks in the Community Land Model (CLM4). *Geophys Res Lett*. 37: L07401.

Booth BBB, Jones CD, Collins M, Totterdell IJ, Cox PM, Sitch S, Huntingford C, Betts RA, Harris GR, Lloyd J (2012) High sensitivity of future global warming to land carbon cycle processes. *Environ Res Lett* 7: 024002.

Brando PM, Balch JK, Nepstad DC, Soares-Filho BS (2014) Abrupt increases in Amazonian tree mortality due to drought-fire interactions. *Proc Natl Acad Sci U.S.A*. 111: 6347–6352.

Brown R, Bouton JH (1993) Physiology and genetics of interspecific hybrids between photosynthetic types. *Annu. Rev. Plant Physiol. Plant Mol Biol* 44: 435–456.

Bruhn D, Mikkelsen TN, Atkin OK (2002) Does the direct effect of atmospheric $CO_2$ concentration on leaf respiration vary with temperature? Responses in two species of Plantago that differ in relative growth rate. *Physiol. Plantarum*. 114: 57–64.

Bucci SJ, Scholz FG, Goldstein G, Meinzer FC, Franco AC, Campanello PI, Villalobos-Vega R, Bustamante M, Miralles Wilhelm F (2006) Nutrient availability constrains the hydraulic architecture and water relations of savannah trees. *Plant Cell Environ*. 29: 2153–2167.

Burrowes PA, Joglar RL, Green DE (2004) Potential causes for amphibian declines in Puerto Rico. *Herpetologica*. 60(2): 141–154.

Carvalho F, de Souza FA, Carrenho R, Moreira FMDS, Jesus EDC, Fernandes GW (2012) The mosaic of habitats in the high-altitude Brazilian rupestrian fields is a hotspot for arbuscularmycorrhizal fungi. *Appl Soil Ecol*. 52: 9–19.

Carvalho F, Godoy EL, Lisboa FJG, Moreira FMDS, de Souza FA, Berbara RLL, Fernandes GW (2014) Relationship between physical and chemical soil attributes and plant species diversity in tropical mountain ecosystems from Brazil. *J Mt Sci*. 11: 875–883.

Cavaleri MA, Reed SC, Smith WK, Wood TE (2015) Urgent need for warming 8 experiments in tropical forests. *Global Change Biol* 21: 2111–2121.

Chambers JQ, Higuchi, N, Ferreira, LV, Melack JM, Schimel JP (2000) Decomposition and carbon cycling of dead trees in tropical forests of the Central Amazon. *Oecologia* 122: 380–388.

Chapin FSI (1980) The mineral nutrition of wild plants. *A Rev Ecol Syst* 21: 423–447.

Chaudhary B (2013) Plant domestication and resistance to herbivory. *Int J Plant Genomics*. 1–14. doi: 10.1155/2013/572784.

Chazdon RL (1992) Photosynthetic plasticity of two rainforest shrubs across natural gap transects. *Oecologia* 92: 586–595.

Chow WS, Liping Q, Goodchild DJ, Anderson JM (1988) Photosynthetic acclimation of Alocasia macrorrhiza (L.) G. Don to growth irradiance: Structure, function and composition of chloroplasts. *Aust J Plant Physiol* 15: 107–122.

Condit R, Hubbell SP, Foster RB (1995) Mortality rates of 205 Neotropical tree and shrub species and the impact of a severe drought. *Ecol Monogr* 65: 419–439.

Corlett RT (2011) Impacts of warming on tropical lowland rainforests. *Trends Ecol Evol* 27: 145–150.

Cornejo FH, Verala A, Wright SJ (1994) Tropical forest litter decomposition under seasonal drought: Nutrient release, fungi and bacteria. *Oikos* 70: 183–190.

Counts JA, Zeldes BM, Lee LL, Straub CT, Adams MWW, Kelly RM (2017) Physiological, metabolic and biotechnological features of extremely thermophilic microorganisms. *Wiley Interdiscip Rev Syst Biol Med* 9(3): e1377. doi: 10.1002/wsbm.1377.

Cunningham S, Read J (2002) Comparison of temperate and tropical rainforest tree species: Photosynthetic responses to growth temperature. *Oecologia* 133: 112–119.

Dale VH. Joyce LA, Mcnulty SG, Neilson RP, Ayres MP, Flannigan M, Hanson PJ, Irland LC, Lugo AE, Peterson CJ, Simberloff D, Swanson FJ, Stocks BJ, Wotton M (2001) Climate change and forest disturbances. *BioScience* 51: 723–734.

De Pinto MC, Locato V, Paradiso A, De Gara L (2015) Role of redox homeostasis in thermo-tolerance under a climate change scenario. *Ann Bot*. 116(4): 487–496. doi: 10.1093/aob/mcv071.

Deligios PA, Rapposelli E, Mameli MG, Baghino L, Mallica GM, Ledda L (2019) Effects of physical, mechanical and hormonal treatments of seed-tubers on bud dormancy and plant productivity. *Agronomy*. 10: 33. doi: 10.3390/agronomy10010033

Deng Y, Shao Q, Li C, Ye X, Tang R (2012) Differential responses of double petal and multi petal jasmine to shading: II. morphology, anatomy and physiology. *Sci Hortic*. 144: 19–28. doi: 10.1016/j.scienta.2012.06.031

Diffenbaugh NS, Scherer M (2011) Observational and model evidence of global emergence of permanent, unprecedented heat in the 20th and 21st centuries. *Clim Change* 107: 615–624.

Dubois M, Van den Broeck L, Inzé D (2018) The pivotal role of ethylene in plant growth. *Trends Plant Sci*. 23: 313–323.

Ellsworth DS, Reich PB (1995) Photosynthesis and leaf nitrogen in five Amazonian tree species during early secondary succession. *Ecology*, 77(2): 581–594.

Enquist CAF (2002) Predicted regional impacts of climate change on the geographical distribution and diversity of tropical forests in Costa Rica. *J Biogeogr*. 29: 519–534.

Faaborg J, Aendt WJ, Kaiser MS (1984) Rainfall correlates of bird population fluctuations in a Puerto Rican dry forest—A nine year study. *Wilson Bullet.* 96(4): 575–593.

Feeley KJ, Davies SJ, Perez R, Hubbell SP, Foster RB (2011) Directional changes in the species composition of a tropical forest. *Ecology* 92: 871–882.

Fernandez D, Fetcher N (1991) Changes in light availability following hurricane Hugo in a subtropical montane forest in Puerto Rico. *Biotropica.* 23: 393–399.

Finzi AC, DeLucia EH, Hamilton JG, Richter DD, Schlesinger WH (2002) The nitrogen budget of a pine forest under free air $CO_2$ enrichment. *Oecologia* 132: 567–578.

Futuyma DJ, Moreno G (1988) The evolution of ecological specialization. *Annu Rev Ecol Syst.* 19: 207–233.

Givnish T (1988) Adaptation to sun and shade: A whole plant perspective. *Aust J Plant Physiol.* 15: 63–92.

Gloor M, Brienen RJW, Galbraith D, Feldpausch TR, Schöngart J, Guyot JL, Espinoza JC, Lloyd J, Phillips OL (2013) Intensification of the Amazon hydrological cycle over the last two decades. *Geophys Res Lett.* 40: 1729–1733.

González AV, Gianoli E (2004) Morphological plasticity in response to shading in three Convolvulus species of different ecological breadth. *Acta Oecol.* 26: 185–190.

Grogan J, Schulze M (2012) The impact of annual and seasonal rainfall patterns on growth and phenology of emergent tree species in Southeastern Amazonia, Brazil. *Biotrop.* 44: 331–340.

Hashimoto S, Carvalhais N, Ito A, Migliavacca M, Nishina K, Reichstein M (2015) Global spatiotemporal distribution of soil respiration modeled using a global database. *Biogeosciences* 12: 4121–4132.

Hobbie JE, Hobbie EA (2013) Microbes in nature are limited by carbon and energy: The starving-survival lifestyle in soil and consequences for estimating microbial rates. *Front Microbiol* 4.

IPCC (2021) Climate change 2021: The physical science basis. In V. Masson-Delmotte, P. Zhai, A. Pirani, SL. Connors, C. Péan, S. Berger, N. Caud, Y. Chen, L. Goldfarb, MI. Gomis, M. Huang, K. Leitzell, E. Lonnoy, JBR. Matthews, TK. Maycock, T. Waterfield, O. Yelekçi, R. Yu, B. Zhou (eds.), *Contribution of Working Group I to the Sixth Assessment Report of the Intergovernmental Panel on Climate Change.* Cambridge University Press, Cambridge, UK and New York, NY, In press. doi: 10.1017/9781009157896

Jackson RB, Lajtha K, Crow SE, Hugelius G, Kramer MG, Pineiro G (2017) The ecology of soil carbon: Pools, vulnerabilities, and biotic and abiotic controls. *Ann Rev Ecol Evol Systematics* 48: 419–445.

Jahnke S, Krewitt M (2002) Atmospheric CO2 concentration may directly affect leaf respiration measurement in tobacco, but not respiration itself. *Pl. Cell Environ.* 25: 641–651.

Janzen DH (1967) Why mountain passes are higher in the tropics. *The American Naturalist* 101: 233–249.

Klay I, Gouia S, Liu M, Mila I, Khoudi H, Bernadac A, Bouzayen M, Pirrello J (2018) Ethylene Response Factors (ERF) are differentially regulated by different abiotic stress types in tomato plants. *Plant Sci.* 274: 137–145.

Larcher W (2006) *Ecofisiologia Vegetal.* RiMa, São Carlos, pp. 525–550.

Laurance WF, Ferreira LV, Rankin-de Merona JM, Laurance SG (1998) Rain forest fragmentation and the dynamics of Amazonian tree communities. *Ecology* 6: 2032–2040.

Laurance WF, Laurance SG, Ferreira LV, Rankin-de Merona JM, Gascon C, Lovejoy TE (1997) Biomass collapse in Amazonian forest fragments. *Science* 278: 1117–1118.

Lawrence D, Vandecar K (2015) Effects of tropical deforestation on climate and agriculture. *Nat Clim Change.* 5: 27–36. doi: 10.1038/nclimate2430

Lee DW, Downum KR (1991) The spectral distribution of biologically active solar radiation at Miami, Florida, USA. *Int J Biometeorol* 35: 48–54.

Lemeur DR, Blad BJ (1974) A critical review of light models for estimating the shortwave radiation regime of plant canopies. *Agric Meteorol* 14: 255–286.

Levitt J (1980) *Responses of Plants to Environmental Stress, Volume 1: Chilling, Freezing, and High Temperature Stresses.* Academic Press, New York, London, 1980.

Lewis SL, Brando PM, Phillips OL, Van der Heijden G, Nepstad DC (2011) The 2010 Amazon drought. *Science* 331: 554.

Li W, Fu R, Dickinson RE (2006) Rainfall and its seasonality over the Amazon in the 21st century as assessed by the coupled models for the IPCC AR4. *J Geophys Res.* 111: D02111.

Locosselli GM, Brienen RJ, de Souza Leite M, Gloor M, Krottenthaler S, de Oliveira AA, Buckeridge M (2020) Global tree-ring analysis reveals rapid decrease in tropical tree longevity with temperature. *Proc Nat Acad Sci.* doi: 10.1073/pnas.2003873117

Lodge DJ, McDowell WH, McSwiney CP (1994) The importance of nutrient pulses in tropical forests. *Trends Ecol Evol* 9: 384–387.

Lombardozzi DL, Bonan GB, Smith NG, Dukes JS, Fisher RA (2015) Temperature acclimation of photosynthesis and respiration: A key uncertainty in the carbon cycle climate feedback. *Geophys Res Lett* 42: 8624–8631.

Lousada JM, Borba EL, Ribeiro KT, Ribeiro LC, Lovato MB (2011) Genetic structure and variability of the endemic and vulnerable *Vellozia gigantea* (Velloziaceae) associated with the landscape in the Espinhaço Range, in southeastern Brazil: Implications for conservation. *Genetica* 139: 431–440.

Lovelock CE, Ball MC, Choat B, Engelbrecht BMJ, Holbrook NM, Feller IC (2006) Linking physiological processes with mangrove forest structure: Phosphorus deficiency limits canopy development, hydraulic conductivity and photosynthetic carbon gain in dwarf Rhizophora mangle. *Plant Cell Environ.* 29: 793–802.

Luo Q (2011) Temperature thresholds and crop production: A review. *Clim Change.* 109(3–4): 583–598. doi: 10.1007/s10584-011-0028-6.

Marengo JA, Souza CJr, Thonicke K, Burton C, Halladay K, Betts R, Alves LM, Soares WR (2018) Changes in climate and land use over the Amazon region: Current and future variability and trends. *Front Earth Sci.* 6: 228. doi: 10.3389/feart.2018.00228

Marengo JA, Tomasella J, Alves LM, Soares WR, Rodriguez DA (2011) The drought of 2010 in the context of historical droughts in the Amazon region. *Geophys Res Lett.* 38: L12703.

Marrs RH, Proctor J, Heaney A, Mouthford MD (1988) Changes in soil nitrogen-mineralization and nitrification along an altitudinal transect in tropical rainforest in Costa Rica. *J Ecol* 76: 466–482.

McGill WB, Cole CV (1981) Comparative aspects of cycling of organic C, N, S and P through soil organic matter. *Geoderma* 26: 267–286.

Medina BMO, Fernandes GW (2007) The potential of natural regeneration of rocky outcrop vegetation on rupestrian field soils in "Serra do Cipó", Brazil. *Rev Bras Bot.* 30: 665–678.

Mercado LM, Medlyn BE, Huntingford C, Oilver RJ, Clark DB, Sitch S, Zelazowski P, Kattge J, Harper AB, Cox PM (2018) Large sensitivity in land carbon storage due to geographical and temporal variation in the thermal response of photosynthetic capacity. *New Phytologist* 218: 1462–1477.

Mittler R (2002) Oxidative stress, antioxidants and stress tolerance. *Trends Plant Sci.* 7: 405–410.

Mooney HA, Winner WE, Pell EJ (1991) *Response of Plants to Multiple Stresses*, 2nd edn. Academic Press, San Diego.

Mora C, Frazier AG, Longman RJ, Dacks RS, Walton MM, Tong EJ, Sanchez JJ, Kaiser LR, Stender YO, Anderson JM, Ambrosino CM, Fernandez-Silva I, Giuseffi LM, Giambelluca TW (2013) The projected timing of climate departure from recent variability. *Nature* 502: 183–187.

Munné-Bosch S, Alegre L (2003) Drought-induced changes in the redox state of alpha-tocopherol, ascorbate, and the diterpene carnosic acid in chloroplasts of Labiatae species differing in carnosic acid contents. *Plant Physiol.* 131: 1816–1825.

Munné-Bosch S, Queval G, Foyer CH (2013) The impact of global change factors on redox signaling underpinning stress tolerance. *Plant Physiol.* 169: 5–19.

Myneni RB, Ross J, Asrar G (1989) A review on the theory of photon transport in leaf canopies. *Agric Forest Meteorol* 45: 1–153.

Noctor G, Mhamdi A, Foyer CH (2014) The roles of reactive oxygen émetabolism in drought: Not so cut and dried. *Plant Physiol.* 164: é1636–1648.

Norby RJ, Gunderson CA, Wullschleger SD, O'neill EG, Mccracken MK (1992) Productivity and compensatory responses of yellow-poplar trees in elevated $CO_2$. *Nature* 357: 322–334.

Norman JM, Welles JM (1983) Radiative transfer in an array of canopies. *Agron J* 75: 481–488.

Pan Y, Birdsey RA, Fang J, Houghton R, Kauppi PE, Kurz WA, Phillips OL, Shvidenko A, Lewis SL, Canadell JG, Ciais P, Jackson RB, Pacala SW, McGuire AD, Piao S, Rautiainen A, Sitch S, Hayes D (2011). A large and persistent carbon sink in the world's forests. *Science* 333: 988–993.

Pan Y, Birdsey RA, Phillips OL, Jackson RB (2013) The structure, distribution, and biomass of the world's forests. *Ann Rev Ecol Evol Syst* 44: 593–622.

Pasquini SC, Santiago LS (2012) Nutrients limit photosynthesis in seedlings of a lowland tropical forest tree species. *Oecologia* 168: 311–319.

Pearcy RW (1990) Sunflecks and photosynthesis in plant canopies. *Ann Rev Plant Physiol Mol Biol* 41: 421–453.

Peñuelas J, Munné-Bosch S, Llusià J, Filella I (2004) Leaf reflectance and photo- and antioxidant protection in field-grown summer-stressed Phillyrea angustifolia. Optical signals of oxidative stress? *New Phytol.* 162: 115–124.

Phillips OL, Aragão LEOC, Lewis SL, Fisher JB, López-González JLG, Malhi Y, Monteagudo A, ... (2009) Drought sensitivity of the Amazon rainforest. *Science* 6;323(5919): 1344–1347. doi: 10.1126/science.1164033

Phillips OL, van der Heijden G, Lewis SL, Lopez-Gonzalez G, Aragao L, Lloyd J, Malhi Y, Monteagudo A, Almeida S, Davila EA, Amaral I, Andelman S, Andrade A, Arroyo L, Aymard G, Baker TR, Blanc L, Bonal D, de Oliveira ACA, Chao KJ, Cardozo ND, da Costa L, Feldpausch TR, Fisher JB, Fyllas NM, Freitas MA, Galbraith D, Gloor E, Higuchi N, Honorio E, Jimenez E, Keeling H, Killeen TJ, Lovett JC, Meir P, Mendoza C, Morel A, Vargas PN, Patino S, Peh KSH, Cruz AP, Prieto A, Quesada CA, Ramirez F, Ramirez H, Rudas A, Salamao R, Schwarz M, Silva J, Silveira M, Slik JWF, Sonke B, Thomas AS, Stropp J, Taplin JRD, Vasquez R, Vilanova E (2010) Drought-mortality relationships for tropical forests. *New Phytol.* 187: 631–646.

PNAS (Proceedings of the National Academy of Sciences) (2020) Global tree ring analysis reveals rapid decrease in tropical tree longevity with temperature. *PNAS.* 117: 33358–33364. doi: 10.1073/pnas.200387311

Raven PH, Evert RF, Eichhorn SE (2005) *Biology of Plants* (Ed. W.H. Freeman). New York and Basingstoke.

Read J (1990) Some effects of acclimation temperature on net photosynthesis in some tropical and extra-tropical Australasian nothofagus species. *J Ecol* 78: 100–112.

Richards PW (1952) *The Tropical Rain Forest*. Cambridge University Press, London.

Rowland L, Harper A, Christoffersen BO, Galbraith DR, Imbuzeiro HMA, Powell TL, Doughty C, Levine NM, Malhi Y, Saleska SR, Moorcroft PR, Meir P, Williams M (2015) Modelling climate change responses in tropical forests: Similar productivity estimates across five models, but different mechanisms and responses. *Geosci Model Dev.* 8: 1097–1110,

Ruelland E, Zachowski A (2010) How plants sense temperature. *Environ Exp Bot.* 69(3): 225–232. doi: 10.1016/j.envexpbot.2010.05.011.

Ruwaimana M, Anshari GZ, Silva LCR, Gavin DG (2020) The oldest extant trop ical peat land in the world: A major carbon reservoir for at least 47000 years. *Environ Res Lett* 15: 114027.

Saatchi SS, Harris NL, Brown S, Lefsky M, Mitchard ETA, Salas W, Zutta BR, Buermann W, Lewis SL, Hagen S, Petrova S, White L, Silman M, Morel A (2011) Benchmark map of forest carbon stocks in tropical regions across three continents. *Proc Natl Acad Sci USA.* 108: 9899–9904.

Saldaña A, Gianoli E, Lusk CH (2005) Ecophysiological responses to light availability in three Blechnum species (Pteridophyta, Blechnaceae) of different ecological breadth. *Oecologia* 145(2): 251–256. doi: 10.1007/s00442-005-0116-2

Saldaña A, Lusk CH (2003) Influencia de las especies del dosel en la disponibilidad de recursos y regeneración avanzada en un bosque templado lluvioso del sur de Chile. *Rev Chil Hist Nat.* 76: 639–650.

Sharkey TD, Monson RK (2014) Estimations of global terrestrial productivity: Converging toward a single number? In Roy J, Saugier B, Mooney HA (eds), *Terrestrial Global Productivity*. Academic Press, New York, pp. 543–557.

Silva CE, Kellner JR, Clark DB, Clark DA (2013) Response of an old-growth tropical rainforest to transient high temperature and drought. *Glob Change Biol.* 19: 3423–3434.

Slik JWF (2004) El Niño droughts and their effects on tree species composition and diversity in tropical rain forests. *Oecologia* 141: 114–120.

Sultan S, Wilczek A, Bell D, Hand G (1998) Physiological response to complex environments in annual Polygonum species of contrasting ecological breadth. *Oecologia* 115: 564–578.

Suzuki N, Koussevitzky S, Mittler R, Miller G (2012) ROS and redox signalling in the response of plants to abiotic stress. *Plant Cell Environ.* 35: 259–270.

Szabó S, Peeters ETHM, Borics G, Nagy PT, Lukács BA (2020) The ecophysiological response of two invasive submerged plants to light and nitrogen. *Front Plant Sci.* doi: 10.3389/fpls.2019.01747

Tanner EVJ, Kapos V, Franco W (1992) Nitrogen and phosphorus fertilization effects on Venezuelan montane forest trunk growth and litterfall. *Ecology* 73: 78–86.

Tanner EVJ, Kapos V, Freskos SRHJ, Theobald AM (1990) Nitrogen and phosphorus fertilization of Jamaican montane forest trees. *J Ecol.* 6: 231–238.

Ter Steege H, Nigel CA, Sabatier D, Baraloto C, Salomao RP, Guevara JE, Phillips OL, Castilho CV, Magnusson WE, Molino JF, Monteagudo A, Vargas PN, Montero JC, Feldpausch TR, Coronado ENH, Killeen TJ, Mostacedo B, Vasquez R, Assis RL, Silman MR (2013) Hyperdominance in the Amazonian tree flora. *Science* 342: 325.

Tewari RK, Kumar P, Sharma PN (2006) Magnesium deficiency induced oxidative stress and antioxidant responses in mulberry plants. *Sci Hortic.* 108: 7–14.

Tinker PB, Nye PH (2000) *Solute Movement in the Rhizosphere. Topics in Sustainable Agronomy.* Oxford University Press, New York.

Tjoelker MG, Oleksyn J, Lee TD, Reich PB (2001) Direct inhibition of leaf dark respiration by elevated CO2 is minor in 12 grassland species. *New Phytol.* 150: 419–424.

Tripathi S, Bhadouria R, Srivastava P, Devi RS, Chaturvedi R, Raghubanshi AS (2020) Effects of light availability on leaf attributes and seedling growth of four tree species in tropical dry forest. *Ecological Processes* 9(1): 1–16.

Uehara G, Gillman GP (1981) *The Mineralogy, Chemistry, and Physics of Tropical Soils with Variable Charge Clays (Westview tropical agriculture series; no. 4).* Westview Press, Boulder, CO.

Vårhammar A, Wallin G, Mclean CM, Dusenge ME, Medlyn BE, Hasper TB, Nsabimana D, Uddling J (2015) Photosynthetic temperature responses of tree species in Rwanda: Evidence of pronounced negative effects of high temperature in montane rainforest climax species. *New Phytologist* 206: 1000–1012.

Veneklaas EJ (1991) Litterfall and nutrient fluxes in two montane tropical rainforests, Colombia. *J Trop Ecol* 7: 319–336.

Vitousek PM (1984) Litterfall, nutrient, cycling, and nutrient limitation in tropical forests. *Ecology* 65: 285–298.

Vitousek PM (1982) Nutrient cycling and nutrient use efficiency. *Am Nat.* 119: 553–572.

Vitousek PM, Denslow JS (1986) Nitrogen and phosphorus availability in treefall gaps of a lowland tropical forest. *J Ecol.* 74: 1167–1178.

Vogel MM, Orth R, Cheruy F, Hagemann S, Lorenz R, van den Hurk BJJM, Seneviratne SI (2017) Regional amplification of projected changes in extreme temperatures strongly controlled by soil moisture-temperature feedbacks. *Geophys Res Lett.* 44: 1511–1519.

Vogelmann TC, Nishio JN, Smith WK (1996) Leaves and light capture: Light propagation and gradients of carbon fixation within leaves. *Trends Plant Sci.* 1: 65–70.

Wagner F, Rossi V, Aubry-Kientz M, Bonal D, Dalitz H, Gliniars R, Stahl C, Trabucco A, Hérault B (2014) Pan-Tropical analysis of climate effects on seasonal tree growth. *PLoS One* 9. doi: 10.1371/journal.pone.0092337

Wagner F, Rossi V, Stahl C, Bonal D, Hérault B (2012) Water availability is the main climate driver of neotropical tree growth. *PLoS One* 7. doi: 10.1371/journal.pone.0034074

Walters MB, Field CB (1987) Photosynthetic light acclimation in two rainforest Piper species with different ecological amplitudes. *Oecologia* 72: 449–456.

Walther GR, Post E, Convey P, Menzel A, Parmesan C, Beebee TJC, Fromentin JM, Hoegh-Guldberg O, Bairlein F (2002) Ecological responses to recent climate change. *Nature* 416: 389–395.

Williams JW, Jackson ST, Kutzbach JE (2007) Projected distributions of novel and 54 disappearing climates by 2100 AD. *Proc Nat Acad Sci* 104: 5738–5742.

Williams M, Malhi Y, Nobre AD, Rastetter EB, Grace J, Pereira MG (1998) Seasonal variation in net carbon exchange and evapotranspiration in a Brazilian rain forest: A modelling analysis. *Plant Cell Environ.* 21: 953–968.

Winter K, Aranda J, Garcia M, Virgo A, Paton SR (2001) Effect of elevated CO2 and soil fertilization on wholeplant growth and water use in seedlings of a tropical pioneer tree. *Ficus Insipida Willd. Flora.* 196: 458–464.

Zang C, Hartl-Meier C, Dittmar C, Rothe A, Menzel A (2014) Patterns of drought tolerance in major European temperate forest trees: Climatic drivers and levels of variability. *Glob Change Biol.* 20: é3767–é3779.

Zeevaart JAD, Creelman R (1988) Metabolism and physiology of abscisic acid. *Annu Rev Plant Physiol Plant Mol Biol.* 39: é439–é473.

Zhu JK (2002) Salt and drought stress signal transduction in plants. *Annu Rev Plant Biol.* 53: 247–273.

# 9 Soil Nutrient Reservoir in a Changing Climate Scenario

*Janaki Subramanyan*

## 9.1 SOIL HEALTH

Terrestrial plants absorb mineral nutrients from the soil solution through their roots. Soil is a complex medium comprising the mineral particles derived from parent rock and organic matter in different stages of decomposition, the soil solution, gases, and soil micro-organisms. Plants take in mineral nutrients in the form of ions from the soil solution. The inorganic and organic soil colloidal particles store nutrients for release into the soil solution. The reservoir of nutrients for plants is made up of ions that have been adsorbed on charged colloidal particles. The health of the soil is determined by a combination of measurable physical, chemical, and biological properties. These attributes are indicators and can be used to evaluate the soil health (Allen et al., 2011).

With soil organic carbon (SOC) and soil inorganic carbon (SIC), soils are the biggest repositories of terrestrial carbon in the world. Soil organic matter (SOM) affects the soil physical and chemical parameters; the physical quality includes soil texture, structure, aggregate quality, tilth, bulk density, porosity, infiltration rates, available water capacity and erodibility, rooting depth, soil surface cover, and soil temperature. The chemical quality includes pH, electrical conductivity, buffer capacity, cation exchange capacity, nutrient retention, and availability. The biological parameters include SOM, soil carbon, the C:N (carbon-to-nitrogen) ratio, soil respiration, soil enzyme activities and soil microbial biomass, and total biomass (Patil & Lamnganbi, 2018). A crucial ecosystem that supports humans, plants, animals, and microbes is the soil. What causes climate change; how climate change is likely to affect soil health, especially the mineralization and nutrient availability; and the importance of soil carbon and soil biodiversity are discussed in the chapter.

## 9.2 CLIMATE CHANGE

Since the earth's evolution began 4–5 billion years ago, the climate has been changing naturally at its own rate. The anthropogenic activities going on at a fast pace have led to the speeding up of global climate change. The result is global warming because of the rise in concentrations of greenhouse gases (GHGs) and the subsequent domino effects on changing rainfall patterns, ice sheet melting, and an increase in sea level (Khan et al., 2009). The soil food webs and soil organisms will be impacted by global warming and altered precipitation patterns, which will then have an impact on plant growth, structure, and physiology (Pritchard, 2011). Climate change will affect plant nutrition by influencing mineralization, decomposition, leaching, and loss of nutrients from soils (Elbasiouny et al., 2022). Rising temperatures, droughts, or increased rainfall owing to global warming will affect soil fertility and mineral nutrition. In turn, soil mineral nutrition is influenced by soil moisture, SOC losses, erosion rates, root growth and function, root-microbe relationships, and plant phenology (St. Clair & Lynch, 2010). Changes to soil variables will likely limit root growth, resulting in nutritional stress. However, the effects of climate change on plant growth and development, yield, water use efficiency, and nutrient uptake from the soil would vary spatiotemporally (Brouder & Volenec, 2008).

Microbes are important for nutrient cycling, soil humus development, and soil structure. A measure of how climate change is affecting soil processes is the microbial quotient, which is the percentage of total organic carbon (C) or nitrogen (N) in the microbial biomass (Patil & Lamnganbi, 2018). The C:N ratio of the soil will decrease when the soil temperature rises along with the atmospheric temperature, enhancing the breakdown of organic matter and releasing carbon dioxide ($CO_2$) (Patil & Lamnganbi, 2018). Compared to temperate regions, tropical regions are more susceptible to climate change because the tropical crops are likely to experience temperature stress, spread of insects and diseases which are already more common in the tropics and because of the prevailing economic and social constraints in the tropics, whereas the temperate $C_3$ plants are more likely to respond to increase in $CO_2$ concentration (Rosenzweig & Liverman, 1992).

## 9.3 EFFECTS OF CLIMATE CHANGE ON THE SOIL

Soil warming stimulates soil microbial respiration and activity, and increases the nitrification and mineralization rates of phosphorus (P) and N. However, these microbial responses may be transient (Pritchard, 2011). Because of global warming, the precipitation patterns will be altered, and some soils will become wetter and others drier. Because of more intense rainfall in the future, soil water is likely to increase. This will favor high microbial diversity, fast microbial turnover rates, fast rates of decomposition, and more robust soil meso- and macrofauna in wet soils compared to dry soils. Wet soils will increase the prevalence of soil-borne diseases (Pritchard, 2011).

### 9.3.1 Mineralization and Nutrient Availability

Global temperature increase poses several uncertainties, such as increased loss of soil minerals because of more rains, increased $CO_2$ concentration resulting in increased uptake of minerals, and stimulation of N fixation by free-living as well as symbiotic micro-organisms. Although the SOM is likely to improve, the availability of mineral nutrients remains unpredictable (Sinclair, 1992). Climate change will have its impact differently in different regions of the earth. Soil nutrient status is affected both in soils facing drought and waterlogging. However, all mineral nutrients will not be affected in the same manner when it comes to concentration and available forms of the mineral. Microbial growth and ion mobility decline in drought conditions, resulting in decreased mineral nutrient availability. In waterlogged soils, the overall nutrient availability is decreased and the reduced forms of iron (Fe), manganese (Mn), N and sulfur (S) are easily available. As a consequence, the mycorrhizae and mineral nutrition of the host plants will also be affected (Kreuzwieser & Gessler, 2010).

There are no reports of tropical legume hosts which form arbuscular mycorrhizal fungal associations under waterlogged conditions. In fact, not many arbuscular mycorrhizal fungi are tolerant to anoxia. Nevertheless, increased uptake of N, P, potassium (K), magnesium (Mg), Fe, Mn, and copper (Cu) by plants has been observed under water stress (Loo et al., 2022). The SOC is closely related to the nutrient cycles. Temperature and rainfall alterations affect the nutrient cycles and nutrient availability. The availability of nutrients in the soil was made more abundant by elevated $CO_2$ and warming. However, it has been demonstrated that elevated atmospheric $CO_2$ concentration decreases the levels of several nutrients in plants, thereby meaning that the uptake of nutrients had been hindered. Studies show that with increase in temperature, mineralization is promoted, nitrification increases, and phosphatase exudation and P uptake are stimulated. Such a condition promotes the ruderal fast-growing plants and limits phytodiversity (Elbasiouny et al., 2022).

The rhizosphere's access to nitrogen is mediated by mineral-associated organic matter (MAOM), which serves as a repository of soil nitrogen. Low molecular weight organic acids found in root exudates either directly or indirectly improve the solubilization and mobilization of MAOM. By directly interacting with the mineral-organic linkages, the exudates can mobilize MAOM from clays. Indirectly, the root exudates can promote the growth of microbes that produce several metabolites and enzymes which destabilize MAOM. By complexing with metals, organic acids like citric acid, malic acid, and oxalic acid produced by microbes can access the MAOM and release the associated organic materials. Organic acids released by roots and microbes exposes new substrates for microbial degradation. The lignin-degrading saprotrophs and ectomycorrhizae produce oxidative enzymes, thus releasing N from organic matter (Jilling et al., 2018).

To understand the effect of increased soil temperature on soil organisms, Ruess et al. (1999) conducted simulation experiments in Swedish Lapland in a heath and a fellfield. At both sites, the altered environmental conditions led to a rise in the density of bacteria- and fungi-eating nematodes. In the heath soil, both the microbial biomass carbon and the active fungal biomass rose. The net effect will be enhanced rates of mineralization of N and P, and plant nutrient availability. The rising $CO_2$ concentration is likely to increase the total soil bacterial biomass together with significant shifts in the composition. The result will be a rise in the number and size of root nodules, as well as an increase in rhizobial populations in the rhizosphere and N-fixing organisms in the bulk soil (Pritchard, 2011).

Selenium (Se) is a trace element essential for human beings. The concentration of Se in the diet depends upon the concentration in the soil. It has been predicted that the climate change scenarios for 2080–2099 will lead to an overall decrease in soil Se concentration, especially in agricultural lands, thereby increasing the prevalence of Se deficiency in the population (Jones et al., 2017).

## 9.4 ROLE OF CARBON SEQUESTRATION

The amount of carbon in the world's soils is 3.3 times more than that of the atmosphere and 4.5 times greater than that of the biotic pool (Pritchard, 2011). Around 2 trillion tonnes of organic carbon are stored in soils, which is more than the sum of the carbon pools in the atmosphere and plants. Microbes regulate the amount of carbon stored in the soil and returned to the atmosphere. The availability of macronutrients—especially N and P, on which productivity largely depends—is influenced by carbon stored in the soil (Cavicchioli et al., 2019). As part of soil carbon sequestration, atmospheric $CO_2$ is transferred to the SOC as humus and to the SIC pool as secondary carbonates with a long mean residence period. Soil carbon management by adopting carbon sequestration strategies such as integrated nutrient management, use of organics, agroforestry for croplands, preserving vulnerable soil carbon pools like those of permafrost and peatlands, and restoring degraded and desertified soils is necessary for the maintenance of the carbon reservoir in the soil (Lal, 2013).

Particulate organic carbon (POC) and mineral-associated organic carbon (MAOC) are two categories under which the impact of climate change on SOC may be examined individually. Compared to the SOC and MAOC, the POC is a superior diagnostic biomarker of short-term soil carbon changes (Rocci et al., 2021). The SOC represents about

50–60% of the SOM. Soil biodiversity is directly related to the SOC. It is the biological activity web of bacteria, fungi, protozoa, worms, insects, and various invertebrates and vertebrates interacting among themselves and with plants and animals which conserves soil fertility (Elbasiouny et al., 2022). As climate change progresses, the rate of soil respiration will increase, whereas $CO_2$ assimilation by terrestrial ecosystems will slow down with time. It is likely that as the turnover of SOM increases, the nutrient cycling increases, hence enhancing soil fertility. In the short term, $CO_2$ intake and plant growth will be improved. When several decades go by, it is likely that there will be a decrease in the rate of absorption of $CO_2$ by terrestrial ecosystems. It has been shown that in free air carbon enrichment (FACE) experiments with trees, the promotion of above-ground growth often declines with time (Millard et al., 2007).

### 9.4.1 Permafrost

The perennially frozen subsurface earth materials remaining below 0°C for more than two consecutive years constitutes the permafrost. It is estimated that up to 100 Pg (petagrams) of carbon could be vulnerable for decay by 2100 from the exposure of thawing permafrost to microbial degradation (Schuur et al., 2008). It is predicted that large carbon reservoirs will become available for microbial respiration and greenhouse gas emissions when permafrost declines by 28–53% compared to the levels in 1960–1990 because of increase in surface temperature by 1.5–2.0°C relative to the global mean surface temperature in 1850–1900 (Cavicchioli et al., 2019).

High elevations where warming has already started to occur contain permafrost soils, as well as a sizable portion of peatland soils. With warming, peatland is exposed to air, allowing oxidase activity by microbes and slowing down the accumulation of phenolic compounds. Thawing permafrost creates a mosaic of flooded and dry areas. In the thawed areas, referred to as thermokarst lakes, anaerobic decomposition of organic matter will proceed slowly, yet the $CH_4$ emissions will be large, constituting loss of carbon, whereas in the drier areas, the carbon losses will be greater (Davidson & Janssens, 2006). Peatlands are being transformed from long-term carbon sinks to carbon sources because of draining and mining by humans; the carbon loss is because of oxidation, leaching of dissolved carbon, and release of $CO_2$, CO, and $CH_4$ from peat fires and combustion of mined peat. Draining of peatlands for land use releases $N_2O$, also contributing to greenhouse gases. Restoration of peatlands through wetting, and at the same time allowing wet agriculture and agroforestry, i.e., paludiculture, will significantly decrease the greenhouse gas emissions (Leifeld & Menichetti, 2018).

## 9.5 SOIL BIODIVERSITY

The biodiversity in soils is far more than that of above-ground ecosystems. The organisms in soils include macroflora, mesofauna, and microfauna. Microflora plays a key role in degradation of SOM and resulting mineralization (Pritchard, 2011). Microbes and microbial processes also play a major role in global fluxes of $CO_2$, $CH_4$, and $N_2O$ (Singh et al., 2010). In fact, the interaction of the soil microflora with the root rhizosphere governs plant nutrition in a big way (Dotaniya & Meena, 2015).

### 9.5.1 Microbiomes

Microbes are vital for ecosystem functioning. The biogeochemical cycling of essential elements vital for plant and animal growth is carried out by soil microbiomes (Jansson & Hofmockel, 2020). Actually, the biosphere is supported by microbiomes, communities of soil bacteria, in the microbial world (Cavicchioli et al., 2019; Naylor et al., 2020). The diversity and activities of microbes will vary as a result of climate change, which will impact other organisms' resistance and responses. Therefore, the overall responses of soil micro-organisms to climate change will determine how well we are able to determine an environmentally sustainable future (Cavicchioli et al., 2019). If the soil temperature rises on account of global climate change, the activity of extracellular enzymes in forest soils has been predicted to increase (Baldrian et al., 2013). Short-term warming increases soil enzyme activity; however, further studies are required to understand what would happen in a long-term scale of warming (Fanin et al., 2022).

### 9.5.2 Rhizosphere Microflora

The approximately 2 mm zone around the root surface is the rhizosphere zone. This zone has a significant role to play in nutrient availability and uptake by roots. The roots exude a variety of substances—including carbohydrates, amino acids, lipids, and vitamins—which stimulate the activities of micro-organisms. The bacteria in the rhizosphere zone are important for the biogeochemical cycling of the micronutrients Cu, Fe, Mn, and zinc (Zn), as well as the macronutrients N and P, which makes the nutrients available to the plants for absorption (Dotaniya & Meena, 2015).

### 9.5.3 Mycorrhizae

Mycorrhizae depend on carbon from plants and in turn benefit the plants by helping in N and P acquisition (Cavicchioli et al., 2019). The hyphae are narrow enough to penetrate pores in the soil which the roots are unable to penetrate, and absorb nutrients in forms which cannot be directly absorbed by the plant root. The mycorrhizae transfer the nutrients to the host in the form which is easily absorbed, and also confer disease resistance to the host. Some plants facing phosphate limitations produce strigolactones that induce growth and branching of arbuscular mycorrhizal fungi, thereby facilitating the proliferation of the mycelium (Pritchard, 2011). Mycorrhizal fungal guilds exhibit genetically based variations in how they obtain N and P from organic materials. The trade-off between nutrient allocation to roots or mycelia, ecophysiological characteristics like root exudation, enzyme production, weathering, plant defense, and the response to climatic change are all altered by mycorrhizal associations (Tedersoo & Bahram, 2019).

## 9.6 CHALLENGES AND THE FUTURE

Climate change models predict a global temperature rise of about 3.7°C by 2100, and the soil microbiomes will increase in their diversity. As a consequence, chemical weathering of rocks will be accelerated by lichens and microbial films, and mycorrhizae; the highly reactive secondary clay minerals will alter the soil properties (Naylor et al., 2020). Understanding the variations in ecophysiology and ecosystem services among the various mycorrhizal species would require a combination of repeatable field trials and controlled laboratory investigations using isotope labeling and -omics approaches (Tedersoo & Bahram, 2019). Global climate change will influence all properties of soils. The soil is a recipient of—as well as a contributor to—the impacts of climate change. Soil management is necessary to restore the SOC so as to improve the soil structure and fertility (Rosenzweig & Hillel, 2000). For effective soil carbon sequestration, both the POM and MAOM should be built up in arable soils (Lugato et al., 2021). How plants will react to scenarios of a changing climate is challenging to anticipate. However, research on root traits which are under quantitative genetic control are of great significance to understand the adaptation to mineral stresses (Lynch & St. Clair, 2004). Climate change encourages plants to accumulate metals and increase biomass production. Suitable high-yielding crop plants that are tolerant to multi-stress conditions and do not accumulate toxic heavy metals should be selected to ensure food security (Rajkumar et al., 2013).

In eutrophic lakes, reservoirs, and estuaries, cyanobacterial blooms are expected to occur more frequently, more intensely, and over longer periods of time. As a result, the waters will contain neurotoxins, hepatotoxins, and dermatoxins, which can be harmful to animals and present a risk for humans who use the water for drinking, fishing, irrigation, and recreation. Moreover, the buoyant cyanobacteria form dense surface blooms, giving them a selective advantage over the nonbuoyant phytoplankton organisms to proliferate further (Cavicchioli et al., 2019).

## 9.7 IMPACT ON FOOD SECURITY AND SAFETY

Understanding how climate change may affect soil fertility and nutrient availability is crucial to foresee the future of global agriculture. According to predictions, food insecurity—especially in developing countries—will become severe because of the adverse effects of climate change on soil fertility and mineral nutrient availability (St. Clair & Lynch, 2010). We can achieve food security by improving soil quality by increasing the soil carbon pool above the critical 1.2–1.5 level in the root zone (Lal, 2013).

Climate change will impact food safety as well; the concentration of micronutrient elements and heavy metals in plant products will change depending on the abundance and availability of the element in the soil. Additionally, it is probable that pathogenic bacteria colonize foods, mycotoxins form on plant products in the field or during storage, marine biotoxins end up in seafood as a result of harmful algal blooms producing phytotoxins, and pesticide residues and polycyclic hydrocarbons are detected in foods and plant products (Miraglia et al., 2009). The estimated increase in the world population to 9.2 billion by 2050, along with climate change, soil degradation, and decreased water availability, pose challenges to food security, thereby necessitating climate-resilient agriculture (Lal, 2013).

## 9.8 MITIGATION

Maintaining the soil tilth, fertility, and productivity by various practices protects the soil aggregates. Soil aggregates that are 0.84 mm in diameter are non-erodible by wind and water. Aggregation is because of the gluey substances synthesized by the biota. The binding gluey substances are also biodegradable, and it is therefore necessary to replenish the SOM. Sowing the seed and adding fertilizers by drilling the soil will help to retain soil aggregates (Patil & Lamnganbi, 2018).

Soil ecosystems would be affected by climate change. Soil micro-organisms can be put to good use for mitigation of the adverse effects of climate change (Jansson & Hofmockel, 2020). Inoculation of field crops with beneficial micro-organisms and genetic modification of N-fixing endosymbiotic bacteria and of mycorrhizae are strategies to mitigate the effects of climate change (Pritchard, 2011). The modification of microbial processes will assist in ameliorating climate change by reducing greenhouse gas emissions from terrestrial ecosystems, and the new molecular tools will aid in assessing the microbial diversity of unculturable taxa (Singh et al., 2010). The influence of climate change on the diversity, distribution, and activity of soil microbes and the consequent effects on plant physiology and root exudation may be mitigated by adding beneficial micro-organisms in cultivated lands and forests (Bhattacharyya et al., 2016).

Suitable site-specific agroforestry systems with trees having a high root biomass–to–above-ground biomass ratio and/or N-fixing trees with the aim of stepping up the SOC can be developed to mitigate the impact of climate change (Lorenz & Lal, 2014). SOC management is the answer for an effective carbon sequestration. There are several methods by which the SOC can be maintained and improved in soils. The sustained release of mineral nutrients in available forms from the SOC controls soil nutrient dynamics (Elbasiouny et al., 2022). For carbon capture and storage in order to restrict the future global temperature rise to 2°C, negative emission technologies need to be adopted. In total, 20 agricultural and tree plantation crops dominate the tropics: Asia dominates in rice, oil palm, seed cotton, coconut, and rubber; the neotropics dominate in soybean, sugarcane, and coffee; Africa dominates in sorghum, millet, cowpea, and cocoa; and Brazil and Indonesia have large plantations of Eucalyptus and Acacia. Targeting the dominant crops for enhanced weathering, spreading crushed reactive silicate rocks (such as basalt) on more than 680 million hectares of tropical agricultural and tree plantations, and restoring depleting biomass carbon sinks could all help sequester carbon and reduce $CO_2$ emissions from fossil fuels (Edwards et al., 2017).

The ecophysiology of the main tropical tree crops such as banana, cashew, cassava, citrus, cocoa, coconut, coffee, mango, papaya, rubber, and tea, demonstrates that the evaporative requirement for a tropical climate is substantially higher than that of temperate trees and annuals. To achieve sustainable agriculture with good yields under changing environment conditions, adoption of intercropping and agroforestry schemes is an effective option (DaMatta, 2007). Several strategies are possible to alleviate the impact of climate change on soil health. However, the ecophysiology of tropical plants with respect to climate change should be a thrust area of research because tropical plants will be affected more than the temperate plants and mankind is dependent on many tropical plants.

## 9.9 CONCLUSIONS

Global climate change is likely to have major impacts on biodiversity—microbial, plant, and animal biodiversity, and ecosystems as a whole. Crop yields can only be maintained by improving soil resilience using beneficial microbiota, especially the arbuscular mycorrhizal fungi which can mitigate biotic and abiotic stresses and help maintain the source-sink balance under elevated $CO_2$ concentrations (Loo et al., 2022, Song et al., 2022). Legumes provide dietary protein, and with the correct soil microbiota, they are able to replenish the organic carbon content of the soil. To circumvent the climate change-associated soil degradation, more research needs to be done on lesser-known tropical legumes such as winged beans and lentils. Emphasis should be on symbiotic microbes, especially on arbuscular mycorrhizal fungi which can withstand anaerobic conditions in waterlogged soils (Loo et al., 2022). Multi-model simulation outputs based on a socioeconomic survey involving over 3,000 households from 72 coastal communities belonging to five Indo-Pacific countries have shown that the impacts of climate change were greater on fisheries than on agriculture; in over 67% of communities, both fisheries and agriculture will be affected simultaneously, and the communities with a lower socioeconomic status will face greater impact of the consequences (Cinner et al., 2022). Future food security lies in international cooperation and support to facilitate research on tropical crops, particularly legumes.

## REFERENCES

Allen, D. E., Singh, B. P. and Dalal, R. C. 2011. Soil health indicators under climate change: A review of current knowledge. In B. Singh, A. Cowie and K. Chan, eds, *Soil Health and Climate Change*, 25–45. Berlin: Springer. https://doi.org/10.1007/978-3-642-20256-8_2

Baldrian, P., Šnajdr, J., Merhautová, V., Dobiášová, P., Cajthaml, T. and V. Valášková. 2013. Responses of the extracellular enzyme activities in hardwood forest to soil temperature and seasonality and the potential effects of climate change. *Soil Biology and Biochemistry* 56: 60–68. https://doi.org/10.1016/j.soilbio.2012.01.020

Bhattacharyya, P. N., Goswami, M. P. and Bhattacharyya, L. H. 2016. Perspective of beneficial microbes in agriculture under changing climatic scenario: A review. *Journal of Phytology* 8: 26–41.

Brouder, S. M. and Volenec, J. J. 2008. Impact of climate change on crop nutrient and water use efficiencies. *Physiologia Plantarum* 133 (4): 705–724. https://doi.org/10.1111/j.1399-3054.2008.01136.x

Cavicchioli, R., Ripple, W. J., Timmis, K. N. et al. 2019. Scientists' warning to humanity: Microorganisms and climate change. *Nature Reviews Microbiology* 17: 569–586.

Cinner, J. E., Caldwell, I. R., Thiault, L., Ben, J., Blanchard, J. L. et al. 2022. Potential impacts of climate change on agriculture and fisheries production in 72 tropical coastal communities. *Nature Communications* 13: 3530. https://doi.org/10.1038/s41467-022-30991-4

DaMatta, F. M. 2007. Ecophysiology of tropical tree crops: An introduction. *Brazilian Journal of Plant Physiology* 19 (4): 239–244. https://doi.org/10.1590/S1677-04202007000400001

Davidson, E. A. and Janssens, I. A. 2006. Temperature sensitivity of soil carbon decomposition and feedbacks to climate change. *Nature* 440: 165–173.

Dotaniya, M. L. and Meena, V. D. 2015. Rhizosphere effect on nutrient availability in soil and its uptake by plants: A review. *Proceedings of the National Academy of Sciences, India, Section B Biological Science* 85: 1–12. https://doi.org/10.1007/s40011-013-0297-0

Edwards, D. P., Lim, F, James, R. H. et al. 2017 Climate change mitigation: Potential benefits and pitfalls of enhanced rock weathering in tropical agriculture. *Biology Letters* 13: 20160715 (7 pp). http://dx.doi.org/10.1098/rsbl.2016.0715

Elbasiouny, H., El-Ramady, H., Elbehiry, F., Rajput, V. D., Minkina, T. and Mandzhieva, S. 2022. Plant nutrition under climate change and soil carbon sequestration. *Sustainability* 14: 914–933. https://doi.org/10.3390/su14020914

Fanin, N., Mooshammer, M., Sauvadet, M., Meng, C., Alvarez, G., Bernard, L., et al. 2022. Soil enzymes in response to climate warming: Mechanisms and feedbacks. *Functional Ecology* 36 (6): 1378–1395.

Jansson, J. K. and Hofmockel, K. S. 2020. Soil microbiomes and climate change. *Nature Reviews Microbiology* 18: 35–46.

Jilling, A., Keiluweit, M., Contosta, A. R. et al. 2018. Minerals in the rhizosphere: Overlooked mediators of soil nitrogen availability to plants and microbes. *Biogeochemistry* 139:103–22. https://doi.org/10.1007/s10533-018-0459-5

Jones, G. D., Droz, B., Greve, P. et al. 2017. Selenium deficiency risk predicted to increase under future climate change. *Proceedings of the National Academy of Sciences PNAS* 114 (11): 2848–2853. https://doi.org/10.1073/pnas.1611576114

Khan, S. A., Kumar, S., Hussain, M. and Kalra, N. 2009. Climate change, climate variability and Indian agriculture: Impacts vulnerability and adaptation strategies. In S. N. Singh, ed, *Climate Change and Crops. Environmental Science and Engineering*, 19–38. Berlin: Springer. https://doi.org/10.1007/978-3-540-88246-6_2

Kreuzwieser, J. and Gessler, A. 2010. Global climate change and tree nutrition: Influence of water availability. *Tree Physiology* 30 (9): 1221–1134. https://doi.org/10.1093/treephys/tpq055

Lal, R. 2013. Food security in a changing climate. *Ecohydrology & Hydrobiology* 13 (1): 8–21. https://doi.org/10.1016/j.ecohyd.2013.03.006

Lal, R. 2013. Soil carbon management and climate change. *Carbon Management* 4 (4): 439–462. https://doi.org/10.4155/cmt.13.31

Leifeld, J. and Menichetti, L. 2018. The underappreciated potential of peatlands in global climate change mitigation strategies. *Nature Communications* 9: 1071–1077. https://doi.org/10.1038/s41467-018-03406-6 |

Loo, W. T., Chua, K.-O., Mazumdar, P., Cheng, A., Osman, N. and Harikrishna, J. A. 2022. Arbuscular mycorrhizal symbiosis: A strategy for mitigating the impacts of climate change on tropical legume crops. *Plants* 11 (21): 2875. https://doi.org/10.3390/plants11212875

Lorenz, K. and Lal, R. 2014. Soil organic carbon sequestration in agroforestry systems. A review. *Agronomy for Sustainable Development* 34: 443–454. https://doi.org/10.1007/s13593-014-0212-

Lugato, E., Lavallee, J. M., Haddix, M. L., Panagos. P. and Francesca Cotrufo, M. 2021. Different climate sensitivity of particulate and mineral-associated soil organic matter. *Nature Geoscience* 14: 295–300.

Lynch, J. P. and St. Clair, S. B. 2004. Mineral stress: The missing link in understanding how global climate change will affect plants in real world soils. *Field Crops Research* 90 (1): 101–115. https://doi.org/10.1016/j.fcr.2004.07.008

Millard, P., Sommerkorn, M. and Grelet, G.-A. 2007. Environmental change and carbon limitation in trees: A biochemical, ecophysiological and ecosystem appraisal. *New Phytologist* 175: 11–28.

Miraglia, M., Marvin, H. J. P., Kleter, G. A. et al. 2009. Climate change and food safety: An emerging issue with special focus on Europe. *Food and Chemical Toxicology* 47 (5): 1009–1021. https://doi.org/10.1016/j.fct.2009.02.005

Naylor, D., Sadler, N., Bhattacharjee, A. et al. 2020. Soil microbiomes under climate change and implications for carbon cycling. *Annual Review of Environment and Resources* 45: 29–59.

Patil, A. and Lamnganbi, M. 2018. Impact of climate change on soil health: A review. *International Journal of Chemical Studies* 6 (3): 2399–2404.

Pritchard, S. G. 2011. Soil organisms and global climate change. *Plant Pathology* 60: 82–99.

Rajkumar, I., Prasad, M. N. V., Swaminathan, S. and Freitas, H. 2013. Climate change driven plant–metal–microbe interactions. *Environment International* 53: 74–86. https://doi.org/10.1016/j.envint.2012.12.009

Rocci, K. S., Lavallee, J. M., Stewart, C. E. and Cotrufo, M. F. 2021. Soil organic carbon response to global environmental change depends on its distribution between mineral-associated and particulate organic matter: A meta-analysis. *Science of The Total Environment* 793 (1): 148569. https://doi.org/10.1016/j.scitotenv.2021.148569

Rosenzweig, C. and Hillel, D. 2000. Soils and global climate change: Challenges and opportunities. *Soil Science* 165 (10): 47–56.

Rosenzweig, C., and Liverman, D. 1992. Predicted effects of climate change on agriculture: A comparison of temperate and tropical regions. In S. K. Majumdar, ed, *Global Climate Change: Implications, Challenges, and Mitigation Measures*, 342–361. Pennsylvania: Philadelphia: The Pennsylvania Academy of Sciences.

Ruess, L., Michelsen, A., Schmidt, I. K. and Jonasson, S. 1999. Simulated climate change affecting microorganisms, nematode density and biodiversity in subarctic soils. *Plant and Soil* 212: 63–73.

Schuur, E. A. G., Bockheim, J., Canadell, J. G. et al. 2008. Vulnerability of permafrost carbon to climate change: Implications for the global carbon cycle. *BioScience* 58 (8): 701–714.

Sinclair, T. R. 1992. Mineral nutrition and plant growth response to climate change. *Journal of Experimental Botany* 43 (8): 1141–1146. https://doi.org/10.1093/jxb/43.8.1141

Singh, B. K., Bardgett, R. D., Smith. P. and Reay, D. S. 2010. Microorganisms and climate change: Terrestrial feedbacks and mitigation options. *Nature Reviews Microbiology* 8: 779–790.

Song, G., Wang, Q. C., Zheng, Y. and He, J. Z. 2022. Responses of arbuscular mycorrhizal fungi to elevated atmospheric $CO_2$ concentration and warming: A review. *The Journal of Applied Ecology* 33 (6): 1709–1718. https://doi.org/10.13287/j.1001-9332.202206.014 PMID: 35729151

St. Clair, S. B. and Lynch, J. P. 2010. The opening of Pandora's Box: Climate change impacts on soil fertility and crop nutrition in developing countries. *Plant and Soil* 335: 101–115.

Tedersoo, L. and Bahram, J. P. 2019. Mycorrhizal types differ in ecophysiology and alter plant nutrition and soil processes. *Biological Reviews* 94 (5): 1857–1880. https://doi.org/10.1111/brv.12538

# 10 Abiotic Stress Responses of a Tropical Plant

## Sugarcane (*Saccharum* species)

*R. Mishra, P. Agarwal, R. Soni, and G. Singh*

## 10.1 INTRODUCTION

Sugarcane belonging to the genus *Saccharum*, tribe Andropogoneae and family Poaceae, is one of the chief commercial cash crops cultivated in tropical and subtropical regions for sugar and biofuel. It is a tall perennial grass that can touch a height of 5–6 meters. The plant body is mainly divided into roots, stalks, leaves, and inflorescence. It produces multiple culms that are distinguished into nodes and internodes, and can be vegetatively propagated. Modern-day commercial sugarcane cultivars have a very complex polyploid or aneuploid genome (>10 gigabases) and are mostly interspecific hybrids formed by combining *S. officinarum* [(x = 10, 2n = 8x = 80), with high sugar content] and *S. spontaneum* (disease tolerant, ratooning display, and robust) (Zhang et al., 2018; Budeguer et al., 2021; Trujillo-Montenegro et al., 2021). It originated in New Guinea. India ranks second in sugarcane production, just after Brazil, and the important sugarcane-producing states of India are Uttar Pradesh, Karnataka, Andhra Pradesh, Maharashtra, Bihar, and Gujarat. Uttar Pradesh has a maximum of 21.99 lakh ha areas under sugarcane cultivation, while Tamil Nadu has the highest sugarcane productivity at 99.70 tons per hectare (https://sugarcane.dac.gov.in/pdf/StatisticsAPY.pdf).

The growth rate of different stages of sugarcane is primarily influenced by the climate. The temperature requirement for sugarcane varies from 25–35°C and differs depending on the different stage of growth namely germination, tillering, ripening, and elongation (https://farmer.gov.in/cropstaticssugarcane.aspx). Sugarcane entails a large amount of water (2000–3000 mm per annum per hectare) and is primarily cultivated as an irrigated crop. Anthropogenic activities—such as the burning of fossil fuels, the production of greenhouse gases, the overuse of fertilizers, and deforestation—have had a significant impact on the climate over the last few decades. Varying climatic conditions have resulted in enhanced $CO_2$ (carbon dioxide) levels, extreme temperatures, altered rainfall frequency, drought, the emergence of new pests and pathogens, floods, etc. Sugarcane plants are subjected to a variety of abiotic stresses such as salt, nutrient deprivation, temperature, and drought. These stresses have a huge undesirable effect on sugarcane growth, development, productivity, and the composition and synthesis of sugar. Depending on the duration and the stage of the plant at which stress is imposed, different varieties of sugarcane respond by modifying the signal transduction, biochemical, metabolic, and molecular pathways (Hussain et al., 2018).

Natural selection, conventional breeding, and modern biotechnological tools have been employed to develop and improve sugarcane. The intricate poly-aneuploid genome, heterogeneity, and vulnerability of sugarcane to a variety of biotic and abiotic stresses, however, present difficulties that call for greater monitoring of the progress of the sugarcane crop. Multi-omics approaches, together with available high-throughput techniques, have contributed to understanding the growth, yield, senescence, and responses to various biotic and abiotic stresses in several plants (Yang et al., 2021). Omics investigations have demonstrated differential regulation of various abiotic stress–related genes in sugarcane. Such studies can shed light on the molecular mechanism and will help in designing strategies for developing abiotic stress-tolerant cultivars.

MicroRNAs (miRNAs) are small and non-coding ribonucleic acid (RNA) molecules 19–24 nucleotides long present in both plants and animals which control the expression of genes via a sequence-specific interaction with the target mRNA. They play a key part in regulating various developmental processes along with stress responses in plants (Bartel, 2009). Several abiotic stress-responsive miRNAs reported in sugarcane play a regulatory role in many metabolic processes.

This chapter principally deliberates the abiotic stress tolerance mechanisms in sugarcane, various omics approaches used to better understand the stress response, different miRNAs up/down-regulated in response to abiotic stresses, and different genes transformed in sugarcane for generating stress-tolerant sugarcane. Information presented in this chapter will aid in the development of sugarcane varieties with increased yield and quality under continuously changing environmental conditions.

## 10.2 ABIOTIC STRESS RESPONSE MECHANISMS IN SUGARCANE

Abiotic stress causes a 50% reduction in agricultural output and productivity, threatening global food security. These abiotic stresses represent a severe challenge to ensure a sustainable food supply for the constantly expanding world population. Plants adapt to unfavorable environmental circumstances by regulating biochemical, physiological, and developmental mechanisms to survive stress and sustain growth (Figure 10.1). Abiotic factors include water stress (drought and flood), salt, high and low temperatures, and toxic metal stress. They can reduce the productivity, lignocellulosic biomass, and downstream products of the sugarcane industry (Srivastava & Kumar, 2020). Any stress during the water demand phase affects growth, photosynthesis, and dry matter build-up in sugarcane (Narwade et al., 2015). Different mechanisms and metabolites—including antioxidants, osmoprotectants, and heat shock proteins—are accumulated by sugarcane plants under various stresses, and many metabolic processes are also triggered (Hussain et al., 2018).

## 10.2.1 Water Stress

Both drought and water logging can have a detrimental impact on the growth and productivity of sugarcane, which is discussed in the following subsections.

### 10.2.1.1 Drought

Water shortage is the main abiotic constraint adversely affecting sugarcane yield. Sugarcane is a high–water-requiring crop, and its growth is susceptible to water scarcity (Lakshmanan & Robinson, 2013). Sugarcane has five primary stages: germination, tillering, stem elongation, maturity, and blooming (Silva et al., 2008). During tillering and stem elongation, drought stress mainly impacts leaf and stem growth, and it also affects root growth (Hoang et al., 2019).

Sugarcane adapts to dryness by rolling its leaves, which leads to a reduction in water loss and absorption of light (Snyman et al., 2016). During water stress, RuBisCo (Ribulose-1,5-bisphosphate carboxylase/oxygenase) and PEPCase (phosphoenolpyruvate carboxylase) activity limit photosynthesis (Lakshmanan & Robinson, 2013). The key processes that inhibit photosynthesis during drought stress are: stomatal closure, which reduces transpiration (Khonghintaisong et al., 2018); reactive oxygen species (ROS)-induced chloroplast damage; and a decrease in relative water content (RWC) (Hoang et al., 2019). During stress, the bundle sheath and mesophyll cells of sugarcane

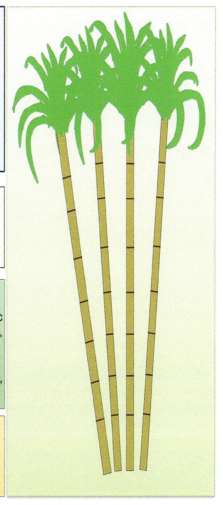

**FIGURE 10.1** Different physiological and biochemical characteristics of sugarcane altered by abiotic stress.

showed a decline in the activity of RuBisCo, PEPCase, FBPase (fructose-1, 6-bisphosphatase), and NADP-ME (NADP-malic enzyme; Sage et al., 2013) involved in photosynthesis. Since photosynthesis is the primary mechanism underlying sugar generation in plants, reduced enzyme activities have a negative impact on sugar build-up and plant biomass (Dlamini, 2021).

#### 10.2.1.2 Waterlogging

The physiological consequences of water logging on sugarcane include the following: (1) decline in transpiration due to closing of stomata; (2) reduction of leaf area, leading to decline in the rate of photosynthesis; (3) drastic lowering of crop growth rates; and (4) submerged organs respire more than leaves. Waterlogging causes a change from aerobic to anaerobic respiration (Gomathi et al., 2015). Oxygen shortage reduces root respiration and cell permeability, slowing absorption of water and nutrients. Under waterlogging circumstances, plants undergo changes in morphology, anatomy, physiology, and biochemistry for their survival (Gomathi & Chandran, 2009). Long-term or short-term flooding influences light interception, chlorophyll degradation, photosynthetic rate and leaf dry matter, enzyme activity (nitrate reductase), nutrient absorption, soluble protein amount, and root and shoot development (Gomathi et al., 2015).

### 10.2.2 Salt Stress

Salinity influences crop production by imposing morphological, physiological, biochemical, and molecular alterations. These alterations may be brought on by the inability of plants to absorb water and the toxicity of certain ions, which restricts plant development, quality, and yield, or results in a loss of total production (Simoes et al., 2020). Investigations show that salt stress is multifactorial; it results in osmotic and ionic stress due to excessive salt concentration (Gupta & Huang, 2014). Osmotic stress reduces root water uptake and induces reduced photosynthetic activity and ROS generation (Munns & Tester, 2008). This occurs due to imbalance in the distribution of ions and water potential that happens at the cellular and whole plant levels.

Ionic stress occurs from increased $Na^+$ (sodium) and $Cl^-$ (chloride) absorption, causing biochemical and physiological harm to the plant. In particular, high levels of $Na^+$ inside the cell block the uptake of $K^+$ (potassium), which is important for many cellular processes (Passamani et al., 2017). The proteome profile of micro-propagated sugarcane shoots has been shown to be altered by salt stress. Salt-induced ionic and oxidative stress can be reduced by the build-up of calcium-dependent protein kinase (CDPK) and phospholipase D (PLD), which maintain normal $Na^+$ in the cytoplasm and limits ROS build-up (Passamani et al., 2017). Salinity alters gas exchange, total soluble sugars, and levels of reducing sugar, which help the plants adapt to salt stress. It affects physiological processes, especially photosynthesis, by reducing leaf area, chlorophyll content, and stomatal conductance and significantly affects plant growth during the reproductive period (Simoes et al., 2020; Srivastava & Kumar, 2020).

### 10.2.3 Temperature

Sugarcane is grown in tropical and subtropical regions and has a $C_4$ metabolism. Sugarcane can withstand high temperatures, but it is susceptible to cold conditions (Zhao & Li, 2015). Depending on the kind, severity, and duration of temperature stress, plants might experience a variety of negative impacts. High temperatures can cause cell death or damage, which can cause the cellular structure to completely collapse (Hu & Xiong, 2014). It affects the membrane integrity and results in increased membrane permeability and solute leakage (Trivedi, 2015).

Plants frequently react when under stress by closing their stomata, increasing their stomatal densities, or altering the composition of membrane lipids (Srivastava et al., 2012). Sugarcane requires 32–33°C for its optimum growth and productivity. It can withstand up to 40°C, but 45°C has negative impact on its development (Wahid, 2007). Temperature stress alters sugarcane physiology and biochemistry, reducing its growth and output (Wahid, 2007). Exposure of sugarcane plants to high temperatures results in shorter internodes, leaf drying, increased tillering, and lower biomass. High temperatures enhances lipid peroxidation, malondialdehyde (MDA) concentration, and damage to cell membranes (Gomathi et al., 2014). It adversely affects metabolic processes by forming ROS, altering the production of primary and secondary metabolites. The production of proline, carotenoids, soluble sugars, glycine betaine (GB), and flavonoids increased following heat stress (40°C) treatment, which improved sugarcane's heat resistance. Castro-Nava et al. (2019) reported that optimal temperature and the phenological stage are used to determine thermal tolerance in sugarcane. Temperature influences the ripening and quality of sugarcane. Low winter temperatures aid natural ripening and slow down the growth of sugarcane, which converts reducing sugars into sucrose (Gawander, 2007).

## 10.3 OMICS APPROACHES TO EXPLICATE ABIOTIC STRESS RESPONSES IN SUGARCANE

Multi "omics" approach in plants have emerged as effective tools over the last few years, and progress in next-generation sequencing (NGS) have further created opportunities for new generations of omics (genomics, transcriptomics, proteomics). Nevertheless, metabolomics, phenomics, and ionomics have been reported for various crop plants (Kaur et al., 2017; Mustafa et al., 2018; Ali et al., 2019). The application of "omics" helps in understanding and unraveling the multipart connections among genes, proteins, and metabolites. In the area of genomics, molecular markers help in finding the genome structure of sugarcane genotypes, as well as deciphering the phylogenetic relationships

(Khueychai et al., 2015; Mustafa et al., 2018). Introgression of genes associated with abiotic stress-resistance traits derived from wild and related species has contributed to the development of various stress-resistant varieties of sugarcane (Meena et al., 2020). In sugarcane, assessment, characterization, and exploitation of germplasm by using various tools like linkage mapping, genomic selection (GS), genome-wide association study (GWAS), and marker-assisted selection (MAS) are suggested to be imperative in breeding programs to develop innovative methods to overcome the deleterious impact of stress. However, in comparison to MAS, GWAS and genomic-assisted selection offer more benefits, and thus play significant roles in sugarcane breeding. NGS jointly with GWAS has improved the resolution for mapping the exact position of QTLs/genes/alleles in the genome (Varshney et al., 2014; Meena et al., 2020).

The transcriptomics study is used to find out the genes that regulate important biological functions. Numerous novel miRNA families that impart a role in providing tolerance to drought, salt, and cold stress have been recognized using the high-throughput sequencing method. For sugarcane transcriptome, ample resources are available, and for various traits, the gene expression (both evident and overlapping gene expression patterns) has been confirmed (Cardoso-Silva et al., 2014; Dharshini et al., 2016). Belesini et al. (2017) performed a transcriptome study in drought-tolerant and susceptible varieties subjected to water-deficient conditions, and genes like aquaporin, *MYB*, *E3*, *SIZ2*, and *SUMO-protein ligase* were found to participate in conferring tolerance to drought stress.

Gene expression studies do not offer whole information on activity or protein synthesis, as function of various proteins is regulated by post-transcriptional/translational modifications (Chen & Harmon, 2006). Proteomics studies help in analyzing protein functions in a multifaceted manner and dynamic applications, like those that are associated with interactions between proteins or cell signaling (Cox & Mann, 2007). The use of proteomics like 2D DIGE (Two-dimensional difference gel electrophoresis) can assist in finding out the differentially expressed proteins and their functions in signaling pathways associated with abiotic and biotic stresses, and isobaric tags can be used for relative and absolute quantitation (iTRAQ) (Barnabas et al., 2015; Su et al., 2016). Proteomics is still at its early stages in sugarcane in comparison to other monocots. Nevertheless, with the increase in genomics research and the introduction of NGS, proteomics studies in sugarcane have expanded. The prominence of sugarcane proteomics has been studied by Barnabas et al. (2015) and Ali et al. (2019). Maranho et al. (2019) used proteomics to study the developmental variations occurring in sugarcane. Micro-propagated sugarcane shoots under salt stress can induce change in its proteomic profile (Passamani et al., 2017). Diverse proteins express under abiotic stresses, and thus, differential proteome studies under varied conditions can facilitate the identification of novel candidate genes that can help in creating new cultivars via the breeding method (Kaur et al., 2022). Chiconato et al. (2021) used proteomics to find out differentially controlled proteins in young plants of two sugarcane cultivars that varied in their response to salt stress.

Lately, metabolomics studies were used to comprehend the complex regulatory processes of metabolites and predict the mode of resistance via the use of high-throughput tools that control metabolic phenotypes (Lee et al., 2016). Vital et al. (2017) used an integrated approach (i.e., transcriptomics, proteomics, and metabolomics) to ascertain the variations at molecular and physiological levels in two associated sugarcane cultivars, including cv. RB867515 under drought conditions. Budzinski et al. (2019) conducted an experiment on sugarcane varieties (CTC15-tolerant; SP90–3414-sensitive) that varied in their response to drought stress to gain an insight into how leaf metabolism adjusts to drought stress. For this, they combined metabolomics and label-free quantitative proteomics data (GC-TOF-MS) by employing a network-based methodology. They applied rCCA (regularized canonical correlation analysis), a variation of standard CCA, to associate protein data and metabolites in sugarcane plants under drought conditions. A study of proteins and metabolite networks indicated that each variety is controlled by diverse biological processes. Perlo et al. (2020) used a combination of trait co-expression and metabolome profiling to identify the association of metabolome with the vital agronomic trait in sugarcane cultivars, and considerable variation was found in the metabolic patterns that showed a specific association with the developmental stage. Thus, these studies suggest that the "omics" approach holds great potential in understanding biological characteristics and can be employed in improving yield, sucrose content, resistance to environmental stresses, etc., in the sugarcane.

## 10.4 ABIOTIC STRESS RESISTANCE IN SUGARCANE

The perception, transmission, and subsequent gene expression govern the abiotic stress responses. Consequently, a deeper understanding of the tolerance mechanisms in sugarcane plants at the molecular, physiological, and cellular levels would help in raising stress-tolerant cultivars. The majority of cultivable hybrids of sugarcane are the result of crossing between *S. spontaneum* and *S. officinarum* (Zhang et al., 2018). Continuous progression in the field of biotechnology has helped in the cloning of several abiotic stress-tolerant genes (Wu et al., 2008) from *S. spontaneum*, a closely related wild variety of sugarcane species which is well studied for its adaptation potential under harsh conditions. It has also been studied as a vital resource for tolerant genes to develop stress-tolerant varieties of sugarcane. In the last few decades, transcriptomics studies in *S. spontaneum* were done under abiotic stress conditions to highlight the stress-tolerant genes and their responsive functions (Park et al., 2015; Wu et al., 2018). The wild relative of sugarcane *Erianthus arundinaceus* is a capable donor of genes for relative biomass and tolerance genes about abiotic stresses. Isolation, characterization, and overexpression of several genes from *E. arundinaceus*, including the heat shock proteins (HSPs), *DREB2* (dehydration-responsive element-binding protein 2), glyoxalase, expansin 1, and chilling tolerant divergence 1 (COLD1)—and their transformation into sugarcane plants via various conventional and biotechnological techniques—will help in developing sugarcane varieties with better agronomic characters.

In the following subsections, studies on sugarcane tolerance mechanisms to abiotic stresses such as salinity, high temperature, and water deficiency are discussed.

### 10.4.1 Salt Stress Tolerance in Sugarcane

Sugarcane is extremely susceptible to salt stress, which negatively affects its growth and development. Salt concentrations as high as 200 mM NaCl are toxic to sugarcane plants. It affects seed germination and causes disbalance in nutrient uptake, resulting in stunted growth which ultimately affects productivity of the plants (Reid, 2013). There is continuous decrease in growth parameters and physiological mechanisms (photosynthetic efficiency, stomatal conductivity, and rate of chlorophyll synthesis) in sugarcane because of the increased value of soil electrical conductivity (EC), which can be overcome by using salt-tolerant varieties (Yunita et al., 2020).

The tolerance level of the sugarcane and its sensitivity to salinity stress determines the screening methodology for screening and developing salt-tolerant sugarcane cultivars. The soil salinity changes slightly over the seasons. In monsoon season, leaching of soluble salts from the soil is well known, which results in a slight decrease in the EC and also the pH of the soil.

To develop salinity-tolerant *Saccharum* species, researchers from ICAR (Indian Council of Agricultural Research) screened approximately 1,200 different genotypes. This study includes the following popular tested varieties—*Co 86011, Co 7717, Co 7219, Co 8208, Co 85004, CoC 671, Co 6806, Co 94008, Co 85019, Co 94012, Co 97008,* and *Co 99004*—that were better adapted for growth in saline soils (Vasantha & Gomathi, 2012).

Genetic manipulations and crop improvement in sugarcane are difficult because the resulting interspecific hybrids of sugarcane have highly complex, large polyploid genomes (10 gigbase pairs [Gbp]) (Vermerris, 2011). The discovery of novel traits and genetic improvement of cultivated crops, such as sugarcane, has been made possible by the decoding of their genome sequences. The study and characterization of quantitative trait loci (QTLs) with respect to salinity tolerance can assist in developing salt-tolerant varieties (Barreto et al., 2019; Shabbir et al., 2021). The responsive genes are associated with proteins linked with photosynthetic mechanisms and vacuolar transport, along with osmoregulatory responsive proteins, membrane channel proteins, ROS activators, and regulatory elements, including TFs (transcription factors). The role of salt overly sensitive (SOS) pathway effective in salt resistance is a signaling pathway that has been well studied in various plants, including sugarcane (Ji et al., 2013).

Membrane-bound receptor proteins like *GPCRs* (G protein-coupled receptors) are important signaling molecules that control plant growth and development. Identification, isolation, and functional characterizing the *GPCR* gene obtained from the hybrid *Saccharum* sp. have revealed that *ShGPCR1*-overexpressed sugarcane plants are more resilient to salinity and drought stresses (Ramasamy et al., 2021). In another report overexpression of *Scdr1* gene (sugarcane drought-related-1; gene coding for a protein induced by water scarcity) in sugarcane resulted in increased resistance to abiotic stresses like salinity (Begcy et al., 2012). Numerous studies have shown that *Scdr2* protects seedlings during germination and its overexpression benefits young plants affected by salinity (Begcy et al., 2019).

### 10.4.2 Heat Stress Tolerance in Sugarcane

Damage to sugarcane crops is proportional to the stage of development at which the plants were subjected to heat stress (Trivedi, 2015). Carbohydrates, amino acids, lipids, and many other metabolic pathways are severely affected under high-temperature stress (Mangrauthia et al., 2016). Different heat shock protein families like HSP20 and HSP70, and other tolerance-responsive proteins (oxidative stress) such as trehalose and glutamate dehydrogenase, play active roles in the regulation of transcription and post-translational modifications during heat stress. According to Park et al. (2009), HSP and thioredoxin expression conferred tolerance against heat stress in various plant species including sugarcane.

Using quantitative real-time PCR (qRT-PCR) and RNA sequencing, temperature-tolerant sugarcane cultivar *Co 99004* with differentially expressed transcripts (DETs) was identified. The selective ten DETs were tested with RNA-Seq using an integrated DNA technology software (IDT) by designing primers of the selective genes and the actin gene as a reference gene., qRT-PCR, and other molecular techniques were used to study differentially expressed genes (DEGs). Testing the selective genes—where phytepsin (vacuole transportation via trafficking through Golgi apparatus), ferredoxin-dependent glutamate synthase (metabolism of nitrogen), and DDR-48, a stress protein (defense regulation)—showed three times up-regulated expression under heat stress (Raju et al., 2020).

In sugarcane, the *HSP70* gene family showed prominent expression among various differentially expressed HSPs. Thus, in order to develop heat-tolerant sugarcane, *HSP70* plays an important target in crop improvement (Augustine et al., 2015a).

### 10.4.3 Drought Tolerance Mechanisms

The ability of sugarcane plants to resist drought stress is a multi-gene complex quantitative trait. The selection of drought-resistant plants (*in vitro*) utilizing polyethylene glycol (PEG) is a frequently adapted method in plant biology advancement (Hernández-Pérez et al., 2021). Given that PEG does not enter plant tissues; it is added to the culture media to induce water stress. There is an urgent need to isolate genes and study their related pathways in order to create sugarcane varieties that can withstand drought. The following subsections discuss various methods to develop drought-tolerant sugarcane varieties.

### 10.4.3.1 Breeding as an Alternative to Screen Drought-Resistant Sugarcane Cultivars

The sugarcane drought tolerance capacity was tested in genotypes and progenies. Prior screening of drought-resistant cultivars was done using an in vitro screening technique utilizing PEG as a stress inducer. Different types of cultivars (cv) such as Mex 69–290, 79–431, CP 72–2086, and MOTZMex 92–207 were exposed to different PEG concentrations (in Murashige and Skoog semi-solid media). After estimating various developmental variables under culture conditions for a month, their total dry matter, net protein, proline, and GB amounts were measured. The results obtained indicated that cv Mex 69–290 was resistant to osmotic stress. With the increase in the concentration of PEG, the total protein content decreased, whereas total dry matter, percentage of proline, and GB increased in the tested varieties. This further suggested that cv Mex 69–290 showed resistance to drought conditions and can be used for better results in areas with little irrigation infrastructure or rainfall. Breeding programs based on MAS were used with the aim to test and enhance screening accuracy (Ali et al., 2019).

### 10.4.3.2 Somaclonal Variation for Developing Drought-Tolerant Sugarcane Cultivars

Somaclonal variation (in vitro genetic variation) was studied by Heinz and Mee (1969) in sugarcane. This method improved performance and yield in sugarcane, reduced fiber percentage, and elongated stems along with internodes, and provided tolerance to drought (Naheed et al., 2018; Manchanda et al., 2018; Naheed et al., 2020). When correlated with CPF-248, the parental genotype, the drought-tolerant somaclonal variant IPSV1 (in vitro selected putative somaclonal variant) displayed high rates of water loss, carbon fixation, osmotic potential, turgor potential, water usage efficiency, and RWC in leaves. The drought-tolerant somaclonal variant (IPSV1) was created by subjecting the parental line to higher 2, 4-D concentrations, which led to high genetic variability in callus cultures. With the aim to develop promising sugarcane cultivars, several studies have explained the induction of mutation in sugarcane via mutagen (Wagih et al., 2004; Khan et al., 2009; Suprasanna et al., 2011; Koch et al., 2012). Because sugarcane is vegetatively propagated, induction of mutations can be carried out in both callus cultures and single cells (Mahlanza et al., 2013). Approximately 13 mutant varieties of sugarcane resistant to different stress were developed via induced mutagenesis as per reports of Mutant Variety Database (Patade & Suprasanna, 2008).

### 10.4.3.3 Genetic Modifications in Sugarcane

The concept of genetic modification (GM) or manipulation entails the introduction of new traits into the recipient plant through the transfer of gene(s) from either the same plant species or a new (completely different) species (Holme et al., 2013). These genetic manipulations are being carried out by utilizing concepts of transgenesis, cisgenesis, and intragenesis (Augustine et al., 2015b). Several traits have been selected for crop improvement, making use of the insertion of drought-tolerant genes in sugarcane (Mbambalala et al., 2021). More research has revealed that the sugarcane gene *Scdr1*, which is evolutionarily conserved and responsive to drought, is expressed more frequently. This overexpression has a positive effect on maintaining high RWC and rate of photosynthesis under drought conditions, but the main effects were recorded during salt stress. Begcy et al. (2012) showed that *Scdr1* and *Scdr2* genes are important for the growth of sugarcane plants and also confer significantly higher tolerance to environmental stresses like salinity and drought. Understanding the function of *Scdr1* and *Scdr2* genes under drought stress is crucial to comprehend and characterizing sugarcane due to the possibility that these regions contain new sources of cisgenic sugarcane plants.

The cisgenic plants are crucial in overcoming problems with conventional breeding methods and regulatory hurdles in the field for the assessment and screening of transgenic lines to gauge drought tolerance (Holme et al., 2013). Applying transgenic technologies, which involve the transfer of HSP coding genes from *Erianthus* to sugarcane to achieve tolerance for drought and salinity stress, showed successful results (Augustine et al., 2015a).

### 10.4.3.4 Differential Gene Expression in Sugarcane

Water scarcity causes the signal transduction network to spread, affecting the functions of various transcription factors, protein kinases, and phosphatases (Rodrigues et al., 2011). A breakthrough in the understanding of sugarcane genes came with the Brazilian sugarcane expressed sequence tags (ESTs) sequencing project (SUCEST). This project aimed to sequence more than two lakhs ESTs from various sugarcane cultivars and different tissues, which were divided into many SAS (sugarcane assembled sequences) (Vettore et al., 2003), which provides ground for beneficial transcriptomics research (Rocha et al., 2007; Rodrigues et al., 2011). The study found that the number of genes whose expression patterns differ depends on the duration and intensity of stress. The results were confirmed by the differential expression of genes, with the tolerant cultivar upregulating the majority of the DEGs (93.3%) in the presence of severe stress. In contrast, 36% of the DEGs, including those involved in photosynthesis and stress responses (such as HSPs), were suppressed in the sensitive cultivars. Differentially expressed functional class genes have intriguing protein-coding functions that highlight how complex the sugarcane genome responds to water scarcity (Lembke et al., 2012).

With the aim to check the ability of the cell to retain water during drought conditions, a balance of water potential equilibrium via modifications in the membrane permeability of the cell is a crucial step. Aquaporin water channels (AQPs) are important for water to move across the cellular membrane (Afzal et al., 2016; de Andrade et al., 2022). It plays a crucial role in maintaining cells' water potential and also in storage of sugar in vacuoles in sugarcane plants. Small basic intrinsic proteins (SIPs), NOD26-like intrinsic proteins (NIPs), and plasma membrane intrinsic proteins (PIPs) are the subfamilies of AQPs. Depending on amino and carboxyl termini lengths, the PIP's form two subgroups, *PIP1* and *PIP2* (Afzal et al., 2016). *PIP2* isoforms (*ShPIP2;1*, *ShPIP2;5*, and *ShPIP2;6*) were found to express in the leaves of sugarcane plants under drought stress (De Andrade et al., 2016). The *PIP2* isoforms were drastically up-regulated after water stress (21 days) and later showed down-regulation after rehydration of soil (De Andrade et al., 2016). Ferreira et al. (2017) argued that *PIP2*'s response to a water shortage illustrates the management and control of the plant's water relationship.

Thus it can be concluded that in sugarcane under drought stress, the tolerance mechanism is mainly dependent on the early perception of stress via several signaling factors resulting in activation of downstream genes, including plant hormonal responses, which play an important role for protection against oxidative stress. In short, the study of targeted gene expression has resulted in the identification of genes that actively contribute to sugarcane's water stress response and plays a necessary role in functional analysis and developing links between the physiological and phenotypic responses.

### 10.4.3.5 Abscisic Acid Signaling as an Outcome of Drought Stress

Abscisic acid (ABA) is the key regulator of abiotic stress responses and provides adaptation to many stresses like drought, salinity, and temperature. Under conditions of water stress, long-term ABA accumulation causes a variety of adaptations, including the closing of stomata, reduced leaf size and stem growth, reorganization of assimilated products, the onset of senescence, the up-regulation of antioxidative enzymes, and enhanced root-to-shoot conductivity potential (Han & Yang, 2015).

In response to environmental stress, ABA-mediated responses involve the activation of cisacting and transacting elements. The dehydration response–related ABA-responsive element (ABRE) participates in ABA-induced gene expression. ABA triggers the activation of the DRE (dehydration-responsive element) and ABRE, which activates the rd29A stress-responsive gene. Numerous additional stress reactions are also controlled by another category of TFs, the basic leucine zipper (bZIP). In sugarcane, expression of six bZIPs was reported to be induced by ABA, with two genes (*ScbZIP29* and *ScbZIP31*) showing up-regulation and four genes (*ScbZIP21*, *ScbZIP24*, *ScbZIP70*, and *ScbZIP79*) showing down-regulation.

Another ABA-regulated gene, *SoNCED*, plays an active role in counteracting drought in sugarcane. Under stress, *SoNCED* expresses in leaves and roots, where it causes an accumulation of ABA (Li et al., 2013). In sugarcane, *SoDip22* (sucrose-phosphate synthase) gene has also been shown to regulate the water content of bundle sheath cells.

## 10.5 ENGINEERING SUGARCANE FOR ABIOTIC STRESS TOLERANCE

Traditional breeding of sugarcane has been practiced for developing varieties with improved agronomic characters. However, the complexity of the genome, limited gene pool, long breeding cycles, and low fertility make the breeding of sugarcane a labor-intensive, expensive, and time-consuming process. In-depth knowledge about the mechanisms underpinning abiotic stress responses has been made possible because of novel omics approaches. These novel tools have helped in explicating the structure, role and interactions of different genes conferring abiotic stress resistance in sugarcane. GM of sugarcane is possible because of the availability of highly efficient in vitro plant regeneration systems. Genetic engineering is a promising tool for generating transgenic sugarcane plants with a higher yield, nutritional value and resistance to different diseases and stress. Different explants—axillary buds, leaf sections, meristem tissue (apical), etc.—have been employed for genetic transformation. Although sugarcane has been genetically modified via particle bombardment/biolistics and electroporation, the most effective method is *Agrobacterium*-mediated transformation because of its simplicity and ability to produce single-copy insertions (Suprasanna et al., 2011). Through genetic engineering, significant advancements in the generation of sugarcane transgenic plants resistant to various abiotic stresses (drought, salt, cold) (Budeguer et al., 2021) have been made (Table 10.1). The reports on genetically modified sugarcane plants that were generated in the past ten years are covered in this chapter.

Water availability is the primary element controlling sugarcane output as it influences tillering, sugar production, and culm height (Sugiharto, 2004). Abiotic stress tolerance in plants is significantly regulated by DREB transcription factor proteins. Sugarcane plants (variety RB855156) transformed with *AtDREB2A CA* gene expressed under the control of *ZmRab17* gene promoter (stress-inducible) displayed drought stress tolerance. Transgenic sugarcane plants displayed enhanced sugar production and better bud sprouting (Reis et al., 2014). Similarly, *DREB2* gene isolated from *E. arundinaceus* when transformed in sugarcane (variety Co 86032) conferred drought tolerance. Co-transformation of *EaDREB2* gene and another stress (salt, dehydration, low temperature, wounding) inducible gene pea DNA helicase 45 (*PDH45*) resulted in drought- and salinity-tolerant transgenic plants (Augustine et al., 2015b). Choline dehydrogenase gene (betA, essential enzyme for the production of GB) isolated from bacteria (*Rhizobium meliloti*) when introduced in sugarcane led to enhanced drought tolerance

### TABLE 10.1
### List of Genes Transformed in Sugarcane for Conferring Abiotic Stress Tolerance

| Name of the Transgene | Source Organism of the Transgene | Effect | Reference |
| --- | --- | --- | --- |
| AtDREB2A CA | Arabidopsis thaliana | Drought tolerance | Reis et al. (2014) |
| P5CS | Vigna aconitifolia | Salinity tolerance | Guerzoni et al. (2014) |
| AVP1 | Arabidopsis thaliana | Drought and salinity tolerance | Kumar et al. (2014); Raza et al. (2016) |
| EaHSP70 | Erianthus arundinaceus | Drought and salinity tolerance | Augustine et al. (2015a) |
| EaDREB2 and PDH45 | Erianthus arundinaceus and Pisum sativum | Drought and salinity tolerance | Augustine et al. (2015b) |
| AtBI-1 | Arabidopsis thaliana | Drought tolerance | Ramiro et al. (2016) |
| betA | Rhizobium meliloti | Drought tolerance | Sugiharto (2017) |
| SoP5CS | Saccharum officinarum | Drought tolerance | Li et al. (2018) |
| BcZAT12 | Brassica carinata | Drought and salinity tolerance | Saravanan et al. (2018) |
| EaGly III | Erianthus arundinaceus | Drought and salinity tolerance | Mohanan et al. (2020); Mohanan et al. (2021) |
| OVP2 | Oryza sativa | Salinity tolerance | Khan et al. (2021) |
| AtBBX29 | Arabidopsis thaliana | Drought tolerance | Mbambalala et al. (2021) |
| TERF1 | Solanum lycopersicum | Drought tolerance | Rahman et al. (2021) |
| EaEXPA1 | Erianthus arundinaceus | Drought tolerance | Narayan et al. (2021) |
| SoTUA | Saccharum officinarum | Cold tolerance | Chen et al. (2021) |

as evidenced by increased osmoprotectant GB and 20%-30% increased sugar production. This variety was commercialized by PT Perkebunan Nusantara XI (University of Jember [East Java, Indonesia]) Ajinomoto and is considered the first drought-tolerant sugarcane variety commercially released in the market (Waltz, 2014; Sugiharto, 2017).

High salinity also negatively affects the growth, yield, and sugar concentration of sugarcane. *Vigna aconitifolia* gene *P5CS* (coding for Δ1-pyrroline-5-carboxylate synthetase) when transformed in sugarcane led to enhanced proline content (25%) and decline in MDA levels and sustained photochemical efficiency under salt stress (Guerzoni et al., 2014). Overexpression of the sugarcane *SoP5CS* gene conferred drought tolerance with transgenic plants showing enhanced proline and ABA concentration, the activity of SOD enzyme, and a decline in MDA levels (Li et al., 2018).

Sugarcane plants transformed with *Arabidopsis* vacuolar pyrophosphatase 1 (*AVP1*) exhibited enhanced drought and salinity tolerance (Kumar et al., 2014). Raza et al. (2016) also generated drought-tolerant transgenic plants by transforming *AVP1* gene. Transgenic plants displayed increased transpiration rate, RWC and stomatal conductance. In another report, the transformation of rice *OVP2* gene coding for vacuolar H⁺-PPase in sugarcane variety CP-77/400 resulted in salt-tolerant transgenic plants as demonstrated by higher proline, sucrose, and chlorophyll levels under salt stress (Khan et al., 2021).

*Erianthus arundinaceus* gene *EaHSP70* (coding for HSPs) conferred drought and salinity stress tolerance to sugarcane plants by enhancing photosynthetic efficacy, chlorophyll levels, cell membrane thermostability, and RWC. Under salt stress, transgenic plants displayed improved germination and chlorophyll retention (Augustine et al., 2015a). *Arabidopsis thaliana* Bax Inhibitor-1 gene (*AtBI-1*; anti-cell death gene) conferred drought stress tolerance to sugarcane plants by enhancing sugar (total and reducing) levels and activity of antioxidant enzymes (Ramiro et al., 2016). Transgenic sugarcane plants (variety Co 86032) transformed with *Brassica carinata* gene *BcZAT12* (reported to confer stress tolerance) displayed enhanced drought and salt tolerance. When compared with control plants, transgenic plants exhibited enhanced vigor, proline, chlorophyll and GB levels, RWC, rate of transpiration, and stomatal conductance (Saravanan et al., 2018).

Enhanced drought and salt tolerance were observed in transgenic sugarcane (variety Co 86032) plants transformed with *Glyoxalase III* gene (*EaGly III*; ROS scavenging enzyme responsible for detoxification of cytotoxic compound methylglyoxal) of *E. arundinaceus* (a wild relative of sugarcane). Transgenic plants exhibited enhanced antioxidant enzyme activities (SOD and peroxidase), soluble sugars, chlorophyll, and proline levels, and improved RWC and photosynthetic efficiency (Mohanan et al., 2020; Mohanan et al., 2021). *Arabidopsis thaliana* B-box gene (*AtBBX29*, shown to regulate stress responses), when introduced in sugarcane plants, conferred increased drought tolerance. Transgenic plants displayed delayed senescence, enhanced proline accumulation, and antioxidant enzyme activity and sustained RWC and photosynthesis (Mbambalala et al., 2021). Transgenic sugarcane (cultivar XintaitangR22) plants harboring *TERF1* (tomato ethylene-responsive factor 1) gene when exposed to water deficit conditions showed enhanced proline, GB and soluble sugar build-up and decreased MDA and hydrogen peroxide ($H_2O_2$) levels

(Rahman et al., 2021). Similarly, *EaEXPA1*, an expansion gene involved in the loosening of the cell was isolated from *E. arundinaceus* and transformed in sugarcane (variety Co 86032). Transgenic plants when subjected to drought stress showed enhanced RWC, chlorophyll levels, photosynthetic efficacy, and cell membrane thermostability when compared with control untransformed plants (Narayan et al., 2021). Recently, Chen et al. (2021) also showed cold tolerance in transgenic sugarcane (cold sensitive cultivar ROC22) plants transformed with the sugarcane alpha (α)-tubulin (*SoTUA*) gene. The transgenic plants displayed enhanced accumulation of soluble sugars and proteins, antioxidant enzyme (peroxidase) activity, up-regulation of genes involved in conferring cold tolerance, and a decline in MDA levels.

Despite the fact that several genes have been genetically transformed into sugarcane to produce abiotic stress-tolerant transgenic plants, there are still numerous challenges to unravel before the commercial tolerant varieties are made available in the market. Recently genome editing (GE) technology (transcription activator-like effector nucleases and clustered regularly interspaced short palindromic repeat [CRISPR]-Associated Nuclease 9 [Cas9]) has made it possible to delete, replace, and insert DNA at specific locations in the genome for the improvement of sugarcane. There are some successful reports in which GE has been used to modify the composition of the cell wall (Jung & Altpeter, 2016), saccharification efficiency (Kannan et al., 2018), and chlorophyll biosynthesis (Eid et al., 2021) in sugarcane. Though sugarcane GE is associated with certain challenges, in the coming years with the availability of whole genome sequences and novel bioinformatic tools for designing gRNA (Mohan, 2016; Mohan et al., 2022), the GE platform can be used to create designer sugarcane by manipulating genes for conferring abiotic stress tolerance. This technique may be particularly effective in nations that have restricted the testing of genetically modified crops.

## 10.6 REGULATORY ROLE OF miRNAs IN ABIOTIC RESPONSES

miRNAs are anticipated to be one of the crucial regulatory factors that are connected with several essential traits in sugarcane. However, recognition of mature miRNAs and their expression studies is insufficient in sugarcane due to unavailability of a well-sequenced complete genome, and moreover, its inherent intricacies further limit the study of miRNA in sugarcane (Vettore et al., 2003). miRNAs play a critical regulatory role in providing tolerance to abiotic stresses and several target genes for miRNA have been detected which impart resistance to various stress in plants. Novel miRNAs for drought and salinity stress resistance have been detected; however the regulatory function of miRNAs—particularly in drought stress tolerance—has been investigated in detail in sugarcane (Ferreira et al., 2012; Thiebaut et al., 2012a; Gentile et al., 2013; Lin et al., 2014; Swapna & Kumar, 2017).

Patade and Suprasanna (2010) examined the expression profile of mature miR159 in leaves of sugarcane subjected to stress for a longer duration (150 mM of NaCl or PEG 8000–20% w/v for 15 days) and shorter duration (200 mM of NaCl or PEG -20% w/v for 24 hours). No substantial change in miRNA transcript level was observed in samples subjected to stress for longer periods, while on the contrary, significant up-regulation was found for samples exposed to stress for a shorter duration in comparison to control. Thiebaut et al. (2012a), based on computational studies, detected 623 candidates for new miRNA (mature) by employing the sequences from eight cultivars of sugarcane, and these were subjected to biotic as well as abiotic stresses. Of these miRNAs, 44 were categorized as high confidence and 67 were precisely detected in water stress (deficit) conditions, with 20 in common to both susceptible and tolerant assays. The predicted targets were zinc-finger, serine/threonine kinases, Myb proteins, etc. Ferreira et al. (2012) reported 18 miRNA families in cultivars of sugarcane (RB867515 with high tolerance; RB855536 with low tolerance) that differed in response to drought tolerance. Out of these, seven showed differential expression during the drought that varied with the duration and degree of stress conditions. Six precursors and targets for the miRNA that showed differential expression were predicted and validated by RT-qPCR. Another study by Gentile et al. (2013) reported 18 miRNA families comprising 30 mature miRNA sequences. They used a small RNA deep sequencing technique and two cultivars of sugarcane whose sensitivity to drought stress varied. Out of these, 13 miRNAs displayed differential expression in plants under drought stress and seven miRNAs were found to differentially express in both cultivars. The target genes for some of the miRNAs (with differential expression) were comprised of transporters, TFs, and proteins linked to flower development and senescence. Lin et al. (2014) examined the expression pattern of the miRNA families in ROC22 leaves (drought-tolerant) subjected to stress; 34 new and 23 conserved miRNA families were detected, and 438 supposed targets for 44 miRNA families were described. From expression studies, 11 miRNAs showed differential expression in control and drought-stressed samples. Selvi et al. (2021) performed deep sequencing to analyze the miRNA regulation in response to drought stress in sugarcane. A total of 1224 conserved miRNAs were identified that belonged to 89 miRNA families; 38% of miRNAs (differentially expressed) were found to be recurrent in both varieties. In addition, 435 novel miRNAs were also detected; 145 miRNAs displayed differential expression in drought-sensitive varieties (V1–31) and 143 in drought-tolerant (V2–31) varieties.

In addition to drought stress, a large number of miRNAs displayed greater expression levels in salt-treated samples. Carnavale Bottino et al. (2013) detected 11 miRNAs that displayed high expression levels in severe salt-stressed samples in comparison to mildly treated samples of sugarcane. Mazalmazraei et al. (2021) examined the expression level of nine miRNAs (miR160, 164, 172, 390, 393, 408, 529, 827,

1432) and their target genes in three cultivars of sugarcane that differed in sensitivity to salt stress (CP-48-sensitive, CP-57-semi-tolerant, CP-69-tolerant). The miRNAs showed differential expression patterns during salinity stress in all three cultivars. Moreover, there was a substantial difference in the expression level of all nine miRNAs and their target genes in salt treated samples than control. All miRNAs were found to be down-regulated, while on the contrary, expression of their target genes increased under stress conditions. The target genes included TFs, genes involved in hormone signaling, and metabolic enzymes, thus helping the plants to overcome salinity stress.

Sugarcane is sensitive to low temperatures, and this sensitivity varies with the varieties of sugarcane. Thiebaut et al. (2012b) studied the role of miRNAs and their corresponding targets in imparting tolerance in sugarcane under low-temperature stress, i.e., 4°C for 24 hours. It was observed that miR319 expression increased in both roots and shoots during cold stress, and on the contrary, the expression of predicted targets for this miRNA decreased. Yang et al. (2017) reported 412 miRNAs in sugarcane that included 261 known and 151 new miRNA. Out of these, 62 displayed considerable variation in expression under cold stress; 34 of these miRNAs were up-regulated, whereas 28 were down-regulated. KEGG and GO study suggested the role of these miRNAs in stress-associated biological pathways. Huang et al. (2021) described a total of 856 miRNAs in *S. spontaneum* that comprised conserved miRNAs (20) and novel miRNAs (836) belonging to 13 families derived from nine libraries. Of these, 109 miRNAs displayed differential expression patterns under cold stress, including 105 novel miRNA and four known miRNAs. In comparison to control samples, the numbers of differentially expressed miRNAs after cold stress for 0.5 hours and one hour, respectively, were 53 and 94.

## 10.7 CONCLUSIONS AND FUTURE PROSPECTS

Sugarcane is an important crop plant, predominantly found in tropical countries. However, its production is largely affected by various abiotic stresses. To date, there are limited studies available on the effect of abiotic stresses on sugarcane. The miRNAs related to abiotic stresses like drought, cold, waterlogging, and salinity tolerance have been documented in sugarcane. Nevertheless, the studies are limited due to lack of availability of genome sequence information. Owing to the complex and polyploid nature of its genome, advanced techniques like NGS, genome selection, and genome editing (GE) offer various advantages to improve its productivity, as well as its resistance to both abiotic and biotic stresses. The advancement in sugarcane monoploid genome sequence has paved way for the researchers in identifying molecular markers and genes related to important agronomic traits that are important for efficient sugarcane breeding programs. Moreover, CRISPR/Cas9 technology—because of its efficacy and exact editing of the genome—has currently appeared as a novel approach to modify the genome of sugarcane and develop new varieties with the desired traits. This system can further help in manipulating several genes that are significant to agronomical traits.

- The miRNA or genes associated with the stress responses are important and can be employed in developing cultivars with the desired traits. The use of GE tools can assist in making effective usage of miRNAs to improve the desired traits in sugarcane.
- The use of sugarcane germplasm and its characterization and assessment via various methods like linkage mapping, MAS, GS, and GWAS can be helpful for breeding programs in developing strategies to reduce the deleterious effects of environmental stress on sugarcane cultivation. These tools further support the detection of valuable molecular markers associated with genes and QTLs by applying methods like fine genetic mapping and bulked segregant analysis.
- In addition to classical pathways, the role of novel loci of miRNAs origin involved in conferring abiotic stress tolerance needs to be decoded so that they are put to efficient use. Thus, genomic-assisted breeding can help in decreasing the negative effect of abiotic stresses on sugarcane cropping, and the usage of CRISPR-based GE tools can help in engineering abiotic stress-resistant sugarcane.

## REFERENCES

Afzal, Zunaira, T. C. Howton, Yali Sun, and M. Shahid Mukhtar. "The roles of aquaporins in plant stress responses." *Journal of Developmental Biology* 4, no. 1 (2016): 9. https://doi.org/10.3390%2Fjdb4010009

Ali, Ahmad, Mehran Khan, Rahat Sharif, Muhammad Mujtaba, and San-Ji Gao. "Sugarcane Omics: An update on the current status of research and crop improvement." *Plants* 8, no. 9 (2019): 344. https://doi.org/10.3390/plants8090344

Augustine, Sruthy Maria, J. Ashwin Narayan, Divya P. Syamaladevi, C. Appunu, M. Chakravarthi, V. Ravichandran, and N. Subramonian. "*Erianthus arundinaceus* HSP70 (EaHSP70) overexpression increases drought and salinity tolerance in sugarcane (*Saccharum* spp. hybrid)." *Plant Science* 232 (2015a): 23–34. https://doi.org/10.1016/j.plantsci.2014.12.012

Augustine, Sruthy Maria, J. Ashwin Narayan, Divya P. Syamaladevi, C. Appunu, M. Chakravarthi, V. Ravichandran, Narendra Tuteja, and N. Subramonian. "Overexpression of EaDREB2 and pyramiding of EaDREB2 with the pea DNA helicase gene (PDH45) enhance drought and salinity tolerance in sugarcane (*Saccharum* spp. hybrid)." *Plant Cell Reports* 34, no. 2 (2015b): 247–263. https://doi.org/10.1007/s00299-014-1704-6

Barnabas, Leonard, Ashwin Ramadass, Ramesh Sundar Amalraj, Malathi Palaniyandi, and Viswanathan Rasappa. "Sugarcane proteomics: An update on current status, challenges, and future prospects." *Proteomics* 15, no. 10 (2015): 1658–1670. http://doi.org/10.1002/pmic.201400463

Barreto, Fernanda Zatti, João Ricardo Bachega Feijó Rosa, Thiago Willian Almeida Balsalobre, Maria Marta Pastina, Renato Rodrigues Silva, Hermann Paulo Hoffmann, Anete

Pereira de Souza, Antonio Augusto Franco Garcia, and Monalisa Sampaio Carneiro. "A genome-wide association study identified loci for yield component traits in sugarcane (*Saccharum* spp.)." *PloS One* 14, no. 7 (2019): e0219843. http://doi.org/10.1371/journal.pone.0219843.

Bartel, David P. "MicroRNAs: Target recognition and regulatory functions." *Cell* 136, no. 2 (2009): 215–233. http://doi.org/10.1016/j.cell.2009.01.002

Begcy, Kevin, Eduardo D. Mariano, Agustina Gentile, Carolina G. Lembke, Sonia Marli Zingaretti, Glaucia M. Souza, and Marcelo Menossi. "A novel stress-induced sugarcane gene confers tolerance to drought, salt and oxidative stress in transgenic tobacco plants." *PloS One* (2012): e44697. https://doi.org/10.1371/journal.pone.0044697. Epub2010

Begcy, Kevin, Eduardo D. Mariano, Carolina G. Lembke, Sonia Marli Zingaretti, Glaucia M. Souza, Pedro Araújo, and Marcelo Menossi. "Overexpression of an evolutionarily conserved drought-responsive sugarcane gene enhances salinity and drought resilience." *Annals of Botany* 124, no. 4 (2019): 691–700. https://doi.org/10.1093/aob/mcz044.

Belesini, A. A., F. M. S. Carvalho, B. R. Telles, G. M. De Castro, P. F. Giachetto, J. S. Vantini, S. D. Carlin, J. O. Cazetta, D. G. Pinheiro, and M. I. T. Ferro. "De novo transcriptome assembly of sugarcane leaves submitted to prolonged water-deficit stress." *Genetics and Molecular Research* (2017). https://doi.org/10.4238/gmr16028845

Budeguer, Florencia, Ramón Atanasio Enrique, María Francisca Perera, Josefina Racedo, Atilio Pedro Castagnaro, Aldo Sergio Noguera, and Bjorn Welin. "Genetic transformation of sugarcane, current status and future prospects." *Frontiers in Plant Science* (2021): 2467. https://doi.org/10.3389/fpls.2021.768609

Budzinski, Ilara Gabriela Frasson, Fabricio Edgar de Moraes, Thais Regiani Cataldi, Lívia Maria Franceschini, and Carlos Alberto Labate. "Network analyses and data integration of proteomics and metabolomics from leaves of two contrasting varieties of sugarcane in response to drought." *Frontiers in Plant Science* 10 (2019): 1524. https://doi.org/10.3389/fpls.2019.01524

Cardoso-Silva, Claudio Benicio, Estela Araujo Costa, Melina Cristina Mancini, Thiago Willian Almeida Balsalobre, Lucas Eduardo Costa Canesin, Luciana Rossini Pinto, Monalisa Sampaio Carneiro, Antonio Augusto Franco Garcia, Anete Pereira de Souza, and Renato Vicentini. "De novo assembly and transcriptome analysis of contrasting sugarcane varieties." *PloS One* 9, no. 2 (2014): e88462. https://doi.org/10.1371/journal.pone.0088462.

Carnavale Bottino, Mariana, Sabrina Rosario, Clicia Grativol, Flávia Thiebaut, Cristian Antonio Rojas, Laurent Farrineli, Adriana Silva Hemerly, and Paulo Cavalcanti Gomes Ferreira. "High-throughput sequencing of small RNA transcriptome reveals salt stress regulated microRNAs in sugarcane." *PloS One* 8, no. 3 (2013): e59423. http://doi.org/10.1371/journal.pone.0059423

Castro-Nava, Sergio, and Enrique López-Rubio. "Thermotolerance and physiological traits as fast tools to heat tolerance selection in experimental sugarcane genotypes." *Agriculture* 9, no. 12 (2019): 251–258. http://doi.org/10.3390/agriculture9120251

Chen, Jiao-Yun, Qaisar Khan, Bo Sun, Li-Hua Tang, Li-Tao Yang, Bao-Qing Zhang, Xing-Yong Xiu, Deng-Feng Dong, and Yang-Rui Li. "Overexpression of sugarcane SoTUA gene enhances cold tolerance in transgenic sugarcane." *Agronomy Journal* 113, no. 6 (2021): 4993–5005. https://doi.org/10.1002/agj2.20618

Chen, Sixue, and Alice C. Harmon. "Advances in plant proteomics." *Proteomics* 6, no. 20 (2006): 5504–5516. https://doi.org/10.1002/pmic.200600143

Chiconato, Denise A., Marília G. de Santana Costa, Tiago S. Balbuena, Rana Munns, and Durvalina MM Dos Santos. "Proteomic analysis of young sugarcane plants with contrasting salt tolerance." *Functional Plant Biology* 48, no. 6 (2021): 588–596. https://doi.org/10.1071/FP20314

Cox, Jürgen, and Matthias Mann. "Is proteomics the new genomics?." *Cell* 130, no. 3 (2007): 395–398. https://doi.org/10.1016/j.cell.2007.07.032

de Andrade, Aline Franciel, Rilner Alves Flores, Derblai Casaroli, Amanda Magalhães Bueno, Marco Aurélio Pessoa-de-Souza, Carlos Cesar Silva Jardim, Klaus de Oliveira Abdala, Eduardo Parra Marques, and Marcio Mesquita. "Biometric and physiological relationships and yield of sugarcane in relation to soil application of potassium." *Sugar Tech* 24, no. 2 (2022): 473–484. https://doi.org/10.1007/s12355-021-01032-z

De Andrade, Larissa Mara, Paula Macedo Nobile, Rafael Vasconcelos Ribeiro, João Felipe Nebó Carlos De Oliveira, Antonio Vargas de Oliveira Figueira, Luis Tadeu Marques Frigel, Daniel Nunes et al. "Characterization of PIP2 aquaporins in *Saccharum* hybrids." *Plant Gene* 5 (2016): 31–37. https://doi.org/10.1016/j.plgene.2015.11.004

Dharshini, S., M. Chakravarthi, V. M. Manoj, M. Naveenarani, Ravinder Kumar, Minturam Meena, Bakshi Ram, and C. Appunu. "De novo sequencing and transcriptome analysis of a low temperature tolerant *Saccharum spontaneum* clone IND 00–1037." *Journal of Biotechnology* 231 (2016): 280–294. https://doi.org/10.1016/j.jbiotec.2016.05.036.

Dlamini, Philani Justice. "Drought stress tolerance mechanisms and breeding effort in sugarcane: A review of progress and constraints in South Africa." *Plant Stress* 2 (2021): 100027. https://doi.org/10.1016/j.stress.2021.100027

Eid, Ayman, Chakravarthi Mohan, Sara Sanchez, Duoduo Wang, and Fredy Altpeter. "Multiallelic, targeted mutagenesis of magnesium chelatase with CRISPR/Cas9 provides a rapidly scorable phenotype in highly polyploid sugarcane." *Frontiers in Genome Editing* 3 (2021): 654996. https://doi.org/10.3389/fgeed.2021.654996

Ferreira, Thaís Helena, Agustina Gentile, Romel Duarte Vilela, Gustavo Gilson Lacerda Costa, Lara Isys Dias, Laurício Endres, and Marcelo Menossi. "microRNAs associated with drought response in the bioenergy crop sugarcane (*Saccharum* spp.)." *PloS One* 7 (2012): e46703. https://doi.org/10.1371/jour-nal.pone.0046703

Ferreira, Thais HS, Max S. Tsunada, Denis Bassi, Pedro Araújo, Lucia Mattiello, Giovanna V. Guidelli, Germanna L. Righetto, Vanessa R. Gonçalves, Prakash Lakshmanan, and Marcelo Menossi. "Sugarcane water stress tolerance mechanisms and its implications on developing biotechnology solutions." *Frontiers in Plant Science* 8 (2017): 1077. https://doi.org/10.3389/fpls.2017.01077.

Gawander, J. "Impact of climate change on sugar-cane production in Fiji." *World Meteorological Organization Bulletin* 56, no. 1 (2007): 34–39.

Gentile, Agustina, Thaís H. Ferreira, Raphael S. Mattos, Lara I. Dias, Andrea A. Hoshino, Monalisa S. Carneiro, Glaucia M. Souza et al. "Effects of drought on the microtranscriptome of field-grown sugarcane plants." *Planta* 237 no. 3 (2013): 783–798. https://doi.org/10.1007/s00425-012-1795-7

Gomathi, R., and K. Chandran. "Effect of water logging on growth and yield of sugarcane clones." *Sugarcane Breeding Institute (SBI-ICAR). Quarterly News Letter* 29, no. 4 (2009): 1–2.

Gomathi, R., P. N. Gururaja Rao, K. Chandran, and A. Selvi. "Adaptive responses of sugarcane to waterlogging stress: An over view." *Sugar Tech* 17, no. 4 (2015): 325–338. http://dx.doi.org/10.21273/HORTSCI13875-19

Gomathi, R., S. Shiyamala, S. Vasantha, and A. Suganya. "Optimization of temperature conditions for screening thermotolerance in sugarcane through temperature induction response (TIR) technique." *International Journal of Sciences* (2014): 5–18. https://ssrn.com/abstract=2573695

Guerzoni, Julia Tufino Silva, Nathalia Geraldo Belintani, Rosangela Maria Pinto Moreira, Andrea Akemi Hoshino, Douglas Silva Domingues, and Luiz Gonzaga Esteves Vieira. "Stress-induced Δ1-pyrroline-5-carboxylate synthetase (P5CS) gene confers tolerance to salt stress in transgenic sugarcane." *Acta Physiologiae Plantarum* 36, no. 9 (2014): 2309–2319. https://doi.org/10.1007/s11738-014-1579-8

Gupta, Bhaskar, and Bingru Huang. "Mechanism of salinity tolerance in plants: Physiological, biochemical, and molecular characterization." *International Journal of Genomics* 2014 (2014): 1–18. https://doi.org/10.1155/2014/701596

Han, Chao, and Pingfang Yang. "Studies on the molecular mechanisms of seed germination." *Proteomics* 15, no. 10 (2015): 1671–1679. https://doi.org/10.1002/pmic.201400375

Heinz, Don J., and Grace WP Mee. "Plant Differentiation from Callus Tissue of *Saccharum* Species 1." *Crop Science* 9, no. 3 (1969): 346–348. https://doi.org/10.2135/cropsci1969.0011183X000900030030x

Hernández-Pérez, César A., Fernando Carlos Gómez-Merino, José L. Spinoso-Castillo, and Jericó J. Bello-Bello. "In vitro screening of sugarcane cultivars (*Saccharum* spp. hybrids) for tolerance to polyethylene glycol-induced water stress." *Agronomy* 11, no. 3 (2021): 598. https://doi.org/10.3390/agronomy11030598

Hoang, Dinh Thai, Takaragawa Hiroo, and Kawamitsu Yoshinobu. "Nitrogen use efficiency and drought tolerant ability of various sugarcane varieties under drought stress at early growth stage." *Plant Production Science* 22, no. 2 (2019): 250–261. https://doi.org/10.1080/1343943X.2018.1540277

Holme, Inger Bæksted, Toni Wendt, and Preben Bach Holm. "Intragenesis and cisgenesis as alternatives to transgenic crop development." *Plant Biotechnology Journal* 11, no. 4 (2013): 395–407. https://doi.org/10.1111/pbi.12055

Hu, Honghong, and Lizhong Xiong. "Genetic engineering and breeding of drought-resistant crops." *Annual Review of Plant Biology* 65 (2014): 715–741. https://doi.org/10.1146/annurev-arplant-050213-040000

Huang, Xing, Yongsheng Liang, Baoqing Zhang, Xiupeng Song, Yangrui Li, Zhengqiang Qin, Dewei Li et al. "Integration of Transcriptional and Post-transcriptional Analysis Revealed the Early Response Mechanism of Sugarcane to Cold Stress." *Frontiers in Genetics* 11 (2021): 581993. https://doi:10.3389/fgene.2020.581993

Hussain, Sadam, Abdul Khaliq, Umer Mehmood, Tauqeer Qadir, Muhammad Saqib, Muhammad Amjed Iqbal, and Saddam Hussain. "Sugarcane production under changing climate: effects of environmental vulnerabilities on sugarcane diseases, insects and weeds" In *Climate Change and Agriculture*, edited by Saddam Hussain. London: IntechOpen, 2018. https://doi: 10.5772/intechopen.81131

Ji, Hongtao, José M. Pardo, Giorgia Batelli, Michael J. Van Oosten, Ray A. Bressan, and Xia Li. "The salt overly sensitive (SOS) pathway: Established and emerging roles." *Molecular Plant* 6, no. 2 (2013): 275–286. https://doi.org/10.1093/mp/sst017

Jung, Je Hyeong, and Fredy Altpeter. "TALEN mediated targeted mutagenesis of the caffeic acid O-methyltransferase in highly polyploid sugarcane improves cell wall composition for production of bioethanol." *Plant Molecular Biology* 92, no. 1 (2016): 131–142. https://doi.org/10.1007/s11103-016-0499-y

Kannan, Baskaran, Je Hyeong Jung, Geoffrey W. Moxley, Sun-Mi Lee, and Fredy Altpeter. "TALEN-mediated targeted mutagenesis of more than 100 COMT copies/alleles in highly polyploid sugarcane improves saccharification efficiency without compromising biomass yield." *Plant Biotechnology Journal* 16, no. 4 (2018): 856–866. https://doi.org/10.1111/pbi.12833

Kaur, Lovejot, S. Dharshini, Bakshi Ram, and C. Appunu. "Sugarcane genomics and transcriptomics." In *Sugarcane Biotechnology: Challenges and Prospects*, pp. 13–32. Cham: Springer, 2017.

Kaur, Navdeep, Shubham Sharma, Mirza Hasanuzzaman, and Pratap Kumar Pati. "Genome editing: A promising approach for achieving abiotic stress tolerance in plants." *International Journal of Genomics* 2022 (2022). https://doi.org/10.1155/2022/5547231

Khan, Imtiaz Ahmed, Muhammad Umar Dahot, Nighat Seema, Shafqat Yasmin, Sajida Bibi, Saboohi Raza, and Abdullah Khatri. "Genetic variability in sugarcane plantlets developed through in vitro mutagenesis." *Pakistan Journal of Botany* 41, no. 1 (2009): 153–166.

Khan, Mohammad Sayyar, Sayed Usman Ali Shah, Mazhar Ullah, Muhammad Zaheer Ahmad, Waqar Ahmad, and Asad Jan. "Genetic engineering of sugarcane with the rice tonoplast H+-ppase (*OVP2*) gene to improve sucrose content and salt tolerance." *Pakistan Journal of Botany* 53, no. 3 (2021): 813–821. http://dx.doi.org/10.30848/PJB2021-3(36)

Khonghintaisong, Jidapa, Patcharin Songsri, B. Toomsan, and Nuntawoot Jongrungklang. "Rooting and physiological trait responses to early drought stress of sugarcane cultivars." *Sugar Tech* 20, no. 4 (2018): 396–406. https://doi.org/10.1007/s12355-017-0564-0

Khueychai, Siriporn, Nisachon Jangpromma, Sakda Daduang, Prasit Jaisil, Khomsorn Lomthaisong, Apisak Dhiravisit, and Sompong Klaynongsruang. "Comparative proteomic analysis of leaves, leaf sheaths, and roots of drought-contrasting sugarcane cultivars in response to drought stress." *Acta Physiologiae Plantarum* 37, no. 4 (2015): 1–16. http://doi: 10.1155/2022/5547231

Koch, Aimée C., Sumita Ramgareeb, R. Stuart Rutherford, Sandra J. Snyman, and M. Paula Watt. "An in vitro mutagenesis protocol for the production of sugarcane tolerant to the herbicide imazapyr." *Vitro Cellular & Developmental Biology-Plant* 48, no. 4 (2012): 417–427. http://dx.doi.org/10.1007/s11627-012-9448-x

Kumar, Tanweer, Muhammad Ramzan Khan, Zaheer Abbas, and Ghulam Muhammad Ali. "Genetic improvement of sugarcane for drought and salinity stress tolerance using *Arabidopsis* vacuolar pyrophosphatase (AVP1) gene." *Molecular Biotechnology* 56, no. 3 (2014): 199–209. https://doi.org/10.1007/s12033-013-9695-z

Lakshmanan, Prakash, and Nicole Robinson. "Stress physiology: Abiotic stresses." *Sugarcane: Physiology, Biochemistry, and Functional Biology* (2013): 411–434. https://doi.org/10.1002/9781118771280.ch16

Lee, Dong-Kyu, Soohyun Ahn, Hae Yoon Cho, Hye Young Yun, Jeong Hill Park, Johan Lim, Jeongmi Lee, and Sung Won Kwon. "Metabolic response induced by parasitic plant-fungus interactions hinder amino sugar and nucleotide sugar metabolism in the host." *Scientific Reports* 6, no. 1 (2016): 1–11. http://doi:10.1038/srep37434

Lembke, Carolina Gimiliani, Milton Yutaka Nishiyama, Paloma Mieko Sato, Rodrigo Fandino de Andrade, and Glaucia Mendes Souza. "Identification of sense and antisense transcripts regulated by drought in sugarcane." *Plant Molecular Biology* 79, no. 4 (2012): 461–477. http://doi.org/10.1007/s11103-012-9922-1

Li, Chang-Ning, Manoj-Kumar Srivastava, Qian Nong, Li-Tao Yang, and Yang-Rui Li. "Molecular cloning and characterization of SoNCED, a novel gene encoding 9-cis-epoxycarotenoid dioxygenase from sugarcane (*Saccharum officinarum* L.)." *Genes & Genomics* 35, no. 1 (2013): 101–109. http://dx.doi.org/10.1007/s13258-013-0065-9

Li, Jian, Thi-Thu Phan, Yang-Rui Li, Yong-Xiu Xing, and Li-Tao Yang. "Isolation, transformation and overexpression of sugarcane SoP5CS gene for drought tolerance improvement." *Sugar Tech* 20, no. 4 (2018): 464–473. https://doi.org/10.1007/s12355-017-0568-9

Lin, Sheng, Ting Chen, Xianjin Qin, Hongmiao Wu, Muhammad Azam Khan, and Wenxiong Lin. "Identification of microrna families expressed in sugarcane leaves subjected to drought stress and the targets thereof." *Pakistan Journal of Agricultural Sciences* 51, no. 4 (2014): 925–934.

Mahlanza, Tendekai, R. Stuart Rutherford, Sandy J. Snyman, and M. Paula Watt. "In vitro generation of somaclonal variant plants of sugarcane for tolerance to *Fusarium sacchari*." *Plant Cell Reports* 32, no. 2 (2013): 249–262. http://dx.doi.org/10.1007/s00299-012-1359-0

Manchanda, Pooja, Ajinder Kaur, and Satbir Singh Gosal. "Somaclonal variation for sugarcane improvement." In *Biotechnologies of Crop Improvement*, Vol. 1, pp. 299–326. Cham: Springer, 2018. https://doi.org/10.1007/978-3-319-78283-6_9

Mangrauthia, Satendra Kumar, Surekha Agarwal, B. Sailaja, N. Sarla, and S. R. Voleti. "Transcriptome analysis of *Oryza sativa* (rice) seed germination at high temperature shows dynamics of genome expression associated with hormones signalling and abiotic stress pathways." *Tropical Plant Biology* 9, no. 4 (2016): 215–228. https://doi.org/10.1007/s12042-016-9170-7

Maranho, Rone C., Mariana M. Benez, Gustavo B. Maranho, Adeline Neiverth, Marise F. Santos, Ana Lúcia O. Carvalho, Adriana Gonela, Claudete A. Mangolin, and P. S. Maria de Fátima. "Proteomic analysis of axillary buds of sugarcane at different cutting stages: Evidence for alterations in axillary bud gene expression." *Crop and Pasture Science* 70, no. 7 (2019): 622–633. https://doi.org/10.1071/CP19115

Mazalmazraei, Tofigh, Khosro Mehdikhanlou, and Daryoosh Nabati Ahmadi. "Comparative analysis of differentially expressed miRNAs in leaves of three sugarcane (*Saacharum Officinarum* L.) cultivars during salinity stress." *Preprint* (2021). https://doi.org/10.21203/rs.3.rs-769558/v1

Mbambalala, Nelisa, Sanjib K. Panda, and Christell van der Vyver. "Overexpression of AtBBX29 improves drought tolerance by maintaining photosynthesis and enhancing the antioxidant and osmolyte capacity of sugarcane plants." *Plant Molecular Biology Reporter* 39, no. 2 (2021): 419–433. https://doi.org/10.1007/s11105-020-01261-8

Meena, Mintu Ram, Ravinder Kumar, Appunu Chinnaswamy, Ramaiyan Karuppaiyan, Neeraj Kulshreshtha, and Bakshi Ram. "Current breeding and genomic approaches to enhance the cane and sugar productivity under abiotic stress conditions." *3 Biotech* 10, no. 10 (2020): 1–18. https://doi.org/10.1007/s13205-020-02416-w

Mohan, Chakravarthi. "Genome editing in sugarcane: Challenges ahead." *Frontiers in Plant Science* 7 (2016): 1542. https://doi.org/10.3389/fpls.2016.01542

Mohan, Chakravarthi, Mona Easterling, and Yuan-Yeu Yau. "Gene editing technologies for sugarcane improvement: Opportunities and limitations." *Sugar Tech* (2022): 1–17. https://doi.org/10.1007/s12355-021-01045-8

Mohanan, Manoj Vadakkenchery, Anunanthini Pushpanathan, Sarath Padmanabhan, Thelakat Sasikumar, Ashwin Narayan Jayanarayanan, Dharshini Selvarajan, Sathishkumar Ramalingam, Bakshi Ram, and Appunu Chinnaswamy. "Overexpression of Glyoxalase III gene in transgenic sugarcane confers enhanced performance under salinity stress." *Journal of Plant Research* 134, no. 5 (2021): 1083–1094. https://doi.org/10.1007/s10265-021-01300-9

Mohanan, Manoj Vadakkenchery, Anunanthini Pushpanathan, Sarath Padmanabhan Thelakat Sasikumar, Dharshini Selvarajan, Ashwin Narayan Jayanarayanan, Sathishkumar Ramalingam, Sathyamoorthy Nagaranai Karuppasamy, Ramanathan Subbiah, Bakshi Ram, and Appunu Chinnaswamy. "Ectopic expression of DJ-1/PfpI domain containing Erianthus arundinaceus Glyoxalase III (EaGly III) enhances drought tolerance in sugarcane." *Plant Cell Reports* 39, no. 11 (2020): 1581–1594. https://doi.org/10.1007/s00299-020-02585-1

Munns, Rana, and Mark Tester. "Mechanisms of salinity tolerance." *Annual Review of Plant Biology* 59 (2008): 651. https://doi.org/10.1146/annurev.arplant.59.032607.092911

Mustafa, Ghulam, Faiz Ahmad Joyia, Sultana Anwar, Aqsa Parvaiz, and Muhammad Sarwar Khan. "Biotechnological interventions for the improvement of sugarcane crop and sugar production." *Sugarcane-Technology and Research*, pp. 113–138. London: IntechOpen, 2018. https://doi.org/10.5772/intechopen.71496

Naheed, Rashda, Muhammad Arfan, Fozia Farhat, Siddra Ijaz, and Hamza Khalid. "Acclimatization of drought tolerance with Somaclonal variants of sugarcane (*Saccharum officinarum* L.)." *Advancements in Life Sciences* 8, no. 1 (2020): 57–62.

Naheed, Rashda, Muhammad Arfan, Siddra Ijaz, and Muhammad Shahbaz. "Induction of somaclonal variation in selected drought sensitive genotype of sugarcane (*Sachharum officinarum*)." *International Journal of Agricultural Biology* 20 (2018): 777–783. https://doi.org/10.17957/IJAB/15.0564

Narayan, J. Ashwin, M. Chakravarthi, Gauri Nerkar, V. M. Manoj, S. Dharshini, N. Subramonian, M. N. Premachandran et al. "Overexpression of expansin EaEXPA1, a cell wall loosening protein enhances drought tolerance in sugarcane." *Industrial Crops and Products* 159 (2021): 113035. https://doi.org/10.1016/j.indcrop.2020.113035

Narwade, Ajay V., Kiran Bhagat, D. V. Patil, Yogeshwar Singh, A. Kumari, Y. G. Ban, H. S. Thakare, and C. Singh. "Abiotic stress responses in sugarcane." In Pasala, R.K., Bhagat, K., & Singh Y. (Eds.), *Challenges and Prospective of Plant Abiotic Stress*, pp. 419–446. Today & Tomorrow's Printers and Publishers, New Delhi, India, 2015.

Park, Jong-Won, Thiago R. Benatti, Thiago Marconi, Qingyi Yu, Nora Solis-Gracia, Victoria Mora, and Jorge A. Da Silva. "Cold responsive gene expression profiling of sugarcane and *Saccharum spontaneum* with functional analysis of a cold inducible *Saccharum* homolog of NOD26-like intrinsic protein to salt and water stress." *PloSone* 10, no. 5 (2015): e0125810. https://doi: 10.1371/journal.pone.0125810

Park, Soo Kwon, Young Jun Jung, Jung Ro Lee, Young Mee Lee, Ho Hee Jang, Seung Sik Lee, Jin Ho Park et al. "Heat-shock and redox-dependent functional switching of an h-type *Arabidopsis* thioredoxin from a disulfide reductase to a molecular chaperone." *Plant Physiology* 150, no. 2 (2009): 552–561. http://dx.doi.org/10.1104/pp.109.135426

Passamani, Lucas Z., Roberta R. Barbosa, Ricardo S. Reis, Angelo S. Heringer, Patricia L. Rangel, Claudete Santa-Catarina, Clícia Grativol, Carlos FM Veiga, Goncalo A. Souza-Filho, and Vanildo Silveira. "Salt stress induces changes in the proteomic profile of micropropagated sugarcane shoots." *PLoS One* 12, no. 4 (2017). https://doi.org/10.1371/journal.pone.0176076.

Patade, Vikas Yadav, and Penna Suprasanna. "Radiation induced in vitro mutagenesis for sugarcane improvement." *Sugar Tech* 10 (2008): 14–19. https://doi.org/10.1007/s12355-008-0002-4

Patade, Vikas Yadav, and Penna Suprasanna. "Short-term salt and PEG stresses regulate expression of MicroRNA, miR159 in sugarcane leaves." *Journal of Crop Science and Biotechnology* 13, no. 3 (2010): 177–182. https://doi.org/10.1007/s12892-010-0019-6

Perlo, Virginie, Frederik C. Botha, Agnelo Furtado, Katrina Hodgson-Kratky, and Robert J. Henry. "Metabolic changes in the developing sugarcane culm associated with high yield and early high sugar content." *Plant Direct* 4, no. 11 (2020): e00276. https://doi.org/10.1002/pld3.276

Rahman, M. Anisur, Wei Wu, Yanchun Yan, and Shamsul A. Bhuiyan. "Overexpression of TERF1 in sugarcane improves tolerance to drought stress." *Crop and Pasture Science* 72, no. 4 (2021): 268–279. https://doi.org/10.1071/CP20161

Raju, Gomathi, Kohila Shanmugam, and Lakshmi Kasirajan. "High-throughput sequencing reveals genes associated with high-temperature stress tolerance in sugarcane." 3 *Biotech* 10, no. 5 (2020): 1–13. https://doi.org/10.1007/s13205-020-02170-z

Ramasamy, Manikandan, Mona B. Damaj, Carol Vargas-Bautista, Victoria Mora, Jiaxing Liu, Carmen S. Padilla, Sonia Irigoyen et al. "A sugarcane G-protein-coupled receptor, ShGPCR1, confers tolerance to multiple abiotic stresses." *Frontiers in Plant Science* (2021). https://doi.org/10.3389/fpls.2021.745891

Ramiro, Daniel Alves, Danila Montewka Melotto-Passarin, Mariana de Almeida Barbosa, Flavio dos Santos, Sergio Gregorio Perez Gomez, Nelson Sidnei Massola Junior, Eric Lam, and Helaine Carrer. "Expression of *Arabidopsis* bax inhibitor-1 in transgenic sugarcane confers drought tolerance." *Plant Biotechnology Journal* 14, no. 9 (2016): 1826–1837. https://doi.org/10.1111/pbi.12540

Raza, Ghulam, Kazim Ali, Muhammad Yasin Ashraf, Shahid Mansoor, Muhammad Javid, and Shaheen Asad. "Overexpression of an H+-PPase gene from *Arabidopsis* in sugarcane improves drought tolerance, plant growth, and photosynthetic responses." *Turkish Journal of Biology* 40, no. 1 (2016): 109–119. https://doi.org/10.3906/biy-1501-100

Reid, Robert L. ed. *The Manual of Australian Agriculture*. Amsterdam: Elsevier, 2013.

Reis, Rafaela Ribeiro, Bárbara Andrade Dias Brito da Cunha, Polyana Kelly Martins, Maria Thereza Bazzo Martins, Jean Carlos Alekcevetch, Antônio Chalfun-Júnior, Alan Carvalho Andrade et al. "Induced over-expression of AtDREB2A CA improves drought tolerance in sugarcane." *Plant Science* 221 (2014): 59–68. https://doi.org/10.1016/j.plantsci.2014.02.003

Rocha, Flávia R., Flávia S. Papini-Terzi, Milton Y. Nishiyama, Ricardo ZN Vencio, Renato Vicentini, Rodrigo DC Duarte, Vicente E. de Rosa et al. "Signal transduction-related responses to phytohormones and environmental challenges in sugarcane." *BMC Genomics* 8, no. 1 (2007): 1–22. https://doi.org/10.1186%2F1471-2164-8-71

Rodrigues, F. A., J. P. Da Graça, M. L. De Laia, A. Nhani-Jr, J. A. Galbiati, M. I. T. Ferro, J. A. Ferro, and S. M. Zingaretti. "Sugarcane genes differentially expressed during water deficit." *Biologia Plantarum* 55, no. 1 (2011): 43–53. https://doi.org/10.1007/s10535-011-0006-x

Sage, Rowan F., Murilo Melo Peixoto, and Tammy L. Sage. "Photosynthesis in sugarcane." In *Sugarcane: Physiology, Biochemistry and Functional Biology*, 1st ed., pp. 121–154. New York: Wiley, 2013. https://dx.doi.org/10.1002/9781118771280.ch6

Saravanan, S., K. K. Kumar, M. Raveendran, D. Sudhakar, L. Arul, and E. Kokiladevi. "Genetic engineering of sugarcane for drought and salt tolerant transgenic plants expressing the BcZAT12 gene." *International Journal of Current Microbiology and Applied Sciences* 7, no. 07 (2018): 2018. https://doi.org/10.20546/ijcmas.2018.707.xx

Selvi, Athiappan, Kaliannan Devi, Ramaswamy Manimekalai, Perumal Thirugnanasambandam Prathima, Rabisha Valiyaparambth, and Kasirajan Lakshmi. "High-throughput miRNA deep sequencing in response to drought stress in sugarcane." *3 Biotech* 11, no. 7 (2021): 1–18. https://doi.org/10.1007/s13205-021-02857-x

Shabbir, Rubab, Talha Javed, Irfan Afzal, Ayman El Sabagh, Ahmad Ali, Oscar Vicente, and Pinghua Chen. "Modern biotechnologies: Innovative and sustainable approaches for the improvement of sugarcane tolerance to environmental stresses." *Agronomy* 11, no. 6 (2021): 1042. https://doi.org/10.3390/agronomy11061042

Silva, Marcelo de Almeida, Jorge Alberto Gonçalves da Silva, Juan Enciso, Vivek Sharma, and John Jifon. "Yield components as indicators of drought tolerance of sugarcane." *Scientia Agricola* 65 (2008): 620–627. https://doi.org/10.1590/S0103-90162008000600008

Simoes, Welson Lima, Daniela Siqueira Coelho, Alessandro Carlos Mesquita, Marcelo Calgaro, and Jucicléia soares da silva. "Physiological and biochemical responses of sugarcane varieties to salt stress." *Revista Caatinga* 32 (2020): 1069–1076. https://doi.org/10.1590/1983-21252019v32n423rc

Snyman, S. J., P. Mhlanga, and M. P. Watt. "Rapid screening of sugarcane plantlets for in vitro mannitol-induced stress." *Sugar Tech* 18, no. 4 (2016): 437–440. https://doi.org/10.1007/s12355-015-0411-0

Srivastava, Sangeeta, and Pavan Kumar. "Abiotic stress responses and tolerance mechanisms for sustaining crop productivity in sugarcane." In *Agronomic Crops*, pp. 29–47. Singapore: Springer, 2020. https://doi.org/10.1007/978-981-15-0025-1_3

Srivastava, Sangeeta, Ashwini Dutt Pathak, Prashant Shekhar Gupta, Ashok Kumar Shrivastava, and Arun Kumar Srivastava. "Hydrogen peroxide-scavenging enzymes impart tolerance to high temperature induced oxidative stress in sugarcane." *Journal of Environmental Biology* 33, no. 3 (2012): 657–661.

Su, Yachun, Liping Xu, Zhuqing Wang, Qiong Peng, Yuting Yang, Yun Chen, and Youxiong Que. "Comparative proteomics reveals that central metabolism changes are associated with resistance against *Sporisorium scitamineum* in sugarcane." *BMC Genomics* 17, no. 1 (2016): 1–21. https://doi.org/10.1186/s12864-016-3146-8

Sugiharto, Bambang. "Biochemical and molecular studies on sucrose-phosphate synthase and drought inducible-protein in sugarcane (*Saccharum officinarum*)." *Jurnal Ilmu Dasar* 5 (2004): 62–67.

Sugiharto, Bambang. "Biotechnology of drought-tolerant sugarcane" In *Sugarcane: Technology and Research*, edited by Alexandre de Oliveira. London: IntechOpen, 2017. https://doi.org/10.5772/intechopen.72436

Suprasanna, P., V. Y. Patade, N. S. Desai, R. M. Devarumath, P. G. Kawar, M. C. Pagariya, A. Ganapathi, M. Manickavasagam, and K. H. Babu. "Biotechnological developments in sugarcane improvement: An overview." *Sugar Tech* 13, no. 4 (2011): 322–335. https://doi.org/10.1007/s12355-011-0103-3

Swapna, M., and Sanjeev Kumar. "MicroRNAs and their regulatory role in sugarcane." *Frontiers in Plant Science* 8 (2017): 997. https://doi.org/10.3389/fpls.2017.00997

Thiebaut, Flávia, Clícia Grativol, Mariana Carnavale-Bottino, Cristian Antonio Rojas, Milos Tanurdzic, Laurent Farinelli, Robert A. Martienssen, Adriana Silva Hemerly, and Paulo Cavalcanti Gomes Ferreira. "Computational identification and analysis of novel sugarcane microRNAs." *BMC Genomics* 13, no. 1 (2012a): 1–14. https://doi.org/10.1186/1471-2164-13-290

Thiebaut, Flavia, Cristian A. Rojas, Karla L. Almeida, Clicia Grativol, Giselli C. Domiciano, Caren Regina C. Lamb, Janice de Almeida Engler, Adriana S. Hemerly, and Paulo CG Ferreira. "Regulation of miR319 during cold stress in sugarcane." *Plant Cell & Environment* 35, no. 3 (2012b): 502–512. https://doi.org/10.1111/j.1365-3040.2011.02430.x

Trivedi, A. K. "Adaptations and mechanisms of heat stress tolerance of plants." *Acadaemic Research Journal of Agricultural Sciences and Research* 3, no. 7 (2015): 151–160. http://dx.doi.org/10.1007/s00497-016-0281-y

Trujillo-Montenegro, Jhon Henry, Maria Juliana Rodriguez Cubillos, Cristian Darío Loaiza, Manuel Quintero, Héctor Fabio Espitia-Navarro, Fredy Antonio Salazar Villareal, Carlos Arturo Viveros Valens et al. "Unraveling the genome of a high yielding colombian sugarcane hybrid." *Frontiers in Plant Science* (2021): 1311. https://doi.org/10.3389/fpls.2021.694859

Varshney, Rajeev K., Ryohei Terauchi, and Susan R. McCouch. "Harvesting the promising fruits of genomics: Applying genome sequencing technologies to crop breeding." *PLoS Biology* 12, no. 6 (2014): e1001883. https://doi.org/10.1371/journal.pbio.1001883.

Vasantha, S., and R. Gomathi. "Growth and development of sugarcane under salinity." *Journal of Sugarcane Research* 2 (2012): 1–10.

Vermerris, Wilfred. "Survey of genomics approaches to improve bioenergy traits in maize, sorghum and sugarcane free access." *Journal of Integrative Plant Biology* 53, no. 2 (2011): 105–119. https://doi.org/10.1111/j.1744-7909.2010.01020.x

Vettore, André L., Felipe R. da Silva, Edson L. Kemper, Glaucia M. Souza, Aline M. da Silva, Maria Inês T. Ferro, Flavio Henrique-Silva et al. "Analysis and functional annotation of an expressed sequence tag collection for tropical crop sugarcane." *Genome Research* 13, no. 12 (2003): 2725–2735. https://doi.org/10.1101/gr.1532103

Vital, Camilo Elber, Andrea Giordano, Eduardo de Almeida Soares, Thomas Christopher Rhys Williams, Rosilene Oliveira Mesquita, Pedro Marcus Pereira Vidigal, Amanda de Santana Lopes et al. "An integrative overview of the molecular and physiological responses of sugarcane under drought conditions." *Plant Molecular Biology* 94, no. 6 (2017): 577–594. https://doi.org/10.1007/s11103-017-0611-y

Wagih, Mohamed E., A. Ala, and Y. Musa. "Regeneration and evaluation of sugarcane somaclonal variants for drought tolerance." *Sugar Tech* 6, no. 1 (2004): 35–40. https://doi.org/10.1007/BF02942615

Wahid, Abdul. "Physiological implications of metabolite biosynthesis for net assimilation and heat-stress tolerance of sugarcane (*Saccharum officinarum*) sprouts." *Journal of Plant Research* 120, no. 2 (2007): 219–228. https://doi.org/10.1007/s10265-006-0040-5

Waltz, Emily. "Beating the heat." *Nature Biotechnology* 32, no. 7 (2014): 610. https://doi.org/10.1038/nbt.2948

Wu, Kai-Chao, Li-Ping Wei, Cheng-Mei Huang, Yuan-Wen Wei, Hui-Qing Cao, Lin Xu, Hai-Bin Luo, Sheng-Li Jiang, Zhi-Nian Deng, and Yang-Rui Li. "Transcriptome reveals differentially expressed genes in *Saccharum spontaneum* GX83-10 leaf under drought stress." *Sugar Tech* 20, no. 6 (2018): 756–764. https://doi.org/10.1007/s12355-018-0608-0

Wu, Y., H. Zhou, Y-X. Que, R-K. Chen, and M-Q. Zhang. "Cloning and identification of promoter Prd29A and its application in sugarcane drought resistance." *Sugar Tech* 10, no. 1 (2008): 36–41. http://dx.doi.org/10.3389/fpls.2017.01077

Yang, Yaodong, Mumtaz Ali Saand, Liyun Huang, Walid Badawy Abdelaal, Jun Zhang, Yi Wu, Jing Li, Muzafar Hussain Sirohi, and Fuyou Wang. "Applications of multi-omics technologies for crop improvement." *Frontiers in Plant Science* (2021): 1846. https://doi.org/10.3389/fpls.2021.563953

Yang, Yuting, Xu Zhang, Yachun Su, Jiake Zou, Zhoutao Wang, Liping Xu, and Youxiong Que. "miRNA alteration is an important mechanism in sugarcane response to low-temperature environment." *BMC Genomics* 18, no. 1 (2017): 1–18. https://doi.org/10.1186/s12864-017-4231-3

Yunita, R., R. S. Hartati, and S. Suhesti. "Response of Bululawang sugarcane variety to salt stress." In IOP Conference Series: Earth and Environmental Science, vol. 418, no. 1, p. 012060. IOP Publishing, 2020. https://doi.org/10.1088/1755-1315/418/1/012060

Zhang, Jisen, Xingtan Zhang, Haibao Tang, Qing Zhang, Xiuting Hua, Xiaokai Ma, Fan Zhu et al. "Allele-defined genome of the autopolyploid sugarcane *Saccharum spontaneum* L." *Nature Genetics* 50, no. 11 (2018): 1565–1573. https://doi.org/10.1038/s41588-018-0237-2

Zhao, Duli, and Yang-Rui Li. "Climate change and sugarcane production: Potential impact and mitigation strategies." *International Journal of Agronomy* (2015). https://doi.org/10.1155/2015/547386

# 11 Effects of Rising Temperature on Flower Production and Pollen Viability in a Widespread Tropical Tree Species, *Muntingia calabura*

*Martijn Slot, Natanja Schuttenhelm, Chinedu E. Eze, and Klaus Winter*

## 11.1 INTRODUCTION

Contemporary tropical forests exist in a narrow temperature range (Janzen, 1967; Wright et al., 2009) close to what may be a high-temperature threshold (Doughty & Goulden, 2008). Further warming will push tropical forests into climate conditions not currently experienced by closed-canopy forests (Wright et al., 2009). This has motivated many studies into warming effects on carbon fluxes in tropical vegetation, particularly photosynthesis and respiration at the leaf level (Slot et al., 2013; Slot & Winter, 2017; Dusenge et al., 2021; Mujawamariya et al., 2021) and stand level (e.g., Tan et al., 2017; Smith et al., 2020). Furthermore, processes with longer timescales such as tree growth and mortality are being studied in relation to changes in environmental factors associated with anthropogenic $CO_2$ (carbon dioxide) emissions (Sullivan et al., 2020; Zuleta et al., 2022). What is typically not considered in natural ecosystems is the effect that elevated temperature can have on plant sexual reproduction (Bogdziewicz, 2022), and therefore, how global warming may affect the next generation of tropical forest trees.

The reproductive phase is one of the most temperature-sensitive parts of a plant's lifecycle. Male reproductive development in particular appears to be temperature sensitive (Dupuis & Dumas, 1990; Lohani et al., 2020), with heat-induced sterility in crops including wheat, corn, barley and rice occurring at 30–39°C (Barnabás et al., 2008; Sage et al., 2015; Ullah et al., 2022), causing worldwide yield loss with global warming (Asseng et al., 2015; Lobell et al., 2011; Cohen et al., 2021). Warm-season crops tend to experience pollen abortion at higher temperatures than cold-season crops (Sage et al., 2015), but even desert plant species such as *Cucurbita palmata* S.Watson, which frequently experiences temperatures in excess of 40°C, have been shown to have significantly reduced pollen viability when exposed to 36°C instead of 28°C daytime temperatures (Sage et al., 2015). Tropical flower temperatures already routinely exceed 30°C (Dobkin, 1985) and as temperatures continue to rise, reproductive failure in tropical forests may lead to profound changes in pollen and seed production, with important consequences for community dynamics (Muller-Landau et al., 2002) and carbon storage (Poorter et al., 2015) in these hyper-diverse and carbon-rich ecosystems. Decreased pollen, seed and fruit production would also have significant downstream effects on pollinator, granivore and frugivore communities. Nonetheless, no experimental research on temperature effects on plant reproduction has been conducted in tropical forests. Even in crops, the mechanisms underlying heat-induced sterility are not completely understood (Sage et al., 2015; Lohani et al., 2020). As a result, we lack the critical understanding required to predict the long-term effects of global warming, and ultimately, to protect natural vegetation—and our food supply—in a changing climate.

Here we report on an experiment designed to test the effects of elevated daytime temperature on flower production and pollen viability of a tropical tree species. Specifically, we tested the hypothesis that moderate daytime warming would have a negative effect on the viability of pollen of the tropical tree species *Muntingia calabura* in lowland Panama.

## 11.2 METHODS AND MATERIALS

### 11.2.1 Study Species

*Muntingia calabura* L. (Muntingiaceae) is a widely distributed small tree species native to tropical central and south America, and introduced as an ornamental or fruit tree in Southeast Asia, Australia and islands of the Pacific Ocean. In Panama, it is a fast-growing species of the lowlands in disturbed areas, at roadsides and in early secondary forests (Carrasquilla, 2005). *M. calabura* is a self-compatible pollination generalist, visited by butterflies, hummingbirds, flies, ants, and beetles, but it is thought to be primarily bee-pollinated (Haber et al., 1981; Bawa & Webb, 1983). English common names for this species include Cotton candy berry, Jamaica Cherry, and strawberry tree, in reference to the edible fruits that can be eaten fresh or used for jams and jellies. In nature, the fruits are eaten by birds, monkeys and bats, which disperse the seeds, thus enabling this species to spread and become invasive (Fleming et al., 1985). In Panama, plants produce flowers year-round, with each flower being open for just one day (Figure 11.1). Most of the flowers are hermaphroditic, but they vary in allocation to male and female components, with some flowers having a prominent pistil and fewer than 40 stamens, others having a reduced pistil and more than 70 stamens, and a third category in between (Bawa & Webb, 1983). In this study, we focus our analyses on the level of the pollen grains and whole flowers, and not on the relative allocation to stamen and pistils. We, therefore, do not distinguish among different flower types.

### 11.2.2 Experimental Setup

The experiments were conducted at the Santa Cruz experimental growth facilities of the Smithsonian Tropical Research Institute in Gamboa, Panama (9°07.214' N, W 79°42.136' W). The Tropical Dome Project consists of six semi-closed geodesic domes with a diameter of 6 m each (Solardomes Industries Limited, Nursling, UK) with

**FIGURE 11.1** Open flower of study species *Muntingia calabura*.

temperature and $CO_2$ control. Three of these were maintained at ambient temperature and three were passively heated during the day, resulting in an average afternoon (noon–5 p.m.) temperature of 36.3°C, which was 3.3°C warmer than the controls. Maximum temperatures were increased by 2.7°C, with the 95th percentile of five-minute mean temperatures reaching 38.3°C in the warmed domes. Afternoon relative humidity was slightly lower in the warmed (51.3%) than in the control domes (56.7%). In both treatments, $CO_2$ concentration was maintained close to ambient (~420 ppm). Nighttime conditions were similar between the treatments, with no warming >0.5°C persisting beyond 1.5 hours after sunset (i.e., by 8 p.m., temperatures were similar between treatments).

For each experiment, we collected six branches from each of three trees in central Panama early in the morning. On average, branches were 180 ± 35 cm long (± standard deviation) and had a basal diameter of 14.5 ± 4.7 mm. The branches were re-cut under water, flowers and large (>4 mm length) flower buds were removed, and one branch from each tree was placed in each of the six domes (i.e., each dome had a bucket with three independent, replicate branches). We replicated the same experiment three times between early February and late March, during the 2022 dry season, sampling a total of 54 branches from nine different parent trees. All parent trees grew naturally (i.e., they were not planted) within a 20 km range in the Panama Canal buffer zone, between the outskirts of Panama City and Parque Nacional Soberanía in Gamboa, Republic of Panama.

### 11.2.3 Sample Collection and Processing

Every morning, starting on day 3 and ending on day 6 of the experiment, we counted all the flowers that were opening that day and collected pollen from up to ten flowers per dome. Pollen was transferred with clean forceps to 8 ml borosilicate scintillation vials with either a few drops of aceto-carmine stain or a modified Alexander triple stain solution. The aceto-carmine consisted of 0.5 grams of carmine dissolved in hot 45% acetic acid, followed by filtering. The triple stain was prepared according to Peterson et al. (2010). The solution contains malachite green, acid fuchsin and orange G as the principal stains, dissolved in a low concentration of ethanol, glycerol and acetic acid. Viable pollen grains are stained purple by the triple stain, whereas non-viable/aborted pollen grains have a blue/green color; in aceto-carmine, viable pollen grains are reddish pink, while aborted pollen grains are colorless (Figure 11.2).

Stained pollen was transferred to a petri dish and one microscope slide was prepared per flower. For each slide, 100 pollen grains were counted under a light microscope (Axioscope 5, Carl Zeiss Microscopy, LLC, White Plains, New York) and the viable and non-viable pollen were scored.

### 11.2.4 Data Analysis

The two stains were compared with Welch's two-sample t-tests of pollen viability scores, using data from the first sampling day when we had equal numbers of control and

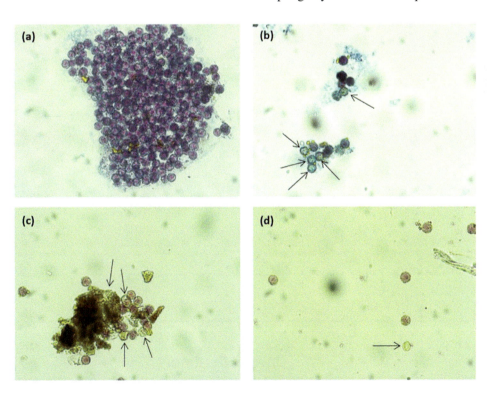

**FIGURE 11.2** Illustration of pollen viability staining used in the current study. Triple stain (a, b) colors viable pollen purple (a) and non-viable pollen blue-green (b). Aceto-carmine (c, d) colors viable pollen pink (c) and non-viable pollen is colorless (d). Arrows indicate non-viable pollen. Microscope magnification in all pictures was 400×.

treatment samples for both stains. For the remaining pollen viability analyses, viability scores obtained with the different stains were pooled together. Temporal trends in pollen viability and flower numbers were analyzed using linear regression by treatment at the level of the parent tree, after establishing that there was no systematic dome effect within treatment using analysis of variance. All data were analyzed in R version 4.0.2 (R Core Team, 2020).

## 11.3 RESULTS

The two stains showed a clear distinction between aborted and non-aborted pollen (Figure 11.2). Samples stained with aceto-carmine recorded on average slightly higher viability than samples stained with triple stain. For example, a comparison of equal sample sizes of the two stains on the first day of sampling revealed a significantly higher viability score in aceto-carmine (Welch two-sample t-test, $t = 2.67$, $df = 92$, $P = 0.009$), but the difference was less than 3.5 percentage points (91% for triple stain, 94.4% for aceto-carmine). The temporal trends in recorded viability and the control vs. treatment contrasts did not differ significantly between the stains, but low sample size after several days of warming (described in what follows) precluded more complete analyses. Viability estimates derived from the two stains were therefore pooled in further analyses. Temperatures and relative humidity were consistent among the treatment replicate domes and among the replicate control domes, and we did not detect significant differences in pollen viability among the three replicate domes per treatment ($F_{297,1} = 0.09$, $P = 0.76$). Because we were interested in knowing the differences in treatment effects among parent trees, we present the viability and flower number results by parent tree in Figure 11.3 and Figure 11.5, respectively.

Across all pooled samples, there was a marginally significant decrease over time in the viability of pollen collected from branches kept in warmed domes ($F_{179,1} = 3.7$, $P = 0.056$). Such a trend was not observed for pollen from flowers kept in control domes ($F_{216,1} = 0.5$, $P = 0.49$). This decreasing trend in viability with duration under treatment conditions was significant ($P < 0.05$) for warmed branches of two trees, and marginally significant ($0.05 < P < 0.1$) for two additional trees; for the remaining five trees, no clear patterns were found (Figure 11.3).

Because of the trend of decreasing viability with greater duration of exposure to elevated temperatures, in one of the

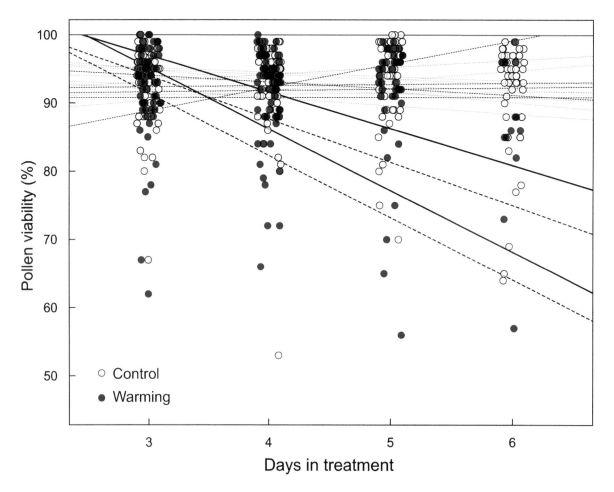

**FIGURE 11.3** Pollen viability of *M. calabura* flowers in relation to the duration under ambient (open symbols) or warmed (closed symbols) conditions. Viability scores from triple stain and aceto-carmine stained flowers are pooled. Solid, dashed and dotted lines represent linear regression trends with time by parent tree with $P < 0.05$, $0.05 < P < 0.1$, and $P > 0.1$, respectively. Data points are jittered on the x-axis to improve clarity.

experiments, we also analyzed pollen viability for buds of flowers collected on day 6 that we anticipated would have opened on day 7 if we had left the branches another day. Consistent with the trend for opened flowers, we found that the viability of pollen from warmed branches was significantly reduced relative to that of control branches (two-sample two-sided t-test with unequal variance. t = 2.23, $P = 0.042$. Figure 11.4a). Figure 11.4b further shows that the size of the buds had a significant positive relationship with the viability in the warming treatment, but not in the control treatment. While this result may simply reflect incomplete development of the small buds (and an absence of such small buds among the controls), a greater part of the development of the smaller buds has occurred under experimental warming conditions, so this might also add to the evidence of warming negatively impacting pollen viability.

From Figure 11.3, it is evident that by day 6, there are very few data points for treatment samples because daily flower production decreased significantly with the duration of the warming treatment (Figure 11.5). For four trees, the decrease was significant (with $P < 0.05$), and for three trees, the decrease was marginally significant ($0.05 < P < 0.1$). For one tree, the decrease was also significant under control conditions, but this branch produced very few flowers to start with (Figure 11.5). On day 3, the median number of new flowers was 3 and 4 in control and warmed branches, respectively (means of 3.4 and 4.8, respectively), while on day 6, the median flower numbers were 1 in control and 0 in treatment branches (means 2.8 and 0.5 flowers).

Among control branches, there was a tendency for larger diameter and longer branches to produce more flowers on day 6 of the experiment ($0.05 < P < 0.10$, both for diameter and length), but this pattern was not observed among warmed branches, which barely produced flowers on day 6 at all.

## 11.4 DISCUSSION

We present results from an experimental warming study aimed at evaluating the effects of elevated daytime temperature on the viability of pollen of a tropical tree species in Panama, Central America. Daytime warming by 3.5°C above the lowland tropical average resulted in a decreasing trend in pollen viability, with a longer duration of warming leading to lower viability, albeit with variation among parent trees. Furthermore, the daily number of flowers produced per branch decreased significantly over time in warmed branches but not in controls. *M. calabura* is a species associated with high-light growth conditions and it has supra-axillary flowers (Webb, 1984). Therefore, it is expected to experience tissue temperatures exceeding air temperature on a regular basis. The negative impact of relatively short-duration warming on the viability of pollen and on flower maturation of this species suggests that sustained warming associated with global environmental change might have a substantial impact on tropical plant fecundity with serious ramifications for the demography of tropical forests in a hotter future.

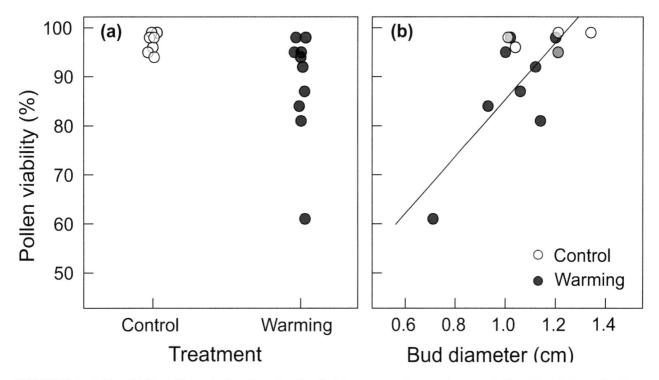

**FIGURE 11.4** Pollen viability of flower buds collected on day 6 of the treatment (a: control vs warming), and in relation to the diameter of the flower buds (b). The line in panel b represents a significant linear regression between pollen viability and flower bud diameter of warmed branches ($F_{7,1} = 10.0$, $P = 0.016$).

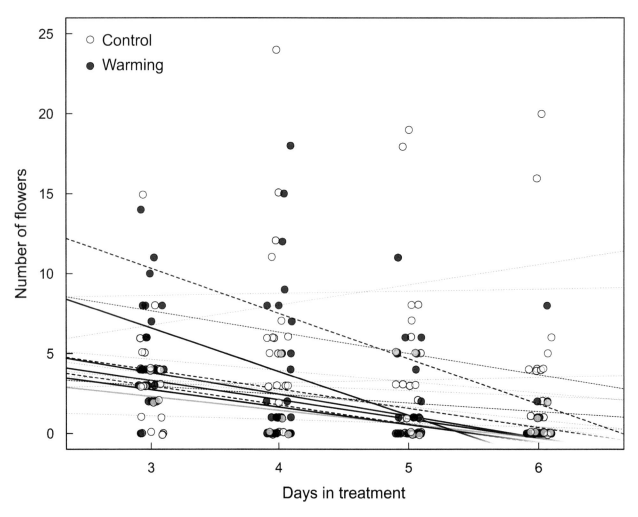

**FIGURE 11.5** The total number of newly-opened *M. calabura* flowers in relation to the duration under ambient (open symbols) or warmed (closed symbols) conditions. Solid, dashed and dotted lines represent linear regression trends with time by parent tree with $P < 0.05$, $0.05 < P < 0.1$, and $P > 0.1$, respectively. Data points are jittered on both axes to improve clarity.

### 11.4.1 Viability Decreases with Duration of Warming

Our observation that five-day exposure to moderate warming caused a decrease in pollen viability is consistent with previous experiments on *Mangifera indica* L. (mango) showing that pollen viability decreases steeply above 33°C (Issarakraisila & Considine, 1994). It also aligns with a recent study on an epiphytic cactus, *Hylocereus undatus* (Haw.) Britton & Rose (pitaya, dragon fruit; the species is now considered *Selenicereus undatus* (Haw.) D.R.Hunt [POWO, 2022]), where the pollen viability decreased significantly ($P < 0.05$) by 91.1% and 99.4%, at temperatures of 35°C and 40°C, respectively (Chu & Chang, 2022). While the decrease in pollen viability with the duration of warming in the current study was significant for several of the parent trees, it was quite moderate in terms of the percentage point reduction. The ex situ experiment we used was not designed for long-term warming, and as such, this study represents a test of the effects of short-term warming. The fact that on days 3–4 of the warming treatment, the viability of pollen collected from flowers opening on those days had not diverged between control and treatment branches suggests that short heatwaves prior to flower opening do not negatively affect the reproduction in this species. Flowers opening on day 6 and later, however, were negatively impacted. At this point we do not know whether this effect of duration of warming indicates that a cumulative warming threshold had been exceeded, or that the steps in early pollen development are particularly sensitive to heat—these steps took place under ambient conditions for flowers that opened in the early days of the experiment, but under experimental conditions for flowers opening on later days.

### 11.4.2 Which Step in Reproduction Is Most Sensitive to Warming?

Male reproductive development involves several steps, each of which may have its own distinct sensitivity to exposure to high temperatures (Lohani et al., 2020). For example, early stages of pollen development include meiotic divisions of microspore mother cells, forming uninucleate microspores, which then go through successive mitotic divisions to form tri-nucleate mature pollen grains, and ultimately, successful anther dehiscence is required to complete the release of the pollen grains. Heat can strongly affect meiosis by changing cross-over rates (Fuchs et al., 2018), but the uninucleate stage of male reproductive development in particular is highly sensitive to temperature (Dolferus et al., 2011; Sage et al., 2015), causing pollen abortion or pollen sterilization. For example, in an experiment with eight species, most of which are tropical crops or otherwise hot-climate species—including *Amaranthus cruentus* L., *Eragrostis tef* (Zuccagni) Trotter, *Capsicum* spp. and *Phaseolus* spp.—pollen abortion at the uninuclear stage was >70% when plants were grown at 36/24°C day/night temperature, whereas pollen abortion was <15% when plants were grown at 28/24°C (Sage et al., 2015).

In the current study, only pollen viability was examined, but even when pollen remains viable following heat exposure, anther dehiscence can be negatively affected by high temperature (Jagadish et al., 2015), as can pollen tube growth, fertilization and ovule and seed development (Bergkamp et al., 2018; Chaturvedi et al., 2021).

Identifying which aspect of reproduction has the lowest temperature threshold could inform management practices aimed at maximizing genetic variation of the most temperature-sensitive traits in vulnerable populations, as short-term exposure during the most sensitive stage of pollen development has the potential to result in complete reproductive failure. *M. calabura* flowers year-round (Figueiredo et al., 2008), so the reproductive loss associated with a short heat wave will not dramatically reduce its annual reproductive output. Most tropical tree species, however, flower at very specific times during the year (Wright et al., 2005)—and some, like *Tachigali versicolor* Standl. & L.O.Williams, flower only once in their entire lifespan (Foster, 1977). For those species, reproductive failure due to short-term heat stress can have serious consequences for their demographic rates. Thus, the timing of heat events and species differences in heat sensitivity could impact the community composition of future forests. Heat tolerance–based conservation efforts may not be realistic in hyperdiverse tropical forests, but less species-rich areas could potentially be managed for reproductive thermal tolerance of keystone species in the near future with technology that is being developed for crops (Driedonks et al., 2016; Tack et al., 2016; Lu et al., 2022).

### 11.4.3 Flower Production

A study based on flowers collected in litter traps reported a positive correlation between air temperature and flower production in Panama and Puerto Rico (Pau et al., 2013), suggesting that a thermal threshold for flower production has not yet been reached for these lowland tropical forests. However, the monthly variation in flower production is unlikely to reflect a true stimulation of flower production by temperature, and instead reflects species-specific floral phenology, triggered by seasonal patterns in solar radiation (Zimmerman et al., 2007), the timing of rainfall events (Wright et al., 2019) and/or accumulation of threshold levels of reserves needed to support a reproductive event. Furthermore, interannual variation in flower production was best explained by atmospheric $CO_2$ concentration due to the directional trends in both (Pau et al., 2018), further suggesting that the observed trend with monthly temperature variation is not in conflict with our observation of temperature effects on flower maturation. The 3.3°C warming in the current study greatly exceeds the <1.5°C seasonal temperature variation in central Panama (Paton, 2019), and the absolute temperatures associated with it (i.e., >36°C), are enough to reduce flower production in some species, e.g., peanut (*Arachis hypogaea* L.) (Prasad et al., 2000). Nonetheless, longer-term studies with free-growing or potted plants across multiple species will be needed to determine how common it is that heat exposure reduces flower production in tropical trees.

### 11.4.4 Long-Term Effects of Atmospheric and Climate Change

The temperature threshold for the fertility of both temperature and tropical crop species appears to be narrow (Prasad et al., 2006; Sage et al., 2015), with little indication of thermal acclimation of reproductive processes (Bagha, 2014). Likewise, Flores-Rentería et al. (2018) found no evidence that pollen germination acclimated to growth temperature in the gymnosperm tree species *Pinus edulis* Engelm. growing along an elevation gradient in Arizona, USA. Tropical trees increase reproduction when more carbohydrates are available, e.g., as a result of increased light availability (Graham et al., 2003) or elevated $CO_2$ concentrations (Pau et al., 2018). In the absence of acclimation of reproductive thermal tolerance, increased carbohydrate availability for flowering could potentially compensate for a reduction in the production of viable pollen or an increase in heat sterilization of flowers. On the other hand, warming could potentially change plant carbohydrate content through changes in metabolism as respiratory $CO_2$ loss increases with temperature (e.g., Slot et al., 2013) and photosynthetic $CO_2$ assimilation decreases above an optimum temperature that corresponds to the average local temperature (Slot & Winter, 2017). Warming could also simply affect total allocation to reproduction through changes in overall plant growth (Bazzaz et al., 2000).

Acclimation—if observed—would have enormous consequences for plant performance in a warmer future world, and assessing the potential for acclimation and the interactive effects of warming and elevated $CO_2$ on reproductive allocation are two important areas of future research.

## 11.5 CONCLUSIONS

Neglecting to consider warming effects on plant reproduction leaves a crucial gap in our understanding of how tropical forests will perform in the future. As is evident from this chapter, very few studies have been conducted on heat effects on the reproductive development of tropical trees. The vast majority of studies are of commercially important (crop) species such as wheat and rice, and while the processes of pollen and flower development are the same for tropical trees and these crops, there are significant differences in lifespan, reproductive allocation and our ability to selectively breed and improve species based on specific environmental challenges. Our study provides a first glance at understanding the implication of rising temperatures for reproduction in tropical forests. More research is needed to evaluate the diversity in heat responses across species and plant functional groups—and the potential importance of reproductive failure for tropical forest survival and management.

## REFERENCES

Asseng, S., Ewert, F., Martre, P., Rötter, R.P., Lobell, D.B., Cammarano, D., Kimball, B.A., Ottman, M.J., Wall, G.W., White, J.W., et al. 2015. Rising temperatures reduce global wheat production. *Nature Climate Change* 5: 143–147.

Bagha, S. 2014. *The Impact of Chronic High Temperatures on Anther and Pollen Development in Cultivated Oryza Species*, Doctoral dissertation. Toronto: University of Toronto.

Barnabás, B., Jäger, K. and Fehér, A. 2008. The effect of drought and heat stress on reproductive processes in cereals. *Plant, Cell & Environment* 31: 11–38.

Bawa, K.S. and Webb, C.J. 1983. Floral variation and sexual differentiation in *Muntingia calabura* (Elaeocarpaceae), a species with hermaphrodite flowers. *Evolution*: 1271–1282.

Bazzaz, F.A., Ackerly, D.D. and Reekie, E.G. 2000. Reproductive allocation in plants. In M. Fenner, Ed, *Seeds: The Ecology of Regeneration in Plant Communities*. Wallingford: CAB International, pp. 1–29.

Bergkamp, B., Impa, S.M., Asebedo, A.R., Fritz, A.K. and Jagadish, S.K. 2018. Prominent winter wheat varieties response to post-flowering heat stress under controlled chambers and field based heat tents. *Field Crops Research* 222: 143–152.

Bogdziewicz, M. 2022. How will global change affect plant reproduction? A framework for mast seeding trends. *New Phytologist* 234: 14–20.

Carrasquilla, R.L.G. 2005. *Árboles y arbustos de Panamá/Trees and Shrubs of Panama*. Panama: Editora Novo Art.

Chaturvedi, P., Wiese, A.J., Ghatak, A., Zaveska Drabkova, L., Weckwerth, W. and Honys, D. 2021. Heat stress response mechanisms in pollen development. *New Phytologist* 231: 571–585.

Chu, Y.C. and Chang, J.C. 2022. Heat stress leads to poor fruiting mainly due to inferior pollen viability and reduces shoot photosystem II efficiency in "Da Hong" Pitaya. *Agronomy* 12: 225.

Cohen, I., Zandalinas, S.I., Huck, C., Fritschi, F.B. and Mittler, R. 2021. Meta-analysis of drought and heat stress combination impact on crop yield and yield components. *Physiologia Plantarum* 171: 66–76.

Dobkin, D.S. 1985. Heterogeneity of tropical floral microclimates and the response of hummingbird flower mites. *Ecology* 66: 536–543.

Dolferus, R., Ji, X. and Richards, R.A. 2011. Abiotic stress and control of grain number in cereals. *Plant Science* 181: 331–341.

Doughty, C.E. and Goulden, M.L. 2008. Are tropical forests near a high temperature threshold? *Journal of Geophysical Research-Biogeosciences* 113: G00B07.

Driedonks, N., Rieu, I. and Vriezen, W.H. 2016. Breeding for plant heat tolerance at vegetative and reproductive stages. *Plant Reproduction* 29: 67–79.

Dupuis, I. and Dumas, C. 1990. Influence of temperature stress on in vitro fertilization and heat shock protein synthesis in maize (*Zea mays* L.) reproductive tissues. *Plant Physiology* 94: 665–670.

Dusenge, M.E., Wittemann, M., Mujawamariya, M., Ntawuhiganayo, E.B., Zibera, E., Ntirugulirwa, B., Way, D.A., Nsabimana, D., Uddling, J. and Wallin, G. 2021. Limited thermal acclimation of photosynthesis in tropical montane tree species. *Global Change Biology* 27: 4860–4878.

Figueiredo, R.A., Oliveira, A.A., Zacharias, M.A., Barbosa, S.M., Pereira, F.F., Cazela, G.N., Viana, J.P. and Camargo, R.A. 2008. Reproductive ecology of the exotic tree Muntingia calabura L. (Muntingiaceae) in southeastern Brazil. *Revista Árvore* 32: 993–999.

Fleming, T.H., Williams, C.F., Bonaccorso, F.J. and Herbst, L.H. 1985. Phenology, seed dispersal, and colonization in *Muntingia calabura*, a neotropical pioneer tree. *American Journal of Botany* 72: 383–391.

Flores-Rentería, L., Whipple, A.V., Benally, G.J., Patterson, A., Canyon, B. and Gehring, C.A. 2018. Higher temperature at lower elevation sites fails to promote acclimation or adaptation to heat stress during pollen germination. *Frontiers in Plant Science* 9: 536.

Foster, R.B., 1977. *Tachigalia versicolor* is a suicidal neotropical tree. *Nature* 268: 624–626.

Fuchs, L.K., Jenkins, G. and Phillips, D.W. 2018. Anthropogenic impacts on meiosis in plants. *Frontiers in Plant Science* 9: 1429.

Graham, E.A., Mulkey, S.S., Kitajima, K., Phillips, N.G. and Wright, S.J. 2003. Cloud cover limits net $CO_2$ uptake and growth of a rainforest tree during tropical rainy seasons. *Proceedings of the National Academy of Sciences* 100: 572–576.

Haber, W.A., Frankie, G.W., Baker, H.G., Baker, I. and Koptur, S. 1981. Ants like flower nectar. *Biotropica* 13: 211–214.

Issarakraisila, M. and Considine, J.A. 1994. Effects of temperature on pollen viability in mango cv. 'Kensington'. *Annals of Botany* 73: 231–240.

Jagadish, S.V.K., Murty, M.V.R. and Quick, W.P. 2015. Rice responses to rising temperatures–challenges, perspectives and future directions. *Plant, Cell & Environment* 38: 1686–1698.

Janzen, D.H. 1967. Why mountain passes are higher in the tropics. *American Naturalist* 101: 233–249.

Lobell, D.B., Schlenker, W. and Costa-Roberts, J. 2011. Climate trends and global crop production since 1980. *Science* 333: 616–620.

Lohani, N., Singh, M.B. and Bhalla, P.L. 2020. High temperature susceptibility of sexual reproduction in crop plants. *Journal of Experimental Botany* 71: 555–568.

Lu, L., Liu, H., Wu, Y. and Yan, G. 2022. Wheat genotypes tolerant to heat at seedling stage tend to be also tolerant at adult stage: The possibility of early selection for heat tolerance breeding. *The Crop Journal* 10: 1006–1013.

Mujawamariya, M., Wittemann, M., Manishimwe, A., Ntirugulirwa, B., Zibera, E., Nsabimana, D., Wallin, G., Uddling, J. and Dusenge, M.E. 2021. Complete or overcompensatory thermal acclimation of leaf dark respiration in African tropical trees. *New Phytologist* 229: 2548–2561.

Muller-Landau, H.C., Wright, S.J., Calderón, O., Hubbell, S.P. and Foster, R.B. 2002. Assessing recruitment limitation: concepts, methods and case-studies from a tropical forest. In D.J. Levey, S.W.R. Silva and M. Galetti, Eds, *Seed Dispersal and Frugivory: Ecology, Evolution and Conservation*. Wallingford, UK: CAB International, pp. 35–53.

Paton, S. 2019. *Monthly Summary_BCI, Vertical*. Panama: Smithsonian Tropical Research Institute (Dataset). https://doi.org/10.25573/data.10059458.v25

Pau, S., Okamoto, D.K., Calderón, O. and Wright, S.J. 2018. Long-term increases in tropical flowering activity across growth forms in response to rising $CO_2$ and climate change. *Global Change Biology* 24: 2105–2116.

Pau, S., Wolkovich, E.M., Cook, B.I., Nytch, C.J., Regetz, J., Zimmerman, J.K. and Wright, S.J. 2013. Clouds and temperature drive dynamic changes in tropical flower production. *Nature Climate Change* 3: 838–842.

Peterson, R., Slovin, J.P. and Chen, C. 2010. A simplified method for differential staining of aborted and non-aborted pollen grains. *International Journal of Plant Biology* 1: e13.

Poorter, L., van der Sande, M.T., Thompson, J., Arets, E.J., Alarcón, A., Álvarez-Sánchez, J., Ascarrunz, N., Balvanera, P., Barajas-Guzmán, G., Boit, A., et al. 2015. Diversity enhances carbon storage in tropical forests. *Global Ecology and Biogeography* 24: 1314–1328.

POWO. 2022. Plants of the world online. *Facilitated by the Royal Botanic Gardens, Kew*. http://www.plantsoftheworldonline.org/ (accessed October 01, 2022).

Prasad, P.V., Boote, K.J. and Allen, L.H. 2006. Adverse high temperature effects on pollen viability, seed-set, seed yield and harvest index of grain-sorghum [*Sorghum bicolor* (L.) Moench] are more severe at elevated carbon dioxide due to higher tissue temperatures. *Agricultural and Forest Meteorology* 139: 237–251.

Prasad, P.V., Craufurd, P.Q., Summerfield, R.J. and Wheeler, T.R. 2000. Effects of short episodes of heat stress on flower production and fruit-set of groundnut (*Arachis hypogaea* L.). *Journal of Experimental Botany* 51: 777–784.

R Core Team. 2020. R: A language and environment for statistical computing. *R Foundation for Statistical Computing, Vienna, Austria*. www.R-project.org/

Sage, T.L., Bagha, S., Lundsgaard-Nielsen, V., Branch, H.A., Sultmanis, S. and Sage, R.F. 2015. The effect of high temperature stress on male and female reproduction in plants. *Field Crops Research* 182: 30–42.

Slot, M. and Winter, K. 2017. In situ temperature response of photosynthesis of 42 tree and liana species in the canopy of two Panamanian lowland tropical forests with contrasting rainfall regime. *New Phytologist* 214: 1103–1117.

Slot, M., Wright, S.J. and Kitajima, K. 2013 Foliar respiration and its temperature sensitivity of trees and lianas: In situ measurements in the upper canopy of a tropical forest. *Tree Physiology* 33: 505–515.

Smith, M.N., Taylor, T.C., van Haren, J., Rosolem, R., Restrepo-Coupe, N., Adams, J., Wu, J., de Oliveira, R.C., da Silva, R., de Araujo, A.C. and de Camargo, P.B. 2020. Empirical evidence for resilience of tropical forest photosynthesis in a warmer world. *Nature Plants* 6: 1225–1230.

Sullivan, M.J., Lewis, S.L., Affum-Baffoe, K., Castilho, C., Costa, F., Sanchez, A.C., Ewango, C.E., Hubau, W., Marimon, B., Monteagudo-Mendoza, A., et al. 2020. Long-term thermal sensitivity of Earth's tropical forests. *Science* 368: 869–874.

Tack, J., Barkley, A., Rife, T.W., Poland, J.A. and Nalley, L.L. 2016. Quantifying variety-specific heat resistance and the potential for adaptation to climate change. *Global Change Biology* 22: 2904–2912.

Tan, Z.H., Zeng, J., Zhang, Y.J., Slot, M., Gamo, M., Hirano, T., Kosugi, Y., Da Rocha, H.R., Saleska, S.R., Goulden, M.L. and Wofsy, S.C. 2017. Optimum air temperature for tropical forest photosynthesis: Mechanisms involved and implications for climate warming. *Environmental Research Letters* 12: 054022.

Ullah, A., Nadeem, F., Nawaz, A., Siddique, K.H. and Farooq, M. 2022. Heat stress effects on the reproductive physiology and yield of wheat. *Journal of Agronomy and Crop Science* 208: 1–17.

Webb, C.J. 1984. Flower and fruit movements in *Muntingia calabura*: A possible mechanism for avoidance of pollinator-disperser interference. *Biotropica* 16: 37–42.

Wright, S.J., Calderón, O. and Muller-Landau, H.C. 2019. A phenology model for tropical species that flower multiple times each year. *Ecological Research* 34: 20–29.

Wright, S.J., Muller-Landau, H.C., Calderón, O. and Hernandéz, A. 2005. Annual and spatial variation in seedfall and seedling recruitment in a Neotropical forest. *Ecology* 86: 848–860.

Wright, S.J., Muller-Landau, H.C. and Schipper, J. 2009. The future of tropical species on a warmer planet. *Conservation Biology* 23: 1418–1426.

Zimmerman, J.K., Wright, S.J., Calderón, O., Pagan, M.A. and Paton, S. 2007. Flowering and fruiting phenologies of seasonal and aseasonal neotropical forests: The role of annual changes in irradiance. *Journal of Tropical Ecology* 23: 231–251.

Zuleta, D., Arellano, G., Muller-Landau, H.C., McMahon, S.M., Aguilar, S., Bunyavejchewin, S., Cárdenas, D., Chang-Yang, C.H., Duque, A., Mitre, D., et al. 2022. Individual tree damage dominates mortality risk factors across six tropical forests. *New Phytologist* 233: 705–721.

# 12 Ecophysiological and Morphological Adaptations of Plants under Temperature Stress
## Influence of Phytohormones

*Ghalia S.H. Alnusairi, Abbu Zaid, Harvinder Kour, Khadiga Alharbi, and Mona H. Soliman*

## 12.1 INTRODUCTION

The threat of global population explosion and the changing climate pattern on agricultural crop production needs is escalating. Crop production needs to be increased in order to feed the larger part of the world's growing population. According to an estimate, agricultural productivity by the mid-century needs to be increased by 70 percent, which could compensate for the need of 2.3 billion human population (Bulgari et al., 2019). In addition, in the natural environment crop plants are challenged by significant biotic and abiotic stress factors, thereby restricting the production to a considerable extent (Zaid et al., 2022; Zulfiqar et al., 2022). These restrictions have become severe and even drastic under the effect of changing climate (Zaid et al., 2021a). One of the principal issues for plant stress physiologists/breeders is to decipher the underlying mechanisms that influence how crop plants mediate response to different stress factors. Salinity, high temperatures, drought, and metal/metalloid are the most universal and imperative exogenous pressures which restrict growth and development of plants (Khan et al., 2020; Altaf et al., 2021; Zaid et al., 2021a, 2022; Mangal et al., 2022). In the context of changing climate, out of a myriad of abiotic pressures, principally high temperature (HT) and heat stress (HS) rigorously affect the quality, quantity, nutritional status, and yield of diverse plant species (Aleem et al., 2021; Chaudhary et al., 2022). Nevertheless, estimates of Waithaka et al. (2013) advocated that increased temperatures (changing climate) can also reveal various new avenues to cultivate crop plants in habitats in which their growth was possible earlier. Also, in water-deficit or HT areas, scenarios of climate change further advocated that the development of crops and selection/preference of particular cultivars could serve as important criteria to trigger adequate yields (Thomas et al., 2007).

High temperature stress limits plant growth and yield by obstructing the plant morphology, vital physiological processes, and activities of enzymatic gadgets of crop plants. In plants, HS orchestrates the phenology and is known to shorten the vegetative and reproductive life stages (Wahid et al., 2007; Waraich et al., 2012; Khajuria et al., 2021). HT is known to reduce quality traits, such as fruit color and texture in tomatoes, cucumbers, and pepper plants (Zipelevish et al., 2000). In a nutshell, HS principally hampers anatomical, physiological, morphological, and biochemical processes of crop plants by altering capacity of photosynthesis, source-sink relationships, water and mineral absorption, enzymatic and non-enzymatic activities, and redox homeostasis (Jenni et al., 2013; Sehar et al., 2022). Crop plants grown under HT stress show disruption of the photosynthetic apparatus by alteration in the functioning of pigment system II (PSII) (Crafts-Brandner & Salvucci, 2000). Various studies advocated the decrement of photosynthesis-related traits, viz., ribulose 1,5-bisphosphate carboxylase (RuBisCo) activity (Perdomo et al., 2017), stomatal conductance (Gs) (Zhou et al., 2018), intercellular $CO_2$ (carbon dioxide) concentration (Ci) (Wang et al., 2010), net photosynthetic rate (Pn), and SPAD value (Sehar et al., 2022) were adversely inhibited under HT stress, resulting in suppression of overall growth and yield in diverse crop plants (Hussain et al., 2019; Poór et al., 2021). Recently, Sehar et al. (2022) exposed a set of wheat plants to 25°C (no stress) and 40°C (HT stress) for 15 days in the environmental growth chamber for six hours daily and reported that HS affected antioxidant metabolism and osmolyte (proline). The involvement of exogenous ethylene (200μL L$^{-1}$ ethephon: 2-chloroethylphosphonic acid) in affecting the biosynthesis of proline and the antioxidant defense system induced changes was noticed under HS. The HS plants showed increased thiobarbituric acid reactive substances (TBARS), hydrogen peroxide ($H_2O_2$), proline biosynthesis, antioxidant enzyme activities, and evolution of ethylene, but inhibited photosynthesis parameters. In another experiment, *Triticum aestivum* L. plants were studied under HT (42°C). It was found that HT stress increased contents of $H_2O_2$ and TBARS but decreased the efficiency of photosynthetic performance, PSII chlorophyll fluorescence efficiency, electron transport ($ET_o$/CS) quantum yield, rate of electron transport fluxes of each reaction centre ($ET_0/RC$) and protein pattern expression profiles of *psbA*, *psbB*, and *psbC*, encoding for D1, CP47, and CP43 proteins, respectively (Fatma et al., 2021). Furthermore, carbohydrate metabolism is inhibited under HT stress conditions, which hamper the development of stamens and pollens, ovule fertilization, fruit sets, and seed development in diverse crop varieties, and exacerbates grain loss (Hasanuzzaman et al., 2013; Schauberger et al., 2017; Poór et al., 2021). Nevertheless, the abandoned generation of reactive oxygen species (ROS) under HT stress disrupts vital functions of cellular organization, accelerates denaturation of DNA and proteins, and accelerates peroxidation of the lipid bilayer—and consequently the death of cells (Chaudhary et al., 2021). In a recent classic research survey, Medina et al. (2021), postulated the signaling and physiological role of ROS biosynthesized in plants under HT stress. It is inferred that various ROS help in HS signaling, trigger the heat shock proteins (HSPs) synthesis, and show regulatory interactions with other different signaling pathways and transcription factors (TFs) for engineering tolerance to HT stress (Lamaoui et al., 2018; Argosubekti, 2020; Kumar et al., 2022). Crop plants are non-motile, when grown under HS conditions, they respond in several complex ways to normalize the altered growth events. However, the response in $C_3$, $C_4$, and CAM plants under HT stress is exaggerated by the stress severity component and its developmental stage (Yamori et al., 2014). Therefore, there is a need for targeted studies to unravel the disastrous effects of HT/HS on the growth, photosynthetic capacity, yield, and quality (flavor, taste, color, nutritional value) of important crop plants, with suitable solution/strategies, engineered/developed to produce/mitigate HS/HT in various cultivars of crop plants. Nonetheless, traditional breeding/engineering techniques have yielded little effectiveness/promising results because stress tolerance characteristics have complexity. Hence, it is momentous to devise new and innovative ideas/means in this area and in the era of changing climate.

Phytohormones could operate as a reasonable choice for yielding climate-resilient crop plants with enhanced production (Figure 12.1). In current times, the influence of various phytohormones has been a substitute and ecofriendly move toward boosting/engineering non-biotic stress resistance including HT stress in crop plants.

## 12.2 PHYTOHORMONES AND TEMPERATURE STRESS: A BURNING ISSUE

Phytohormones are naturally occurring organic elicitors that act at very low concentrations and control various regulatory physiological processes in diverse crop plants (Ahmad et al., 2019; Zaid et al., 2021b; Altaf et al., 2022). Phytohormones are biosynthesized at a particular site and are transported to the site of their requirement in plants (Went & Thimann, 1937). Phytohormones are of various kinds in plants, viz., auxins, cytokinins (CKs), gibberellins (GAs), brassinosteroids (BRs), abscisic acid (ABA), and ethylene

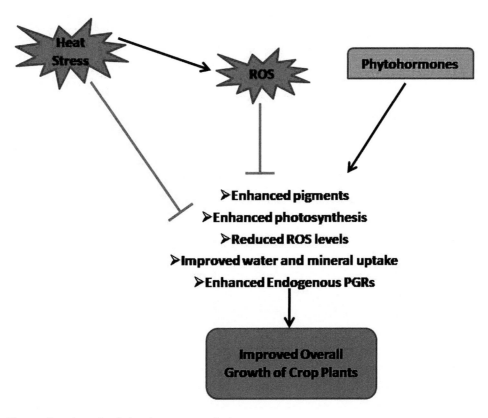

**FIGURE 12.1** The ameliorative role of phytohormones under heat stress.

(ET). According to Wani et al. (2016), phytohormones exert an imperative task during underlying abiotic stress signaling responses by involving complex tolerance trade-offs between signal transduction mechanisms. Phytohormones regulate a myriad of in-house and peripheral stimuli that exert foremost plant developmental switch (Kazan, 2015). During evolution, plants have adapted to various mechanisms to minimize/cope with the HS-induced negative effects, including intrinsic HS tolerance mechanisms under which plants switch on their integral ability to resist HS and acquire tolerance enabling them to optimize their metabolism under HS following proper acclimatization (Song et al., 2019). Plants also undergo physical changes recurrently by elevating signals for adjusting metabolism to re-establish the cellular redox changes and homeostasis to protect themselves from HT-induced oxidative stress (Park & Seo, 2015). In view of this background, the present chapter primarily sheds light on important conceptual improvements on the task of various phytohormones in shifting plants headed for adaption by principally domineering ecophysiological and morphological responses under HT stress conditions.

## 12.3 ADAPTIVE ROLE OF PHYTOHORMONES UNDER TEMPERATURE STRESS

As was already noted, phytohormones significantly contribute to the development of temperature stress tolerance in a variety of crop plants by altering numerous physiological and morphological mechanisms. The potential contribution of specific phytohormones to the development of HT stress tolerance in a variety of crop plants will be covered in the sections that follow.

## 12.4 SALICYLIC ACID

In a variety of crop plants, salicylic acid (SA) plays a fundamental function in thermo-tolerance. Salicylic acid, also known as ortho-hydroxy benzoic acid, is a powerful phytohormone that is found across the entire plant kingdom (Raskin et al., 1990) and is thought to help crop plants adapt to HT stress. Actually, SA is a powerful natural inducer of heat generation in arum lilies (Raskin et al., 1987). Due to its role in the control of epigenetic variables, SA controls plant growth characteristics and other crucial developmental processes under HT stress situations. In a variety of crops, including *Arabidopsis thaliana*, mustard, potato, tobacco, tomato, and bean, SA-induced signaling systems increase HT stress tolerance (Horváth et al., 2007; Li et al., 2010; Khan et al., 2013; Jahan et al., 2019). In a study, HT stress was administered to hyacinth bean plants (40–42°C), along with exogenous applications of various SA concentrations (0.50, 1.0, 1.5, and 2.0 mM), in a field setting. Exogenous SA administration during HT synchronized growth and biophysical processes by impressively controlling the activities of antioxidant enzymes, modulating the mRNA levels of certain enzymes, boosting photosynthesis, and optimizing water relations. According to both increases and decreases in distinct profile bands, methylation and de-methylation patterns as evidenced by the coupled restriction enzyme digestion-random amplification (CRED-RA) technique, it was discovered that HT stress considerably increased DNA damage in the examined plants (Rai et al., 2018). In a pilot experiment, the impact of SA (100 mM) on two wheat cultivars grown under HS (38°C for two hours) and their thermo-tolerance levels was investigated by 2-DE, plants grown under HS and receiving SA treatment during post-anthesis stage were shown to have a significant number of protein spots. Additionally, the MALDI-TOF-TOF/MS analysis was used to identify signaling molecules, heat-responsive transcription factors (HSFs), and heat-shock proteins (HSPs) as differentially expressed proteins. The abundance of transcripts for HSFs, HSPs, CDPK, SOD, and RCA was seen in the specific wheat cultivar under the HT and SA treatments. Last but not least, it was proposed that SA treatment decreased the negative effects of HS on the activity of soluble starch synthase and starch granule synthesis in both wheat cultivars (Kumar et al., 2015). The traditional function of SA in shade avoidance was figured out in *Arabidopsis thaliana*. It was discovered that SA pathway genes play a significant role in preventing petiole shadow (Nozue et al., 2018). The ability of externally applied SA to mediate HS tolerance in genotypes of chosen bread wheat from Pakistan was examined (Munir et al., 2018). SA is applied by first pre-soaking seeds in a solution of $10^{-4}$ M (SA) for 24 hours, and then applying it topically to leaves. The findings showed that HS enhanced proline content, soluble proteins, and soluble sugar but decreased grain yield, thousand kernel weights, and chlorophyll content. However, under HS conditions, the treatment of SA by both techniques increased the amount of chlorophyll, soluble protein, proline, sugar build-up, and net yield. Wheat plants were subjected to HS (40°C for six hours) (Khan et al., 2013) in an environmental growth chamber. The findings suggested that whereas proline and stress ethylene levels rose in HS growing plants, RuBisCo activity, net photosynthetic rate, and photosynthetic nitrogen use efficiency (NUE) declined. By enhancing γ-glutamyl kinase and lowering proline oxidase activities, two enzymes involved in the synthesis of proline, the SA (0.5 mM) therapy had a calming effect. Under HS, coordination of the processes was facilitated by SA, which boosted osmotic and water potential. SA treatment limited the production of stress ethylene under HS (Khan et al., 2013). The reduction of the negative effects of HS in the examined plants was caused by an improvement in proline metabolism, N absorption, and photosynthesis in plants receiving SA and its interaction with ethylene production. Rice plants were grown in a greenhouse equipped with an autonomous temperature control system in research by Zhang et al. (2017). Plants of rice were exposed to HT (40°C for 10 days). The SA concentrations that were utilized were 0.01, 0.1, 1, 10, and 50 mmol L$^{-1}$. It was noticed that rice plants exposed to HT had spikelets with damage. Superoxide dismutase (SOD), peroxidase (POD), catalase (CAT), and ascorbate peroxidase

(APX)—antioxidant enzymes of which SA mediates increased levels—were important for preventing spikelet degeneration brought on by HS. After subjecting *Digitalis trojana* plants to HS, Cingoz and Gurel (2016) examined the subsequent modifications brought on by SA pre-treatment. However, SA pre-treatment had an additive effect on their activities, elevated proline levels, and overall phenolic and flavonoid contents. Plants subjected to HT showed decreased SOD and CAT activities. However, SA also significantly increased the amount of cardenolides that the studied plants produced, up to 472.28g/g DW. In three separate tri-genomic hybrids of *Brassica napus* L., Ghani et al. (2021) investigated the ameliorative effects of SA under HS. Before being exposed to HS, the seedlings were sprayed with SA (0.13 mM) (Figure 12.1). After 30 days of seeding, the plants were harvested for biochemical examination and growth. It was discovered that plant growth, chlorophyll content, and enzyme activity significantly decreased in HS-grown plants. However, SA applied topically under HS increased leaf and root biomass as well as antioxidative enzyme levels. In *Arabidopsis thaliana*, SA-deficient *NahG* transgenic plants showed the lowest tolerance to T stress, whereas SA-accumulating *cpr5* (constitutive expressor of PR genes) mutants showed the best tolerance ability to T stress, demonstrating the role of endogenous SA in granting thermo-tolerance (Ahammed et al., 2015). HS had a significant negative impact on the SA mutants *npr1-1* and *npr1-5* (Ahammed et al., 2015). According to Clarke et al. (2009), SA-mediated functional interaction with jasmonates gave HS in *Arabidopsis thaliana* baseline tolerance. According to Zhao et al. (2011), SA pre-treatment in wheat plants increases protein kinase activity and prevents the degradation of the D1 protein when both high light and heat are present. This suggests that SA is involved in the D1 protein's accelerated recovery under HS. SA was used by Wang et al. (2010) in *Vitis vinifera* L. to reduce the photosynthetic inhibition caused by HS. The SA pre-treatment increased the effectiveness of photosystem II, maintained a greater activation state of RuBisCo, and caused *HSP21* to accumulate during the recovery phase. Similar to *WRKY*, the TF superfamily is important for the way plants react to HS. Numerous *WRKY* genes have been linked to HS response, but only a handful of them play a significant part in phytohormone-mediated HS tolerance. For instance, *WRKY39* belongs to *WRKY* protein group II. This TF in *Arabidopsis* exhibits regulatory interaction in the cross-talk of activated signaling pathways mediated by SA and JA under HS (Li et al., 2010). Under HS circumstances (42°C), Jahan et al. (2019) exogenously administered SA (1mM) to *Solanum lycopersicum* L. plants. The effects of HS on photosynthetic characteristics, chlorophyll pigment, carotenoid content, leaf water potential, osmotic potential, and antioxidant enzyme activities were reported to limit plant growth and development. The results showed that SA application increased the leaf gas-exchange traits, the maximum quantum yield of PS II (Fv/Fm), water use efficiency, and antioxidant enzyme activities such as SOD, POD, CAT, and APX that help reduce oxidative damage by scavenging of HS-induced excess reactive oxygen species (ROS), but decreased the accumulation of TBARS, $H_2O_2$, and electrolyte leakage (EL). In order to improve the intake of water and minerals under HT stress, SA also boosted the proline content, recovering the osmotic potential that had been disturbed by HS. Under HT stress (40°C for 10 days), the impact of SA on the reproductive organs of rice plants was determined (Feng et al., 2018). The genes *EAT1* (Eternal Tapetum 1), *MIL2* (Microsporeless 2), and *DTM1* (Defective Tapetum and Meiocytese 1) were engaged in this mechanism, which lowered the formation of ROS in anthers to stop tapetum degeneration and death. As a result of inhibiting tapetum cell death, $H_2O_2$ served as a downstream signaling molecule in the SA-induced suppression of pollen abortion under HS. In order to evaluate the changes in essential oil (EO) composition and antioxidant enzyme activity, Haydari et al. (2019) applied SA (2, 3, and 4mM) in a growth chamber under HS to *Mentha piperita* and *Mentha arvensis* L. plants. The HS significantly altered the EO composition and antioxidant enzyme activity in both cultivars. Exogenous SA was used by Wassie et al. (2020) to protect *Medicago sativa* L. plants from the harm caused by HS in terms of growth and photosynthetic efficiency. Alfalfa seedlings that were four weeks old were given SA (0.25mM or 0.5mM) treatment, followed by HS five days later. The findings showed that SA increased plant height, biomass, leaf morphology, chlorophyll content, and photosynthetic efficiency under HT stress by lowering the concentration of EL and TBARS and by modifying the activities of CAT, SOD, and POD. More pronounced effects were seen at lower SA concentrations (0.25 mM). SA (0.1mmol/L) was used by Fan et al. (2022) in a recent two-year pot culture experiment to lessen yield loss in wheat plants under HT stress. The experiment involved applying SA from the heading stage (SAH), anthesis stage (SAA), five days after anthesis (DAA; SA5), and ten days after anthesis (DAA; SA10), then exposing plants to HT stress. The yield was significantly reduced by the HS at the grain-filling stage, which also encourages photosynthetic capacity suppression. In descending order, SAAG > SA5G > SA10G and SAHG, the SA application reduced yield loss and lowered photosynthetic capacity inhibition under HS. The leaf area, stomatal density, chlorophyll content, soluble protein content, maximum photochemical efficiency (Fv/Fm), actual photochemical efficiency (ΦPSII), and sucrose phosphate synthase (SPS) activity were all improved by exogenous SA administration. The reaction of *Capsicum annuum* L. under HT stress was noticed (Zhang et al., 2020), as well as the effectiveness of exogenous SA in reducing the effects of HT on seed germination and seedling growth. The findings suggested that SA (0.01, 0.1 mM) improved HS tolerance by speeding up and perhaps enhancing germination and decreasing oxidative damage. Under HT stress, SA spraying (0.1mM) decreased yellow leaf dwarfing, prevented water loss, preserved high root vigor and the integrity of cell structure, activated the antioxidant defense system, reduced production of various HT-induced ROS, increased protective enzyme activities, and increased the content of non-enzymatic ROS scavengers.

*Brassica juncea* L. plants were subjected to HT (30°C or 40°C) for 24 hours (Hayat et al., 2009). The follow-up treatment included SA ($10^{-5}$M) at the eight-day seedling stage and after plants were sampled 30 days after seeding. According to the findings, HS exposure caused substantial ($p < 0.05$) decreases in growth, chlorophyll content, nitrate reductase (NR), and carbonic anhydrase (CA) activity, and leaf gas exchange parameters in plants. However, SA spray increased HT stress tolerance by notably enhancing the aforementioned factors, where antioxidative enzymes and proline levels significantly played a critical role. He et al. (2005) investigated the effects of SA (0, 0.1, 0.25, 0.5, 1, and 1.5 mmol) in a growth chamber while HS (46°C for 72 hours) was applied to Kentucky bluegrass (*Poa pratensis* L.) to understand the function of antioxidant metabolism. The most effective SA for increasing HS tolerance in Kentucky bluegrass was determined to be SA (0.25 mmol), which also improved re-growth potential and maintained leaf hydration status.

Khanna et al. (2016) further investigated the role of SA (400 µM)-induced enhancement of the antioxidant response in conferring HT stress resistance in two spring maize genotypes of *Zea mays* L. The findings demonstrated that HT decreased dry biomass in genotypes while increasing APX, CAT, and SOD activities, as well as the levels of $H_2O_2$, proline, and TBARS. Under HT stress conditions, SA treatment enhanced dry biomass, proline, and Halliwell-Asada pathway efficiency, and decreased $H_2O_2$ levels. Thus, the body of research indicates that SA functions as a significant adaptation mechanism in a variety of crop plants under HT circumstances.

## 12.5 NITRIC OXIDE

Nitric oxide (NO) is recognized as a free radical signaling elicitor involved in the complex array of functions in crop plants grown under physiological and undesirable environmental stress factors. NO controls principal morpho-physiological, anatomical, developmental, and biochemical processes under biotic and abiotic pressures in crop plants. NO exerts beneficial—as well as detrimental—effects which depend on its application and site of production. NO—and/or its derivatives called reactive nitrogen species (RNS)—exert a benign task in redox signaling and modification in the activation of plants' antioxidant defense mechanisms and adaptation under abiotic stress (Corpas et al., 2011; Sahay & Gupta, 2017; Hasanuzzaman et al., 2020). Increasing experimental evidence indicates that NO as a key player imparts stress tolerance under HT stress conditions by modulating metabolic processes and adaptation to HT stress. In order to achieve HS resistance, NO acts as a signaling molecule under heat stress and influences photosynthesis, oxidative defense reactions, osmolyte accumulation, gene expression, protein changes, and interactions with other signaling molecules and phytohormones. Although evidence suggests that the concentration of NO is increased in various crop plants under HT stress conditions (Bouchard & Yamasaki, 2008; Yu et al., 2014), its biosynthesis is reliant on the extent and severity of HS treatment. Gould et al. (2003) noted that in tobacco leaf cells, NO is produced at HS (45°C). Song et al. (2013) observed that NO release has a functional specificity in providing thermo-tolerance in rice plants as scavenging of endogenously produced NO by 2–4-carboxphenyl-4,4,5,5-teramethyllimidazoline-1-oxyl-3-oxide (cPTIO) palliates the advantageous effect of NO under heat stress. However, Beard et al. (2012) found that NO production is reduced under HS for auxin-mediated gene expression. The effectiveness of exogenous NO (0.5mM) in reducing HS (38°C)-induced oxidative stress in *Triticum aestivum* L. seedlings was investigated (Hasanuzzaman et al., 2012). The findings showed that HS significantly reduced the amounts of chlorophyll and ascorbate (AsA) while increasing the levels of TBARS and $H_2O_2$, glutathione (GSH), glutathione disulfide (GSSG), and antioxidant enzymes APX, GR, GPX, and GST, as well as enzymes connected to the glyoxalase system. Under HS, NO dramatically decreased MDA and $H_2O_2$ levels while increasing Chl, AsA, and GSH concentrations as well as the GSH/GSSG ratio and enhancing the activities of APX, MDHAR, DHAR, GR, GST, CAT, and Gly I. The detoxifying system for methylglyoxal (MG) and NO-mediated up-regulation of antioxidant defense were determined to be the mechanisms defending wheat plants against HS-induced oxidative stress. Li et al. (2013) studied the involvement of $H_2S$ in the acquisition of NO-induced HS forbearance in *Zea mays* L. seedlings. NO in presence of $H_2S$ enhanced the thermo-tolerance of seedlings but $H_2S$ biosynthesis inhibitors-DL-propargylglycine, aminooxyacetic acid, potassium pyruvate, and hydroxylamine, as well as the scavenger hypotaurine, allayed these effects. In conclusion, it was urged that $H_2S$ acts as a downstream signaling elicitor in NO-induced HS resistance of maize plants. Litvinovskaya et al. (2022) considered the involvement of NO in brassinosteroids (epicastasterone) and SA (EC 2-monosalicylate)-induced HS tolerance in wheat. Both conjugates (50nM) increased the seedling survival after being exposed to HS and showed increased levels of NO content in roots. The responses were not effective with NO scavenger PTIO (2-phenyl-4,4,5,5-tetramethylimidazoline-1-oxyl-3-oxide) which allayed the stress-protective effects of conjugates. In *Arabidopsis* seedlings, mutants deficient in NO showed reduced thermo-tolerance (Xuan et al., 2010). Hou et al. (2021) applied HS to *Pleurotu sostreatus*, NO application decreased the ATP and ROS production, and it stimulated the alternative oxidation pathway, thereby sustaining HS tolerance. Thus, the research suggests that NO makes plant mechanisms adjustable to adapt HS stress-induced oxidative stress.

## 12.6 BRASSINOSTEROIDS

Brassinosteroids (BRs), found in all plant species, are polyhydroxylated sterol–derived phytohormones and have structural resemblance with hormones of animal and insect steroids. Mitchell and his colleagues from the U.S. Department of Agriculture found BRs in pollen extracts from over 30 different species (Siddiqui et al., 2018). Then, a *Brassica napus* extract with growth-promoting properties was obtained and named "brassins." Later, the "brassinolide" was identified as its bioactive component and is thought to be the most physiologically active form of BRs (Grove et al., 1979). Under stress, particularly HS in crop plants, BRs can play a variety of ameliorative roles in plant growth and development (Siddiqui et al., 2018; Planas-Riverola et al., 2019; Ahammed et al., 2020; Kothari & Lachowiec, 2021; Wang et al., 2022). According to reports, HT stress has a significant impact on photosynthesis (Berry & Bjorkman, 1980, Rath et al., 2022; Yang et al., 2022). It has been noted that crop plants acquire thermo-tolerance when exposed to 24-epibrassinolide (EBL) and 28-homobrassinolide (HBR) (Hayat et al., 2010; Geetika et al., 2014; Sirhindi et al., 2017; Kaur et al., 2018; Wang et al., 2022). EBL treatment increases thermo-tolerance in tomato and rapeseed plants (Dhaubhadel et al., 1999; Singh & Shono, 2005). By comparing 7,8-dihydro-8a-20-hydroxyecdysone DHECD, a BR mimic, to EBL at lethal HS (47°C for two hours), Thussagunpanit et al. (2015) investigated the effects of BRs on rice plant photosynthesis, lipid peroxidation (LPOX), and seed set. Increases in shoot fresh weight, leaf area, chlorophyll content, and carotenoid content were achieved with the application of EBL and DHECD, while high levels of net $CO_2$ assimilation rate and photochemical quenching were maintained. Both the net $CO_2$ assimilation rate and LPOX were lower in the plants exposed to HS. The plants that had been exposed to EBL and DHECD also displayed lower levels of TBARS and $H_2O_2$, as well as higher total soluble sugar concentrations. The impact of EBL was investigated in *Lycopersicon esculentum* Mill. cv. 9021 plants for eight days at 40/30°C. The maximum potential rate of electron transport that contributed to RuBP, the Pn, Gs, and relative quantum efficiency of PSII photochemistry (APSII), as well as photochemical quenching (qP), were all shown to be decreased by HS, whereas non-photochemical quenching was increased (NPQ). By raising the pool of antioxidant activities such as SOD, APX, guaiacol peroxidase (GPOX), and CAT on the one hand and reducing $H_2O_2$ and TBARS contents on the other, the EBL pre-treatment considerably eased HS-induced photosynthetic capacity inhibition. Additionally, under HS, the EBL supply greatly boosted shoot weight (Ogweno et al., 2008). In *Brassica juncea* L., Sirhindi et al. (2017) investigated the scavenging capacity of HBR ($10^{-9}$ M) in reducing the HS-induced formation of free radicals ($O_2^{?-}$, $OH^{?-}$, $H_2O_2$) and NO. The plants raised in HS produced more TBARS, $H_2O_2$, and NO, leading to oxidative stress and up-regulated levels of the antioxidant enzymes SOD, CAT, GPOX, and APX. However, the priming treatment of HBR decreased the production of MDA, $H_2O_2$, and NO, while increasing growth characteristics, membrane stability, and the enzymatic activity of SOD, CAT, GPOX, and APX. Kaur et al. (2018) looked at the effects of seed priming on *Brassica juncea* exposed to HS (44°C) and salinity (180mM) using HBR (0, $10^{-6}$, $10^{-9}$, $10^{-12}$ M). However, the $H_2O_2$ concentration and the activity of SOD, CAT, APOX, GR, dehydroascorbate reductase (DHAR), and monodehydroascorbatereductase (MDHAR) rose due to the combined heat and salt stress. These activities were affected by the HBR supply in an additive manner, and the reaction was dose-dependent. By controlling ROS generation and enhancing the redox state of antioxidants, the HBR seed priming enhanced the tested plant's capacity to mitigate the harmful effects caused by heat and salt. According to Singh and Shono (2005), tomato seedlings treated with EBL under HT stress showed greater Pn than untreated tomato plants. In order to improve plants' ability to withstand heat, various phytohormones interact significantly. For instance, abscisic acid (ABA), a plant stress hormone, is a signaling elicitor that affects a variety of physiological processes. Since the production of ABA is required for plants to endure the stress conditions created by the induction of HSPs, HT stress causes a rise in ABA levels (Maestri et al., 2002; Pareek et al., 1998). According to observations made under HT, BRs increased the production of ABA (Bajguz, 2009), pointing to a potential regulatory relationship between ABA and BRs. Pn, Gs, Ci, the relative quantum efficiency of PSII photochemistry (APSII), and qP were reduced under HT stress in tomatoes (Ogweno et al., 2008). However, through improving antioxidant systems and carboxylation efficiency, the BRs treatment reduced heat-stress-induced photosynthetic capacity inhibition. In a two-year field experiment, Li et al. (2022) investigated the effects of 24-epicastasterone and $KH_2PO_4$ on wheat during the grain filling stage. Greater grain weight under HS was seen as a result of maintaining leaf greenness and better photoassimilates partitioning. However, the spray treatments with $KH_2PO_4$ or 24-epicastasterone preserved even a high level of photosynthetic activity by increasing antioxidant enzymes and lowering damage to lipid membranes and chlorophyll. To investigate the use of BRs to prevent spikelet degeneration, Zhang et al. (2022) gave *Oryza sativa* L. HT treatments during meiosis. The root activity, photosynthetic capacity, activity of the tricarboxylic acid cycle, and antioxidant capacity were all increased by the BRs. On the other hand, it was found that BRs levels at HT had decreased, which was explained by the breakdown of BRs. Additionally, interaction between BRs and HT was discovered in mutant rice lines that lack BRs, transgenic rice, and chemical regulators (Zhang et al., 2022). In order to evaluate growth, chlorophyll, photosynthesis, photosystem II, antioxidant system, and proline concentration under HT stress (30°C or 40°C for 48 hours), Qazi et al. (2014) administered HBR (0.01μM) to Indian mustard. The heat-grown plants displayed decreased growth, chlorophyll, photosynthetic rate, and CA and NR activities. The HBR increased proline and antioxidative enzyme levels, which enhanced growth and photosynthesis. In a different study, Harpreet et al. (2014) primed *Brassica juncea* L. with HBR ($10^{-6}$, $10^{-9}$, and $10^{-12}$ M) to examine its therapeutic effects during HT stress (44°C). All of the HBR doses utilized in the study were found to have a variety of effects on morphology and light-quenching pigments, as well as to stimulate growth and light-quenching pigments. But $10^{-9}$M HBR produced the best outcomes. In a greenhouse, Yadava et al. (2016) investigated the physiological and biochemical effects of EBL on *Zea mays* L. hybrid PMH3 plants during heat stress (48°C). Application of EBL preserved the stability of cell membranes and stopped protein deterioration. Nevertheless, it was discovered that EBL therapy dynamically increased the antioxidant enzymes CAT, SOD, and POX. Dhaubhadel et al. (2002) applied BRs to study their protecting effects under thermal stress in *Brassica napus*. The seedlings grown under EBL showed a higher buildup of four major classes of HSPs. Under HT stress, EBL prevented loss of the translational mechanism and increased its expression levels which corresponded to regained cellular protein synthesis. Therefore, it is evident from the described research that these phytohormones cooperate a fundamental adaptive task in shifting plants toward adaptation under the changing climate.

## 12.7 CONCLUSION

The complicated regulatory interplay between various phytohormones under temperature stress is presumed to be ameliorative in diverse crop plants. The high temperature stress under changing climate in plants causes oxidative stress by orchestrating the rate of ROS biosynthesis. The endogenous modulation of the levels of various phytohormone—either by genetic engineering or external supplements—could mitigate the heat/temperature-induced oxidative stress state by harmonizing the advanced plant physiology through modulation of essential processes viz-stimulation of the antioxidant and osmoprotectant defense system. Nevertheless, the associated phytohormone-mediated underlying mechanisms and their cross-talk need to be exploited further so that the current knowledge for adaptive plant ecophysiological and morphological processes under temperature stress in the changing climate could be utilized for better understanding.

## REFERENCES

Ahammed, G. J., Li, X., Liu, A., & Chen, S. (2020). Brassinosteroids in plant tolerance to abiotic stress. *Journal of Plant Growth Regulation*, *39*, 1451–1464.

Ahammed, G. J., Li, X., Yu, J., & Shi, K. (2015). *NPR1*-dependent salicylic acid signaling is not involved in elevated $CO_2$-induced heat stress tolerance in Arabidopsis thaliana. *Plant Signaling and Behavior*, *10*(6), e1011944.

Ahmad, B., Zaid, A., Sadiq, Y., Bashir, S., & Wani, S. H. (2019). Role of selective exogenous elicitors in plant responses to abiotic stress tolerance. In *Plant Abiotic Stress Tolerance* (pp. 273–290). Springer, Cham.

Aleem, S., Sharif, I., Tahir, M., Najeebullah, M., Nawaz, A., Khan, M. I., & Arshad, W. (2021). Impact of heat stress on cauliflower (*Brassica Oleracea* var. Botrytis): A physiological assessment. *Pakistan Journal of Agricultural Research*, *34*, 479–486.

Altaf, M. A., Shahid, R., Kumar, R., Altaf, M. M., Kumar, A., Khan, L. U., & Naz, S. (2022). Phytohormones mediated modulation of abiotic stress tolerance and potential crosstalk in horticultural crops. *Journal of Plant Growth Regulation*, 1–27. https://doi.org/10.1007/s00344-022-10812-0

Altaf, M. A., Shahid, R., Ren, M. X., Mora-Poblete, F., Arnao, M. B., Naz, S., & Chen, J. T. (2021). Phytomelatonin: An overview of the importance and mediating functions of melatonin against environmental stresses. *Physiologia Plantarum*, *172*, 820–846.

Argosubekti, N. (2020, April). A review of heat stress signaling in plants. In *IOP Conference Series: Earth and Environmental Science* (Vol. 484, No. 1, p. 012041). IOP Publishing, Bristol.

Bajguz, A. (2009). Brassinosteroid enhanced the level of abscisic acid in Chlorella vulgaris subjected to short-term heat stress. *Journal of Plant Physiology*, *166*, 882–886.

Beard, R. A., Anderson, D. J., Bufford, J. L., & Tallman, G. (2012). Heat reduces nitric oxide production required for auxin-mediated gene expression and fate determination in tree tobacco guard cell protoplasts. *Plant Physiology*, *159*, 1608–1623.

Berry, J., & Bjorkman, O. (1980). Photosynthetic response and adaptation to temperature in higher plants. *Annual Review of Plant Physiology*, *31*, 491–543.

Bouchard, J. N., & Yamasaki, H. (2008). Heat stress stimulates nitric oxide production in *Symbiodinium microadriaticum*: A possible linkage between nitric oxide and the coral bleaching phenomenon. *Plant and Cell Physiology*, *49*, 641–652.

Bulgari, R., Franzoni, G., & Ferrante, A. (2019). Biostimulants application in horticultural crops under abiotic stress conditions. *Agronomy*, *9*(6), 306.

Chaudhary, C., Sharma, N., & Khurana, P. (2021). Decoding the wheat awn transcriptome and overexpressing *TaRca1β* in rice for heat stress tolerance. *Plant Molecular Biology*, *105*, 133–146.

Chaudhary, S., Devi, P., Hanumantha Rao, B., Jha, U. C., Sharma, K. D., Prasad, P. V., & Nayyar, H. (2022). Physiological and molecular approaches for developing thermotolerance in vegetable crops: A growth, yield and sustenance perspective. *Frontiers in Plant Science*, *13*.

Cingoz, G. S., & Gurel, E. (2016). Effects of salicylic acid on thermotolerance and cardenolide accumulation under high temperature stress in *Digitalis trojana* Ivanina. *Plant Physiology and Biochemistry*, *105*, 145–149.

Clarke, S. M., Cristescu, S. M., Miersch, O., Harren, F. J., Wasternack, C., & Mur, L. A. (2009). Jasmonates act with salicylic acid to confer basal thermotolerance in *Arabidopsis thaliana*. *New Phytologist*, *182*, 175–187.

Corpas, F. J., Leterrier, M., Valderrama, R., Airaki, M., Chaki, M., Palma, J. M., & Barroso, J. B. (2011). Nitric oxide imbalance provokes a nitrosative response in plants under abiotic stress. *Plant Science*, *181*, 604–611.

Crafts-Brandner, S. J., & Salvucci, M. E. (2000). Rubiscoactivase constrains the photosynthetic potential of leaves at high temperature and $CO_2$. *Proceedings of the National Academy of Sciences*, *97*, 13430–13435.

Dhaubhadel, S., Browning, K. S., Gallie, D. R., & Krishna, P. (2002). Brassinosteroid functions to protect the translational machinery and heat-shock protein synthesis following thermal stress. *The Plant Journal*, *29*, 681–691.

Dhaubhadel, S., Chaudhary, S., Dobinson, K. F., & Krishna, P. (1999). Treatment with 24-epibrassinolide, a brassinosteroid, increases the basic thermotolerance of *Brassica napus* and tomato seedlings. *Plant Molecular Biology*, *40*, 333–342.

Fan, Y., Lv, Z., Li, Y., Qin, B., Song, Q., Ma, L., & Huang, Z. (2022). Salicylic acid reduces wheat yield loss caused by high temperature stress by enhancing the photosynthetic performance of the flag leaves. *Agronomy*, *12*(6), 1386.

Fatma, M., Iqbal, N., Sehar, Z., Alyemeni, M. N., Kaushik, P., Khan, N. A., & Ahmad, P. (2021). Methyl jasmonate protects the PSII system by maintaining the stability of chloroplast D1 protein and accelerating enzymatic antioxidants in heat-stressed wheat plants. *Antioxidants*, *10*(8), 1216.

Feng, B., Zhang, C., Chen, T., Zhang, X., Tao, L., & Fu, G. (2018). Salicylic acid reverses pollen abortion of rice caused by heat stress. *BMC Plant Biology*, *18*, 1–16.

Geetika, S., Harpreet, K., Renu, B., Spal, K. N., & Poonam, S. (2014). Thermo-protective role of 28-homobrassinolide in *Brassica juncea* plants. *American Journal of Plant Sciences*, *2014*.

Ghani, M. A., Abbas, M. M., Ali, B., Ziaf, K., Azam, M., Anjum, R., & Jillani, U. (2021). Role of salicylic acid in heat stress tolerance in tri-genomic *Brassica napus* L. *Bioagro*, *33*, 13–20.

Gould, K. S., Lamotte, O., Klinguer, A., Pugin, A., & Wendehenne, D. (2003). Nitric oxide production in tobacco leaf cells: A generalized stress response? *Plant, Cell and Environment*, *26*, 1851–1862.

Grove, M. D., Spencer, G. F., Rohwedder, W. K., Mandava, N., Worley, J. F., Warthen, J. D., & Cook, J. C. (1979). Brassinolide, a plant growth-promoting steroid isolated from Brassica napuspollen. *Nature*, *281*, 216–217.

Harpreet, K., Geetika, S., Renu, B., Poonam, S., & Mir, M. (2014). 28-homobrassinolide modulate antenna complexes and carbon skeleton of *Brassica juncea* L. under temperature stress. *Journal of Stress Physiology & Biochemistry*, *10*, 186–196.

Hasanuzzaman, M., Bhuyan, M. B., Zulfiqar, F., Raza, A., Mohsin, S. M., Mahmud, J. A., & Fotopoulos, V. (2020). Reactive oxygen species and antioxidant defense in plants under abiotic stress: Revisiting the crucial role of a universal defense regulator. *Antioxidants*, *9*(8), 681.

Hasanuzzaman, M., Nahar, K., Alam, M. M., & Fujita, M. (2012). Exogenous nitric oxide alleviates high temperature induced oxidative stress in wheat (*Triticum aestivum* L.) seedlings by modulating the antioxidant defense and glyoxalase system. *Australian Journal of Crop Science*, *6*, 1314–1323.

Hasanuzzaman, M., Nahar, K., Alam, M. M., Roychowdhury, R., & Fujita, M. (2013). Physiological, biochemical, and molecular mechanisms of heat stress tolerance in plants. *International Journal of Molecular Sciences*, *14*, 9643–9684.

Hayat, S., Hasan, S. A., Yusuf, M., Hayat, Q., & Ahmad, A. (2010). Effect of 28-homobrassinolide on photosynthesis, fluorescence and antioxidant system in the presence or absence of salinity and temperature in *Vigna radiata*. *Environmental and Experimental Botany*, *69*, 105–112.

Hayat, S., Masood, A., Yusuf, M., Fariduddin, Q., & Ahmad, A. (2009). Growth of Indian mustard (*Brassica juncea* L.) in response to salicylic acid under high-temperature stress. *Brazilian Journal of Plant Physiology*, *21*, 187–195.

Haydari, M., Maresca, V., Rigano, D., Taleei, A., Shahnejat-Bushehri, A. A., Hadian, J., & Basile, A. (2019). Salicylic acid and melatonin alleviate the effects of heat stress on essential oil composition and antioxidant enzyme activity in Mentha× piperita and Mentha arvensis L. *Antioxidants*, *8*(11), 547.

He, Y., Liu, Y., Cao, W., Huai, M., Xu, B., & Huang, B. (2005). Effects of salicylic acid on heat tolerance associated with antioxidant metabolism in Kentucky bluegrass. *Crop Science*, *45*, 988–995.

Horváth, E., Szalai, G., & Janda, T. (2007). Induction of abiotic stress tolerance by salicylic acid signaling. *Journal of Plant Growth Regulation*, *26*, 290–300.

Hou, L., Zhao, M., Huang, C., He, Q., Zhang, L., & Zhang, J. (2021). Alternative oxidase gene induced by nitric oxide is involved in the regulation of ROS and enhances the resistance of *Pleurotus ostreatus* to heat stress. *Microbial Cell Factories*, *20*, 1–20.

Hussain, H. A., Men, S., Hussain, S., Chen, Y., Ali, S., Zhang, S., & Wang, L. (2019). Interactive effects of drought and heat stresses on morpho-physiological attributes, yield, nutrient uptake and oxidative status in maize hybrids. *Scientific Reports*, *9*, 1–12.

Jahan, M. S., Wang, Y., Shu, S., Zhong, M., Chen, Z., Wu, J., & Guo, S. (2019). Exogenous salicylic acid increases the heat tolerance in Tomato (*Solanum lycopersicum* L) by enhancing photosynthesis efficiency and improving antioxidant defense system through scavenging of reactive oxygen species. *Scientia Horticulturae*, *247*, 421–429.

Jenni, S., Truco, M. J., & Michelmore, R. W. (2013). Quantitative trait loci associated with tipburn, heat stress-induced physiological disorders, and maturity traits in crisphead lettuce. *Theoretical and Applied Genetics*, *126*, 3065–3079.

Kaur, H., Sirhindi, G., Bhardwaj, R., Alyemeni, M. N., Siddique, K. H., & Ahmad, P. (2018). 28-homobrassinolide regulates antioxidant enzyme activities and gene expression in response to salt-and temperature-induced oxidative stress in Brassica juncea. *Scientific Reports*, *8*(1), 1–13.

Kazan, K. (2015). Diverse roles of jasmonates and ethylene in abiotic stress tolerance. *Trends in Plant Science*, *20*, 219–229.

Khajuria, M., Jamwal, S., Ali, V., Rashid, A., Faiz, S., & Vyas, D. (2021). Temperature mitigation strategies in Lepidium latifolium L., a sleeper weed from Ladakh himalayas. *Environmental and Experimental Botany*, *184*, 104352.

Khan, M. I. R., Iqbal, N., Masood, A., Per, T. S., & Khan, N. A. (2013). Salicylic acid alleviates adverse effects of heat stress on photosynthesis through changes in proline production and ethylene formation. *Plant Signaling & Behavior*, *8*(11), e26374.

Khan, N., Bano, A., Ali, S., & Babar, M. (2020). Crosstalk amongst phytohormones from planta and PGPR under biotic and abiotic stresses. *Plant Growth Regulation*, *90*, 189–203.

Khanna, P., Kaur, K., & Gupta, A. K. (2016). Salicylic acid induces differential antioxidant response in spring maize under high temperature stress. *Indian Journal of Experimental Biology*, *54*, 386–393.

Kothari, A., & Lachowiec, J. (2021). Roles of brassinosteroids in mitigating heat stress damage in cereal crops. *International Journal of Molecular Sciences*, *22*(5), 2706.

Kumar, R. R., Rai, G. K., Kota, S., Watts, A., Sakhare, A., Kumar, S., & Praveen, S. (2022). Fascinating dynamics of silicon in alleviation of heat stress induced oxidative damage in plants. *Plant Growth Regulation*, 1–15. https://doi.org/10.1007/s10725-022-00879-w

Kumar, R. R., Sharma, S. K., Goswami, S., Verma, P., Singh, K., Dixit, N., & Rai, R. D. (2015). Salicylic acid alleviates the heat stress-induced oxidative damage of starch biosynthesis pathway by modulating the expression of heat-stable genes and proteins in wheat (*Triticum aestivum*). *Acta Physiologiae Plantarum, 37*(8), 1–12.

Lamaoui, M., Jemo, M., Datla, R., & Bekkaoui, F. (2018). Heat and drought stresses in crops and approaches for their mitigation. *Frontiers in Chemistry, 6,* 26.

Li, M., Wei, Q., Zhu, Y., Li, J., Ullah, N., & Song, Y. (2022). 24-Epicastasterone and $KH_2 PO_4$ protect grain production of wheat crops from terminal heat impacts by modulating leaf physiology. *Archives of Agronomy and Soil Science,* 1–14.

Li, S., Zhou, X., Chen, L., Huang, W., & Yu, D. (2010). Functional characterization of *Arabidopsis thaliana* WRKY39 in heat stress. *Molecules and Cells, 29,* 475–483.

Li, Z. G., Yang, S. Z., Long, W. B., Yang, G. X., & Shen, Z. Z. (2013). Hydrogen sulphide may be a novel downstream signal molecule in nitric oxide-induced heat tolerance of maize (*Zea mays* L.) seedlings. *Plant, Cell & Environment, 36,* 1564–1572.

Litvinovskaya, R. P., Shkliarevskyi, M. A., Kolupaev, Y. E., Kokorev, A. I., & Khripach, V. A. (2022). Involvement of nitric oxide in implementation of a protective effect of epicastasterone and its monosalicylate on wheat seedlings under heat stress. *Applied Biochemistry and Microbiology, 58,* 368–374.

Maestri, E., Klueva, N., Perrotta, C., Gulli, M., Nguyen, H. T., & Marmiroli, N. (2002). Molecular genetics of heat tolerance and heat shock proteins in cereals. *Plant Molecular Biology, 48,* 667–681.

Mangal, V., Lal, M. K., Tiwari, R. K., Altaf, M. A., Sood, S., Kumar, D., & Aftab, T. (2022). Molecular insights into the role of reactive oxygen, nitrogen and sulphur species in conferring salinity stress tolerance in plants. *Journal of Plant Growth Regulation,* 1–21.

Medina, E., Kim, S. H., Yun, M., & Choi, W. G. (2021). Recapitulation of the function and role of ROS generated in response to heat stress in plants. *Plants, 10*(2), 371.

Munir, M. A. H., & Shabbir, G. (2018). Salicylic acid mediated heat stress tolerance in selected bread wheat genotypes of Pakistan. *Pakistan Journal of Botany, 50,* 2141–2146.

Nozue, K., Devisetty, U. K., Lekkala, S., Mueller-Moulé, P., Bak, A., Casteel, C. L., & Maloof, J. N. (2018). Network analysis reveals a role for salicylic acid pathway components in shade avoidance. *Plant Physiology, 178,* 1720–1732.

Ogweno, J. O., Song, X. S., Shi, K., Hu, W. H., Mao, W. H., Zhou, Y. H., & Nogués, S. (2008). Brassinosteroids alleviate heat-induced inhibition of photosynthesis by increasing carboxylation efficiency and enhancing antioxidant systems in Lycopersicon esculentum. *Journal of Plant Growth Regulation, 27,* 49–57.

Pareek, A., Singla, S. L., & Grover, A. (1998). Protein alterations associated with salinity, desiccation, high and low temperature stresses and abscisic acid application in Lal nakanda, a drought-tolerant rice cultivar. *Current Science,* 1170–1174.

Park, C. J., & Seo, Y. S. (2015). Heat shock proteins: A review of the molecular chaperones for plant immunity. *The Plant Pathology Journal, 31*(4), 323.

Perdomo, J. A., Capó-Bauçà, S., Carmo-Silva, E., & Galmés, J. (2017). Rubisco and rubisco activase play an important role in the biochemical limitations of photosynthesis in rice, wheat, and maize under high temperature and water deficit. *Frontiers in Plant Science, 8,* 490.

Planas-Riverola, A., Gupta, A., Betegón-Putze, I., Bosch, N., Ibañes, M., & Caño-Delgado, A. I. (2019). Brassinosteroid signaling in plant development and adaptation to stress. *Development, 146*(5), dev151894.

Poór, P., Nawaz, K., Gupta, R., Ashfaque, F., & Khan, M. I. R. (2021). Ethylene involvement in the regulation of heat stress tolerance in plants. *Plant Cell Reports,* 1–24.

Qazi, F., Mohammad, Y., Mahmooda, B., & Aqil, A. (2014). 28-homobrassinolide protects photosynthetic machinery in Indian mustard under high temperature stress. *Journal of Stress Physiology & Biochemistry, 10,* 181–194.

Rai, K. K., Rai, N., & Rai, S. P. (2018). Salicylic acid and nitric oxide alleviate high temperature induced oxidative damage in *Lablab purpureus* L. plants by regulating bio-physical processes and DNA methylation. *Plant Physiology and Biochemistry, 128,* 72–88.

Raskin, I., Ehmann, A., Melander, W. R., & Meeuse, B. J. (1987). Salicylic acid: A natural inducer of heat production in *Arum lilies. Science, 237,* 1601–1602.

Raskin, I., Skubatz, H., Tang, W., & Meeuse, B. J. (1990). Salicylic acid levels in thermogenic and non-thermogenic plants. *Annals of Botany, 66,* 369–373.

Rath, J. R., Pandey, J., Yadav, R. M., Zamal, M. Y., Ramachandran, P., Mekala, N. R., & Subramanyam, R. (2022). Temperature-induced reversible changes in photosynthesis efficiency and organization of thylakoid membranes from pea (*Pisum sativum*). *Plant Physiology and Biochemistry, 185,* 144–154.

Sahay, S., & Gupta, M. (2017). An update on nitric oxide and its benign role in plant responses under metal stress. *Nitric Oxide, 67,* 39–52.

Schauberger, B., Archontoulis, S., Arneth, A., Balkovic, J., Ciais, P., Deryng, D., & Frieler, K. (2017). Consistent negative response of US crops to high temperatures in observations and crop models. *Nature Communications, 8,* 1–9.

Sehar, Z., Gautam, H., Masood, A., & Khan, N. (2022). Ethylene and proline-dependent regulation of antioxidant enzymes to mitigate heat stress and boost photosynthetic efficacy in wheat plants. *Journal of Plant Growth Regulation.* https://doi.org/10.1007/s00344-022-10737-8.

Siddiqui, H., Hayat, S., & Bajguz, A. (2018). Regulation of photosynthesis by brassinosteroids in plants. *Acta Physiologiae Plantarum, 40,* 1–15.

Singh, I., & Shono, M. (2005). Physiological and molecular effects of 24-epibrassinolide, a brassinosteroid on thermotolerance of tomato. *Plant Growth Regulation, 47,* 111–119.

Sirhindi, G., Kaur, H., Bhardwaj, R., Sharma, P., & Mushtaq, R. (2017). 28-Homobrassinolide potential for oxidative interface in *Brassica juncea* under temperature stress. *Acta Physiologiae Plantarum, 39,* 1–10.

Song, L., Zhao, H., & Hou, M. (2013). Involvement of nitric oxide in acquired thermotolerance of rice seedlings. *Russian Journal of Plant Physiology, 60,* 785–790.

Song, Q., Yang, F., Cui, B., Li, J., Zhang, Y., Li, H., & Gao, J. (2019). Physiological and molecular responses of two Chinese cabbage genotypes to heat stress. *Biology Plant, 63,* 548–555.

Thomas, D. S. G., Twyman, C., Osbahr, H., & Hewitson, B. (2007). Adaptation to climate change and variability: Farmer responses to intraseasonal precipitation trends in South Africa. *Climate Change, 83,* 301–322.

Thussagunpanit, J., Jutamanee, K., Kaveeta, L., Chai-arree, W., Pankean, P., Homvisasevongsa, S., & Suksamrarn, A. (2015). Comparative effects of brassinosteroid and brassinosteroid mimic on improving photosynthesis, lipid peroxidation, and rice seed set under heat stress. *Journal of Plant Growth Regulation, 34,* 320–331.

Wahid, A., Gelani, S., Ashraf, M., & Foolad, M. R. (2007). Heat tolerance in plants: An overview. *Environmental and Experimental Botany, 61,* 199–223.

Waithaka, M., Nelson, G. C., Thomas, T. S., & Kyotalimye, M. (2013). *East African Agriculture and Climate Change: A Comprehensive Analysis*. Washington, DC: IFPRI, p. 387.

Wang, L. J., Fan, L., Loescher, W., Duan, W., Liu, G. J., Cheng, J. S., & Li, S. H. (2010). Salicylic acid alleviates decreases in photosynthesis under heat stress and accelerates recovery in grapevine leaves. *BMC Plant Biology*, 10, 1–10.

Wang, W., Xie, Y., Liu, C., & Jiang, H. (2022). The exogenous application of brassinosteroids confers tolerance to heat stress by increasing antioxidant capacity in soybeans. *Agriculture*, 12(8), 1095.

Wani, S. H., Kumar, V., Shriram, V., & Sah, S. K. (2016). Phytohormones and their metabolic engineering for abiotic stress tolerance in crop plants. *The Crop Journal*, 4, 162–176.

Waraich, E. A., Ahmad, R., Halim, A., & Aziz, T. (2012). Alleviation of temperature stress by nutrient management in crop plants: A review. *Journal of Soil Science and Plant Nutrition*, 12, 221–244.

Wassie, M., Zhang, W., Zhang, Q., Ji, K., Cao, L., & Chen, L. (2020). Exogenous salicylic acid ameliorates heat stress-induced damages and improves growth and photosynthetic efficiency in alfalfa (*Medicagosativa* L.). *Ecotoxicology and Environmental Safety*, 191, 110206.

Went, F. W., & Thimann, K. V. (1937). *Phytohormones*. New York: Macmillan, p. 3.

Xuan, Y., Zhou, S., Wang, L., Cheng, Y., & Zhao, L. (2010). Nitric oxide functions as a signal and acts upstream of AtCaM3 in thermotolerance in *Arabidopsis* seedlings. *Plant Physiology*, 153, 1895–1906.

Yadava, P., Kaushal, J., Gautam, A., Parmar, H., & Singh, I. (2016). Physiological and biochemical effects of 24-epibrassinolide on heat-stress adaptation in maize (*Zea mays* L.). *Natural Science*, 8, 171–179.

Yamori, W., Hikosaka, K., & Way, D. A. (2014). Temperature response of photosynthesis in C3, C4, and CAM plants: Temperature acclimation and temperature adaptation. *Photosynthesis Research*, 119, 101–117.

Yang, X., Han, Y., Hao, J., Qin, X., Liu, C., & Fan, S. (2022). Exogenous spermidine enhances the photosynthesis and ultrastructure of lettuce seedlings under high-temperature stress. *Scientia Horticulturae*, 291, 110570.

Yu, M., Lamattina, L., Spoel, S. H., & Loake, G. J. (2014). Nitric oxide function in plant biology: A redox cue in deconvolution. *New Phytologist*, 202, 1142–1156.

Zaid, A., Ahmad, B., & Wani, S. H. (2021b). Medicinal and aromatic plants under abiotic stress: A crosstalk on phytohormones' perspective. *Plant Growth Regulators*, 115–132.

Zaid, A., Mohammad, F., & Siddique, K. H. (2022). Salicylic acid priming regulates stomatal conductance, trichome density and improves cadmium stress tolerance in mentha arvensis L. *Frontiers in Plant Science*, 13.

Zaid, A., Mushtaq, M., & Wani, S. H. (2021a). Interactions of phytohormones with abiotic stress factors under changing climate. In *Frontiers in Plant-Soil Interaction*. Cambridge: Academic Press, pp. 221–236.

Zhang, C. X., Feng, B. H., Chen, T. T., Zhang, X. F., Tao, L. X., & Fu, G. F. (2017). Sugars, antioxidant enzymes and IAA mediate salicylic acid to prevent rice spikelet degeneration caused by heat stress. *Plant Growth Regulation*, 83, 313–323.

Zhang, W., Huang, H., Zhou, Y., Zhu, K., Wu, Y., Xu, Y., & Yang, J. (2022). Brassinosteroids mediate moderate soil-drying to alleviate spikelet degeneration under high temperature during meiosis of rice. *Plant, Cell & Environment*. https://doi.org/10.1111/pce.14436.

Zhang, Z., Lan, M., Han, X., Wu, J., & Wang-Pruski, G. (2020). Response of ornamental pepper to high-temperature stress and role of exogenous salicylic acid in mitigating high temperature. *Journal of Plant Growth Regulation*, 39, 133–146.

Zhao, H. J., Zhao, X. J., Ma, P. F., Wang, Y. X., Hu, W. W., Li, L. H., & Zhao, Y. D. (2011). Effects of salicylic acid on protein kinase activity and chloroplast D1 protein degradation in wheat leaves subjected to heat and high light stress. *Acta Ecologica Sinica*, 31, 259–263.

Zhou, R., Wu, Z., Wang, X., Rosenqvist, E., Wang, Y., Zhao, T., & Ottosen, C. O. (2018). Evaluation of temperature stress tolerance in cultivated and wild tomatoes using photosynthesis and chlorophyll fluorescence. *Horticulture, Environment, and Biotechnology*, 59, 499–509.

Zipelevish, E., Grinberge, A., Amar, S., Gilbo, Y., & Kafkafi, U. (2000). Eggplant dry matter composition fruit yield and quality as affected by phosphate and total salinity caused by potassium fertilizers in the irrigation solution. *Journal of Plant Nutrition*, 23, 431–442.

Zulfiqar, F., Nafees, M., Chen, J., Darras, A., Ferrante, A., Hancock, J. T., Ashraf, M., Zaid, A., Latif, N., Corpas, F. J., & Altaf, M. A. (2022). Chemical priming enhances plant tolerance to salt stress. *Frontiers in Plant Science*, 13.

#  13  Ecophysiological Response of Dipterocarp Seedlings to Ectomycorrhizal Colonisation
## A Fungicide Addition Study

*Francis Q. Brearley*

## 13.1 INTRODUCTION

Mycorrhizas are a mutualistic association between specialised root-inhabiting fungi and the plant roots, found in around 95% of all plant species in virtually all terrestrial ecosystems. Benefits are accrued by both partners, with the plant providing the fungus with carbon derived from its photosynthesis and receiving a number of benefits from the fungus in return, such as improvement of nutrient uptake, growth, water relations and metal resistance (van der Heijden & Sanders, 2002; Smith & Read, 2008, and references therein). Most tropical tree species form arbuscular mycorrhizas but a key minority form ectomycorrhizas (EcM) and often dominate the ecosystems in which they are found (Corrales et al., 2018). Dipterocarps are the dominant family in the lowland evergreen rain forests of Southeast Asia (Brearley et al., 2016), and nearly all the species examined have been confirmed as being EcM (*e.g.* Singh, 1966; Hong, 1979; Smits, 1992; Pampolina et al., 1995; Lee, 1998; Hoang & Tuan, 2008; Brearley, 2012). It is important to study dipterocarp seedlings, as their growth and performance influences regeneration processes that are key for the replacement of adult trees and maintenance of diversity in the forest ecosystem. Effective uptake of mineral nutrients is important for growth of seedlings of tropical forest trees, even in highly shaded forest understoreys (Holste et al., 2011; Bhadouria et al., 2020), and mycorrhizas have a role to play in this key ecophysiological process.

Ectomycorrhizal colonisation can increase the growth and nutrient uptake of dipterocarp seedlings, although studies have been mostly conducted under controlled pot conditions (reviewed in Brearley, 2011). Presently, one of the major problems facing dipterocarp mycorrhizal research is the ability to manipulate EcM colonisation of seedlings. This problem is confounded by the irregular fruiting of most dipterocarps (Ashton et al., 1988; Brearley et al., 2007a), leading to an unpredictable supply of seeds—and hence, experimental material. To date, most studies examining the responses of dipterocarp seedlings to the presence or absence of EcM fungi have inoculated seedlings or cuttings with a single EcM fungal species (Brearley, 2011), confining them to tightly controlled conditions. Under more natural conditions, where there will be multiple fungal species on a single tree, the only way to manipulate mycorrhizal colonisation is by fungicide application. Controlling EcM colonisation is problematic, as there are no fungicides specific to EcM fungi (predominantly in the Basidiomycota, with some Ascomycota), but a number of fungicides have shown varying degrees of success (Trappe et al., 1984; Teste et al., 2006). Preliminary studies (Brearley, 2003) showed that the dithiocarbamate fungicide, Mancozeb, has good potential for controlling EcM colonisation of dipterocarp seedlings without having phytotoxic effects. It does, however, have notable effects on the soil nutrient status which it is essential to bear in mind when interpreting the results of the experiments presented here. The main effects are an increase in nitrogen (N) mineralisation which may last for up to four months and an increase in soil zinc and manganese (which are part of the active ingredient of the fungicide). However, there is no evidence that soil pH, phosphorus (P) concentrations, or cation exchange capacity are affected (Brearley, 2003).

This chapter describes an experiment addressing the question of how the addition of Mancozeb fungicide reduces EcM colonisation on dipterocarp seedlings and how this reduction in EcM colonisation affects the ecophysiology of two contrasting dipterocarp species.

## 13.2 METHODS

### 13.2.1 Study Species

Two species that had contrasting ecologies were used in this study. *Parashorea tomentella* (Symington) Meijer (Urat mata beludu) is a relatively light-demanding species and common in the lowland forests (< 200 m) of north-east Borneo. It is a very large light hardwood tree and has been used as a popular timber tree. *Hopea nervosa* King (Selangan jangkang), in contrast, is a shade-tolerant medium hardwood that is locally common in northern Borneo although rarely used for timber due to its small stature.

## 13.2.2 Experimental Set-Up

The study was carried out in the nursery of the Sabah Forestry Department's Forest Research Centre which is situated adjacent to Kabili-Sepilok Forest Reserve at 5°52' N, 117°56' E. One-and-a-half year old seedlings of both species were obtained from the INFAPRO nursery, Danum Valley, Sabah, and planted in plastic pots (1.2 litres) in a mixture of alluvial forest soil from Kabili-Sepilok Forest Reserve (sieved to c. 1 cm) and river sand in a ratio of 3:1 (Brearley et al., 2007b). All seedlings had EcM when they were planted in the pots. They were placed on four shade tables (nominally receiving 50% of full sunlight) with c. 1200 cm$^2$ for every seedling. The irradiance within the tables was recorded using a SKP 215 quantum sensor and DataHog2 data-logger (both from Syke Instruments Ltd., Llandrindod Wells, Powys, Wales, UK). The values were recorded on four consecutive sunny days towards the end of the experiment with the irradiance received up to 12.8 mol m$^{-2}$ day$^{-1}$. Mean annual precipitation is around 3000 mm and the mean daily temperature range was from 23.8–31.3°C (measurements taken at Sandakan Airport about 11 km to the east) (Brearley et al., 2003).

Two treatments were applied weekly:

1. Control: 100 ml water.
2. Fungicide: 100 ml of Mancozeb fungicide (Dithane GR, Rohm & Haas, Paris, France) as 0.2 g l$^{-1}$ soil in 100 ml of water, i.e., 0.24 g per pot per week.

The experiment was conducted for six months, during which time the seedlings received natural rainfall; they were also given supplemental water on days without rain until field capacity was reached, and the seedling positions were re-randomised within the shade tables monthly.

## 13.2.3 Seedling Measurements

Measurements of height and leaf number were taken at the beginning of the experiment. At that time, the seedling biomass was determined by harvesting, dividing and drying (80°C, 48 hours) a further 30 (*P. tomentella*) or 31 (*H. nervosa*) individuals and then weighing them and using a best subsets regression to create the equations:

*P. tomentella* biomass (g) = –7.99 + [0.0823 × height (cm)] + [0.121 × number of leaves] + [1.79 × basal diameter (mm)] ($r^2$ = 74.4%, $p < 0.001$)

*H. nervosa* biomass (g) = –3.64 + [0.0451 × height (cm)] + [0.0597 × number of leaves] + [0.0281 × total branch length (cm)] + [1.20 × basal diameter (mm)] ($r^2$ = 89.3%, $p < 0.001$)

After six months, at the end of the experiment, the height and leaf number were re-measured. The seedlings were harvested and divided into four fractions (leaves, main stem and branches, tap root, and fine roots [< 2 mm diameter]), which were dried at 80°C for 48 hours and each then weighed. Mass ratios were calculated to determine the proportion of the seedling's biomass allocated to each fraction as:

Mass ratio = biomass of plant fraction/total biomass of seedling

The relative height growth rate (RGR) and relative leaf production rate (RPR) were calculated according to Hunt (1990) as:

RGR or RPR = (loge $M_t$ – loge $M_0$)/t months

where $M_0$ and $M_t$ are the Measurements at the beginning of the experiment and after t months, respectively.

The youngest fully expanded leaf was taken from each of the seedlings and nitrogen (N), phosphorus (P), potassium (K), calcium (Ca) and magnesium (Mg) concentrations were measured following digestion in a salicylic in sulphuric acid mix (33 g l$^{-1}$) with a lithium sulphate/copper sulphate catalyst (in a 10:1 ratio). N was analysed using the gas diffusion method (Tecator Ltd., 1984) and P was analysed using the ammonium molybdate-stannous chloride method (Tecator Ltd., 1983) on a Tecator Flow Injection Analyser 5042 Detector and 5012 Analyser (Foss UK Ltd., Didcot, Oxfordshire, UK). Potassium, Ca and Mg were measured by atomic absorption spectrophotometry with an air-acetylene flame (Perkin-Elmer 2100 Atomic Absorption Spectrophotometer, Beaconsfield, Buckinghamshire, UK). Leaf mass per area (LMA) was recorded by tracing the area and then measuring the tracing using a leaf area meter (Mark 2, Delta-T Devices Ltd., Burwell, Cambridgeshire, UK); the leaf was then weighed. In addition, samples of c. 1 mg were analysed for $\delta^{15}$N (PDZ Europa ANCA-GSL preparation module connected to a 20–20 isotope ratio mass spectrometer, Northwich, Cheshire, UK). Isotope ratios were calculated as:

$\delta^{15}$N (‰) = (Rsample/Rstandard) × 1000

where R is the isotope ratio of $^{15}$N/$^{14}$N of either the sample or the standard (atmospheric N).

## 13.2.4 Ectomycorrhizal Colonisation

Percentage EcM colonisation (% EcM) of the fine roots was calculated as the number of EcM root tips out of a total of c. 100–200 root tips examined from every seedling. Root tips were examined from at least four sub-samples chosen randomly from the whole fine root system.

## 13.2.5 Statistical Analyses

Analyses were carried out using t-tests. Box-Cox transformations were carried out as required and all analyses were done using Minitab 12.2 (Minitab Inc., State College, Pennsylvania, USA).

## 13.3 RESULTS

### 13.3.1 ECTOMYCORRHIZAL COLONISATION

As anticipated, total % EcM decreased with fungicide addition from c. 60–70% to c. 10–20% (Figure 13.1).

### 13.3.2 GROWTH RATES, BIOMASS AND BIOMASS ALLOCATION

The relative biomass and height growth rates and relative leaf production rate of *H. nervosa* were all reduced following fungicide addition, although the reduction was only significant for the leaf production rate (Figure 13.2). There were no consistent or significant effects on the growth of *P. tomentella* following fungicide addition (Figure 13.2). *Hopea nervosa* showed a significant reduction in root-to-shoot ratio, root-to-mass ratio and fine root–to-mass ratio following fungicide addition (Table 13.1). *Parashorea tomentella* showed similar patterns of biomass reallocation following fungicide addition (with the exception of fine root-to-mass ratio), but the changes were not significant; *P. tomentella* did, however, show a significantly reduced LMA with fungicide addition (Table 13.2).

### 13.3.3 FOLIAR NUTRIENTS

Foliar N concentration increased with fungicide addition in both species; foliar P and Ca concentrations were reduced in *P. tomentella*, whereas only foliar Ca concentration was reduced in *H. nervosa* (Figure 13.3). Fungicide addition led to an increase in foliar Mg concentration in *P. tomentella* (the increase was not significant in *H. nervosa*)

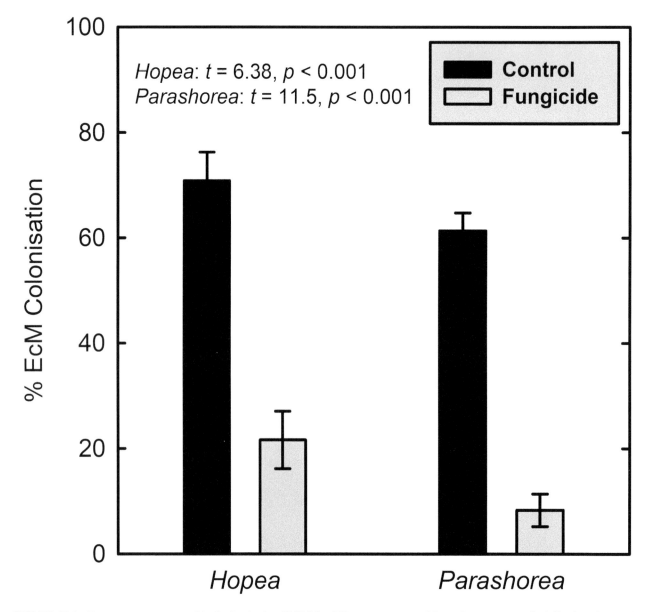

**FIGURE 13.1** Percentage ectomycorrhizal colonisation (% EcM) of *Hopea nervosa* and *Parashorea tomentella* following six months growth in the nursery given fungicide additions weekly and (all values are mean ± SE).

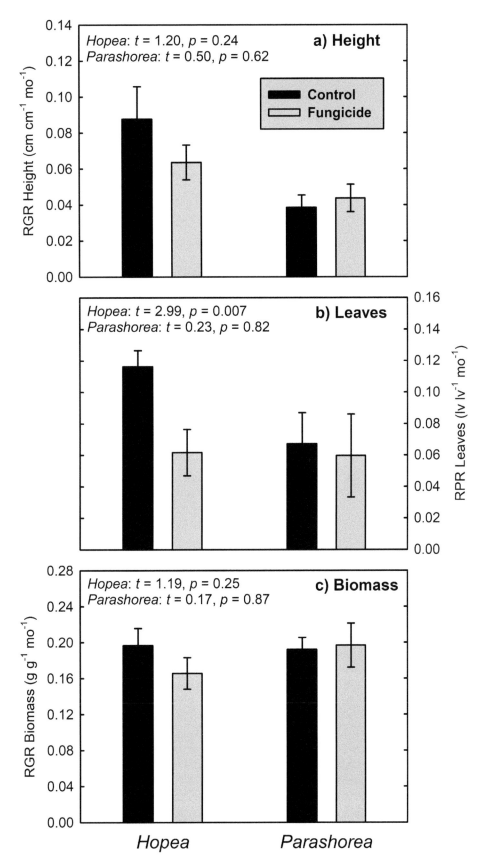

**FIGURE 13.2** Relative height growth rate (a), relative leaf production rate (b) and relative biomass growth rate (c) of *Hopea nervosa* and *Parashorea tomentella* following six months growth in the nursery given fungicide additions weekly (all values are mean ± SE).

### TABLE 13.1
**Biomass Allocation Patterns of *Hopea nervosa* and *Parashorea tomentella* Following Six Months Growth in the Nursery Given Fungicide Additions Weekly**

| | Control | Fungicide | Statistics |
|---|---|---|---|
| ***Hopea nervosa*** | | | |
| Leaf mass per area (g m$^{-2}$) | 62.7 ± 3.3 | 62.7 ± 1.7 | t = 0.01, p = 0.99 |
| Root-to-shoot ratio | 0.68 ± 0.03 | 0.52 ± 0.04 | t = 3.14, p = 0.005 |
| Leaf-to-mass ratio | 0.26 ± 0.01 | 0.30 ± 0.02 | t = 1.67, p = 0.11 |
| Stem-to-mass ratio | 0.36 ± 0.02 | 0.38 ± 0.02 | t = 1.00, p = 0.33 |
| Root-to-mass ratio | 0.38 ± 0.01 | 0.31 ± 0.01 | t = 3.55, p = 0.002 |
| Fine root–to-mass ratio | 0.15 ± 0.01 | 0.10 ± 0.01 | t = 3.48, p = 0.002 |
| ***Parashorea tomentella*** | | | |
| Leaf mass per area (g m$^{-2}$) | 72.3 ± 2.0 | 65.0 ± 1.9 | $t_{22}$ = 2.69, p = 0.013 |
| Root-to-shoot ratio | 0.70 ± 0.04 | 0.62 ± 0.03 | $t_{22}$ = 1.59, p = 0.13 |
| Leaf-to-mass ratio | 0.19 ± 0.01 | 0.19 ± 0.01 | $t_{22}$ = 0.29, p = 0.78 |
| Stem-to-mass ratio | 0.41 ± 0.01 | 0.43 ± 0.01 | $t_{22}$ = 1.19, p = 0.25 |
| Root-to-mass ratio | 0.41 ± 0.01 | 0.38 ± 0.01 | $t_{22}$ = 1.53, p = 0.14 |
| Fine root–to-mass ratio | 0.10 ± 0.01 | 0.12 ± 0.02 | $t_{22}$ = 0.88, p = 0.39 |

*Note:* All values are mean ± SE.

(Figure 13.3). *Hopea nervosa* showed a non-significant increase in foliar $\delta^{15}N$ with the addition of fungicide, whereas in contrast, *P. tomentella* showed a marginally significant decrease (Figure 13.3).

### 13.4 DISCUSSION

The responses of the two dipterocarp seedling species to the reduction in EcM colonisation with fungicide addition were different. A reduction in % EcM from 60–70% to 10–20% led to a reduction in the growth (most notably for leaf production) of *H. nervosa*. In contrast, reductions in % EcM in *P. tomentella* led to more obvious changes in LMA and reductions in foliar nutrient concentrations. The results give some indication that EcMs may be more important for *H. nervosa* than for *P. tomentella*, as *H. nervosa* grew more slowly in the nursery with the reduction in % EcM. However, foliar nutrients were more strongly influenced by the reduction in % EcM in *P. tomentella*, as shown by reductions in P, Ca and $\delta^{15}N$ that may take longer to translate to increased growth rates.

However, there was some evidence that *P. tomentella* took up extra N resulting from fungicide addition,- leading to a reduced LMA and minor increases in growth, so the effect of fungicide addition on factors other than EcMs must be carefully considered. Although the $\delta^{15}N$ values of the N sources in the nursery experiment were not measured, the contrasting results exhibited by *H. nervosa* and *P. tomentella* suggest that fungicide addition may be having different effects on the two species. A reduction in % EcM led to a (non-significant) increase in $\delta^{15}N$ for *H. nervosa*, suggesting that this species was obtaining a considerable proportion of its N *via* EcMs (Hobbie, 2005). In contrast, the addition of fungicide led to a decrease in the $\delta^{15}N$ values of *P. tomentella*, suggesting alternative processes may have been occurring such as the uptake of more isotopically lighter N obtained following its release from senescing micro-organisms or from the fungicide directly.

Other nursery studies have shown an increase in the growth/biomass of dipterocarp seedlings when inoculated with a single fungal species (Hadi & Santoso, 1988; Santoso, 1988; Yazid et al., 1994, 1996; Omon, 1996; Turjaman et al., 2005, 2011; Kaewgrajang et al., 2013) or unidentified species through inoculation with root fragments (Lee & Alexander, 1994; Yap, 1998). These growth increases were very variable depending on the seedling species and fungus combinations under consideration. For example, Lee and Alexander (1994) found a c. 20% increase in *Hopea helferi* (Dyer) Brandis seedling dry weight when inoculated with root fragments which showed a marked contrast of the results of Hadi and Santoso (1988), who found a nearly 30-fold increase in *Shorea compressa* Burck (syn. *Shorea pinanga* Scheff.) seedling height when inoculated with *Scleroderma* sp. The most common increases in growth are between 50% and 300% (Hadi & Santoso, 1988; Santoso, 1988; Lee & Alexander, 1994; Yazid et al., 1994, 1996; Omon, 1996; Turjaman et al., 2005; 2011; Kaewgrajang et al., 2013). However, one distinction of this current study from most of these other studies is that a more natural complement of EcM fungi was present on the roots of the 'control' EcM plants (Brearley et al., 2003, 2007b) in contrast to a single inoculated species or an unknown range of unidentified species.

Foliar N concentration was increased with fungicide addition in the nursery experiment, but this was assumed to be due

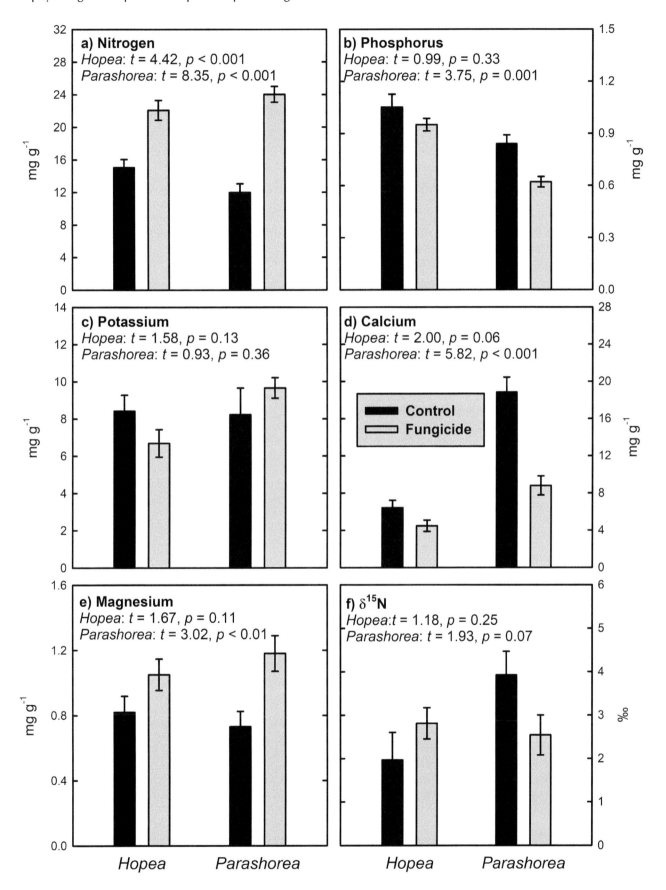

**FIGURE 13.3** Foliar nutrient concentrations and $\delta^{15}N$ values of *Hopea nervosa* and *Parashorea tomentella* following six months growth in the nursery given fungicide additions weekly (all values are mean ± SE).

to the death of other soil micro-organisms and the subsequent release of nutrients from them. There may also have been some effect of the N within the fungicide itself (which contained around 10% N; Francis Q. Brearley, unpublished data). Foliar P and Ca concentrations were reduced in one or both species with fungicide addition, with the reduction in foliar Ca particularly notable (> 50%) for *P. tomentella*. Increased P uptake is considered to be the 'classic' benefit of mycorrhizal colonisation (Smith & Read, 2008, and references therein), particularly as it is one of the most poorly mobile nutrients within tropical soils. Lee and Alexander (1994) also tentatively concluded that EcMs could be important for Ca uptake in dipterocarp seedlings. The increase in foliar Mg concentration with fungicide addition is difficult to explain but may be related to the high concentrations of manganese (Mn) in Mancozeb fungicide (16%; Thomson, 1978), which may affect Mg uptake, as Mg and Mn are closely related metals and have similar uptake pathways (Marshner, 1995).

P is often considered to be a key limiting nutrient in soils under lowland evergreen rain forests due its low concentrations and potential for fixation with other soil minerals (Wright, 2019). However, the reduction in foliar P concentration in *P. tomentella* with fungicide addition with no concomitant reduction in growth suggests that P may not, in fact, be limiting *P. tomentella* seedling growth, as it can gain sufficient P *via* its symbiotic EcM partners. Reductions in P and Ca concentrations with fungicide addition indicate the importance of EcM associations for accessing a number of soil nutrients and suggest that *H. nervosa* also did not appear to be nutrient limited. These results are important, as even under poorly lit understory conditions, nutrients still play a role in seedling growth (Holste et al., 2011; Santiago et al., 2012).

## 13.5 CONCLUSIONS

The addition of Mancozeb fungicide was successful in reducing EcM colonisation on seedlings of *H. nervosa* and *P. tomentella*, both members of the Dipterocarpaceae family. The reduction in EcM colonisation led to reductions in phosphorus and calcium uptake. However, reductions in seedling growth were less marked, with small growth reductions shown by *H. nervosa*. Future work to examine the importance of EcM fungi under more natural conditions would be valuable. Bearing in mind the limitations imposed by a fungicide addition study (particularly increases in nitrogen availability due to microbial senescence), we show here contrasting responses of two dipterocarp species to reductions in EcM colonisation that may influence their regeneration under more natural conditions.

## ACKNOWLEDGEMENTS

The British Ecological Society is thanked for financially supporting this work (through the grant "Biotic Interactions in Tropical Rain Forests"), and the academic contributions of Malcolm Press and Julie Scholes are gratefully acknowledged.

## REFERENCES

Ashton, P. S., T. J. Givnish and S. Appanah. 1988. Staggered flowering in the Dipterocarpaceae: New insights into floral induction and the evolution of mast fruiting in the aseasonal tropics. *American Naturalist* 132:44–66.

Bhadouria, R., P. Srivastava, R. Singh, S. Tripathi, P. Verma and A. S. Raghubanshi. 2020. Effects of grass competition on tree seedlings growth under different light and nutrient availability conditions in tropical dry forests in India. *Ecological Research* 35:807–18.

Brearley, F. Q. 2003. *The Role of Ectomycorrhizas in the Regeneration of Dipterocarp Seedlings*. PhD Thesis. Sheffield: University of Sheffield.

Brearley, F. Q. 2011. The importance of ectomycorrhizas for the growth of dipterocarps and the efficacy of ectomycorrhizal inoculation schemes. In *Diversity and Biotechnology of Ectomycorrhizae*, eds. M. Rai and A. Varma, 3–17. Berlin, Germany: Springer-Verlag.

Brearley, F. Q. 2012. Ectomycorrhizal associations of the Dipterocarpaceae in tropical forests. *Biotropica* 44:637–48.

Brearley, F. Q., L. F. Banin and P. Saner. 2016. The ecology of the Asian dipterocarps. *Plant Ecology and Diversity* 9:429–36.

Brearley, F. Q., M. C. Press and J. D. Scholes. 2003. Nutrients obtained from leaf litter can improve the growth of dipterocarp seedlings. *New Phytologist* 160:101–10.

Brearley, F. Q., J. Proctor, Suriantata, L. Nagy, G. Dalrymple and B. C. Voysey. 2007a. Reproductive phenology over a 10-year period in a lowland evergreen rain forest of central Borneo. *Journal of Ecology* 95:828–39.

Brearley, F. Q., J. D. Scholes, M. C. Press and G. Palfner. 2007b. How does light and phosphorus fertilisation affect the growth and ectomycorrhizal community of two contrasting dipterocarp species? *Plant Ecology* 192:237–49.

Corrales, A., T. W. Henkel and M. E. Smith. 2018. Ectomycorrhizal associations in the tropics—Biogeography, diversity patterns, and ecosystem roles. *New Phytologist* 220:1076–91.

Hadi, S. and E. Santoso. 1988. Effect of *Russula* spp., *Scleroderma* sp. and *Boletus* sp. on the mycorrhizal development and growth of five dipterocarp species. In *Agricultural and Biological Research Priorities in Asia, Proceedings of the IFS Symposium of Science Asia 87*, ed. M. Mohinder Singh, 183–5. Kuala Lumpur, Malaysia: International Foundation for Science & Malaysian Scientific Association.

Hoang, P. N. D. and D. L. A. Tuan. 2008. Investigating the ectomycorrhizal appearance of seedlings in the Tan Phu forest enterprise's nursery, Dong Nai Province. *Science and Technology Development Journal* 11:96–100.

Hobbie, E. A. 2005. Using isotopic tracers to follow carbon and nitrogen cycling of fungi. In *The Fungal Community: Its Organisation and Role in the Ecosystem*, 3rd Ed, eds. J Dighton, J. F. White and P. Oudemans, 361–81. Boca Raton, FL, USA: CRC Press, Taylor & Francis Group.

Holste, E. K., R. K. Kobe and C. F. Vriesendorp. 2011. Seedling growth responses to soil resources in the understory of a wet tropical forest. *Ecology* 92:1828–38.

Hong, L. T. 1979. A note on dipterocarp mycorrhizal fungi. *Malaysian Forester* 42:280–3.

Hunt, R. 1990. *Basic Growth Analysis*. London: Unwin Hayman.

Kaewgrajang, T., U. Sangwanit, K. Iwase, M. Kodama and M. Yamato. 2013. Effects of ectomycorrhizal fungus *Astraeus odoratus* on *Dipterocarpus alatus* seedlings. *Journal of Tropical Forest Science* 5:200–5.

Lee, S. S. 1998. Root symbiosis and nutrition. In *A Review of Dipterocarps Taxonomy, Ecology and Silviculture*, eds. S. Appanah and J. M. Turnbull, 99–114. Bogor: Centre for International Forestry Research.

Lee, S. S. and I. J. Alexander. 1994. The response of seedlings of two dipterocarp species to nutrient additions and ectomycorrhizal infection. *Plant and Soil* 163:299–306.

Marshner, H. 1995. *Mineral Nutrition of Higher Plants*, 2nd Ed. London: Academic Press.

Omon, R. M. 1996. Pengaruh beberapa jamur mikoriza dan media terhadap pertumbuhan stek Shorea leprosula Miq. *Bulletin Penelitian Hutan* 603:27–36.

Pampolina, N. M., R. E. de la Cruz and M. U. Garcia. 1995. Ectomycorrhizal roots and fungi of Philippine dipterocarps. In *Mycorrhizas for Plantation Forestry in Asia* (ACIAR Proceedings No. 62), eds. M. C. Brundrett, B. Dell, N. Malajczuk and M. Q. Gong, 47–50. Canberra: Australian Centre for International Agricultural Research.

Santiago, L., S. J. Wright, K. E. Harms et al. 2012. Tropical tree seedling growth responses to nitrogen, phosphorus and potassium addition. *Journal of Ecology* 100: 309–16.

Santoso, E. 1988. Pengaruh mikoriza terhadap diameter batang dan bobot kering anakan Dipterocarpaceae. *Bulletin Penelitian Hutan* 504:11–21.

Singh, K. G. 1966. Ectotrophic mycorrhiza in equatorial rain forests. *Malay Forester* 29:13–8.

Smith, S. E. and D. J. Read. 2008. *Mycorrhizal Symbiosis*, 3rd Ed. San Diego, CA: Academic Press.

Smits, W. T. M. 1992. Mycorrhizal studies in dipterocarp forests in Indonesia. In *Mycorrhizas in Ecosystems*, eds. D. J. Read, D. H. Lewis, A. H. Fitter and I. J. Alexander, 283–92. Wallingford: CAB International.

Tecator Ltd. 1983. *Determination of Orthophosphate Based on the Stannous Chloride Method Using Flow Injection Analysis* (Application Note AN 60/83). Didcot, Oxfordshire: Tecator Ltd.

Tecator Ltd. 1984. *Determination of Ammonium Nitrogen in Water by Flow Injection Analysis of Gas Diffusion* (Application Note ASN 50–02/84). Didcot, Oxfordshire: Tecator Ltd.

Teste, F. P., J. Karst, M. D. Jones, S. W. Simard and D. M. Durall. 2006. Methods to control ectomycorrhizal colonization: Effectiveness of chemical and physical barriers. *Mycorrhiza* 17:51–65.

Thomson, W. T. 1978. *Agricultural Chemicals: Book IV Fungicides*. Fresno, CA: Thomson Publications.

Trappe, J. M., R. Molina, R. and M. A. Castellano. 1984. Reactions of mycorrhizal fungi and mycorrhiza formation to pesticides. *Annual Review of Phytopathology* 22:331–59.

Turjaman, M., E. Santoso, A. Susanto et al. 2011. Ectomycorrhizal fungi promote growth of *Shorea balangeran* in degraded peat swamp forest. *Wetlands Ecology and Management* 19:331–339.

Turjaman, M., Y. Tamai, H. Segah et al. 2005. Inoculation with the ectomycorrhizal fungi *Pisolithus arhizus* and *Scleroderma* sp. improves early growth of *Shorea pinanga* nursery seedlings. *New Forests* 30:67–73.

van der Heijden, M. G. A. and I. R. Sanders (eds). 2002. *Mycorrhizal Ecology* (Ecological Studies 157). Berlin, Germany: Springer-Verlag.

Yap, S. W. 1998. *Large Scale Enrichment Planting of Dipterocarps in Logged-Over Forest in Sabah, Malaysia: Potential Role of Fertiliser Application*. Ph.D. Thesis. London: Wye College, University of London.

Yazid, S. M., S. S. Lee and F. F. Lapeyrie. 1994. Growth stimulation of *Hopea* spp. (Dipterocarpaceae) seedlings following mycorrhizal inoculation with an exotic strain of *Pisolithus tinctorius*. *Forest Ecology and Management* 67:339–43.

Yazid, S. M., S. S. Lee and F. F. Lapeyrie. 1996. Mycorrhizal inoculation of *Hopea odorata* (Dipterocarpaceae) in the nursery. *Journal of Tropical Forest Science* 9:276–8.

Wright, S. J. 2019. Plant responses to nutrient addition experiments conducted in tropical forests. *Ecological Monographs* 89:e01382.

# 14 Ecophysiological Responses of Tropical Plants to Changing Concentrations of Carbon Dioxide

*Rupali Jandrotia, Ipsa Gupta, Riya Raina and Daizy R. Batish*

## 14.1 INTRODUCTION

The ongoing noise around the accelerating trends of climate change has exposed the threats and hazards being imposed by the ongoing climate crisis that are affecting the quality and quantity of resources (van Kessel, 2020). The harms caused by climate change are inter-related directly or indirectly—for example, in addition to droughts and poor crop yields, increased temperature due to poor air quality leads to startling variation in the patterns of rainfall and increases the melting of glaciers and mountain snow which in turn causes rise in sea level, thereby affecting the coastal life forms and island communities (Martin & Watson, 2016). Although the major influence of climate change is registered in the Arctic and Antarctic, recent studies have confirmed the climate crisis also in the tropics. Subtropical regions are experiencing profound changes due to expansion of the tropical belt and poleward movement of circulation of atmospheric systems including jet streams and storm tracks (Dunning et al., 2018).

Tropical ecosystems differ from other ecosystems in numerous key ways, as high biodiversity has also increased the functional diversity of these ecosystems. In comparison to temperate ecosystems, oxygen is scarcer in the tropics because of lower saturation levels of dissolved oxygen in saltwater and higher rates of aerobic metabolism, which use up and deplete oxygen (Altieri et al., 2021). Tropical ecosystems are important to the global carbon cycle, owing to their high size and biomass. Tropical latitudes constitute around 40% of the surface of earth and are home to around 75% of terrestrial plant species (Barlow et al., 2018). The primary structural element of the environment in the majority of ecosystems is vegetation. Biodiversity and ecosystem services may be facilitated by this complexity. It is crucial to distinguish between horizontal and vertical vegetation structure because some taxa, such as birds and bees, may be more dependent on abrupt changes in horizontal structures (such as trees for nesting and open areas for foraging) than other taxa (Altieri et al., 2021). In tropical forests with usually homogeneous closed canopies, tree-fall gaps encourage variation. At the moment, tropical forests are dealing with both positive and negative observable regional changes and undetermined universal changes brought on mostly by the use of fossil fuels and changes in remote land cover imposed by the changing climate.

## 14.2 WHY THE TROPICS ARE IMPORTANT FROM A CLIMATE CHANGE PERSPECTIVE

Tropical environments possess specific features like lack of seasonality of temperature, much less fluctuation in the mean monthly temperature throughout the year than other terrestrial environments, variable rainfall producing distinct and localized ecosystems, and seasons (Sheldon, 2019). The tropical areas are generally very rich in biodiversity and the species have very small elevational ranges, resulting in highly specialized species and making the species particularly vulnerable to the changing climate (Forister et al., 2015). Also, tropical climates have registered wide variations in climatic variables due to climate change. Wang and Dillon (2014) showed that the tropics have registered a daily temperature increase of 0.29°C and 1.0 ± 0.8% per decade decrease in precipitation (Malhi & Wright, 2004), and these changes in parameters are still going to shift. Contrary to other regions, a small warming of a magnitude of 1°C is projected to give rise to extreme consequences in the tropical regions (Nottingham et al., 2020), although the predictions for precipitation are less clear. Also, the episodes and potency of extreme weather events like El Niño are going to increase. Changes in climatic conditions are known to impart a number of biotic changes to the plant species in the tropics. The almost stable climate of the tropics makes its species more vulnerable to climate changes as they come from a stable environment and are already present at their upper thermal limits (Deutsch et al., 2008). Thus, the warming of the atmosphere due to elevated $CO_2$ (carbon dioxide) concentrations are believed to make the surroundings too hot for many species to survive. Therefore, climate change is pushing tropical species towards their limits—and it hence is thought to pose a serious threat to millions of organisms found in the tropics (Sentinella et al., 2020). The key features of tropical ecosystems are illustrated in Figure 14.1. The chapter encapsulates the status of understanding of elevated $CO_2$ effects in the plant species inhabiting the tropics.

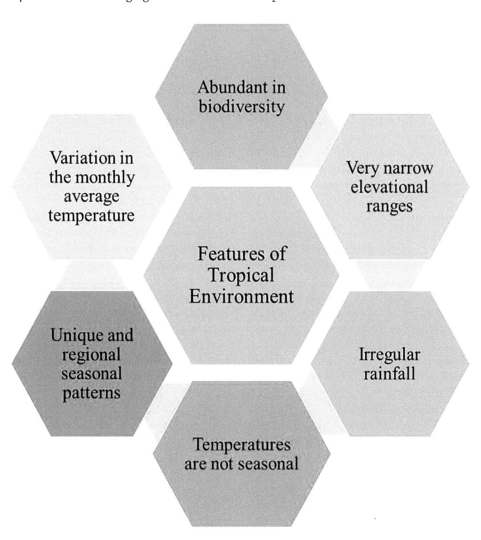

FIGURE 14.1 Features of tropical ecosystems.

Different modes of plant adjustment to elevated $CO_2$ concentrations and the importance of degree of genetic differentiation vs. phenotypic plasticity and the transplant experiments as the key to address these fundamental questions of genotype environment interactions determining fitness to the changing climate and minimizing the chances of extinction is also highlighted in the later sections.

## 14.3 SPECIES RESPONSES TO ELEVATED $CO_2$ CONCENTRATIONS

### 14.3.1 Ecological Responses of Plant Species to Increasing $CO_2$ Concentrations

Climate has long been acknowledged by biologists as a key factor in the dispersion of organisms (Fourcade et al., 2021). Whether tropical species can adapt to the changes caused by climate change is a crucial subject in this field of study. Research on species of tropical plants and animals have given robust evidence of uphill range migrations with the warming climate conditions. Morueta-Holme et al. (2015) found upward change in distribution of vegetation zones and 51 plant species in Chimborazo in 2012. Also, creation of new habitats has been experienced in the Andes due to significant changes in the abiotic factors (Diaz & Graham, 1996).

Organization of the community and variation in biodiversity is dependent upon the elevational range alterations. Warm-adapted, low-elevation species are becoming more abundant relative to other species in tropical montane communities (Fadrique et al., 2018). If the replacement of the up-migrating species does not take place by the other species, lowland biotic attrition may occur. Given the high endemism in tropical highlands, species may lose their preferred habitat and become extinct at higher elevations (Freeman et al., 2018). The upslope migration of species from low elevations may result in decrease in population of the species, decrease in fitness of the species and increase in their vulnerability due to the presence of new rivals since competition can limit species' elevational ranges in the tropics. According to Sheldon et al. (2011), the changing climate is expected to cause more tropical community fractionation than in temperate mountain communities. This prediction is supported by data showing that turnover of species, a gauge of changing community, has been higher in tropical regions due to range shifts triggered by climate change (Gibson-Reinemer & Rahel, 2015).

Because temperatures consistently decrease with elevation, range adjustments as feedback to warming are predicted to happen predominantly along the elevational gradients in mountains. In locations without topographic relief, it will be impossible to escape heat by moving upslope (Sheldon, 2019). Tropical species will have to migrate to colder, higher latitudes in order to track thermal niches in lowland areas with flat terrain. The mean annual temperature from 25°N to 25°S latitude is quite high and stable, and it subsequently decreases linearly with corresponding increasing latitude. Because of the weak temperature gradient in the tropics, tropical species need to move significantly farther north than do temperate species to reach cool refuges (Wright et al., 2009). Although distributions of lowland species may change as a result of heat stress imposed by the changing climate in the warmest regions of their range, near to the geographic limits of the tropics, warming-induced latitudinal shifts are anticipated to occur. Changes in moisture will change which habitats are suitable for which species. However, changes in rainfall patterns vary among different regions and on small geographical scales, in contrast to directional increases in temperature across the tropics. As a result, it is harder to forecast range alterations in response to precipitation. In the end, it may be concluded that the magnitude and direction of the species range shifts is probably influenced by a number of species; however, more mechanistic knowledge is required to pinpoint the contribution of climate change to these distributional shifts (Sheldon, 2019).

Given the significant natural fluctuations in tropical species' populations and the dearth of abiding data for major taxa, it is challenging to establish a connection between climate change and abundance of the species (Sheldon, 2019). Numerous tropical species also experience various challenges, making it difficult to pinpoint the cause of an issue. However, research shows that to induce population reduction, changes in temperature and precipitation can either work alone or in a complicated association with other factors. These pantropical, unidirectional forces shrink in abundance across multiple species are in accordance with the climate change projections (Cohen et al., 2019).

The tropics have seen global extinctions linked to climate change, as well. According to Wiens (2016), the climate-related local extinctions have been found to be significantly common in tropical species (55% versus 39% of species surveyed) than in temperate species. These extinctions may be attributed to the constricted geographic ranges; small population sizes, escalated specialization and reduced chances of dispersal are a number of characteristics that render tropical species particularly vulnerable to extinction due to the changing climate (Urban, 2015) and are anticipated to be more conspicuous in mountains as populations become constrained at the mountain's peak.

### 14.3.2 Whole Plant Responses of Plant Species to Elevated $CO_2$

The plant species may show variable of responses at elevated $CO_2$ concentration. The responses of whole plants to elevated $CO_2$ is shown in Figure 14.2.

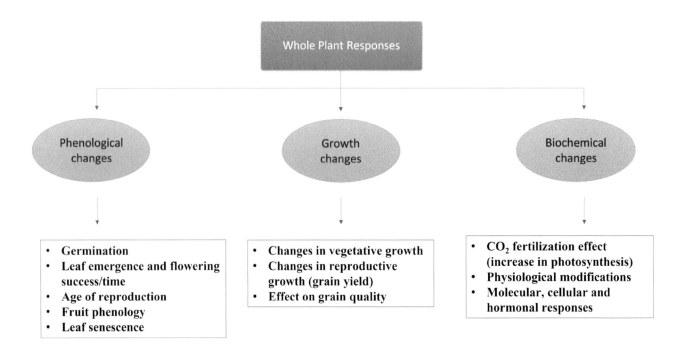

**FIGURE 14.2** Flow chart showing whole plant responses to elevated carbon dioxide ($CO_2$) concentrations.

## 14.3.2.1 Phenological Changes

Phenology—or the timing of botanical events over the annual life cycle of plants—can be correlated directly to the environmental factors sensed by the plants, and therefore can be considered crucial in determining the fitness of a species in a specific environment. As climatic parameters (rising atmospheric $CO_2$ concentrations and temperature) strongly affects phenological events like germination success (Walck et al., 2011), growth (Davies et al., 2006), flowering success (Springer & Ward, 2007), fruiting phenology (Gallinat et al., 2015) and leaf senescence (Fu et al., 2018) of plants; therefore, changes in phenological patterns may serve as bioindicators of changing climate. The seedling may not receive specific temperature and moisture levels for breaking of the dormancy which in result affects the success and timing of seed germination (Parmesan & Hanley, 2015). Although increased $CO_2$ is believed to augment the production of seeds (Li et al., 2018), elevated temperature is believed to reduce the viability of seeds (Ooi et al., 2012). Cleland et al. (2007) elaborated that changes in schedule of many crucial phenological activities—i.e., leaf emergence and flowering has advanced in the spring and events like leaf coloration and leaf fall have been delayed in fall—has occurred in response to the changing climate. Elevated temperatures enhance the development of flowers in tropical species (Pau et al., 2013), but disruption in the development of pollens and reduction in pollen viability is also observed (Pacini & Dolferus, 2019). Also, since age of reproduction is altered, plant fitness is hence also affected, as it is known to affect the distribution of resources between roots and leaves, which further alters potential of plants to acquire water and nutrients (Nord & Lynch, 2009). Other issues related to altered phenology are mistiming of the event and/or pollinators and flowering of different species at the same time, which intensifies the competition between the species, to induce ecological mismatch between the species (Tachiki et al., 2010). Climate-induced change also alters the composition of fruits, particularly fruit sugars which pose resistance to pathogens, thereby making them more prone to attack by pathogens (van Leeuwen & Darriet, 2016). The alteration in phenological events like life cycles of a species/populations of species has a number of implications for plant performance and survival. Altered phenology has been less explored in natural ecosystems, particularly for its consequences related to climate change (Sherry et al., 2007). Numata et al. (2022) performed the most comprehensive assessment of the reproductive phenology of 210 Malaysian tropical plant species for over 35 years and showed a decline magnitude of flowering and fruiting events in the species due to cooler conditions. Therefore, it may be concluded that rising $CO_2$ concentrations will change the timing of these botanical events—which in turn may affect the survival of the species in the tropics.

## 14.3.2.2 Changes in Plant Growth

### 14.3.2.2.1 Changes in Vegetative Growth

Growth of plants is dependent upon water availability which is in turn affected by temperature. Therefore, growth of plants in cool climates could experience elevated growth under climate change (Reich et al., 2018; Dusenge et al., 2020). However, increased temperatures lead to poor plant growth in the tropics as the temperature in the tropics is already equivalent to their thermal optimum (Sheldon, 2019), and water deficit and nitrogen limitation negates the fertilizing effect of $CO_2$ (Terrer et al., 2019).

Additionally, the declining moisture content connected with warming induces drought stress in the plants (Hatfield & Prueger, 2015). Therefore, it may be concluded that the photosynthetic capacity of plants is reduced in response of changing climate which thereby is going to affect the growth of plants in the tropics. Ironically, the growth models proposed to date have focused on particular regions and ignored the species from the tropics (Sheldon, 2019). Also, in these models, the response of community to the climate change is ignored (Korell et al., 2020). Therefore, there is a need to develop specific models to forecast regional scale plant growth which will enhance our understanding of these multi-factorial manipulations of distant future effects of climate change.

### 14.3.2.2.2 Changes in Reproductive Growth and Yield of Plants

Seed establishment is the first and the most crucial parameter in the success of the species in a particular environment. In the tropics, the seedling establishment speed is already quite slow and variable due to the involvement of a variety of interacting factors (like soil, climatic conditions, etc.) and is going to deteriorate under the changing climatic regime (Bhadouria et al., 2017). Hence, a multi-factorial interaction study under different regimes of resources and disturbances is recognized as an important aspect of studying the fate of future plant communities during changing climate (Bhadouria et al., 2017).

A small increase in the temperature of a range of 2–4°C is observed to affect the grain quality to a greater extent than elevated $CO_2$ (Williams et al., 1995). The rate of various phenological stages of plants advances is determined by the temperature; therefore, in regions like the tropics, where the temperature is already towards the higher side of the scale, the further increase will surely affect yields and quality (Wheeler et al., 2000). This increase in the average temperature is going to constrain the accumulation of the biomass by limiting the availability of nutrients and moisture to the plant. The elevated temperature progresses the maturity of plants. This leads to the conclusion that although per day yield is elevated in response to the elevated $CO_2$ concentration, the total growth season is shortened, which is going to decrease the overall yield. The temperature extremes may be damaging and induce high-temperature injuries to the plants that may be due to suppression of seedling emergence, grain filling and flowering (Mederski, 2019; Shaw, 2019).

Also, since the elevated $CO_2$ is proven to increase the photosynthetic capacity, and hence, biomass, growth, and yield of the plant (Ainsworth & Long, 2005), its effect on the quality of the seeds cannot be defined (Hampton et al., 2013) and the quality of the seeds may either decrease, increase or remain unchanged (Li et al., 2018; Thinh et al., 2017; Lamichaney et al., 2019). Therefore, it may be concluded that the beneficial direct effects of enhanced $CO_2$ on the grain yield may be invalidated by the non-beneficial indirect effects of the same.

#### 14.3.2.2.3 Effect on Grain Quality

Tropics are the center of origin and development of a number of important edible crops cultivated in the world. Therefore, yield and wholesomeness of grains in tropical crops are particularly important. Elevated $CO_2$ is believed to lower the nutrient quality of the grains because the seed filling is dependent upon both carbon assimilation rates and the nutrient status of the plants (Bhargava & Mitra, 2021). Plants grown under elevated $CO_2$ showed 6% increase in carbon content and 15% reduction in nitrogen content of the grains (Loladze, 2014). The "diluting effect of carbohydrates" decreases the nitrogen content under enhanced conditions of $CO_2$ (Broberg et al., 2017) further reducing the protein and amino acid content of the grains. Also, the phosphorus, sulfur and mineral content of the seeds are shown to decrease under elevated $CO_2$ (Loladze, 2014), thereby declining the overall nutritional value of the grains.

### 14.3.2.3 Effect of Elevated $CO_2$ on Crucial Biological Processes of Tropical Plants

#### 14.3.2.3.1 $CO_2$ Fertilization Effect

The photosynthetic rates are believed to increase with increase in the availability of $CO_2$, which acts as a substrate to the plants. Upsurge in the net photosynthetic rates and corresponding decline in the stomatal conductance is the key response of the plants to elevation in ambient $CO_2$ concentrations. Due to enhancement of photosynthetic efficiency, growth and corresponding yield enhancement is known as "$CO_2$-fertilization" effect (Chen et al., 2022). Most of the vascular plant species uses $C_3$ fixation, approximately 3% uses $C_4$ fixation, and the remaining 6–7% use Crassulacean acid metabolism (CAM) mode of fixation (Heyduk, 2022). Different types of plants show differential responses to enhanced $CO_2$. Enhanced $CO_2$ gives an initial kick to the photosynthesis in $C_3$ plants (Hussain et al., 2021). Since, $C_4$ plants already possess the concentrating mechanism for $CO_2$ in the mesophyll cells of their leaves, these plants are not known to be affected by the elevation in $CO_2$ to a larger extent (Habermann et al., 2019). Also, this change cannot be continued in the prolonged exposure to elevated $CO_2$ as a plant undergoes photosynthetic acclimation (Seneweera & Norton, 2011), since the regulation of $CO_2$ fertilization effects by the nutrients is reported in many climatic zones (Walker et al., 2021). Tropical vegetation predominantly grows in soil having poor nutrient quality; therefore, response to elevated $CO_2$ is strongly limited by the nutrient availability (Fleischer & Terrer, 2022). Reduction in the $CO_2$ fertilization effect on the carbon cycle due to climatic changes in tropical regions is reported by Liu et al. (2017). However, we still cannot predict the quantum elevated $CO_2$ on $CO_2$ fertilization, as it is a complex interplay of various factors.

#### 14.3.2.3.2 Physiological and Biochemical Modification

The enhanced $CO_2$ level stimulates plant growth and has a good impact on the photosynthetic process in C3 plants. Increased $CO_2$ at the location of RuBisCo, which restricts the availability of adenosine triphosphate (ATP) and nicotinamide adenine dinucleotide phosphate (NADPH) for photorespiration, is thought to be the cause of the increase in photosynthesis with elevated $CO_2$ levels. This makes plants readily accessible for the assimilation of $CO_2$ (Hay et al., 2017). According to Kaiser et al. (2017), higher ambient $CO_2$ at 800 bar enhanced photosynthesis by 17% and reduced the induction of photosynthesis by 14% in *Solanum lycopersicum* in contrast to 400 bar due to a faster carboxylation rate and a slower rate of RuBisCo oxygenation. Increased $CO_2$ has been linked to enhanced levels of carboxylation efficiency, apparent quantum efficiency, and light-saturated maximal photosynthetic rate in *S. lycopersicum* (Pan et al., 2020). The biomass, height, number of flowers and number of leaves of *Parthenium hysterophorus* are all positively impacted by the raised $CO_2$ level, stipulating improved growth and reproductive capacity of plants (Bajwa et al., 2019). Through satellite imaging, it is possible to see worldwide forest productivity in the form of numerous green patches of vegetation due to the increasing $CO_2$ concentration. Increased $CO_2$ causes trees to accumulate more carbon, which increases the amount of wood they can store. It also causes plants to require less water and their stomata to close (Paudel et al., 2018). However, as a result of drought and rising mean temperatures brought on by climate change, terrestrial ecosystems' morphological and physiological functions would suffer significant harm (Ainsworth et al., 2020).

Tropical forests are highly productive and are therefore crucial in determining the global carbon budget. Plant responses to increased temperature and elevated $CO_2$ could be different in different geographic locations of the world, e.g., rises in temperature enhance growth in temperate environments but may be detrimental in tropical environments (Cernusak et al., 2013). There are numerous approaches deployed for studying the effect of elevated temperature and $CO_2$ which include studies within controlled growth chambers, open top chambers (OTC), etc. Of these approaches, free air carbon dioxide enhancement (FACE) and free air temperature increase (FATI) facilities are those which provide near natural future conditions of the two parameters (Aeschlimann et al., 2005; Ainsworth et al., 2003a, 2003b). In addition to studying response of single or few species, network studies like no-crop FACE studies are imperative in generating the network information of different ecosystems (Nowak et al., 2004). Despite numerous advantages, the high cost of operation has limited the use FACE technology in developing nations (Ainsworth & Long, 2005).

The main effect of increase in atmospheric $CO_2$ concentration at leaf level is mainly believed to influence physiological processes. According to studies, increased $CO_2$ concentrations have an undeviating impact on the development of plant activities such as photosynthesis and gas exchange (Zhao et al., 2021). In the short term, increasing carboxylation by RuBisCo and decreasing stomatal openings is expected to be a direct response of elevated $CO_2$ concentration, followed by a number of indirect effects (Long et al., 2004). An initial gain in the photosynthesis rate after short-term exposure to enhanced $CO_2$ is believed to be due to elevation in onsite substrate availability. However, the initial stimulation of growth and photosynthesis due to enhanced $CO_2$ may disappear in the longer run. Factor responsible for it could be reduced carboxylation efficiency at enhanced $CO_2$ (Drake et al., 1997) due to reduction in RuBisCo protein and down-regulation in the expression of *rbcS* and other photosynthetic genes.

A variety of other factors, including both biotic and abiotic factors such as temperature (Ruiz-Vera et al., 2013), nitrogen availability (Markelz et al., 2014), genetic variation (Ainsworth et al., 2004), water availability (Xu et al., 2014) and age of leaves (Liu et al., 2016) may also be associated with the decline in the photosynthetic efficiency. This may lead to photosynthetic down-regulation due to reduction in chlorophyll content, reduced activity and content of RuBisCo, decrease in the per-unit mass concentration of leaf nitrogen, and reduced regeneration of ribulose bisphosphate (Ward & Strain, 1999). The downregulation of photosynthesis is more profound in $C_3$ species as compared to $C_4$ species because $C_3$ species do not have any carbon concentrating mechanisms and hence undergo severe N dilution of exposure to $CO_2$ enrichment (Duarte et al., 2014). However, whether such a response is common to all tropical species is little known.

Mid-day depression is a very pronounced phenomenon in tropical species of plants to minimize transpiration losses, and this depression is found to be independent of the water content of the soil (Kosugi et al., 2008). Therefore, in tropical species, the rise in leaf temperature due to elevation in ambient $CO_2$ concentration would cause longer and more frequent episodes of mid-day depression (Cernusak et al., 2013).

Variation in leaf stomatal conductance and net photosynthetic rate of $CO_2$ assimilation substantially alters water-use efficiency (WUE) of plants. Increases in plant growth and corresponding increases in the ecosystem productivity is expected in reaction to enhanced $CO_2$, based on decreased availability of water. Enhanced atmospheric $CO_2$ generally improve WUE in plants that may enhance their productivity in drier/desert areas. However, the assumption that there will be enhancement of productivity in the drier areas has a very limited support from the FACE studies from four different ecosystems of the world (Nowak et al., 2004). Constraints imposed on photorespiration due to enhanced $CO_2$ may reduce oxidation stress to the plants, therefore safeguarding the photosynthetic apparatus (Watanabe et al., 2014). However, another dilemma in the issue is that enhanced photoperoxidation in chloroplast due to inhibition of photorespiration under surging $CO_2$ levels may induce the destruction and disassembly of the chloroplast, resulting in the decline in the photosynthesis (Moroney et al., 2013). Therefore, it may be concluded that the major effects of enhanced $CO_2$ to plants are invigoration of growth and photosynthesis, particularly in $C_3$ species; decline in stomatal conductance; and reduced content of RuBisCo (Ainsworth & Long, 2005; Ainsworth & Rogers, 2007), which is believed to downregulate the antioxidant metabolism (Mishra et al., 2019). However, many species like *Beta vulgaris*, *Lolium perenne*, *Medicago lupulina*, etc., showed stimulation in the antioxidant system on exposure to elevated $CO_2$ (Farfan-Vignolo & Asard, 2012; Kumari et al., 2013). It may be, therefore, presumed that the effect of $CO_2$ on antioxidant system of tropical plants is yet not clear and its modulation depends upon species type and abiotic stress imposed (Singh & Agrawal, 2015).

#### 14.3.2.3.3 Molecular, Cellular and Hormonal Responses

The photosynthesis capacity of plants is dependent upon the developmental stages and the environmental conditions to which plant are exposed. Exposure to elevated $CO_2$ elevates the sugar content of the plants, which in turn results in the production of sink organs such as seeds, leaves, tillers, etc. The activity of sink is determined by the carbon metabolism–associated enzymes. Plants exposed to elevated $CO_2$ showed up-regulation of sucrose phosphate synthase and sucrose synthase, two key enzymes of sucrose metabolism in most of the plant species (Li et al., 2008), propounding that sucrose metabolism is elevated in response to elevated $CO_2$. The elevated sucrose content in the leaves of plants in turn leads to feedback inhibition of photosynthesis at elevated $CO_2$.

The increased carbon content is followed by lowering of the nitrogen content of plant at elevated $CO_2$ (Takatani et al., 2014), particularly in the leaf tissues which may be attributed to dilution of the plant nitrogen by extra sugars produced in the leaves. It may also be attributed to the downregulation of important enzymes associated with nitrogen metabolism like *aspartate synthetase*, *ferredoxin-nitrate reductase*, *glutamate dehydrogenase* and *low-affinity nitrate transporter*s, etc., as reported in various plant species. The increased carbon nitrogen ratio of plants' biomass initially helps in the elevation of photosynthesis. However, in the long run, the photosynthetic capacity declines due to nitrogen limitation. The turnover number of RuBisCo, a key enzyme of photosynthesis, is dependent upon the plant nitrogen content. Prolonged exposure to enhanced $CO_2$ suppresses the synthesis of RuBisCo and accelerates the signaling pathway that suppresses the transcription of *rbcL* and *rbcS* genes involved in encoding the large and small subunit of RuBisCo, respectively (Cheng et al., 1998).

The cell cycle is known to be influenced by the enhanced carbon supply at elevated $CO_2$ levels, thereby promoting premature growth and development of the plants in the tropics (Thilakarathne et al., 2014). Enhanced $CO_2$ affects cell wall loosening and cell cycle genes. It also up-regulates ribosomal protein genes regulating the growth of cytoplasm (Wei et al., 2013) and downregulation of secondary wall construction enzymes and hemicelluloses metabolism enzymes (Kontunen-Soppela et al., 2010). Gene modification in regulating these cell cycle–related enzymes account for the increased growth, division and expansion of cells noticed under elevated $CO_2$. However, still there is a dearth of information available on the transcriptomic responses of plants to elevated $CO_2$, which will impart us with understanding a cut above of the responses of the cell cycle to elevated $CO_2$.

Plant hormones are very important in altering the development of plants at elevated $CO_2$ levels (Wei et al., 2013). Plants exposed to elevated $CO_2$ levels are shown to produce higher levels of hormones like ethylene (Seneweera et al., 2003), dihydrozeatin riboside, gibberellic acid, indole-3-acetic acid, isopentenyl adenosine and zeatin riboside (Teng et al., 2006), playing a pivotal role in the development of the plant. Also, elevated $CO_2$ lowers down the concentration of the abscisic acid (ABA), which promotes senescence and lowers the stomatal conductance, which in turn improves the WUE of the plants (Finkelstein, 2013).

### 14.3.3 Elevated $CO_2$ and Plant Biotic Factors

It is estimated that the concentration of $CO_2$ by the end of the twenty-first century will reach an elevated value of around 730–1000 ppm, which will result in severe changes to the planet's climate (Chaudhry & Sidhu, 2022). This will impact the following biotic factors associated with the plants.

#### 14.3.3.1 Pests

According to studies, increased $CO_2$ modifies the components of nutrients and defensive molecules in leaves, which has a substantial impact on insect infestation (Li et al., 2019). Due to impact on the biomass and plant nutrients, elevation in the concentration of $CO_2$ affects the growth and development of insects and animal species dependent upon them (Kazan, 2018). In general, increased $CO_2$ causes populations of sap-sucking insects to increase, including aphids, whiteflies and planthoppers like *Nilaparvata lugens*, *Myzus persicae* and *Bemisia tabaci* (Jiang et al., 2016; Ahammed et al., 2020). In addition, increased $CO_2$ inhibits the survival, growth and population density of most leaf-chewing insects, presumably as a result of the potential decline in the nutritious content of their diets. Recently, Li et al. (2019) showed that feeding on tea seedlings cultivated in high-$CO_2$ environments considerably boosted the abundance of the population of tea aphids (*Toxoptera aurantii*). Higher $CO_2$ concentrations are probably going to have an impact on plant physiology by boosting photosynthetic activity, which will lead to more growth and higher plant output. This increase in the growth and output would in turn affect the quantity and quality of the plant foliage, therefore indirectly affecting the insects. A frequent trait of plants growing in high-$CO_2$ environments is an alteration in the chemical makeup of leaves. The alteration in the chemistry may impact the nutritional value and palatability of the foliage to the insect feeding on these leaves. Additionally, these plants frequently build up sugars and starches in their leaves, which alters the carbon-to-nitrogen ratio and decreases palatability. Because there is an important role of nitrogen in the growth of insects, some pest groups consume more plants when $CO_2$ concentrations are higher. To compensate for the nitrogen deficiency, the pests need to increase their quantity of food—thereby increasing plant damage. When nitrogen levels drop as predicted by $CO_2$ fertilization and compensatory feeding, foliage feeders like caterpillars, miners and chewers frequently increase their consumption rates (Skendžić et al., 2021).

#### 14.3.3.2 Diseases and Soil Microbes

Higher levels of $CO_2$ can also increase respiration rates and decrease cooling due to transpiration. Reactions of crops to high $CO_2$ depend on the availability of nitrogen; higher levels of $CO_2$ with insufficient levels of nitrogen and lack of sinks of carbon might restrict photosynthetic capability and growth. The number of the host population for a pathogen population increases with escalation in crop biomass and canopy at high $CO_2$ concentrations. A greater concentration of agricultural wastes also promotes the survival of the pathogens and provides a boost of inoculums for nearby fields and succeeding crops. Decline in nutritional value, increase in carbohydrate and fiber content, elevation in leaf wax, mesophyll and epidermal cells, etc., are some of the physiological and anatomical modifications in the plants due to elevated $CO_2$ affecting plant diseases. Reduced silicon concentration in sensitive rice types under increasing $CO_2$ levels has been linked to sheath blight severity and increase in leaf blast and sheath blight (Kobayashi et al., 2006). Increase in leaf wax and epidermal thickness in rice leads to increases in physical sensitivity to diseases besides varying responses like increase in virulence, abundance, distribution, fecundity and activity of the pathogens (Asibi et al., 2019).

Studies on the effects of global changes on soil organisms have provided an explanation for the variability in soil organism responses by pointing to resources connected to plant inputs or to soil water and temperature. For instance, in conditions of low soil water content, heat and decreased precipitation interact and leads to decline in feeding activity of soil invertebrates (Thakur et al., 2018). The beneficial impact on soil organisms due to rising $CO_2$ may be due to higher carbon inputs into soils through increased plant root formation and/or increased moisture content of soil at enhanced $CO_2$ (Nie et al., 2013). Studies have also repeatedly demonstrated that increased nitrogen mineralization and soil respiration due to warming occur at ambient precipitation and decrease in moisture deficient soil settings, which consequently increases soil nitrogen (Thakur et al., 2018). As a result, elevated nitrogen in warm, dry conditions of soil can make up for the nitrogen shortfall there, whereas elevated nitrogen in moist and warm soil conditions may worsen the impact on microbial activities in soil. Additionally, when the soil is adequately moist compared to soil with lower nitrogen levels and drier conditions, plant inputs into the soil at elevated $CO_2$ may likewise multiply elevated nitrogen (Thakur et al., 2019).

## 14.4 PHENOTYPIC PLASTICITY VS. GENETIC DIFFERENTIATION

The ability of plastic and adaptive species to changing conditions of climate determines its fitness to the changing climate and minimizes the chances of its extinction (Williams et al., 2017). A limited number of studies have evaluated the plasticity in plant species from different origins. These studies suggest that plants from low-elevation climates showed greater plastic responses as compared the plant species from higher elevations (Frei et al., 2014). The plasticity of plant species to elevated $CO_2$ in terms of life span of seeds and leaves, phenology and metabolic processes are all well documented. However, the plastic capacity of a tropical plant species to changing temperature is variable. The temperature fluctuation in the tropical regions is limited, and therefore, physiological adjustment to temperature variations is believed to be limited in tropical climates. Acclimatization of respiration and photosynthesis rates to temperature were observed in the seedlings and leaves of tropical trees provided water, light and nutrients (Slot et al., 2014). Temperate *Eucalytpus* sp. Showed enhanced growth (20–60%) at elevated temperatures as compared to tropical species which showed reduced growth (10%) at the elevated temperatures (Drake et al., 2015). Therefore, plasticity may be considered as a rapid response of an individual that will enable the organism to survive under changing climate conditions. However, if phenotypic responses cannot overcome the environmental variation, the absence of plasticity may be advantageous for the species' survival. Also, sometimes the plasticity may suppress the genetic modifications in the species, thereby preventing the natural selection. Due to these reasons, phenotypic plasticity may be useful in case of short term, random and unpredictable environmental changes. In order to cope with the extensive and directional selection pressures imposed by changing climate, genetic changes in the populations corresponding to the natural selection is required (Nicotra et al., 2010). Therefore, to understand the species' response to changing climatic conditions, assessment of degree of genetic differentiation vs. phenotypic plasticity is important (Münzbergová et al., 2017), and the transplant experiments are the key to address these fundamental questions of genotype–environment interactions (Franks et al., 2014).

## 14.5 MICROCLIMATES AS BUFFERS OF CLIMATE CHANGE

Forest microclimates are naturally occurring regions where the organisms living in the understory face distinct climatic conditions that vary considerably from the external forests area (De Frenne et al., 2019). Under a changing climate conditions, forest canopies may protect species in the understory by promoting microclimates that can act as feedback to macroclimate warming (De Frenne et al., 2019). Species have been significantly buffered against regional extirpations due to current climate change by microclimatic variation (Suggitt et al., 2018). Microclimates are isolated from the macroclimates due to decreased seasonal and annual fluctuations represented by minimum temperatures rising and maximum temperatures getting lower. Forests have an inherent ability to protect their internal microclimates from extremes. Macroclimatic circumstances is diminished by fragmentation of forest and the development of forest margins, which expose segments of the forest to the extremes of external climatic conditions (Ewers & Leite, 2013). The alteration in the microclimatic conditions near the forest edges are routinely reported from forests. Factors such as gradients in temperature and humidity around edges of forest have largely been hypothesized to mediate the responses of species within (Ewers & Leite, 2013). The biophysical factors that support and keep up microclimatic buffering and its continuity across time remain mainly unexplained, however. Studies have indicated the critical role of topographic variations in microrefugia formation. Suggitt et al. (2018) collected five million distribution records belonging to 430 species and concluded that England's population losses were found to be lesser in locations where topography produced more variance in the microclimate. Microclimate has a significant impact on how forest ecosystems function, having a direct impact on a variety of processes. The species inhabiting tropical forests are most susceptible to changing climates and the forest protects their microclimates by buffering them from the unpredictable climatic conditions on the environment outside (Ewers & Banks-Leite, 2013). Therefore, the probability of extinction due to climate change can be significantly decreased by microclimatic heterogeneity.

## 14.6 CONCLUSION AND PRIORITIES FOR FUTURE RESEARCH

Atmospheric concentration of $CO_2$ is continuously on the rise and is expected to impact plant performance, especially in the tropics. The responses of tropical plants can be assessed on different levels (ecosystem, phenological, physiological and molecular and biochemical levels). However, major gaps still exist in the understanding of the tropical plant species, communities and ecosystem responses at elevated $CO_2$ concentrations. Therefore, rigorous research is required to provide a spontaneous understanding of the aftermath of elevated $CO_2$ in the tropics. Future scope of work may include the following.

1. Studies to track the same in habitat of representative populations, providing further insight in face of climate change.
2. Study of the species under simulated conditions of relevant environmental factors.
3. Study of interactive effects of elevated $CO_2$, higher temperature and other important biotic and abiotic parameters essential for establishment, growth and survival of plants.
4. Studies of better understanding of the magnitude, drivers and implications of microclimatic variations in a warmer climate.

## ACKNOWLEDGMENTS

Rupali Jandrotia is thankful to Department of Health Research (DHR), India, for research funding. Ipsa Gupta is thankful to University Grants Commission (UGC), India for research fellowship.

## REFERENCES

Aeschlimann, U., Nösberger, J., Edwards, P.J., Schneider, M.K., Richter, M. and Blum, H., 2005. Responses of net ecosystem $CO_2$ exchange in managed grassland to long-term $CO_2$ enrichment, N fertilization and plant species. *Plant, Cell & Environment*, 28, 823–833.

Ahammed, G.J., Li, X., Liu, A. and Chen, S., 2020. Physiological and defense responses of tea plants to elevated $CO_2$: A review. *Frontiers in Plant Science*, 11, 305.

Ainsworth, E.A., Davey, P.A., Hymus, G.J., Osborne, C.P., Rogers, A., Blum, H., Nösberger, J. and Long, S.P., 2003a. Is stimulation of leaf photosynthesis by elevated carbon dioxide concentration maintained in the long term? A test with Lolium perenne grown for 10 years at two nitrogen fertilization levels under Free Air $CO_2$ Enrichment (FACE). *Plant, Cell & Environment*, 26, 705–714.

Ainsworth, E.A., Lemonnier, P. and Wedow, J.M., 2020. The influence of rising tropospheric carbon dioxide and ozone on plant productivity. *Plant Biology*, 22, 5–11.

Ainsworth, E.A. and Long, S.P., 2005. What have we learned from 15 years of free-air $CO_2$ enrichment (FACE)? A meta-analytic review of the responses of photosynthesis, canopy properties and plant production to rising $CO_2$. *New Phytologist*, 165, 351–372.

Ainsworth, E.A. and Rogers, A., 2007. The response of photosynthesis and stomatal conductance to rising $CO_2$: Mechanisms and environmental interactions. *Plant, Cell & and Environment*, 30, 258–270.

Ainsworth, E.A., Rogers, A., Blum, H., NoÈsberger, J. and Long, S.P., 2003b. Variation in acclimation of photosynthesis in Trifolium repens after eight years of exposure to Free Air $CO_2$ Enrichment (FACE). *Journal of Experimental Botany*, 54, 2769–2774.

Ainsworth, E.A., Rogers, A., Nelson, R. and Long, S.P., 2004. Testing the "source–sink" hypothesis of down-regulation of photosynthesis in elevated [$CO_2$] in the field with single gene substitutions in Glycine max. *Agricultural and Forest Meteorology*, 122, 85–94.

Altieri, A.H., Johnson, M.D., Swaminathan, S.D., Nelson, H.R. and Gedan, K.B., 2021. Resilience of tropical ecosystems to ocean deoxygenation. *Trends in Ecology & Evolution*, 36(3), 227–238.

Asibi, A.E., Chai, Q. and Coulter, J.A., 2019. Rice blast: A disease with implications for global food security. *Agronomy*, 9, 451.

Bajwa, A.A., Wang, H., Chauhan, B.S. and Adkins, S.W., 2019. Effect of elevated carbon dioxide concentration on growth, productivity and glyphosate response of parthenium weed (*Parthenium hysterophorus* L.). *Pest Management Science*, 75, 2934–2941.

Barlow, J., França, F., Gardner, T.A., Hicks, C.C., Lennox, G.D., Berenguer, E., Castello, L., Economo, E.P., Ferreira, J., Guénard, B. and Gontijo Leal, C., 2018. The future of hyperdiverse tropical ecosystems. *Nature*, 559, 517–526.

Bhadouria, R., Srivastava, P., Singh, R., Tripathi, S., Singh, H. and Raghubanshi, A.S., 2017. Tree seedling establishment in dry tropics: An urgent need of interaction studies. *Environment Systems and Decisions*, 37, 88–100.

Bhargava, S. and Mitra, S., 2021. Elevated atmospheric $CO_2$ and the future of crop plants. *Plant Breeding*, 140(1), 1–11.

Broberg, M.C., Högy, P. and Pleijel, H., 2017. $CO_2$-induced changes in wheat grain composition: Meta-analysis and response functions. *Agronomy*, 7, 32.

Cernusak, L.A., Ubierna, N., Winter, K., Holtum, J.A., Marshall, J.D. and Farquhar, G.D., 2013. Environmental and physiological determinants of carbon isotope discrimination in terrestrial plants. *New Phytologist*, 200, 950–965.

Chaudhry, S. and Sidhu, G.P.S., 2022. Climate change regulated abiotic stress mechanisms in plants: A comprehensive review. *Plant Cell Reports*, 41, 1–31.

Chen, C., Riley, W.J., Prentice, I.C. and Keenan, T.F., 2022. $CO_2$ fertilization of terrestrial photosynthesis inferred from site to global scales. *Proceedings of the National Academy of Sciences*, 119(10), e2115627119.

Cheng, S.H., Moore, B.D. and Seemann, J.R., 1998. Effects of short-and long-term elevated $CO_2$ on the expression of ribulose-1, 5-bisphosphate carboxylase/oxygenase genes and carbohydrate accumulation in leaves of Arabidopsis thaliana (L.) Heynh. *Plant Physiology*, 116, 715–723.

Cleland, E.E., Chuine, I., Menzel, A., Mooney, H. A. and Schwartz, M.D., 2007. Shifting plant phenology in response to global change. *Trends in Ecology & Evolution*, 22, 357–365.

Cohen, J.M., Civitello, D.J., Venesky, M.D., McMahon, T.A. and Rohr, J.R., 2019. An interaction between climate change and infectious disease drove widespread amphibian declines. *Global Change Biology*, 25, 927–937.

Davies, Z. G., Wilson, R.J., Coles, S. and Thomas, C.D., 2006. Changing habitat associations of a thermally constrained species, the silver-spotted skipper butterfly, in response to climate warming. *Journal of Animal Ecology*, 75(1), 247–256.

De Frenne, P., Zellweger, F., Rodríguez-Sánchez, F., Scheffers, B.R., Hylander, K., Luoto, M. . . . and Lenoir, J., 2019. Global buffering of temperatures under forest canopies. *Nature Ecology & Evolution*, 3(5), 744–749.

Deutsch, C.A., Tewksbury, J.J., Huey, R.B., Sheldon, K.S., Ghalambor, C.K., Haak, D.C. and Martin, P.R., 2008. Impacts of climate warming on terrestrial ectotherms across latitude. *Proceedings of the National Academy of Sciences*, 105(18), 6668–6672.

Diaz, H.F. and Graham, N.E., 1996. Recent changes in tropical freezing heights and the role of sea surface temperature. *Nature*, 383, 152–155.

Drake, B.G., González-Meler, M.A. and Long, S.P., 1997. More efficient plants: A consequence of rising atmospheric $CO_2$. *Annual Review of Plant Physiology and Plant Molecular Biology*, 48, 609–639.

Drake, J.E., Aspinwall, M.J., Pfautsch, S., Rymer, P.D., Reich, P.B., Smith, R.A. and Tjoelker, M.G., 2015. The capacity to cope with climate warming declines from temperate to tropical latitudes in two widely distributed Eucalyptus species. *Global Change Biology*, 21, 459–472.

Duarte, B., Santos, D., Silva, H., Marques, J.C. and Cacador, I., 2014. Photochemical and biophysical feedbacks of C3 and C4 Mediterranean halophytes to atmospheric $CO_2$ enrichment confirmed by their stable isotope signatures. *Plant Physiology and Biochemistry*, 80, 10–22.

Dunning, C.M., Black, E. and Allan, R.P., 2018. Later wet seasons with more intense rainfall over Africa under future climate change. *Journal of Climate*, 31(23), 9719–9738.

Dusenge, M.E., Madhavji, S. and Way, D.A., 2020. Contrasting acclimation responses to elevated $CO_2$ and warming between an evergreen and a deciduous boreal conifer. *Global Change Biology*, 26, 3639–3657.

Ewers, R.M. and Banks-Leite, C., 2013. Fragmentation impairs the microclimate buffering effect of tropical forests. *PLOS One*, 8, e58093.

Fadrique, B., Báez, S., Duque, Á., Malizia, A., Blundo, C., Carilla, J., Osinaga-Acosta, Q., Malizi, L., Silman, M., Farfán-Ríos, W., Malhi, Y., Young, K.R., Cuesta, C.F., Homeier, J., Peralvo, M., Pinto, E., Jadan, O., Aguirre, N., Aguirre, Z. and Feeley, K.J., 2018. Widespread but heterogeneous responses of Andean forests to climate change. *Nature*, 564, 207–212.

Farfan-Vignolo, E.R. and Asard, H., 2012. Effect of elevated $CO_2$ and temperature on the oxidative stress response to drought in Lolium perenne L. and *Medicago sativa* L. *Plant Physiology and Biochemistry*, 59, 55–62.

Finkelstein, R., 2013. Abscisic acid synthesis and response. *The Arabidopsis Book/American Society of Plant Biologists*, 11, e0166.

Fleischer, K. and Terrer, C., 2022. Estimates of soil nutrient limitation on the $CO_2$ fertilization effect for tropical vegetation. *Global Change Biology*, 28, 6366–6369.

Forister, M.L., Novotny, V., Panorska, A.K., Baje, L., Basset, Y., Butterill, P.T. and Dyer, L.A., 2015. The global distribution of diet breadth in insect herbivores. *Proceedings of the National Academy of Sciences*, 112, 442–447.

Fourcade, Y., WallisDeVries, M.F., Kuussaari, M., van Swaay, C.A., Heliölä, J. and Öckinger, E., 2021. Habitat amount and distribution modify community dynamics under climate change. *Ecology Letters*, 24, 950–957.

Franks, S.J., Weber, J.J. and Aitken, S.N., 2014. Evolutionary and plastic responses to climate change in terrestrial plant populations. *Evolutionary Applications*, 7, 123–139.

Freeman, B.G., Lee-Yaw, J.A., Sunday, J.M. and Hargreaves, A.L., 2018. Expanding, shifting and shrinking: The impact of global warming on species' elevational distributions. *Global Ecology and Biogeography*, 27, 1268–1276.

Frei, E.R., Ghazoul, J., Matter, P., Heggli, M. and Pluess, A.R., 2014. Plant population differentiation and climate change: Responses of grassland species along an elevational gradient. *Global Change Biology*, 20, 441–455.

Fu, Y.H., Piao, S., Delpierre, N., Hao, F., Hänninen, H., Liu, Y. and Campioli, M., 2018. Larger temperature response of autumn leaf senescence than spring leaf-out phenology. *Global Change Biology*, 24, 2159–2168.

Gallinat, A.S., Primack, R.B. and Wagner, D.L., 2015. Autumn, the neglected season in climate change research. *Trends in Ecology & Evolution*, 30, 169–176.

Gibson-Reinemer, D.K. and Rahel, F.J., 2015. Inconsistent range shifts within species highlight idiosyncratic responses to climate warming. *PloS One*, 10, e0132103.

Habermann, E., Dias de Oliveira, E.A., Contin, D.R., Delvecchio, G., Viciedo, D.O., de Moraes, M. A. . . . and Martinez, C.A., 2019. Warming and water deficit impact leaf photosynthesis and decrease forage quality and digestibility of a C4 tropical grass. *Physiologia Plantarum*, 165(2), 383–402.

Hampton, J.G., Boelt, B., Rolston, M.P. and Chastain, T.G., 2013. Effects of elevated $CO_2$ and temperature on seed quality. The Journal of Agricultural Science, 151, 154–162.

Hatfield, J.L. and Prueger, J.H., 2015. Temperature extremes: Effect on plant growth and development. *Weather and Climate Extremes*, 10, 4–10.

Hay, W.T., Bihmidine, S., Mutlu, N., Le Hoang, K., Awada, T., Weeks, D.P. and Long, S.P. (2017). Enhancing soybean photosynthetic $CO_2$ assimilation using a cyanobacterial membrane protein, ictB. *Journal of Plant Physiology*, 212, 58–68.

Heyduk, K., 2022. Evolution of Crassulacean acid metabolism in response to the environment: Past, present, and future. *Plant Physiology*, 190(1), 19–30.

Hussain, S., Ulhassan, Z., Brestic, M., Zivcak, M., Zhou, W., Allakhverdiev, S.I. . . . and Liu, W., 2021. Photosynthesis research under climate change. *Photosynthesis Research*, 150, 5–19.

Jiang, S., Liu, T., Yu, F., Li, T., Parajulee, M.N., Zhang, L. and Chen, F., 2016. Feeding behavioral response of cotton aphid, *Aphis gossypii*, to elevated $CO_2$: EPG test with leaf microstructure and leaf chemistry. *Entomologia Experimentalis et Applicata*, 160, 219–228.

Kaiser, E., Zhou, D., Heuvelink, E., Harbinson, J., Morales, A. and Marcelis, L.F., 2017. Elevated $CO_2$ increases photosynthesis in fluctuating irradiance regardless of photosynthetic induction state. *Journal of Experimental Botany*, 68, 5629–5640.

Kazan, K., 2018. Plant-biotic interactions under elevated $CO_2$: A molecular perspective. *Environmental and Experimental Botany*, 153, 249–261.

Kobayashi, T., Ishiguro, K., Nakajima, T., Kim, H.Y., Okada, M. and Kobayashi, K., 2006. Effects of elevated atmospheric $CO_2$ concentration on the infection of rice blast and sheath blight. *Phytopathology*, 96, 425–431.

Kontunen-Soppela, S., Parviainen, J., Ruhanen, H., Brosche, M., Keinänen, M., Thakur, R. C. and Vapaavuori, E., 2010. Gene expression responses of paper birch (Betula papyrifera) to elevated $CO_2$ and $O_3$ during leaf maturation and senescence. *Environmental Pollution*, 158, 959–968.

Korell, L., Auge, H., Chase, J.M., Harpole, S. and Knight, T.M., 2020. We need more realistic climate change experiments for understanding ecosystems of the future. *Global Change Biology*, 26, 325–327.

Kosugi, Y., Takanashi, S., Ohkubo, S., Matsuo, N., Tani, M., Mitani, T. and Nik, A.R., 2008. $CO_2$ exchange of a tropical rainforest at Pasoh in Peninsular Malaysia. *Agricultural and Forest Meteorology*, 148, 439–452.

Kumari, S., Agrawal, M. and Tiwari, S., 2013. Impact of elevated $CO_2$ and elevated $O_3$ on Beta vulgaris L.: Pigments, metabolites, antioxidants, growth and yield. *Environmental Pollution*, 174, 279–288.

Lamichaney, A., Swain, D.K., Biswal, P., Kumar, V., Singh, N.P. and Hazra, K.K., 2019. Elevated atmospheric carbon–dioxide affects seed vigour of rice (*Oryza sativa* L.). *Environmental and Experimental Botany*, 157, 171–176.

Li, L., Wang, M., Pokharel, S.S., Li, C., Parajulee, M.N., Chen, F. and Fang, W., 2019. Effects of elevated $CO_2$ on foliar soluble nutrients and functional components of tea, and population dynamics of tea aphid, Toxoptera aurantii. *Plant Physiology and Biochemistry*, 145, 84–94.

Li, P., Ainsworth, E.A., Leakey, A.D., Ulanov, A., Lozovaya, V., Ort, D.R. and Bohnert, H.J., 2008. Arabidopsis transcript and metabolite profiles: Ecotype-specific responses to open-air elevated [$CO_2$]. *Plant, Cell & Environment*, 31, 1673–1687.

Li, Y., Yu, Z., Jin, J., Zhang, Q., Wang, G., Liu, C. and Liu, X., 2018. Impact of elevated $CO_2$ on seed quality of soybean at the fresh edible and mature stages. *Frontiers in Plant Science*, 9, 1413.

Liu, C., Colón, B.C., Ziesack, M., Silver, P.A. and Nocera, D.G., 2016. Water splitting–biosynthetic system with $CO_2$ reduction efficiencies exceeding photosynthesis. *Science*, 352, 1210–1213.

Liu, H., Qin, Q., Zhang, R., Ling, L. and Wang, B., 2017. Insights into the mechanism of the capture of $CO_2$ by $K_2CO_3$ sorbent: A DFT study. *Physical Chemistry Chemical Physics*, 19, 24357–24368.

Loladze, I., 2014. Hidden shift of the ionome of plants exposed to elevated $CO_2$ depletes minerals at the base of human nutrition. *Elife*, 3, e02245.

Long, S.P., Ainsworth, E.A., Rogers, A. and Ort, D.R., 2004. Rising atmospheric carbon dioxide: Plants FACE the future. *Annual Review of Plant Biology*, 55, 591.

Malhi, Y. and Wright, J., 2004. Spatial patterns and recent trends in the climate of tropical rainforest regions. *Philosophical Transactions of the Royal Society of London. Series B: Biological Sciences*, 359, 311–329.

Markelz, R.C., Vosseller, L.N. and Leakey, A.D., 2014. Developmental stage specificity of transcriptional, biochemical and $CO_2$ efflux responses of leaf dark respiration to growth of A rabidopsis thaliana at elevated [$CO_2$]. *Plant, Cell & Environment*, 37, 2542–2552.

Martin, T. and Watson, J., 2016. Intact ecosystems provide best defence against climate change. *Nature Climate Change*, 6, 122–124.

Mederski, H.J., 2019. Effects of water and temperature stress on soybean plant growth and yield in humid, temperate climates. In *Crop Reactions to Water and Temperature Stresses in Humid, Temperate Climates* (pp. 35–48). CRC Press.

Mishra, A.K., Agrawal, S.B. and Agrawal, M., 2019. Rising atmospheric carbon dioxide and plant responses: Current and future consequences. In Choudhary, K.K., Kumar A., Singh, A.K. (Eds.), *Climate Change and Agricultural Ecosystems* (pp. 265–306). Woodhead Publishing.

Moroney, J.V., Jungnick, N., DiMario, R.J. and Longstreth, D.J., 2013. Photorespiration and carbon concentrating mechanisms: Two adaptations to high $O_2$, low $CO_2$ conditions. *Photosynthesis Research*, 117, 121–131.

Morueta-Holme, N., Engemann, K., Sandoval-Acuña, P., Jonas, J.D., Segnitz, R.M. and Svenning, J.C., 2015. Strong upslope shifts in Chimborazo's vegetation over two centuries since Humboldt. *Proceedings of the National Academy of Sciences*, 112, 12741–12745.

Münzbergová, Z., Hadincová, V., Skálová, H. and Vandvik, V., 2017. Genetic differentiation and plasticity interact along temperature and precipitation gradients to determine plant performance under climate change. *Journal of Ecology*, 105, 1358–1373.

Nicotra, A.B., Atkin, O.K., Bonser, S.P., Davidson, A.M., Finnegan, E.J., Mathesius, U. and van Kleunen, M., 2010. Plant phenotypic plasticity in a changing climate. *Trends in Plant Science*, 15, 684–692.

Nie, M., Lu, M., Bell, J., Raut, S. and Pendall, E., 2013. Altered root traits due to elevated $CO_2$: A meta-analysis. *Global Ecology and Biogeography*, 22, 1095–1105.

Nord, E.A. and Lynch, J.P. 2009. Plant phenology: A critical controller of soil resource acquisition. *Journal of Experimental Botany*, 60, 1927–1937.

Nottingham, A.T., Meir, P., Velasquez, E., & Turner, B.L., 2020. Soil carbon loss by experimental warming in a tropical forest. *Nature*, 584(7820), 234–237.

Nowak, R.S., Ellsworth, D.S. and Smith, S.D., 2004. Functional responses of plants to elevated atmospheric CO2–do photosynthetic and productivity data from FACE experiments support early predictions? *New Phytologist*, 162, 253–280.

Numata, S., Yamaguchi, K., Shimizu, M., Sakurai, G., Morimoto, A., Alias, N. and Satake, A., 2022. Impacts of climate change on reproductive phenology in tropical rainforests of Southeast Asia. *Communications Biology*, 5, 1–10.

Ooi, M.K., Auld, T.D. and Denham, A.J., 2012. Projected soil temperature increase and seed dormancy response along an altitudinal gradient: Implications for seed bank persistence under climate change. *Plant and Soil*, 353, 289–303.

Pacini, E. and Dolferus, R., 2019. Pollen developmental arrest: Maintaining pollen fertility in a world with a changing climate. *Frontiers in Plant Science*, 679.

Pan, T., Wang, Y., Wang, L., Ding, J., Cao, Y., Qin, G. and Zou, Z., 2020. Increased CO2 and light intensity regulate growth and leaf gas exchange in tomato. *Physiologia Plantarum*, 168, 694–708.

Parmesan, C. and Hanley, M.E., 2015. Plants and climate change: Complexities and surprises. *Annals of Botany*, 116, 849–864.

Pau, S., Wolkovich, E.M., Cook, B.I., Nytch, C.J., Regetz, J., Zimmerman, J.K. and Joseph Wright, S., 2013. Clouds and temperature drive dynamic changes in tropical flower production. *Nature Climate Change*, 3, 838–842.

Paudel, I., Halpern, M., Wagner, Y., Raveh, E., Yermiyahu, U., Hoch, G. and Klein, T., 2018. Elevated $CO_2$ compensates for drought effects in lemon saplings via stomatal down-regulation, increased soil moisture, and increased wood carbon storage. *Environmental and Experimental Botany*, 148, 117–127.

Reich, P.B., Hobbie, S.E., Lee, T.D. and Pastore, M.A., 2018. Unexpected reversal of C3 versus C4 grass response to elevated $CO_2$ during a 20-year field experiment. *Science*, 360, 317–320.

Ruiz-Vera, U.M., Siebers, M., Gray, S.B., Drag, D.W., Rosenthal, D.M., Kimball, B.A. and Bernacchi, C.J., 2013. Global warming can negate the expected $CO_2$ stimulation in photosynthesis and productivity for soybean grown in the Midwestern United States. *Plant Physiology*, 162, 410–423.

Seneweera, S., Aben, S.K., Basra, A.S., Jones, B. and Conroy, J.P., 2003. Involvement of ethylene in the morphological and developmental response of rice to elevated atmospheric $CO_2$ concentrations. *Plant Growth Regulation*, 39, 143–153.

Seneweera, S. and Norton, R.M., 2011. Plant responses to increased carbon dioxide. *Crop Adaptation to Climate Change*, 364, 198–217.

Sentinella, A.T., Warton, D.I., Sherwin, W.B., Offord, C.A. and Moles, A.T., 2020. Tropical plants do not have narrower temperature tolerances, but are more at risk from warming because they are close to their upper thermal limits. *Global Ecology and Biogeography*, 29(8), 1387–1398.

Shaw, R.H., 2019. Estimates of yield reductions in corn caused by water and temperature stress. In *Crop Reactions to Water and Temperature Stresses in Humid, Temperate Climates* (pp. 49–65). CRC Press.

Sheldon, K.S., 2019. Climate change in the tropics: Ecological and evolutionary responses at low latitudes. *Annual Review of Ecology, Evolution, and Systematics*, 50, 303–333.

Sheldon, K.S., Yang, S. and Tewksbury, J.J., 2011. Climate change and community disassembly: Impacts of warming on tropical and temperate montane community structure. *Ecology Letters*, 14, 1191–1200.

Sherry, R.A., Zhou, X., Gu, S., Arnone III, J.A., Schimel, D.S., Verburg, P.S. and Luo, Y., 2007. Divergence of reproductive phenology under climate warming. *Proceedings of the National Academy of Sciences*, 104, 198–202.

Singh, A. and Agrawal, M., 2015. Effects of ambient and elevated $CO_2$ on growth, chlorophyll fluorescence, photosynthetic pigments, antioxidants, and secondary metabolites of Catharanthus roseus (L.) G Don. grown under three different soil N levels. *Environmental Science and Pollution Research*, 22, 3936–3946.

Skendžić, S., Zovko, M., Živković, I. P., Lešić, V. and Lemić, D., 2021. The impact of climate change on agricultural insect pests. *Insects*, 12, 440.

Slot, M., Rey-Sánchez, C., Gerber, S., Lichstein, J.W., Winter, K. and Kitajima, K., 2014. Thermal acclimation of leaf respiration of tropical trees and lianas: Response to experimental canopy warming, and consequences for tropical forest carbon balance. *Global Change Biology*, 20, 2915–2926.

Springer, C.J. and Ward, J.K., 2007. Flowering time and elevated atmospheric $CO_2$. *New Phytologist*, 176, 243–255.

Suggitt, A.J., Wilson, R.J., Isaac, N.J., Beale, C.M., Auffret, A.G., August, T., Bennie, J.J., Crick, H.Q., Duffield, S., Fox, R. and Hopkins, J.J., 2018. Extinction risk from climate change is reduced by microclimatic buffering. *Nature Climate Change*, 8, 713–717.

Tachiki, Y., Iwasa, Y. and Satake, A., 2010. Pollinator coupling can induce synchronized flowering in different plant species. *Journal of Theoretical Biology*, 267, 153–163.

Takatani, N., Ito, T., Kiba, T., Mori, M., Miyamoto, T., Maeda, S.I. and Omata, T., 2014. Effects of high $CO_2$ on growth and metabolism of Arabidopsis seedlings during growth with a constantly limited supply of nitrogen. *Plant and Cell Physiology*, 55, 281–292.

Teng, N., Wang, J., Chen, T., Wu, X., Wang, Y. and Lin, J., 2006. Elevated $CO_2$ induces physiological, biochemical and structural changes in leaves of Arabidopsis thaliana. *The New Phytologist*, 172, 92–103.

Terrer, C., Jackson, R.B., Prentice, I.C., Keenan, T.F., Kaiser, C., Vicca, S. and Franklin, O., 2019. Nitrogen and phosphorus constrain the $CO_2$ fertilization of global plant biomass. *Nature Climate Change*, 9, 684–689.

Thakur, M.P., Del Real, I.M., Cesarz, S., Steinauer, K., Reich, P.B., Hobbie, S., Eddy C.E. and Eisenhauer, N., 2019. Soil microbial, nematode, and enzymatic responses to elevated $CO_2$, N fertilization, warming, and reduced precipitation. *Soil Biology and Biochemistry*, 135, 184–193.

Thakur, M.P., Reich, P.B., Hobbie, S.E., Stefanski, A., Rich, R., Rice, K.E., Eddy C.E. and Eisenhauer, N., 2018. Reduced feeding activity of soil detritivores under warmer and drier conditions. *Nature Climate Change*, 8, 75–78.

Thilakarathne, C.L., Tausz-Posch, S., Cane, K., Norton, R.M., Fitzgerald, G.J., Tausz, M. and Seneweera, S., 2014. Intraspecific variation in leaf growth of wheat (Triticum aestivum) under Australian Grain Free Air $CO_2$ Enrichment (AGFACE): Is it regulated through carbon and/or nitrogen supply? *Functional Plant Biology*, 42, 299–308.

Thinh, N.C., Kumagai, E., Shimono, H. and Kawasaki, M., 2017. Effects of elevated $CO_2$ concentration on bulbil germination and early seedling growth in Chinese yam under different air temperatures. *Plant Production Science*, 20, 313–322.

Urban, M.C., 2015. Accelerating extinction risk from climate change. *Science*, 348(6234), 571–573.

van Kessel, C., 2020. Teaching the climate crisis: Existential considerations. *Journal of Curriculum Studies Research*, 2(1), 129–145.

Van Leeuwen, C. and Darriet, P., 2016. The impact of climate change on viticulture and wine quality. *Journal of Wine Economics*, 11, 150–167.

Walck, J.L., Hidayati, S.N., Dixon, K.W., Thompson, K.E.N. and Poschlod, P., 2011. Climate change and plant regeneration from seed. *Global Change Biology*, 17, 2145–2161.

Walker, A.P., De Kauwe, M.G., Bastos, A., Belmecheri, S., Georgiou, K., Keeling, R.F. and Zuidema, P.A., 2021. Integrating the evidence for a terrestrial carbon sink caused by increasing atmospheric $CO_2$. *New Phytologist*, 229, 2413–2445.

Wang, G. and Dillon, M.E., 2014. Recent geographic convergence in diurnal and annual temperature cycling flattens global thermal profiles. *Nature Climate Change*, 4, 988–992.

Ward, J.K. and Strain, B.R., 1999. Elevated $CO_2$ studies: Past, present and future. *Tree Physiology*, 19, 211–220.

Watanabe, C.K., Sato, S., Yanagisawa, S., Uesono, Y., Terashima, I. and Noguchi, K., 2014. Effects of elevated $CO_2$ on levels of primary metabolites and transcripts of genes encoding respiratory enzymes and their diurnal patterns in Arabidopsis thaliana: Possible relationships with respiratory rates. *Plant and Cell Physiology*, 55, 341–357.

Wei, H., Gou, J., Yordanov, Y., Zhang, H., Thakur, R., Jones, W. and Burton, A., 2013. Global transcriptomic profiling of aspen trees under elevated $[CO_2]$ to identify potential molecular mechanisms responsible for enhanced radial growth. *Journal of Plant Research*, 126, 305–320.

Wheeler, T.R., Craufurd, P.Q., Ellis, R.H., Porter, J.R., & Prasad, P.V., 2000. Temperature variability and the yield of annual crops. *Agriculture, Ecosystems & Environment*, 82(1–3), 159–167.

Wiens, J.J., 2016. Climate-related local extinctions are already widespread among plant and animal species. *PLoS Biology*, 14, e2001104.

Williams, M., Shewry, P.R., Lawlor, D.W. and Harwood, J.L., 1995. The effects of elevated temperature and atmospheric carbon dioxide concentration on the quality of grain lipids in wheat (Triticum aestivum L.) grown at two levels of nitrogen application. *Plant, Cell & Environment*, 18, 999–1009.

Williams, R.G., Roussenov, V., Goodwin, P., Resplandy, L. and Bopp, L., 2017. Sensitivity of global warming to carbon emissions: Effects of heat and carbon uptake in a suite of Earth system models. *Journal of Climate*, 30, 9343–9363.

Wright, S.J., Muller-Landau, H.C. and Schipper, J.A.N., 2009. The future of tropical species on a warmer planet. *Conservation Biology*, 23, 1418–1426.

Xu, Z., Shimizu, H., Ito, S., Yagasaki, Y., Zou, C., Zhou, G. and Zheng, Y., 2014. Effects of elevated $CO_2$, warming and precipitation change on plant growth, photosynthesis and peroxidation in dominant species from North China grassland. *Planta*, 239, 421–435.

Zhao, H., Guan, J., Liang, Q., Zhang, X., Hu, H. and Zhang, J., 2021. Effects of cadmium stress on growth and physiological characteristics of sassafras seedlings. *Scientific Reports*, 11, 1–11.

# Section 4

*Ecophysiological responses of tropical plants to disturbance events*

# 15 Fire and Ecophysiological Responses of Tropical Plants

*Shikha Singh and Tanu Kumari*

## 15.1 INTRODUCTION

Natural processes like fire have greatly shaped our ecosystems and preserved biodiversity all around the globe (Bond & Midgley, 2012). Fire controls organismal attributes, population trends, community interactions and structure, carbon and nutrient cycles, and ecosystem dynamics as a dominant ecological and evolutionary factor (Juárez-Orozco et al., 2017, Mclauchlan et al., 2020). For instance, studies on tropical savannas and grasslands demonstrate that fire is a common phenomenon that has long been used as a management tool for these ecosystems (Sankaran et al., 2004; Bhadouria et al., 2018). Fire incorporates intricate feedbacks among biological, social, and geophysical processes as an ecological phenomenon, necessitating collaboration across several disciplines and scales of research. Most tropical environments are among the 20% of the world's ecosystems that are classified as being sensitive to fire; these habitats are dominated by species that did not primarily develop in areas that experience frequent fires (Shlisky et al., 2007). Currently, around 3% of the earth's surface continues to burn per year (Van der Werf et al., 2017). Humans have a significant impact on the current fire regimes through sources of ignition, inhibition and alterations in land use, and fuels (Bistinas et al., 2013; Kelley et al., 2019). As a result of shifting climate, unpredictable weather, land use, population increase, and changes in vegetation patterns, fire regimes are changing constantly at a faster rate (Jolly et al., 2015; Andela et al., 2017; Veraverbeke et al., 2017). However, at a global scale, fire regimes have changed in more than 60% of the terrestrial environments (Shlisky et al., 2009). The predicted continuation of these changes over the next few decades might have significant effects on ecosystems, biodiversity, habitat, nutrient cycles, climate, and population (Rogers et al., 2020). Similarly, fire can be considered as a major natural disturbance which influences vegetation worldwide. The heat transfers that occur on particular plant components as a result of different fire behavior characteristics, such as fire-line intensity and residence time, impact the level of vegetation damage by fire (Michaletz & Johnson, 2007; O'Brien et al., 2018). For example, Saha and Howe (2003) discovered that low-intensity fires occurring annually in a deciduous savanna in Central India which supports a relatively small number of tree species to regrow, projecting that there will be significant decreases in variety of trees for these communities in the near future (Bhadouria et al., 2018). Heat damage from a fire can result in a chain reaction of complicated mechanisms that impact the physiology of trees after fires. Thus, appropriate ecosystem functioning in the current scenario of global environmental changes requires an understanding of plant ecophysiology. Plant ecophysiology is crucial in analyzing forest responses to global change at broader spatial and temporal scales. For instance, many of the functional linkages that were discovered through experimentation are applied in global change studies, such as the relationship between temperature and respiration (Heskel et al., 2016), the light response, and stomatal restriction of leaf gas exchange (Buchmann, 2002; Vico et al., 2011). The current focus of intense scientific research is on determining the precise physiological pathways and linking specific damage to whole-plant and ecosystem functioning. Some recent studies have made significant progress in our understanding of the physiological processes that occur in trees after fires, also showing possible interactions between fire damage and other stresses including drought, insects, and pathogens (Bär et al., 2019; Daniels & Throop, 2022).

Global climate change, which has an impact on all aspects of fire management, is the result of human activity linked with atmospheric and landscape processes (Vitousek, 1994). It is well known that since pre-industrial times, anthropogenic greenhouse gas emissions have exacerbated the frequency and severity of several weather and climate events, notably for extreme temperatures. According to the sixth assessment report (AR6) of IPCC (2021), even at 1.5°C of global warming, the frequency of such severe events will increase as global warming continues. Biomass burning and industrial activities are responsible for changes in the global atmospheric processes. The changes in the chemical composition of the atmosphere are expected to have a considerable impact on biogeochemical processes and the earth's radiation balance, known as the "greenhouse effect." Such changes in the atmosphere and the earth's energy balance are projected to alter precipitation, temperature, humidity, and vegetation development, which will further exacerbate the phenomena of fire and have an impact on its management (Ryan, 2000). Every year, wildfires destroy 350 M ha, or 2.3% of the total land area (Chuvieco et al., 2016; Giglio et al., 2013).

Forest fires play a major role in highly flammable ecosystems such as shrublands, grasslands, and forests, and they contribute to climate change. Thus, forest fire and climate change are concepts that are inextricably linked (Singh, 2022). Since, it is well known that the carbon stored in vegetation is released as soon as it starts burning. This is the primary reason why large-scale forest fires emit atmospheric carbon dioxide ($CO_2$) and thus make a significant contribution to the rate of climate change. The increased occurrences of fire events in temperate and Mediterranean ecosystems has been studied extensively where plants are well adapted to fire (Ganteaume et al., 2013), but there is still a lack of information on fire susceptibility of tropical forests. Scientists and foresters are now investigating the causes and consequences of tropical forest fires (Juárez-Orozco, 2017). The tropics occupy 33.7% (more than one-third) of the earth's land area. They support more than 35% of the world's population, more than 40% of the world's forests, and a significant number of flora and fauna (Cochrane, 2009a). Fire poses a hazard to tropical ecosystems as it limits their potential for scientific and socioeconomic growth. However, in comparison to temperate or seasonally dry forests, less attention has been paid to fires in these ecosystems. Global climate change, deforestation, and land use change are all variables that increase the danger and extent of fires in these ecosystems. As there is a close connection between fires and tropical ecosystems dating back to prehistoric times, it is imperative to understand how fires affect the ecophysiology of plants (Bowman et al., 2009; Roberts et al., 2017). To manage these human-influenced landscapes sustainably, a critical information gap regarding the response of the variety of tropical plants to fire regimes over long timescales must be addressed.

Thus, the objective of this chapter is to explore how plants respond ecophysiologically to fires and how they have adapted to survive in tropical ecosystems. The knowledge gaps concerning fire in tropical ecosystems are discussed in the chapter, which also identifies the main causes of forest fires and the different types of fire regimes. We investigate how much an understanding of plant ecophysiological responses and function can help to fill the gap between fire behavior and plant ecology. Plant physiological ecology and fire science both have an extensive research history; however, they have generally evolved in parallel. These gaps in the study on forest fires could serve as a guide for decision-making to prevent further fire spread or at least reduce its detrimental impacts in tropical ecosystems.

## 15.2 FIRE IN TROPICAL ECOSYSTEMS

The tropics are inhabited by extensive savanna, alpine grasslands, dry deciduous and thorn forests, mangrove swamps, deserts, and a number of other ecosystems, the majority of which burns frequently. According to satellite observation of thermal anomalies, the tropics have more fires per year than any other region on the earth (Cochrane, 2009a, 2011). The world's largest repository of biodiversity is found in the tropics, where fire predominates throughout many different landcover types. The tropical rainforests in this region also appear to be practically fire-immune. The most intense solar activity occurs in the tropics. This results in significant land surface heating, moisture evaporation, cloud formation, and heavy rainfall. The resulting rainforests, which cover a considerable percentage of the area, have grown dense vegetation that nearly totally blocks the sunlight from reaching the forest floor. Only around 2% of the sunlight's intensity may be lowered beneath these canopies. Since there is a struggle for light, these woods have evolved tight canopies with several leaf layers. More than 6 square meters of vegetation may cover every square meter of land. As a result, compared to non-forest land cover types, the subcanopy climate is significantly colder and more humid. The tropical rainforests experience fires despite their moist environment, which has a significant impact on both environmental and monetary values. But compared to temperate or seasonally dry forests, fires in these ecosystems have received less attention. A huge portion of tropical forests have declined despite conservation efforts over the past 25 years, regardless of the lack of precise statistics showing their loss (Keenan et al., 2015). Rainforests are seriously threatened by agricultural activities, cattle grazing, deforestation, mining, construction of roads, and the development of bioenergy plants (Prince's Charities, 2015). In many regions of the world, poor farmers practice slash-and-burn farming on small portions of land, as compared to intensive and extensive agriculture, which often requires many hectares. Such human actions have a huge effect, changing both legal and illicit land utilization (Zhang & Pearse, 2011; Jakovac et al., 2016).

### 15.2.1 Causes of Forest Fires in the Tropics

Agriculture and ranching are two common human activities that directly contribute to the underlying causes of fires in tropical rainforests (Cochrane, 2009b; Butler, 2012; Van Vliet et al., 2012). Deforestation has an impact on the tropical ecosystem by boosting the frequency of agricultural fires, changing the microclimate, and modifying the plant structure, which creates a fire-friendly feeedback loop (Juárez-Orozco et al., 2017). For instance, the tropical rainforest of the Congo Basin is surrounded by savannahs and deforested areas as a result of logging and land clearing practices, which increase the risk of fires, particularly under the influence of global climate change (Bucini & Lambin, 2002). However, the conflicts over land tenure and land-use allocation are also influenced by other socioeconomic forces (Dennis et al., 2005). The consequences of shifting cultivation have been extensively debated in the research. Plant ecology, crop productivity and management, and soil science are the primary research areas in this practice (Mukul & Herbohn, 2016). On the other hand, shifting cultivation is a method of farming linked to extensive environmental degradation and deforestation. Fire regimes in the tropics could vary significantly as a result of anthropogenic greenhouse gas emissions (Goldammer & Price, 1998). The increased occurrence and intensity of extreme drought conditions is among the most important factors influencing such events (Thonicke et al., 2001).

### 15.2.1.1 Climate Change

The sensitivity of the tropics to climate change is still largely unknown, but some data suggests that potential drops in rainfall and rises in temperature due to global warming in the tropics are currently being discussed by scientists. According to IPCC (2001), tropical regions can expect to experience rising temperatures in the future. The geographical distribution of precipitation is likely to shift due to future warming in the tropics, but the likelihood of decreased precipitation is less certain. Additionally, recurring and strong El Niño/Southern Oscillation events are predicted to cause global temperatures to rise, which could speed up evapotranspiration and exacerbate the effects of droughts (Foley et al., 2003). The tropical ecosystems are highly susceptible to anthropogenic greenhouse gas (GHG)-induced climate change (Goldammer & Price, 1998). Future droughts may be more severe and frequent, increasing the risk of land clearing by facilitating the spread of fire and decreasing its return period in tropical forests.

### 15.2.1.2 Agriculture and Livestock Farming

The tropical region experiences a specific type of fire that is commonly used to clear land and alter land use (Nepstad et al., 1999a, 1999b). Humans have been using fire to clear the land since ancient times, turning the native vegetation into farming lands and pastures which are the most typical land uses in tropical areas (Walker et al., 2000; Chazdon, 2003; Sorrensen, 2004). The link between tropical agricultural productivity and land clearing, which considerably aids in the development of fire-prone conditions, is primarily responsible for the expansion of anthropogenic sources of fire in the tropics.

### 15.2.1.3 Rural and Urban Expansion

The change in fire regimes in the tropics has been largely attributed to the growth of rural and urban development (Shlisky et al., 2009). Road infrastructure, for example, has a significant impact on rural and urban development in tropical regions of South America (Laurance et al., 2001). Investment in transportation encourages population growth and rural-to-urban migration. Large-scale land clearing and subsequent fire use are encouraged by this process.

### 15.2.1.4 Energy Production

Energy production can influence changes in tropical fire regimes on a variety of scales. One of the primary energy sources that significantly influences the occurrence and severity of fires locally is the production of charcoal for thermoelectric plants. For instance, logging and the spread of forest fires are linked to the production of charcoal in eastern Amazonia (Alencar et al., 2004).

### 15.2.1.5 Fire Exclusion and Suppression

The suppression and exclusion of fires are the factors that most significantly and regionally alter the fire regimes. In the tropics, the lack of fire in savanna habitats near patches of forest may stimulate forest species to migrate toward savanna ecosystems (Hoffmann et al., 2003).

### 15.2.1.6 Exotic Species

The rapid spread of exotic species and their special feature of benefiting from past disturbances have an impact on both their establishment and their resistance to fire occurrences (Shlisky et al., 2009). As a result, the presence of such species may alter the frequency, season, and intensity of fires.

### 15.2.1.7 Arson

Arson is a significant contributor to unchecked forest fires in the tropics. Conflicts over natural resources are common in areas where this type of fire occurs (Vayda, 2006). Arson is the most complex type of human-induced fire that cannot be precisely detected and defined due to its illicit nature. Typically, areas where arson has taken place are classified as accidental or escaped fires.

In order to understand the intricate responses of forest ecosystems to climate change, it may be useful to combine the disciplines, because many aspects of plant ecophysiology and ecosystem functioning are complementary (Figure 15.1). Predicting upcoming disturbance interactions introduces another factor of uncertainty on top of probable changes in regional weather patterns, vegetation responses, and fire regimes. Even while disturbance interactions are normal and frequent landscape processes (such as fires that follow outbreaks of bark beetles, debris flows, blowdown events that follow fires, and drought followed by insect or pest outbreaks), they have the potential to cause unanticipated disturbance behaviors and vegetation responses due to nonlinear, cascading, or cumulative impacts.

## 15.2.2 Fire Regime Types in the Tropics

A key concept in fire research is the fire regime, which includes both the temporal and spatial distribution of fire. When studying the impacts of fires on different ecosystems and their consequences for land management, the fire regime should be taken into account. This means that trends in fire frequency, intensity, size distribution, and season vary significantly between different biomes, locations, and ecosystem types (Archibald et al., 2013). A distinctive fire regime dynamic is a result of the interactions between the atmosphere, geography, soils, and plant species. A general classification of ecosystems can be done based on how they interact with particular fire regime characteristics in a specific environment, such as fuels, flammability, ignitions, and fire distribution conditions. Three different forms of fire dependency—*fire sensitive*, *fire dependent*, and *fire independent*—have been recognized by scientists to define the fire ecology of terrestrial ecosystems.

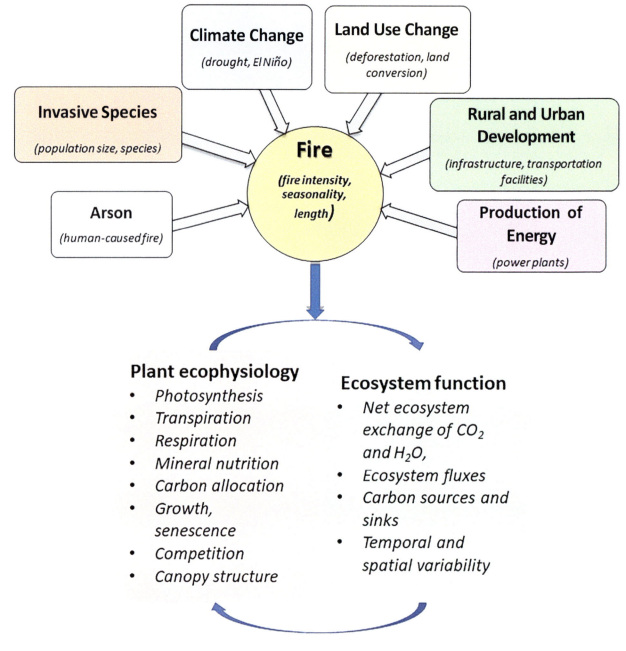

**FIGURE 15.1** Effects of fire on plant ecophysiology and ecosystem functioning.

- *Fire-dependent ecosystems* are those in which the majority of organisms have their origins in fire and where the presence of fire is an essential mechanism for preserving species diversity (e.g., savannas, temperate forests). If fire is eliminated or introduced in a manner that is ecologically inappropriate, such as at an incorrect frequency, intensity, or seasonal timing, these systems may experience significant changes (Trejo, 2008; Shlisky et al., 2009).
- *Fire-sensitive ecosystems* include the majority of species that have not experienced considerable changes as a result of the presence of fire (Kraus & Goldammer, 2007). In fire-sensitive systems, fire may only play a limited role in maintaining the structure and functionality of natural ecosystems, but the presence of environmentally unsuitable fire can show large negative effects on biodiversity—for example, in tropical moist broadleaf forests. A positive response loop that results from excessive burning in fire-prone areas can also be established, making these areas more vulnerable to subsequent fires and rapidly degrading one of the most intact forest ecosystems (Shlisky et al., 2009).
- *Fire-independent ecosystems* are those that are inherently incapable of supporting fires as an evolutionary force due to inadequate supply of fuel or ignition inputs (e.g., deserts, tundra).

The three types of fire regimes are distributed globally, and this data suggests that the tropics are where the majority of ecosystems are most vulnerable to fire (Shlisky et al., 2007). Large regions of tropical and subtropical dry and moist broadleaf forests, which experience few natural fires because of high precipitation, can be used to explain this spatial distribution (Goldammer, 2012). However, as human activity increases, more forests are being converted to agricultural or grazing pastures, which weakens the ability of forests to resist wildfires.

## 15.3 ENVIRONMENTAL CHANGES AND PLANT RESPONSES

Stress is defined as an environmental element that causes some physiological processes (such growth or photosynthesis) to occur at a rate that is lower than what the plant could normally support (Lambers et al., 2008). Both biotic and abiotic processes have the potential to cause stress. Some of the abiotic stresses that adversely affect plant growth, development, or crop productivity are drought, extreme heat or cold, heavy metals, or high salinity (Wani et al., 2016; Hossain et al., 2016; Sah et al., 2016). The immediate reaction of plant to stress is a decline in performance. Depending on the type of stress and the physiological processes that are impacted, plants counteract the negative effects of stress using a variety of mechanisms that function on various time frames. These compensatory responses work together to allow the plant to continue its physiological activities at a fairly constant rate, despite periodic stresses that would otherwise cause it to perform relatively poorly. A certain level of stress tolerance is necessary for a plant to thrive in a stressful environment. The mechanisms by which different species survive stress vary greatly. Different physiological mechanisms react differently to stress. However, in order to comprehend how plants respond, we must take into account how certain mechanisms react on a smaller scale (e.g., the response of photosynthesis or of light-harvesting pigments to a change in light intensity). There are at least the following three different time scales by which plants react to stress.

- 15.3.1 The initial negative impact of a stress on a plant process is known as the stress response. This often happens over a period of seconds to days, leading to deterioration in the performance of a process. A number of complicated physiological, structural, and morphological adaptations are a part of both biotic and abiotic stress responses in plants (Rejeb et al., 2014).
- 15.3.2 Acclimation is the process by which individual plants change their morphology and physiological functions to make up for the performance reduction that occurs after the initial stress reaction (Lambers et al., 2008). Plant acclimation is a result of terrestrial plants' sessile behavior, which forces them to live in a dynamic environment where many important factors are subject to fast change (Kleine et al., 2021). Acclimation occurs as a reaction to environmental change by altering the activity or synthesis of new biochemical components like enzymes, which are frequently linked to the growth of new tissue. These biochemical alterations then initiate a chain reaction that manifests itself at various levels, including changes in the rate or environmental sensitivity of a particular process (like photosynthesis), the rate at which entire plants grow, and the morphology of individual organs or the entire plant. Stress acclimation happens at some point during an individual's lifespan, usually within a few days or weeks. Comparing plants that are genetically similar but are growing in various conditions can explain how acclimation occurs.
- 15.3.3 Rapid environmental changes require organisms to adapt in order to survive. Depending on the type of response, the trait, the population, and the environmental conditions, adaptation can take many forms and occur at different rates (Kristensen et al., 2020). Adaptation is the evolved response arising from genetic variations in populations that make up for the performance reduction caused by the stress (Lambers et al., 2008). The physiological mechanisms of response and acclimation are frequently comparable because both include modifications to the activity or synthesis of biochemical components and result in adjustments to the rates of specific physiological processes, growth rate, and morphology. The ability of plants to adapt to transient environmental variation may even change as a result of adaptation. In order to deal with the environmental stresses, plants create effective coping mechanisms to prevent or tolerate the stresses, enabling them to adjust and defend themselves. These adaptation techniques operate at the molecular, morphological, anatomical, and biochemical levels (Lamalakshmi Devi et al., 2017). Comparing genetically different plants raised in similar environments can help in understanding adaptation.

## 15.4 ECOPHYSIOLOGICAL RESPONSES OF PLANTS TO FIRE

The interplay between physiological and biological processes as a result of diverse environmental pressures is referred to as "ecophysiology" (Anderson, 2003; Sadeghifar et al., 2020). Since many elements of plant ecophysiological response are paired with the ecosystem functioning at various levels of organization, fire regimes and resistivity (Figure 15.2), combining the disciplines may be a beneficial step toward unraveling key functional information toward any environmental change or shift (Buchmann, 2002).

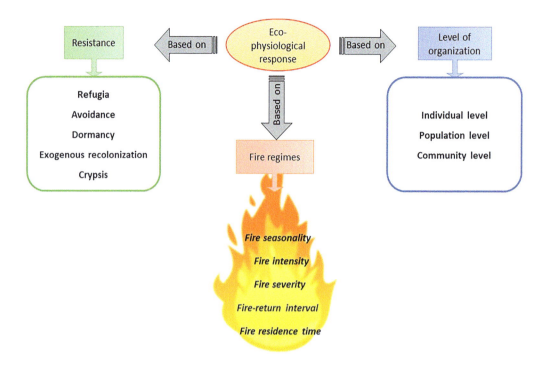

**FIGURE 15.2** Overview of plant ecophysiological responses to fire based on different ecological aspects.

The most prevalent environmental disturbances in the last several years have been changes in fire regimes in savannas and forests of tropical ecosystems (Trumbore et al., 2015; Pellegrini & Jackson, 2020). Owing to the critical role of tropical ecosystem in global carbon cycle, it becomes crucial to understand their post-fire reactions required for a proper "forest management action plan" for a shorter time scale and a "fire-risk reduction plan" on a long-term basis (Battipaglia et al., 2016; Niccoli et al., 2019).

### 15.4.1 Responses Based on Fire Regimes

Variable plant responses to fire are frequently dictated by the intricate interactions provided by external influences including temperature, soil moisture, heat duration, and fire characteristics. It is widely known that the response of plants to fire depends on the species (Catry et al., 2010) and is influenced by its various regimes. The components of fire regime that affect plant responses are season, intensity, severity, residence time, and fire-return interval (Tangney et al., 2022). The following discussion explains how fire regimes control the plant's response.

- *Fire seasonality:* The plants show response to fire seasonality in terms of changes in phenology, seed dormancy, vegetative and reproductive stage. For instance, if fire incidence takes place during a crucial stage in phenological cycle of a plant community, then it becomes very difficult for the post-fire plant populations to recover from the damage (Tangney et al., 2022). One such key stage is represented by seedling establishment. A number of studies have reported the detrimental effects of fire on tree seedling survival and development, which are especially pronounced in the case of ground-laid seedlings (Bhadouria et al., 2016). Seedling growth following a fire event is highly reliant on the availability of moisture. Therefore, insufficient moisture in drier seasons results in seedling death (Bhadouria et al., 2017). The recurring fire incidents in African woodlands during the late dry season, which results in the death of seedlings, is one such illustration of the effect of fire seasonality (Khurana & Singh, 2001).
- *Fire intensity:* The amount of heat emitted per unit area is known as fire intensity, which is assessed by the fuel load, i.e., quality and quantity of fuel (Savadogo et al., 2007). The flow of nutrients is regulated by the intensity of the fire, which affects overall survival (Saha, 2002).
- *Fire severity:* The influence a fire has had on a significant property indicates the fire's severity. The terms "fire severity" and "fire intensity," although not synonymous, are often used interchangeably. This may be understood from instances of ground fire when the fire intensity is relatively low yet tree mortality is substantial owing to high fire severity. Hence, tree mortality rate or biomass loss is seen as one of the quantifiable metrics and indicators of fire severity (Keeley, 2009; Resco de Dios, 2020).

- **Fire residence time:** The term "fire residence time" refers to the overall amount of time a plant or area of interest is exposed to fire. The higher the residence period, the more severe are the impacts on plants. For example, based on the residence time of the fire, some plants burn entirely, while others partially (Saha, 2002).
- **Fire-return interval:** The duration of time flanked by two successive fire events in a unit area is known as the fire-return interval (Saha, 2002). Fire frequency is inversely related to fire-return interval. Because of frequent fire occurrences, sapling growth is stifled by the top kill. Therefore, a prolonged period of fire-free condition is necessary for the recruitment of seedlings to a specific adult size at which they are no longer sensitive to fire (Hoffmann et al., 2012). This may be seen in the tropical savanna, where with rising fire frequency, tree density increased rather than decreased. This can be attributed to the resilience of tree species showing post-fire resprouting from root stocks. It was observed that mortality rates were greater for small stems (<2 m height) than large stems (>2 m height), thereby resulting in Oskar syndrome (promoting growth of small sized plants as a response to repeated exposure to fire) and Gulliver syndrome (encouraging plant growth after it escapes fire) (Figure 15.3) (Higgins et al., 2007).

### 15.4.2 Responses Based on Level of Organization

#### 15.4.2.1 Individual Level

The ecophysiological response of plants can be studied at various sub-levels, described as follows.

- **Leaf-level ecophysiology:** After fire, high photosynthetic rate has been reported in many cases. This can be linked to greater availability of water as a result of decreased vegetation cover and minimized plant uptake due to foliage loss. Further high post-fire photosynthesis can be attributed to high root-to-shoot ratio, enabling higher water uptake capacity (Renninger et al., 2013). However, as time passes post-fire, this ratio is maintained due to shoot regrowth. This property was reported in a plant species in South Africa, where high photosynthesis just after fire was observed which then declined after six months, the reason being rapid revival of the plant population (Schutz et al., 2009; Daniels & Throop, 2022).
- **Process-level ecophysiology:** According to Teixeira et al. (2022), fire encourages both above- and below-ground biomass, and helped increase the allocation of below-ground biomass. Additionally, fire boosted functional variety, as well as dispersion. Although soil respiration in burned areas was lower than in unburned areas, NEE (net ecosystem carbon dioxide exchange) and total SOC (soil organic carbon) was higher in burned areas. ET (ecosystem transpiration) and eWUE (ecosystem water use efficiency) were reported to be unaffected by the fire.
- **Biochemical-level ecophysiology:** The production of secondary metabolites enables plants to be biochemically ready to respond to environmental stress. Fire is such a stress, encourage the production of these metabolites (e.g., terpenes, tannins, carotenoids, phenolic compounds, etc.). This improved biosynthesis gives plant resilience, but makes it more vulnerable to future fire occurrences. Due to the considerable susceptibility of phenolic compounds to different fire regimes, these metabolites that are thought to be stress indicators (Santacruz-García et al., 2021). In *Pinus laricio*, sudden increase in phenolic compound was reported following fire occurrences (Santacruz-García et al., 2021).

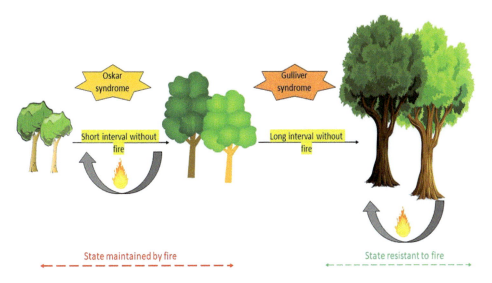

**FIGURE 15.3** Effects of fire-return interval on tree density and size.

- *Phenology-level ecophysiology:* The level of damage caused by fire depends on the stage of growth for various plants. For example, the majority of grasses are more vulnerable to fire damage during their active growth than during dormancy (Trollope, 1982). Another response is stature-based, whereby smaller plants survival is more due to less fuel content than larger ones. In addition, temperatures at the ground's surface are often lower during a wildfire, which is favorable for little plants to survive (Bond & Van Wilgen, 2012).

### 15.4.2.2 Population Level

Plant population response to various fire regimes have been discussed in many hypotheses. One such is "fire-interval hypothesis," which states that recurring fire restricts the increase of plant population size (Keeley et al., 2005). Another hypothesis, "event-dependent effect," states that if a population is facing same fire-interval even then their response can be quite different due to difference in any component of the fire regime such as intensity, seasonality or residence time (Bond & Van Wilgen, 2012).

Some of the population attributes governing plant population response to fire are discussed in what follows.

- Life history is a key feature to understand the response of a plant population to fire. It comprises of post-fire mortality and germination rate, establishment of seedlings, and their subsequent post-fire survival (Werner & Prior, 2013). Understanding of these life history characteristics, makes the prediction of population response to fire better, which is a key step in formulation of fire policies at regional level.
- In order to forecast how a population would respond to a fire, Noble and Slatyer (1980) formulated a concept known as "vital attributes." According to this, functional groups are categorized based on following four processes.

  1. Mode of arrival at fire-prone site
  2. Method of persistence within site
  3. Sequence of new species establishment
  4. Period of attaining adult stage, plant mortality, and propagule mortality

### 15.4.2.3 Community Level

Certain species may survive and continue to grow after a fire, while others are more susceptible and die. With the extinction of certain species and the emergence of others, the structure of a community may occasionally change, i.e., a successional change (He et al., 2019).

The time between fires is currently regarded to be the main factor influencing changes in community composition, with succession occurring along a variety of paths depending on fire history (Qin et al., 2020). The seral stage most frequently seen on the landscape depends on the fire intensities and fire intervals unique to a place (Burrows, 2008).

According to direct succession, areas with less frequent fire have longer odds of having disturbance-free periods long enough to support the emergence of later seral species. Therefore, these locations would have a larger quantity of seed supply for the later seral species than would the more often burned sites when fire burns through such places. As a result, the subsequent seral species would colonize these areas much more quickly, giving rise to the name "direct succession."

According to the theory of accelerated succession, disturbances like fire eliminate overstory seral species and free up understory climax species, speeding up the successional process (Cannon et al., 2017).

### 15.4.3 Response Based on Resistance

Some plant species manage to resist the fire and survive, while others are killed or have varied responses, i.e., they can grow depending upon environmental condition and plant stage. The survival of plants to withstand fire owes to the resistive nature of the plant species. This resistive response can be achieved by various mechanisms (Pausas, 2019)

- *Refugia:* Some species inhabit fire-protected microhabitats or have been known to migrate there. For example, in areas where fire is prevalent, plant species that are sensitive to fire are solely found in microsites such as gullies or pockets of rainforest (Woinarski et al., 2004) that are far from the reach of fire impacts. Such phenomenon is evident in *Pinus* species where refugia is used as a survival strategy (Keeley et al., 2011). During fire-free periods, *Pinus* species colonize from refugia. However, as a part of evolutionary processes, populations from refugia may differentiate from the original populations on a long-term basis (Leonard et al., 2018).
- *Avoidance:* In this mechanism, unlike refugia, plant species do not migrate to other places; rather, they either evolve reduced flammability in directly exposed parts or grow them away from fire. Low flammability is due to thick and sparse leaves and twigs in some species, while property of self-pruning of lower branches (fire-tolerator syndrome) in other species such as in pines (Pausas & Keeley, 2014). This strategy is often used for survival against surface fires.
- *Dormancy:* When some species are not able to withstand fire in its adult stage, they enter into a latent or dormant phase, allowing their population to survive. This is called dormancy, which is one of the most common tactics used by post-fire seeders as fire is required for breaking the dormant seed (Keeley et al., 2011). Intense fire regimes with extended fire intervals are favorable for this mechanism. Due to less competition, this strategy leads to increase in plant populations size (Pausas & Keeley, 2014).

- **Exogenous recolonization:** In this mechanism, recolonization occurs quickly after fire but the recolonizing species are the individuals from neighboring populations and not from the burned one. Post-fire environments with less competition are conducive to such fast colonization. Plants lacking crown fire adaptations but comparatively strong dispersal ability are characteristic examples of exogenous colonization (Rodrigo et al., 2004; Owen et al., 2017).
- **Crypsis:** This mechanism favors post-fire circumstances over pre-fire conditions for plant species. It has been discovered that numerous plant species develop darker seeds to make them less visible or cryptic to predators in the environment following a fire (Lev-Yadun & Ne'eman, 2013).

## 15.5 PLANT ADAPTATIONS TO FIRE

Plant species are not "fire-adapted"; rather, they evolve adaptations to fit in a specific fire habitat (Keeley et al., 2011). Therefore, even a slight shift in the fire regime leads to remarkable changes in the fire-prone environment. The plants in the fire-affected ecosystems are either susceptible types (filtered out of community) or resistant types (survive during or after fire) due to various adaptive traits and post-fire strategies (Clarke et al., 2015) (Figure 15.4). The adaptations owing to pre-fire condition is collectively related to avoidance mechanism and that of during fire and post-fire to resistance and tolerance mechanisms respectively. The tolerance mechanism (post-fire regeneration) can be further grouped as resprouters and recruiters.

In the case of recruiters, only post-fire recruitment (P+) through a seed bank allows populations to regenerate as they are fire-susceptible (R−). Hence, recruiters are also called as "obligate seeders" (R−P+). Resprouters can withstand fire (R+) and establish by vegetative regrowth from insulated parts regardless of germination of seed induced by fire. Species that are considered "obligate resprouters" are those in which establishment and possible growth exclusively take place in the absence of fire, with seedling recruitment not occurring after a fire (Cowling et al., 2018).

### 15.5.1 Fire-Adaptive Traits Associated with Resistance

- **Fire-resistant bark:** One of the oldest reported fire adaptations is thought to have been bark thickening for bud and meristem insolation. The types of bark on eucalypts vary greatly (Wrigley & Fagg, 2010). Bark's capacity to operate as an insulator may be impacted by a number of factors, including the amount of moisture present, its thickness, and its density (Vines, 1968; Hoffmann et al., 2009; Lawes et al., 2011). The insolation capacity of bark saves the cambium from heat of the fire.
- **Bud position:** In grasses, perennating organs (such as buds) are near the soil surface, but if the position of the bud is below the soil surface, they are protected from fire (Archibald et al., 2019).
- **Culm orientation:** Vertical culm growth enhances plant height and light absorption, and keeps combustibles away from the delicate buds; parallel growth, on the other hand, promotes lateral expansion while keeping appetizing material out of grazers' access. Within a species and over the course of an individual's lifespan, this characteristic is remarkably adaptable. The orientation of the culm can be vertical or horizontal. In vertical growth, sensitive buds are protected from

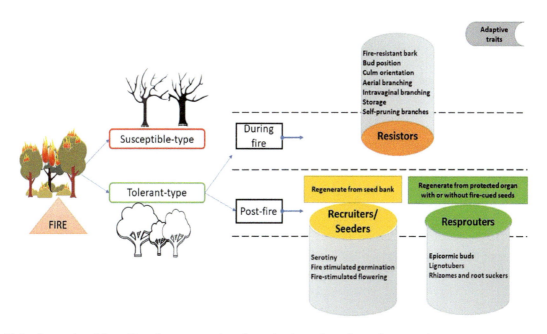

**FIGURE 15.4** Categories of fire-affected ecosystems based on adaptive traits and post-fire strategies.

fire due to increased height. In the horizontal orientation, the buds are protected due to their lateral growth and position away from the grazers (Kellogg, 2015).
- **Aerial branching:** In the grasses, tillering at distant position—i.e., aerial branching—enables branched growth of grasses which increases light capture, as well as biomass required for survival in fire-prone condition (Archibald et al., 2019).
- **Intravaginal branching:** The morphology of tillers developed by intravaginal branching allows for height increase and shields underlying buds behind several leaf coverings. The mat-grass stolons grow from extravaginal tillers, which are likewise able to fill gaps and expand horizontally but incur the risk of exposing seedlings (Schulte, 2011).
- **Storage:** Most species' subterranean portions, such as roots, rhizomes, etc., contain ample storage necessary for resprouting swiftly after a single defoliation. On the other hand, a robust plant that could withstand repeated defoliation needs to sustain photosynthetic activity and cannot just solely rely on storage that has been accumulated (Brodie et al., 2021). Therefore, storage organs help in the species persistence.
- **Self-pruning branches:** Another ability of plants to withstand flames is self-pruning. Because ascending fires are eliminated by self-pruning, branches—and the likelihood that a surface fire may spread to the canopy—are greatly decreased. Self-pruning branches are frequent in low-severity or surface fire regimes (Pausas et al., 2017).

### 15.5.2 Fire-Adaptive Traits Associated with Resprouters

- **Epicormic buds:** Epicormic buds, i.e., buds present on the stem, are adaptive to various fire regimes. They are the protected or latent buds that resprout after defoliation due to fire. As after crown fires, the whole tree is not lost—the epicormic buds resprout and enable quick recovery of the canopy. The two ways by which epicormic buds are protected are development of thick bark and deep burial of buds in the bark. In the case of deep burial, the bark can be thin (e.g., *Eucalyptus*) (Burrows, 2002).
- **Lignotubers:** Not every plant has epicormic buds and thick bark. However, certain trees and plants have underground buds that can reappear even after a fire kills the stems. Lignotubers, which are woody structures around a plant's roots that store numerous latent buds, are essential for post-fire plant recovery. In crown or high-severity fire regimes, species with lignotubers are frequently observed.
- **Rhizomes and root suckers:** In addition to lignotubers, certain eucalypts also contain a range of additional subterranean structures that aid in vegetative regeneration following fire, such as rhizomes and root suckers (Daniels & Throop, 2022). Three species of *Eucalyptus* found to have rhizomes, suckers, and stolons as the regenerative structure (Table 15.1).

### 15.5.3 Fire-Adaptive Traits Associated with Recruiters

- **Serotiny:** One of the earliest reported fire-adaptive phenomena, serotiny refers to the release of accumulated seeds in the cone or fruit in response to fire (Causley et al., 2016). While structural characteristics may differ between species, serotiny frequently necessitates that enclosed scales of cone soften due to fire-induced temperature, enabling seed dispersal (Pastor, 2016).
- **Fire-stimulated germination:** The ability of certain species to store seeds in soil and create soil seed banks to germinate those seeds is yet another regrowth property that may have arisen as a result of fire. These species' seeds go into dormancy and need physical and chemical stimulation to revive their ecophysiological mechanism. Refractory seeds, also known as fire recruiters, are seeds that are revived from dormancy as a result of a wildfire. Since fire promotes seed germination and eases establishment, *Tectona grandis* (teak) is known to have enhanced regeneration in burned regions as opposed to in fire-protected areas (Khurana & Singh, 2001).

In general, the species showing fire-stimulated generation possess a pathway whereby the thick testa (outer covering of seed) is scarified by the heat of fire. The heat or smoke probably induces changes in the seed coat due to production of the chemicals. Other changes, such as removal of water permeability barrier and growth regulator production, are also thought to be the result of heat-mediated signals.

**TABLE 15.1**

**Different Types of Regenerative Parts in Different Species of *Eucalyptus***

| Type of *Eucalyptus* Species | Type of Regenerative Structure | References |
| --- | --- | --- |
| *E. porrecta* | Lignotubers + rhizome | Lacey (1974) |
| *E. tetrodonta* | Lignotuber + root sucker | Lacey & Whelan (1976) |
| *E. moluccana* | Lignotuber + stolon | Gillison et al. (1980) |

- **Fire-stimulated flowering:** The initial post-fire flowering evidence dates to around 80 million years ago, making it one of the most phenomenal fire-related adaptative features (Lamont et al., 2019). The majority that blossoms after a fire are pollinated by wind, giving them a comparative benefit due to enhanced plants recruitment (He et al., 2019). When nutrients are abundant and plant cover is minimal after a fire, these species may quickly colonize new areas, which helps wind-dispersed species (Fidelis & Zirondi, 2021).

## 15.6 CONCLUSION

Forest fires caused by human activity pose a threat to the world's remaining tropical forests, even in protected regions. Despite the fact that several studies have been conducted in various countries, data show that there have not been enough studies to assess the global situation. Therefore, it is critical to determine the precise number and extent of fires that occur in tropical rainforests around the world. Global climate change has the potential to increase or alleviate fire hazards in particular tropical forest regions, depending on how precipitation amounts and seasonality change. The source, severity, and frequency of fires in the tropics should be the subject of more study because these factors determine the impact on soil and plants. The extent of the effects of fire on tropical ecosystems can range from low to high. Plant ecophysiology is a critical component of research on forest responses to climate change. It is critical to explain the function of ecophysiological research in clarifying plant community responses to changing environments. Thus, the current discussion will aid in decision-making, recognizing the significance of plant ecophysiological changes in relation to forest fires and identifying gaps in fire ecology research. We attempted to investigate how much an understanding of plant ecophysiological responses and function can help to fill the gap between fire behavior and plant ecology. Plant physiological ecology and fire science both have an extensive research history, but they have generally evolved in parallel. The study focuses on the causes of fires and the different types of fire regimes. These gaps in the study on forest fires could serve as a guide for decision-making to prevent further fire spread or at least reduce its detrimental impacts in tropical ecosystems. More quantitative investigations on physiological-level responses are needed to increase knowledge of the mechanisms involved in plant responses and adaptations. Species respond differently to different fire intervals and fire intensities due to differences in fire tolerance, maturation period, and reproductive strategies. Some species respond well to regular fires, while others benefit from slightly longer fire-free intervals. Therefore, extensive research is required to improve our understanding of the underlying biophysical processes, as well as how various species react to fire injuries under various post-fire climatic and ecological conditions.

## REFERENCES

Alencar, A. A., Solórzano, L. A., & Nepstad, D. C. (2004). Modeling forest understory fires in an eastern Amazonian landscape. *Ecological Applications, 14*(suppl. 4), 139–149.

Andela, N., Morton, D. C., Giglio, L., Chen, Y., van der Werf, G. R., Kasibhatla, P. S., . . . & Randerson, J. T. (2017). A human-driven decline in global burned area. *Science, 356*(6345), 1356–1362.

Anderson, T. H. (2003). Microbial eco-physiological indicators to asses soil quality. *Agriculture, Ecosystems & Environment, 98*(1–3), 285–293.

Archibald, S., Hempson, G. P., & Lehmann, C. (2019). A unified framework for plant life-history strategies shaped by fire and herbivory. *New Phytologist, 224*(4), 1490–1503.

Archibald, S., Lehmann, C. E., Gómez-Dans, J. L., & Bradstock, R. A. (2013). Defining pyromes and global syndromes of fire regimes. *Proceedings of the National Academy of Sciences, 110*(16), 6442–6447.

Bär, A., Michaletz, S. T., & Mayr, S. (2019). Fire effects on tree physiology. *New Phytologist, 223*(4), 1728–1741.

Battipaglia, G., Savi, T., Ascoli, D., Castagneri, D., Esposito, A., Mayr, S., & Nardini, A. (2016). Effects of prescribed burning on ecophysiological, anatomical and stem hydraulic properties in *Pinus pinea* L. *Tree Physiology, 36*(8), 1019–1031.

Bhadouria, R., Singh, R., Srivastava, P., & Raghubanshi, A. S. (2016). Understanding the ecology of tree-seedling growth in dry tropical environment: A management perspective. *Energy, Ecology and Environment, 1*(5), 296–309.

Bhadouria, R., Srivastava, P., Singh, R., Tripathi, S., Singh, H., & Raghubanshi, A. S. (2017). Tree seedling establishment in dry tropics: An urgent need of interaction studies. *Environment Systems and Decisions, 37*(1), 88–100.

Bhadouria, R., Tripathi, S., & Rao, K. S. (2018). Understanding plant community assemblage, functional diversity and soil attributes of Indian savannas through a continuum approach. *Tropical Ecology, 59*(4), 545–554.

Bistinas, I., Oom, D., Sá, A. C., Harrison, S. P., Prentice, I. C., & Pereira, J. M. (2013). Relationships between human population density and burned area at continental and global scales. *PLoS One, 8*(12), e81188.

Bond, W. J., & Midgley, G. F. (2012). Carbon dioxide and the uneasy interactions of trees and savannah grasses. *Philosophical Transactions of the Royal Society B: Biological Sciences, 367*(1588), 601–612.

Bond, W. J., & Van Wilgen, B. W. (2012). *Fire and Plants* (Vol. 14). Springer Science & Business Media.

Bowman, D. M., Balch, J. K., Artaxo, P., Bond, W. J., Carlson, J. M., Cochrane, M. A., . . . & Pyne, S. J. (2009). Fire in the Earth system. *Science, 324*(5926), 481–484.

Brodie, E. G., Miller, J. E., & Safford, H. D. (2021). Productivity modifies the effects of fire severity on understory diversity. *Ecology, 102*, e03514.

Buchmann, N. (2002). Plant ecophysiology and forest response to global change. *Tree Physiology, 22*(15–16), 1177–1184.

Bucini, G., & Lambin, E. F. (2002). Fire impacts on vegetation in Central Africa: A remote-sensing-based statistical analysis. *Applied Geography, 22*(1), 27–48.

Burrows, G. E. (2002). Epicormic strand structure in Angophora, Eucalyptus and Lophostemon (Myrtaceae): Implications for fire resistance and recovery. *New Phytologist*, 111–131.

Burrows, N. D. (2008). Linking fire ecology and fire management in south-west Australian forest landscapes. *Forest Ecology and Management, 255*(7), 2394–2406.

Butler, R. (2012). Types of rainforests. Retrieved from http//www.rainforests. mongabay. com/0103. htm on, 3(02), 2017.

Cannon, J. B., Peterson, C. J., O'Brien, J. J., & Brewer, J. S. (2017). A review and classification of interactions between forest disturbance from wind and fire. *Forest Ecology and Management*, *406*, 381–390.

Catry, F. X., Rego, F., Moreira, F., Fernandes, P. M., & Pausas, J. G. (2010). Post-fire tree mortality in mixed forests of central Portugal. *Forest Ecology and Management*, *260*(7), 1184–1192.

Causley, C. L., Fowler, W. M., Lamont, B. B., & He, T. (2016). Fitness benefits of serotiny in fire-and drought-prone environments. *Plant Ecology*, *217*(6), 773–779.

Chazdon, R. L. (2003). Tropical forest recovery: Legacies of human impact and natural disturbances. *Perspectives in Plant Ecology, Evolution and Systematics*, *6*(1–2), 51–71.

Chuvieco, E., Yue, C., Heil, A., Mouillot, F., Alonso-Canas, I., Padilla, M., . . . & Tansey, K. (2016). A new global burned area product for climate assessment of fire impacts. *Global Ecology and Biogeography*, *25*(5), 619–629.

Clarke, P. J., Lawes, M. J., Murphy, B. P., Russell-Smith, J., Nano, C. E., Bradstock, R., . . . & Gunton, R. M. (2015). A synthesis of postfire recovery traits of woody plants in Australian ecosystems. *Science of the Total Environment*, *534*, 31–42.

Cochrane M. A. (2009a). Fire in the tropics. In Cochrane M. A. (Ed.), *Tropical Fire Ecology* (pp. 1–23). Springer Praxis Books. Springer, Berlin, Heidelberg. https://doi.org/10.1007/978-3-540-77381-8_1

Cochrane, M. A. (2009b). *Tropical Fire Ecology: Climate Change, Land Use, and Ecosystem Dynamics* (p. 682). Springer.

Cochrane M. A. (2011). The past, present, and future importance of fire in tropical rainforests. In Bush M., Flenley J., & Gosling W. (Eds.), *Tropical Rainforest Responses to Climatic Change* (pp. 213–240). Springer Praxis Books. Springer, Berlin, Heidelberg. https://doi.org/10.1007/978-3-642-05383-2_7

Cowling, R. M., Gallien, L., Richardson, D. M., & Ojeda, F. (2018). What predicts the richness of seeder and resprouter species in fire-prone Cape fynbos: Rainfall reliability or vegetation density? *Austral Ecology*, *43*(6), 614–622.

Daniels, Q. F., & Throop, H. L. (2022). Ecophysiological responses of Terminalia sericea to fire history in a semi-arid woodland savanna, central Namibia. *South African Journal of Botany*, *146*, 205–212.

Dennis, R. A., Mayer, J., Applegate, G., Chokkalingam, U., Colfer, C. J. P., Kurniawan, I., . . . & Tomich, T. P. (2005). Fire, people and pixels: Linking social science and remote sensing to understand underlying causes and impacts of fires in Indonesia. *Human Ecology*, *33*(4), 465–504.

Fidelis, A., & Zirondi, H. L. (2021). And after fire, the Cerrado flowers: A review of post-fire flowering in a tropical savanna. *Flora*, *280*, 151849.

Foley, J. A., Costa, M. H., Delire, C., Ramankutty, N., & Snyder, P. (2003). Green surprise? How terrestrial ecosystems could affect earth's climate. *Frontiers in Ecology and the Environment*, *1*(1), 38–44.

Ganteaume, A., Camia, A., Jappiot, M., San-Miguel-Ayanz, J., Long-Fournel, M., & Lampin, C. (2013). A review of the main driving factors of forest fire ignition over Europe. *Environmental Management*, *51*(3), 651–662.

Giglio, L., Randerson, J. T., & Van Der Werf, G. R. (2013). Analysis of daily, monthly, and annual burned area using the fourth-generation global fire emissions database (GFED4). *Journal of Geophysical Research: Biogeosciences*, *118*(1), 317–328.

Gillison, A. N., Lacey, C. J., & Bennett, R. H. (1980). Rhizostolons in Eucalyptus. *Australian Journal of Botany*, *28*(3), 299–304.

Goldammer, J. G. (Ed.). (2012). *Fire in the Tropical Biota: Ecosystem Processes and Global Challenges* (Vol. 84). Springer Science & Business Media.

Goldammer, J. G., & Price, C. (1998). Potential impacts of climate change on fire regimes in the tropics based on MAGICC and a GISS GCM-derived lightning model. *Climatic Change*, *39*(2), 273–296.

He, T., Lamont, B. B., & Pausas, J. G. (2019). Fire as a key driver of Earth's biodiversity. *Biological Reviews*, *94*(6), 1983–2010.

Heskel, M. A., O'sullivan, O. S., Reich, P. B., Tjoelker, M. G., Weerasinghe, L. K., Penillard, A., . . . & Atkin, O. K. (2016). Convergence in the temperature response of leaf respiration across biomes and plant functional types. *Proceedings of the National Academy of Sciences*, *113*(14), 3832–3837.

Higgins, S. I., Bond, W. J., February, E. C., Bronn, A., Euston-Brown, D. I., Enslin, B., . . . & Trollope, W. S. (2007). Effects of four decades of fire manipulation on woody vegetation structure in savanna. *Ecology*, *88*(5), 1119–1125.

Hoffmann, W. A., Adasme, R., Haridasan, M., T. de Carvalho, M., Geiger, E. L., Pereira, M. A., . . . & Franco, A. C. (2009). Tree topkill, not mortality, governs the dynamics of savanna–forest boundaries under frequent fire in central Brazil. *Ecology*, *90*(5), 1326–1337.

Hoffmann, W. A., Geiger, E. L., Gotsch, S. G., Rossatto, D. R., Silva, L. C., Lau, O. L., . . . & Franco, A. C. (2012). Ecological thresholds at the savanna-forest boundary: How plant traits, resources and fire govern the distribution of tropical biomes. *Ecology Letters*, *15*(7), 759–768.

Hoffmann, W. A., Orthen, B., & Do Nascimento, P. K. V. (2003). Comparative fire ecology of tropical savanna and forest trees. *Functional Ecology*, 720–726.

Hossain, M. A., Wani, S. H., Bhattacharjee, S., Burritt, D. J., & Tran, L. S. P. (2016). *Drought Stress Tolerance in Plants*. Springer. https://doi.org/10.1007/978-3-319-28899-4

IPCC. (2001). Climate change 2001: The scientific basis. In Houghton, J. T., Ding, Y., Griggs, D. J., Noguer, M., van der Linden, P. J., Dai, X., Maskell, K., & Johnson C. A. (eds.), *Contribution of Working Group I to the Third Assessment Report of the Intergovernmental Panel on Climate Change* (p. 881). Cambridge University Press.

IPCC. (2021). Climate change 2021: The physical science basis. In Masson-Delmotte, V., Zhai, P., Pirani, A., Connors, S. L., Péan, C., Berger, S., Caud, N., Chen, Y., Goldfarb, L., Gomis, M. I., Huang, M., Leitzell, K., Lonnoy, E., Matthews, J. B. R., Maycock, T. K., Waterfield, T., Yelekçi, O., Yu, R., & B. Zhou, B. (eds.), *Contribution of Working Group I to the Sixth Assessment Report of the Intergovernmental Panel on Climate Change*. Cambridge University Press, In press. https://doi.org/ 10.1017/9781009157896

Jakovac, C. C., Peña-Claros, M., Mesquita, R. C., Bongers, F., & Kuyper, T. W. (2016). Swiddens under transition: Consequences of agricultural intensification in the Amazon. *Agriculture, Ecosystems & Environment*, *218*, 116–125.

Jolly, W. M., Cochrane, M. A., Freeborn, P. H., Holden, Z. A., Brown, T. J., Williamson, G. J., & Bowman, D. M. (2015). Climate-induced variations in global wildfire danger from 1979 to 2013. *Nature Communications*, *6*(1), 1–11.

Juárez-Orozco, S. M., Siebe, C., & Fernández y Fernández, D. (2017). Causes and effects of forest fires in tropical rainforests: A bibliometric approach. *Tropical Conservation Science*, *10*, 1940082917737207.

Keeley, J. E. (2009). Fire intensity, fire severity and burn severity: A brief review and suggested usage. *International Journal of Wildland Fire*, *18*(1), 116–126.

Keeley, J. E., Fotheringham, C. J., & Baer-Keeley, M. (2005). Determinants of postfire recovery and succession in Mediterranean-climate shrublands of California. *Ecological Applications*, *15*(5), 1515–1534.

Keeley, J. E., Pausas, J. G., Rundel, P. W., Bond, W. J., & Bradstock, R. A. (2011). Fire as an evolutionary pressure shaping plant traits. *Trends in Plant Science*, *16*(8), 406–411.

Keenan, R. J., Reams, G. A., Achard, F., de Freitas, J. V., Grainger, A., & Lindquist, E. (2015). Dynamics of global forest area: Results from the FAO Global Forest Resources Assessment 2015. *Forest Ecology and Management*, *352*, 9–20.

Kelley, D. I., Bistinas, I., Whitley, R., Burton, C., Marthews, T. R., & Dong, N. (2019). How contemporary bioclimatic and human controls change global fire regimes. *Nature Climate Change*, *9*(9), 690–696.

Kellogg, E. A. (2015). Flowering plants. Monocots. *Cham: Springer International Publishing*, *10*, 978–3.

Khurana, E., & Singh, J. S. (2001). Ecology of tree seed and seedlings: Implications for tropical forest conservation and restoration. *Current Science*, 748–757.

Kleine, T., Nägele, T., Neuhaus, H. E., Schmitz-Linneweber, C., Fernie, A. R., Geigenberger, P., . . . & Leister, D. (2021). Acclimation in plants–the Green Hub consortium. *The Plant Journal*, *106*(1), 23–40.

Kraus D., & Goldammer J. G. (2007). Fire regimes and ecosystems: an overview of fire ecology in tropical ecosystems. In Schmerbeck J., Hiremath A., & Ravichandran C. (Eds.), *Forest Fires in India–Workshop Proceedings* 2007 Feb 19. ATREE, Bangalore, India and Institute of Silviculture, Freiburg, Germany (pp. 9–13).

Kristensen, T. N., Ketola, T., & Kronholm, I. (2020). Adaptation to environmental stress at different timescales. *Annals of the New York Academy of Sciences*, *1476*(1), 5–22.

Lacey, C. J. (1974). Rhizomes in tropical eucalypts and their role in recovery from fire damage. *Australian Journal of Botany*, *22*(1), 29–38.

Lacey, C. J., & Whelan, P. I. (1976). Observations on the ecological significance of vegetative reproduction in the Katherine-Darwin region of the Northern Territory. *Australian Forestry*, *39*(2), 131–139.

Lamalakshmi Devi, E., Kumar, S., Basanta Singh, T., Sharma, S. K., Beemrote, A., Devi, C. P., . . . & Wani, S. H. (2017). Adaptation strategies and defence mechanisms of plants during environmental stress. In Ghorbanpour M., & Varma A. (Eds.), *Medicinal Plants and Environmental Challenges* (pp. 359–413). Springer, Cham. https://doi.org/10.1007/978-3-319-68717-9_20

Lambers, H., Chapin, F. S., & Pons, T. L. (2008). *Plant Physiological Ecology* (Vol. 2, pp. 11–99). Springer.

Lamont, B. B., He, T., & Yan, Z. (2019). Evolutionary history of fire-stimulated resprouting, flowering, seed release and germination. *Biological Reviews*, *94*(3), 903–928.

Laurance, W. F., Cochrane, M. A., Bergen, S., Fearnside, P. M., Delamônica, P., Barber, C., . . . & Fernandes, T. (2001). The future of the Brazilian Amazon. *Science*, *291*(5503), 438–439.

Lawes, M. J., Richards, A., Dathe, J., & Midgley, J. J. (2011). Bark thickness determines fire resistance of selected tree species from fire-prone tropical savanna in north Australia. *Plant Ecology*, *212*(12), 2057–2069.

Leonard, J., West, A. G., & Ojeda, F. (2018). Differences in germination response to smoke and temperature cues in 'pyrophyte'and 'pyrofuge'forms of Erica coccinea (Ericaceae). *International Journal of Wildland Fire*, *27*(8), 562–568.

Lev-Yadun, S., & Ne'eman, G. 2013. Bimodal colour pattern of individual *Pinus* halepensis Mill seeds: A new type of crypsis.—Biol. *Botanical Journal of the Linnean Society*, *109*, 271–278.

McLauchlan, K. K., Higuera, P. E., Miesel, J., Rogers, B. M., Schweitzer, J., Shuman, J. K., . . . & Watts, A. C. (2020). Fire as a fundamental ecological process: Research advances and frontiers. *Journal of Ecology*, *108*(5), 2047–2069.

Michaletz, S. T., & Johnson, E. A. (2007). How forest fires kill trees: A review of the fundamental biophysical processes. *Scandinavian Journal of Forest Research*, *22*(6), 500–515.

Mukul, S. A., & Herbohn, J. (2016). The impacts of shifting cultivation on secondary forests dynamics in tropics: A synthesis of the key findings and spatio temporal distribution of research. *Environmental Science & Policy*, *55*, 167–177.

Nepstad, D. C., Moreira, A. G., & Alencar, A. A. (1999a). *Flames in the Rain Forest: Origins, Impacts and Alternatives to Amazonian Fires* (p. 190). Pilot Program to Conserve the Brazilian Rain Forest.

Nepstad, D. C., Verssimo, A., Alencar, A., Nobre, C., Lima, E., Lefebvre, P., . . . & Brooks, V. (1999b). Large-scale impoverishment of Amazonian forests by logging and fire. *Nature*, *398*(6727), 505–508.

Niccoli, F., Esposito, A., Altieri, S., & Battipaglia, G. (2019). Fire severity influences ecophysiological responses of Pinus pinaster Ait. *Frontiers in Plant Science*, *10*, 539.

Noble, I. R., & Slatyer, R. O. (1980). The use of vital attributes to predict successional changes in plant communities subject to recurrent disturbances. *Vegetatio*, *43*(1), 5–21.

O'Brien, J. J., Hiers, J. K., Varner, J. M., Hoffman, C. M., Dickinson, M. B., Michaletz, S. T., . . . & Butler, B. W. (2018). Advances in mechanistic approaches to quantifying biophysical fire effects. *Current Forestry Reports*, *4*(4), 161–177.

Owen, S. M., Sieg, C. H., Meador, A. J. S., Fulé, P. Z., Iniguez, J. M., Baggett, L. S., . . . & Battaglia, M. A. (2017). Spatial patterns of ponderosa pine regeneration in high-severity burn patches. *Forest Ecology and Management*, *405*, 134–149.

Pastor, J. (2016). Fire regimes and the correlated evolution of serotiny and flammability. In *What Should a Clever Moose Eat?* (pp. 227–235). Island Press.

Pausas, J. G. (2019). Generalized fire response strategies in plants and animals. *Oikos*, *128*(2), 147–153.

Pausas, J. G., & Keeley, J. E. (2014). Evolutionary ecology of resprouting and seeding in fire-prone ecosystems. *New Phytologist*, *204*(1), 55–65.

Pausas, J. G., Keeley, J. E., & Schwilk, D. W. (2017). Flammability as an ecological and evolutionary driver. *Journal of Ecology*, *105*(2), 289–297.

Pellegrini, A. F., & Jackson, R. B. (2020). The long and short of it: A review of the timescales of how fire affects soils using the pulse-press framework. *Advances in Ecological Research*, *62*, 147–171.

Prince's Charities (Great Britain). (2015). *Tropical Forests: A Review*. Prince's Charities' International Sustainability Unit.

Qin, C., Zhu, K., Chiariello, N. R., Field, C. B., & Peay, K. G. (2020). Fire history and plant community composition outweigh decadal multi-factor global change as drivers of microbial composition in an annual grassland. *Journal of Ecology*, *108*(2), 611–625.

Rejeb, I. B., Pastor, V., & Mauch-Mani, B. (2014). Plant responses to simultaneous biotic and abiotic stress: Molecular mechanisms. *Plants*, *3*(4), 458–475.

Renninger, H. J., Clark, K. L., Skowronski, N., & Schäfer, K. V. (2013). Effects of a prescribed fire on water use and photosynthetic capacity of pitch pines. *Trees*, *27*(4), 1115–1127.

Resco de Dios, V. (2020). Effects of fire on plant performance. In *Plant-Fire Interactions* (pp. 117–132). Springer.

Roberts, P., Hunt, C., Arroyo-Kalin, M., Evans, D., & Boivin, N. (2017). The deep human prehistory of global tropical forests and its relevance for modern conservation. *Nature Plants*, *3*(8), 1–9.

Rodrigo, A., Retana, J., & Picó, F. X. (2004). Direct regeneration is not the only response of Mediterranean forests to large fires. *Ecology*, *85*(3), 716–729.

Rogers, B. M., Balch, J. K., Goetz, S. J., Lehmann, C. E., & Turetsky, M. (2020). Focus on changing fire regimes: Interactions with climate, ecosystems, and society. *Environmental Research Letters*, *15*(3), 030201.

Ryan, K. C. (2000). Global change and wildland fire [Chapter 8]. In Brown, J. K. & Smith, J. K. (eds.), *Wildland Fire in Ecosystems: Effects of Fire on Flora* (Vol. 2, pp. 175–183). USDA Forest Service, Rocky Mountain Research Station, General Technical Report RMRS-GTR-42.

Sadeghifar, M., Agha, A. B. A., & Pourreza, M. (2020). Comparing soil microbial eco-physiological and enzymatic response to fire in the semi-arid Zagros woodlands. *Applied Soil Ecology*, *147*, 103366.

Sah, S. K., Kaur, G., & Wani, S. H. (2016). Metabolic engineering of compatible solute trehalose for abiotic stress tolerance in plants. In Iqbal, N., Nazar, R., & Khan, N. A. (eds.), *Osmolytes and Plants Acclimation to Changing Environment: Emerging Omics Technologies* (pp. 83–96). Springer, New Delhi.

Saha, S. (2002). Anthropogenic fire regime in a deciduous forest of central India. *Current Science*, 1144–1147.

Saha, S., & Howe, H. F. (2003). Species composition and fire in a dry deciduous forest. *Ecology*, *84*(12), 3118–3123.

Sankaran, M., Ratnam, J., & Hanan, N. P. (2004). Tree–grass coexistence in savannas revisited–insights from an examination of assumptions and mechanisms invoked in existing models. *Ecology Letters*, *7*(6), 480–490.

Santacruz-García, A. C., Bravo, S., del Corro, F., García, E. M., Molina-Terrén, D. M., & Nazareno, M. A. (2021). How do plants respond biochemically to fire? The role of photosynthetic pigments and secondary metabolites in the post-fire resprouting response. *Forests*, *12*(1), 56.

Savadogo, P., Zida, D., Sawadogo, L., Tiveau, D., Tigabu, M., & Odén, P. C. (2007). Fuel and fire characteristics in savanna-woodland of West Africa in relation to grazing and dominant grass type. *International Journal of Wildland Fire*, *16*(5), 531–539.

Schulte, J. R. (2011). *The Fire Ecology of Kentucky Bluegrass (Poa Pratensis)*. Master of Science Dissertation. North Dakota State University.

Schutz, A. E. N., Bond, W. J., & Cramer, M. D. (2009). Juggling carbon: Allocation patterns of a dominant tree in a fire-prone savanna. *Oecologia*, *160*(2), 235–246.

Shlisky, A., Alencar, A. A. C., Nolasco, M. M., & Curran, L. M. (2009). Overview: Global fire regime conditions, threats, and opportunities for fire management in the tropics. In Cochrane M. A. (Ed.), *Tropical Fire Ecology* (pp. 65–83). Springer Praxis Books. Springer, Berlin, Heidelberg. https://doi.org/10.1007/978-3-540-77381-8_3

Shlisky, A., Waugh, J., Gonzalez, P., Gonzalez, M., Manta, M., Santoso, H., ... & Fulks, W. (2007). Fire, ecosystems and people: Threats and strategies for global biodiversity conservation. In *Global Fire Initiative Technical Report 2*. The Nature Conservancy.

Singh, S. (2022). Forest fire emissions: A contribution to global climate change. *Frontiers in Forests and Global Change*, *5*, 925480.

Sorrensen, C. (2004). Contributions of fire use study to land use/cover change frameworks: Understanding landscape change in agricultural frontiers. *Human Ecology*, *32*(4), 395–420.

Tangney, R., Paroissien, R., Le Breton, T. D., Thomsen, A., Doyle, C. A., Ondik, M., ... & Ooi, M. K. (2022). Success of post-fire plant recovery strategies varies with shifting fire seasonality. *Communications Earth & Environment*, *3*(1), 1–9.

Teixeira, J., Souza, L., Le Stradic, S., & Fidelis, A. (2022). Fire promotes functional plant diversity and modifies soil carbon dynamics in tropical savanna. *Science of The Total Environment*, *812*, 152317.

Thonicke, K., Venevsky, S., Sitch, S., & Cramer, W. (2001). The role of fire disturbance for global vegetation dynamics: Coupling fire into a Dynamic Global Vegetation Model. *Global Ecology and Biogeography*, *10*(6), 661–677.

Trejo, D. A. R. (2008). Fire regimes, fire ecology, and fire management in Mexico. *AMBIO: A Journal of the Human Environment*, *37*(7), 548–556.

Trollope, W. S. W. (1982). Ecological effects of fire in South African savannas. In *Ecology of Tropical Savannas* (pp. 292–306). Springer.

Trumbore, S., Brando, P., & Hartmann, H. (2015). Forest health and global change. *Science*, *349*(6250), 814–818.

Van Der Werf, G. R., Randerson, J. T., Giglio, L., Van Leeuwen, T. T., Chen, Y., Rogers, B. M., ... & Kasibhatla, P. S. (2017). Global fire emissions estimates during 1997–2016. *Earth System Science Data*, *9*(2), 697–720.

Van Vliet, N., Mertz, O., Heinimann, A., Langanke, T., Pascual, U., Schmook, B., ... & Ziegler, A. D. (2012). Trends, drivers and impacts of changes in swidden cultivation in tropical forest-agriculture frontiers: A global assessment. *Global Environmental Change*, *22*(2), 418–429.

Vayda, A. P. (2006). Causal explanation of Indonesian forest fires: Concepts, applications, and research priorities. *Human Ecology*, *34*(5), 615–635.

Veraverbeke, S., Rogers, B. M., Goulden, M. L., Jandt, R. R., Miller, C. E., Wiggins, E. B., & Randerson, J. T. (2017). Lightning as a major driver of recent large fire years in North American boreal forests. *Nature Climate Change*, *7*(7), 529–534.

Vico, G., Manzoni, S., Palmroth, S., & Katul, G. (2011). Effects of stomatal delays on the economics of leaf gas exchange under intermittent light regimes. *New Phytologist*, *192*(3), 640–652.

Vines, R. G. (1968). Heat transfer through bark, and the resistance of trees to fire. *Australian Journal of Botany*, *16*(3), 499–514.

Vitousek, P. M. (1994). Beyond global warming: Ecology and global change. *Ecology*, *75*(7), 1861–1876.

Walker, R., Moran, E., & Anselin, L. (2000). Deforestation and cattle ranching in the Brazilian Amazon: External capital and household processes. *World Development*, *28*(4), 683–699.

Wani, S. H., Sah, S. K., Hossain, M. A., Kumar, V., & Balachandran, S. M. (2016). Transgenic Approaches for Abiotic Stress Tolerance in Crop Plants. In Al-Khayri, J., Jain, S., & Johnson, D. (Eds.), *Advances in Plant Breeding Strategies: Agronomic, Abiotic and Biotic Stress Traits* (pp. 345–396). Springer, Cham. https://doi.org/10.1007/978-3-319-22518-0_10

Werner, P. A., & Prior, L. D. (2013). Demography and growth of subadult savanna trees: Interactions of life history, size, fire season, and grassy understory. *Ecological Monographs*, *83*(1), 67–93.

Woinarski, J. C. Z., Risler, J., & Kean, L. (2004). Response of vegetation and vertebrate fauna to 23 years of fire exclusion in a tropical Eucalyptus open forest, Northern Territory, Australia. *Austral Ecology*, *29*(2), 156–176.

Wrigley, J., & Fagg, M. (2010). *Eucalypts: A Celebration*. Allen & Unwin.

Zhang, D., & Pearse, P. H. (2011). *Forest Economics*. UBC Press.

# 16 Climate Change-Mediated Fire Effects on Community Structure and the Physio-Anatomical Adaptations of Plants in Tropical Savannas

*S. Oyedeji, C.O. Ogunkunle, S.A. Adeniran, O.O. Agboola, and P.O. Fatoba*

## 16.1 INTRODUCTION

Tropical savannas are biomes characterized by tropical wet and dry climatic conditions with a typical continuous cover of perennial grasses that are often from 3–6 feet tall, either with an open canopy or not. They are strongly seasonal, with frequent fires, and their ecological dynamics and distinct vegetation physiognomy are determined mainly by climatic factors and fire (Bond & Keeley, 2005; Bowman et al., 2009). Tropical savannas may have fire-resistant or browse-resistant tree species or open shrub layers but are usually distinguished by the dominant taxon of tree stratum which varies in cover and are usually poorly understood (Lehmann et al., 2009).

Fire is widely employed as a management option for maintaining grassland or rangeland, and for the utilization of savannas for productive use. It has been generally agreed upon that fire has been a regular phenomenon in savanna landscapes over many decades, and this has great influence on the landscape and in creating adaptations for plants and animals (Ramos-Neto & Pivello, 2000). In fact, fire has been opined as a tool to shift the balance of tree–grass co-existence in savannas towards tree dominance, thereby leading to a grass-dominated or woodland type of savanna (van Langevelde et al., 2003; Bond & Keeley, 2005). Tropical savanna fire is a regular phenomenon that usually confers several environmental benefits to the ecosystem such as the stimulus to seeding and dispersal among several plant species, as well as the recycling of nutrients in accumulated biomass. However, fires in tropical savannas are usually of high frequency and low heat (low intensity) and appear to be a different phenomenon than fire in other biomes (Lonsdale & Braithwaite, 1991). According to some authors (e.g. Bond & Midgley, 2012; Murphy & Bowman, 2012), tropical savanna fires are mostly surface-type fires that are fueled by accumulated herbaceous biomass. There may occur crown fires in tropical savannas, but due to the fact that trees are mostly scattered, crown fires are rare (Kahiu & Hanan, 2018).

Fire has evolved as an integral component of savanna vegetation (Oyedeji et al., 2021b), but the growing impact of climate change on the global ecosystem is exacerbating hot and dry conditions, especially in seasonally dry regions, which is increasing drying up of vegetation and the risk of wildfires (WEF, 2022). According to the United Nations Environment Program (UNEP 2022), climate change and wildfire are jointly impacting vegetation—the changing climate is increasing fire impacts by increasing temperatures, drought, and strong winds, while reducing relative humidity; these effects in turn lengthen fire season, frequency of fires, and the extent of burned areas. There is therefore the need to expose the determinants of fire events in savannas, the impact on the flora, and their adaptive traits/strategy to surviving recurring fire events in the vegetation.

## 16.2 FIRE DISTURBANCES IN TROPICAL SAVANNAS

Fire incidence, one of the important features of savannas, is a more frequent occurrence in the biome than in other biomes (Huntley & Walker, 1982). This is due to the fact that fire can either be naturally occurring or anthropogenic as a 'use and control' strategy in the management of savannas. It can have significant impacts on the savanna ecosystem, depending on several factors like tree height, level of fuel loads, and climatic conditions (Govender et al., 2006). Fires in tropical savannas are not mainly extrinsically induced but plant species have also been reported to act as inducers and influencers of fire frequency (Beckage et al., 2003; 2009). This is probably attributed to the tendency of fire to be initiated or impacted by the species abundance and distribution, as well as to the state of the savanna (Mermoz et al., 2005).

Fire can either be natural or human-caused. Natural fires in tropical savannas are usually initiated by lightning that may result at the last part of the dry season in the presence of the first thunderstorms, as opined by Bloesch (1999). Ancient fires have also been documented as natural and caused by volcanic activities and friction between rocks (Ramos-Neto & Pivello, 2000) but are seldom documented or perhaps occur less often. However, natural fires have been altered or modified by human activities through suppression and ignition (Archibald, 2016; Harrison et al., 2010; Kahiu & Hanan, 2018).

Human-caused or deliberate burning is one of the causes of fire in a tropical savanna, and has been in existence for ages. In most tropical savannas, anthropogenic activities are primarily the causative agent of fire. For instance, the bulk of fires in the African savanna are reportedly initiated by human activities (Cahoon et al., 1992). In the tropical savanna of South America, especially in most Cerrado parks and reserves of the Brazillian savanna, human-caused fires are reportedly frequent and major causes of savanna fires (Pivello & Coutinho, 1992; Ramos-Neto & Pivello, 2000). Bloesch (1999) has listed pasture burning, charcoal production, the harvest of honey, fire as a political protest, and prescribed fire as the major cause of human-caused fires in tropical savannas.

- *Pasture burning:* This is a type of fire created by pastoralists to renew and maintain grazing land by subjecting old tussocks to combustion in order for new ones to sprout (Johansson et al., 2021). For instance, Bloesch (1999) reported that fire is usually set up by pastoralists at the concluding part of the dry season on the Ankarafantshika plateau in Madagascar. In fact, the author asserted that induced fire in this area as a management strategy is seasonal with a frequency of 2–3 years.
- *Charcoal production*: The making of charcoal from woody species is another cause of human-caused fires in the savanna, which are usually occasioned by activities of peasants for economic means. Aside from the direct harvesting of trees to produce charcoal, fire may spread and burn up the vegetation, especially during dry and windy conditions.
- *Harvest of honey:* The deployment of fire to harvest honey from bee hives in tropical savannas can also create unintended fire in the vegetation. Higher occurrences of bushfires have been reported in areas where the honey gathering has been practiced (Snook et al., 2015; Ribeiro et al., 2019).
- *Political protest*: In some regions, people present grievances at governments or government policies by inducing fires. For instance, peasants in Madagascar showed disapproval toward the government over the loss of their savanna land to FaMaMa Project, Mevadoany by setting fires on the newly installed cashew plantations in the acquired savanna land (Bloesch, 1999).
- *Prescribed fire*: This type of fire is employed in managing ecosystems in order to avoid uncontrolled fire events and protect fire-sensitive species/vegetation while maintaining the biodiversity. This type of fire is usually practiced in most tropical and sub-tropical savannas for productive reasons, as well as for conservative purposes by managers of ecosystems not only to minimize fuel load and control the incidence, intensity, and severity of fires (Govender et al., 2006), but also to preserve the existing local biodiversity and structure of the vegetation (Govender et al., 2006; van Wilgen et al., 2007; Rissi et al., 2017).

### 16.2.1 Fire Regimes in Tropical Savannas

A fire regime relates to the frequency, seasonality, intensity, severity, type, and spread patterns of fires that exist in a particular area (Laris, 2013). The frequency and intensity of fires occurring in a specified biome (e.g., savanna) are important components of the fire regime of that biome, and are sometimes loosely referred to as the 'fire regime' (van Wilgen et al., 1990; Rodrigues et al., 2021). In tropical savannas, there is high frequency of fires, typically owing to the high litter production of grasses which results in large fuel accumulation, but they are usually less intense (Bloesch, 1999; Rodrigues et al., 2021). According to Russell-Smith et al. (2007), the dynamics of biomass development/fuel load accumulation and propensity of the biomass to burn (e.g., dryness of biomass) determine the trends of the fire regime. Fire regimes describe the occurrence and how fires at a particular biome/site are characterized by their frequency, intensity, season (time of the year), and type of fire, whether surface or canopy (Keeley, 2009). Fire regimes are shaped by the drivers of fire—including climate, season, ignition and fuel, and land management methods (Marlon et al., 2008). The link between the drivers and the fundamental biophysical processes elucidate the unpredictability in the regimes of fire between or among biomes (Archibald, 2016). The variation or changes in fire regime in a tropical savanna, for instance—the frequency changes—could lead to detrimental effects on the vegetation/biome as such changes could lead to alterations in behavior parameters of fire (Rodrigues et al., 2021; Teixeira et al., 2022).

### 16.2.2 Determinants of Fire Frequency and Intensity in Tropical Savanna

Fire frequency and intensity are among the major attributes of a fire regime that are of particular importance to ecologists and land managers who monitor these changes and trends in natural events (Sugihara et al., 2006; Lutz et al., 2011). Fire greatly determines spatiotemporal patterns of biodiversity via its impact on the composition and abundance of plant and wildlife communities (Sugihara et al., 2006). Ecological processes—including nutrient cycling, carbon storage, soil structure and moisture—are impacted by the frequency and intensity of fire (Steel et al., 2015). Fire frequency refers to the number of occurrences of fires over a specified period (Curt, 2018). Fire intensity, on the other hand, refers to the heat energy per unit of time released by the fire. It is directly related to the amount of biomass or fuel load available for the fire to burn (Rossi et al., 2018). Fire regime is influenced by environmental factors including seasonal dryness, period of ignition, species composition, and availability of fuel load, among other factors.

### 16.2.2.1 Seasonal or Environmental Dryness

The season of occurrence of fire is an important factor influencing the intensity and regularity in a tropical savanna, and usually influences the dynamics of vegetation. This is connected to the fact that the availability of fuel load and the amount of moisture in fuel load varies among seasons and influence fires (Whelan & Tait, 1995). There is the general assertion that fires in tropical savannas occurring just at the start of the dry season when fuel (dry biomasses) still contains or holds significant moisture tend to be less extreme or severe than fires during the late dry season (the time when there is a larger amount of dry matter), mostly owing to the accumulation of fuel load (Rissi et al., 2017). Fire season seems to be the least understood element of a fire regime, despite the fact that it significantly affects the behavior of fire in savannas (Gill et al., 2002; Knapp et al., 2009; Gomes et al., 2018). Plants growing in infertile soils of savannas are also prone to fire incidence. This is connected to the suppressed rate of development of tree and shrub canopies that allows fast growth of grasses that serve as fuel for the fire and create a persistence for fire episodes (Kellman, 1984).

### 16.2.2.2 Ignition Timing

The influence of season on ignition timing in tropical savannas is great, and causes major or significant changes to fire intensity, burned area, and fire regime (Gill et al., 2000; Russell-Smith et al., 2007). The time of the year or ignition time actually influences or varies the intensity of tropical savanna fires. Schmidt et al. (2016) have found that fires during the later part of the dry season in Brazilian savanna tended to be more severe than that occurring during the early dry season; however, this difference was actually noted to be not significant. Lonsdale & Braithwaite (1991) and Cheney (1981) also reported that in Australian savannas, early-season fires are usually patchy with low scorch height and mostly are naturally extinguished within a few hours after ignition when perennial vegetation is moist or patchy. Conversely, the fires during the driest periods (August–September) are often intense and may sometimes burn for weeks.

### 16.2.2.3 Presence of 'Fire-Fighter' or 'Fire-Sensitive' Species

The incidence (frequency) and severity (intensity) of savanna fires can also be influenced by the level of abundance of species that are fire-sensitive, and are susceptible to fire damage. Once species are readily susceptible to damages by fire, their abundance tend to decline as fire incidence increases (Keeley et al., 2005). 'Fire-sensitive' woody species are common in the savannas biome (Frost, 1985) and mostly form the dominant component of vegetation despite recurrent fires—for instance, the *Brachystegia* and *Julbernardia* spp. in Central African Miombo (Trapnell et al., 1976). Plant species with potentials that can inhibit the spread of fire are referred to as 'fire-fighting' species (Platt & Schwartz, 1990; Cole et al., 1992). Therefore, any increase in the abundance of fire-fighting/fire-inhibiting species could influence the incidence and severity of the fire in tropical savanna.

### 16.2.2.4 Litter or Fuel Load

Fuel load is one of the major drivers of fire regimes in the tropical savanna. Schmidt et al. (2016) reported that fuel loads conferred significant influence on the intensity of the fire in the Cerrado Brazillian tropical savanna, where early fire with low fuel load presented the lowest temperature and the site with high fuel load was characterized by very high temperatures. The high amount of fuel load accumulates from seasonal growth in African savannas and is responsible for the recurrent fires in the region. African savannas constitute the most widely and regularly burned regions on the earth and account for nearly 64% of the global land area burned yearly (Laris, 2013).

### 16.2.2.5 Determinants of Fire Intensity in Tropical Savannas

The damaging effect of fire on flora in tropical savanna is directly proportional to the fire intensity (De Ronde et al., 2004; Groen et al., 2008), and the intensity of the fire is determined by several factors but most importantly the grass biomass/fuel load—the more grass biomass available, the more intense the fires (van Langevelde et al., 2003).

Table 16.1 presents various fuel loads and corresponding fire intensities in some savanna environments. On other hand, the intensity of the fire in the savanna was reported to be determined by temperature and the duration of fire, which according to Bloesch (1999) relates to the speed of fire spread. The speed of spread of fires in savannas actually varies based on biomass density, fuel type, and wind velocity. For instance, the speed of fire in the tropical savanna of Ivory Coast was documented as ca. 500 m/h (Gillon, 1983), and this is usually dictated by wind speed. Head fires that are burning in the direction of the wind will proceed faster than fires burning in opposing direction to the wind, and this will naturally affect both the duration of burning and temperature—and consequently, the fire's intensity.

According to some authors (e.g., Bloesch, 1999; Bradstock, 2010), the intensity of fire in savannas is

### TABLE 16.1
### Fuel Loads and Fire Intensity in Some Tropical Savanna Environments

| Region | Intensity (kW/m) | Fuel Load (kg/m²) | References |
|---|---|---|---|
| Central Brazil | 240–1083 | 0.43–1.04 | Schmidt et al. (2016) |
| Pantanal, Brazil | Not available | 0.3–0.5 | Cardoso et al. (2000) |
| North Australia | 500–18,000 | 1.5–13 | Williams et al. (1998) |
| Central Brazil | 1200–20,393 | 0.2–1.6 | Miranda et al. (2009) |
| Roraima, Brazil | 1256–16,394 | 0.2–0.4 | Barbosa and Fearnside (2005) |

generally depending on the factors such as the: (1) quantity and type of fuel influenced by the precipitation, previous fire events, and intensity of grazing; (2) moisture content of the fuel load (which depends on the season and the daytime; (3) prevailing climatic conditions; (4) fire spread; and (5) relief/slope: uphill fire is much more severe and more damaging than downhill fire.

## 16.3 FIRE-MEDIATED MODIFICATIONS IN SAVANNA ECOSYSTEMS

Fire is a vital component of savanna ecosystems, and has been shaping the biomes for centuries (Neary & Leonard, 2020). Fire has been known to cause alteration in ecosystems by influencing the abundance and structural patterns of flora in savannas. It also influences the diversity of fauna, as well as soil, air, and water resources that are important determinants of ecosystem structure and functions (Neary et al., 2005). Despite the various impacts of fire that have been reported, the ecological impact of fire events is a function of both the regime and the occurrence of single fires. Fire regime expresses the spatiotemporal impact of fire events and is regulated by multiple factors including climate, biomass properties, and frequency of burning. It describes properties such as type of fire, mean occurrence of fire, and variance in fire intensity, frequency, pattern, and season, as well as the extent of a burn. Fire types relate to those that consume the organic layers of the soil (ground fires), those that consume the above-ground components of the vegetation (surface fires), and those that affect the canopies of trees (crown fires). A mixed fire regime could occur where evidence of ground, surface, and crown fires are visible (Keane et al., 2002; Bond & Keane, 2017). Ground fires take place predominantly in areas with organic soils such as tropical regions and cause damaging effects to roots through direct destruction or complete alteration of soil properties on which root growth depends. Crown fires are high-severity fires characteristic of low-productivity landscapes such as boreal forests and Mediterranean shrublands of the United States. Surface fires are widespread in many wooded vegetation areas—including forests—where litter deposition is the primary fuel. Surface fires occur primarily in regions where tree litter and/or herbaceous fuels accumulate—as found in savannas—and can burn at annual or even sub-annual intervals, depending on the yield of the plant components (Keane et al., 2002). Fire induces a huge number of changes in the environment which in turn affect the structure and dynamics of vegetation (Keeley & Fotheringham, 2000). Fire impact on vegetation can be broadly classified as direct or indirect—depending on whether vegetation cover is directly impacted by burn incidents or those effects that arise from fire impacts on other components of the vegetation.

### 16.3.1 Direct Fire Impact on Savanna Vegetation

Fire can alter site conditions directly through seedling recruitment and thus offer the selective drive for fire-dependent germination of seeds (Bhadouria et al., 2016; 2017). Apart from stimulating the germination of seeds, the removal of vegetation cover by fire creates the gaps that increase irradiance at ground level (Figure 16.1) and reduce competition for the germinating seedlings (Keeley & Fotheringham, 2000). For some species, one of a number of disturbances may activate the germination of seeds and subsequent recruitment of seedlings, whereas fire-dependent species rely on fire stimulus for the recruitment of their seedlings. Fire may also trigger seed regeneration directly by breaking open serotinous cones/fruits or by triggering germination of dormant seeds in the soil seed bank. Fire may also indirectly influence the recruitment of seedling by opening up gaps in closed vegetation, thus allowing for conditions suitable for colonization. In this condition, the seeds are expected to both endure fire conditions and also utilize the gaps created by fire to begin germination. In fire-triggered germination, seeds are unconstrained from dormancy by heat and/or smoke, which also help in priming the seeds to germinate. In physiologically dormant seeds, heat and definite smoke chemicals help to release the seed from dormancy and trigger germination. These chemicals may be wrapped up in the dry seeds or imbibed from soil solution during the wet season (Pausas & Lamont, 2022).

Fire could also suppress or trigger flowering, seed release, germination, and resprouting. Suppressive impact happens when unprotected structures of plants are damaged by heat or flame (Lamont & He, 2017). Tree–grass coexistence in tropical savannas is a reflection of varied interacting factors, including fire disturbances, that help to maintain a lower tree-to-grass ratio by pruning/reducing trees to smaller size classes and promoting grass population

**FIGURE 16.1** A fire incident: (a) opened up the canopy of the trees, (b) stimulated the growth of the grasses, and (c) and increased their dominance (Photograph courtesy of S. Oyedeji).

(Bond & van Wilgen, 1996; Scholes & Archer, 1997; van Langevelde et al., 2003). There is no consensus on the factors that determine the structure and function of tropical savanna; however, the influence of disturbances (fire and/or grazing) in regulating the growth and recruitment of new species or stands in savannas cannot be overemphasized (Sankaran et al., 2004). Studies from the African savanna have shown fire to be a critical regulator that helps in shaping the vegetation (Swaine et al., 1992; Smit et al., 2010). The fire effect is critical for vegetation structure, especially in highly wooded landscapes compared to more open ones. Likewise, drier ecosystems are also likely to be more impacted than wetter ones (Smit et al., 2010). Exclusion of fire in tropical savannas may lead to a change in the vegetation structure by increasing the recruitment of trees (Keane et al., 2002), thus leading to close-up through thicket formation, especially in regions that border forest ecosystem (Swaine et al., 1992; Staver & Levin, 2012).

According to Simon and Pennington (2012), fire disturbance suppresses woody plants and favors herbaceous growth, especially grasses. Fire regimes in tropical savannas essentially pose critical challenges to perennial plants which have to endure the impact in order to survive. The diverse set of morphological traits in plants that allow them to survive in fire-prone tropical savannas includes below-ground biomass allocation coupled with resprout potential, thick corky bark that provides insulation and protection of internal tissues against heat, heat-resistant terminal branches (that are usually thick), dense leafy bases to reduce burning, and increased stipules formation to protect the shoot apices from fire. Some species also exhibit specific phenological and reproductive adaptive strategies such as early maturity and fire-triggered flowering to escape fire or cushion its impact (Hoffmann et al., 2003; Simon & Pennington, 2012). Fires may accelerate plant reproductive phenology by altering the surrounding vegetation (Paritsis et al., 2006). For example, fire synchronizes reproduction in *Echinacea angustifolia*, leading to an increased flower set that improves mating opportunities and eventual fruit set (Wagenius et al., 2020). This change in reproductive phenology is more likely to result from spatial changes in environmental conditions that alter microclimate—particularly temperature (Paritsis et al., 2006).

### 16.3.2 Indirect Fire Impact on Savanna Vegetation

Tree–grass co-existence in tropical savannas is the result of different interacting factors (Bond & van Wilgen, 1996; Scholes & Archer, 1997). Fire can alter these factors and cause a shift in the dynamics of the vegetation. One of the factors influenced by fire events is soil moisture availability—a major determinant for the development of trees and grasses in these systems (van Langevelde et al., 2003). Wildfires can trigger evaporative losses through the removal of the absorbent layers of deposited and decaying plant litter that cover the soil surface. These layers, called litter and duff, facilitate the storage of excess water after heavy rain, thereby reducing erosion. Fires can also harden the soil surfaces, thereby increasing runoff and sediment transport (Parlak, 2015). Burn incidents can also remove vegetation cover, leaving soils exposed and effortlessly erodible by precipitation.

Fire events can also create water-repellent soils. High surface heat from fire consumes organic matter and creates vapors that descend into the inner soil layer in response to a temperature gradient which later condenses on soil particles to create a water-repellent layer. Water-repellent soils have high water entry pressure, low infiltration, increase surface runoff, and nutrient and sediment transport and deposition. Fire-induced water-repellent soil could lead to low rate of establishment of grasses in the savanna ecosystem, as most of their seeds are washed off from the soil. Seed germination may also be impaired, as the soil will lack enough soil water to trigger the process (Letey, 2001).

Furthermore, fire can accelerate the nutrient release from organic matter and increase the availability of inorganic nutrients (Dunn & Debano, 1977; Oyedeji et al., 2021a). Low-intensity fire can increase soil fertility by facilitating the chemical conversion of nutrients locked up in plant litters and the soil surface, making it readily available for plants to uptake, or the fire indirectly triggers the rate of mineralization through its impacts on micro-organisms in the soil (Schoch & Binkley, 1986). High-force fire sterilizes the soil by altering microbial populations and plummeting pathogens that influence seed germination and development (Keeley & Fotheringham, 2000). Fire impact also reduces the population of herbivores directly by 'heat killing' and indirectly by exposing their hideouts (habitat) and making them more vulnerable to attack by predators (Quinn, 1994). The pulses of recruitment of predators reduce seedling predation and thus increase post-fire germination (Keeley & Fotheringham, 2000).

## 16.4 PLANT DIVERSITY IN FIRE-EXPOSED SAVANNAS

Fire influences ecosystems in many ways and ecosystems themselves strongly affect fire, chiefly through its feedback on the biomass, its condition, and its distribution. Fire is a disturbance because its impact and regularity is dependent on, among other things, the accumulated fuel load, often determined by the growth and decomposition of plants that it consumes (Steel et al., 2015). Wildfires have played a crucial role in shaping the composition and functioning of a number of ecosystems across the globe (Bond & Keeley, 2005; Bowman et al., 2009). Despite its importance in maintaining natural ecosystems, neither ecologists nor land managers have a full understating about the influences of wildfires on biodiversity and ecosystem services across complex landscapes with varying environmental conditions (Burkle et al., 2015).

It has been well known that the open canopy in savanna vegetation is maintained by fire (Bond & Keeley, 2005), and scientists have posited that fire regimes in the vegetation may behave accounted for the evolution of $C_4$ grasses (Beerling & Osborne, 2006). It is possible that fire may have also accounted for the richness and distribution of other plants and plant communities such as ferns and primitive flowering plants, as many of these have been found in sites associated with frequent fires (Friis et al., 2006). Fire can alter species composition, diversity, and structure of savanna vegetation through its influence on growth, recruitment, and mortality (Reis et al., 2015). In a post-fire study of savanna vegetation in Brazil, Reis et al. (2015) observed that some species (such as *Alchornea discolor*, *Andira cujabensis*, *Byrsonima coccolobifolia*, *Guapira graciliflora*, *Mezilaurus crassiramea*, and *Qualea grandiflora*) disappeared from the burned region while new ones (such as *Machaerium acutifolium*, *Vochysia haenkeana*, and *Tabebuia aurea*) appeared. Likewise, even though some species could escape fire impact, site conditions at post-fire favor others. Species such as *Senna obtusifolia*, *Oldenlandia herbacea*, and *Physalis micrantha* have been associated with fire events in the savanna vegetation of Nigeria (Oyedeji et al., 2021b). Features presented by fire-exposed ecosystems often include the following: (1) some species that are insensitive to fire impact often benefit from fire impact by increasing growth, diameter, or density after the fire. Other have new individuals recruited at post-fire; (2) plant traits often decrease after fire, as some characteristics are lost to the disturbance; and (3) higher mortality of particularly smaller-sized species occur. Wildfires greatly impact closed formations compared with more open formations, and the feature is common with savannas on the African continent (Smit et al., 2010).

### 16.4.1 Species Richness and Diversity

Disturbance—especially wildfire—is a major driver of diversity (Huston, 2014) because it facilitates occurrence of multiple species by delaying competitive exclusion. However, high-intensity fire may limit the recruitment of many species into the community. This is achieved as the heat from such fires reduces or completely destroys the seed bank of certain species in the soil, and in some situations, eliminates both aerial and subsurface organs used for resprouting/regeneration (dos Santos et al., 2020). Fire disturbance can also steer phenotypic and genetic divergence within species (Pausas, 2015) by delaying competitive exclusion, increasing heterogeneity within the landscape, and generating new niches for the diversity of species (Pausas & Ribeiro, 2017). Durigan et al. (2020) observed that even a single prescribed fire event after a long period without fire was enough to induce major changes in the structure of the vegetation—with a reduction in tree density mostly due to top-killing of adult trees by 23% in grasslands and 6% in savannas, and a reduction of tree basal-area by up to 30% in grasslands and 9% in savannas.

The frequency of fires has also been reported to impact on species diversity, as it can cause a reduction in both abundance and richness by influencing the quantity of deposited seeds (seed rain), accumulated seeds (seed bank), seedling emergence and survival, and resprout potentials of stands (dos Santos et al., 2020), and these effects are higher at vegetation boundaries that are often exposed to recurring fire events and spread. Apart from the direct fire impact on plants, fire influences diversity by affecting soil nutrients and other environmental conditions leading to the selection of plant species in different communities (Silva et al., 2013). Attributes of plant community that are influenced by soil nutrients include biomass, cover, and floristic composition—which directly affect diversity (Perroni-Ventura et al., 2006; Long et al., 2018). A post-fire study in Nigeria's savanna showed increases in organic carbon and exchangeable calcium and concomitant decreases in magnesium and total nitrogen (Oyedeji et al., 2021b). Jones et al. (2013) explained that nitrogen losses via volatilization occurred because of its increased solubility as ammonia at higher temperatures and subsequent loss of ammonia as gas. Studies, especially in the tropics, have shown species diversity to increase with soil organic matter and nutrients including nitrogen (N), phosphorus (P), and potassium (K) (Long et al., 2018; Oyedeji et al., 2021b). A shift in the soil nutrient status could therefore alter the species diversity over time by influencing species composition, especially for the dominant species (Long et al., 2018).

### 16.4.2 Wildfire and Plant Traits

Wildfires have also been reported to influence adaptive traits in plants (Pausas et al., 2004; Burkle et al., 2019). Adaptive traits are features that provide a fitness advantages to species in a given environment. Plants exhibit many features that help them to survive in the face of recurrent fire, and these features are very distinct with fire regimes. According to Pausas et al. (2004), the major plant traits associated with post-fire persistence are the resprouting capacity (a feature that shows the persistence of an individual plant species) and propagule-persistent capacity (a feature characteristic of a persistent population and linked to the soil seed bank). The existence of functional types among plants in ecosystems susceptible to fire also suggests the existence of some fundamental constraints or tradeoffs—e.g., vegetative vs. sexual regeneration (Keeley, 1986)—that border the potential blending of life-history traits (Pausas & Lavorel, 2003). Fire may serve as a filter preventing some plant traits (Diaz et al., 1998), causing different blends of traits to exist in vegetations with varying fire history (Keeley & Zedler, 1978), more especially as fire regimes are far from being constant (Pausas et al., 2004). No plant is adapted species to fire; however, adaptation to fire regimes exists. Traits such as resprouting (survival by bud regrowth), serotiny (seed preservation in cones and other protective structures until after fire), and germination induced by heat and smoke from fires are survival strategies for plants in fire-prone environments. Fire-adaptive traits are the products of different evolutionary pathways. However, it is difficult to separate the traits that originate due to adaptation in response to fire from those that are exaptations (which is used to describe adaptation[s] that originate in response to other factors) (Keeley et al., 2011).

Resprouting in response to fire episodes is an essential functional character among plants—especially woody species and tall grasses—in fire-prone ecosystems, and it serves as the basis for their persistence in disturbed ecosystems, such as savannas where recurring fire events occur. The potential of a plant to resprout is dictated by the production, safeguarding, and nourishment of a viable bud bank—from which regrowth can occur (Clarke et al., 2012). Classification for re-sprouts has focused mainly on the disturbance severity (Bellingham & Sparrow, 2000), or location of the bud bank in trees and shrubs (Del Tredici, 2001; Klimešová & Klimeš, 2003), and in herbs (Klimešová & Klimeš, 2003). The re-sprouted buds are located above ground (aerial or basal) or below ground (as found in geophytes) (Clarke et al., 2012). It is thought that bark protection alone is responsible for the distinct mosaic patterns of savanna and even forest vegetation in fire-prone regions of Africa. However, recent research in the Australian savanna ecosystem now shows that bud position, not bark protection alone, determines the resilience of plants to fire by means of resprouting (Charles-Dominique et al., 2015).

Fire could also alter plant traits such as the sizes, shapes, and thicknesses of organs. For example, Burkle et al. (2019) observed the plants in fire-ravaged plots to be taller and with higher numbers of flowers per flower head than in the unburned plots. The plants in the latter were with bigger flowers and higher specific leaf areas. Their findings imply that post-wildfire conditions encourage tall plants and flowering. These conditions were facilitated because wildfire cleared the canopy and created gaps for sunlight and penetration, thus allowing for the formation of thick and diverse understory plants (Swanson et al., 2010). The dense population of understory plants established to compete with each other for resources, thus accounting for the reduction in their traits. Gómez-González et al. (2011) observed that seed pubescence, shape, and pericarp thickness are key persistent population traits that are strongly associated with fire frequency. This happens because fire-selected individuals with features such as pubescent seeds, thick pericarp, and less rounded seeds from within a population pass on the traits to their offspring, as these traits have been shown to be heritable.

## 16.5 ADAPTATION STRATEGIES OF PLANT SPECIES TO FIRE EFFECTS

Fire has existed organically for ages, controlling the dimensional dispersion of biomes and maintaining the composition and functioning of smoke-sensitive environments (Scott, 2000; Bond & Keeley, 2005). As a result, there has evolved a huge mutative force for species residing lush grass-grown biomes (Keeley et al., 2011; Simon & Pennington, 2012) that has developed in excellent diversity and an entire reliance on burning in many places (Fidelis et al., 2019; Pilon et al., 2020).

Plant species respond to disturbances with varying degrees of adaptation. The degree of disturbances can strongly affect the survival rate and direction of plant species. Plant species have developed adaptive properties to fire disturbances. Plant responses to fire vary depending on several factors, but most importantly fire regime and plant adaptive properties, which determine their survival after a fire. Under the same condition of fire, some plants may die completely, while others may only be partly affected (Figure 16.2). The tolerance or susceptibility of plants in fire-ravaged ecosystems triggers a change in the vegetation (Wang, 1961; Jefferies et al., 1994). Plants exhibit varied responses in fire-ravaged ecosystems which are adaptive for survival. Resprouting offers an important mechanism for recruitment, especially in seasonally dry ecosystems like tropical savannas. Resprouters are plant species that endure fire and regrow through the activation of dormant vegetative buds. Resprout responses in plants may be apical, basal, soil (clonal), or epicormic. Apical sprouters are typically grow shorter than fire resisters (except for some palms), but their water-conducting tissues and apical buds are well sheltered to resist fire damage. They guard their apical meristems with firm clustered leaf primordia, as well as the bases of the mature leaves (Lamont et al., 2004; Clarke et al., 2012). After a fire event, most small plants resprout from the base. For instance, most palms exhibit clonal resprouting in savannas, although they may exhibit apical sprouting in forests. Woody plants exhibit two main strategies to escape from fire. Some plants escape by fast growth to reach a height above the reach of flame before the next fire. This way, the plants have a propensity to exhibit pole-like or lanky architecture and the inability to attain sufficient growth above the flame zone during the interfire period will lead to top-killing and subsequent growth is possible only from basal resprouting. Woody resprouters rejuvenate from buds located at or below ground surface (Bond & Midgley, 2001; Moreira et al., 2008), and their ability to resprout varies with species, fire regime, and region (Clarke et al., 2012). Woody basal resprouters are dominant in African savannas, while other savannas contain more of a combination of stem and basal resprouters (Hoffmann et al., 2003; Clarke et al., 2012).

In fire-prone ecosystems, the absence of basal buds does not limit the ability of plants to resprout, as frequent elimination of up-and-coming buds in lignotuberous species does not weaken the bud bank of the adult plants (Canadell & López-Soria, 1998; Wildy & Pate, 2002). Seedlings of lignotuberous plants quickly produce buds above what they require for resprouting (bud bank)—an adaptation against recurring fire events (Fidelis et al., 2010). Plant species adapted to frequently burned environments thus exhibit life-history features that guard against fire injury, allowing them to endure fires and re-establish themselves in large numbers—compared to species devoid of such traits—as fire frequency intensifies (Beckage & Stout, 2000). Fire-adapted species may also possess traits that amplify the chances and degree of fire events compared with fire-sensitive species (Ellair & Platt, 2012). For example, high levels of volatile organic compounds in the foliage of some plant species can increase combustibility. Also, the low leaf area index of many trees species in savannas can support grasses and other fine fuels that aid fire (Ratnam et al., 2011; Beckage et al., 2019).

### 16.5.1 Physiological Strategies

Plant physiological responses after fires often relate to conditions of adaptation to low soil moisture. This is due to the damaging consequences of fire on the soil structure, including

**FIGURE 16.2** Fire killed the entire plant (a), and damaged portions of the shoot of the plant (b) (Photograph courtesy of Sunday Adeniran).

those of reducing the permeability to water and deterioration of nutrient and organic matter levels, which ultimately affect the normal growth and development of the plant (Filipović, 2021). According to Beckage et al. (2019), water limitation in soil shifts the tree–grass competitive balance toward grasses because of the increase in water demand of trees species due to their higher leaf area index and low water-use efficiency, particularly compared with $C_4$ grasses dominant in tropical savannas. This water limitation on tree species leads to an indirect positive impact on grasses. Water limitations that reduce the formation of a closed woody canopy allow for adequate understory irradiance to sustain the growth of understory grasses—thus maintaining the savanna structure. The severity of post-fire water shortage in the soil is often dependent on the seasonal distribution of rainfall, including those of other environmental conditions (Scholes & Archer, 1997; Sankaran et al., 2004). For example, sandy soils allow rapid water infiltration into deeper soil horizons and thus facilitate tree–grass niche partitioning—grasses, because of their shallow roots, compete mainly for water in the upper portions of the soil layer, while trees will take up water from a deeper soil layer using their taproots (Walter, 1971; Casper & Jackson, 1997; Beckage et al., 2019). Conditions of low soil moisture may favor succulent plants—which have specialized water-use strategies to survive. Succulent plants exhibit different life forms—including annuals, perennials, shrubs, and trees (Filipović, 2021).

The resources (such as N and P) available in post-fire soil can also determine the magnitude at which plants can regrow (Chapin et al., 1990). In grasses (and other small plants), the resprouting pattern is determined by the available below-ground biomass that escapes fire (Figure 16.3). In tree species, thicker bark lowers the risk of fire damage to the inner tissue, and this will protect the buds and cambium from the ravaged fire. Post-fire responses of plants will be dependent on the extent of bud damage to twigs and branches, and the level of devastation to the shoot and root cambial cells, against the quantity of subversive reserves—that is, the amount of stored carbohydrates and additional mineral resources that are available and that can be mobilized to re-establish the destroyed biomass (Chapin et al., 1990; Bond & van Wilgen, 1996; Iwasa & Kubo, 1997; Bellingham & Sparrow, 2000). At moderate severity levels, two types of reaction may be seen. First, if the extent of damage is great, the buds may be destroyed either straightforwardly by heat or obliquely by the devastation of the vascular cambium in the stem—although at below-ground level, the physiological responsibility of carbohydrates remains to facilitate the development of anatomical structures (Del Tredici, 2001). Furthermore, apical dominance will be restrained straightforwardly through the demolition of buds by heat or obliquely through cambium damage (Miller, 2000; Kozlowski & Pallardy, 2002), and the seedling will accordingly need to counter this by way of basal resprouting.

Plants also require a high amount of energy in the form of carbon store during the post-fire regrowth stage (De Moraes et al., 2016). Under this condition, most plants rely upon the mobilization of food reserves in different storage organs such as underground stems, roots, leaves, and seeds. Studies have shown that reserve compounds contain carbohydrates (including monosaccharides [glucose and fructose], disaccharides [sucrose], and polysaccharides [fructans]) and nitrogen compounds such as proteins (Figueiredo-Ribeiro et al., 1986). This accumulation is of great importance, particularly in plants in which the catabolism of previously accumulated reserves rather than current assimilates is the major driver for growth (Hendry, 1993). Also, the rapid conversion of water-soluble carbohydrates, especially fructans, via polymerization and depolymerization, serve as an alternative mechanism for osmoregulation in plants (De Moraes et al., 2016). Storage substances are therefore important substances for the growth and survival of plants, even in stress-induced environments (Fujiwara et al., 2002). This is evident in a lot of resprouting savanna herbaceous species, where the accumulated reserves are enough for them to survive the top-killing effects of fire and achieve rapid post-fire succession (Diaz-Toribio & Putz, 2021).

### 16.5.2 Anatomical Strategies

In surface fire regimes such as that typical of tropical savannas, the first requirement for trees to survive fires is to develop a thick bark, especially at juvenile stages (Lawes et al., 2011). These traits have not been demonstrated in the forest, except for a few cases when shrubs and smaller trees regrow (Bond & Midgley, 2003). The stem/trunk cambium becomes less vulnerable to fire damage due to the protection offered by the bark, and plant species might survive fires. Excessive fire will cause deterioration of the bark, which will eventually affect the trunk cambium. In this instance, damage to the crown buds results either by direct heat killing or indirectly through the destruction of the vascular cambium in the stem (Del Tredici, 2001). The adaptation and survival of consumed foliage, buds, and twigs is dependent upon the intensity of the fire and the phase of fire prevention at the individual level (Bond & van Wilgen, 1996). When the fire damage level is reduced and the stem cambium is invulnerable because of the dense bark, the plant will necessarily re-emerge from the fire-ravaged disturbance, but extreme damage occurs where the cambium is most severed and the likely outcome is plant death. Furthermore, apical growth will be minimized directly due to bud damage resulting from direct heat injury or indirectly through cambium damage (Kozlowski, 1971; Kozlowski et al., 1991; Miller, 2000). The tree will in turn respond through regrowth from basal parts (basal resprouting). On the other hand, if the extent of damage is not so severe (e.g., fractional injury of the crown buds and cambium), apical growth is only weakened (Kozlowski, 1971) and there is at least some accessibility to underground reserves that will facilitate simultaneous resprouting of the crown and cambium.

Durigan et al. (2020) reported that younger savanna trees possess thicker bark compared to forest tree species for a

**FIGURE 16.3** Resprouting of grasses from below-ground organs after a fire (Photograph courtesy of Sunday Adeniran).

given trunk diameter. This strategy ensures that the xylem is insulated from the heat of fires (Balfour & Midgley, 2006; Michaletz, 2018). Plants use a variety of schemes to cope with fire output. Wood varieties such as pitch pine and chestnut oak produce a dense bark that insulates the wood and protects it against surface fires. Trees with finer bark, such as highbush cranberry and walnut trees, are more susceptible to damage from surface fires. Grasses, aspen, and scrub oaks use a rescue policy following a fire accident. Although the above-ground part of the plant is ruptured, the roots and stems can grow quickly after a fire and relocate the affected portion. Other classes of plant species help to stabilize scorched areas, e.g., fire provokes the germination of mesquite and ceanothus seeds. Dickison (2000) identified the following modifications in the anatomical features: development of denticles on the midribs of leaves, increases in the density of vascular tissues in the leaf and stem, enhanced lignification of cell walls, and relative presence of sclerenchymatous tissues.

Of all the modes of disturbances, however, none pose a greater impact on shaping the evolution of plant species and the changes in the environment like fire. Even after the maximum trenchant fire damage, plants can re-establish their communities. Ground fires are feasible above ground, frequently blaze fast, and are not remarkably hot along the ground, consuming leaf mud, bushes, and various small plants. The soil insulates and stops the deep heating of surface fires and other negative impacts on roots, tubers, bulbs, or plant organs that are concealed underground. While crown fires are moving, the directionality is predictable and the impacts are visible. With large amounts of fuel on the ground, surface fires can blaze vehemently enough to burn branches and add up to crown fires. Crown fires blaze remarkably hot and can even produce strong winds that quickly blaze the vegetation.

The commonness and force of fires vary among ecosystems. In grasslands and coniferous woods, decomposition is slow and minerals are held up in the ground cover. Fires quickly help to loosen the nutrients and make them accessible to the plants. Fire poses damaging effects to plants and even kills inactive seeds in the soil, thus limiting the potential for seedling recruitment. Fire impacts in dry tropical ecosystems, such as savannas, have been worsened by drought-related stress imposed global warming and climate change. These conditions have increasingly induced stand mortality, vegetation dryness, and flammable conditions (Allen et al., 2010, 2015) across savannas. While change in fire regimes can independently compel significant transformation in plant communities and ecological systems, the direct impact of warm weather conditions (climate change) can compel additional stress on major demographic processes (including survival, growth, and reproductive rates) both before and after fires. Environmental conditions of droughts and heat waves can pose serious threats to the preservation of plant diversity in fire-adapted vegetations, thus raising concerns of ecosystem alteration and transformation (Harvey & Enright, 2022).

## 16.6 CONCLUSION

Fire is a common feature in savanna vegetations across different continents of the world, and the disturbance has been responsible for shaping the biome. The fire regime in savannas is linked to conditions imposed by climate, soil, and/or plant communities. The impact of fire on plants in savannas could be direct (including effects on plant structures, germination, growth, resprouting, flowering, and seed development) or indirect (such as the effect on soil moisture, structure, nutrients, temperature, and other microclimatic conditions). Fire impacts in savannas have been worsened by increasing hot and dry conditions imposed by the changing climate. Although plants in savannas possess diverse strategies (anatomical, physiological, and phenological) to survive recurring fire disturbances, increasing heat waves and droughts brought about by the changing climate are compelling modification of plant adaptive traits/strategies in the biome. Such modifications may alter post-fire survival, growth, and regeneration potentials of plant populations, communities, and ecosystems, while raising serious concerns for the diversity and conservation of flora in savannas.

## REFERENCES

Allen, C. D., Breshears, D. D., & McDowell, N. G. (2015). On underestimation of global vulnerability to tree mortality and forest die-off from hotter drought in the anthropocene. *Ecosphere, 6*, art129. https://doi.org/10.1890/ES15-00203.1

Allen, C. D., Macalady, A. K., Chenchouni, H., Bachelet, D., McDowell, N., Vennetier, M., Kitzberger, T., Rigling, A., Breshears, D. D., Hogg, E. H., Gonzalez, P., Fensham, R., Zhang, Z., Castro, J., Demidova, N., Lim, J.-H., Allard, G., Running, S. W., Semerci, A., & Cobb, N. (2010). A global overview of drought and heat-induced tree mortality reveals emerging climate change risks for forests. *Forest Ecology and Management, 259*, 660–684. https://doi.org/10.1016/j.foreco.2009.09.001

Archibald, S. (2016). Managing the human component of fire regimes: Lessons from Africa. *Philosophical Transactions of the Royal Society of London. Series B, Biological Sciences, 371*(1696), 20150346. https://doi.org/10.1098/rstb.2015.0346

Balfour, D. A., & Midgley, J. J. (2006). Fire induced stem death in an African acacia is not caused by canopy scorching. *Austral Ecology, 31*(7), 892–896. https://doi.org/10.1111/j.1442-9993.2006.01656.x

Barbosa, R. I., & Fearnside, P. M. (2005). Above-ground biomass and the fate of carbon after burning in the savannas of Roraima, Brazilian Amazonia. *Forest Ecology and Management, 216*(1–3), 295–316. https://doi.org/10.1016/j.foreco.2005.05.042

Beckage, B., Bucini, G., Gross, L. J., Platt, W. J., Higgins, S. I., Fowler, N. L., Slocum, M. G., & Farrior, C. (2019). Water limitation, fire, and savanna persistence. *Savanna Woody Plants and Large Herbivores*, 643–659. https://doi.org/10.1002/9781119081111.ch19

Beckage, B., Platt, W. J., & Gross, L. J. (2009). Vegetation, fire, and feedbacks: A disturbance-mediated model of savannas. *The American Naturalist, 174*(6), 805–818. https://doi.org/10.1086/648458

Beckage, B., Platt, W. J., Slocum, M. G., & Panko, B. (2003). Influence of the el niño southern oscillation on fire regimes in the Florida everglades. *Ecology, 84*(12), 3124–3130. https://doi.org/10.1890/02-0183

Beckage, B., & Stout, I. J. (2000). Effects of repeated burning on species richness in a Florida pine savanna: A test of the intermediate disturbance hypothesis. *Journal of Vegetation Science, 11*(1), 113–122. https://doi.org/10.2307/3236782

Beerling, D. J., & Osborne, C. P. (2006). The origin of the savanna biome. *Global Change Biology, 12*, 2023–2031.

Bellingham, P. J., & Sparrow, A. D. (2000). Resprouting as a life history strategy in woody plant communities. *Oikos, 89*(2), 409–416. https://doi.org/10.1034/j.1600-0706.2000.890224.x

Bhadouria, R., Singh, R., Srivastava, P., & Raghubanshi, A. S. (2016). Understanding the ecology of tree-seedling growth in dry tropical environment: A management perspective. *Energy, Ecology and Environment, 1*(5), 296–309. https://doi.org/10.1007/s40974-016-0038-3

Bhadouria, R., Srivastava, P., Singh, R., Tripathi, S., Siingh, H., & Raghubanshi, A. S. (2017). Tree seedling establishment in dry tropics: An urgent need for interaction studies. *Environment, Systems and Decisions, 37*(1), 88–100. https://doi.org/10.100710669-017-9625-x

Bloesch, U. (1999). Fire as a tool in the management of a savanna/dry forest reserve in Madagascar. *Applied Vegetation Science, 2*(1), 117–124. https://doi.org/10.2307/1478888

Bond, W. J., & Keane, R. E. (2017). Fires, ecological effects of ☆. In *Reference Module in Life Sciences*. Elsevier. https://doi.org/10.1016/b978-0-12-809633-8.02098-7

Bond, W. J., & Keeley, J. E. (2005). Fire as a global "herbivore": The ecology and evolution of flammable ecosystems. *Trends in Ecology and Evolution, 20*(7), 387–394. https://doi.org/10.1016/j.tree.2005.04.025

Bond, W. J., & Midgley, G. F. (2012). Carbon dioxide and the uneasy interactions of trees and savannah grasses. *Philosophical Transactions of the Royal Society of London. Series B, Biological Sciences, 367*(1588), 601–612. https://doi.org/10.1098/rstb.2011.0182

Bond, W. J., & Midgley, J. J. (2001). Ecology of sprouting in woody plants: The persistence niche. *Trends in Ecology & Evolution, 16*(1), 45–51. https://doi.org/10.1016/s0169-5347(00)02033-4

Bond, W. J., & Midgley, J. J. (2003). The evolutionary ecology of sprouting in woody plants. *International Journal of Plant Sciences, 164*(S3), S103–S114. https://doi.org/10.1086/374191

Bond, W. J., & van Wilgen, B. W. (1996). *Fire and Plants*. Springer. https://doi.org/10.1007/978-94-009-1499-5

Bowman, D. M. J. S., Balch, J. K., Artaxo, P., Bond, W. J., Carlson, J. M., Cochrane, M. A., D'Antonio, C. M., DeFries, R. S., Doyle, J. C., Harrison, S. P., Johnston, F. H., Keeley, J. E., Krawchuk, M. A., Kull, C. A., Marston, J. B., Moritz, M. A., Prentice, I. C., Roos, C. I., Scott, A. C., . . . & Pyne, S. J. (2009). Fire in the earth system. *Science, 324*(5926), 481–484. https://doi.org/10.1126/science.1163886

Bradstock, R. A. (2010). A biogeographic model of fire regimes in Australia: Current and future implications. *Global Ecology and Biogeography, 19*(2), 145–158. https://doi.org/10.1111/j.1466-8238.2009.00512.x

Burkle, L. A., Myers, J. A., & Belote, R. T. (2015). Wildfire disturbance and productivity as drivers of plant species diversity across spatial scales. *Ecosphere, 6*(10), art202. https://doi.org/10.1890/es15-00438.1

Burkle, L. A., Simanonok, M. P., Durney, J. S., Myers, J. A., & Belote, R. T. (2019). Wildfires influence abundance, diversity, and intraspecific and interspecific trait variation of native bees and flowering plants across burned and unburned landscapes. *Frontiers in Ecology and Evolution, 7*. https://doi.org/10.3389/fevo.2019.00252

Cahoon, D. R., Stocks, B. J., Levine, J. S., Cofer, W. R., & O'Neill, K. P. (1992). Seasonal distribution of African savanna fires. *Nature, 359*(6398), 812–815. https://doi.org/10.1038/359812a0

Canadell, J., & López-Soria, L. (1998). Lignotuber reserves support regrowth following clipping of two Mediterranean shrubs. *Functional Ecology, 12*(1), 31–38. https://doi.org/10.1046/j.1365-2435.1998.00154.x

Cardoso, E. L., Crispim, S. M. A., Rodrigues, C. A. G., & Barioni Junior, W. (2000). Biomassa aérea e produção primária do estrato herbáceo em campo de Elyonurus muticus submetido à queima anual, no Pantanal. *Pesquisa Agropecuária Brasileira, 35*(8), 1501–1507. https://doi.org/10.1590/s0100-204x2000000800001

Casper, B. B., & Jackson, R. B. (1997). Plant competition underground. *Annual Review of Ecology and Systematics, 28*(1), 545–570. https://doi.org/10.1146/annurev.ecolsys.28.1.545

Chapin, F. S., Schulze, E., & Mooney, H. A. (1990). The ecology and economics of storage in plants. *Annual Review of Ecology and Systematics, 21*(1), 423–447. https://doi.org/10.1146/annurev.es.21.110190.002231

Charles-Dominique, T., Beckett, H., Midgley, G. F., & Bond, W. J. (2015). Bud protection: A key trait for species sorting in a forest–savanna mosaic. *New Phytologist, 207*(4), 1052–1060. https://doi.org/10.1111/nph.13406

Cheney, N. P. (1981). Fire behaviour. In A. M. Gill, R. H. Groves, & I. R. Noble (Eds.), *Fire and the Australian Biota*. Australian Academy of Science.

Clarke, P. J., Lawes, M. J., Midgley, J. J., Lamont, B. B., Ojeda, F., Burrows, G. E., Enright, N. J., & Knox, K. J. E. (2012). Resprouting as a key functional trait: How buds, protection and resources drive persistence after fire. *New Phytologist, 197*(1), 19–35. https://doi.org/10.1111/nph.12001

Cole, K. L., Klick, K. F., & Pavlovic, N. B. (1992). Fire temperature monitoring during experimental burns at Indiana Dunes National Lakeshore. *Natural Areas Journal, 12*, 177–183.

Curt, T. (2018). Fire frequency. In *Encyclopedia of Wildfires and Wildland-Urban Interface (WUI) Fires* (pp. 1–5). Springer International Publishing. https://doi.org/10.1007/978-3-319-51727-8_110-1

De Moraes, M. G., De Carvalho, M. A. M., Franco, A. C., Pollock, C. J., & Figueiredo-Ribeiro, R. D. C. L. (2016). Fire and drought: Soluble carbohydrate storage and survival mechanisms in herbaceous plants from the Cerrado. *BioScience, 66*(2), 107–117. https://doi.org/10.1093/biosci/biv178

De Ronde, C., Trollope, W. S. W., Parr, C. L., Brockett, B. H., & Geldenhuys, C. J. (2004). Fire effects on flora and fauna. In J. G. Goldhammer, & C. De Ronde (Eds.), *Wildland Fire Management: Handbook for Sub-Sahara Africa* (pp. 60–87). Global Fire Monitoring Centre.

Del Tredici, P. (2001). Sprouting in temperate trees: A morphological and ecological review. *The Botanical Review, 67*(2), 121–140. https://doi.org/10.1007/bf02858075

Diaz, S., Cabido, M., & Casanoves, F. (1998). Plant functional traits and environmental filters at a regional scale. *Journal of Vegetation Science, 9*(1), 113–122. https://doi.org/10.2307/3237229

Diaz-Toribio, M. H., & Putz, F. E. (2021). Underground carbohydrate stores and storage organs in fire-maintained longleaf pine savannas in Florida, USA. *American Journal of Botany, 108*(3), 432–442. https://doi.org/10.1002/ajb2.1620

Dickison, W. C. (2000). Ecological anatomy. In *Integrative Plant Anatomy* (pp. 295–337). Elsevier. https://doi.org/10.1016/b978-012215170-5/50009-1

dos Santos, C. R. T., Montibeller-Santos, C., Balch, J. K., Brando, P. M., & Torezan, J. M. D. (2020). Effects of fire frequency on seed sources and regeneration in southeastern Amazonia. *Frontiers in Forests and Global Change, 3*. https://doi.org/10.3389/ffgc.2020.00082

Dunn, P. H., & Debano, L. F. (1977). Fire's effect on biological and chemical properties of chaparral soils. In H. A. Mooney, & C. E. Conrad (Eds.), *Symposium on the Environmental Consequences of Fire and Fuel Management in Mediterranean Ecosystems* (pp. 75–84). ESDA Forest Service General Technical Report.

Durigan, G., Pilon, N. A. L., Abreu, R. C. R., Hoffmann, W. A., Martins, M., Fiorillo, B. F., Antunes, A. Z., Carmignotto, A. P., Maravalhas, J. B., Vieira, J., & Vasconcelos, H. L. (2020). No net loss of species diversity after prescribed fires in the Brazilian savanna. *Frontiers in Forests and Global Change, 3*. https://doi.org/10.3389/ffgc.2020.00013

Ellair, D. P., & Platt, W. J. (2012). Fuel composition influences fire characteristics and understorey hardwoods in pine savanna. *Journal of Ecology, 101*(1), 192–201. https://doi.org/10.1111/1365-2745.12008

Fidelis, A., Müller, S. C., Pillar, V. D., & Pfadenhauer, J. (2010). Population biology and regeneration of forbs and shrubs after fire in Brazilian Campos grasslands. *Plant Ecology, 211*(1), 107–117. https://doi.org/10.1007/s11258-010-9776-z

Fidelis, A., Rosalem, P., Zanzarini, V., Camargos, L. S., & Martins, A. R. (2019). From ashes to flowers: A savanna sedge initiates flowers 24 h after fire. *Ecology, 100*(5), e02648. https://doi.org/10.1002/ecy.2648

Figueiredo-Ribeiro, R. C. L., Dietrich, S. M., Chu, E. P., Carvalho, M. A. M., Vieira, C. C. J., & Graziano, T. (1986). Reserve carbohydrates in underground organs of native Brazilian plants. *Revista Brasileira de Botânica, 9*, 159–166.

Filipović, A. (2021). Water plant and soil relation under stress situations. In *Soil Moisture Importance*. IntechOpen. https://doi.org/10.5772/intechopen.93528

Friis, E. M., Pedersen, K. R., & Crane, P. R. (2006). Cretaceous angiosperm flowers: Innovation and evolution in plant reproduction. *Palaeogeography, Palaeoclimatology, Palaeoecology, 232*(2–4), 251–293. https://doi.org/10.1016/j.palaeo.2005.07.006

Frost, P. G. H. (1985). Organic matter and nutrient dynamics in a broadleafed African savanna. In J. C. Tothill, & J. J. Mott (Eds.), *Ecology and Management of the World's Savannas* (pp. 200–206). Australian Academy of Science.

Fujiwara, T., Nambara, E., Yamagishi, K., Goto, D. B., & Naito, S. (2002). Storage proteins. In *The Arabidiopsis Book* (Vol. e0020). American Society of Plant Biologists. https://doi.org/10.1199/tab.0020

Gill, A. M., Bradstock, R. A., & Williams, J. E. (2002). Fire regimes and biodiversity: Legacy and vision. In *Flammable Australia: The Fire Regimes and Biodiversity of a Continent* (pp. 429–446). Cambridge University Press.

Gill, A. M., Ryan, P. G., Moore, P. H. R., & Gibson, M. (2000). Fire regimes of world heritage kakadu national park, Australia. *Austral Ecology, 25*(6), 616–625. https://doi.org/10.1046/j.1442-9993.2000.01061.x

Gillon, D. (1983). The fire problem in tropical savannas. In F. Bourlière (Ed.), *Tropical Savannas. Ecosystems of the World* (Vol. 13, pp. 617–641). Elsevier.

Gomes, L., Miranda, H. S., & Bustamante, M. M. da C. (2018). How can we advance the knowledge on the behavior and effects of fire in the Cerrado biome? *Forest Ecology and Management, 417*, 281–290. https://doi.org/10.1016/j.foreco.2018.02.032

Gómez-González, S., Torres-Díaz, C., Bustos-Schindler, C., & Gianoli, E. (2011). Anthropogenic fire drives the evolution of seed traits. *Proceedings of the National Academy of Sciences of the United States of America, 108*(46), 18743–18747. https://doi.org/10.1073/pnas.1108863108

Govender, N., Trollope, W. S. W., & van Wilgen, B. W. (2006). The effect of fire season, fire frequency, rainfall and management on fire intensity in savanna vegetation in South Africa. *Journal of Applied Ecology, 43*(4), 748–758. https://doi.org/10.1111/j.1365-2664.2006.01184.x

Groen, T. A., van Langevelde, F., van de Vijver, C. A. D. M., Govender, N., & Prins, H. H. T. (2008). Soil clay content and fire frequency affect clustering in trees in South African savannas. *Journal of Tropical Ecology, 24*(3), 269–279. https://doi.org/10.1017/s0266467408004872

Harrison, S. P., Marlon, J. R., & Bartlein, P. J. (2010). Fire in the earth system. In Dodson, J. (Ed.), *Changing Climates, Earth Systems and Society* (pp. 21–48). Springer. https://doi.org/10.1007/978-90-481-8716-4_3

Harvey, B. J., & Enright, N. J. (2022). Climate change and altered fre regimes: Impacts on plant populations, species, and ecosystems in both hemispheres. *Plant Ecology, 223*, 699–709. https://doi.org/10.1007/s11258-022-01248-3

Hendry, G. A. F. (1993). Evolutionary origins and natural functions of fructans: A climatological, biogeographic, and mechanistic appraisal. *New Phytologist, 123*, 3–14.

Hoffmann, W. A., Orthen, B., & Nascimento, P. K. V. do. (2003). Comparative fire ecology of tropical savanna and forest trees. *Functional Ecology, 17*(6), 720–726. https://doi.org/10.1111/j.1365-2435.2003.00796.x

Huntley, B. J., & Walker, B. H. (1982). Conclusion: Characteristic Features of Tropical Savannas. In Huntley, B. J., & Walker, B. H. (Eds.), *Ecology of Tropical Savannas* (pp. 657–660). *Ecological Studies*, vol 42. Springer, Berlin, Heidelberg. https://doi.org/10.1007/978-3-642-68786-0_30

Huston, M. A. (2014). Disturbance, productivity, and species diversity: Empiricism vs. logic in ecological theory. *Ecology, 95*, 2382–2396. https://doi.org/10.1890/13-1397.1

Iwasa, Y., & Kubo, T. (1997). Optimal size of storage for recovery after unpredictable disturbances. *Evolutionary Ecology, 11*(1), 41–65. https://doi.org/10.1023/a:1018483429029

Jefferies, R. L., Klein, D. R., & Shaver, G. R. (1994). Vertebrate herbivores and northern plant communities: Reciprocal influences and responses. *Oikos, 71*(2), 193. https://doi.org/10.2307/3546267

Johansson, M. U., Abebe, F. B., Nemomissa, S., Bekele, T., & Hylander, K. (2021). Alien plant dynamics following fire in mediterranean-climate California shrublands. *Ambio, 50*(1), 190–202. https://doi.org/10.1007/s13280-020-01343-7

Jones, C., Brown, B. D., Engel, R., Horneck, D., & Olson-Ru, K. (2013). Factors affecting nitrogen fertilizer volatilization. In *Management to Minimize Nitrogen Fertilizer Volatilization, Montana State University, Extension Publication EB0802*. Montana State University. http://landresources.montana.edu/soilfertility/

Kahiu, M. N., & Hanan, N. P. (2018). Fire in sub-Saharan Africa: The fuel, cure and connectivity hypothesis. *Global Ecology and Biogeography, 27*(8), 946–957. https://doi.org/10.1111/geb.12753

Keane, R. E., Ryan, K. C., Veblen, T. T., Allen, C. D., Logan, J., & Hawkes, B. (2002). *Cascading Effects of Fire Exclusion in the Rocky Mountain Ecosystems: A Literature Review*. U.S. Department of Agriculture, Forest Service, Rocky Mountain Research Station. https://doi.org/10.2737/rmrs-gtr-91

Keeley, J. E. (1986). Resilience of mediterranean shrub communities to fires. In *Resilience in Mediterranean-Type Ecosystems* (pp. 95–112). Springer. https://doi.org/10.1007/978-94-009-4822-8_7

Keeley, J. E. (2009). Fire intensity, fire severity and burn severity: A brief review and suggested usage. *International Journal of Wildland Fire, 18*(1), 116. https://doi.org/10.1071/wf07049

Keeley, J. E., Baer-Keeley, M., & Fotheringham, C. J. (2005). Alien plant dynamics following fire in mediterranean-climate California shrublands. *Ecological Applications, 15*(6), 2109–2125. https://doi.org/10.1890/04-1222

Keeley, J. E., & Fotheringham, C. J. (2000). Role of fire in regeneration from seed. In *Seeds: The Ecology of Regeneration in Plant Communities* (pp. 311–330). CABI Publishing. https://doi.org/10.1079/9780851994321.0311

Keeley, J. E., Pausas, J. G., Rundel, P. W., Bond, W. J., & Bradstock, R. A. (2011). Fire as an evolutionary pressure shaping plant traits. *Trends in Plant Science, 16*(8), 406–411. https://doi.org/10.1016/j.tplants.2011.04.002

Keeley, J. E., & Zedler, P. H. (1978). Reproduction of chaparral shrubs after fire: A comparison of sprouting and seedling strategies. *American Middle Naturalist, 99*, 142–161.

Kellman, M. (1984). Synergistic relationships between fire and low soil fertility in neotropical savannas: A hypothesis. *Biotropica, 16*(2), 158. https://doi.org/10.2307/2387850

Klimešová, J., & Klimeš, L. (2003). Resprouting of herbs in disturbed habitats: Is it adequately described by Bellingham-Sparrow's model? *Oikos*, *103*(1), 225–229. https://doi.org/10.1034/j.1600-0706.2003.12725.x

Knapp, E. E., Estes, B. L., & Skinner, C. N. (2009). *Ecological Effects of Prescribed Fire Season: A Literature Review and Synthesis for Managers*. U.S. Department of Agriculture, Forest Service, Pacific Southwest Research Station. https://doi.org/10.2737/psw-gtr-224

Kozlowski, T. T. (1971). Structral and growth characteristics of trees. In *Seed Germination, Ontogeny, and Shoot Growth*. Elsevier. https://doi.org/10.1016/b978-0-12-424201-2.50007-9

Kozlowski, T. T., Kramer, P. J., & Pallardy, S. G. (1991). *The Physiological Ecology of Woody Plants*. Academic Press. https://doi.org/10.1016/b978-0-12-424160-2.50004-5

Kozlowski, T. T., & Pallardy, S. G. (2002). Acclimation and adaptive responses of woody plants to environmental stresses. *The Botanical Review*, *68*(2), 270–334. https://doi.org/10.1663/0006-8101(2002)068[0270:aaarow]2.0.co;2

Lamont, B. B., & He, T. (2017). Fire-proneness as a prerequisite for the evolution of fire-adapted traits. *Trends in Plant Science*, *22*(4), 278–288. https://doi.org/10.1016/j.tplants.2016.11.004

Lamont, B. B., Wittkuhn, R., & Korczynskyj, D. (2004). Ecology and ecophysiology of grasstrees. *Australian Journal of Botany*, *52*(5), 561. https://doi.org/10.1071/bt03127

Laris, P. (2013). Integrating land change science and savanna fire models in West Africa. *Land*, *2*(4), 609–636. https://doi.org/10.3390/land2040609

Lawes, M. J., Richards, A., Dathe, J., & Midgley, J. J. (2011). Bark thickness determines fire resistance of selected tree species from fire-prone tropical savanna in north Australia. *Plant Ecology*, *212*(12), 2057–2069. https://doi.org/10.1007/s11258-011-9954-7

Lehmann, C. E. R., Prior, L. D., & Bowman, D. M. J. S. (2009). Fire controls population structure in four dominant tree species in a tropical savanna. *Oecologia*, *161*(3), 505–515. https://doi.org/10.1007/s00442-009-1395-9

Letey, J. (2001). Causes and consequences of fire-induced soil water repellency. *Hydrological Processes*, *15*(15), 2867–2875. https://doi.org/10.1002/hyp.378

Long, C., Yang, X., Long, W., Li, D., Zhou, W., & Zhang, H. (2018). Soil Nutrients Influence Plant Community Assembly in Two Tropical Coastal Secondary Forests. *Tropical Conservation Science*, *11*, 194008291881795. https://doi.org/10.1177/1940082918817956

Lonsdale, W. M., & Braithwaite, R. W. (1991). Assessing the effects of fire on vegetation in tropical savannas. *Austral Ecology*, *16*(3), 363–374. https://doi.org/10.1111/j.1442-9993.1991.tb01064.x

Lutz, J. A., Key, C. H., Kolden, C. A., Kane, J. T., & van Wagtendonk, J. W. (2011). Fire frequency, area burned, and severity: A quantitative approach to defining a normal fire year. *Fire Ecology*, *7*(2), 51–65. https://doi.org/10.4996/fireecology.0702051

Marlon, J. R., Bartlein, P. J., Carcaillet, C., Gavin, D. G., Harrison, S. P., Higuera, P. E., Joos, F., Power, M. J., & Prentice, I. C. (2008). Climate and human influences on global biomass burning over the past two millennia. *Nature Geoscience*, *1*(10), 697–702. https://doi.org/10.1038/ngeo313

Mermoz, M., Kitzberger, T., & Veblen, T. T. (2005). Landscape influences on occurrence and spread of wildfires in patagonian forests and shrublands. *Ecology*, *86*(10), 2705–2715. https://doi.org/10.1890/04-1850

Michaletz, S. T. (2018). Xylem dysfunction in fires: Towards a hydraulic theory of plant responses to multiple disturbance stressors. *New Phytologist*, *217*(4), 1391–1393. https://doi.org/10.1111/nph.15013

Miller, M. (2000). Fire autecology. In J. K. Brown, & J. K. Smith (Eds.), *Wildland Fire in Ecosystems: Effects of Fire on Flora. Gen Tech Rep RMRS-GTR-42* (Vol. 2, pp. 9–34). Department of Agriculture, Forest Service, Rocky Mountain Research Station.

Miranda, H. S., Sato, M. N., Neto, W. N., & Aires, F. S. (2009). Fires in the cerrado, the Brazilian savanna. In *Tropical Fire Ecology* (pp. 427–450). Springer Berlin Heidelberg. https://doi.org/10.1007/978-3-540-77381-8_15

Moreira, F., Catry, F., Duarte, I., Acácio, V., & Silva, J. S. (2008). A conceptual model of sprouting responses in relation to fire damage: An example with cork oak (Quercus suber L.) trees in Southern Portugal. *Plant Ecology*, *201*(1), 77–85. https://doi.org/10.1007/s11258-008-9476-0

Murphy, B. P., & Bowman, D. M. J. S. (2012). What controls the distribution of tropical forest and savanna? *Ecology Letters*, *15*(7), 748–758. https://doi.org/10.1111/j.1461-0248.2012.01771.x

Neary, D. G., Ryan, K. C., & DeBano, L. F. (2005). *Wildland Fire in Ecosystems: Effects of Fire on Soils and Water*. U.S. Department of Agriculture, Forest Service, Rocky Mountain Research Station. https://doi.org/10.2737/rmrs-gtr-42-v4

Neary, G. D., & Leonard, M. J. (2020). Effects of fire on grassland soils and water: A review. In *Grasses and Grassland Aspects*. IntechOpen. https://doi.org/10.5772/intechopen.90747

Oyedeji, S., Agboola, O. O., Animasaun, D. A., Ogunkunle, C. O., & Fatoba, P. O. (2021a). Organic carbon, nitrogen and phosphorus enrichment potentials from litter fall in selected greenbelt species during a seasonal transition in Nigeria's savanna. *Tropical Ecology*, *62*(4), 580–588. https://doi.org/10.1007/s42965-021-00172-3

Oyedeji, S., Agboola, O. O., Oriolowo, T. S., Animasaun, D. A., Fatoba, P. O., & Isichei, A. O. (2021b). Early-season effects of wildfire on soil nutrients and weed diversity in two plantations. *Scientia Agriculturae Bohemica*, *52*(1), 1–10. https://doi.org/10.2478/sab-2021-0001

Paritsis, J., Raffaele, E., & Veblen, T. (2006). Vegetation disturbance by fire affects plant reproductive phenology in a shrubland community in northwestern Patagonia, Argentina. *New Zealand Journal of Ecology*, *30*(3), 387–395.

Parlak, M. (2015). Effects of wildfire on runoff and soil erosion in the southeastern Marmara region, Turkey. *Ekoloji*, *24*(94), 43–48. https://doi.org/10.5053/ekoloji.2015.946

Pausas, J. G. (2015). Alternative fire-driven vegetation states. *Journal of Vegetation Science*, *26*, 4–6.

Pausas, J. G., Bradstock, R. A., Keith, D. A., & Keeley, J. E. (2004). Plant functional traits in relation to fire in crown-fire ecosystems. *Ecology*, *85*(4), 1085–1100. https://doi.org/10.1890/02-4094

Pausas, J. G., & Lamont, B. B. (2022). Fire-released seed dormancy—a global synthesis. *Biological Reviews of the Cambridge Philosophical Society*, *97*(4), 1612–1639. https://doi.org/10.1111/brv.12855

Pausas, J. G., & Lavorel, S. (2003). A hierarchical deductive approach for functional types in disturbed ecosystems. *Journal of Vegetation Science*, *14*(3), 409–416. https://doi.org/10.1111/j.1654-1103.2003.tb02166.x

Pausas, J. G., & Ribeiro, E. (2017). Fire and plant diversity at the global scale. *Global Ecology and Biogeography*, *26*(8), 889–897. https://doi.org/10.1111/geb.12596

Perroni-Ventura, Y., Montaña, C., & García-Oliva, F. (2006). Relationship between soil nutrient availability and plant species richness in a tropical semi-arid environment. *Journal of Vegetation Science, 17*(6), 719–728. https://doi.org/10.1111/j.1654-1103.2006.tb02495.x

Pilon, N. A. L., Cava, M. G. B., Hoffmann, W. A., Abreu, R. C. R., Fidelis, A., & Durigan, G. (2020). The diversity of post-fire regeneration strategies in the cerrado ground layer. *Journal of Ecology, 109*(1), 154–166. https://doi.org/10.1111/1365-2745.13456

Pivello, V. R., & Coutinho, L. M. (1992). Transfer of macronutrients to the atmosphere during experimental burnings in an open cerrado (Brazilian savanna). *Journal of Tropical Ecology, 8*(4), 487–497. https://doi.org/10.1017/s0266467400006829

Platt, W. J., & Schwartz, M. W. (1990). Temperate hardwood forests. In R. L. Myers, & J. J. Ewel (Eds.), *Ecosystems of Florida* (pp. 194–229). University of Central Florida Press.

Quinn, R. D. (1994). Animals, fire, and vertebrate herbivory in California chaparral and other Mediterranean-type ecosystems. In J. M. Moreno, & W. C. Oechel (Eds.), *The Role of Fire in Mediterranean-Type Ecosystems* (pp. 46–78). Springer-Verlag.

Ramos-Neto, M. B., & Pivello, V. R. (2000). Lightning fires in a brazilian savanna national park: Rethinking management strategies. *Environmental Management, 26*(6), 675–684. https://doi.org/10.1007/s002670010124

Ratnam, J., Bond, W. J., Fensham, R. J., Hoffmann, W. A., Archibald, S., Lehmann, C. E. R., Anderson, M. T., Higgins, S. I., & Sankaran, M. (2011). When is a 'forest' a savanna, and why does it matter? *Global Ecology and Biogeography, 20*, 653–660. https://doi.org/10.1111/j.1466-8238.2010.00634.x

Reis, S. M., Lenza, E., Marimon, B. S., Gomes, L., Forsthofer, M., Morandi, P. S., Marimon Junior, B. H., Feldpausch, T. R., & Elias, F. (2015). Post-fire dynamics of the woody vegetation of a savanna forest (Cerradão) in the Cerrado-Amazon transition zone. *Acta Botanica Brasilica, 29*(3), 408–416. https://doi.org/10.1590/0102-33062015abb0009

Ribeiro, N. S., Snook, L. K., Nunes de Carvalho Vaz, I. C., & Alves, T. (2019). Gathering honey from wild and traditional hives in the Miombo woodlands of the Niassa National Reserve, Mozambique: What are the impacts on tree populations? *Global Ecology and Conservation, 17*, e00552. https://doi.org/10.1016/j.gecco.2019.e00552

Rissi, M. N., Baeza, M. J., Gorgone-Barbosa, E., Zupo, T., & Fidelis, A. (2017). Does season affect fire behaviour in the Cerrado? *International Journal of Wildland Fire, 26*(5), 427. https://doi.org/10.1071/wf14210

Rodrigues, C. A., Zirondi, H. L., & Fidelis, A. (2021). Fire frequency affects fire behavior in open savannas of the Cerrado. *Forest Ecology and Management, 482*, 118850. https://doi.org/10.1016/j.foreco.2020.118850

Rossi, J. L., Chatelon, F. J., & Marcelli, T. (2018). Fire Intensity. In *Encyclopedia of Wildfires and Wildland-Urban Interface (WUI) Fires* (pp. 1–6). Springer International Publishing. https://doi.org/10.1007/978-3-319-51727-8_51-1

Russell-Smith, J., Yates, C. P., Whitehead, P. J., Smith, R., Craig, R., Allan, G. E., Thackway, R., Frakes, I., Cridland, S., Meyer, M. C. P., & Gill, A. M. (2007). Bushfires "down under": Patterns and implications of contemporary Australian landscape burning. *International Journal of Wildland Fire, 16*(4), 361. https://doi.org/10.1071/wf07018

Sankaran, M., Ratnam, J., & Hanan, N. P. (2004). Tree-grass coexistence in savannas revisited—insights from an examination of assumptions and mechanisms invoked in existing models. *Ecology Letters, 7*(6), 480–490. https://doi.org/10.1111/j.1461-0248.2004.00596.x

Schmidt, I. B., Fidelis, A., Miranda, H. S., & Ticktin, T. (2016). How do the wets burn? Fire behavior and intensity in wet grasslands in the Brazilian savanna. *Brazilian Journal of Botany, 40*(1), 167–175. https://doi.org/10.1007/s40415-016-0330-7

Schoch, P., & Binkley, D. (1986). Prescribed burning increased nitrogen availability in a mature loblolly pine stand. *Forest Ecology and Management, 14*(1), 13–22. https://doi.org/10.1016/0378-1127(86)90049-6

Scholes, R. J., & Archer, S. R. (1997). Tree-Grass Interactions in Savannas. *Annual Review of Ecology and Systematics, 28*(1), 517–544. https://doi.org/10.1146/annurev.ecolsys.28.1.517

Scott, A. C. (2000). The Pre-Quaternary history of fire. *Palaeogeography, Palaeoclimatology, Palaeoecology, 164*(1–4), 281–329. https://doi.org/10.1016/s0031-0182(00)00192-9

Silva, D. M., Batalha, M. A., & Cianciaruso, M. V. (2013). Influence of fire history and soil properties on plant species richness and functional diversity in a neotropical savanna. *Acta Botanica Brasilica, 27*(3), 490–497. https://doi.org/10.1590/S0102-33062013000300005

Simon, M. F., & Pennington, T. (2012). Evidence for adaptation to fire regimes in the tropical savannas of the Brazilian Cerrado. *International Journal of Plant Sciences, 173*(6), 711–723. https://doi.org/10.1086/665973

Smit, I. P. J., Asner, G. P., Govender, N., Kennedy-Bowdoin, T., Knapp, D. E., & Jacobson, J. (2010). Effects of fire on woody vegetation structure in African savanna. *Ecological Applications, 20*(7), 1865–1875. https://doi.org/10.1890/09-0929.1

Snook, L., Alves, T., Sousa, C., Loo, J., Gratzer, G., Duguma, L., Schrotter, C., Ribeiro, N., Mahanzule, R., Mazuze, F., Cuco, E., & Elias, M. (2015). Relearning traditional knowledge to achieve sustainability: Honey gathering in the miombo woodlands of northern Mozambique. *XIV World Forestry Congress, Durban, South Africa*, 7–11 September 2015. http://foris.fao.org/wfc2015/api/file/552e8b8e9e00c2f116f8eac2/contents/028c3429-30dd-4b71-856b-22ea44d4849c.pdf

Staver, A. C., & Levin, S. A. (2012). Integrating theoretical climate and fire effects on savanna and forest systems. *The American Naturalist, 180*(2), 211–224. https://doi.org/10.1086/666648

Steel, Z. L., Safford, H. D., & Viers, J. H. (2015). The fire frequency-severity relationship and the legacy of fire suppression in California forests. *Ecosphere, 6*(1), 1–23. https://doi.org/10.1890/es14-00224.1

Sugihara, N. G., Van Wagtendonk, J. W., & Fites-Kaufman, J. (2006). Fire as an ecological process. In *Fire in California's Ecosystems* (pp. 58–74). University of California Press. https://doi.org/10.1525/california/9780520246058.003.0004

Swaine, M. D., Hawthorne, W. D., & Orgle, T. K. (1992). The effects of fire exclusion on savanna vegetation at Kpong, Ghana. *Biotropica, 24*(2), 166. https://doi.org/10.2307/2388670

Swanson, M. E., Franklin, J. F., Beschta, R. L., Crisafulli, C. M., DellaSala, D. A., Hutto, R. L., Lindenmayer, D. B., & Swanson, F. J. (2010). The forgotten stage of forest succession: Early-successional ecosystems on forest sites. *Frontiers in Ecology and the Environment, 9*(2), 117–125. https://doi.org/10.1890/090157

Teixeira, J., Souza, L., Le Stradic, S., & Fidelis, A. (2022). Fire promotes functional plant diversity and modifies soil carbon dynamics in tropical savanna. *Science of The Total Environment, 812*, 152317. https://doi.org/10.1016/j.scitotenv.2021.152317

Trapnell, C. G., Friend, M. T., Chamberlain, G. T., & Birch, H. F. (1976). The effects of fire and termites on a Zambian woodland soil. *The Journal of Ecology, 64*(2), 577. https://doi.org/10.2307/2258774

UNEP. (2022). Number of wildfires to rise by 50% by 2100 and governments are not prepared, experts warn. *United Nations Environment Programme.* www.unep.org/news-and-stories/press-release/number-wildfires-rise-50-2100-and-governments-are-not-prepared

van Langevelde, F., van de Vijver, C. A. D. M., Kumar, L., van de Koppel, J., de Ridder, N., van Andel, J., Skidmore, A. K., Hearne, J. W., Stroosnijder, L., Bond, W. J., Prins, H. H. T., & Rietkerk, M. (2003). Effects of fire and herbivory on the stability of savanna ecosystems. *Ecology, 84*(2), 337–350. https://doi.org/10.1890/0012-9658(2003)084[0337:eofaho]2.0.co;2

van Wilgen, B. W., Everson, C. S., & Trollope, W. S. W. (1990). Fire management in Southern Africa: Some examples of current objectives, practices, and problems. In *Ecological Studies* (pp. 179–215). Springer Berlin Heidelberg. https://doi.org/10.1007/978-3-642-75395-4_11

van Wilgen, B. W., Govender, N., & Biggs, H. C. (2007). The contribution of fire research to fire management: A critical review of a long-term experiment in the Kruger National Park, South Africa. *International Journal of Wildland Fire, 16*(5), 519. https://doi.org/10.1071/wf06115

Wagenius, S., Beck, J., & Kiefer, G. (2020). Fire synchronizes flowering and boosts reproduction in a widespread but declining prairie species. *Proceedings of the National Academy of Sciences, 117*(6), 3000–3005. https://doi.org/10.1073/pnas.1907320117

Walter, H. (1971). *Ecology of Tropical and Subtropical Vegetation.* Oliver & Boyd.

Wang, C. W. (1961). The forests of China, with a survey of grassland and desert vegetations. In *Harvard University Maria Moors Cabot Foundation No. 5* (pp. 171–187). Harvard University Press.

WEF. (2022). Wildfire risk has increased, but we can still influence where and how fires strike. *World Economic Forum.* www.weforum.org/agenda/2022/07/climate-change-wildfire-risk-has-grown-nearly-everywhere-but-we-can-still-influence-where-and-how-fires-strike

Whelan, R. J., & Tait, I. (1995). Responses of plant populations to fire: Fire season as an understudied element of fire regime. *CALMscience, 4*(Supplement), 147–150.

Wildy, D. T., & Pate, J. S. (2002). Quantifying above- and belowground growth responses of the western Australian oil mallee, Eucalyptus kochii subsp. Plenissima, to contrasting decapitation regimes. *Annals of Botany, 90*(2), 185–197. https://doi.org/10.1093/aob/mcf166

Williams, R. J., Gill, A. M., & Moore, P. H. R. (1998). Seasonal changes in fire behaviour in a tropical savanna in northern Australia. *International Journal of Wildland Fire, 8*(4), 227. https://doi.org/10.1071/wf9980227

# Section 5

*Ecophysiology and geoclimatic factors*

# 17 Adaptation of Fruit Trees to Different Elevations in the Tropical Andes

*Gerhard Fischer, Helber Enrique Balaguera-López, Alfonso Parra-Coronado, and Stanislav Magnitskiy*

## 17.1 INTRODUCTION

Climate change affects microclimates—and, thus, canopy conditions—in orchards, impacting the suitability of a region to grow specific fruit species (Atkinson et al., 2013). Global climate change is a new constraint for tropical fruit producers that alters normal climatic variables (Gutiérrez-Villamil et al., 2022). As a result of climate change and global biodiversity problems, tropical ecosystems are an important object of study; however, interest is focused, in many cases, on lowland humid tropics, and little is known about tropical ecosystems at higher elevations (Rundel et al., 1994). The general pattern of solar radiation results in small seasonal changes in temperature in the tropics, with the climate regime of the wet tropics being noticeable at high elevations (Körner, 2007; Grabherr et al., 2010), but very large climatic changes occur diurnally in the mountains (Rundel, 1994). That is, diurnal temperature ranges in the mountains can be 3–10 times greater than the annual seasonal changes in tropical alpine environments (Rundel, 1994).

Since about 25% of the earth's surface is covered by mountains, elevation gradients are among the most powerful "natural experiments" (Körner, 2007, p. 569). For a comparative ecology, the climatic data collected within elevation gradients always include a combined effect between the phenomena of elevation and regional particularities (Körner, 2007), taking into account the fact that local climatic conditions often differ from regional ones (Janicot et al., 2015).

According to Ramírez and Kallarackal (2014), tropical regions refer to the areas between 23.5°N and 23.5°S latitude, which include mountains or tropical highlands at elevations greater than 900 m above sea level. According to Guerrero et al. (2011), the South American Andes consist of a chain of mountains that is 8,500 km long and that is found in Chile, northern Argentina, Bolivia, Peru, Ecuador, Colombia and Venezuela, with an average elevation of 3,000–4,000 m above sea level (m a.s.l.), bordering the Pacific Ocean. They are between 250 km and 750 km wide and have an area of about 2,870,000 km² (Orme, 2007). With 15% of the total richness of the world's plants, the tropical Andes constitute one of the main biodiversity hotspots on the planet (Peyre et al., 2019).

Many tropical countries have great potential for the production of fruits because they have a zone of primary diversity and ecological niches that are suitable for fruit tree cultivation, such as the highlands (Ligarreto, 2012). The adaptation of fruit plants to higher elevations is generally related to climatic conditions, such as lower temperatures and partial pressure of gases and, importantly, higher solar radiation—especially ultraviolet (UV) radiation—as elevation increases (Fischer et al., 2022a). Decreases in temperature, about 0.6–0.7°C per 100 m of tropical elevation (Benavides et al., 2017), and increases in phenological cycles (Ramírez & Kallarackal, 2014) facilitate the introduction of fruit species from subtropical and temperate climates to the inner tropics.

Ecophysiology is vital to find adaptation strategies in fruit plants subjected to changing environmental conditions (Sánchez-Reinoso et al., 2019). Several climatic factors affect crops in terms of excessive heat, intense UV light, prolonged drought, strong winds, and other environmental stress factors (Zandalinas et al., 2021). These climatic factors affect crop performance and physiology in combination and not alone (Fischer et al., 2016). All aspects of plant growth and development are directly or indirectly influenced by environmental variables (Fischer & Orduz-Rodríguez, 2012). In general, fruit species that are not growing within their optimal elevation range do not generate economic profitability (Fischer & Parra-Coronado, 2020).

UV-B radiation (280–315 nm), which has the highest energy in the daylight spectrum with the potential to damage macromolecules including DNA (deoxyribonucleic acid) (Jenkins et al., 2009), increases with elevation and exerts various effects on morphology, biochemical composition, and molecular responses in plant species (Terfa et al., 2014). Many fruit tree cultivars from the tropics have developed mechanisms of tolerance to excessive UV-B radiation. For example, foliar adaptations (size, leaf/stem thickness, and trichomes, among others) are very common in fruit species from tropical, mountainous regions (Fischer & Melgarejo, 2020). Fruits are also protected from UV-B rays via the production of antioxidants in the epidermis, as reported by Kumar et al. (2019), who observed that the total phenol content was positively correlated with increasing elevation in apples. On the other hand, the pubescence of the aerial parts of plants better counteract nocturnal cold in the highlands (Fischer et al., 2021a).

Species that adapt to conditions at tropical elevation are generally from the same Andean zone, such as the Solanaceae species cape gooseberry or goldenberry (*Physalis peruviana*), lulo or naranjilla (*Solanum quitoense*) and tree tomato (*Solanum betaceum*); Passifloraceae species granadilla (*Passiflora ligularis*), purple passionfruit or gulupa (*Passiflora edulis*), and banana passionfruit or curuba (*Passiflora tripartita* var. *mollissima*); Myrtaceae feijoa or pineapple guava (*Acca sellowiana*); and Ericaceae Andean (or Colombian) blueberry (*Vaccinium meridionale*), among others, but deciduous fruit trees have also been introduced from mountainous regions in the temperate zones. Ligarreto (2012) commented that these Andean fruit crops have a very low environmental impact.

For "mountain farmers" in the Colombian Andes, cultivating these fruit trees is a great challenge but has great benefits socioeconomically (Fischer, 1995) because several of these exotic species enjoy large-scale production and export at the global level, such as the purple passionfruit, cape gooseberry, and "Hass" avocado. However, there is little knowledge of how these species are affected by environmental conditions or which adaptations and defense mechanisms they have, especially in terms of high UV radiation and drastic temperature changes.

## 17.2 CLIMATIC CHARACTERISTICS WITH INCREASING TROPICAL ELEVATION

Given the thermal uniformity of the tropics, the largest fluctuations occur during the 24-hour cycle of a day, such that the climate in the highlands can be described as "summer" during the day and "winter" at night time (Rundel, 1994). As tropical elevation increases, there are characteristic climatic changes (Table 17.1) that affect the growth and development of fruit trees and fruit quality (Fischer et al., 2022a).

Tropical, low latitudes have constant net heat gains and relatively constant levels of solar irradiation throughout the year (Rundel, 1994). In the high-elevation tropics, where sun rays are less filtered, the atmosphere layer is thinner, causing an increase in UV radiation by 10–12% for each 1,000 m increase in elevation (Benavides et al., 2017). The greater incidence of solar radiation in the highlands, with a stronger effect when the sky is clear, not only increases the temperature of plant tissues but also the soil temperature (Fischer et al., 2022b). On the contrary, a very low soil temperature at high elevations is a critical environmental element for plants (Beck, 1994).

These climate factors can vary, e.g., temperature decreases with elevation (Table 17.1) and is modified by place (latitude), time of day (Benavides et al., 2017) and, according to Paull and Duarte (2011), cloudiness, precipitation, and wind pattern.

The inverse temperature/elevation relationship turns tropical zones into thermal floors, as described by Paull and Duarte (2011) for the equatorial region: (1) hot zone, 0–1,000 m a.s.l.; (2) temperate zone, 1,000–2,000 m a.s.l.; and (3) cold zone, >2,000 m a.s.l. Notably, frosts occur that affect all fruit crops, depending on their stage of development and phenology. For example, in the Eastern Cordillera of Colombia, frosts occur above 2,400 m a.s.l. The temperature drops as elevation increases in all tropical highlands, about 0.6°C per 100 m of elevation worldwide (Nagy & Grabherr, 2009).

Parra-Coronado et al. (2015a, 2015b) recorded the climatic conditions on two feijoa (*Acca sellowiana* [O. Berg] Burret)-producing farms in the Department of Cundinamarca (Colombia) at extreme production elevations. They found that the precipitation accumulated during the reproductive phase of the crop (from flower bud to

### TABLE 17.1
### Climate Changes as Tropical Elevation Increases

| Climatic Factors | Changes with Increasing Elevation | Author(s) |
| --- | --- | --- |
| Air temperature | Temperature reduction of about 0.6–0.7°C per 100 m of elvation. | Benavides et al. (2017) |
| Sunlight | Increased visible light, UV, and IR. | Fischer and Orduz-Rodríguez (2012) |
| Solar radiation | Increase in solar radiation accumulated during a time period. | Parra-Coronado et al. (2015a, 2017) |
| Atmospheric pressure and partial gas pressure | Reduction of atmospheric and partial pressure of $CO_2$, $O_2$, $N_2$, and water vapor. | Körner (2007), Fischer and Orduz-Rodríguez (2012) |
| Precipitation | Reduction, with an inverse relationship with radiation and elevation. | Benavides et al. (2017), Parra-Coronado et al. (2015a, 2015b, 2017) |
| Wind | In continental climates, there is often no increase with elevation, but in mountains near the sea, there may be an increase. | Körner (2007) |

harvest) decreased as elevation increased, with mean values of 987 mm at 1,800 m a.s.l. but only 304 mm at 2,580 m a.s.l. In addition, they also detected that the accumulated solar radiation increased with elevation (8.918 W m$^{-2}$ at 1,800 m vs. 11.082 W m$^{-2}$ at 2,580 m a.s.l.).

While Körner (2007) classified humidity as not specifically dependent on elevation, the importance of water in characterizing the regional climate should not be left because Marengo et al. (2011) predicted that the rate of precipitation in the tropical Andes will increase by 20–25%. Mendoza et al. (2017) studied the phenology of fructification in the American neotropics and found that rainfall is the climatic factor that most influences the reproductive phase (73.4%), with much less influence attributed to air temperature (19.3%), and only 3.2% to solar radiation or photoperiod.

Fischer and Lüdders (2002) recorded very different agroclimatic conditions on cape gooseberry (*Physalis peruviana*) plantations at two elevations in the Boyaca region of Colombia. It was noted that soil temperature decreases on a smaller gradient than the air temperature. The mean soil temperature was 2.1°C higher than that of the air at the lower site (2,300 m a.s.l.) but 4.3°C higher at the elevated site (2,690 m a.s.l.), most likely because of higher solar radiation at the higher elevation, with the same hours of sunshine (5.3 h day$^{-1}$) at both sites (Table 17.2).

Parra-Coronado et al. (2015a) determined the thermal time for the reproductive phenological states of feijoa (*Acca sellowiana* [O. Berg] Burret), from flower bud to fruit harvest, during four different reproductive periods in two locations in the Department of Cundinamarca (Colombia), at extreme elevations of production, indicating that phenological cycles increased at higher elevations (lower temperature and less precipitation), with a greater number of calendar days from one phenological state to another than at the lower elevation (Table 17.3).

Parra-Coronado et al. (2015a) found that, although fewer calendar days are required for the reproductive phase—from flower bud to fruit at harvest—of feijoa at lower elevations, the required thermal time is greater because of a higher temperature. Thus, the thermal time in degree days of growth or development (GDD) required to go from the phenological stage of a flower bud to fruit at harvest was 2,965 ± 113 GDD for the location at 1,800 m a.s.l. and 2,337 ± 124 GDD for the location at 2,580 m a.s.l.

For atmospheric pressure, including the partial pressure of gases (pO$_2$, pCO$_2$, pN$_2$), Körner (2007) reported that a decrease of about 11% (at 20°C) for each increase of 1,000 m, which can vary somewhat with temperature and humidity.

## 17.3 TROPICAL ELEVATION RANGES FOR DIFFERENT FRUIT SPECIES

As seen in Figure 17.1, the elevation ranges in Colombia for the best growth and production of different fruit trees vary according to the species, going from sea level up to 3,200 m a.s.l. For each species, if there are different varieties or genotypes, these ranges can be specific, such as, in the Japanese plum "Beauty", the range is 2,600–2,800 m a.s.l., and the pear "Triunfo de Viena" has a range of 2,400–2,800 m a.s.l. (Fischer, 2000), while the apple "Anna" ("Adassin Red" × "Golden Delicious"), the most adapted to Colombian highland conditions, covers the entire range for the different varieties, from 1,700–2,800 m

**TABLE 17.2**
**Agro-Climatic Conditions on Experimental Cape Gooseberry Plantations at Two Elevations during 44 Weeks of the Study in the Boyaca Region (Colombia)**

| Climatic Conditions | Unit | Villa de Leyva (2,300 m a.s.l.) | Tunja (2,690 m a.s.l.) |
|---|---|---|---|
| Latitude | N | 5°39' | 5°34' |
| Longitude | W | 73°32' | 73°22' |
| Mean air temperature | °C | 17.4 | 12.5 |
| Max. air temperature | °C | 20.5 | 15.1 |
| Min. air temperature | °C | 14.3 | 9.9 |
| Mean soil temperature | °C | 19.5 | 16.8 |
| Mean relative humidity (RH) | % | 66.6 | 79.0 |
| Mean max. RH | % | 78.4 | 93.1 |
| Mean min. RH | % | 54.8 | 64.9 |
| Precipitation | mm year$^{-1}$ | 837 | 302 |
| Evaporation | mm year$^{-1}$ | 1,438 | 1,095 |
| Direct sunlight | h day$^{-1}$ | 5.3 | 5.3 |
| UV-B radiation | mW m$^{-2}$ | 1,294 | 1,399 |
| Atmospheric pressure | Mb | 776 | 736 |

*Source:* Data from Fischer & Lüdders (2002)

### TABLE 17.3
### Calendar Time (Days ± S.D.) for the Reproductive Phenological Phase of Feijoa in Two Locations at Extreme Production Elevations

| Phenological Phase | San Francisco (1,800 m a.s.l.) T: 12.4°C, P: 476 mm | Tenjo (2,580 m a.s.l.) T: 20.5°C, P: 722 mm |
|---|---|---|
| Floral bud to anthesis | 11.75 ± 0.43 | 18.75 ± 2.28 |
| Anthesis to fruit set | 6.75 ± 0.43 | 11.75 ± 0.83 |
| Fruit set to harvest maturity | 149.00 ± 5.83 | 179.75 ± 12.21 |
| Floral bud to harvest maturity | 167.5 ± 4.97 | 210.25 ± 10.08 |

*Source:* Adapted from Parra-Coronado et al. (2015a)
*Notes:* T = Mean temperature; P = Accumulated mean precipitation.

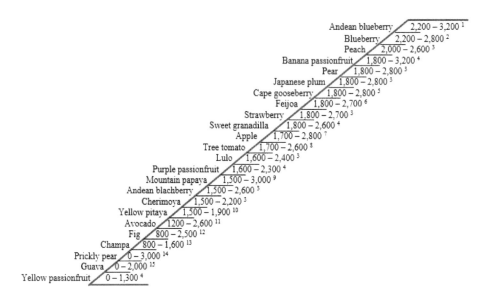

**FIGURE 17.1** Increasing elevation ranges in commercial fruit grown in Colombia, South America.

*Sources:* Adapted from Fischer et al., (2022a). [1]Medina et al. (2015); [2]Cleves (2021); [3]Fischer & Orduz-Rodríguez (2012); [4]Fischer & Miranda (2021); [5]Fischer & Melgarejo (2020); [6]Parra-Coronado et al. (2016); [7]Gutiérrez-Villamil et al. (2022); [8]Bonnet & Cárdenas (2012); [9]Reyes (2012); [10]Corredor (2012); [11]Carvalho et al. (2015); [12]Fischer et al. (2012); [13]Balaguera-López et al. (2022); [14]Almanza-Merchán & Fischer (2012); [15]Fischer & Melgarejo (2021)

a.s.l. (Gutiérrez-Villamil et al., 2022). Because the ranges for some species are very large (Table 17.4), there are more specific ones for the best performance of plants, such as the elevation ranges from 1,700–2,500 m a.s.l. for the prickly pear (Almanza-Merchán & Fischer, 2012), 1,700–2,000 m a.s.l. for purple passionfruit (Ocampo et al., 2020), and 2,200–2,400 m a.s.l. for the cape gooseberry (Fischer et al., 2021a). Also, in other countries, banana passionfruit crops have been cultivated at higher elevations, that is, up to 3,400 m a.s.l. in Cuzco, Peru (National Research Council, 1989). Cultivation sites closer to the equator have increased temperatures, allowing fruit trees to be grown at higher elevations (Fischer et al., 2022c). Carillo et al. (2015) reported cape gooseberry crops in Ecuador up to 3,300 m a.s.l.

As can be seen in Table 17.4, many fruit species come from mountainous areas, meaning that they are adapted to elevation conditions, not only those from the tropics and subtropics, but also the deciduous fruit trees apple, pear, peach, and plum, which originate in temperate climates and require chilling hours. Thus, Caldwell et al. (1980) suggested that species that originate from regions near the equator and/or higher elevations, as is the case of Andean fruit crops, have developed a greater tolerance to UV-B radiation than fruit species originating in temperate climates.

## 17.4 ADAPTATION OF FRUIT TREES TO CONDITIONS IN TROPICAL HIGHLANDS

### 17.4.1 MORPHOLOGICAL ADAPTATIONS

Plants adapted to tropical high-elevation conditions have developed various strategies for all plant organs. Limited root growth is one strategy in fruit trees for conditions in the

## TABLE 17.4
### Elevation Ranges in Commercial Fruit Grown in Colombia (South America), and Their Origins

| Common Name | Scientific Name | Elevation (m a.s.l.) | Origin |
|---|---|---|---|
| Andean blueberry | *Vaccinium meridionale* | 2,200–3,200[1] | Andes of Venezuela, Colombia, and Peru, mountain forests of Jamaica[16] |
| Blueberry | *Vaccinium corymbosum* | 2,200–2,800[2] | North America[17] |
| Peach | *Prunus persica* | 2,000–2,600[3] | China[18] |
| Banana passionfruit | *Passiflora tripartita* var. *mollissima* | 1,800–3,200[4] | The mountainous area of South America (Colombia, Ecuador, Venezuela, Bolivia, Peru)[19] |
| Pear | *Pyrus communis* | 1,800–2,800[3] | Caucasic mountains and the Caspian Sea coast[20] |
| Japanese plum | *Prunus salicina* | 1,800–2,800[3] | China[18] |
| Cape gooseberry | *Physalis peruviana* | 1,800–2,800[5] | Andes of Peru, Brazil, Chile, Ecuador, and Colombia[5] |
| Feijoa | *Acca sellowiana* | 1,800–2,700[6] | Southern Brazil, Uruguay, upper region of western Paraguay, and northeastern Argentina[21] |
| Strawberry | *Fragaria* × *ananassa* | 1,800–2,700[3] | USA and Chile[22] |
| Sweet granadilla | *Passiflora ligularis* | 1,800–2,600[4] | From Bolivia to Central America[23] |
| Apple | *Malus domestica* | 1,700–2,800[7] | Caucasic Mountains[18] |
| Tree tomato | *Solanum betaceum* | 1,700–2,600[8] | Andes of Colombia, Ecuador, Peru, Bolivia, and Chile[23] |
| Lulo | *Solanum quitoense* | 1,600–2,400[3] | Humid forests of Colombia, Ecuador, and Peru[24] |
| Purple passionfruit | *Passiflora edulis* | 1,600–2,300[4] | South of Brazil, Paraguay, and North of Argentina[25] |
| Mountain papaya | *Vasconcellea pubescens* | 1,500–3,000[9] | Andes of América Meridional, Colombia, and Perú[9] |
| Andean blackberry | *Rubus glaucus* | 1,500–2,600[3] | Tropical mountains between Ecuador and Mexico[26] |
| Cherimoya | *Annona cherimola* | 1,500–2,200[3] | Inter-Andean valleys of Ecuador, Colombia, and Bolivia[23] |
| Yellow pitaya | *Hylocereus megalanthus* | 1,500–1,900[10] | Northern and Eastern South America[23] |
| Avocado | *Persea americana* | 1,200–2,600[11] | Mexico and Central America[27] |
| Fig | *Ficus carica* | 800–2,500[12] | Warm areas of Occidental Asia[12] |
| Champa | *Campomanesia lineatifolia* | 800–1,600[13] | Brazilian, Peruvian, Colombian and Bolivian Amazon[28] |
| Prickly pear | *Opuntia ficus-indica* | 0–3,000[14] | Central America[14] |
| Guava | *Psidium guajava* | 0–2,000[15] | Central America and southern Mexico[23] |
| Yellow passionfruit | *Passiflora edulis* f. *flavicarpa* | 0–1,300[4] | Amazon region[23] |

*Source:* Modified from Fischer et al., 2022a). [1]Medina et al., 2015; [2] Cleves, 2021; [3] Fischer & Orduz-Rodríguez, 2012; [4] Fischer & Miranda, 2021; [5] Fischer & Melgarejo, 2020; [6] Parra-Coronado et al., 2016; [7] Gutiérrez-Villamil et al., 2022; [8] Bonnet & Cárdenas, 2012; [9] Reyes, 2012; [10] Corredor, 2012; [11] Carvalho et al., 2015;[12] Fischer et al., 2012;[13] Balaguera-López et al., 2022; [14] Almanza-Merchán & Fischer, 2012; [15] Fischer & Melgarejo, 2021; [16] Ehlenfeldt & Luteyn, 2021; Tineo et al., 2022; [17] Retamales & Hancock, 2012; [18] Cárdenas & Fischer, 2013; [19] Fischer et al., 2020; [20] Campos, 2013; [21] Parra-Coronado & Fischer, 2013; [22] Ruiz & Piedrahita, 2012; [23] Blancke, 2016; [24] Lobo et al., 2007; [25] Ocampo et al., 2020; [26] Morales & Villegas, 2012; [27] Hoyos, 1989; [28] Villachica, 1996.

high tropics (Fischer, 2000). For example, cape gooseberry that grows in the Boyaca region (Colombia) at an elevation of 2,690 m a.s.l. possesses more superficial root systems than plants grown at 2,300 m a.s.l. (Fischer et al., 2007). Similarly, Walter (1990) showed that, at increased tropical elevations, plant roots become shallower to utilize the short daytime heating in the upper soil layer. Importantly, this region had an average soil temperature of 2,690 m a.s.l. of 16.8°C, which was 2.7°C lower than at 2,300 m a.s.l. (Fischer et al., 2007). At the same time, Beck (1994) reported that low soil temperatures are a critical factor in tropical highlands, which limit plant growth. In addition, superficial root systems are susceptible to very dry soil conditions that are common in tropical highlands (Fischer, 2000). One practical strategy for strawberries in high-elevation tropics (up to 2,700 m a.s.l., Table 17.4) could be covering furrows with black plastic (Fischer et al., 2022b).

For stem growth in higher areas, many researchers have reported decreasing longitudinal growth (e.g., Neugart & Schreiner, 2018). Thus, the cape gooseberry developed shorter main stems and lateral shoots in higher orchards (2,690 m a.s.l.) (Fischer & Lüdders, 2002). This stunting effect was found to be more marked under conditions with low photosynthetic active radiation (PAR) (Fischer, 2000). Supposedly, the higher UV light affects the synthesis of auxins (Fischer & Melgarejo, 2014; Kulandaivelu et al., 1989) and can decrease the production of gibberellins in the internodes, as reported by Buchanan et al. (2015), contributing to the stunting effect. On the other hand, the decrease in atmospheric pressure as elevation increases could have contributed to the decrease in the length of stem internodes, as found by Daunicht and Brinkjans (1996) in tomato plants kept at low atmospheric pressure. In addition, according to Larcher (2003), night temperature generally affects stem growth more than daytime temperature. Thus, stems become shorter at higher elevations, which have decreased night temperatures (Fischer & Lüdders, 2002).

Generally, plants develop a series of foliar adaptations, including changes in leaf anatomy and morphology, with changes in elevation (Fischer, 2000). These changes with increasing elevation manifest as: (1) decreased leaf area (Figure 17.2) inhibiting leaf expansion in many species because of increased UV-B light incidence (Nobel, 2020); and (2) increased leaf thickness (Figure 17.2) because of the increased number of mesophyll layers and a thicker cuticle that better resists UV light (Fischer & Miranda, 2021) and that better protects against foliar dehydration in the drier climate at elevation (Fischer & Lüdders, 2002). Tranquilini (1964) also observed an increase in leaf thickness in high mountain plants with 5–6 layers of palisade parenchyma tissue.

In the cape gooseberry, Fischer and Melgarejo (2020) reported that this plant can increase the number of leaf stomata at higher elevations (Figure 17.2) to compensate for the reduced concentration of $pCO_2$ and $pO_2$ at these sites, where the partial pressure is about 21% lower at 2,000 m a.s.l. than at sea level (Körner, 2007). The latter author argued that the declined $pCO_2$ is not always the primary cause for the increased number of stomata at high elevations because this trend was not observed in cloudy tropical mountains.

The observed pubescence (epidermal hairs) of the aerial green part of cape gooseberry plants (Figure 17.3) growing in tropical elevation zones (Fischer & Melgarejo, 2020) was described by Meinzer et al. (1994) as a key factor in the regulation of the foliar thermal balance in many *Espeletia* species, which results in reduced absorbance of solar radiation, lower leaf temperature, and less water loss by transpiration (Nobel, 2020). Walter (1990) defined the formation of a dense pubescence in the aerial organs of plants to counteract high UV radiation and high nocturnal cooling of the atmosphere (Fischer et al., 2021a). Also, in the genus *Rubus* (Andean blackberry), the particular anatomical features found by Moreno-Medina and Casierra-Posada (2021) in species cultivated in various locations in Boyaca (Colombia) included the presence of glandular trichomes and simple hairs, which can defend plants from the effects of UV radiation and, in other cases, store or secrete compounds for plant defense.

### 17.4.2 Physiological Adaptations

When lulo plants (*Solanum quitoense*) are grown at elevations greater than 2,400 m a.s.l., the flowers, leaves, and stems synthesize more anthocyanins (Fischer & Orduz-Rodríguez, 2012) (Figure 17.4).

Larcher (2003) reported that, in recently expanded leaves, anthocyanins act as an absorption filter and shade the mesophyll, a reaction that occurs especially in the tropics. Also, the contents of protective pigments induced by high light intensities, such as carotene and lutein, increase in chloroplasts. Likewise, plants can form flavonoids that absorb UV-B radiation (Mphahlele et al., 2014) and epicuticular waxes to protect themselves from high solar radiation (Larcher, 2003).

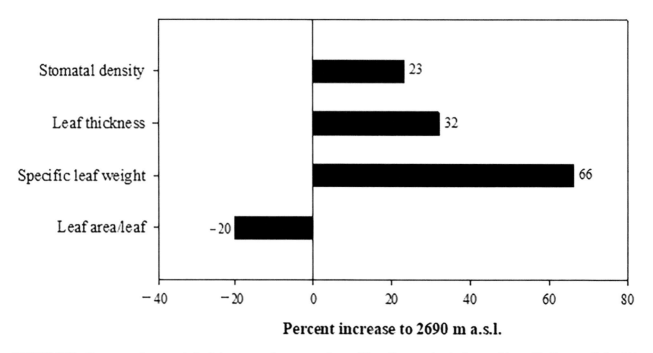

**FIGURE 17.2** Percentage increase in leaf characters of cape gooseberry (*Physalis peruviana*) plants cultivated in Boyaca (Colombia) at an elevation of 2,690 m a.s.l. compared to plants cultivated at 2,300 m a.s.l.

# Adaptation of Fruit Trees to Elevations in the Tropical Andes 199

**FIGURE 17.3** High pubescence in all green parts of cape gooseberry plants growing in the Colombian highlands (1,800–2,700 m a.s.l.) to protect against UV-B radiation and low night temperatures.

*Source:* From Miranda and Fischer (2021), with permission

**FIGURE 17.4** Lulo plant at 2,650 m a.s.l., near Bogota (Colombia), with anthocyanins in aerial plant parts, which act as an absorption filter of UV-B radiation.

Varieties/ecotypes with higher photosynthetic performance and lower susceptibility to photoinhibition are better adapted to high-elevation belts (Fischer et al., 2016). It is important that the crops are cultivated within the recommended elevation range (Table 17.3). For example, in an ecophysiological study of sweet granadilla in the Department of Huila (Colombia) planted at 2,060 and 2,270 m a.s.l., the values before dawn (predawn) of the maximum photochemical efficiency of photosystem II (Fv/Fm >0.86) confirmed that the plants were not exposed to stress conditions, and the two elevations were suitable for the cultivation of this passionfruit crop in this region (Fernández et al., 2014).

### 17.4.3 Adaptation of Fruits to the High Incidence of UV Light and Low Temperatures at High Elevations

In recent decades, radiation has increased steadily, especially UV-B (280–320 nm) (Van Leeuwen & Darriet, 2016). UV-B radiation can increase plant frost tolerance under extremely high-temperature environments. While extreme temperatures can directly determine the UV-B radiation sensitivity of different plants (Caldwell et al., 2007), plants growing at high-elevation areas and/or near the equator—as is the case with Andean fruit crops—have developed a greater tolerance to UV-B radiation (Caldwell et al., 1998). Photoprotection mechanisms include anthocyanins, the xanthophyll cycle, and the antioxidant system (Chen et al., 2013). It is possible that the high-elevation environment alters defense metabolism (Karagiannis et al., 2020).

The phytochemical compounds in fruits are considerably affected by the climatic conditions of the growth site (Zeng et al., 2020). Higher elevations and high light intensities can increase soluble phenolic, flavanols, antioxidant capacity (Pereira et al., 2006; Martz et al., 2010), and total flavonoids (Mphahlele et al., 2014).

Karagiannis et al. (2020) reported the accumulation of some polyphenolic compounds, mainly quercetin-3-O-rhamnoside, quercetin-3-O-glucoside+ galactoside, quercetin-3-O-rutinoside, and chlorogenic acid in apple peels in high-elevation environments. Voronkov et al. (2019) compared elevations from 300–1,200 m a.s.l. in the Caucasus and found that the concentrations of phenolic substances and polyunsaturated fatty acids increased in the skin of apples with increased elevations. This increase in phenols protects fruits from high UV radiation because of their powerful antioxidant effects, while unsaturated fatty acids retain the fluidity of the fruit's cell membranes within the physiological range (Voronkov et al., 2019). In avocado, elevation also induced an increase in the percentage of oil, oleic, palmitoleic and linoleic acids (Carvalho et al., 2015).

The content of total phenols and flavonoids in the epidermis of peach fruits (Karagiannis et al., 2016) increased at higher elevations, as did anthocyanins and tannins in grapes (Oliveira et al., 2019), confirming the importance of these protective substances in the skin of fruits that reduce the incidence of high UV radiation. In the bilberry (*Vaccinium myrtillus*), a direct relationship was found between antioxidant capacity and elevation in the range from 900–1,450 m a.s.l.: the relationship between elevation and total anthocyanins was inversely proportional, while from 1,500 m a.s.l., the dependence was proportional (Papanov et al., 2021). In peaches from different elevations, Wang et al. (2021) reported genetic differentiation and genes related to signaling in response to light stress, mainly UV stress signaling. They also found an increased expression of these genes in plants grown at high elevations. Through profiling the metabolomes, these authors determined that 379 metabolites had genetic correlations with elevation and that phenylpropanoids compounds were important metabolites associated with high-elevation adaptations (Wang et al., 2021).

For flavanol accumulation, the temperature has less influence than light exposure. UV-B radiation stimulates flavanol biosynthesis (Oliveira et al., 2019), so high elevations favor these compounds in fruits. The epicatechin was not affected by elevation in grape seeds, while the resveratrol in the peel was higher at low elevations, apparently because there was a greater influence from temperature than radiation (Oliveira et al., 2019).

Another important antioxidant, ascorbic acid, increased in banana passionfruit when they were grown at higher elevations (2,498 vs. 2,006 m a.s.l.) (Mayorga et al., 2020). A similar result was reported in strawberries (Pérez de Camacaro et al., 2017). In the cape gooseberry, the increasing elevation did not affect this acid (Fischer et al., 2000), while in apple trees, from 1,400–1,800 m a.s.l., ascorbic acids decreased in fruits, but antioxidant activity and total phenols increased (Kumar et al., 2019).

For plant breeders, climatic adaptation is required for a new variety, and fruit trees are continuously modified through genetic selection to adapt to new environments, such as light intensity, photoperiod, cold and heat, type of soil, and moisture conditions. Many of these requirements change with elevation and latitude (Sherman & Beckman, 2003). Interestingly, Wang et al. (2021) stated that the use of retrotransposons allows a promising approach for the future characterization of genes and metabolites, both for increasing our understanding of the mechanisms that contribute to adaptation to high elevations and the genetic improvement of fruit tree crops through breeding.

Light is a determining stimulus in the accumulation of pigments (carotenoids and anthocyanins) in fruits because it affects the expression of genes associated with the synthesis of fruit pigments (Yahia & Carillo-López, 2019).

Carotenoids are an important factor in adaptations to low temperatures and high sunlight (Kato et al., 2019). Carotenoids protect chlorophyll-like photosynthetic pigments from photodamage (Yahia & Carillo-López, 2019). In peach fruits, Zheng et al. (2023) found that carotenoids are affected by light and temperature and that the contents of nine kinds of carotenoid pigments were associated with elevation. The pigments *PSY*, *CCD4*, and *BCH* were differentially expressed between the different elevations (low elevation and high elevation). Therefore, carotenoids also have determinant functions in the mountainous region adaptation of peach fruits (Zheng et al., 2023).

The intensity of the red color in strawberry (Pérez de Camacaro et al., 2017) and pomegranate (Al-Kalbani et al., 2021) fruits, resulting from increased concentrations of anthocyanins, has also been found in peaches (Karagiannis et al., 2016), blueberries (Zeng et al., 2020) and grapes (Oliveira et al., 2019). In apples, Bai et al. (2014) found that UV-B induces the expression of the Bbox (BBX) gene *MdCOL11* in apples, and also that the *MdCOL11* promoter is targeted by MdHY5 to stimulate anthocyanin biosynthesis in apple peels.

In apple fruits, high-elevation red coloration was stimulated by inducing anthocyanin (Cy-3-O-galactoside and cyanidin-3-O-sambubioside) synthesis. This suggests that most of the color-related phenotypic variability apparently can be related with the different UV levels and air temperature, both associated with elevation (Karagiannis et al., 2020). So, UV light plays an important role in anthocyanin and flavanol synthesis via regulation of *HY5* and *MYB10/MYB22* (Henry-Kirk et al., 2018). A low temperature was necessary to elicit an important increase in the transcription of *MYB10*, and in anthocyanin abundance in apples (Lin-Wang et al., 2011). Therefore, fruits growing in high-elevation environments may develop strategies to adapt to these conditions (Karagiannis et al., 2020). However, at high elevations, the reduction of photosynthetic proteins was observed in parallel with the degradation of chlorophylls and the loss of green color in the epidermis of apple fruits during maturity transition, indicating that these processes are associated with the anthocyanin content of high-elevation fruits (Karagiannis et al., 2020). These authors suggested the presence of a tight regulatory control between proteins related to photosynthesis and anthocyanin production (Karagiannis et al., 2020), whereby anthocyanins protect photosynthetic tissues at high elevations and reduce photobleaching and photoinhibition in the apple peel (Smillie & Hetherington, 1999; Spinardi et al., 2019, Karagiannis et al., 2020).

Thus, high elevations can be a very important condition because they improve the color of fruits and increase the content of pigments. A more intense and more uniform coloration of fruits generates greater acceptance among consumers (Rodríguez-Mena et al., 2023). So, pigments provide health benefits that reduce the probability of acquiring diseases such as type-1 diabetes, obesity, and coronary diseases (Rodríguez-Mena et al., 2023).

For more details on the effect of high elevation on the physicochemical characteristics of fruits, such as sugars, acids, and firmness, among others, the review by Fischer et al. (2022a) is recommended. These authors stated that the firmness of the fruits decreased with increasing elevation. In most cases, the total soluble solids of fruits increased with elevation, which may have been due to the increase in photosynthesis with greater solar radiation, while the total acidity did not show a clear trend when increasing the elevation (Fischer et al., 2022a).

Although several species of fruit trees have different mechanisms to adapt to high elevations, it is possible that when they are grown at elevations above their optimal range, they suffer damage from abiotic stress, mainly because of low temperatures (chilling and freezing stress) and high UV radiation, which can cause significant production losses (Madani et al., 2019).

On clear nights in areas above 2,200 m a.s.l., frost commonly occurs, and, because of the thermal exchange on the surface of the outer tissues of plants, cooling below the air temperature leads to damage in cell membranes and intracellular compartments (Voronkov et al., 2019). Also, sudden changes in temperature between day and night at elevation can generate physiological disorders such as fruit cracking in gooseberry and granadilla, among others (Fischer et al., 2021b).

The high and prolonged radiation, especially UV-B, at high elevations can cause sunburn on the epidermis of fruits (Figure 17.5), which occurs in different species (Fischer et al., 2022a). Excessive radiation could produce damage in tissues as a result of reactive oxygen species (ROS) production. Photo-oxidative sunburn leads to symptomatic pigmentations because of chlorophyll loss and brown pigment accumulation (Rustioni et al., 2020; Rustioni, 2017; Rustioni et al., 2015). Sun injury symptoms differ in species, varieties, and environmental conditions, from light to dark brown tissue areas and necrosis, with or without the presence of tissue bleaching (Morales-Quintana et al., 2020; Felicetti & Schrader, 2008). Discoloration, irregular ripening, and cell rupture may also occur, while internally there is a release of electrolytes, reduction of $Fv/Fm$, damage to DNA, proteins and lipids, and reduction in $\Psi$s, among others (Morales-Quintana et al., 2020).

Direct solar radiation increases temperature and evapotranspiration through the fruit surface, causing high moisture loss and increases in the susceptibility of fruits to cracking (Ikram et al., 2020; Fischer et al., 2021b). In Japanese plums (*Prunus salicina* Lindl.), excessive radiation manifests as a brown to yellow discoloration on the surface of the fruits, which can even result in necrotic spots and cracking of the fruit epidermis (Makeredza et al., 2018). In sweet granadilla, sunburn appears as brown spots on the epidermis that can cover a large part of the surface (Fischer et al., 2022a).

## 17.5 FIRST STUDIES ON THE ANDEAN BLUEBERRY AT HIGH ELEVATION

In the Andean blueberry (*Vaccinium meridionale* Swartz) (Figure 17.6), temperature—which starts to produce leaf damage from the cold that is registered by increased electrolyte leakage from cells—is equal to −10.5°C in the Andes of Venezuela (Cáceres & Rada, 2011), making this species cold tolerant. According to these authors, low temperatures are not a limiting factor for this species' natural distribution since the plants were sampled at 3,420 m a.s.l., close to the upper limit of the species distribution, where

**FIGURE 17.5** Sunburn symptoms on feijoa fruits al 2,600 m a.s.l. in the Boyaca highlands of Colombia, with chlorophyll loss and reddish-brown pigment accumulation. Parra-Coronado et al. (2015b) observed that the color of healthy fruits was not affected by different elevation conditions (1,800 m a.s.l. *vs.* 2,580 m a.s.l.).

**FIGURE 17.6** Ripe Andean blueberry fruits grown at 3,000 m a.s.l. in the Department of Santander, Colombia.

the air temperatures were well above −10.5°C. The average temperatures for the optimum photosynthesis of this species were reported at 15.7°C, while about 80% of carbon assimilation by the plants occurred from 9.3–22.2°C (Cáceres & Rada, 2011).

In the Colombian Andes (Department of Cundinamarca), the juvenile Andean blueberry produced smaller plants with shortened internodes and less ramification as the elevation increased from 1,885 m a.s.l. to 3,300 m a.s.l. (Becerra Rojas et al., 2022). The height of the plants had a direct correlation with the number of primary branches and the number of leaves. Therefore, the Andean blueberry grown at 3,300 m a.s.l. produced a smaller leaf area than the plants grown at 2,556 m a.s.l. (Becerra Rojas et al., 2022).

## 17.6 DECIDUOUS FRUIT TREES IN THE HIGH TROPICS: A SPECIAL SITUATION

Deciduous fruit trees, such as apple, pear, quince, peach, Japanese plums, grapes, and blueberry, among others, are adapted to certain conditions of the climate in the high tropics, between latitudes 0° and 15°, maintaining photosynthetically active leaves for 11 months (Ramírez & Kallarackal, 2014). For this reason, the cultivation of deciduous fruit trees is possible in nearly all regions near the equator (Pio et al., 2019). In the adaptation of deciduous fruits to the tropics, the relatively uniform temperature and photoperiod throughout the year in low latitudes are not part of the natural physiology (Fischer, 2013). But these plants, native to climatic zones with cold winters, enter a state of dormancy that allows them to tolerate low temperatures (<0°C) and renew their growth when the cold season ends (Luedeling, 2012).

Fruit farmers, in order to meet the chilling requirements for breaking rest in vegetative and reproductive buds in these species, selected sites in Colombia from 1,700–2,800 m a.s.l. (Fischer & Orduz-Rodríguez, 2012) (Table 17.4), with minimum annual temperatures of 12–13°C (Fischer, 2013). However, because of global warming, these species—in most cases—grew abnormally at these elevations and presented erratic bud breaking, with asynchronous fruiting and, in extreme cases, a prolonged rest period.

In Colombia, the cultural adaptations of these deciduous species include—in most cases—continuous growth cycles to avoid endodormancy (deep rest period), because their requirement for chilling hours (<7.2°C or <7°C) is not met under the conditions of the inner tropics (Fischer, 2000). Edwards (1987) observed that deciduous fruit trees can grow without any winter cold, with artificially induced dormancy through defoliation. This forced or continuous cropping system in high tropical areas requires a uniform temperature throughout the year, without a cold season (Fischer, 2013; Pinzón et al., 2014). In the equatorial tropics, day lengths are not short enough and temperatures are not sufficiently low to induce the natural abscission of leaves (which contain the inhibitor of bud breaking, abscisic acid). So, manual or chemical defoliation of leaves is necessary to activate gibberellin and cytokinin synthesis in bud breaking (Edwards, 1987).

In Colombia, fruit growers apply a sequence of techniques that start after the harvest: suppression of the vegetative growth of trees by withholding irrigation for one month, then defoliation, pruning, irrigation, fertilization, and, in the end, applying of rest-avoiding chemicals (e.g., hydrogen cyanamide or potassium nitrate) if the tree has poor bud breaking (Fischer, 2013). There is a bimodal rain pattern, making two crops per year possible (using this rest avoidance technique) in varieties with short fruit development, e.g., the apple "Anna" with a low chilling requirement (250–300 hours), and fruit development of only 100–120 days (Gutiérrez-Villamil et al., 2022). Varieties with longer fruit development or excessive forced reproductive growth (risk of causing dwarfing effects) can achieve three harvests in two years. Only cases of a colder period (compared to a mild "winter"), such as in Nuevo Colón (Boyaca, Colombia) that presents an elongated unimodal rain pattern, can have only one harvest per year, where a rest-breaking chemical is applied that compensates for the missing chilling hours.

In all cases, much research is still required to understand and handle dormancy, bud breaking, flowering, and fructification of deciduous fruit species in the tropics (Ramírez & Kallarackal, 2014; Pio et al., 2019). In addition, the introduction of new varieties can alleviate some of these difficulties because the response of genotypes depends on environmental conditions (Fischer et al., 2013).

## 17.7 INFLUENCE OF ELEVATION ON FEIJOA, A FRUIT INTRODUCED INTO A TROPICAL REGION

Research carried out by Parra-Coronado et al. (2016) on feijoa crops (*Acca sellowiana* [O. Berg] Burret) at elevations of 1,800 m a.s.l. and 2,580 m a.s.l. with similar soil conditions and crop management showed that the fruits produced at higher elevations required more calendar days and less thermal time or GDD from anthesis to harvest than those produced at lower elevations. In addition, the authors developed mathematical models for the weight based on the size and growth of the fruits (increase in fruit weight) in relation to the elevation and GDD of the orchard locations in the study. These models showed that the weight and size of ripe feijoa fruits had a direct relationship with the elevation of the plantation (Parra-Coronado et al., 2016).

In the studies carried out on the growth and development of feijoa fruits, with similar soil conditions and crop management, Parra-Coronado et al. (2015b) indicated that the weight and size of the ripe fruits were directly associated to the elevation of the production site. The study showed that the fruits produced at the highest elevation (2,580 m a.s.l.)—where climatic conditions had lower temperatures, less precipitation, lower relative humidity and greater solar radiation accumulated during the fruit development—had a higher total soluble solids (TSS) value than fruits produced at lower elevations (1,800 m a.s.l.). Similar behaviors were reported for different fruit species grown at low-temperature conditions (Benkeblia & Tennant, 2011).

The total titratable acidity (TTA) of the ripe feijoa fruit was not determined by the prevailing climatic conditions in the two studied sites (Parra-Coronado et al., 2015b), as found for cape gooseberry fruits grown at 2,300 m a.s.l. and 2,690 m a.s.l. by Fischer et al. (2007).

Parra-Coronado et al. (2017) proposed mathematical models (equations) to predict the quality of feijoa fruits during their growth and development as a function of elevation and climatic conditions of cultivation in two sites in the Colombian Department of Cundinamarca: San Francisco de Sales (1,800 m a.s.l.; mean air temperature 20.6°C; RH from 63–97%; mean annual precipitation of 1,493 mm) and Tenjo (2,580 m a.s.l.; mean air temperature 12.5°C; RH from 74–86%; mean annual precipitation of 765 mm). Equations were given for the variation in fruit weight, TTA, TSS, skin and pulp firmness, and skin color (hue) (Parra-Coronado et al., 2017). Regression analyses showed that the models sufficiently predicted various fruit attributes during growth for both sites. The results illustrated that elevation, GDD, and accumulated precipitation had more impact on the physicochemical attributes during fruit growth. The models for fresh weight, TTA, and skin firmness of fruits best predicted quality behavior during their growth and development (Parra-Coronado et al., 2017).

## 17.8 CONCLUSIONS

- The adaptation of fruit plants to higher elevations is generally related to climatic conditions, such as lower temperatures and partial pressure of gases, and importantly, higher solar radiation—especially UV-B—as elevation increases.
- Decreases in temperature, about 0.6–0.7°C per 100 m of tropical elevation, increase plant phenological cycles and facilitate the introduction of fruit species from subtropical (e.g., feijoa) and temperate climates (e.g., deciduous fruits) to the inner tropics.
- In the Colombian Andes, many fruit species are grown up to 2,600–2,800 m a.s.l., but the Andean blueberry and banana passionfruit withstand up to 3,200 m a.s.l.
- The ecophysiological responses of fruit species to tropical elevation conditions include morphological, anatomical, and biochemical adaptations, especially in the leaves and fruits.
- Many fruit species and cultivars from the tropics have developed mechanisms of tolerance to UV-B radiation, particularly in leaves that are smaller and thicker, with a more resistant cuticle, a dense pubescence, and an increased number of stomata at higher elevation (e.g., the cape gooseberry).
- Other physiological adaptations of fruits to increased elevation and UV-B radiation include an increase in anthocyanin contents, carotenoid synthesis, the xanthophyll cycle, and the antioxidant system.
- Deciduous fruits, which grow up to 2,800 m a.s.l. in the Colombian Andes, in most cases, do not fulfill their chilling requirements for bud breaking and, therefore, are managed without deep dormancy in continuous cropping.
- Research results for well-adapted feijoa fruits have shown that elevation, GDDs and accumulated precipitation influence the physicochemical characteristics of fruits more.

Generally, many studies are needed to understand the adaptation of fruit trees to tropical elevations, where the microclimate has been affected by recent climate change conditions. Plant breeding and the selection of new varieties offer new possibilities for the cultivation of fruit species in the high tropics with new genotype × environment interactions.

## REFERENCES

Al-Kalbani, B. S., R. A. Al-Yahyai, A. M. Al-Sadi, and A. G. H. Al-Mamari. 2021. Physical and chemical fruit quality attributes of two pomegranate cultivars grown at varying altitudes of Al-Hajar Mountains in Oman. *Journal of Agricultural and Marine Sciences* 26(2):42–50. https://doi.org/10.53541/jams.vol26iss2pp42-50

Almanza-Merchán, P. J., and G. Fischer. 2012. Tuna (*Opuntia ficus-indica* (L.) Miller). In *Manual para el cultivo de frutales en el trópico*, ed. G. Fischer, 1014–1023. Bogota: Produmedios.

Atkinson, C. J., R. M. Brennan, and H. G. Jones. 2013. Declining chilling and its impact on temperate perennial crops. *Environmental and Experimental Botany* 91:48–62. https://doi.org/10.1016/j.envexpbot.2013.02.004

Bai, S., T. Saito, C. Honda, Y. Hatsuyama, A. Ito, and T. Moriguchi. 2014. An apple B-box protein, MdCOL11, is involved in UV-B-and temperature-induced anthocyanin biosynthesis. *Planta* 240:1051–1062. https://doi.org/10.1007/s00425-014-2129-8

Balaguera-López, H. E., G. Fischer, and A. Herrera-Arévalo. 2022. Postharvest physicochemical aspects of Campomanesia lineatifolia R. & P. fruit, a Myrtaceae with commercial potential. *Revista Colombiana de Ciencias Hortícolas* 16(2):e14185. https://doi.org/10.17584/rcch.2022v16i2.14185

Becerra Rojas, A. D., S. Quevedo Rubiano, S. Magnitskiy, and H. O. Lancheros Redondo. 2022. Morphological responses of Andean blueberry (*Vaccinium meridionale* Swartz) plants growing in three environments different in altitude. *Revista Colombiana de Ciencias Hortícolas* 16(3):e15034. https://doi.org/10.17584/rcch.2022v16i3.15034

Beck, E. 1994. Cold tolerance in tropical alpine plants. In *Tropical Alpine Environments—Plant Form and Function*, eds. P. W. Rundel, A. P. Smith, and F. C. Meinzer, 77–110. Cambridge: Cambridge University Press.

Benavides, H. O., O. Simbaqueva, and H. J. Zapata. 2017. *Atlas de radiación solar, ultravioleta y ozono de Colombia*. Bogota: Instituto de Hidrología, Meteorología y Estudios Ambientales (IDEAM). www.andi.com.co//Uploads/RADIACION.compressed.pdf (accessed August 10, 2022).

Benkeblia, N., and P. F. Tennant. 2011. Preharvest and harvest factors influencing the postharvest quality of tropical and subtropical fruits. In *Postharvest Biology and Technology of Tropical and Subtropical Fruits* (Vol. 1), ed. E. M. Yahia, 112–141. Cambridge: Woodhead Publishing.

Blancke, R. 2016. *Tropical Fruits and Other Edible Plants of the World*. Ithaca and London: Cornell University Press.

Bonnet, J. G., and J. F. Cárdenas. 2012. Tomate de árbol (*Cyphomandra betacea* Sendt.). In *Manual para el cultivo de frutales en el trópico*, ed. G. Fischer, 825–850. Bogota: Produmedios.

Buchanan, B. B., W. Gruissem, and R. L. Jones. 2015. *Biochemistry and Molecular Biology of Plants*, 2nd ed. New York: John Wiley & Sons.

Cáceres, Y., and F. Rada. 2011. ¿Cómo responde la especie leñosa *Vaccinum meridionale* a la temperatura en su límite altitudinal de distribución en los andes tropicales? *Ecotropicos* 24(1):80–91.

Caldwell, M. M., L. O. Bjorn, J. F. Bornman et al. 1998. Effects if increased solar ultraviolet radiation on terrestrial ecosystems. *Journal of Photochemistry and Photobiology B: Biology* 46(1–3):40–52. https://doi.org/10.1016/S1011-1344(98)00184-5

Caldwell, M. M., J. F. Bornman, C. L. Ballaré, S. D. Flint, and G. Kulandaivelu. 2007. Terrestrial ecosystems, increased solar ultraviolet radiation, and interactions with other climate change factors. *Photochemical & Photobiological Sciences* 6(3):252–66. https://doi.org/10.1039/b700019g

Caldwell, M. M., R. Robberecht, and W. D. Billings. 1980. A steep latitudinal gradient of solar ultraviolet-B radiation in the Arctic-Alpine life zone. *Ecology* 61(3):600–611. https://doi.org/10.2307/1937426.

Campos, T. J. 2013. Especies y variedades de hoja caduca en Colombia. In *Los frutales caducifolios en Colombia—Situación actual, sistemas de cultivo y plan de desarrollo*, eds. D. Miranda, G. Fischer, and C. Carranza, 47–66. Bogota: Sociedad Colombiana de Ciencias Hortícolas.

Cárdenas, J., and G. Fischer. 2013. Clasificación botánica y morfología de manzano, peral, duraznero y ciruelo. In *Los frutales caducifolios en Colombia—Situación actual, sistemas de cultivo y plan de desarrollo*, eds. D. Miranda, G. Fischer, and C. Carranza, 21–30. Bogota: Sociedad Colombiana de Ciencias Hortícolas.

Carillo, E., A. Aller, S. M. Cruz-Quintana, F. Giampieri, and J. M. Alvarez-Suarez, 2015. Andean berries from Ecuador: A review on botany, agronomy, chemistry and health potential. *Journal of Berry Research* 5(2):49–69. https://doi.org/10.3233/JBR-140093

Carvalho, C. P., J. Bernal, M. A. Velásquez, and J. R. Cartagena. 2015. Fatty acid content of avocados (*Persea americana* Mill. cv. Hass) in relation to orchard altitude and fruit maturity stage. *Agronomía Colombiana* 33(2):220–227. https://doi.org/10.15446/agron.colomb.v33n2.49902

Chen, C., H. Li, D. Zhang, P. Li, and F. Ma. 2013. The role of anthocyanin in photoprotection and its relationship with the xanthophyll cycle and the antioxidant system in apple peel depends on the light conditions. *Physiologia Plantarum* 149(3):354–66. https://doi.org/10.1111/ppl.12043.

Cleves, J. A. 2021. *Fundamentos técnicos del cultivo del arándano (Vaccinium corymbosum L.) en la región central de Colombia*. Tunja: Editorial de la Universidad Pedagógica y Tecnológica de Colombia-UPTC.

Corredor, D. 2012. Pitahaya amarilla (Hylocereus megalanthus [K. Schum. ex Vaupel] Ralf Bauer). In *Manual para el cultivo de frutales en el trópico*, ed. G. Fischer, 802–824. Bogota: Produmedios.

Daunicht, H.-J., and H. J. Brinkjans. 1996. Plant responses to reduced air pressure: Advanced techniques and results. *Advances in Space Research* 18(4–5):273–281. https://doi.org/10.1016/0273-1177(95)00889-M

Edwards, G. R. 1987. Producing temperate zone fruit at low latitudes: Avoiding rest and the chilling requirement. *HortScience* 22(6):1236–1240.

Ehlenfeldt, M. K., and J. L. Luteyn. 2021. Fertile intersectional F1 hybrids of 4x Vaccinium meridionale (section Pyxothamnus) and Highbush blueberry, V. corymbosum (section Cyanococcus). *HortScience* 56(3):318–323.

Felicetti, D. A., and L. E. Schrader. 2008. Photooxidative sunburn of apples: Characterization of a third type of apple sunburn. *International Journal of Fruit Science* 8(3):160–172.

Fernández, G. E. M., L. M. Melgarejo, and N. A. C. Rodríguez. 2014. Algunos aspectos de la fotosíntesis y potenciales hídricos de la granadilla (*Passiflora ligularis* Juss.) en estado reproductivo en el Huila, Colombia. *Revista Colombiana de Ciencias Hortícolas* 8(2):206–216. https://doi.org/10.17584/rcch.2014v8i2.32142014

Fischer, G. 1995. *Effect of root zone temperature and tropical altitude on the growth, development and fruit quality of cape gooseberry (Physalis peruviana L.)*. PhD thesis. Berlin: Humboldt-Universität zu Berlin.

Fischer, G. 2000. Ecophysiological aspects of fruit growing in tropical highlands. *Acta Horticulturae* 531:91–98. https://doi.org/10.17660/ActaHortic.2000.531.13

Fischer, G. 2013. Comportamiento de los frutales caducifolios en el trópico. In *Los frutales caducifolios en Colombia—Situación actual, sistemas de cultivo y plan de desarrollo*, eds. D. Miranda, G. Fischer, and C. Carranza, 31–46. Bogota: Sociedad Colombiana de Ciencias Hortícolas.

Fischer, G., P. J., Almanza-Merchán, and W. Piedrahíta. 2012. Brevo (*Ficus carica* L.). In *Manual para el cultivo de frutales en el trópico*, ed. G. Fischer, 943–952. Bogota: Produmedios.

Fischer, G., H. E. Balaguera-López, and J. Álvarez-Herrera. 2021b. Causes of fruit cracking in the era of climate change. A review. *Agronomía Colombiana* 39(2):196–207. https://doi.org/10.15446/agron.colomb.v39n2.97071

Fischer, G., H. E. Balaguera-López, and S. Magnitskiy. 2021a. Review on the ecophysiology of important Andean fruits: Solanaceae. *Revista UDCA Actualidad & Divulgación Científica* 24(1):e1701. http://doi.org/10.31910/rudca.v24.n1.2021.1701

Fischer, G., J. A. Cleves-Leguizamo, and H. E. Balaguera-López. 2022b. Impact of soil temperature on fruit species within climate change scenarios. *Revista Colombiana de Ciencias Hortícolas* 16(1):e12769. http://doi.org/10.19053/rcch.2022v16i1.12769

Fischer, G., G. Ebert, and P. Lüdders. 2000. Provitamin A carotenoids, organic acids and ascorbic acid content of cape gooseberry (*Physalis peruviana* L.) ecotypes grown at two tropical altitudes. *Acta Horticulturae* 531:263–267. https://doi.org/10.17660/ActaHortic.2000.531.43

Fischer, G., G. Ebert, and P. Lüdders. 2007. Production, seeds and carbohydrate contents of cape gooseberry (*Physalis peruviana* L.) fruits grown at two contrasting Colombian altitudes. *Journal of Applied Botany and Food Quality* 81:29–35.

Fischer, G., and P. Lüdders. 2002. Efecto de la altitud sobre el crecimiento y desarrollo vegetativa de la uchuva (*Physalis peruviana* L.). *Revista Comalfi* 29(1):1–10.

Fischer, G., and L. M. Melgarejo. 2014. Ecofisiología de la uchuva (*Physalis peruviana* L.). In *Physalis peruviana: Fruta andina para el mundo*, eds. C. P. Carvalho, and D. A. Moreno, 31–47. Limencop: Programa Iberoamericano de Ciencia y Tecnología para el Desarrollo—CYTED.

Fischer, G., and L. M. Melgarejo. 2020. The ecophysiology of cape gooseberry (*Physalis peruviana* L.)—an Andean fruit crop. A review. *Revista Colombiana de Ciencias Hortícolas* 14(1):76–89. https://doi.org/10.17584/rcch.2020v14i1.10893

Fischer, G., and L. M. Melgarejo. 2021. Ecophysiological aspects of guava (*Psidium guajava* L.). A review. *Revista Colombiana de Ciencias Hortícolas* 15(2):e12355. https://doi.org/10.17584/rcch.2021v15i2.12355

Fischer, G., and D. Miranda. 2021. Review on the ecophysiology of important Andean fruits: *Passiflora* L. *Revista Facultad Nacional de Agronomía Medellín* 74(2):9471–9481. https://doi.org/10.15446/rfnam.v74n2.91828

Fischer, G., L. M. Melgarejo, and H. E. Balaguera-López. 2022c. Review on the impact of elevated $CO_2$ concentrations on fruit species in the face of climate change. *Ciencia y Tecnología Agropecuaria* 23(2):e2475. https://doi.org/10.21930/rcta.vol23_num2_art:2475

Fischer, G., and J. O. Orduz-Rodríguez. 2012. Ecofisiología en frutales. In *Manual para el cultivo de frutales en el trópico*, ed. G. Fischer, 54–72. Bogota: Produmedios.

Fischer, G., and A. Parra-Coronado. 2020. Influence of environmental factors on the feijoa (*Acca sellowiana* [Berg] Burret): A review. *Agronomía Colombiana* 38(3):388–397. https://doi.org/10.15446/agron.colomb.v38n3.88982

Fischer, G., A. Parra-Coronado, and H. E. Balaguera-López. 2022a. Altitude as a determinant of fruit quality with emphasis on the Andean tropics of Colombia. A review. *Agronomía Colombiana* 40(2). https://doi.org/10.15446/agron.colomb.v40n2.101854

Fischer, G., O. C. Quintero, C. P. Tellez, and L. M. Melgarejo. 2020. Curuba: Passiflora tripartita var. mollissima y Passiflora tarminiana. In Pasifloras—especies cultivadas en el mundo, eds. A. Rodríguez, F. G. Faleiro, M. Parra, and A. M. Costa, 105–121. Brasilia: ProImpress.

Fischer, G., F. Ramírez, and F. Casierra-Posada. 2016. Ecophysiological aspects of fruit crops in the era of climate change. A review. *Agronomía Colombiana* 34(2):190–199. https://doi.org/10.15446/agron.colomb.v34n2.56799

Grabherr, G., M. Gottfried, and H. Pauli. 2010. Climate change impacts in alpine environments. *Geography Compass* 4(8):1133–1153. https://doi.org/10.1111/j.1749-8198.2010.00356.x

Guerrero, A. L., S. Gallucci, P. Michalijos, and S. M. Visciarelli. 2011. Países Andinos: Aportes teóricos para un abordaje integrado desde las perspectivas geográfica y turística. *Huellas* 15:121–138.

Gutiérrez-Villamil, D. A., J. G. Ávarez-Herrera, and G. Fischer. 2022. Performance of the 'Anna' apple (*Malus domestica* Borkh.) in tropical highlands: A review. *Revista de Ciencias Agrícolas* 39(1):123–141. https://doi.org/10.22267/rcia.223901.175

Henry-Kirk, R. A., B. Plunkett, M. Hall et al. 2018. Solar UV light regulates flavonoid metabolism in apple (*Malus × domestica*). *Plant Cell Environment* 41:675–688. https://doi.org/10.1111/pce.13125

Hoyos, J. 1989. *Frutales en Venezuela*. Caracas: Sociedad de Ciencias Naturales La Salle.

Ikram, S., W. Shafqat, M. A. Qureshi, et al. 2020. Causes and control of fruit cracking in pomegranate: A review. *Journal of Global Innovations in Agricultural and Social Sciences* 8(4):183–190. https://doi.org/10.22194/JGIASS/8.920

Janicot, S., C. Aubertin, M. Bernoux et al. 2015. Chapter 11. Highland zones: The rapid change of Andean environments. In *Climate Change. What Challenges for the South?* 129–143. Marseille: IRD Editions. https://doi.org/10.4000/books.irdeditions.34209

Jenkins, G. I. 2009. Signal transduction in responses to UV-B radiation. *Annual Review of Plant Biology* 60:407–431. http://dx.doi.org/10.1146/annurev.arplant.59.032607.092953

Karagiannis, E., M. Michailidis, G. Tanou, et al. 2020. Decoding altitude-activated regulatory mechanisms occurring during apple peel ripening. *Horticulture Research* 7:120. https://doi.org/10.1038/s41438-020-00340-x

Karagiannis, E., G. Tanou, M. Samiotaki, et al. 2016. Comparative physiological and proteomic analysis reveal distinct regulation of peach skin quality traits by altitude. *Frontiers in Plant Science* 7:1689. https://doi.org/10.3389/fpls.2016.01689

Kato, S., Y. Tanno, S. Takaichi, and T. Shinomura. 2019. Low temperature stress alters the expression of phytoene desaturase genes (crtP1 and crtP2) and the zeta-carotene desaturase gene (crtQ) together with the cellular carotenoid content of Euglena gracilis. *Plant and Cell Physiology* 60(2):274–284. https://doi.org/10.1093/pcp/pcy208.

Körner, C. 2007. The use of 'altitude' in ecological research. *Trends in Ecology and Evolution* 22(11):569–574. https://doi.org/10.1016/j.tree.2007.09.006

Kulandaivelu, G., S. Maragatham, and N. Nedunchezhian. 1989. On the possible control of ultraviolet-B induced response in growth and photosynthetic activities in higher plants. *Physiologia Plantarum* 76:398–404.

Kumar, P., S. Sethi, R. R. Sharma, et al. 2019. Influence of altitudinal variation on the physical and biochemical characteristics of apple (*Malus domestica*). *Indian Journal of Agricultural Sciences* 89(1):145–152.

Larcher, W. 2003. *Physiological Plant Ecology: Ecophysiology and Stress Physiology of Functional Groups*, 4th ed., New York: Springer. http://dx.doi.org/10.1007/978-3-662-05214-3

Ligarreto, G. A. 2012. Recursos genéticos d especies frutícolas en Colombia. In *Manual para el cultivo de frutales en el trópico*, ed. G. Fischer, 35–53. Bogota: Produmedios.

Lin-Wang, K., D. Micheletti, J. Palmer, et al. 2011. High temperature reduces apple fruit colour via modulation of the anthocyanin regulatory complex. *Plant Cell Environment* 34:1176–1190.

Lobo, M., C. I. Medina, O. A. Delgado, and A. Bermeo. 2007. Variabilidad morfologica de la colección colombiana de lulo (*Solanum quitoense* Lam.) y especies elacionadas de la sección Lasiocarpa. *Revista Facultad Nacional de Agronomía Medellín* 60(2):3939–3964.

Luedeling, E. 2012. Climate change impacts on winter chill for temperate fruit and nut production: A review. *Scientia Horticulturae* 144:218–229. http://dx.doi.org/10.1016/j.scienta.2012.07.011

Madani, B., A. Mirshekari, and Y. Imahori. 2019. Physiological responses to stress. In *Postharvest physiology and biochemistry of fruits and vegetables*, eds. E. Yahía, and A. Carrillo-López, 405–423. Elsevier. https://doi.org/10.1016/B978-0-12-813278-4.00020-8

Makeredza, B., M. Jooste, E. Lötze, M. Schmeisser, and W. J. Steyn. 2018. Canopy factors influencing sunburn and fruit quality of Japanese plum (*Prunus salicina* Lindl.). *Acta Horticulturae* 1228:121–128. https://doi.org/10.17660/ActaHortic.2018.1228.18

Marengo, J. A., J. D. Pabón, A. Díaz et al. 2011. Climate change: Evidence and future scenarios for the Andean region. In *Climate Change and Biodiversity in the Tropical Andes*, eds. S. Herzog, R. Martinez, P. M. Jorgensen, and H. Tiessen, 110–127. Paris: IAI-SCOPE-UNESCO.

Martz, F., L. Jaakola, R. Julkunen-Tiitto, and S. Stark. 2010. Phenolic composition andantioxidant capacity of bilberry (*Vaccinium myrtillus*) leaves in Northern Europe following foliar development and along environmental gradients. *Journal of Chemical Ecology* 36:1017–1028.

Mayorga, M., G. Fischer, L. M. Melgarejo, and A. Parra-Coronado. 2020. Growth, development and quality of Passiflora tripartita var. Mollissima fruits under two environmental tropical conditions. *Journal of Applied Botany and Food Quality* 93:66–75. https://doi.org/10.5073/JABFQ.2020.093.009

Medina, C. I., M. Lobo, A. A. Castaño, and L. E. Cardona. 2015. Análisis del desarrollo de plantas de mortiño (*Vaccinium meridionale* Swart.) bajo dos sistemas de propagación: clonal y sexual. *Corpoica Ciencia y Tecnología Agropecuaria* 16(1):65–77.

Meinzer, F. C., G. Goldstein, and F. Rada. 1994. Páramo microclimate and leaf thermal balance of Andean giant rosette plants. In *Tropical Alpine Environments. Plant form and Function*, eds. P. W. Rundel, A. P. Smith, and F. C. Meinzer, 45–60. Cambridge: Cambridge University Press.

Mendoza, M., C. A. Peres, and L. P. C. Morellato. 2017. Continental-scale patterns and climatic drivers of fruiting phenology: A quantitative neotropical review. *Global Planetary Change* 148:227–241. http://dx.doi.org/10.1016/j.gloplacha.2016.12.001

Miranda, D. and G. Fischer. 2021. Avances tecnológicos en el cultivo de la uchuva (*Physalis peruviana* L.) en Colombia. In *Avances en el cultivo de las berries en el trópico*, eds. G. Fischer, D. Miranda, S. Magnitskiy, H. E. Balaguera-López, and Z. Molano, 14–36. Bogota: Sociedad Colombiana de Ciencias Hortícolas. https://doi.org/10.17584/IBerries

Morales, C. S. and B. Villegas. 2012. Mora (*Rubus glaucus* Benth.). In *Manual para el cultivo de frutales en el trópico*, ed. G. Fischer, 728–754. Bogota: Produmedios.

Morales-Quintana, L., J. M. Waite, L. Kalcsits, C. A. Torres, and P. Ramos. 2020. Sun injury on apple fruit: Physiological, biochemical and molecular advances, and future challenges. *Scientia Horticulturae* 260:108866.

Moreno-Medina, B. L. and F. Casierra-Posada. 2021. Caracterización de especies de mora (*Rubus* sp.) cultivadas en los altiplanos tropicales. In *Avances en el cultivo de las berries en el trópico*, eds. G. Fischer, D. Miranda, S. Magnitskiy, H. E. Balaguera-López, and Z. Molano, 102–112. Bogota: Sociedad Colombiana de Ciencias Hortícolas. https://doi.org/10.17584/IBerries

Mphahlele, R. M., M. A. Stander, O. A. Fawole, and U. L. Opara. 2014. Effect of fruit maturity and growing location on the postharvest contents of flavonoids, phenolic acids, vitamin C and antioxidant activity of pomegranate juice (cv. Wonderful). *Scientia Horticulturae* 179:36–45. https://doi.org/10.1016/j.scienta.2014.09.007

Nagy, L., and G. Grabherr. 2009. *The Biology of Alpine Habitats*. Oxford: Oxford University Press.

National Research Council. 1989. *Lost Crops of the Incas*. Washington, DC: National Academic Press.

Neugart, S. and M. Schreiner. 2018. UVB and UVA as eustressors in horticultural and agricultural crops. *Scientia Horticulturae* 234:370–381. https://doi.org/10.1016/j.scienta.2018.02.021

Nobel, P. S. 2020. *Physicochemical and Environmental Plant Physiology*, 5th ed. London: Elsevier.

Ocampo, J., A. Rodríguez, and M. Parra. 2020. Gulupa: *Passiflora edulis* f. *edulis* Sims. In *Pasifloras—especies cultivadas en el mundo*, eds. A. Rodríguez, F. G. Faleiro, M. Parra, and A. M. Costa, 139–157. Brasilia: ProImpress.

Oliveira, J. B., R. Egipto, O. Laureano, R. Castro, G. E. Pereira, and J. M. Ricardo-da-Silva. 2019. Climate effects on physicochemical composition of Syrah grapes at low and high altitude sites from tropical grown regions of Brazil. *Food Research International* 121:870–879. https://doi.org/10.1016/j.foodres.2019.01.011

Orme, A. R. 2007. The tectonic framework of South America. In *The Physical Geography of South America*, eds. T. T. Veblen, K. R. Young, and A. R. Orme, 23–44. Oxford: Oxford University Press.

Papanov, S. I., E. G. Petkova, and I. G. Ivanov. 2021. Polyphenols content and antioxidant activity of bilberry juice obtained from different altitude samples. *Journal of Pharmaceutical Research International* 33(29A):218–223. https://doi.org/10.9734/JPRI/2021/v33i29A31581

Parra-Coronado, A., and G. Fischer. 2013. Maduración y comportamiento poscosecha de la feijoa (*Acca sellowiana* (O. Berg) Burret). *Revista Colombiana de Ciencias Hortícolas* 7(1):98–110. https://doi.org/10.17584/rcch.2013v7i1.2039

Parra-Coronado, A., G. Fischer, and B. Chavez-Cordoba. 2015a. Tiempo térmico para estados fenológicos reproductivos de la feijoa (*Acca sellowiana* (O. Berg) Burret). *Acta Biológica Colombiana* 20(1):167–177. http://dx.doi.org/10.15446/abc.v20n1.43390

Parra-Coronado, A., G. Fischer, and J. H. Camacho-Tamayo. 2015b. Development and quality of pineapple guava fruit in two locations with different altitudes in Cundinamarca, Colombia. *Bragantia* 74(3):359–366. http://dx.doi.org/10.1590/1678-4499.0459

Parra-Coronado, A., G. Fischer, and J. H. Camacho-Tamayo. 2016. Growth model of the pineapple guava (*Acca sellowiana* (O. Berg) Burret), as a function of thermal time and tropical altitude. *Ingeniería e Investigación* 36(3):6–14. https://doi.org/10.15446/ing.investig.v36n3.52336

Parra-Coronado, A., G. Fischer, and J. H. Camacho-Tamayo. 2017. Model of pre-harvest quality of pineapple guava fruits (*Acca sellowiana* (O. Berg) Burret) as a function of weather conditions of the crops. *Bragantia* 76(1):177–186. http://dx.doi.org/10.1590/1678-4499.652

Paull, R. E. and O. Duarte. 2011. *Tropical Fruits*, 2nd ed. CAB International.

Pereira, G. E., J. P. Gaudillere, P. Pieri et al. 2006. Microclimate influence on mineral and metabolic profiles of grape berries. *Journal of Agriculture and Food Chemistry* 54:6765–6775.

Pérez de Camacaro, M., M. Ojeda, A. Giménez, M. González, and A. Hernández. 2017. Atributos de calidad en frutos de fresa 'Capitola' cosechados en diferentes condiciones climáticas en Venezuela. *Bioagro* 29(3):163–174.

Peyre, G., H. Balslev, X. Font, and J. S. Tello. 2019. Fine-scale plant richness mapping of the Andean paramo according to macroclimate. *Frontiers in Ecology and Evolution* 7:377. https://doi.org/10.3389/fevo.2019.00377

Pinzón, E. H., A. C. Morillo, and G. Fischer. 2014. Aspectos fisiológicos del duraznero (*Prunus persica* [L.] Batsch.) en el trópico alto. Una revisión. *Revista UDCA Actualidad & Divulgación Científica* 17(2):401–411.

Pio, R., F. B. M. d. Souza, L. Kalcsits, R. B. Bisi, and D. d. H. Farias. 2019. Advances in the production of temperate fruits in the tropics. *Acta Scientiarum. Agronomy* 41:e39549.

Ramírez, F., and J. Kallarackal. 2014. Chapter 6: Ecophysiology of temperate fruit trees in the tropics. In *Advances in Environmental Research* (Vol. 31), ed. J. A. Daniels, 1–12. New York: Nova Science Publishers.

Retamales, J. B., and J. F. Hancock. 2012. *Blueberries*. Wallingford: CAB International.

Reyes, C. 2012. Papayuela (*Vasconcellea pubescens*) Lenné et Koch) Badillo). In *Manual para el cultivo de frutales en el trópico*, ed. G. Fischer, 1001–1006. Bogota: Produmedios.

Rodríguez-Mena, A., L. A. Ochoa-Martínez, S. M. González-Herrera, O. M. Rutiaga-Quiñones, R. F. González-Laredo, and B. Olmedilla-Alonso. 2023. Natural pigments of plant origin: Classification, extraction and application in foods. *Food Chemistry* 398:133908. https://doi.org/10.1016/j.foodchem.2022.133908

Ruiz, R. y W. Piedrahita. 2012. Fresa (*Fragaria × ananassa*). In *Manual para el cultivo de frutales en el trópico*, ed. G. Fischer, 474–495. Bogota: Produmedios

Rundel, P. W. 1994. Tropical alpine climates. In *Tropical Alpine Environments. Plant Form and Function*, eds. P. W. Rundel, A. P. Smith, and F. C. Meinzer, 21–44. Cambridge: Cambridge University Press.

Rundel, P. W., Smith, A. P., and Meinzer, F. C. (eds.). 1994. *Tropical Alpine Environments. Plant Form and Function*. Cambridge: Cambridge University Press. https://doi.org/10.1017/CBO9780511551475

Rustioni, L. 2017. Oxidized polymeric phenolics: Could they be considered photoprotectors? *Journal of Agriculture and Food Chemistry* 65(36):7843–7846. https://doi.org/10.1021/acs.jafc.7b03704

Rustioni, L., D. Fracassetti, B. Prinsi et al. 2020. Oxidations in white grape (*Vitis vinifera* L.) skins: Comparison between ripening process and photooxidative sunburn symptoms. *Plant Physiology and Biochemistry* 150:270–278. https://doi.org/10.1016/j.plaphy.2020.03.003

Rustioni, L., C. Milani, S. Parisi, and O. Failla. 2015. Chlorophyll role in berry sunburn symptoms studied in different grape (*Vitis vinifera* L.) cultivars. *Scientia Horticulturae* 185:145–150.

Sánchez-Reinoso, A. D., Y. Jiménez-Pulido, J. P. Martínez-Pérez, C. S. Pinilla, and G. Fischer. 2019. Chlorophyll fluorescence and other physiological parameters as indicators of waterlogging and shadow stress in lulo (*Solanum quitoense* var. *septentrionale*) seedlings. *Revista Colombiana de Ciencias Hortícolas* 13(3):325–335. https://doi.org/10.17584/rcch.2019v13i3.100171

Sherman, W. B. and T. G. Beckman. 2003. Climatic adaptation in fruit crops. *Acta Horticulturae* 622:411–428. https://doi.org/10.17660/ActaHortic.2003.622.43

Smillie, R. M. and S. E. Hetherington. 1999. Photoabatement by anthocyanin shields photosynthetic systems from light stress. *Photosynthetica* 36:451–463.

Spinardi, A., G. Cola, C. S. Gardana, and I. Mignani. 2019. Variation of anthocyanin content and profile throughout fruit development and ripening of highbush blueberry cultivars grown at two different altitudes. *Frontiers in Plant Science* 10:1–14.

Terfa, M. T., A. G. Roro, J. E. Olsen, and S. Torre. 2014. Effects of UV radiation on growth and postharvest characteristics of three pot rose cultivars grown at different altitudes. *Scientia Horticulturae* 178:184–191. http://dx.doi.org/10.1016/j.scienta.2014.08.021

Tineo, D., D. E. Bustamante, M. S. Calderon, and E. Huaman. 2022. Exploring the diversity of Andean berries from northern Peru based on molecular analyses. *Heliyon* 8(2):e08839.

Tranquilini, W. 1964. The physiology of plants at high altitudes. *Annual Review of Plant Physiology* 15:345–362.

Van Leeuwen, C., and P. Darriet. 2016. The impact of climate change on viticulture and wine quality. *Journal of Wine Economics* 11(1):150–167. https://doi.org/10.1017/jwe.2015.21

Villachica, H. 1996. *Frutales y hortalizas promisorios del Amazonas*. Lima: Tratado de Cooperación Amazónica, Secretaría Pro Tempore.

Voronkov, A. S., T. V. Ivanova, E. I. Kuznetsova, and T. K. Kumachova. 2019. Adaptations of *Malus domestica* Borkh. (Rosaceae) fruits grown at different altitudes. *Russian Journal of Plant Physiology* 66(6):922–931. https://doi.org/10.1134/S1021443719060153

Walter, H. 1990. *Vegetation und Klimazonen*, 6th ed. Stuttgart: Ulmer Verlag.

Wang, X., S. Liu, H. Zuo et al. 2021. Genomic basis of high-altitude adaptation in Tibetan Prunus fruit trees. *Current Biology* 31(17):3848–3860; e3848. https://doi.org/10.1016/j.cub.2021.06.062

Yahia, E. M., and A. Carillo-López (eds.). 2019. *Postharvest Physiology and Biochemistry of Fruits and Vegetables*. Cambridge: Elsevier. https://doi.org/10.1016/C2016-0-04653-3

Zandalinas, S. I., F. B. Fritschi, and R. Mittler. 2021. Global warming, climate change, and environmental pollution: Recipe for a multifactorial stress combination disaster. *Trends in Plant Science* 26(6):588–599. https://doi.org/10.1016/j.tplants.2021.02.011

Zeng, Q., G. Dong, L. Tian et al. 2020. High altitude is beneficial for antioxidant components and sweetness accumulation of rabbiteye blueberry. *Frontiers of Plant Science* 11:573531. https://doi.org/10.3389/fpls.2020.573531

Zheng, W., Y. Shiqi, W. Zhang et al. 2023. The content and diversity of carotenoids associated with high-altitude adaptation in Tibetan peach fruit. *Food Chemistry* 398:133909. https://doi.org/10.1016/j.foodchem.2022.133909

# 18 Impact of Altitudinal Shifts and Climatic Changes on Ecophysiological Responses of Tropical Plants

*Nagaraj Nallakaruppan, Kalaivani Thiagarajan, and Rajasekaran Chandrasekaran*

## 18.1 INTRODUCTION

In the last few decades, the ecosystems are experiencing climatic change at a rapid rate, which is evident from increased temperatures, altered precipitation patterns, decreased sea and snow levels in the artic zone, and an upsurge in the incidence and extremity of incidents occurring in ecosystems (IPCC, 2014). These variations have a tremendous impact on living organisms, ranging from individual levels to species, ecosystems, and biomes. At the species level, three possible reactions to climate change are adaptation, range changes, and inhabitant or worldwide extinction (IPBES, 2019). Since the quaternary period, mountains have protected species from climatic fluctuations, and this role will likely be reinforced throughout the Anthropocene as substantial anthropogenic disruptions eradicate many lowland species (Loarie et al., 2009; Sandel et al., 2011; Rahbek et al., 2019).

The infrastructures and functions of ecosystems may be dramatically impacted by species redistribution due to climate change, including biological invasions, the mismatch between pollinators and host plants, and local biodiversity loss (Mantyka-Pringle et al., 2015), but not all species will react to alterations in a similar way as they respond to environmental factors (i.e., each plant species will uniquely respond according to the impact of environmental factors). Globally distributed organisms are probably genetically diverse enough to be able to acclimatize to a wide variety of environmental and climatic gradients in a extremely flexible and changing environment (Schierenbeck, 2017). Still, maximum attempts to evaluate species transitions to date have engrossed on climate modeling or plant experiments that are not likely to have a significant impact on community diversity or biome services (Chapman et al., 2014; Springate & Kover, 2014). More likely, species with substantial genetic variability and inheritable phenotypical variations (i.e., cosmopolitan species) have the characteristic elasticity to develop in response to climatic alterations when compared to species with limited intraspecies diversity and narrow geographic distributions (Lavergne & Molofsky, 2007). Furthermore, genotypes with increased phenotypic adaptability (i.e., the increased capability if a genotype to generate diverse phenotypes as a response to ecological alterations) are often able to acclimatize to transformed ecological conditions better than species with less adaptability (Franks et al., 2014). Several factors like genetic and environmental stresses influence the ecophysiological response of tropical plants to altitude. Generally, tropical plants at high altitudes experience decreased vapor pressure, temperature, and partial pressure of air. Plants experience many physiological, biochemical, and morphological changes with an increase of altitude like a decrease in the height of the plant (Friend & Woodward, 1990). Additionally, the species richness of tropical plants has declined with increasing altitudes (Körner, 2019).

Global biodiversity varies because climatic and geographical conditions have an impact on the development and diversity of living organisms. Since the 1992 Earth Summit (the United Nations Conference on Environment and Development [UNCED] in Rio de Janeiro), conservation and management of biodiversity has been prioritized to lead a safe and healthy lifestyle (Thapa, 2010). Climate change is known to have a adverse impact such as a reduction in the mountain species along the altitudinal gradient, which could enhance the consequence of local extinctions of indigenous species (Rowe et al., 2015). There is an uprising surge to know if climate change is affecting mountain species along the elevational ranges. The shrinkage of mountain species along elevational shifts may be due to the upward movement of lower limits or the downward movement of higher limits, or even both. A recent report revealed that there is a significant decline in the elevational ranges of several mountain plants and animal species (Freeman et al., 2018). The variation in the distribution of plant communities along different altitudes of the mountain is depicted in Figure 18.1.

Hence, in this chapter, the diversity of various tropical plants along the major biodiversity hotspots—the Western Ghats and the Eastern Ghats—has been discussed, in

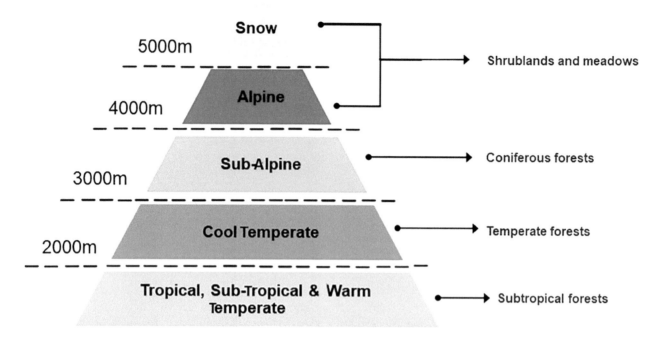

**FIGURE 18.1** Influence of altitude on the structure of the plant community.

addition to the alterations in the climatic characteristics along the altitudes. Additionally, the effect of various environmental factors like temperature, precipitation, atmospheric pressure, primary productivity, and biotic interactions on the ecophysiological response of the plants is discussed. Furthermore, a brief note is provided on the impact of altitudinal shifts on the plant's secondary metabolites.

## 18.2 PLANT DIVERSITY OF TROPICAL REGIONS IN INDIA

The evaluation of biological diversity has become a part of conservational ecology, owing to the increasing interest in today's conservational practice (Chazdon et al., 2009). Tropical forests are considered to have the most highly diverse ecosystem worldwide (Kothandaraman et al., 2020). India, being one of the 17 mega-diverse nations in the world, holds the 10th-largest forest area (329 million ha) globally. It is home to 47,513 different plant species in total, 28% of which are endemic, making up 11.4% of the world's flora (Singh & Dash, 2013: Nachiappan et al., 2022).

Despite taking up only 7% of the earth's surface, tropical regions are reported to be home to greater than half of the living species (Armenteras et al., 2009; Cusack et al., 2016, Sullivan et al., 2017; Subashree et al., 2021). The number of tree varieties in tropical forests may range from 40,000–53,000 (Slik et al., 2015). Tropical forests are under constant anthropogenic pressure, which has led to their existence as fragments in many regions of the world, in part because of their abundant biodiversity (Bhadouria et al., 2019). About 15% of tropical forests are given conservational priority due to their tremendous biological diversity and the numerous dangers they face (Köhl & Marchetti, 2016).

The tropical forests of the Western Ghats are among the world's 36 biodiversity hotspots and are geographically enriched with flora and fauna due to their abundance in biodiversity (Kothandaraman et al., 2020). Significantly more endemic species can be found in the southern portion of the Western Ghats, which are also regarded to have the region's richest floristic composition (Subashree et al., 2021).

The Central Eastern Ghats support local inhabitants' security of livelihood while also containing a wide variety of trees. The Eastern Ghats, which stretch about 1,750 km and have an average length of roughly 100 km throughout peninsular India, is situated between latitudes 11°30'N and 22°34'N and longitudes 72°22'E and 87°29'E. The varied topography and other physical characteristics of the Eastern Ghats have sculpted the region to support a rich and diverse flora. The area has significant phytogeographic features and is thought to represent a plant migration route from the Himalayas to the southern peninsula, and vice versa (Naidu & Kumar, 2015; Naidu et al., 2018).

A study conducted by Chandrashekara (2016) in the Shola Forest of the Western Ghats revealed that the growth and photosynthetic rates of tropical plants increased with increasing altitude. This suggests that plants in high altitudes may be genetically adapted to such environmental conditions which are depicted by high chlorophyll content in leaves and stomatal conductance, but very few plants showed reduced transpiration rate and stomatal conductance at high altitudes besides a high photosynthetic rate. They predict that the increased photosynthetic rate may be the result of the enhanced efficiency of tropical plants to utilize water at high altitudes.

A study conducted by Sureshkumar et al. (2020) evaluated the species richness of tropical plants in the Eastern Ghats region of India. It depicts that the impact of altitude on species richness could not be endorsed to a single factor but several environmental and anthropogenic factors were involved. This study reveals that as elevation increases, the richness of species decreases. This may be due to combination of many factors like dimensional constraints and climatic variations including fluctuations in rainfall, temperature, and humidity.

## 18.3 CLIMATE CHARACTERISTICS IN TROPICAL HIGHLANDS

Most tropical mountains and highlands are situated roughly between 20°S and 20°N latitudes. Tropical climates can be found as far north as 30°N latitude in Southeast Asia and eastern India (Henry, 2005). In tropical mountainous places, the zonation of the vegetation varies greatly. While there is significantly less daily temperature variation in tropical highlands than in mid-latitude mountains, there exists very little seasonal variation, as well. In tropical highlands, the daily temperature swings lead to a lot of freeze-thaw days (Kidane et al., 2019).

The growth of plants is frequently hampered by low air and soil temperatures in highly elevated tropical zones (Rundel et al., 1994). The three temperature zones that make up the altitudinal habitats at the equator are the hot zone between sea level and 1,000 meters, the temperate zone between 1,000 and 2,000 meters, and the cool zone above 2,000 meters, with frost occurring at roughly 4,000 meters at the equator (Nakasone & Paull, 1998). Plants that thrive at sea level will grow more slowly in mountainous areas because their growth rate is influenced by temperature (Aldenderfer, 2008).

Because of the sun's more vertical position relative to the latitude zones, high solar radiation occurs at low altitudes. Temperature decreases with altitude due to increased radiation losses, which are a 20% increase in solar energy from 300 m to 3,000 m. The steep rise in solar UV-B (ultraviolet) radiation (280–320 nm) from high to low latitudes is the result of a natural gradient caused by the thinning of the ozone layer and the shortening of the sun's solar paths as it approaches the equator (Krause & Winter, 2020).

Due to DNA and protein absorption, UV-B radiation exerts significant biological consequences in plants. Anatomy, biomass, and glucose status are all affected when sensitive plants are exposed to UV-B light. Elevation affects UV-B radiation levels, in addition to the place of residence. Although Rehm (1988) thinks that the cutin and cell walls in the epidermal tissue absorb UV radiation, Tranquillini (1964) believes that UV resistance of plants at high altitudes is caused by the filtering activity of the cell sap.

Characterization of the main climatic alternations with increasing tropical altitudes can be presented as follows.

- Decrease of Temperature (0.6°C per 100 m elevation)
- Radiation (UV, visible, and infrared [IR] light)
- Decrease of partial gas pressure ($CO_2$, $N_2$, $O_2$, and $H_2O$)
- Decreased rainfall beginning at 1,300–1,500 m
- Increase in intensity of wind

## 18.4 EFFECT OF ALTITUDINAL SHIFT AND CLIMATE CHANGE FACTORS ON THE ECOPHYSIOLOGY OF TROPICAL PLANTS

The global climate has warmed by around 0.6°C throughout the 20th century. Particularly in the past 50 years, high-mountain regions and regions at high latitudes have seen even larger temperature increases (Salinger, 2005). The global distribution of plants and plant communities, and in particular their altitudinal distribution, are predicted to vary due to changing climatic conditions, and as a result, it is anticipated that ecosystems and species that are delicate to climatic variations would respond (Walther et al., 2005). Stunted aerial parts growth and rigorous growth of underground parts like roots, rhizomes, bulbs, etc., can occur, as well as anomalous biomass accumulation, flowering, fruiting, seed formation, and seed setting patterns, and even influence on seed viability and seed germination rate. Overall, the phenology of tropical plants is hampered by abiotic components of tropical environmental factors.

Any living organism finds it difficult to survive in high altitudinal environments, as well as in elevational gradients, due to their rapid physical and ecological alterations. This concept was expected to possess an ideal situation to study concerns regarding how life histories alter in response to adaptation. Along with elevation, organisms' growth, development, reproduction, and survival prototypes are very diverse, yet they also tend to converge in comparable environmental conditions (Laiolo & Obeso, 2017). Major variations in living organisms along the altitudinal gradient are depicted in Figure 18.2, and the impact of altitude and climate on the ecophysiology of tropical plants was tabulated in Table 18.1.

### 18.4.1 Temperature

Temperature is the most important abiotic component that differs along elevation sites for plants because of its wide range and significant impact on biochemical or physiological processes. It drops by 0.54–0.65°C per 100 m of ascent; however, significant variance is brought on by the weather, local geography, and altitude (Körner, 2021). The assimilation of energy and all metabolic activity slows down in colder temperatures, as do physical and chemical reactions. The effects of this process on plants are particularly

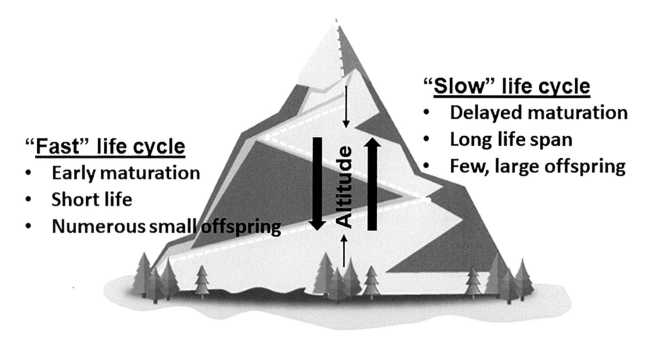

**FIGURE 18.2** Major variations in living organisms along the altitudinal gradient.

**TABLE 18.1**
**Case Studies on the Effect of Altitudinal Shift and Climate Change Factors on the Ecophysiology of Tropical Plants**

| Plant | Geographic Location | Factor | Impact | References |
|---|---|---|---|---|
| *Rhododendron campanulatum* | Western Himalayas | Altitudinal gradient | Changes in leaf morphology | Jamloki et al. (2022) |
| *Rhododendron lepidotum* | Nepal Himalayas | Elevation gradient | Variability in morphological features | Pandey et al. (2021) |
| *Polylepis rugulosa* | Dry tropical Andes | Altitudinal shift | Changes in morphology and ecophysiology | Macek et al. (2009) |
| *Polylepis tarapacana* | Dry tropical Andes | Altitudinal shift | Changes in morphology and ecophysiology | Macek et al. (2009) |
| *Fagus sylvatica* | Catalonia (northeastern Spain) | Increasing temperature | Elevated mortality and reproductive decline | Penuelas et al. (2007) |
| *Populus tremuloides* | Colorado, USA | Increasing temperature | Elevated mortality | Worrall et al. (2008) |
| *Picea rubens* | Green Mountains of Vermont, USA | Increasing temperature and drought stress | Increased mortality and decline in reproductivity | Beckage et al. (2008) |
| *Machilus thumbergii* | Japan | High temperature | Photosynthetic rates were lower | Hara et al. (2021) |
| *Abies pinsapo* Boiss | Ronda Mountains (southern Spain) | Altitudinal shift | Overlap of offspring production | González-Hernández et al. (2021) |
| *Anthyllis vulneraria* | Central Europe | Altitudinal and latitudinal gradient | Plant size and reproduction decreased | Daco et al. (2021) |
| *Larix sibirica* | Altai Mountains, China | Altitudinal variation | Affected radial growth | Zhou et al. (2021) |
| *Polylepis tomentella* | Dry tropical Andes | Altitudinal shift | Changes in morphology and ecophysiology | Macek et al. (2009) |
| *Larix gmelinii* Rupr. | Xing'an Mountains, northeast China | Temperature/precipitation | Decline of growth and radial growth | Bai et al. (2019) |
| *Ferula jaeschkeana* Vatke | Himalayan region | Habitat variability and altitudinal gradient | Reduction in the number of umbels and increase in the number of stigma and anthers in flowers | Yaqoob & Nawchoo (2017) |

striking because they either result in decreased activity or extensive homeostatic reactions to counteract the passive response to freezing and decreasing temperature (Barry & Chorley, 2009; Rolland, 2003). Growth, development, and physiological activity are temporarily stopped during a time of dormancy (including diapause and hibernation), which is a defense strategy against the winter's cold or resource scarcity. Temperature directly affects phenology, growth and reproductive patterns, and life histories by limiting the production rates throughout ontogeny and reproduction. The following processes are controlled by temperature, which is also closely related to seasonality and productivity (Laiolo & Obeso, 2017).

### 18.4.2 Atmospheric Pressure

Gas exchanges in plants are impacted by atmospheric pressure, air moisture content, and the partial pressures of physiologically appropriate gases, like oxygen and carbon dioxide, which all decrease consistently with elevation. However, impacts of pressure on photosynthetic rate are less pronounced than would be expected from a fall in ambient $pCO_2$ alone because the enhanced rate of molecular diffusivity brought on by thinner air is offset by a decrease in environmental temperature, which slows diffusion. Furthermore, the liquid phase, which is unaffected by pressure, accounts for more than 80 percent of the overall carbon dioxide transfer resistance between air and chloroplasts (Körner, 2007).

### 18.4.3 Precipitation

Precipitation has non-linear correlations with elevation and regional patterns as opposed to global ones, unlike temperature and air pressure. At low elevations, precipitation tends to gradually rise with altitude but shows no pattern or reduces at very high elevations or in tropical regions. It is predicted that precipitation decreases above the cloud zone (Nagy & Grabherr, 2009).

In addition to providing thermally protective snow cover, orographic precipitation of snow is often characterized in high altitudes and determines—together with seasonal variations in temperature and photoperiod—the extent of the growing seasons of tropical plants. The balance of plant water resources is rarely a serious issue because evapotranspiration decreases with elevation. However, the need to limit transpiration may have an impact on the photosynthetic rates of carbon dioxide uptake and nutritional assimilation at times of increased evaporative demand or protracted soil moisture depletion (Schulze & Chapin, 1987; Körner, 2007).

### 18.4.4 Primary Productivity

The environmental surroundings of mountains generally exhibit high differences in the fertility of the soil which in turn depends on various factors like climate, rocky layers, soil composition and age, topographic variations of individual plants, and fertilization of soil by herbivores (primary consumers). Generally, at high altitudes, reduced or mild temperature has a negative impact on the enzymatic activity of the soil, rate of biodegradation of minerals, and their relevant turnover, which in turn decreases the amount of nutrients available for the primary producers and also has a consequent impact on the latitudinal region above the tropic zone. Lower productivity is also caused in some mountainous regions by shorter grazing seasons, fewer nutrients supplied by animals, and shorter nutrient permanence in steep and shallow soils (Mariotti et al., 1980, Laiolo et al., 2015). Although there are some exceptions, plants may adapt by shrinking and increasing the concentration of their mineral nutrients. It is also important to note that the concentration of nutrients in the soil may not necessarily represent their availability to organisms because some nutrients may be present in indigestible forms. Undoubtedly, the availability of nutrients has an impact on how well plants function overall up to higher trophic levels and has a significant impact on the evolution of life history (Stearns, 1992).

### 18.4.5 Biotic Interactions

The drop in species richness seen in many taxa at high altitudes, as well as the decline in biotic interactions, are attributed mostly to the aforementioned causes, along with land size and history (McCain & Grytnes, 2010). According to the Stress Gradient Hypothesis, severe environments should result in a drop in negative interactions and a rise in positive ones (Brooker, 2006). Numerous studies have actually found negative trends in competition, predation, and parasitism with altitude, but they have also found some beneficial associations (e.g., Abbate & Antonovics, 2014). There is more clarity regarding the improvement of plant facilitation with elevation, which has been detected in floral communities throughout the world (Callaway et al., 2002). Extrinsic lethality variance caused by interactions, together with productivity, is a significant driver of an organism's life strategy, affecting growth, maturity, extent, and developmental patterns.

## 18.5 VARIATION OF PLANT METABOLITES UNDER ALTITUDINAL SHIFT AND CLIMATE CHANGE

Along with altering plant phenology, climate change also affects species distribution, habitat loss, and species composition. In recent years, much attention has been placed on comprehending this phenomenon in research oriented on climate change. Among other living organisms, much less attention was given to examining how tropical plants' production of secondary metabolites is impacted by climatic variations (Wookey et al., 1993). According to Galen and Stanton (1993), climate change may have an effect on blooming phenology, which in turn may have an impact on the fitness of plants. Through shifting species ranges and varying phenology, plants are adapted to recent climatic changes (Parmesan, 2006).

Temperature fluctuations and climatic changes may affect the ability of the region's ecosystem to function and may cause long-term disturbances during interspecific interactions. Reports suggest that climate change has impacted the biochemical components and survival of some tropical plants in high-altitude regions. Since the growth of tropical plants is typically constrained at higher elevations, photosynthesis and carbon fixation have an adverse impact on primary metabolites. Consequently, there is an increase in the production of secondary metabolites when plants are under stress (Mooney et al., 1991).

The hard and complicated climatic conditions at high altitudes restrict the dispersal and variety of plants. Reduced $O_2$ and $CO_2$, high sun irradiation, strong winds, shallow rocky soils, lower temperatures, and less water and nutrient contents are some of these challenging environmental factors. Despite the fact that these stress factors could slow the growth of plants, most of the species have endured and evolved a variety of adaptation systems in response to them (Wonsick & Pinker, 2014). In order to be tolerant of various environmental biotic and abiotic challenges, secondary metabolite accumulation has been shown to be crucial. These secondary metabolites actually aid in the adaptability of stressed plants to various environmental conditions, such as the over-accumulation of antioxidants like flavonoids and phenols, an upsurge in the content proline and soluble protein to fend off drought, and an increase in carotenoids relative to chlorophyll to lessen the damage due to intensified light by upregulating the xanthophyll-cycle pigment pool. Plant secondary metabolite production is influenced by a number of variables, including age, season, and nutritional status. Ecological factors such as elevation, increased/decreased temperature, drought, and light conditions also greatly affect the production of secondary compounds in plants and population groups. These variables include quantitative and qualitative changes in secondary compound production (Hashim et al., 2020).

The impact of environmental stressors and factors associated with global change on plant secondary metabolites appears to depend on the chemical type, plant species, genotype, and maybe other factors as well. However, it is predicted that the warming temperatures and increasing $CO_2$ levels will disrupt the growth cycles of tropical plants, and the active components of the plants may vary due to physiological variations (Chaturvedi et al., 2007).

## 18.6 CONCLUSION

Global warming is one of the major threats to the adaptation of plants along the altitudinal gradient. Changes in temperature specifically influence the physiology of tropical plants by affecting their growth, development, and productivity. Consequently, excessive levels of these factors only will show a negative response. Solar radiation enhances photosynthesis, and when this level becomes excessive, it leads to photoinhibition in plants. On the other hand, the variation in temperature could also contribute to the acceleration of the plant life cycle and enables the survival of plants at high altitudes—but also intensifies the negative effects of other abiotic parameters on the plant system such as soil sub-zero temperature, drought, variation in mean air temperature, elevated $CO_2$, precipitation, and UV radiation.

Among abiotic factors, temperature plays a vital role in limiting the species' availability along latitudinal and altitudinal edges. Additionally, a shift in the biotic interactions also has an impact on the diversity of tropical plants. Hailing weather/climate along the altitudinal regions influences the upward shifts usually indicating the biotic adoption in the landscape of tropical regions. For the conservation of floristic diversity, a proper understanding of climate change, ecological alterations, and adaptation among different taxa along altitude is mandatory.

## ACKNOWLEDGMENTS

The authors would like to extend our profound sense of gratitude to our management, Vellore Institute of Technology for their constant support and encouragement in all our endeavors. Also, we extend our sincere thanks to DST—SPLICE—Climate Change Programme (DST/CCP/NCC&CV/133/2017 (G), Govt. of India for providing financial support in carrying over this project successfully to date.

## REFERENCES

Abbate JL, Antonovics J (2014) Elevational disease distribution in a natural plant–pathogen system: Insights from changes across host populations and climate. *Oikos* 123(9):1126–1136.

Aldenderfer MS (2008) High elevation foraging societies. In *The Handbook of South American Archaeology*, 131–143, Springer.

Armenteras D, Rodríguez N, Retana J (2009) Are conservation strategies effective in avoiding the deforestation of the Colombian Guyana Shield? *Biol. Conserv.* 142(7):1411–1419.

Bai X, Zhang X, Li J, Duan X, Jin Y, Chen Z (2019) Altitudinal disparity in growth of Dahurian larch (*Larix gmelinii* Rupr.) in response to recent climate change in northeast China. *Sci. Total Environ.* 670:466–477.

Barry RG, Chorley RJ (2009) *Atmosphere, Weather and Climate*. Routledge.

Beckage B, Osborne B, Gavin DG, Pucko C, Siccama T, Perkins T (2008) A rapid upward shift of a forest ecotone during 40 years of warming in the Green Mountains of Vermont. *Proc. Natl. Acad. Sci.* 105(11):4197–4202.

Bhadouria R, Tripathi S, Srivastava P, Singh P (2019) *Handbook of Research on the Conservation and Restoration of Tropical Dry Forests*. IGI Global.

Brooker RW (2006) Plant–plant interactions and environmental change. *New Phytol.* 171(2):271–284.

Callaway RM, Brooker RW, Choler P, Kikvidze Z, Lortie CJ, Michalet R, Paolini L, Pugnaire FI, Newingham B, Aschehoug ET, Armas C (2002) Positive interactions among alpine plants increase with stress. *Nature* 417(6891):844–848.

Chandrashekara UM (2016) Growth and physiological responses to an elevation gradient by co-occurring tree species in a shola forest of Kerala, India. *Curr. Sci.* 110(10):1900–1901.

Chapman DS, Haynes T, Beal S, Essl F, Bullock JM (2014) Phenology predicts the native and invasive range limits of common ragweed. *Glob. Change Biol.* 20:192–202.

Chaturvedi AK, Vashistha RK, Prasad P, Nautiyal MC (2007) Need of innovative approach for climate change studies in alpine region of India. *Curr. Sci.* 93(12):1648–1649.

Chazdon RL, Harvey CA, Komar O, Griffith DM, Ferguson BG, Martínez-Ramos M, Morales H, Nigh R, Soto-Pinto L, Van Breugel M, Philpott SM (2009) Beyond reserves: A research agenda for conserving biodiversity in human-modified tropical landscapes. *Biotropica* 41(2):42–153.

Cusack DF, Karpman J, Ashdown D, Cao Q, Ciochina M, Halterman S, Lydon S, Neupane A (2016) Global change effects on humid tropical forests: Evidence for biogeochemical and biodiversity shifts at an ecosystem scale. *Rev. Geophys.* 54(3):523–610.

Daco L, Colling G, Matthies D (2021) Altitude and latitude have different effects on population characteristics of the widespread plant Anthyllis vulneraria. *Oecologia* 197(2):537–549.

Franks SJ, Weber JJ, Aitken SN (2014) Evolutionary and plastic responses to climate change in terrestrial plant populations. *Evol. Appl.* 7:123–139.

Freeman BG, Scholer MN, Ruiz-Gutierrez V, Fitzpatrick JW (2018) Climate change causes upslope shifts and mountaintop extirpations in a tropical bird community. *P. Natl. Acad. Sci USA*. 115:11982–11987.

Friend AD, Woodward FI (1990) Evolutionary and ecophysiological responses of mountain plants to the growing season environment. *Adv. Ecol. Res.* 20:59–124 (Academic Press).

Galen C, Stanton ML (1993) Short-term responses of alpine buttercups to experimental manipulations of growing season length. *Ecology* 74(4):1052–1058.

González-Hernández A, Nieto-Lugilde D, Peñas J, Alba-Sánchez F (2021) Lean pattern in an altitude range shift of a tree species: Abies pinsapo Boiss. *Forests.* 12(11):1451.

Hara C, Inoue S, Ishii HR, Okabe M, Nakagaki M, Kobayashi H (2021) Tolerance and acclimation of photosynthesis of nine urban tree species to warmer growing conditions. *Trees* 35(6):1793–1806.

Hashim AM, Alharbi BM, Abdulmajeed AM, Elkelish A, Hozzein WN, Hassan HM (2020) Oxidative stress responses of some endemic plants to high altitudes by intensifying antioxidants and secondary metabolites content. *Plants* 9(7):869.

Henry J (2005) Tropical and equatorial climates. *Encyclopedia of World Climatology*:742–750.

IPBES (2019) Global assessment report of the Intergovernmental science- policy platform on biodiversity and ecosystem services. In: Brondízio ES, Settele J, Díaz S, Ngo HT (eds), IPBES secretariat, Bonn. ISBN: 978-3-947851-20-1

IPCC (2014) Climate change 2014: Synthesis report. In Core Writing Team, Pachauri RK, Meyer LA (eds), *Contribution of Working Groups I, II and III to the Fifth Assessment Report of the Intergovernmental Panel on Climate Change*, 151. IPCC.

Jamloki A, Singh A, Chandra S, Shukla V, Nautiyal MC, Malik ZA (2022) Population structure, regeneration potential and leaf morphological traits of Rhododendron campanulatum D. Don along an altitudinal gradient in Western Himalaya. *Plant Biosystems-An International Journal Dealing with all Aspects of Plant Biology*:1–16.

Kidane YO, Steinbauer MJ, Beierkuhnlein C (2019) Dead end for endemic plant species? A biodiversity hotspot under pressure. *Global Ecol. Conserv.* 19:e00670.

Köhl M, Marchetti M (2016) Measurements and assessments on field plots. In Pancel, L., & Köhl, M. (Eds.), *Tropical Forest Handbook*, 1–51. Springer, Berlin, Heidelberg.

Körner C (2007) The use of 'altitude'in ecological research. *Trends Ecol. Evol.* 22(11):569–574.

Körner C (2019) *Mountain Biodiversity, Its Causes and Function: An Overview*, 3–20. Mountain Biodiversity.

Körner C (2021) *Alpine Plant Life: Functional Plant Ecology of High Mountain Ecosystems*. Springer Nature.

Kothandaraman S, Dar JA, Sundarapandian S, Dayanandan S, Khan ML (2020) Ecosystem-level carbon storage and its links to diversity, structural and environmental drivers in tropical forests of Western Ghats, India. *Sci. Rep.* 10(1):1–15.

Krause GH, Winter K (2020) The photosynthetic system in tropical plants under high irradiance and temperature stress. *Progress in Botany* 82:131–169.

Laiolo P, Illera JC, Meléndez L, Segura A, Obeso JR (2015) Abiotic, biotic, and evolutionary control of the distribution of C and N isotopes in food webs. *The American Naturalist* 185(2):169–182.

Laiolo P, Obeso JR (2017) Life-history responses to the altitudinal gradient. In *High Mountain Conservation in a Changing World*, 253–283. Springer.

Lavergne S, Molofsky J (2007) Increased genetic variation and evolutionary potential drive the success of an invasive grass. *Proc. Natl. Acad. Sci. U.S.A.* 104: 3883–3888.

Loarie SR, Duffy PB, Hamilton H, Asner GP, Field CB, Ackerly DD (2009) The velocity of climate change. *Nature* 462: 1052–1055.

Macek P, Macková J, de Bello F (2009) Morphological and ecophysiological traits shaping altitudinal distribution of three Polylepis treeline species in the dry tropical Andes. *Acta Oecol.* 35(6):778–785.

Mantyka-Pringle CS, Visconti P, Di Marco M, Martin TG, Rondinini C, Rhodes JR (2015) Climate change modifies risk of global biodiversity loss due to land-cover change. *Biol. Conserv.* 187:103–111.

Mariotti A, Pierre D, Vedy JC, Bruckert S, Guillemot J (1980) The abundance of natural nitrogen 15 in the organic matter of soils along an altitudinal gradient (Chablais, Haute Savoie, France). *Catena* 7(1):293–300.

McCain CM, Grytnes JA (2010) *Elevational Gradients in Species Richness*. ELS.

Mooney HA, Winner WE, Pell EJ (1991) *Response of Plants to Multiple Stresses*. Academic Press.

Nachiappan K, Nallakaruppan N, Alphonse M, Sekaran M, Veluchamy C, Ramamoorthy S, Thaigarajan K, Chandrasekaran R (2022) Status, conservation, and sustainability on medicinal plant resources of India. In *Plant Genetic Resources, Inventory, Collection and Conservation*, 351–387. Springer.

Nagy L, Grabherr G (2009) *The Biology of Alpine Habitats*. Oxford University Press on Demand.

Naidu MT, Kumar OA (2015) Tree species diversity in the Eastern Ghats of northern Andhra Pradesh, India. *JoTT*. 7(8): 7443–7459.

Naidu MT, Premavani D, Suthari S, Venkaiah M (2018) Assessment of tree diversity in tropical deciduous forests of Northcentral Eastern Ghats, India. *Geol. Ecol. Landscapes.* 2(3):216–227.

Nakasone HY, Paull RE (1998) *Tropical Fruits*. Cab International.

Pandey M, Pathak ML, Shrestha BB (2021) Morphological and wood anatomical traits of Rhododendron lepidotum Wall ex G. Don along the elevation gradients in Nepal Himalayas. *AAAR.* 53(1):35–47.

Parmesan C (2006) Ecological and evolutionary responses to recent climate change. *Annu. Rev. Ecol. Syst.* 37:637–639.

Penuelas J, Ogaya R, Boada M, Jump A (2007) Migration, invasion and decline: Changes in recruitment and forest structure in a warming-linked shift of European beech forest in Catalonia (NE Spain). *Ecography* 30(6):829–837.

Rahbek C, Borregaard MK, Colwell RK, Dalsgaard B, Holt BG, Morueta-Holme N, . . . Fjeldså J (2019) Humboldt's enigma: What causes global patterns of mountain biodiversity? *Science* 365:1108–1113.

Rehm S (1988) Fundamentals of crop production in the tropics and subtropics. *Handbook of Agriculture and Food in Developing Countries* 2(3):478.

Rolland C (2003) Spatial and seasonal variations of air temperature lapse rates in Alpine regions. *J. Clim.* 16(7): 1032–1046.

Rowe KC, Rowe KMC, Tingley MW, Koo MS, Moritz C (2015) Spatially heterogeneous impact of climate change on small mammals of montane California. *P. Roy. Soc. B-Biol. Sci.* 282 (1799):20141857.

Rundel PW, Smith AP, Meinzer FC (1994) *Tropical Alpine Environments: Plant Form and Function.* Cambridge University Press.

Salinger MJ (2005) Climate variability and change: Past, present and future—an overview. *Increasing Climate Variability and Change*:9–29.

Sandel B, Arge L, Dalsgaard B, Davies RG, Gaston KJ, Sutherland WJ, Svenning JC (2011) The influence of late Quaternary climate-change velocity on species endemism. *Science* 334:660–664.

Schierenbeck KA (2017) Population-level genetic variation and climate change in a biodiversity hotspot. *Ann. Bot.* 119(2):215–228.

Schulze ED, Chapin FS (1987) Plant specialization to environments of different resource availability. In *Potentials and Limitations of Ecosystem Analysis*, 120–148. Springer.

Singh P, Dash SS (2013–2014) *Plant Discoveries—New Genera, Species and New Records.* Botanical Survey of India.

Slik JF, Arroyo-Rodríguez V, Aiba SI, Alvarez-Loayza P, Alves LF, Ashton P, Balvanera P, Bastian ML, Bellingham PJ, Van Den Berg E, Bernacci L (2015) An estimate of the number of tropical tree species. *Proc. Natl. Acad. Sci.* 112(24):7472–7477.

Springate DA, Kover PX (2014) Plant responses to elevated temperatures: A field study on phenological sensitivity and fitness responses to simulated climate warming. *Glob. Change Biol.* 20:456–465.

Stearns SC (1992) *The Evolution of Life Histories.* Oxford University Press, p. 249.

Subashree K, Dar JA, Karuppusamy S, Sundarapandian S (2021) Plant diversity, structure and regeneration potential in tropical forests of Western Ghats, India. *Acta Ecol. Sin.* 41(4):259–284.

Sullivan MJ, Talbot J, Lewis SL, Phillips OL, Qie L, Begne SK, Chave J, Cuni-Sanchez A, Hubau W, Lopez-Gonzalez G, Miles L (2017) Diversity and carbon storage across the tropical forest biome. *Sci. Rep.* 7(1):1–12.

Sureshkumar J, Ayyanar M, Silambarasan R (2020) Pteridophyte species richness along elevation gradients in Kolli Hills of the Eastern Ghats, India. *J. Asia-Pac. Biodivers.* 13(1):92–106.

Thapa D (2010) Bio-diversity conservation in Nepal. *Bibechana* 6:31–36.

Tranquillini W (1964) The physiology of plants at high altitudes. *Annu. Rev. Plant Physiol.* 15(1):345–362.

Walther GR, Beißner S, Pott R (2005) Climate change and high mountain vegetation shifts. In *Mountain Ecosystems*, 77–96. Springer.

Wonsick MM, Pinker RT (2014) The radiative environment of the Tibetan Plateau. *Int. J. Climatol.* 34(7):2153–2162.

Wookey PA, Parsons AN, Welker JM, Potter JA, Callaghan TV, Lee JA, Press MC (1993) Comparative responses of phenology and reproductive development to simulated environmental change in sub-arctic and high arctic plants. *Oikos*:490–502.

Worrall JJ, Egeland L, Eager T, Mask RA, Johnson EW, Kemp PA, Shepperd WD (2008) Rapid mortality of Populus tremuloides in southwestern Colorado, USA. *For. Ecol. Manage.* 255(3–4):686–696.

Yaqoob U, Nawchoo IA (2017) Impact of habitat variability and altitude on growth dynamics and reproductive allocation in Ferula jaeschkeana Vatke. *J King Saud Univ Sci.* 29(1):19–27.

Zhou P, Huang JG, Liang H, Rossi S, Bergeron Y, Shishov VV, Jiang S, Kang J, Zhu H, Dong Z (2021) Radial growth of Larix sibirica was more sensitive to climate at low than high altitudes in the Altai Mountains, China. *Agric. For. Meteorol.* 304:108392.

# Section 6

*Emerging techniques in Ecophysiological Research*

# 19 An Overview of Emerging Techniques in Ecophysiological Research

*Surbhi Sharma, Joat Singh, Neeru Bala, Priyanka Sharma, Shalini Bahel, and Jatinder Kaur Katnoria*

## 19.1 INTRODUCTION

Plants are vulnerable to harsh geographical and climatic conditions like high altitudes, extreme variations in temperatures, salt, drought, flood and environmental pollution because of their sessile nature (Mahan et al., 1995; Haggag et al., 2015; Beltran-Peña et al., 2020). As a result of their inability to escape from their natural environment, plants develop internal "tolerance" as well as "resistance" mechanisms to cope with varieties of stresses. Therefore, plants are well known to exhibit an extensive range of ecophysiological responses, functional variability, growth rates and productivity, along with dynamics of population and communities. In comparison to temperate forests, tropical forests have far larger levels of genetic, species and ecological richness. They also have a greater impact on the global carbon cycle. Only 7% of the earth's land surface is covered by the tropical forests but they are habitat to 50% of all plant species. Moreover, these forests provide a number of ecosystem services for the welfare of people and other animals, in addition to having a higher species diversity (Bhadouria et al., 2016). Development and productivity of plants are documented to be severely affected by various environmental causes such as severe soil salinity, insufficient mineral content and high amounts of phytotoxic substances in the soil (Sadak, 2016). Stress is a term that describes sub-optimal growth conditions that have a deleterious impact on a plant's yield level and production (Orcutt & Nilsen, 2000). The study relating to the behavior of plants in response to stress is termed "ecophysiology." Thus, ecophysiologists study the interaction of plants with their biotic and abiotic environments, which are the key factors to influence their processes such as survival, growth, reproduction, abundance and geographic distribution of plants. Understanding the ecophysiological trends and pathways facilitates comprehension of the evolutionary history and functional importance of particular plant features (Fitter & Robert, 2012). Ecophysiological techniques are documented to improve various functions of plants, viz., the relations of plants and water, respiration, photosynthesis and different mechanisms in response to stresses on timescales ranging from immediate to evolutionary (Verdeguer & Mercedes, 2018).

The adoption of an ecophysiological perspective is beneficial for many significant societal concerns like agriculture, climate change and nature protection. The study of environmental plant physiology and ecology has produced techniques for analyzing the survival, dispersion and abundance of plant species in the world's various climates. Plants have evolved to thrive in a staggering variety of habitats (Ainsworth et al., 2016). Huge ecophysiological diversity is present in plants, and this diversity contributes to variance in development rates, yield level, dynamics of population, community and the functioning of the ecosystem (Ackerly et al., 2000). Tropical forests are very rich in flora and fauna. This chapter explains the ecophysiological mechanisms of tropical plants to varied stresses. Furthermore, this chapter also explains the central role of different plant ecophysiological techniques, viz., thermal imaging, metabolomics, proteomics, in vitro and in vivo assays and isotope techniques in applying basic research to address present and future hurdles to plant ecosystems.

## 19.2 ECOPHYSIOLOGICAL RESPONSE OF TROPICAL PLANTS AGAINST VARIOUS STRESSES

A stressful environment often leads to the production of varied oxygen species which are very reactive due to the presence in them of unpaired electrons, and they are often known as ROS (reactive oxygen species). They are categorized as hydrogen peroxide ($H_2O_2$), hydroxyl radicals ($OH^{\cdot-}$) and superoxide radicals ($O^{2\cdot-}$) in plants which in turn have the potential to initiate harmful oxidative reactions involving lipid peroxidation (LPO), protein oxidation, chlorophyll bleaching and nucleic acid impairment. To combat the toxicity induced by ROS, various enzymatic antioxidant systems—consisting of glutathione reductase (GR), catalase (CAT), superoxide dismutase (SOD), dehydroascorbate reductase (DHAR), monodehydroascorbate reductase (MDHAR) and ascorbate peroxidase (APX)—and metabolites like ascorbate (ASC), glutathione (GSH), tochopherol and carotenoids play a great roles in the protection of plant system (Gupta et al., 2015). Modulation in the enzymes activities has been documented as a significant factor in the tolerance of various plants against stressful environmental conditions (Das & Roychoudhury, 2014). Demiralay et al. (2013) stated that *C. setosa* showed variations in antioxidant enzymes activities during leaf rolling. The antioxidant enzymes—viz., CAT, GR, APX, DHAR and MDHAR—have shown an increase in their activities in *C. setosa* as a result of drought stress which induced leaf rolling, while SOD activity showed no appreciable change.

Plants have evolved defense strategies for adapting toward various abiotic and biotic stresses which were the consequences of exposure to the frequently unfavorable environment (Bhadouria et al., 2019). Because of the complexity of the biology of a plant cell, as well as plant tissues, each stimulus activates several cellular signaling pathways as result of intricate interactions among them (Desikan et al., 2005). According to a study by Sanders et al. (1999), plant cells responded to a variety of harsh environmental conditions by increasing their cytosolic calcium ($Ca^{2+}$) levels. These conditions included an attack by pathogens and stress during the osmosis process. With the upsurge of $Ca^{2+}$ concentration in the cellular space, some simultaneous pathways are stimulated by calcium-interacting proteins such as $Ca^{2+}$-dependent protein kinases (CDPKs), calmodulin and calcineurin B–like proteins (CBLs). When biotic and abiotic stress is sensed, signaling cascades are triggered, activating ion channels and kinase cascades, and producing ROS. Environmental stresses—viz., temperature, humidity and pathogens—can transcriptionally activate defense response genes in plants. It has further been suggested that a complicated signaling network exists that allows a plant to recognize and defend itself against the attacks of pathogens and environmental stresses by dependence on the induction of defensive gene expression in the response against specific infections (Ali et al., 2018). It was shown that NADPH oxidases generated $O_{2-}$ which was rapidly converted to hydrogen peroxide. Recent evidence concludes that the NADPH oxidases were activated by $Ca^{2+}$ signatures (Quan et al., 2008). A brief discussion about the different types of stress is given in the following sections, and a few reports are highlighted in Table 19.1.

## 19.2.1 Drought

Drought stress is a period of time when a land experiences below-normal precipitation. The plant's turgor pressure drops abruptly when water loss occurs. Stress caused by drought has a detrimental effect on cell growth, which results in membrane protein separation, lowers chlorophyll concentration and reduces seed germination (Farooq et al., 2012). The typical symptoms include reduced leaf morphology (size reduction and rolling), weakened grain forms and inadequate size development in many plant sections, including the stem, leaves and apical buds. Brassinosteroids (BRs), auxins and gibberellins encourage the development and maintenance of the balance between a plant tolerance to water stress, whereas ethylene and abscisic acid (ABA) restrict growth and close the stomata during transpiration (Naqvi, 1999; Kumawat et al., 2018). By doing this, ABA stops water from evaporating and delays the growth

**TABLE 19.1**
**Ecophysiological Response of Tropical Plants against Environmental Stresses**

| S. No. | Types of Environmental Stresses | Ecophysiological Response of Tropical Plants | References |
|---|---|---|---|
| 1. | Temperature | (1) Heat: reduced water potential, reduced leaf area and premature leaf senescence<br>(2) Cold: formation of ice crystals in hydrated cells | Impe et al. (2022)<br>Wahid et al. (2007) |
| 2. | Drought | Development of hairy roots, reduction in the size of the leaves, flowering time and height<br>Dropping of plant's turgor pressure<br>Reduction in seed germination | Bhadouria et al. (2016)<br>Burgess et al. (2016) |
| 3. | Environmental pollution | Thickening of cell walls, storage of pheolics and structural changes in the vegetative organs | Mishra et al. (2016)<br>Gupta et al. (2015) |
| 4. | Heavy metals | Physiological and metabolic alterations, reduction in plant growth, ultrastructural, biochemical and molecular changes | Baig et al. (2020)<br>Dubey (2011) |
| 5. | Pathogens | Breaching of structural barriers<br>Activation of innate immunity | Lacava et al. (2022)<br>Walling (2009) |
| 6. | Pesticides | Reduction of plant diversity<br>Oxidative stress | Shahzad et al. (2018)<br>Sharma et al. (2015)<br>Goh et al. (2011) |
| 7. | Salinity | Slow-down of protein synthesis and decrease of cell membrane permeability | Santini et al. (2022)<br>Parihar et al. (2015) |
| 8. | Radiation | Generation of reactive oxygen species and increase in the activities of antioxidant enzymes, viz., APX, CAT, GR, GST, POD and SOD<br>Changes in the physiology due to the variations in DNA | Lung et al. (2016)<br>Baulcombe and Dean (2014) |

of apical organs, permitting roots to use more water by increasing their ability to locate water in the soil. Water stress brought on by water shortage or waterlogging causes a variety of responses that includes varied ecophysiological and biochemical responses, reduction in size and distinctive structural changes, especially in the leaves (Burgess et al., 2016). The environment in dry tropical ecosystems is harsh and unpredictable, which makes plants more vulnerable to external stress as the plant community evolves. Variations in soil water availability were shown to have a significant impact on species diversity, distribution and community composition because they alter niche differentiation (Bhadouria et al., 2017). With the primary goal of conserving or maintaining the water availability for their growth during the vegetative period, plants grow a "deep root system" during pre-drought and following rainy periods. The protective and stress resistance strategies of xerophytic plants include hairiness, acanthaceous and waxiness, which play a significant role during harsh environmental conditions. During stressful environments, the plants excrete different compounds, viz., ABA, amino acids, proline, polyamines, sorbitol, raffinose and mannitol (Beltran-Peña et al., 2020).

### 19.2.2 Atmospheric Carbondioxide ($CO_2$)

Community, structure and ecosystem of plant functions are influenced by their exposure to environmental changes. Evaluating those changes is essential for predicting future ecosystem functions because plants can adapt to global climate change and have significant major effects on ecosystem functioning. The concentration of carbon dioxide ($CO_2$) in the atmosphere has drastically increased in the post-industrial era, and the present concentration of $CO_2$ concentration has reached around 400 ppm. Both ecophysiological characteristics such as respiration and photosynthesis are impacted by global variations in composition of atmospheric gases, ambient temperature, precipitation and changes in the concentration of pollutants. Several studies have documented that seedling growth in response to atmospheric $CO_2$ in dry tropical environments induces variability in relative growth rate (RGR), leaf area ratio (LAR) and total dry weight (Hungate, 2014; Cowles et al., 2016; Langley & Bhadouria et al., 2016). The various processes of a plant's body respond differently to environmental changes—ambient climate modifications may change the equilibrium between them, affecting carbon distribution within ecosystems and net carbon flux (Avolio et al., 2015).

### 19.2.3 Temperature

Temperature is among the major elements in our surroundings that affect plant development and yield. Field-grown plants are frequently exposed to heat fluctuations that have substantial impacts on plant metabolism. Both low and high temperatures affect plants. The ideal temperature for plant growth ranges between 12°C and 22°C, and if exposure to temperatures above that range results in heat stress, plant output can be quickly and dramatically decreased. Heat stress also leads to a reduction in water holding capacity, reduction in the area of the leaf and premature leaf senescence in the plant system (Gulen et al., 2004; Wahid et al., 2007).

When the temperature drops, ice crystals form in hydrated cells as a result of the sudden changes in the structure of cytoplasm and cellular lipids. "Desiccation stress" is brought on or triggered by this extracellular ice production. When the temperature drops, it causes the cell membrane to become rigid because it readily breaks the links between molecules, weakening the structure of the membrane. In general, plants grown between temperatures of 0°C and 15°C in the tropical zones and semi-tropical zones should never be subjected to low-temperature stress. The degree of damage is mostly determined by the thickness and norm of the plant, age, ecophysiological responses and viscosity of the water that is readily available (Impe et al., 2022).

### 19.2.4 Environmental Pollution

The ecosystem and living things are seriously threatened by pollution on a global scale. It greatly affects regional and worldwide atmospheric problems such as ozone layer depletion, acidification and global warming. Plants are greatly influenced by increased pollution levels.

Air pollution was considered as a local problem earlier and it has gained great attention due to the widespread of pollutants that can cause ill health effects in distant places. The pollutants like sulfur dioxide, ground-level ozone, carbon monoxide, lead, nitrogen dioxides and particulate matter have been shown to increase as a result of acid rain, smog, industrialization and long-distance transfer of pollutants. Numerous researchers have looked into the structural changes in the vegetative organs of various tropical plant species as a result of air pollution (Rai et al., 2011). A typical action of plant leaves to stress brought on by the air contaminants which include heavy metals, is to thicken their cell walls. Alvarez et al. (1998) reported necrosis involving thickening and pigmentations of the cell walls of palisade parenchyma. In response to the physiological stress of pollution, plants can manufacture and store a wide range of phenolics (Møller & Swaddle, 1997). Plants are capable of absorbing toxins through their roots when chemical contaminants accumulate in terrestrial or aquatic habitats. Phytotoxicity is the poisoning of plants by harmful substances. When plants are subjected to phytotoxicity, they have shown poor growth, dying seedlings and dead leaf tissue (Gupta et al., 2015)

Due to the high degree of chemical stability of many pesticides, notably chlorinated hydrocarbons like DDT stay in the soil for a very long time. For the protection of crops from pests and fungi, the inappropriate and persistent use of herbicides, pesticides and fungicides changes the fundamental constitution of the soil and renders it poisonous for plant growth. Due to the contamination of soil, the ecological system's delicate equilibrium is compromised. Most plants are not able to rapidly respond to the fluctuations in soil chemistry (Mishra et al., 2016).

### 19.2.5 HEAVY METALS

Naturally, plants are subjected to many adverse environmental stresses like temperature, humidity and various anthropogenic activities. Among them, heavy metal stress poses a distinguished adverse effects on crop productivity and development. Heavy metals—viz., lead, cadmium, nickel, cobalt, chromium and arsenic—trigger altered responses in plants that may include biochemical responses to crop production. Heavy metals are largely found in dispersed forms in rock formations. The main sources of heavy metal are soil and water-originated toxic substances. Numerous physiological and metabolic changes are induced in plants when they are exposed to hazardous heavy metals (Dubey, 2011; Villiers et al., 2011).

Metal contamination is linked to a decrease in plant development (Sharma & Dubey, 2007) including leaf chlorosis, necrosis, turgor loss, a reduction in the rate of seed percent germination and a crippled photosynthetic apparatus, which is directly linked with continuing senescence processes or with plant death (Carrier et al., 2003; Dalcorso et al., 2010). All these impacts are a result of the ultrastructural, biochemical and molecular modifications that heavy metal presence has on plant tissues and cells. It is worth noting that metals play important roles in cellular functions like homeostasis and photosynthesis in plants, where they also suppress root growth and cell division (Baig et al., 2020).

### 19.2.6 SALINITY

Salinity results in the reduction of plant's height, root development and formation of root, stem, leaf and bud. The degradation of fruit color, ion balance of the breakdown of the cell leading to stress when plants are subjected to excessive salt stress. Under saline conditions, protein synthesis reduces the permeability of the cell membrane reduces and whole organelles are injured. Ion toxicity, water scarcity, carbon-hydrogen shortage and associated symptoms emerge in the plant leaf that are subjected to long-term salinity stress (Parihar et al., 2015).

### 19.2.7 AGRONOMIC PROCESSES

Pesticides are applied to give protection to crops both in the field and throughout storage following harvest to reduce crop loss. Pesticides are primarily used to manage the various pests that attack crop plants, viz., soil insects, cutworms, leaf rollers, aphids, etc. (Goh et al., 2011). Utilizing biopesticides and creating pest-resistant transgenic cultivars are the two major alternative options available for controlling insect pests. The most extensively used method of protecting crops against pests, however, is still the use of chemical pesticides that further produces crops with a high yield. The application of pesticides also leads to the toxicity of plants that may be seen in the form of necrosis, stunting, discoloration of leaves and burns. The extreme use of pesticides is one of the key factors in the decrease of plant diversity (Donald, 2004). The toxicity of plants depends on the use of pesticides, application rate and use of different techniques during spraying, climatic conditions, humidity and properties of the soil (Shahzad et al., 2018). The application of pesticides subjects plants to oxidative stress as a result of the production of ROS. Oxidative stress causes proteins and pigments in chlorophyll to degrade, which reduces the ability of plants to absorb sunlight efficiently (Xia et al., 2006; Sharma et al., 2015). For combating this kind of stress, the antioxidative defense system of plants is stimulated, which involves enzymatic and non-enzymatic antioxidants that further decrease the stress in plants produced by pesticide toxicity (Xia et al., 2009; Sharma et al., 2015).

### 19.2.8 RADIATIONS

The world is moving toward growth of the telecommunication field, which is producing increased non-ionizing radiation in the surrounding environment that ultimately affects all living beings and plants standing in the vicinity of communication base stations. Radio waves are non-ionizing radiation that holds low quantum energy and does not manage the ionization of atoms. When such waves interact with living matter, they get absorbed and the transfer of energy happens within the living system, which further increases the frequency of collisions that leads to incremental change in heat termed as thermal effects.

Plants are severely affected by non-ionizing radiation due to continuous exposure. When exposed to non-ionizing electromagnetic radiation (EMR), plants experience changes in their physiology that may be caused by variations in their deoxyribonucleic acid (DNA) and gene expression patterns (Baulcombe & Dean, 2014). The photosystems and photosynthetic pigment concentrations of plants are both impacted by electromagnetic waves, which further effect plant growth. The ROS content is increased by electromagnetic fields, which has a negative impact on tissue health. As a result, the plant defense mechanism is stimulated to produce more ROS scavengers such as antioxidants and secondary metabolites. The initiation of plant defense mechanism and increased generation of secondary metabolites were documented by few researchers upon employment of extremely low-frequency EMRs (Lung et al., 2016). The two phytohormones, jasmonic acid (JA) and ABA, work as electrical signal transducers which further facilitate long-term communications and are accountable for systemic responses. The plant system functions as a biological antenna that changes its electrical properties when exposed to EMF radiation. Radiation changes the electric polarity of plant tissue by impacting charged ions and molecules that further change the mobility of ions and spread incorrect signals that are responsible for numerous metabolic, morphological and physiological processes in plants.

### 19.2.9 Pathogens

Plants encounter continuous threats from bacteria, fungi, oomycetes, viruses and herbivores that inhabit their environment. The plant's resistance toward specific invaders is observed by its survival rate, which depends on its defense mechanism. Meanwhile, pathogens have developed mechanisms to break structural barriers and avoid induced chemical defenses of their host plants. On the other hand, plants have also developed certain mechanisms to perceive pathogens and activate innate immune responses upon encountering non-adapted microbes, virulent microbes and insects. To limit pathogen and pest damage, plants have evolved various mechanisms. Innate immunity is triggered by the mechanical and chemical signals which are associated with injury, as well as pathogen-associated molecular responses (PAMPs). Using mitogen-activated protein kinase (MAPK) cascades, PAMP-triggered immunity integrates with the defense pathways regulated by salicylic acid (SA), JAs and ethylene (ET). This network provides effective, inducible defenses to resist pathogens and herbivores (Glazebrook, 2005; Walling, 2009).

### 19.2.10 Ozone

Ozone ($O_3$) gas is one of the major factors that lead to stress. Ozone is considered one of the potent destructive agent to plants which reduces photosynthesis and enhances up senescence. When plants react with ozone they exhibit a wide range of physiological responses (Apel & Hirt, 2004; Tiwari & Agarawal, 2018). It is well recognized that many natural environmental factors, including UV light, heavy metals, fertilizers, pesticides, drought, floods, heat stress, cold stress and other anthropogenic factors induce oxidative stress in plants. Stress in plants results from conditions that led to the production of ROS, which further causes damage to plant cells and plays a significant role in endonuclease activation in the plant cell, leading to DNA damage (Mehla et al., 2017). The majority of abiotic stressors in plants result in oxidative cell structural damage, which in turn reduces cellular activity.

## 19.3 VARIOUS EMERGING TECHNIQUES USED IN ECOPHYSIOLOGICAL RESEARCH

Different methods have been developed to explore the ecophysiological responses of plants toward various types of abiotic stress. Some of the emerging techniques used in ecophysiological research are as follow.

### 19.3.1 Hybridization Method

Hybridization is defined as the breeding of two different organisms from genetically dissimilar groups or species. It is one of the old techniques that has been used to increase genetic variability among the population. The goal of traditional hybridization methods was to create a genotype with desirable characteristics, such as insect resistance and high plant flowering potential to boost their commercial worth. Due to negative selection on advanced generation hybrids, the introduction of advantageous alleles across species barriers and speciation, hybridization is a mechanism for evolutionary change that is concerned with the conservation of species barriers because of negative selection on advanced generation hybrids (Christe et al., 2016; Goulet et al., 2017; Chhatre et al., 2019). Hybrid populations provide a rare chance to reset the clock on selection and examine possibly harmful trait variations before selection eliminates them. In addition to identifying the effects of hybridization on trait variance, trade-offs and conflicts in selection were also seen, as well as the development of growth-defense mechanisms through the selection of stomatal characteristics on the upper leaf surface. These findings highlighted the significant role that stomata play in controlling growth-defense mechanisms in organisms that are susceptible to foliar diseases.

In a study conducted by Goulet et al. (2017), significant observations regarding the impact of hybridization on phenotypic traits, particularly through the manifestation of hybrid vigor, were made. Hybridization not only leads to the formation of new hybrid species but also facilitates local adaptation by introducing novel alleles and exhibiting transgressed segregation. The field of plant biology has a rich history of investigating the prevalence of hybridization and its implications for evolution. The authors highlighted that recent advancements in genetic tools revolutionized our ability to document and analyze the occurrence and outcomes of hybridization in plants, providing profound insights into the underlying mechanisms driving phenomena such as heterosis and transgressive segregation.

### 19.3.2 Thermal Imaging

Thermal imaging is a technique for increasing the visibility of an object in a dark environment by detecting the infrared radiation emitted by the object and creating an image from that data. Thermography offers a very powerful tool for the study and investigation of spatial variation in plant and canopy temperatures, with many potential applications in plant physiology and ecology. Presently, thermal imaging is mostly utilized in medicine (Jones et al., 2004; Kumar et al., 2022).

Jones et al. (2004) studied the application of thermal imaging and infrared sensing in plant physiology and ecophysiology, and the authors recognized that the great majority of its applications in the physiology of plant and agronomy depend to an extent on variations in plant–water relationships. Although many applications of thermography in the areas of plant response to pollutants, chemicals, pests or pathogens have been proposed, in nearly all cases, any observed responses can be traced to changes in leaf surface conductance mainly due to changes in stomatal aperture. In addition to the significance of plant–water relations, it has been suggested that such imagery also gives information on differences in photosynthetic rate (Inoue, 1990), although this relationship only exists because photosynthetic gas exchange depends on stomatal conductance and would not detect non-stomata regulation of photosynthesis (Jones, 1999). The principal uses of thermal sensing and thermal photography to understand plant physiology are discussed in the following subsections.

#### 19.3.2.1 Water Loss, Transpiration, Stomatal Conductance and Stress Indices

Thermal imaging has clear advantages over conventional gas exchange and porometry approaches to the study of stomatal conductance, due to its non-destructive use and lack of direct leaf interference. Moreover, thermal imaging has the potential to obtain numerous values rapidly or to follow dynamic changes without affecting the leaf's natural environment. There has been significant interest in using airborne and satellite-based thermal imagery and photography for the study of water stress for irrigation scheduling reasons, as well as for the estimation of evaporation rates from plant canopies. The following subsections deal with the various physiological and ecological aspects which contribute to stress studies. Thermal imaging is particularly well adapted in the investigation of stomatal response spatial patterns, as well as in the stomatal conductance spatial diversity over and in between plant leaves, including stomatal "patchiness" (Weyers & Lawson, 1997). This study thus showed synchronous and asynchronous patterns of stomatal behavior in *Avena* leaves (Jones, 1999; Prytz et al., 2003).

The development of agricultural water stress indices as references for irrigation scheduling has always been of primary interest in the use of thermal sensing. In water-scarce conditions, such indices are supposed to mostly show results in evaporation rate and changes in the opening of stomatal activity. Because of the great temporal variability of stomata due to various environmental factors, raw temperature measurements are of relatively little use as indicators of crop water stress. The primary useful step in methods and techniques advancements using temperature information in irrigation scheduling was to normalize canopy temperature regarding air temperature and to calculate stress degree day (SDD), which is defined as the canopy-air temperature difference measured soon after midday, significant elevation of canopy temperature above air temperature was indicative of stomatal closure and crop water stress in a study (Jackson et al., 1977). An alternative approach studied by Clawson and Blad (1982) and Fuchs and Tanner (1966) to determine the temperature difference between the experimental canopy and a comparable well-irrigated crop was referred to as temperature stress day (TSD). This method required a well-watered reference crop, which is rather a limited application of the approach, and both these methods were found unsatisfactory in the studies due to the magnitude of both indices such as SDD and TSD varied as a function of climatic factors (atmospheric humidity). In more humid climates, cloudiness also becomes a serious factor; therefore, a new index called crop water stress index (CWSI) was developed in the advancement of thermal sensing for irrigation management motives (Idso, 1982; Jackson, 1982). This index took into account variations in atmospheric humidity. It was observed that (T$canopy$–T$air$) exhibits a linear relationship with atmospheric vapor pressure loss for well-watered crops and was defined as CWSI.

$$CWSI = (Tcanopy-Tnws)/(Tdry-Tnws) \quad (19.1)$$

Where $Tnws$ stands for temperature predicted for a well-watered crop under the same vapor pressure deficit conditions and $Tdry$ is the temperature of a non-transpiring crop

As it was not possible to have an actual non-watered crop adjacent to any field to provide this "non-water-stressed baseline," a standard relationship between (T$canopy$–T$air$) and vapor pressure deficit, Jackson et al. (1981) derived an analytical derivation for CWSI which was based on the energy balance equations demonstrated that CWSI was proportional to (1–E/Eo), where Eo is the potential evaporation rate for a well-watered crop (Jackson et al., 1981; Hipps et al., 1985).

#### 19.3.2.2 Temperature Variability

It was proposed that the variability of canopy temperature within a plot would provide useful information as an alternate method for the construction of stress indices (Clawson & Blad, 1982; Inoue, 1986). Because of the underlying soil heterogeneity and the fact that certain areas dry out more quickly than others, it was reasoned that soil water loss could be expected to result in increased variability across a field.

#### 19.3.2.3 Irrigation Scheduling

Thermal infrared information for irrigation scheduling has been widely used. Recent developments in imaging and control systems are now starting to bring up the possibilities of thermal imaging–based irrigation control systems for crop water stress, wherein the following two main strategies are anticipated.

1. Hand-held thermal imagers which are used as rapid monitoring systems to assess the efficacy of current irrigation scheduling systems on a farm, as well as direct the application of supplemental irrigation.
2. Imagery may be incorporated into an automated irrigation control system that provides nearly continuous monitoring and the capacity to supply water whenever it is needed—for example, in standard trickle irrigation systems. In this case, a standard amount of irrigation would be applied whenever the stress index surpasses a specific threshold. To create a robust irrigation schedule that could change the amount of irrigation water applied at any moment, expert systems that integrate data from various sources could be modified to incorporate temperature data with a water budget calculation.

### 19.3.2.4 Biophysical and Aerodynamic Properties

Infrared thermography has been documented to provide biophysical information regarding factors determining the thermal properties of leaves and plants. An early study conducted by Clark (1975) using thermography of model and *Phaseolus vulgaris* leaves in a wind tunnel investigated the convective transfer processes between leaves and their environment. Kummerlen et al. (1999) described two methods which were based on active thermography (studying responses to imposed environmental changes) for determining the spatial distribution over the leaf surface of thermal capacitance called constant flux approach and periodic flux method.

### 19.3.2.5 Metabolic Processes

Any aspect of plant metabolism results in the production of heat because no metabolic process is fully thermodynamically efficient. Oxidative respiration is one such procedure that organisms used to produce reducing equivalents and adenosine triphosphate (ATP) and to regulate the metabolic processes. Plants, however, have a second respiratory pathway whereby electron transport is not linked to ATP synthesis, directly releasing more free energy as heat. This alternative respiratory pathway, also known as the alternate oxidase (AOX) pathway, is resistant to cyanide and is characteristic of plant mitochondria (Vanlerberghe & McIntosh, 1997). Because of the high heat production in the AOX pathway, there has been specific interest in its effect on the regulation of the thermal balance of plant tissues. The burst of heat production at flowering—seen in many plant species, particularly of the Araceae family (including *Arum maculatum*, *Symplocarpus foetidus* and *Philodendron selloum*) and in custard apples, cycads, palms, lotus, Ravelesia and water lilies—is the most well-known example of thermogenesis in plants (Ito et al., 2003).

It has been demonstrated that the increased rates of respiration in the inflorescences of these thermogenic species caused a temperature increase of 20°C (Knutson, 1974). Even though the use of thermocouples is a perfectly adequate method for tracking the temperature changes of specific parts of the inflorescence during the metabolic burst, thermography allows for rapid visualization and quantification of the spatial differences in temperature dynamics and was used as a particular effect in the study of the dynamics of thermogenesis in all the various parts of an *Arum* inflorescence (Bermadinger-Stabentheiner & Stabentheiner, 1995).

### 19.3.2.6 Disease and Infection

Infrared thermometry was initially used to show the potential sensitivity of temperature measurements for the study of plant disease. For instance, an early study conducted by Pinter et al. (1979) showed that root infections of cotton (by *Phymatotrichum omnivorum*) and sugarbeet (by *Pythium aphanidermatum*) could raise leaf temperatures by as much as 3–8°C above healthy plants, even though visual symptoms were not apparent without excavating the roots.

Laboratory thermographic studies could have demonstrated that thermal effects can even be utilized to predict future cell death events on infection by tobacco mosaic virus (TMV) and in bacterioopsin (bO) transgenic tobacco plants (Chaerle et al., 1999, Chaerle & Van Der Straeten, 2001). With TMV inoculation, there was a close correlation between visible disease symptoms and localized temperature rises of particular significance; however, it was observed that clear thermal responses could be detected at least eight hours before the onset of visible symptoms of hypersensitive response (HR) symptoms (Chaerle et al., 1999). Catena (2003) has reported that the application of thermography reveals cavities or rotten tissue in trees, based on the differential thermal behavior of rotten tissue.

### 19.3.2.7 Pollution and Agronomic Effects

The potential to use thermal imagery together with reflectance and fluorescence photography has been used for the diagnosis and analysis of plants with various agronomic treatments, especially for the studying the effects of pollution. Omasa et al. (1981) used thermal imaging to evaluate the consequences of pollution. These investigations showed that exposure to pollutants like $NO_2$, $SO_2$ and $O_3$ caused stomatal closure. The development of hand-held thermal cameras has made it possible to use thermal imaging to diagnose the physiological status of both individual urban trees and wild trees (Omasa et al., 1993; Catena, 2003).

Omasa et al. (1993) reported a successful diagnosis of the physiological activity of specific trees using a thermal camera carried by helicopter. These studies supported the hypothesis that leaves are spatially heterogeneously impaired by biotic and abiotic stresses. Thermography is currently considered the best method that provides sufficient information on the spatial variation of stomatal conductance for such studies.

## 19.3.3 Metabolomics, Proteomics and QSARS/SARS

### 19.3.3.1 Metabolomics

Metabolomics is the study of the set of metabolites in a cell, tissue or organism at any specified time. The creation of secondary metabolites makes metabolomics particularly significant. Metabolites are obtained from tissues, sorted and thoroughly analyzed in the metabolite profiling research (Arbona et al., 2013; Barik & Mishra, 2018). Using metabolic fingerprinting, samples are differentiated according to their phenotype or biological relevance. These studies generally rely on crude data collected from detecting instruments like gas and liquid chromatography–mass spectrometry or nuclear magnetic resonance to analyze and calculate metabolites from samples. The study of metabolomics plays a significant role in the networking of molecules to connect the gap between the genotypes and phenotypes (Katam et al., 2022).

Metabolic pathways are expected to be active and evolve continuously in response to changes in energy level and physiological status. Metabolomics is widely applicable and realistic to many species in their respective ecophysiological modifications (Barik & Mishra, 2018).

#### 19.3.3.2 Proteomics

Proteomics is a branch of molecular biology that investigates living organisms from the perspective of their protein structures and biomolecules that carry out information about encoded genes, along with the goal of understanding the translation of genotype to phenotype (Barik & Mishra, 2018; Borrajo et al., 2018). The goal of these investigations will outline the identification and classification of the protein species, as well as the qualitative and quantitative assessment of two or more biological samples. The knowledge of proteomics has also been for both fundamental and applied research associated with plant studies. Proteomics research has been conducted on different stages of plants (Barik & Mishra, 2018; Sánchez et al., 2018).

#### 19.3.3.3 QSARS/SARS

Studies of structure-activity relationships (SARs) and quantitative structure-activity relationships (QSARs) were utilized extensively for toxicological chemistry and medicinal chemistry to promote a drug's discovery, as well as in the study and research of hazardous and toxic compounds (Sánchez et al., 2018). QSAR has also been used as an instrument to learn the behavior of biological systems, proving that a compound's physiological effects are a concern of its chemical composition (Sánchez et al., 2018). SAR investigations aim to identify chemical information in a series of synthetic compounds with similar biological activities. SAR research should take into account any changes to the compound's chemical structure that may affect its solubility in water, transport through membranes, binding to receptors and other kinetic aspects (Shanmugam & Jeon, 2017). One of the fundamental presumptions of SAR analysis is that active and comparable chemical compounds interact with or combine with biological targets through a similar mode of action. These studies can be categorized into two methods: structure-based or ligand-based methods. Ligand-based methods are associated with the properties of ligands, whereas structure-based linked with the biological target structures. Since many of these properties are similar in geomorphology, the metamorphic term "activity landscape" can be used to conceptualize many features of SARs of large series of chemical substances. The analysis of structural similarity and power differences between compounds that exhibit the same biological activity is integrated into a representation known as the "activity landscape". According to Borrajo et al. (2018) and Stumpfe and Bajorath (2012), it is challenging to characterize the activity landscapes that represent particular SARs of a group of compounds in a systematic. The studies that evaluate or count the relationships between the chemical structure and the activity of several chemicals, that use a statistical approximation to build mathematical equations, are known by the term QSAR.

### 19.3.4 Isotopes Techniques for Plant and Soil Studies

Over the past two decades, research into how plants and their environments interact has significantly increased. This pattern will persist as more environmental changes affect how ecosystems function. Examining the interactions between plants and their environments involves concentrating on how plants react to stresses, such as biochemical communication among plants, as well as between plants and micro-organisms (allelopathy), and plant reaction to pollutants. Isotopes are two or more types of atoms with the same atomic number but different neutron counts. The stable isotopes of light elements—especially carbon, hydrogen, nitrogen and oxygen—which all have more than one isotope are used in many ecological investigations. In addition to the fact that these elements are more frequently used in the biological system, these elements show a greater relative percent mass change than heavier ones for light isotopes with variations of just one or two atomic mass numbers (Chiapusio et al., 2018).

Primarily carbon radiochemicals ($^{14}C$) and stable carbon isotope tracers ($^{13}C$) are effective tools in plant ecophysiological studies to explain the impacts of allelochemicals and contaminants in plants and their surroundings (Chiapusio et al., 2018). Carbon-12 ($^{12}C$), carbon-13 ($^{13}C$) and carbon-14 ($^{14}C$) are the three most prevalent isotopes of carbon. In nature, $^{12}C$ makes up the majority of the carbon, followed by $^{13}C$ at 1.1% and $^{14}C$ at very low concentrations. In other words, neither $^{12}C$ nor $^{13}C$ changes over time. In contrast, $^{14}C$ which has a half-life of 5700 years, is unstable (Michener & Lajtha, 2007).

#### 19.3.4.1 Isotope Technique Using Carbon

These effective methods enable the quantitative and qualitative description of the transmission of molecules that are radiolabeled in mesocosms by tracing the $^{14}C$ compound in plants and soil.

Stable $^{13}C$ isotope techniques are the following.

*19.3.4.1.1 Elemental Analyzer (EA)*

Elemental analyzers are used to determine the organic carbon of a sample. Here, only small amounts of samples are required. Before analysis, the liquid samples should be freeze-dried or vacuum-dried (organic solvent), and the $^{13}C$ label quantities must be significant enough to be detectable.

*19.3.4.1.2 Liquid or Gas Chromatography*

By separating particular biomolecules to obtain their isotopic signature, liquid or gas chromatography is used to quantify the quantities of different molecules. The time needed to assess each sample in liquid or gas chromatography is long. It is difficult to couple an isotope ratio mass spectrometer (IRMS) since the most abundant molecule is not always the one with the highest concentration of $^{13}C$. If this is the case, the dilution factor of the sample is challenging to identify, and a fraction collector is frequently required to distinguish between minor and significant molecules.

#### 19.3.4.1.3 Isotopic Ratio Mass Spectrometer (IRMS)

IRMS is used to determine the isotope ratio $\partial^{13}C$ (ratio $^{13}CO_2/^{12}CO_2$).

#### 19.3.4.1.4 Isotopic Ratio Infrared Spectroscopy (IRIS)

To estimate the levels of $^{13}CO_2$ and $^{12}CO_2$ appropriate in the field for estimation of $CO_2$ (interference with non-target gases: $CH_4$, $H_2O$, $H_2S$, volatile organic compounds [VOCs]).

IRIS is used in evaluating drought-tolerant plant species using carbon stable isotopes:

> Carbon, the main element of proteins and carbohydrates in plant tissues, contains both $^{12}C$ and $^{13}C$ isotopes (Gouveia et al., 2019). In order to choose and assess plant cultivars that can tolerate drought, it is becoming more and more common to measure inherent variations in the quantity of $^{13}C$ and $^{12}C$ in plant matter (Bhat & Bhat, 2010). Because stressed plants absorb less carbon from the environment, notably $^{13}C$, during a drought, there is a significant change in the natural isotopic ratios of $^{13}C$ and $^{12}C$ in plant matter. Plant variety that resists water scarcity ought to exhibit less $^{13}C$ depletion than a variety that is susceptible to it (Dercon et al., 2006; Bhat & Bhat, 2010). Drought-resistant barley, wheat, rice, and peanut have been effectively selected using this discrimination against $^{13}C$ in plant tissues.
>
> *(Wright et al., 1994)*

### 19.3.4.2 Isotopes Techniques Using Hydrogen and Oxygen

#### 19.3.4.2.1 Water and Evaporative Enrichment

Although studies on hydrogen and oxygen isotope ratios originated outside of biology, these elements' relevance in plant biology is growing quickly. Early researchers in Europe and the United States measured the $^{18}O$ and $^{2}H$ content of various freshwater samples in the 1950s (Epstein & Mayeda, 1953), but it was Dansgaard's (1961, 1964) systematic investigations in Denmark that provided the blueprint for the mechanistic understanding of the variables causing spatial isotopic variance in meteoric fluids. According to Harold Urey's (1947) study, seawater should have 6% more $^{18}O$ than atmospheric oxygen at equilibrium. Early studies focused on the isotopic composition of leaf water that is increased during transpiration since diatomic oxygen is produced by photosynthesis and evaporative processes are known to enrich rivers and streams (Dongmann et al., 1974). Relative humidity is one of the key factors influencing the enrichment of the heavier isotopes, which can be used to simulate the amount of the leaf water evaporative enrichment (Dongmann et al., 1974; Zundel et al., 1978). Early in these attempts, Farris and Strain (1978) proposed that water stress might affect the degree of leaf-water enrichment, although it now seems like a large percentage of the isotopic enrichment in water-stressed leaves may be brought on by elevated leaf temperatures in plants exposed (Zundel et al., 1978).

However, it has recently become clearer how models relating carbon isotope fractionation and transpiration efficiency work (Farquhar & Lloyd, 1993). One of the ultimate aims of the growing interest has been to determine whether or not isotopic evaporative enrichment of leaf water may be used as a primary determinant of the leaf-to-air water vapor gradient, the driving force behind transpiration.

### 19.3.4.3 Isotopes Techniques Using Hydrogen and Nitrogen

Since plants get a substantial percentage of their nitrogen from the soil, it is important to know the processes which cause variations in the soil's $\delta^{15}N$ (Handley & Raven, 1992; Dawson et al., 2002). This can help to identify the sources of nitrogen that plants use. The main soil nitrogen conversions—which are driven by micro-organisms, including mineralization, nitrification, and denitrification—yield nitrogen compounds deficient in $^{15}N$ relative to the substrates from which they were produced (Dawson et al., 2002). If the amount of $NO_3$ released after nitrification is greater than the quantity needed by the soil and plants, the pool of remaining organic matter sometimes become increasingly $^{15}N$ rich (Natelhoffer & Fry, 1988; Dawson et al., 2002).

Natural abundance measurements of $^{15}N$ are increasingly being used as a measure of variation in the forest nitrogen cycle. Using enriched $^{15}N$ tracers to monitor the fate of nitrogen in ecosystems has a wide range of uses. In greenhouse trials, $^{15}N$ tracers have also been employed to evaluate if foliage uptake and assimilation of nitrogen happens and to estimate the overall contribution to the nitrogen needs of each plant species (Dawson et al., 2002).

## 19.3.5 Flow Cytometry

The technology known as "flow cytometry" allows for the fast investigation of individual cells or particles that travel past one or more lasers while suspended in a buffered salt solution. Each particle is assessed for a scattering of visible light and one or more fluorescence properties. To govern a cell's internal complexity or granularity, visible light scatter is measured in two separate directions: frontward, which might reveal the cell's relative size, and at 90 degrees. Numerous fields—including immunology, molecular biology, cancer biology, virology, and infectious disease surveillance—use the potent technique of flow cytometry (Macey, 2010). For instance, it works wonders when researching the immune system, as well as how plants' immune systems react to infectious illnesses and cancer. The tools that are utilized for flow cytometry have evolved during the last few decades. Common pieces of equipment include systems with 96-well loaders for bead analysis, systems that combine microscopy and flow cytometry, and systems that combine mass spectrometry and flow cytometry. Multiple laser systems are also often used (McKinnon, 2018).

### 19.3.5.1 Instrumentation in Flow Cytometers

#### *19.3.5.1.1 Conventional Flow Cytometers*

Conventional or traditional flow cytometers are composed of three structures: fluidics, optics and electronics. The fluidics system employs sheath fluid to transport and concentrate the sample to the laser intercept or interrogation point, where the sample is assessed (often a buffered saline solution). The excitation optics (lasers) and collection that provide the visible and fluorescent light signals needed to examine the sample make up the optical system. To detect and quantify each distinct fluorochrome, a sequence of dichroic filters direct luminous light to certain detectors, and bandpass filters choose the light wavelengths that are read. In more detail, dichroic filters are those that let in the light of a different wavelength and then imitate the light that remains at an angle (Manohar et al., 2021).

#### *19.3.5.1.2 Acoustic Focusing Cytometers*

To well concentrate cells for laser examination, this cytometer employs ultrasonic vibrations. Excessive sample input and reduced sample congestion are both possible with this kind of acoustic concentrating. This cytometer has 14 fluorescence channels and up to four lasers (Ward et al., 2009; Piyasena et al., 2012).

#### *19.3.5.1.3 Other Types of Instrumentation in Flow Cytometers*

The cell sorter is a specific type of traditional flow cytometer that may gather and filter samples for extra investigation. An individual can choice (gate) a population of cells or particles that is positive (or negative) for the necessary parameters and then direct those cells into a collecting vessel using a cell sorter. The cell sorter produces drops by rapidly swinging the sample stream of liquid to separate the cells. The drops are then given a positive or negative charge and sent through metal deflection plates where—depending on their charge—they are guided to a certain collection receptacle (Orfao & Ruiz-Argüelles, 1996). Fluorescence microscopy and conventional flow cytometry are combined in imaging flow cytometers (IFC). This enables fast analysis of a sample's structure and multi-parameter fluorescence, both at the scale of a single cell and a population (Barteneva et al., 2012). IFC has the ability to process several cells similarly to a flow cytometer while also tracking protein dispersals inside specific cells like a confocal or fluorescence microscopy. They are especially helpful in various type of applications, including studies on the co-localization of cell signaling, cell-to-cell communication, and DNA damage and repair, along with any other where it is necessary to be able to organize the location of cells with their fluorescence expression in huge numbers of cells. In mass cytometers, time-of-flight mass spectrometry and flow cytometry are integrated. As an alternative to fluorescent antibodies, heavy metal ion-tagged antibodies (often from the lanthanide series) are used to mark cells, and time-of-flight mass spectrometry is used to identify them (Leipold et al., 2015). Additionally, since reagents used in mass cytometry lack the emission spectrum overlap associated with fluorescent labels and cellular autofluorescence signals, no adjustment is required. Cell sorting is not feasible since the sample is destroyed during investigation, and the attainment rate is substantially lower (1,000 cells per second as opposed to 10,000 cells per second) than with a conventional flow cytometer (Mei et al., 2016).

### 19.3.5.2 The Use of Flow Cytometry in Ecophysiological Research

Flow cytometry can detect apoptotic cells quickly and quantitatively. For the most accurate findings, methods for identifying cells that are experiencing apoptosis (programmed cell death) ought to identify apoptotic features of cells quickly and permit them to maintain their original form. The human proto-oncogene Bcl-2 is found in the outer membranes of mitochondria, the nuclear envelope and the endoplasmic reticulum. Bcl-2 is an anti-apoptotic intracellular protein, and studies have shown that its level is lowered as apoptosis progresses (Wlodkowic et al., 2012). The knowledge of ecophysiology techniques in research field helps to understand the physiological responses of tropical plants to the changing climate, $CO_2$ level and development of invasiveness in plants. We still are waiting for new molecular biology, biochemistry and genetics technologies to be discovered in this area of research.

## 19.4 CONCLUSION

Ecophysiology attempts to understand the potential limits placed on plants by their physiology and how plant species respond to particular environmental challenges. The past two decades have seen a renaissance in ecophysiological research. The fact of global climate change that includes the reduction in the level of oxygen in our environment is indisputable. Undeniably, flora and fauna are being effected by the changes in the natural environment due to severe anthropogenic activities. This chapter aimed to give an insight into the techniques used in ecophysiological research and the ecophysiological response of plants to various types of stresses. This chapter serves as a catalyst for more work in this significant and difficult area of research, in addition to addressing some recent developments in the ecophysiology of tropical plants. However, for substantial benefit to the farmer, the performance of several plants must also be assessed in the context of the naturally changing tropical environment. The chapter further highlighted the mechanisms of adaptations of plant species in stressful environments

## CONFLICT OF INTEREST

The authors state that there are no conflicts of interest in the publication of this chapter.

# REFERENCES

Ackerly, David D., Susan A. Dudley, Sonia E. Sultan, Johanna Schmitt, James S. Coleman, C. Randall Linder, Darren R. Sandquist et al. "The evolution of plant ecophysiological traits: Recent advances and future directions: New research addresses natural selection, genetic constraints, and the adaptive evolution of plant ecophysiological traits." *Bioscience* 50, no. 11 (2000): 979–995.

Ainsworth, Elizabeth A., Carl J. Bernacchi, and Frank G. Dohleman. "Focus on ecophysiology." *Plant Physiology* 172, no. 2 (2016): 619–621.

Ali, Muhammad, Zhihui Cheng, Husain Ahmad, and Sikandar Hayat. "Reactive oxygen species (ROS) as defenses against a broad range of plant fungal infections and case study on ROS employed by crops against Verticillium dahliae wilts." *Journal of Plant Interactions* 13, no. 1 (2018): 353–363.

Alvarez, D., G. Laguna, and I. Rosas. "Macroscopic and microscopic symptoms in Abies religiosa exposed to ozone in a forest near Mexico City." *Environmental Pollution* 103, no. 2–3 (1998): 251–259.

Apel, Klaus, and Heribert Hirt. "Reactive oxygen species: Metabolism, oxidative stress, and signaling transduction." *Annual Review of Plant Biology* 55 (2004): 373.

Arbona, Vicent, Matías Manzi, Carlos de Ollas, and Aurelio Gómez-Cadenas. "Metabolomics as a tool to investigate abiotic stress tolerance in plants." *International Journal of Molecular Sciences* 14, no. 3 (2013): 4885–4911.

Avolio, Meghan L., Kimberly J. La Pierre, Gregory R. Houseman, Sally E. Koerner, Emily Grman, Forest Isbell, David Samuel Johnson, and Kevin R. Wilcox. "A framework for quantifying the magnitude and variability of community responses to global change drivers." *Ecosphere* 6, no. 12 (2015): 1–14.

Baig, Mohammad Affan, Sadia Qamar, Arlene Asthana Ali, Javed Ahmad, and M. Irfan Qureshi. "Heavy metal toxicity and tolerance in crop plants." In *Contaminants in agriculture*, pp. 201–216. Springer, 2020.

Barik, Bibhuti Prasad, and Amarendra Narayan Mishra. "Computational approach to study ecophysiology." In *Advances in plant ecophysiology techniques*, pp. 483–497. Springer, 2018.

Barteneva, Natasha S., Elizaveta Fasler-Kan, and Ivan A. Vorobjev. "Imaging flow cytometry: Coping with heterogeneity in biological systems." *Journal of Histochemistry & Cytochemistry* 60, no. 10 (2012): 723–733.

Baulcombe, David C., and Caroline Dean. "Epigenetic regulation in plant responses to the environment." *Cold Spring Harbor Perspectives in Biology* 6, no. 9 (2014): a019471.

Beltran-Peña, Areidy, Lorenzo Rosa, and Paolo D'Odorico. "Global food self-sufficiency in the 21st century under sustainable intensification of agriculture." *Environmental Research Letters* 15, no. 9 (2020): 095004.

Bermadinger-Stabentheiner, Edith, and Anton Stabentheiner. "Dynamics of thermogenesis and structure of epidermal tissues in inflorescences of Arum maculatum." *New Phytologist* 131, no. 1 (1995): 41–50.

Bhadouria, Rahul, Pratap Srivastava, Rishikesh Singh, Sachchidanand Tripathi, Hema Singh, and A. S. Raghubanshi. "Tree seedling establishment in dry tropics: An urgent need of interaction studies." *Environment Systems and Decisions* 37, no. 1 (2017): 88–100.

Bhadouria, Rahul, Rishikesh Singh, Pratap Srivastava, and Akhilesh Singh Raghubanshi. "Understanding the ecology of tree-seedling growth in dry tropical environment: A management perspective." *Energy, Ecology and Environment* 1, no. 5 (2016): 296–309.

Bhadouria, Rahul, Sachchidanand Tripathi, Pratap Srivastava, and Pardeep Singh, eds. *Handbook of research on the conservation and restoration of tropical dry forests*. IGI Global, 2019.

Bhat, M. I., and M. A. Bhat. "Applications of stable and radioactive isotopes in soil science." *Current Science* (2010): 1458–1471.

Borrajo, C. I., A. M. Sánchez-Moreiras, and M. J. Reigosa. "Morpho-physiological responses of tall wheatgrass populations to different levels of water stress." *PLoS One* 13, no. 12 (2018): e0209281.

Burgess, Patrick, and Bingru Huang. "Mechanisms of hormone regulation for drought tolerance in plants." In *Drought stress tolerance in plants*, Vol. 1, pp. 45–75. Springer, 2016.

Carrier, P., A. Baryla, and M. Havaux. "Cadmium distribution and microlocalization in oilseed rape (Brassica napus) after long-term growth on cadmium-contaminated soil." *Planta* 216, no. 6 (2003): 939–950.

Catena, Alessandra. "Thermography reveals hidden tree decay." *Arboricultural Journal* 27, no. 1 (2003): 27–42.

Chaerle, Laury, and Dominique Van Der Straeten. "Seeing is believing: Imaging techniques to monitor plant health." *Biochimica et Biophysica Acta (BBA)-Gene Structure and Expression* 1519, no. 3 (2001): 153–166.

Chaerle, Laury, Wim Van Caeneghem, Eric Messens, Hans Lambers, Marc Van Montagu, and Dominique Van Der Straeten. "Presymptomatic visualization of plant–virus interactions by thermography." *Nature Biotechnology* 17, no. 8 (1999): 813–816.

Chhatre, Vikram E., Karl C. Fetter, Andrew V. Gougherty, Matthew C. Fitzpatrick, Raju Y. Soolanayakanahally, Ronald S. Zalesny, and Stephen R. Keller. "Climatic niche predicts the landscape structure of locally adaptive standing genetic variation." *BioRxiv* (2019): 817411.

Chiapusio, Geneviève, Dorine Desalme, Philippe Binet, and François Pellissier. "Carbon radiochemicals (14 C) and stable isotopes (13 C): Crucial tools to study plant-soil interactions in ecosystems." In *Advances in plant ecophysiology techniques*, pp. 419–437. Springer, 2018.

Christe, Camille, Kai N. Stölting, Luisa Bresadola, Barbara Fussi, Berthold Heinze, Daniel Wegmann, and Christian Lexer. "Selection against recombinant hybrids maintains reproductive isolation in hybridizing populus species despite F1 fertility and recurrent gene flow." *Molecular Ecology* 25, no. 11 (2016): 2482–2498.

Clark, J. A. "Heat and mass transfer from real and model leaves." *Heat and Mass Transfer in the Biosphere* (1975): 413–422.

Clawson, Kirk L., and Blaine L. Blad. "Infrared thermometry for scheduling irrigation of corn 1." *Agronomy Journal* 74, no. 2 (1982): 311–316.

Cowles, Jane M., Peter D. Wragg, Alexandra J. Wright, Jennifer S. Powers, and David Tilman. "Shifting grassland plant community structure drives positive interactive effects of warming and diversity on aboveground net primary productivity." *Global Change Biology* 22, no. 2 (2016): 741–749.

Dalcorso, G., Farinati, S. and Furini, A. "Regulatory networks of cadmium stress in plants." *Plant Signaling and Behavior* 5, no. 6 (2010): 1–5.

Dansgaard, Willi. "Stable isotopes in precipitation." *Tellus* 16, no. 4 (1964): 436–468.

Dansgaard, Willi. *The Isotopic Composition of Natural Waters with Special Reference to the Greenland Ice Cap.* Oersted Institue, 1961.

Das, Kaushik, and Aryadeep Roychoudhury. "Reactive oxygen species (ROS) and response of antioxidants as ROS-scavengers during environmental stress in plants." *Frontiers in Environmental Science* 2 (2014): 53.

Dawson, Todd E., Stefania Mambelli, Agneta H. Plamboeck, Pamela H. Templer, and Kevin P. Tu. "Stable isotopes in plant ecology." *Annual Review of Ecology and Systematics* (2002): 507–559.

Demiralay, Mehmet, Aykut Sağlam, and Asim Kadioğlu. "Salicylic acid delays leaf rolling by inducing antioxidant enzymes and modulating osmoprotectant content in Ctenanthe setosa under osmotic stress." *Turkish Journal of Biology* 37, no. 1 (2013): 49–59.

Dercon, G., E. Clymans, Jan Diels, Roel Merckx, and J. Deckers. "Differential 13C isotopic discrimination in maize at varying water stress and at low to high nitrogen availability." *Plant and Soil* 282, no. 1 (2006): 313–326.

Desikan, Radhika, John Hancock, and Steven Neill. "Reactive oxygen species as signalling molecules." *Antioxidants and Reactive Oxygen Species in Plants* (2005): 169–196.

Donald, Paul F. "Biodiversity impacts of some agricultural commodity production systems." *Conservation Biology* 18, no. 1 (2004): 17–37.

Dongmann, G., H. W. Nürnberg, H. Förstel, and K. Wagener. "On the enrichment of H2 18O in the leaves of transpiring plants." *Radiation and Environmental Biophysics* 11, no. 1 (1974): 41–52.

Dubey, R. S. "Metal toxicity, oxidative stress and antioxidative defense system in plants." In S. D. Gupta, ed., *Reactive oxygen species and antioxidants in higher plants*, pp. 177–203. CRC Press, 2011.

Epstein, Samuel, and Toshiko Mayeda. "Variation of O18 content of waters from natural sources." *Geochimica et cosmochimica acta* 4, no. 5 (1953): 213–224.

Farooq, Muhammad, M. Hussain, Abdul Wahid, and K. H. M. Siddique. "Drought stress in plants: An overview." *Plant responses to drought stress* (2012): 1–33.

Farquhar, Graham D., and Jon Lloyd. "Carbon and oxygen isotope effects in the exchange of carbon dioxide between terrestrial plants and the atmosphere." In *Stable isotopes and plant carbon-water relations*, pp. 47–70. Academic Press, 1993.

Farris, F., and B. R. Strain. "The effects of water-stress on leaf H2 18O enrichment." *Radiation and Environmental Biophysics* 15, no. 2 (1978): 167–202.

Fitter, Alastair H. and Robert KM. Hay. *Environmental physiology of plants.* Academic press, 2012.

Fuchs, M. and C. B. Tanner. "Infrared thermometry of vegetation 1." *Agronomy Journal* 58, no. 6 (1966): 597–601.

Glazebrook, Jane. "Contrasting mechanisms of defense against biotrophic and necrotrophic pathogens." *Annual Review of Phytopathology* 43 (2005): 205.

Goh, Wei-Ling, Pang-Hung Yiu, Sie-Chuong Wong, and Amartalingam Rajan. "Safe use of chlorpyrifos for insect pest management in leaf mustard (Brassica juncea L. Coss)." *Journal of Food, Agriculture and Environment* 9 (2011): 1064–1066.

Goulet, Benjamin E., Federico Roda, and Robin Hopkins. "Hybridization in plants: Old ideas, new techniques." *Plant Physiology* 173, no. 1 (2017): 65–78.

Gouveia, Carla SS, José FT Gananca, Jan Slaski, Vincent Lebot, and Miguel ÂA Pinheiro de Carvalho. "Stable isotope natural abundances ($\delta$13C and $\delta$15N) and carbon-water relations as drought stress mechanism response of taro (Colocasia esculenta L. Schott)." *Journal of Plant Physiology* 232 (2019): 100–106.

Gulen, Hatice, and Atilla Eris. "Effect of heat stress on peroxidase activity and total protein content in strawberry plants." *Plant Science* 166, no. 3 (2004): 739–744.

Gupta, Dharmendra K., José M. Palma, and Francisco J. Corpas, eds. *Reactive oxygen species and oxidative damage in plants under stress.* Springer, 2015.

Haggag, Wafaa M., H. F Abouziena, F. Abd-El-Kreem, and S. El Habbasha. "Agriculture biotechnology for management of multiple biotic and abiotic environmental stress in crops." *Journal of Chemical and Pharmaceutical Research* 7, no. 10 (2015): 882–889.

Handley, L. L., and John Albert Raven. "The use of natural abundance of nitrogen isotopes in plant physiology and ecology." *Plant, Cell & Environment* 15, no. 9 (1992): 965–985.

Hipps, L. E., G. Asrar and E. T. Kanemasu. "A theoretically-based normalization of environmental effects on foliage temperature." *Agricultural and Forest Meteorology* 35, no. 1–4 (1985): 113–122.

Idso, Sherwood B. "Non-water-stressed baselines: A key to measuring and interpreting plant water stress." *Agricultural Meteorology* 27, no. 1–2 (1982): 59–70.

Impe, Daniela, Daniel Ballesteros, and Manuela Nagel. "Impact of drying and cooling rate on the survival of the desiccation-sensitive wheat pollen." *Plant Cell Reports* 41, no. 2 (2022): 447–461.

Inoue, Yoshio. "Remote detection of physiological depression in crop plants with infrared thermal imagery." *Japanese Journal of Crop Science* 59, no. 4 (1990): 762–768.

Inoue, Yoshio. "Remote-monitoring of function and state of crop community: I analysis of thermal image of crop canopy." *Japanese Journal of Crop Science* 55, no. 2 (1986): 261–268.

Ito, K., Y. Onda, T. Sato, Y. Abe, and M. Uemura. "Structural requirements for the perception of ambient temperature signals in homeothermic heat production of skunk cabbage (Symlocarpus foetidus)." *Plant, Cell and Environment* 26, no. 6 (2003): 783–788.

Jackson, Ray D. "Canopy temperature and crop water stress." In *Advances in Irrigation*, vol. 1, pp. 43–85. Elsevier, 1982.

Jackson, Ray D., S. B. Idso, R. J. Reginato, and P. J. Pinter Jr. "Canopy temperature as a crop water stress indicator." *Water Resources Research* 17, no. 4 (1981): 1133–1138.

Jackson, Ray D., R. J. Reginato, and Serwood B. Idso. "Wheat canopy temperature: A practical tool for evaluating water requirements." *Water Resources Research* 13, no. 3 (1977): 651–656.

Jones, Hamlyn G. "Use of thermography for quantitative studies of spatial and temporal variation of stomatal conductance over leaf surfaces." *Plant, Cell and Environment* 22, no. 9 (1999): 1043–1055.

Jones, Hamlyn G., Nicole Archer, and Eyal Rotenberg. "Temperature and evaporation from forest canopies." *Forests at the Land-atmosphere Interface* (2004): 123.

Katam, Ramesh, Chuwei Lin, Kirstie Grant, Chaquayla S. Katam, and Sixue Chen. "Advances in plant metabolomics and its applications in stress and single-cell biology." *International Journal of Molecular Sciences* 23, no. 13 (2022): 6985.

Knutson, Roger M. "Heat production and temperature regulation in eastern skunk cabbage." *Science* 186, no. 4165 (1974): 746–747.

Kumar, Prasoon, Ankit Gaurav, Rajesh Kumar Rajnish, Siddhartha Sharma, Vishal Kumar, Sameer Aggarwal, and Sandeep Patel. "Applications of thermal imaging with infrared thermography in Orthopaedics." *Journal of Clinical Orthopaedics and Trauma* 24 (2022): 101722.

Kummerlen, B., S. Dauwe, D. Schmundt, and U. Schurr. "Thermography to measure water relations of plant leaves." In B. Jahne, ed., *Handbook of computer vision and applications (Systems and Applications)*, Vol. 3, pp. 763–781. Academic Press, 1999.

Lacava, Paulo Teixeira, Andréa Cristina Bogas, and Felipe de Paula Nogueira Cruz. "Plant growth promotion and biocontrol by endophytic and rhizospheric microorganisms from the tropics: A review and perspectives." *Frontiers in Sustainable Food Systems* 6 (2022): 796113.

Langley, J. Adam, and Bruce A. Hungate. "Plant community feedbacks and long-term ecosystem responses to multi-factored global change." *AoB Plants* 6 (2014).

Leipold, Michael D., Evan W. Newell, and Holden T. Maecker. "Multiparameter phenotyping of human PBMCs using mass cytometry." In *Immunosenescence*, pp. 81–95. Humana Press, New York, NY, 2015.

Lung, Ildikó, Maria-Loredana Soran, Ocsana Opriş, Mihail Radu Cătălin Truşcă, Ülo Niinemets, and Lucian Copolovici. "Induction of stress volatiles and changes in essential oil content and composition upon microwave exposure in the aromatic plant Ocimum basilicum." *Science of the Total Environment* 569 (2016): 489–495.

Macey, M. G. *Principles of flow cytometry. Flow cytometry: Principles and applications* (Macey MG, ed.), Vol. 1, p. 15. Humana Press, 2010.

Mahan, J. R., B. L. McMichael, and D. F. Wanjura. "Methods for reducing the adverse effects of temperature stress on plants: A review." *Environmental and Experimental Botany* 35, no. 3 (1995): 251–258.

Manohar, Sonal M., Prachi Shah, and Anusree Nair. "Flow cytometry: Principles, applications and recent advances." *Bioanalysis* 13, no. 3 (2021): 181–198.

McKinnon, Katherine M. "Flow cytometry: An overview." *Current Protocols in Immunology* 120, no. 1 (2018): 5–1.

Mehla, Neeti, Vinita Sindhi, Deepti Josula, Pooja Bisht, and Shabir H. Wani. "An introduction to antioxidants and their roles in plant stress tolerance." In *Reactive oxygen species and antioxidant Systems in Plants: Role and regulation under abiotic stress*, pp. 1–23. Springer, 2017.

Mei, Henrik E., Michael D. Leipold, and Holden T. Maecker. "Platinum-conjugated antibodies for application in mass cytometry." *Cytometry Part A* 89, no. 3 (2016): 292–300.

Mishra, Rajesh Kumar, Naseer Mohammad, and Nilanjan Roychoudhury. "Soil pollution: Causes, effects and control." *Van Sangyan* 3, no. 1 (2016): 1–14.

Møller, Anders Pape, and John P. Swaddle. *Asymmetry, developmental stability and evolution*. Oxford University Press, 1997.

Naqvi, Syed Shamshad Mehdi. "Plant hormones and stress phenomena." *Handbook of Plant and Crop Stress* 43 (1999): 709–730.

Natelhoffer, K. J., and B. Fry. "Controls on natural nitrogen-15 and carbon-13 abundances in forest soil organic matter." *Soil Science Society of America Journal* 52, no. 6 (1988): 1633–1640.

Omasa, Kenji, Yasushi Hashimoto, and Ichiro AIGA. "A quantitative analysis of the relationships between O3 sorption and its acute effects on plant leaves using image instrumentation." *Environment Control in Biology* 19, no. 3 (1981): 85–92.

Omasa, Kenji, Hideyuki Shimizu, Kazuo Ogawa and Akihisa Masuki. "Diagnosis of trees from helicopter by thermographic system." *Environment Control in Biology* 31, no. 3 (1993): 161–168.

Orcutt, David M. and Erik T. Nilsen. *Physiology of plants under stress: Soil and biotic factors*, Vol. 2. John Wiley and Sons, 2000.

Orfao, Alberto, and Alejandro Ruiz-Argüelles. "General concepts about cell sorting techniques." *Clinical Biochemistry* 29, no. 1 (1996): 5–9.

Parihar, P., S. Singh, R. Singh, V. P. Singh, and S. M. Prasad. "Effect of salinity stress on plants and its tolerance strategies: A review." *Environmental Science and Pollution Research* 22, no. 6 (2015): 4056–4075.

Pinter Jr, P. J., M. E. Stanghellini, R. J. Reginato, S. B. Idso, A. D. Jenkins, and R. D. Jackson. "Remote detection of biological stresses in plants with infrared thermometry." *Science* 205, no. 4406 (1979): 585–587.

Piyasena, Menake E., Pearlson P. Austin Suthanthiraraj, Robert W. Applegate Jr, Andrew M. Goumas, Travis A. Woods, Gabriel P. López, and Steven W. Graves. "Multinode acoustic focusing for parallel flow cytometry." *Analytical Chemistry* 84, no. 4 (2012): 1831–1839.

Prytz, Gunnar, Cecilia M. Futsaether, and Anders Johnsson. "Thermography studies of the spatial and temporal variability in stomatal conductance of Avena leaves during stable and oscillatory transpiration." *New Phytologist* 158, no. 2 (2003): 249–258.

Quan, Li-Juan, Bo Zhang, Wei-Wei Shi, and Hong-Yu Li. "Hydrogen peroxide in plants: A versatile molecule of the reactive oxygen species network." *Journal of Integrative Plant Biology* 50, no. 1 (2008): 2–18.

Rai, Richa, Madhu Rajput, Madhoolika Agrawal, and S. B. Agrawal. "Gaseous air pollutants: A review on current and future trends of emissions and impact on agriculture." *Journal of Scientific Research* 55, no. 771 (2011): 1.

Sadak, Mervat S. "Physiological role of signal molecules in improving plant tolerance under abiotic stress." *International Journal of ChemTech Research* 9, no. 7 (2016): 46–60.

Sanders, Dale, Colin Brownlee, and Jeffrey F. Harper. "Communicating with calcium." *The Plant Cell* 11, no. 4 (1999): 691–706.

Santini, Rachel, Jéssica Pacheco de Lima, Priscila Lupino Gratão, and Antonio Fernando Monteiro Camargo. "Evaluation of growth and oxidative stress as indicative of salinity tolerance by the invasive tropical aquatic macrophyte tanner grass." *Hydrobiologia* 849, no. 5 (2022): 1261–1271.

Shahzad, Babar, Mohsin Tanveer, Zhao Che, Abdul Rehman, Sardar Alam Cheema, Anket Sharma, He Song, Shams ur Rehman, and Dong Zhaorong. "Role of 24-epibrassinolide (EBL) in mediating heavy metal and pesticide induced oxidative stress in plants: A review." *Ecotoxicology and Environmental Safety* 147 (2018): 935–944.

Shanmugam, Gnanendra, and Junhyun Jeon. "Computer-aided drug discovery in plant pathology." *The Plant Pathology Journal* 33, no. 6 (2017): 529.

Sharma, Isha, Renu Bhardwaj, and Pratap Kumar Pati. "Exogenous application of 28-homobrassinolide modulates the dynamics of salt and pesticides induced stress responses in an elite rice variety Pusa Basmati-1." *Journal of Plant Growth Regulation* 34, no. 3 (2015): 509–518.

Sharma, P. and R. S. Dubey. "Involvement of oxidative stress and role of antioxidative defense systemin growing rice seedlings exposed to toxic concentrations of aluminum." *Plant Cell Reports* 26, no. 11 (2007): 2027–2038.

Stumpfe, Dagmar, and Jürgen Bajorath. "Methods for SAR visualization." *RSC Advances* 2, no. 2 (2012): 369–378.

Tiwari, Supriya, and Madhoolika Agrawal. "Effect of ozone on physiological and biochemical processes of plants." In *Tropospheric Ozone and its Impacts on Crop Plants*, pp. 65–113. Springer, 2018.

Urey, Harold C. "The thermodynamic properties of isotopic substances." *Journal of the Chemical Society (Resumed)* (1947): 562–581.

Vanlerberghe, Greg C. and Lee McIntosh. "Alternative oxidase: From gene to function." *Annual Review of Plant Biology* 48, no. 1 (1997): 703–734.

Verdeguer, Mercedes. "In Vitro and In Vivo Bioassays." In *Advances in plant ecophysiology techniques*, pp. 1–13. Springer, 2018.

Villiers, F. Ducruix, C. and Hugouvieux, V. "Investigating the plant response to cadmium exposure by proteomic and metabolomic approaches." *Proteomics* 11, no. 9 (2011): 1650–1663.

Wahid, Abdul, Saddia Gelani, M. Ashraf, and Majid R. Foolad. "Heat tolerance in plants: An overview." *Environmental and Experimental Botany* 61, no. 3 (2007): 199–223.

Walling, Linda L. "Adaptive defense responses to pathogens and insects." *Advances in Botanical Research* 51 (2009): 551–612.

Ward, Michael, Patrick Turner, Marc DeJohn, and Gregory Kaduchak. "Fundamentals of acoustic cytometry." *Current Protocols in Cytometry* 49, no. 1 (2009): 1–22.

Weyers, Jonathan D. B. and Tracy Lawson. "Heterogeneity in stomatal characteristics." In *Advances in botanical research*, Vol. 26, pp. 317–352. Academic Press, 1997.

Wlodkowic, Donald, Joanna Skommer, and Z. Darzynkiewicz. "Cytometry of apoptosis. Historical perspective and new advances." *Experimental Oncology* 34, no. 3 (2012): 255.

Wright, G. C., R. C. Nageswara Rao, and G. D. Farquhar. "Water-use efficiency and carbon isotope discrimination in peanut under water deficit conditions." *Crop Science* 34, no. 1 (1994): 92–97.

Xia, Xiao Jian, Yue Yuan Huang, Li Wang, Li Feng Huang, Yun Long Yu, Yan Hong Zhou, and Jing Quan Yu. "Pesticides-induced depression of photosynthesis was alleviated by 24-epibrassinolide pretreatment in Cucumis sativus L." *Pesticide Biochemistry and Physiology* 86, no. 1 (2006): 42–48.

Xia, Xiao Jian, Yun Zhang, Jing Xue Wu, Ji Tao Wang, Yan Hong Zhou, Kai Shi, Yun Long Yu, and Jing Quan Yu. "Brassinosteroids promote metabolism of pesticides in cucumber." *Journal of Agricultural and Food Chemistry* 57, no. 18 (2009): 8406–8413.

Zundel, G., W. Miekeley, Breno M. Grisi, and H. Förstel. "The $H_2$ $^{18}O$ enrichment in the leaf water of tropic trees: Comparison of species from the tropical rain forest and the semi-arid region in Brazil." *Radiation and Environmental Biophysics* 15, no. 2 (1978): 203–212.

# 20 A Critical Review of Different Methods of Estimation of the Above-Ground Biomass and Carbon Stocks in India

*Dipti Karmakar, Srimanta Gupta, and Pratap Kumar Padhy*

## 20.1 INTRODUCTION

In global energy balance, greenhouse gases play a very important role and a small change in concentration affects the climatic conditions on the earth. A major greenhouse gas, $CO_2$ (carbon dioxide), is increasing in the atmosphere since the preindustrial periods (from 280 ppmv in 1800, ~ 315 ppmv in 1957, ~358 ppmv in 1996, 400 ppmv in 2013, to surpassed levels of ~ 418 ppmv in 2022) (Lal & Singh, 2000; NASA, 2018, 2022). Instantaneous attention is required to save the life on the earth from rising levels of greenhouse gases due to anthropogenic activities (Raupach et al., 2007; Rathore et al., 2018).

A forest is defined as an ecosystem where trees and other woody plants predominate and have a land area greater than 0.5 ha, have a canopy cover of at least 10%, and are not used for agriculture or any other non-forest land use (FAO, 2001). As a reservoir of biological diversity, tropical forests play a very important role in maintaining global biodiversity (Naidu & Kumar, 2016; Bhadouria et al., 2016; Bhadouria et al., 2017). Tropical forests are rapidly disappearing, as about 1–4% of the current area is reduced annually (Naidu & Kumar, 2016; Khan et al., 2017).

Terrestrial ecosystems and oceans are the two natural carbon sinks, and the forest is the most important of terrestrial ecosystems. Vegetation takes away the $CO_2$ from the atmosphere and stores it in plant tissue as carbon. Forests store about 86% of the terrestrial above-ground carbon and 73% of earth's soil carbon (Rodger, 1993; Vashum & Jayakumar, 2012), and about 62–78% of terrestrial carbon is stored in the forests (Dixon et al., 1994; Ray et al., 2011). Forests accounts for about 27%, followed by tropical savannas and grasslands totaling approximately 23% together, and croplands about 10% (Ravindranath & Ostwald, 2008).

The identification of vegetation carbon biomass in tropical regions has garnered considerable interest from REDD+ (reducing emissions from deforestation and forest degradation and enhancing carbon stocks), which seeks to preserve and boost the carbon sink in the forests of developing nations. The Conference of Parties (COP) of the United Nations Framework Convention on Climate Change (UNFCCC) has proposed REDD+ as a means of offering financial incentives to reduce carbon dioxide emissions into the atmosphere. To qualify for REDD+ incentives, developing nations often need to keep an eye on the carbon biomass store in their land covers (Mertz et al., 2012; Day et al., 2014; Yuen et al., 2016).

According to Intergovernmental Panel on Climate Change (IPCC), the five carbon pools of above-ground biomass (AGB), below-ground biomass (BGB), dead wood, litter, and soil organic matter (SOM) contribute to the total biomass in the forest biomass (Suganthi et al., 2017). Carbon concentration is strongly interconnected with the stand age over the life cycle of forest ecosystems. AGB, BGB, and dead biomass are the main carbon sinks with stand development. The ecologist primarily relies on the mature forest to recognize the global pattern of carbon sequestration since old forests are carbon neutral and in a constant state, and mature forests have a high and stable carbon density (Ma et al., 2017, Karmakar, 2018).

Among the five carbon pools (AGB, BGB, deadwood, litter, and soil carbon), the most essential, noticeable, and predominate carbon resource is AGB. Above the earth are living plants, both woody and herbaceous (stems, stumps, bark, branches, seeds and foliage). Live root biomass is a type of BGB that can be kept for very extended periods of time. Only around 6% of the forests' ability to store carbon comes from deadwood and other woody areas. It includes naturally dead (standing or fallen) forest and the effects of pest infestations, wind degradation, and human meddling, but it excludes naturally fallen woody and non-woody litter or biomass. Deadwood generally occurs in natural forests, but it is occasional in new plantations, agroforestry systems, savanna, grasslands and croplands (Bhadouria et al., 2018). Biomass that is not alive, an organic detritus layer, discarded or fallen plant matter, and plant components that are not linked to living plants are all considered to be litter, which may even contribute as little as 6% of the biomass of the plants. The live and dead fine roots in the soil that are included wherever they cannot be empirically differentiated from the soil organic matter are considered soil carbon (of less than the recommended requirement for BGB). Dead organic materials decompose and become soil organic matter (Ravindranath & Ostwald, 2008, Karmakar, 2018).

## 20.2 VARIOUS METHODS FOR ESTIMATION OF CARBON STOCKS AND ABOVE-GROUND TREE BIOMASS OF FORESTS

The forest biomass is invaded by various factors, like stand age, the composition of species, topography, environmental heterogeneity, and natural and anthropogenic disturbances (Fang et al., 2006; Singh et al., 2011; Chaturvedi & Raghubanshi, 2013). The estimation of an individual tree is needed for the estimation of total carbon dioxide ($CO_2$) which has been stored from the atmosphere to plant tissue (Rathore et al., 2018), and also to deepen our comprehension of the processes and mechanisms underlying the global C (carbon) cycle (Ma et al., 2017). For the management of different ecosystems (forest, scrubland, grassland) and mapping of different properties of forest fuels through different models, the quantification of AGB and other variables is very essential. Strategic values, large-scale evaluation, and mapping are crucial for obtaining an accurate model of the dynamics of ecosystems (Mandal & Joshi, 2015, Karmakar, 2018).

### 20.2.1 DESTRUCTIVE METHOD

This is a direct method to quantify the AGB. This method includes harvesting all the trees in a sample plot and weighing different parts (e.g., tree trunks, branches, leaves, flowers, fruits) of harvested trees (Mandal & Joshi, 2015; Feyisa et al., 2016; Hossain et al., 2016). This method provides very accurate quantification of biomass but the conventional approach to calculating the carbon biomass in the forest is removing or uprooting, and then weighing, complete components. The entire process is cumbersome, costly, unworkable, occasionally illegal, and goes against the goals of safeguarding forests (Mandal & Joshi, 2015; Feyisa et al., 2016; Yuen et al., 2016; López-López et al., 2017). This method is not allowable in a degraded forest containing threatened species (Montès et al., 2000, Karmakar, 2018).

### 20.2.2 NON-DESTRUCTIVE METHOD

This method is indirect but ultimately based on tree harvesting and weighing (Montès et al., 2000; Chave et al., 2005). This non-destructive—or least destructive—method also includes harvesting of trees to validate the estimated biomass. This method involves some measurable independent variables (diameter at breast height or girth at breast height, tree height, tree volume, crown diameter, and some other components) (Chave et al., 2005; Kale et al., 2004; Singh et al., 2011; Hossain et al., 2016; Rathore et al., 2018). This method also involves some common regression equations (linear, quadratic, exponential, and logarithmic) with different parameters (tree height, basal area, tree volume, etc.) for the estimation of biomass (Husch et al., 2003). Development of a generalized allometric model for different tree species and forests in large-scale global and regional comparisons were tried by various researchers (Montès et al., 2000; Chave et al., 2005; Hossain et al., 2016; Rathore et al., 2018), but for getting precise and dependable results of biomass, the use of species and site-specific allometric equations are desirable (Khan et al., 2005; Smith & Whelan, 2006; Hossain et al., 2016; Yuen et al., 2016; Djomo & Chimi, 2017, Karmakar, 2018).

### 20.2.3 REMOTE SENSING OR SATELLITE METHOD

Aerial photography, optical parameters, and radar are a few examples of remote sensing techniques that can be used to monitor changes in land usage in a project region (Ravindranath & Ostwald, 2008). This is the best method for the quantification of biomasses at a regional level, where field data is scarce (Anaya et al., 2009). We are unable to get a direct estimation of the biomass of the forest through remote sensing. To estimate the AGB, remote sensing data are integrated with other empirical data (either directly using allometric relationships or indirectly using attributes like canopy cover) (Ravindranath & Ostwald, 2008; Vashum & Jayakumar, 2012). The predictive and allometric equations are developed from the field measurements and the data is validated with remotely sensed data. After validation, these remotely sensed data can be used in the measurement of carbon stock in a larger area (Vashum & Jayakumar, 2012). The information on biomass and stand attributes is mostly obtained using three forms of remotely sensed data: optical, SAR (microwave), and LiDAR (light detection and ranging) (Sinha et al., 2015). In optical remote sensing, the reflected energy is dependent on different vegetation characteristics (leaf structure, pigmentation, moisture, etc.). The size, density, orientation, and dielectric characteristics of materials with dimensions similar to the radar wavelength determine the scattered microwave energy. Through optical remote sensing, different information on land properties (normalized difference vegetation index [NDVI]) is presented because of the availability of data in several bands (visible to infrared wavelengths) (Sinha et al., 2015; Joshi et al., 2016; Behera et al., 2016). The wavelength of optical remote sensing is too short to penetrate through clouds and atmospheric dust, and due to the dependency on solar illumination, it cannot operate during light (Kumar, 2009), so optical remote sensing is not suitable to use in tropical regions due to cloud cover (Jha et al., 2006).

Wavelengths of radar sensors can penetrate through clouds, atmospheric moisture, and the upper vegetation layer of forest canopy due to its longer length (1 mm–1 m). Synthetic aperture radar (SAR) systems have different bands, like X-band (2.4–3.75 cm), C-band (3.75–7.5 cm), L-band (10–30 cm), and P-band (30–100 cm) (Kumar, 2009). Radar backscatter values are positively correlated with forest biomass, tree height, and basal area, and the intensity of correlation is more for P-band (Hussin et al., 1991; Dobson et al., 1992; Kumar, 2009; Behera et al., 2016). A weak correlation was observed between the C-band backscatter of SAR and AGB (Le Toan et al., 1992). For each sensor, single-wavelength radar signals are generated which interact with structural land properties in a characteristic way (e.g., backscattered energy from active radar signals is returned primarily from canopies and stems of forests, and differences in the roughness and moisture content of these surfaces may be extracted, depending on the wavelength and incidence angle of the radar pulses used). However, polarization combinations of the signals sent to and returned from land surfaces can make up different bands of SAR backscatter (e.g., horizontal send and horizontal receive [HH], vertical send and vertical receive [VV], and horizontal send and vertical receive [HV]) (Sinha et al., 2015; Joshi et al., 2016; Behera et al., 2016). For the measurement of biomass, remote sensing technologies are a better substitute conventional method (Lu, 2006). Remote sensing canopy reflectance models can be used to estimate foliage, biomass content, and productivity (Roy, 1989; Franklin & Hiernaux, 1991). The integrated vegetation index is mutually interrelated with AGB and primary productivity (Tucker et al., 1983; Goward & Dye, 1987). Backscattering coefficients increase with increasing biomass levels. A significant correlation was there only in the backscattering coefficient HH, whereas VH and VV are not well correlated. This relationship can be utilized for the estimation of the biophysical parameters of a study area (Jha et al., 2006). The radar backscatter at P- and L-bands (higher wavelengths) is significantly correlated with tree density, biomass, and volume due to their penetration ability through the crown. And radar backscatter at C- and X-bands (lower wavelengths) can affect only leaves, young twigs, and branches, due to their inability to penetrate through crown layers (Alappat et al., 2011). Wide swath satellite data from NOAA-AVHRR show the distribution of different types of forest cover (Roy & Kumar, 1986), while broad vegetation type distribution is provided by satellites like Landsat, SPOT, and IRS based on the principal species composition, canopy density, and site circumstances in India (Roy & Ravan, 1996).

The total area of forest cover in different states of India is represented in Figure 20.1 (India State of Forest Report (ISFR), 2011).

This chapter mainly deals with the various methods in India to quantify the AGB in different zonal councils in India and to compare among them. Many authors have developed various species and site-specific allometric models by using different methods in an area, so it will become helpful in further studies to use these models in the estimation of AGB without felling the trees, conserving biodiversity.

## 20.3 ESTIMATION OF BIOMASS IN DIFFERENT ZONAL COUNCILS OF INDIA

### 20.3.1 NORTH ZONAL COUNCIL

#### 20.3.1.1 Haryana

The AGB predictions at Yamunanagar and Panchkula districts in Haryana was done by correlating field biomass with the reflectance value of different bands (red band, infrared band and NDVI) and indices of MODIS satellite data to develop combined regression models and the best-fitted model was derived from the red band (MODIS) (Kumar, Gupta et al., 2011; Karmakar, 2018) (Table 20.1).

#### 20.3.1.2 Himachal Pradesh

Estimates of Haryana's AGB were made using multispectral high-resolution data from Landsat-5 TM and IRS P-6 LISS-III, as well as various allometric equations. NDVI retrieved for the same site from the 2010 LISS-III picture and the biomass derived from sample plot data were shown against one another. The model that best defined the link between NDVI and AGB among the other models was called "Power model" (Sharma et al., 2013). Temporal variation in carbon stock (1956–2011) in *Pinus roxburghii* of Solan and Dharampur forest ranges of Himachal Pradesh was estimated by Shah et al. (2014) by using allometric equations. With a mean value of 189.93 tons/ha, the above-ground carbon store was calculated in *Cedrus deodara* (CD) forests in the Mandi district in Himachal Pradesh (Thakur & Verma, 2019). The AGB was also recorded in the silvipasture land-use system in the Kinnaur district of Himachal Pradesh (Chisanga et al., 2018) (Table 20.1).

#### 20.3.1.3 Jammu and Kashmir

Above-ground and below-ground biomass of the trees of seven forests in temperate forests of the Kashmir Himalaya was estimated by using species-specific volume equations using diameter at breast height (dbh), basal area, or height as a factor (Dar & Sundarapandian, 2015, Karmakar, 2018). The highest and lowest amount of AGB density was recorded in *Cedrus* and *Acacia* forest types of Kashmir Himalaya (Sajad et al., 2021). In another study, the mean values of AGB carbon were measured with the values of 63.70 t ha$^{-1}$, 63.24 t ha$^{-1}$, 68.81 t ha$^{-1}$, and 51.39 t ha$^{-1}$ in *Cedrus deodara*, mixed I (*Cedrus deodara-Pinus wallichiana*), mixed II (*Abies pindrow-Picea smithiana*), and *Pinus wallichiana*, respectively, in north Kashmir region of Himalayas. The AGB carbon stock was higher in mixed II followed by *Cedrus deodara*, mixed I, and *Pinus wallichiana*, respectively (Wani et al., 2019). The mean AGB values were measured about 328.6±281.1 Mg ha$^{-1}$, 358.4±276.5 Mg ha$^{-1}$, 393.7±221.3 Mg ha$^{-1}$, 306.4±239.5 Mg ha$^{-1}$, and 253.3±165.9 Mg ha$^{-1}$ in *Pinus wallichiana*, *Abies pindrow*, *Cedrus deodara*, *Picea simithiana*, and *Betula utilis*, respectively, in Gulmarg forest range of northern Kashmir

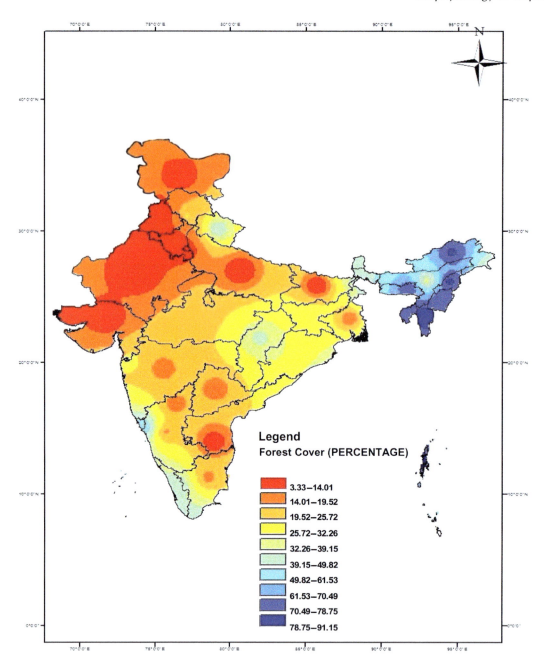

**FIGURE 20.1** State-wise forest cover (percentage) in India.

*Source:* India State of Forest Report (ISFR) (2011)

Himalaya (Dar & Sahu, 2018). In another study, the AGB of trees in an agriculture land-use system was measured using non-destructive methods in the sub-tropics of Jammu and Kashmir (Mahajan et al., 2021) (Table 20.1).

### 20.3.2 North-Eastern Council

#### 20.3.2.1 Assam

India's bamboo covers about 10 million hectares of forest land, homestead land, and private plantations, and has a naturally growing capacity in heights from sea level to over 3,500 m (Verwijst & Telenius, 1999). Nath et al. (2009) measured the AGB (121.51 t ha$^{-1}$) of bamboos of Barak Valley, North-East India by harvesting method and allometric equations were also developed. In another previous work, the C pool in the AGB was measured in three bamboo species (*B. vulgaris* Schrad. ex Wendl. [Jai borua], *Bambusa cacharensis* R. Majumder [Betua], and *B. balcooa* Roxb. [Sil borua]) in the Cachar district of Assam (Nath & Das, 2012). Borah et al. (2015) have also estimated the AGB in the Kholahat reserve forest and Gibbon wildlife sanctuary of Assam by using some allometric models. AGB of tree species were measured in Kholahat reserve forest and Gibbon wildlife sanctuary, respectively. The biomass estimation of *Barringtonia acutangula* forests in Chatla floodplain in Assam was done by harvesting method and developing of the allometric equation (Nath et al., 2017, Karmakar, 2018) (Table 20.1).

### 20.3.2.2 Arunachal Pradesh

Using 2016 ASTER elevation datasets and Landsat OLI optical remote sensing datasets, the AGB of the main land-use sector of the Arunachal Himalaya was calculated. The AGB of selected land use was about 332.28 t ha$^{-1}$, 246.63 t ha$^{-1}$, 145.36 t ha$^{-1}$, 179.31 t ha$^{-1}$, 149.63 t ha$^{-1}$, 55.40 t ha$^{-1}$, and 16.84 t ha$^{-1}$ in dense forest, moderately dense forest, open forest, plantations, jhum >5 years, jhum <5 years, and current jhum, respectively (Das et al., 2021, Karmakar, 2018) (Table 20.1).

### 20.3.2.3 Manipur

The AGB of two forest stands in the Chandel district of Manipur was estimated by the harvesting method (Devi & Yadava, 2009, Karmakar, 2018). In a different study, the largest sacred grove in Manipur, Northeast, had its AGB measured (Waikhom et al., 2017) (Table 20.1).

### 20.3.2.4 Meghalaya

The Nongkhyllem wildlife sanctuary in Meghalaya used IRS LISS-III data and a biomass regression model to quantify the AGB of the native semi-evergreen forest and sal (*Shorea robusta*) planted forest Baishya et al. (2009). By using a similar biomass regression model, Upadhaya et al. (2015) also estimated the AGB in a wildlife sanctuary and five reserve forests at Meghalaya and the wildlife sanctuary had the highest total AGB (Karmakar, 2018) (Table 20.1).

## 20.3.3 Central Zonal Council

### 20.3.3.1 Chhattisgarh

To quantify the existing land use, volume, biomass, and carbon storage at Balamdi watershed, a part of Barnawapara Sanctuary, Raipur, Chhattisgarh was carried out by using some species-specific allometric equations (FSI, 1996). To localize the variation in biomass distribution across several physiographic divisions of the research area, an NDVI map (created using data from IRS 1D LISS-III) was used. The mixed forest has higher AGB and carbon storage than teak and sal mixed forest, as also proved by having maximum NDVI value in mixed forest (Bijalwan et al., 2010). Similar kinds of estimation were also done by Thakur and Swamy (2010) at Barnawpara Sanctuary, Raipur, Chhattisgarh by using species-specific regression equations and IRS 1D LISS-III data. NDVI value was more in dense mixed forests. Also, the AGB was measured at Katghora forest, Korba, Chhattishgarh (Pawar et al., 2014). In another study, the quantification of AGB at Barnawapara Wildlife Sanctuary, Raipur, Chhattisgarh was done by Sahu et al. (2013) by using allometric equations (Karmakar, 2018) (Table 20.1).

### 20.3.3.2 Madhya Pradesh

The estimation of tree biomass in dry deciduous and mixed-deciduous forests in seven different districts of Madhya Pradesh was carried out by using non-destructive allometric site- and species-specific equations (Salunkhe et al., 2016). For the evaluation of AGB of an agro-pastoral ecology in Shivpuri Tehsil in Madhya Pradesh, support vector machines (SVM) and conventional regression models were used Deb et al. (2020). This study demonstrates that for the estimation of AGB, SVM performed better than conventional regression models. The prediction of AGB of central Indian deciduous forests in Madhya Pradesh was done using an integrated set of advanced land observing satellite phased array type L-band synthetic aperture radar (ALOS PALSAR) L-band data and field-based data (Thumaty et al., 2015). By combining field-based data with space-borne QuadPOL ALOS PALSAR-2 L-band and dual-pol Sentinel 1A C-band synthetic aperture radar (SAR) data, the AGB of the Satpura Tiger Reserve in the Satpura mountain ranges, Madhya Pradesh, was estimated (Lal et al., 2021) (Table 20.1).

### 20.3.3.3 Uttarakhand

For the retrieval of forest biophysical parameters (height and AGB) in Barkot and Thano forest ranges, Uttarakhand, using polarimetric space-borne C-band data with Polarimetric SAR Interferometry (PolInSAR) techniques (RADARSAT-2) has been done. The backscatter value has been increased with AGB. PolInSAR coherence-based modeling approach is the most accurate method to retrieve forest biophysical parameters (Kumar et al., 2017; Karmakar, 2018). Agarwal and Purwar (2017) have estimated the AGB of *Bambusa balcooa*, *B. nutans*, and *B. tulda* in the mid-Himalayan region of Uttarakhand by using a regression equation. Kumar et al. (2014) measured the AGB in five randomly selected sites of forest in the hydroelectric project of Srinagar Garhwal region of Uttarakhand. Mukhopadhyay et al. (2021) used integrated polarimetric (PolSAR) and interferometric (PolInSAR) data for measuring the AGB in Malhan Forest Range, Dehradun Forest Division, located in the southwesternmost part of Uttarakhand. The estimation of AGB using these data also allowed comparisons and evaluations of several linear regression and random forest regression models. In another study, the AGB was measured in different age groups of teak plantations in Kotdwara and Kothari Range of Kotdwara Forest Division, Pauri Garhwal, Uttarakhand (Nirala et al., 2018). Sentinel-1A and Landsat 8 Operational Land Imager spectral and textural characteristics, as well as the random forest technique, were used to forecast the spatial distribution of AGB in the various forest types in the Doon Valley, located in the Himalayan foothills (Purohit et al., 2021) (Table 20.1).

### 20.3.3.4 Uttar Pradesh

An allometric equation was used to estimate the AGB in a mono-specific teak, mono-specific sal, multi-specific teak, multi-specific sal, and multi-specific teak and sal forest in the Sonebhadra district of Uttar Pradesh (Chaturvedi & Raghubanshi, 2015). Behera et al. (2016) estimated the AGB and carbon content in two forest types (*Shorea robusta* and *Tectona grandis*) in Katerniaghat Wildlife Sanctuary, Uttar Pradesh by using QuadPOL ALOS PALSAR data, water cloud model and species-specific allometric equations. The AGB ranges from 6.5–38.44 Mg/0.04 ha and 6.10–35.9 Mg/0.04 ha in *Shorea robusta* and *Tectona grandis* forests on the basis of in situ measurements. A significant correlation has been found between backscatter coefficient and field-measured AGB, and the backscatter coefficient value was maximum in HV of *Shorea robusta* and minimum in HH for *Tectona grandis*. Using allometric equations, Behera et al. (2017) also estimated the AGB in the sal mixed, dry mixed and teak plantations of the Katerniaghat Wildlife Sanctuary in Uttar Pradesh. Another study used advanced land observing satellite (ALOS) data from the fully polarimetric phased array type L-band synthetic aperture radar (PALSAR) to retrieve the AGB of the forest in Dudhwa National Park (Kumar et al., 2019) (Table 20.1).

### 20.3.4 EASTERN ZONAL COUNCIL

#### 20.3.4.1 Bihar

In order to estimate the AGB and BGB of *Populus deltoids* at Pusa, Samastipur, Bihar, Ajit et al. (2011) created models (triplicate from 1–9 years of age). The net biomass was 10.32 t ha$^{-1}$ yr$^{-1}$ at the age of nine years, and each of the five developed models (allometric, logistic, Chapman-Richards, linear, Gompertz) are equally efficient ($R^2 = 0.99$) for the estimation of biomass of *Populus deltoids* (Karmakar, 2018). A regression model for forecasting above-ground bole biomass across tropical mixed-deciduous forests in Munger, Bihar, was constructed in a separate study by Sinha et al. (2019). It has been accomplished to employ field inventory data in conjunction with space-borne L-band ALOS PALSAR and X-band COSMO-Skymed SAR data. The average estimated AGB was about 73.31 Mg ha$^{-1}$ (Table 20.1).

#### 20.3.4.2 Jharkhand

The carbon sequestration potential of *Eucalyptus* hybrid in Dhalbhum, Jharkhand, was measured by destructive method (Bala et al., 2012). Three models were developed to estimate the carbon sequestration potential (Karmakar, 2018). Panda et al. (2020) have estimated the AGB of tree layer at two different sites of dry sal forest in Ranchi by using allometric equations. At both the sites, *Shorea robusta* contributed more in total biomass and it contributes about 71.5% and 76.65% to the total biomass in site 1 and site 2, respectively (Table 20.1).

#### 20.3.4.3 Odisha

AGB of a tropical dry forest in Deogarh district, Odisha was estimated by using some allometric equation by Sahu et al. (2016). No significant relationship has been found between AGB and species diversity, but a positive relationship was there between AGB and basal area (Karmakar, 2018). Ghosh and Behera (2021) have used machine learning algorithms and Sentinel-1 C-band SAR data for the prediction of AGB of tropical mangrove forests, Bhitarkanika Wildlife Sanctuary along the eastern Indian coast. Interferometric water cloud model (IWCM) was found to be an unsuitable method for AGB estimation because the deep learning algorithm could precisely translate the relationship between predictor variables and mangrove AGB in Bhitarkanika Wildlife Sanctuary due to low canopy penetration power of C-band SAR, high temporal decorrelation caused by longer time gaps between interferometric image pairs, and high spatial heterogeneity of mangrove forests. In another study, with the use of several allometric equations, six different secondary tropical deciduous forests (e.g., planted teak forest, peninsular sal forest [PSF], semi-evergreen forest [SEF], moist mixed-deciduous forest, Bamboo Brakes [BB], and degraded thorny shrubby forest) of the Chandaka Wildlife Sanctuary, Eastern Ghats of Odisha was chosen for the measurement of AGB. The highest AGB was estimated in the bamboo brakes (BB) and lowest for the degraded thorn shrub forest (DTSF) (Pattnayak et al., 2020). Two different forests (Debrigarh Wildlife Sanctuary and Andhari Sacred Forest) were selected to quantify the AGB by using allometric equations. The presence of more old trees in the wildlife sanctuary compared to the sacred forest may be the cause of the increased biomass content in the wildlife sanctuary (Pradhan et al., 2019). In another study, the AGB of 18 selected woody mangrove species in Bhitarkanika Wildlife Sanctuary was estimated by using some allometric equations (Banerjee et al., 2018). To measure the AGB of various dominating tree species, Mohanta et al. (2020) chose two tropical moist deciduous forest sites (namely, *Xylia*-dominated forest and sal-dominated forest) of the Similipal Biosphere Reserve, Odisha. The study revealed that at both the sites *Shorea robusta* Gaertn. contributes more in biomass and it contributed about 21.56% in *Xylia*-dominated forest and about 47.8% in sal-dominated forest, respectively (Table 20.1).

#### 20.3.4.4 West Bengal

Using allometric equations, Shukla and Chakravarty (2018) quantified the AGB of the Chilapatta Reserve Forest, which is part of the Cooch Behar Wildlife Division and is situated in the Terai zone of West Bengal state, at the northern edge of the state in the foothills of the sub-Himalayan mountain belts. Stems made up the majority of the estimated AGB of the trees (58.74%), followed by branches (28.04%), roots (12.90%), and leaves (0.32%). In another study, Sentinel-1A C-band SAR data was used to quantify the AGB in a *Shorea robusta* (sal)-dominated forest cover in sub-tropical region of Jhargram, West Bengal. Also, field measurement data was collected to estimate AGB by using allometric equations, and the data were statistically linked with satellite data. The AGB was derived from field measurement data and satellite-based data. The majority of the estimated AGB was in the range of 100–200 tons/ha, which may indicate that the forest cover in this area is uniform, as

**TABLE 20.1**

**Different Methods of Estimation of Above-Ground Biomass in India**

| Place | Method/Sensor | Model Used or Developed | Above-Ground Biomass | References |
|---|---|---|---|---|
| **Northern Zonal Council** | | | | |
| Haryana | Allometric equations, IKONOS (MODIS) | $Y = 162046 \times$ basal area$^{-1.2506}$ (D) | 30.46–310 Mg ha$^{-1}$ | Kumar, Gupta et al. (2011) |
| Himachal Pradesh | Landsat-5 TM and IRS P-6 LISS-III | AGB $=1169.8 \times$ NDVI$^{4.2043}$ (D) | – | Sharma et al. (2013) |
| Himachal Pradesh | Allometric equations | – | 39.23–38.62 t ha$^{-1}$ | Shah et al. (2014) |
| Himachal Pradesh | Allometric equations | – | 189.93 ton/ha | Thakur & Verma (2019) |
| Himachal Pradesh | Allometric equations | | 84.65 t ha$^{-1}$ | Chisanga et al. (2018) |
| Jammu and Kashmir | Regression equation | – | 79–237 Mg ha$^{-1}$, | Dar & Sundarapandian (2015) |
| Kashmir | – | – | 275.63 ± 52.12 Mg ha$^{-1}$ | Sajad et al. (2021) |
| Kashmir | Allometric equation | – | 63.70 t ha$^{-1}$, 63.24 t ha$^{-1}$, 68.81 t ha$^{-1}$ and 51.39 t ha$^{-1}$ | Wani et al. (2019) |
| Kashmir | Allometric equation | – | 282±99.67 Mg ha$^{-1}$ | Dar & Sahu (2018) |
| Jammu and Kashmir | Non-destructive methods | – | 9.06 Mg ha$^{-1}$ | Mahajan et al. (2021) |
| **North-Eastern Council** | | | | |
| Assam | Harvesting method Allometric equation | log (dry weight in g) = a + b log (dbh in cm) (D) | – | Nath et al. (2009) |
| Assam | Allometric equation | log (dry weight in g) = a + b log (dbh in cm) (U) | 21.69–76.55 Mg ha$^{-1}$ | Nath & Das (2012) |
| Assam | Allometric equation | AGB $= 0.0509 \times \rho \times D^2 \times H$ (U) | 135.30 and 146.42 Mg ha$^{-1}$ | Borah et al. (2015) |
| Assam | Harvesting method Allometric equation | Y= aDBH$^b$ (D) | 552 ± 23 Mg ha$^{-1}$ | Nath et al. (2017) |
| Arunachal Pradesh | Landsat OLI optical remote sensing and ASTER elevation data | - | 16.84 t ha$^{-1}$–332.28 t ha$^{-1}$ | Das et al. (2021) |
| Manipur | Harvesting method | - | 15.60 and 15.84 t ha$^{-1}$ | Devi & Yadava (2009) |
| Manipur | Regression equation | - | 962.94–1130.79 Mg ha$^{-1}$ | Waikhom et al. (2017) |
| Meghalaya | IRS LISS III, Regression model | Y= exp [−0.37 + 0.33 ln(D) + 0.933 ln(D)$^2$ 0.122 ln(D)$^3$] (U) | 406 Mg ha$^{-1}$ and 324 Mg ha$^{-1}$ | Baishya et al. (2009) |
| Meghalaya | Regression model | Y = exp [- 0.37 + 0.33 ln (D) + 0.933 ln (D)$^2$-0.122 ln (D)$^3$] (U) | 204–314 Mg ha$^{-1}$ | Upadhaya et al. (2015) |
| **Central Zonal Council** | | | | |
| Chhattisgarh | Species-specific allometric equations, remote sensing and GIS (IRS 1D LISS III) | – | 45.94–78.31 Mg ha$^{-1}$ | Bijalwan et al. (2010) |
| Chhattisgarh | Species-specific allometric equations, remote sensing and GIS (IRS 1D LISS III) | – | 20.25–103.43 Mg ha$^{-1}$ | Thakur & Swamy (2010) |
| Chhattisgarh | Allometric equations | | 111.20–199.42 (t ha$^{-1}$) | Pawar et al. (2014) |
| Chhattisgarh | Allometric equations | – | 99.08–197.88 t ha$^{-1}$ | Prasad et al. (2013) |
| Madhya Pradesh | Landsat TM, Regression equation | y = 73709·9241–48420·44 NDVI + 67242·43 MIR index value (D) | 3·31–129·951 t ha$^{-1}$ | Roy & Ravan (1996) |

*(Continued)*

**TABLE 20.1**
**Continued**

| Place | Method/Sensor | Model Used or Developed | Above-Ground Biomass | References |
|---|---|---|---|---|
| Madhya Pradesh | Harvesting method | – | 28.1–85.3 t ha$^{-1}$ | Pande (2005) |
| Madhya Pradesh | Allometric equations | – | 31.8 and 20.7 t ha$^{-1}$ | Salunkhe et al. (2016) |
| Madhya Pradesh | Regression models and support vector machine (SVM) | – | 82.99 Mg ha$^{-1}$ | Deb et al. (2020) |
| Madhya Pradesh | ALOS-PALSAR data and field-based data | – | 58 t/ha | Thumaty et al. (2015) |
| Madhya Pradesh | ALOS PALSAR-2 L-band, Sentinel 1A C-band and field-based data | $\sigma°$ (dB) = Constant + Slope X ln (AGB) | 186.87 Mg ha$^{-1}$ ±7.04 | Lal et al. (2021) |
| Uttarakhand | Allometric equation | $\log_{10} Y = \beta_1 + \beta_2 \log_{10}$ crown cover (D) Where, $\beta_1$ = Intercept, $\beta_2$ = slope | – | Tiwari & Singh (1984) |
| Central Himalaya | Landsat, aerial photography | – | – | Singh et al. (1985) |
| Uttarakhand | Regression equation | – | – | Chaturvedi & Singh (1987) |
| Uttarakhand | Harvesting method, allometric equation | – | 26–237.7 t ha$^{-1}$ | Garkoti & Singh (1995) |
| Uttarakhand | Harvesting method, regression model | – | 52.5–118.1 t ha$^{-1}$ | Lodhiyal & Lodhiyal (2003) |
| Uttarakhand | Allometric equations | – | 101.42–434.43 Mgha$^{-1}$ | Sharma et al. (2010) |
| Uttarakhand | Allometric equations | – | 171.9 and 380.3 Mg ha$^{-1}$ | Gairola et al. (2011) |
| Uttarakhand | Allometric regression equations | – | 77.75 and 377.93 t ha$^{-1}$ | Gautam et al. (2011) |
| Uttarakhand | Allometric equations | – | 134.1 ± 10.6–518.2 ± 44.8 Mg ha$^{-1}$ | Sharma et al. (2011) |
| Uttarakhand | MODIS, allometric equation | Predicted phytomass = −267.85 × ln (NDVI) + 302.18 (D) | | Patil et al. (2015) |
| Uttarakhand | Allometric equation | – | 202.26, 347.75, and 335.27 Mg ha$^{-1}$ | Shahid & Joshi (2015) |
| Uttarakhand | Harvesting method | – | – | Kaushal et al. (2016) |
| Uttarakhand | PolInSAR | – | – | Kumar et al. (2017) |
| Uttarakhand | Regression equation | – | – | Kumar & Sharma (2015) |
| Uttarakhand | Forest inventory, IRS P6 LISS-III satellite data, and geostatistical techniques | – | 143–421 Mg ha$^{-1}$ | Yadav & Nandy (2015) |
| Uttarakhand | Regression equation | – | 4.9–53 t ha$^{-1}$ | Agarwal and Purwar (2017) |
| Uttarakhand | Allometric equation | – | 121.05–549.27 t ha$^{-1}$ | Kumar et al. (2014) |
| Uttarakhand | polarimetric (PolSAR) and interferometric (PolInSAR) data, random forest regression, and multiple linear regression model | – | 158.14–384.84 Mg/ha | Mukhopadhyay et al. (2021) |
| Uttarakhand | Allometric equation | – | 205.40–687.07 t ha$^{-1}$ | Nirala et al. (2018) |

# Estimation of Above-Ground Biomass and Carbon Stocks in India

| Place | Method/Sensor | Model Used or Developed | Above-Ground Biomass | References |
|---|---|---|---|---|
| Uttarakhand | Landsat 8 and Sentinel-1A data, and random forest regression algorithm | – | 49.36–596.15 Mg ha$^{-1}$ | Purohit et al. (2021) |
| Uttar Pradesh | Harvesting method, allometric equations | – | 42–78 t ha$^{-1}$ | Singh & Singh (1991) |
| Uttar Pradesh | Landsat ETM, EnviSat ASAR, ETM NDVI | – | – | Kumar (2007) |
| Uttar Pradesh | EnviSat ASAR, regression equation | Water cloud model, interferometric water cloud model | 41.07 t ha$^{-1}$ | Kumar (2009) |
| Uttar Pradesh | Allometric equation | – | – | Chaturvedi et al. (2011) |
| Uttar Pradesh and Haryana | Allometric equation | – | 65.62 t ha$^{-1}$ | Rizvi et al. (2011) |
| Uttar Pradesh | Landsat TM, EnviSat ASAR, allometric equation | Water cloud model, interferometric water cloud model | | Kumar et al. (2012) |
| Uttar Pradesh | Allometric equation | – | 247 ± 28.2–392 ± 42.3 t ha$^{-1}$ | Chaturvedi & Raghubanshi (2015) |
| Uttar Pradesh | QuadPOL ALOS PALSAR, | Water cloud model (WCM) | – | Behera et al. (2016) |
| Uttar Pradesh | Allometric equation | – | 290.82–455.99 Mg ha$^{-1}$ | Behera et al. (2017) |
| Uttar Pradesh | PALSAR | Extended water cloud model (EWCM) | 92.90 t ha$^{-1}$ | Kumar et al. (2019) |
| Bihar | Harvesting method Allometric equations | $Y= 1.79 \times (dbh)^{1.25}$<br>$Y= -261.01/1+exp(5.54-(-0.01) \times dbh)$<br>$Y=4510.04 *[1-exp(-0.0021 \times dbh)]^{[1/(1-0.2249)]}$<br>$Y= 300.21 \times exp(-3.84 \times exp[-0.050 \times dbh])$<br>$Y=-11.923+4.717 \times x (D)$ | – | Ajit et al. (2011) |
| **Eastern Zonal Council** | | | | |
| Bihar | ALOS PALSAR, COSMO-Skymed SAR data, and field inventory data | $AGB = 88.5536 \times e^{(0.0442*\sigma^o_{X\_VV})} + 988.3198 \times e^{(0.2765*\sigma^o_{L\_HH})} - 53.1242 (D)$ | 73.31 Mg ha$^{-1}$ | Sinha et al. (2019) |
| Jharkhand | Destructive method | $y = 4.57-162.05x + 1761.80x^2$<br>$y = 14.25 - 372.53x + 3160.37x^2 - 2831.88x^3$<br>$y = 0.00458465 x^{3.1785456} (D)$ | 13.07 tC ha$^{-1}$ (6 yrs) | Bala et al. (2012) |
| Jharkhand | Allometric equations | – | 333.31 t ha$^{-1}$ and 254.84 t ha$^{-1}$ | Panda et al. (2020) |
| Orissa | Harvesting method | – | 30.12–261.08 Mg ha$^{-1}$ | Behera & Misra (2006) |
| Orissa | Allometric equation | $AGB= \rho \times exp(-0.667 + 1.784 \times ln(D) + 0.207 \times (ln(D))^2 - 0.0281 \times (ln(D))^3)$ (U) | 6.98–257.25 Mg C ha$^{-1}$. | Sahu et al. (2016) |
| Orissa | Sentinel-1 and machine learning algorithms | – | 70–666 t/ha | Ghosh & Behera (2021) |
| Orissa | Allometric equations | $AGB = \rho * exp(-0.667 + 1.784 * ln(D))^2 - 0.0281 * (lnD))^3$ where, AGB is above-ground biomass, $\rho$ = wood specific gravity (grams/cm$^3$), ln is natural logarithm, $D$ is DBH (cm). (D) | 4.3–145.5 Mg ha$^{-1}$ | Pattnayak et al. (2020) |

*(Continued)*

**TABLE 20.1**
**Continued**

| Place | Method/Sensor | Model Used or Developed | Above-Ground Biomass | References |
|---|---|---|---|---|
| Orissa | Allometric equations | $\ln(Y1) = -0.37 + 0.333(\ln D) + 0.933 (\ln D)^2 - 0.122 (\ln D)^3$ Y1 = biomass value per tree; and D = diameter at breast height (dbh) for each tree measured at 1.37 m above the surface (U) | 223.49 and 246.79 Mg/ha | Pradhan et al. (2019) |
| Orissa | Allometric equations | – | 263.90 ± 30.63–1050.12 ± 224.43 t ha$^{-1}$ | Banerjee et al. (2018) |
| Orissa | Allometric equations | AGB (in kg) = $\rho \times \exp(-1.499 + 2.148 \times \ln(D) + 0.207 \times (\ln(D))^2 - 0.0281 \times (\ln(D))^3$ where D is the diameter at breast height (in cm), $\rho$ is the wood specific gravity (in g cm$^{-3}$) (U) | 337.7 and 410.7 Mg ha$^{-1}$ | Mohanta et al. (2020) |
| West Bengal | Allometric equation | – | – | Jana et al. (2009) |
| West Bengal | Allometric equation | – | – | Mitra et al. (2011) |
| West Bengal | Harvest method, allometric equation | AGB = 1.0471 (d)$^{0.864}$ (H)$^{0.635}$ ($\rho$)$^{-1.37}$ AGB = 1.3799 (H)$^{0.687}$ (d)$^{0.955}$ (D) | 39.93 ± 14.05 t C ha$^{-1}$ | Ray et al. (2011) |
| West Bengal | Allometric equation | Y = exp. {−2.4090 + 0.9522 ln (D$^2$HS)} where 'Exp.' denotes 'e' to the power of . . .", 'D' is dbh in meters, 'H' is height of the tree (m) and 'S' is density of wood (t/m$^2$) assumed as 0.5 for tropical woods | 1733.13 Mg ha$^{-1}$ | Shukla & Chakravarty (2018) |
| West Bengal | Sentinel-1A and Field measurement data | AGB = 0.0673 ($\rho \times D^2 \times H$)$^{0.976}$ where, $\rho$ = Wood density in gcm$^3$, D = DBH in cm, and H = Tree height in m | 88.56–170.29 tonnes/ha, 44.1–249 tonnes/ha | Roy et al. (2021) |
| West Bengal | Allometric equations | – | 673.89 and 185.14 Mg ha$^{-1}$ | Karmakar et al. (2019) |
| West Bengal | Allometric equations | – | 71.08–102.85 t ha$^{-1}$ and 51.02–90.09 t ha$^{-1}$ | Banerjee et al. (2013) |
| **Western Zonal Council** | | | | |
| Maharashtra | Landsat and IRS LISS III, allometric equations | – | 140.76 t ha$^{-1}$ (1989) and 149.98 t ha$^{-1}$ (2005) | Kale et al. (2009) |
| Maharashtra | DLR ESAR | – | – | Alappat et al. (2011) |
| Maharashtra | Allometric equations | – | 82.8- t ha$^{-1}$ | Chavan & Rasal (2012) |
| Maharashtra | Allometric equations | – | 0.626 t/tree | Suryawanshi et al. (2014) |
| Maharashtra | Landsat TM and field measurement data | – | – | Das & Singh (2016) |
| Maharashtra | LISS-III and MODIS and field measured data | – | 10.12–84.62 t/ha and 8.75–27.75 t/ha | Joshi et al. (2020) |
| Gujarat | LISS-III and MODIS and field measured data | – | 5.534 t/ha –134.082 t/ha | Patil et al. (2011) |
| Gujarat | Semi harvest method | – | 0.09–168.28 Mg ha_1 | Mehta et al. (2014) |

| Place | Method/Sensor | Model Used or Developed | Above-Ground Biomass | References |
|---|---|---|---|---|
| Rajasthan | Harvesting method, regression equation | Y= a + bx (D) | – | Kumar, Sajish et al. (2011) |
| **Southern Zonal Council** | | | | |
| Andhra Pradesh | Allometric equation | Aboveground biomass = $34.4703 - 8.0671 \times dbh + 0.6589 \times (dbh)^2$ (U) | 58.04–368.39 Mg/ha | Srinivas & Sundarapandian (2019) |
| Andhra Pradesh | Allometric equations | – | 172.32 t ha$^{-1}$ | Manickam et al. (2014) |
| Karnataka | Destructive method, regression equation | $\log_e T = -0.435 + 2.12 \log_e dbh$ (T=Total above-ground biomass) (D) | 420–649 t ha$^{-1}$ | Rai & Proctor (1986) |
| Karnataka | Destructive method, allometric equation | Biomass = $\exp(-2.997 + \ln(\text{wood density} \times (dbh)^2 \times H))$ (U) | – | Osuri et al. (2014) |
| Karnataka | ENVISAT-ASAR, regression equation | – | – | Jha et al. (2006) |
| Karnataka | LISS-IV and forest inventory data | – | 280 (±72.5) and 297.6 (±55.2) Mg ha$^{-1}$ | Madugundu et al. (2008) |
| Karnataka | Destructive method | – | 216.19 and 41.09 Mg/ha | Subbanna and Viswanath (2021) |
| Karnataka | Allometric equations | – | 43.86 and 95.87 Mg ha$^{-1}$ | Khaple et al. (2016) |
| Kerala | Destructive sampling method | – | 241.7 Mg ha$^{-1}$ | Kumar et al. (2005) |
| Kerala | Allometric equation | W top = $0.251 \rho D^{2.46}$ W top is above-ground biomass ρ is the wood density of the respective species D is the diameter (D) | 80.23 ± 15.95 t/ha | Harishma et al. (2020) |
| Kerala | Allometric equation | W top = $0.251 \rho D^{2.46}$ W top is above-ground biomass ρ is the wood density of the respective species D is the diameter (D) | 189.26 ± 97.80 t ha$^{-1}$ | Vinod et al. (2019) |
| Tamil Nadu | Regression equation | AGB = 12.05 + 0.876(BA), AGB = 11.27 + 6.03(BA) + 1.83 (H) (U) | 39.69–170.02 Mg ha$^{-1}$ and 73.06–173.10 Mg ha$^{-1}$ | Mani & Parthasarathy (2007) |
| Tamil Nadu | IRS 1D LISS III, regression equations | – | | Ramachandran et al. (2007) |
| Tamil Nadu | Allometric equation | – | 15.61–597.13 t ha$^{-1}$ | Mohanraj et al. (2010) |
| Tamil Nadu | Allometric equation | – | 58.43–102.76 Mg/ha | Sundarapandian et al. (2013) |
| Tamil Nadu | Allometric formula | AGB dry = $\exp(2.2014 \times LN(DBH) - 1.0615)$ where, AGB dry = above-ground dry biomass of tree (kg); dbh = diameter at breast height (cm); 2.2014 and −1.0615 are constants (D) | 117.77 t ha$^{-1}$ | Udayakumar et al. (2016) |
| Tamil Nadu | Allometric formula | AGB dry = $\exp(2.2014 \times LN(DBH) - 1.0615)$; where, AGB dry = Above-ground dry biomass of tree (kg); dbh = diameter at breast height (cm); 2.2014 and −1.0615 are constants (D) | 274.07 tonne/ha | Manikandan et al. (2019) |

*(Continued)*

**TABLE 20.1**
**Continued**

| Place | Method/Sensor | Model Used or Developed | Above-Ground Biomass | References |
|---|---|---|---|---|
| Tamil Nadu | Allometric formula | AGB dry = exp (1.9724 × LN (dbh)−1.0717); where, AGB dry is above-ground dry biomass of tree (kg); dbh is stem diameter at breast height (cm); LN is natural logarithm; 1.9724 and 1.0717 are constants (D) | 36.8 ± 18.9 Mg ha$^{-1}$ | Udayakumar et al. (2018) |
| Tamil Nadu | Non-harvesting method | – | 99–216 Mg/ha | Naveenkumar et al. (2017) |

*Notes:* Dbh = diameter at breast height, H = tree height, ρ = density of tree species, (D) = developed model, (U) = used model

suggested by the data on tree height and tree density (Roy et al., 2021). A non-destructive sampling technique and allometric equations were used in a different study to estimate and compare the AGB between two forests (Barjora Forest, Bankura; and Ballavpur Wildlife Sanctuary, Bolpur, West Bengal), each of which had a different level of air pollution. The reason behind the decrease in AGB at Barjora forest was increasing levels of air pollution. The findings also showed that *Shorea robusta* produced greater biomass overall in both forests, and that all tree species with dbh>30 cm supplied roughly 40.89% and 69.97% CSP at Barjora and BWLS, respectively (Karmakar et al., 2019). *Avicennia alba*, *Excoecaria agallocha*, and *Sonneratia apetala*, three even-aged dominating mangrove species in two different salinity regimes of the Indian Sundarbans, were calculated to have AGB by Banerjee et al. (2013). In the western region compared to the central region, all three of the selected species had higher AGB (Table 20.1).

### 20.3.5 Western Zonal Council

#### 20.3.5.1 Maharashtra

Suryawanshi et al. (2014) also measured the above and below ground biomass of tree species present at North Maharashtra University campus, Jalgaon, Maharashtra by using allometric equations (Karmakar, 2018). The integrated use of field measured data and satellite remote sensing (Landsat TM satellite) and GIS were used in prediction of AGB in tropical forest in the Western Ghats of Maharashtra. The estimated AGB ranged from 30.2–51.1 ton/ha in moist deciduous forest, 9.2–99.1 ton/ha in dry deciduous forest, 42.1–158.6 ton/ha in semi-evergreen forest, and 160.9–271 ton/ha in evergreen forest, respectively (Das & Singh, 2016). Joshi et al. (2020) have also measured the AGB by using field-measured data and LISS-III and MODIS data in tropical dry deciduous forest, Dhule District, Maharashtra. In a different study, Singh and Das (2014) evaluated AGB in the Ratnagiri district of Maharashtra using geostatistical modeling, remote sensing (Landsat TM satellite data), and field inventory data. The estimated value ranged from 0.05–271 t-dry wt ha$^{-1}$ (Table 20.1).

#### 20.3.5.2 Gujarat

Patil et al. (2011) have used the aggregation of LISS-III and MODIS data and field measurement data to estimate the AGB phytomass in moist deciduous forests of Surat district, and bamboo contributed more in total AGB in these areas. In another study, the semi-harvest method was used to measure the AGB in forest covers of Gujarat (Mehta et al., 2014) (Table 20.1).

#### 20.3.5.3 Rajasthan

Regression equations and the harvesting method were used to determine the biomass content of three different age groups (five, ten, and 15 years old) in the *Butea monosperma* forest in Rajasthan. The tree biomass ranges between 183.7 ± 3.21 t ha$^{-1}$ and 298.3 ± 3.57 t ha$^{-1}$ (5–15 years) (Kumar, Sajish et al., 2011; Karmakar, 2018) (Table 20.1).

### 20.3.6 Southern Zonal Council

#### 20.3.6.1 Andhra Pradesh

In order to calculate the AGB of trees using an allometric equation, the locations were chosen in the East Godavari region of the Eastern Ghats. The highest AGB was estimated in site I and the lowest value was observed in site III. *Xylia xylocarpa* contributed more in AGB in site I and II while, *Terminalia arjuna* contributed more in site III (Srinivas & Sundarapandian, 2019). In another study, the AGB of 13 tree species was predicted in three different districts (Khammam, Kurnool, and Anantapur) of Andhra Pradesh by using allometric equations (Manickam et al., 2014) (Table 20.1).

#### 20.3.6.2 Karnataka

Variations in carbon storage at contiguous and fragmented forests in Karnataka were estimated by Osuri et al. (2014). It has been that relatively large, well-protected fragment forests store 40% less carbon per hectare than contiguous forests. The AGB of the deciduous forests in the Western Ghats of Karnataka was estimated by Madugundu et al. (2008) using LISS-IV data and forest inventory data. In another study, the AGB was estimated in five different bamboo species

(*Bambusa balcooa, Bambusa bambos, Dendrocalamus asper, Dendrocalamus stocksii,* and *Dendrocalamus strictus*) in tropical humid and semiarid regions of Karnataka by using destructive sampling method. *B. bambos* and *D. strictus* performed better in semiarid regions than in tropical humid regions, and vice-versa for the remaining three species; the estimated AGB was more in tropical humid regions as compared to semiarid regions (Subbanna & Viswanath, 2021). Khaple et al. (2016) have estimated and compared the AGB in two different vegetation types (moist deciduous forest and dry deciduous forest) of Dharwad district in Karnataka by using some allometric equations. The calculated AGB was greater in moist deciduous forest and lesser in dry deciduous forest. Also, it has been observed that taller trees and trees with larger diameter contributes more in total AGB (Table 20.1).

### 20.3.6.3 Kerala

Kumar et al. (2005) have estimated the AGB of bamboo clumps by using destructive sampling method in Vellanikkara, Thrissur. In Thalassery Estuarine Wetland in Kerala, South-West Coast, Vinod et al. (2019) also used an allometric equation to quantify the AGB of eight species of mangroves. Among all the eight species, *Avicennia officinalis* recorded the highest AGB with the value of 63.53 ± 34.26 t ha$^{-1}$ (Table 20.1).

### 20.3.6.4 Tamil Nadu

Allometric equations were used to calculate the AGB in distinct forest types (evergreen, deciduous, mixed, open scrub, and plantation forest) in the Kolli forest, Eastern Ghat, Tamil Nadu. The mean AGB were 412.52 ± 96.67, 348.08 ± 68.75, 290.34 ± 124.99, 293.72 ± 260.30, and 361.91 ± 41.88 t ha$^{-1}$ in evergreen, deciduous, mixed, open scrub and plantation forest, respectively (Mohanraj et al., 2010) (Karmakar, 2018). Sundarapandian et al. (2013) selected four different sites of dry forests (sacred groves) in the Sivagangai district of Tamil Nadu to calculate the AGB by using some allometric equations. Greater AGB was calculated in site III, followed by site I, site II and site IV, respectively. Pachaimalai in Tamil Nadu's Southern Eastern Ghats was the site of a variety of studies. To establish an allometric equation and determine the AGB of trees, destructive harvesting of trees was used in a study (Udayakumar et al., 2016).

By creating an allometric formula and destructively sampling trees from a variety of tropical dry forests, Manikandan et al. (2019) have determined the AGB of trees in the tropical forests of the Pachamalai Hills, Southern Eastern Ghats of Tamil Nadu. In Chennai Metropolitan City, the AGB was calculated by developing allometric formula through destructively sampled healthy individuals of *Albizia saman* (Udayakumar et al., 2018). In a tropical dry forest in the Javadi Hills, Eastern Ghats, Naveenkumar et al. (2017) conducted a new study in three different elevations (high-elevation forest, mid-elevation forest, and low-elevation forest). Using a non-harvesting method, the AGB of adult trees (≥30 cm girth at breast height), juvenile trees (≥10 to <30 cm girth at breast height), and lianas (≥3 cm girth at 1.37 m from the roots point) was estimated (Table 20.1).

## 20.4 COMPARISON OF BIOMASS AMONG THE ZONAL COUNCILS OF INDIA

The AGB of six different zones has been represented in Figure 20.2. The lowest AGB value was found in Western Zonal Council with a median value of 85.5075 and the highest AGB value was found in Eastern Zonal Council with the median value of 223.4900. The dispersion of AGB are nearly similar in North-East Council and Eastern Zonal Council (interquartile ranges are 311.17 and 278.75, respectively). The overall range of AGB dataset is greater in Eastern Zonal Council and lower in Western Zonal Council.

## 20.5 TRENDS IN MEASUREMENT OF TREE BIOMASS IN INDIA

Trends in measurement of tree biomass in different zonal councils of India were reviewed of 107 articles. The measurement method, location, and the different models which were either used or developed are represented in Table 20.1. The study revealed that the highest percentage (51%) for the measurement of biomass is covered through non-destructive means (Figure 20.3). The use of remote sensing or satellite method is being covered about 31%, followed by destructive method (about 18%).

## 20.6 ADVANTAGES AND LIMITATIONS OF DIFFERENT METHODS

In destructive method, the plant species is being removed or uprooted and have to weigh the entire components. In addition, the entire process is time-consuming, expensive, impractical, occasionally unlawful, and goes against the goals of safeguarding forests (López-López et al., 2017). Furthermore, this method is not appropriate in degraded forests that contain threatened species (Montès et al., 2000). But among these three measurement methods, the destructive method provides the most accurate results. The non-destructive method also ultimately requires tree harvesting and weighing—and also, for getting precise and dependable results of biomass, use of species and site-specific allometric equations are desirable (Djomo & Chimi, 2017). But at a regional level, for which field data is not available, remote sensing or satellite data is very useful. Although biomass estimation directly through remote sensing is not possible, the combination of empirical data and remotely sensed data are used to quantify the biomass (Vashum & Jayakumar, 2012). The predictive and allometric equations are developed from the field measurements and the data is validated with remotely sensed data. So, these remotely sensed data

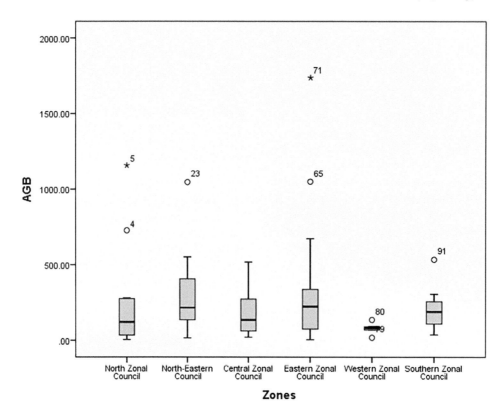

**FIGURE 20.2** Box-plot diagrams representing the AGB data in six different zones in India (North Zonal Council, North-Eastern Council, Central Zonal Council, Eastern Zonal Council, Western Zonal Council, and Southern Zonal Council).

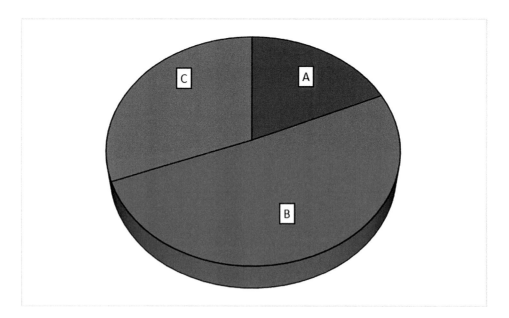

**FIGURE 20.3** Different measurement methods for estimation of biomass in different zonal councils of India: (A) destructive method; (B) non-destructive method; and (C) remote sensing or satellite method.

after validation can be used in a greater area to predict the biomass (Vashum & Jayakumar, 2012).

Estimating the AGB has primarily relied on non-destructive techniques, and these techniques have typically only been employed in certain regions of India. Use of remote sensing and satellite data are limited in the available research. Using of integrated geospatial techniques (like optical, microwave, and LiDAR) and also using different models (like IWCM and water cloud model [WCM]) the estimation method can become quicker, more economical, and easier. Due to the climatic, geographical, and biophysical peculiarities of the research location, it has become increasingly difficult to transfer a single established model from one area to another (Sinha et al., 2015). Use of these techniques is encouraged. Incorporation of remote sensing data and field inventory data can reduce the uncertainties in the estimation of AGB (Sharma & Chaudhry, 2013).

## 20.7 CONCLUSIONS

Owing to its high dynamic nature and tedious accessibility conditions of tropical forests, the estimation of aboveground biomass (AGB) is very difficult. Although the conventional or direct method of biomass estimation is the accurate way to quantify AGB, this method has some limitations due to cost and time concerns, and inaccessibility of some areas—but using remote sensing technology, the AGB can be estimated rapidly at lower cost and on a broader scale. Despite the complexity of all estimate techniques, there are more benefits to be gained from combining various datasets for model creation and using separate datasets for model validation. AGB estimation method is improving progressively in India. There is a lack of complete information on these aspects from India, and this chapter will help researchers, policy makers, and other people who work on forests in assessing and choosing the best method/model for estimating AGB without felling the trees—which is key for the conservation of biodiversity—in future studies.

## DECLARATION ON CONFLICTS OF INTEREST

The ethical standard is maintained and there is no conflict of interest between the authors.

## ACKNOWLEDGMENTS

One of the authors (DK) was given a fellowship by the Science and Engineering Research Board (NPDF-SERB), Government of India, in the form of a post-doctoral fellowship (File No: PDF/2020/001009).

## REFERENCES

Agarwal A, Purwar JP (2017) Carbon sequestration and above ground biomass produced by Bambusa spp. in the mid Himalayan region of Uttarakhand, India. *Indian Forester* 143(4):303–6.

Ajit, Das DK, Chaturvedi OP, Jabeen N, Dhyani SK (2011) Predictive models for dry weight estimation of above and below ground biomass components of *Populus deltoides* in India: Development and comparative diagnosis. *Biomass and Bioenergy* 35(3):1145–52.

Alappat VO, Joshi AK, Krishnamurthy YV (2011) Tropical dry deciduous forest stand variable estimation using SAR data. *Journal of the Indian Society of Remote Sensing* 39(4):583–9. http://dx.doi.org/10.1007/s12524-011-0118-y

Anaya JA, Chuvieco E, Palacios-Orueta A (2009) Aboveground biomass assessment in Colombia: A remote sensing approach. *Forest Ecology and Management* 257(4):1237–46.

Baishya R, Barik SK, Upadhaya K (2009) Distribution pattern of aboveground biomass in natural and plantation forests of humid tropics in northeast India. *Tropical Ecology* 50(2):295–304

Bala S, Biswas S, Mazumdar A (2012) Potential of carbon benefits from eucalyptus hybrid in dry-deciduous coppice forest of Jharkhand. *Journal of Engineering and Applied Sciences* 7:1614–22.

Banerjee K, Bal G, Paul R (2018) Total biomass and carbon estimates in mangrove species of Bhitarkanika Wildlife Sanctuary (BWLS), Odisha. *International Journal of Plant and Environment* 4(02):27–34.

Banerjee K, Sengupta K, Raha A, Mitra A (2013) Salinity based allometric equations for biomass estimation of Sundarban mangroves. *Biomass and Bioenergy* 56:382–91.

Behera MD, Tripathi P, Mishra B, Kumar S, Chitale VS, Behera SK (2016) Above-ground biomass and carbon estimates of Shorea robusta and Tectona grandis forests using Quadpol Alos Palsar data. *Advances in Space Research* 57(2):552–61. http://dx.doi.org/10.1016/j.asr.2015.11.010

Behera SK, Misra MK (2006) Aboveground tree biomass in a recovering tropical sal (Shorea robusta Gaertn. f.) forest of Eastern Ghats, India. *Biomass and Bioenergy* 30(6):509–21.

Behera SK, Sahu N, Mishra AK, Bargali SS, Behera MD, Tuli R (2017) Aboveground biomass and carbon stock assessment in Indian tropical deciduous forest and relationship with stand structural attributes. *Ecological Engineering* 99:513–24. http://dx.doi.org/10.1016/j.ecoleng.2016.11.046

Bhadouria R, Singh R, Srivastava P, Raghubanshi AS (2016) Understanding the ecology of tree-seedling growth in dry tropical environment: A management perspective. *Energy, Ecology and Environment* 1(5):296–309.

Bhadouria R, Srivastava P, Singh R, Tripathi S, Singh H, Raghubanshi AS (2017) Tree seedling establishment in dry tropics: An urgent need of interaction studies. *Environment Systems and Decisions* 37(1):88–100.

Bhadouria R, Tripathi SA, Rao KS (2018) Understanding plant community assemblage, functional diversity and soil attributes of Indian savannas through a continuum approach. *Tropical Ecology* 59(4):545–54.

Bijalwan A, Swamy SL, Sharma CM, Sharma NK, Tiwari AK (2010) Land-use, biomass and carbon estimation in dry tropical forest of Chhattisgarh region in India using satellite remote sensing and GIS. *Journal of Forestry Research* 21(2):161–70. http://dx.doi.org/10.1007/s11676-010-0026-y

Borah M, Das D, Kalita J, Boruah HP, Phukan B, Neog B (2015) Tree species composition, biomass and carbon stocks in two tropical forest of Assam. *Biomass and Bioenergy* 78:25–35.

Chaturvedi OP, Singh JS (1987) The structure and function of pine forest in Central Himalaya. II. Nutrient dynamics. *Annals of Botany* 60:53–67.

Chaturvedi RK, Raghubanshi AS (2013) Above ground biomass estimation of small diameter woody species of tropical dry forest. *New Forests* 44(4):509–19. http://dx.doi.org/10.1007/s11056-012-9359-z

Chaturvedi RK, Raghubanshi AS (2015) Assessment of carbon density and accumulation in mono-and multi-specific stands in Teak and Sal forests of a tropical dry region in India. *Forest Ecology and Management* 339:11–21. http://dx.doi.org/10.1016/j.foreco.2014.12.002

Chaturvedi RK, Raghubanshi AS, Singh JS (2011) Carbon density and accumulation in woody species of tropical dry forest in India. *Forest Ecology and Management* 262(8):1576–88. http://dx.doi.org/10.1016/j.foreco.2011.07.006

Chavan B, Rasal G (2012) Total sequestered carbon stock of Mangifera indica. *Journal of Environment and Earth Science* 2(1).

Chave J, Andalo C, Brown S, Cairns MA, Chambers JQ, Eamus D, Fölster H, Fromard F, Higuchi N, Kira T, Lescure JP (2005) Tree allometry and improved estimation of carbon stocks and balance in tropical forests. *Oecologia* 145(1):87–99.

Chisanga K, Bhardwaj DR, Pala NA, Thakur CL (2018) Biomass production and carbon stock inventory of high-altitude dry temperate land use systems in North Western Himalaya. *Ecological Processes* 7(1):1–3. https://doi.org/10.1186/s13717-018-0134-8

Dar DA and Sahu P (2018) Assessment of biomass and carbon stock in temperate forests of Northern Kashmir Himalaya, India. *Proceedings of the International Academy of Ecology and Environmental Sciences* 1;8(2):139–50.

Dar JA, Sundarapandian S (2015) Variation of biomass and carbon pools with forest type in temperate forests of Kashmir Himalaya, India. *Environmental Monitoring and Assessment* 187(2):55. http://dx.doi.org/10.1007/s10661-015-4299-7

Das B, Bordoloi R, Deka S, Paul A, Pandey PK, Singha LB, Tripathi OP, Mishra BP, Mishra M (2021) Above ground biomass carbon assessment using field, satellite data and model based integrated approach to predict the carbon sequestration potential of major land use sector of Arunachal Himalaya, India. *Carbon Management* 8;12(2):201–14.

Das S, Singh TP (2016) Forest type, diversity and biomass estimation in tropical forests of Western Ghat of Maharashtra using geospatial techniques. *Small-scale Forestry* 15(4):517–32.

Day M, Baldauf C, Rutishauser E, Sunderland TC (2014) Relationships between tree species diversity and aboveground biomass in Central African rainforests: Implications for REDD. *Environmental Conservation* 41(1):64–72.

Deb D, Deb S, Chakraborty D, Singh JP, Singh AK, Dutta P, Choudhury A (2020) Aboveground biomass estimation of an agro-pastoral ecology in semi-arid Bundelkhand region of India from Landsat data: A comparison of support vector machine and traditional regression models. *Geocarto International* 29:1–6.

Devi LS, Yadava PS (2009) Aboveground biomass and net primary production of semi-evergreen tropical forest of Manipur, north-eastern India. *Journal of Forestry Research* 20(2):151–5. http://dx.doi.org/10.1007/s11676-009-0026-y

Dixon RK, Solomon AM, Brown S, Houghton RA, Trexier MC, Wisniewski J (1994) Carbon pools and flux of global forest ecosystems. *Science* 263(5144):185–90.

Djomo AN, Chimi CD (2017) Tree allometric equations for estimation of above, below and total biomass in a tropical moist forest: Case study with application to remote sensing. *Forest Ecology and Management* 391:184–93. https://doi.org/10.1016/j.foreco.2017.02.022.

Dobson MC, Ulaby FT, LeToan T, Beaudoin A, Kasischke ES, Christensen N (1992) Dependence of radar backscatter on coniferous forest biomass. *IEEE Transactions on Geoscience and Remote Sensing* 30(2):412–5.

Fang J, Brown S, Tang Y, Nabuurs GJ, Wang X, Shen H (2006) Overestimated biomass carbon pools of the northern mid- and high latitude forests. *Climatic Change* 74(1–3):355–68.

FAO (2001) Global forest resources assessment 2000—main report. *FAO Forestry Paper* 140, Food and Agriculture Organization of the United Nations, Rome, p. 363.

Feyisa K, Beyene S, Megersa B, Said MY, Angassa (2016) Allometric equations for predicting above-ground biomass of selected woody species to estimate carbon in East African rangelands. *Agroforestry Systems* 1–23.

Franklin J, Hiernaux PH (1991) Estimating foliage and woody biomass in Sahelian and Sudanian woodlands using a remote sensing model. *International Journal of Remote Sensing* 12(6):1387–404.

FSI (1996) *Volume Equations for forest of India, Nepal and Bhutan*. Dehradun: Forest Survey of India, pp. 249.

Gairola S, Sharma CM, Ghildiyal SK, Suyal S (2011) Live tree biomass and carbon variation along an altitudinal gradient in moist temperate valley slopes of the Garhwal Himalaya (India). *Current Science* 1862–70.

Garkoti SC, Singh SP (1995) Variation in net primary productivity and biomass of forests in the high mountains of Central Himalaya. *Journal of Vegetation Science* 6(1):23–8.

Gautam MK, Tripathi AK, Manhas RK (2011) Assessment of critical loads in tropical sal (Shorea robusta Gaertn. f.) forests of Doon Valley Himalayas, India. *Water, Air, & Soil Pollution* 218(1–4):235–64. http://dx.doi.org/10.1007/s11270-010-0638-z

Ghosh SM, Behera MD (2021) Aboveground biomass estimates of tropical mangrove forest using Sentinel-1 SAR coherence data-The superiority of deep learning over a semi-empirical model. *Computers & Geosciences* 150:104737.

Goward SN, Dye DG (1987) Evaluating North American net primary productivity with satellite observations. *Advances in Space Research* 7(11):165–74.

Harishma KM, Sandeep S, Sreekumar VB (2020) Biomass and carbon stocks in mangrove ecosystems of Kerala, southwest coast of India. *Ecological Processes* 9(1):1–9.

Hossain M, Saha C, Abdullah SR, Saha S, Siddique MR (2016) Allometric biomass, nutrient and carbon stock models for Kandelia candel of the Sundarbans, Bangladesh. *Trees* 30(3):709–17.

Husch B, Beers TW, Kershaw JA (2003) *Forest mensuration*, 4th edn. New Jersey: Wiley.

Hussin YA, Reich RM, Hoffer RM (1991) Estimating splash pine biomass using radar backscatter. *IEEE Transactions on Geoscience and Remote Sensing* 29(3):427–31.

India State of Forest Report (ISFR) (2011) Published by Forest Survey of India (Ministry of Environment & Forests).

Jana BK, Biswas S, Majumder M, Roy PK, Mazumdar A (2009) Carbon sequestration rate and aboveground biomass carbon potential of four young species. *Journal of Ecology and Natural Environment* 1(2):15–24.

Jha CS, Rangaswamy M, Vyjayanthi N, Murthy MS (2006) Estimation of forest biomass using Envisat-ASAR data. In *Microwave Remote Sensing of the Atmosphere and Environment* (Vol. 6410, p. 641002). International Society for Optics and Photonics. http://dx.doi.org/10.1117/12.6933040

Joshi B, Rane G, Singh S (2020) Distribution of biomass in dry deciduous forest from Dhule District, Maharashtra, India. *BIOINFOLET-A Quarterly Journal of Life Sciences* 17(1b):93–5.

Joshi N, Baumann M, Ehammer A, Fensholt R, Grogan K, Hostert P, Jepsen MR, Kuemmerle T, Meyfroidt P, Mitchard ET, Reiche J (2016) A review of the application of optical and radar remote sensing data fusion to land use mapping and monitoring. *Remote Sensing* 8(1):70. http://dx.doi.org/10.3390/rs8010070

Kale M, Singh S, Roy PS, Deosthali V, Ghole VS (2004) Biomass equations of dominant species of dry deciduous forest in Shivpuri district, Madhya Pradesh. *Current Science*:683–7.

Kale MP, Ravan SA, Roy PS, Singh S (2009) Patterns of carbon sequestration in forests of Western Ghats and study of applicability of remote sensing in generating carbon credits through afforestation/reforestation. *Journal of the Indian Society of Remote Sensing* 37(3):457–71.

Karmakar D (2018) *Effects of Air Pollution on Forest and Its Consequence on Ecosystem Services a Case Study of Barjora Forest Bankura.* Ph.D thesis, Visva-Bharati Santiniketan—731235 India. http://hdl.handle.net/10603/273359

Karmakar D, Ghosh T, Padhy PK (2019) Effects of air pollution on carbon sequestration potential in two tropical forests of West Bengal, India. *Ecological Indicators* 98:377–88.

Kaushal R, Subbulakshmi V, Tomar JM, Alam NM, Jayaparkash J, Mehta H, Chaturvedi OP (2016) Predictive models for biomass and carbon stock estimation in male bamboo (Dendrocalamus strictus L.) in Doon valley, India. *Acta Ecologica Sinica* 36(6):469–76.

Khan M, Khan SM, Ilyas M, Alqarawi AA, Ahmad Z, Abd_Allah EF (2017) Plant species and communities assessment in interaction with edaphic and topographic factors; an ecological study of the mount Eelum District Swat, Pakistan. *Saudi Journal of Biological Sciences* 24(4):778–86. https://doi.org/10.1016/j.sjbs.2016.11.018

Khan MNI, Suwa R, Hagihara A, Ogawa K (2005) Allometric relationships for estimating the above ground phytomass and leaf area of mangrove Kandelia candel (L.) Druce trees in the Manko Wetland, Okinawa Island, Japan. *Trees* 19:266–272.

Khaple AK, Puttaiah ET, Devagiri GM, Mishra SB (2016) Vegetation structure and above ground biomass carbon in dry and moist deciduous forests of Dharwad district in Karnataka, India. *International Journal of Environmental Sciences* 5(2):89–97.

Kumar A, Sharma MP (2015) Estimation of carbon stocks of Balganga reserved forest, Uttarakhand, India. *Forest Science and Technology* 11(4):177–81.

Kumar BM, Rajesh G, Sudheesh KG (2005) Aboveground biomass production and nutrient uptake of thorny bamboo [*Bambusa bambos* (L.) Voss] in the homegardens of Thrissur, Kerala. *Journal of Tropical Agriculture* 43:51–6.

Kumar JN, Sajish PR, Kumar RN, Patel K (2011) Biomass and net primary productivity in three different aged Butea forest ecosystems in Western India, Rajasthan. *Our Nature* 9(1):73–82.

Kumar M, Sheikh MA, Saleem S (2014) Carbon stock in submergence forest of Srinagar hydroelectric project, Uttarakhand, India. *Forest Science and Technology* 10(2):61–6.

Kumar NR (2007) *Forest Cover, Stand Volume and Biomass Assessment in Dudhwa National Park Using Satellite Remote Sensing Data (optical and EnviSat ASAR).* Doctoral Dissertation, Andhra University, India.

Kumar R, Gupta SR, Singh S, Patil P, Dadhwal VK (2011) Spatial distribution of forest biomass using remote sensing and regression models in northern Haryana, India. *International Journal of Ecology and Environmental Sciences* 37(1):37–47.

Kumar S (2009) *Retrieval of Forest Parameters from Envisat ASAR Data for Biomass Inventory in Dudhwa National Park, UP Indiana.* Kolkata: ITC.

Kumar S, Garg RD, Govil H, Kushwaha SP (2019) PolSAR-decomposition-based extended water cloud modeling for forest aboveground biomass estimation. *Remote Sensing* 11(19):2287.

Kumar S, Khati UG, Chandola S, Agrawal S, Kushwaha SP (2017) Polarimetric SAR Interferometry based modeling for tree height and aboveground biomass retrieval in a tropical deciduous forest. *Advances in Space Research* 60(3):571–86. http://dx.doi.org/10.1016/j.asr.2017.04.018

Kumar S, Pandey U, Kushwaha SP, Chatterjee RS, Bijker W (2012) Aboveground biomass estimation of tropical forest from Envisat advanced synthetic aperture radar data using modeling approach. *Journal of Applied Remote Sensing* 6(1):063588.

Lal M, Singh R (2000) Carbon sequestration potential of Indian forests. *Environmental Monitoring and Assessment* 60(3):315–27.

Lal P, Kumar A, Saikia P, Das A, Patnaik C, Kumar G, Pandey AC, Srivastava P, Dwivedi CS, Khan ML (2021) Effect of vegetation structure on above ground biomass in tropical deciduous forests of Central India. *Geocarto International* 29(just-accepted):1–5.

Le Toan T, Beaudoin A, Riom J, Guyon D (1992) Relating forest biomass to SAR data. *IEEE Transactions on Geoscience and Remote Sensing* 30(2):403–11. http://dx.doi.org/10.1109/36.134089.

Lodhiyal N, Lodhiyal LS (2003) Biomass and net primary productivity of Bhabar Shisham forests in Central Himalaya, India. *Forest Ecology and Management* 176:217–235.

López-López SF, Martínez-Trinidad T, Benavides-Meza H, Garcia-Nieto M, Héctor M (2017) Non-destructive method for above-ground biomass estimation of *Fraxinus uhdei* (Wenz.) Lingelsh in an urban forest. *Urban Forestry & Urban Greening* 24:62–70. https://doi.org/10.1016/j.ufug.2017.03.025.

Lu D (2006) The potential and challenge of remote sensing-based biomass estimation. *International Journal of Remote Sensing* 27(7):1297–328.

Ma L, Shen C, Lou D, Fu S, Guan D (2017) Patterns of ecosystem carbon density in edge-affected fengshui forests. *Ecological Engineering* 107:216–23. https://doi.org/10.1016/j.ecoleng.2017.07.037.

Madugundu R, Nizalapur V, Jha CS (2008) Estimation of LAI and above-ground biomass in deciduous forests: Western Ghats of Karnataka, India. *International Journal of Applied Earth Observation and Geoinformation* 10(2):211–9.

Mahajan V, Choudhary P, Raina NS, Sharma P (2021) Carbon sequestration potential of trees in arable land-use and allometric modelling for dominant tree species in sub-tropics of Jammu and Kashmir. *Journal of Environmental Biology* 1;42:414–9.

Mandal G, Joshi SP (2015) Estimation of above-ground biomass and carbon stock of an invasive woody shrub in the subtropical deciduous forests of Doon Valley, western Himalaya, India. *Journal of Forestry Research* 26(2):291–305.

Mani S, Parthasarathy N (2007) Above-ground biomass estimation in ten tropical dry evergreen forest sites of peninsular India. *Biomass and Bioenergy* 31(5):284–90. https://doi.org/10.1016/j.biombioe.2006.08.006.

Manickam V, Krishna IV, Shanti SK, Radhika R (2014) Biomass calculations for carbon sequestration in forest ecosystem. *Journal of Energy and Chemical Engineering* 2(1):30–8.

Manikandan S, Udayakumar M, Sekar T (2019) Woody stem density and above-ground biomass in Pachaimalai Hills of Southern Eastern Ghats, Tamil Nadu, India. *International Journal for Research in Applied Science and Engineering Technology* 7(1):151–8.

Mehta N, Pandya NR, Thomas VO, Krishnayya NS (2014) Impact of rainfall gradient on aboveground biomass and soil organic carbon dynamics of forest covers in Gujarat, India. *Ecological Research* 29(6):1053–63.

Mertz O, Müller D, Sikor T, Hett C, Heinimann A, Castella JC, Lestrelin G, Ryan CM, Reay DS, Schmidt-Vogt D, Danielsen F, Theilade I, van Noordwijk M, Verchot LV, Burgess ND, Berry NJ, Pham TT, Messerli P, Xu J, Fensholt R, Hostert P, Pflugmacher D, Bruun TB, Neergaard AD, Dons K, Dewi S, Rutishauser E, Sun Z (2012) The forgotten D: Challenges of addressing forest degradation in complex mosaic landscapes under REDD+. *Geografisk Tidsskrift-Danish Journal of Geography* 112(1):63–76.

Mitra A, Sengupta K, Banerjee K (2011) Standing biomass and carbon storage of above-ground structures in dominant mangrove trees in the Sundarbans. *Forest Ecology and Management* 261(7):1325–35. https://doi.org/10.1016/j.foreco.2011.01.012.

Mohanraj R, Saravanan J, Dhanakumar S (2010) Carbon stock in Kolli forests, Eastern Ghats (India) with emphasis on aboveground biomass, litter, woody debris and soils. *iForest-Biogeosciences and Forestry* 4(2):61.

Mohanta MR, Mohanta A, Mohapatra U, Mohanty RC, Sahu SC (2020) Carbon stock assessment and its relation with tree biodiversity in Tropical Moist Deciduous Forest of Similipal Biosphere Reserve, Odisha, India. *Tropical Ecology* 61(4):497–508.

Montès N, Gauquelin T, Badri W, Bertaudiere V, Zaoui EH (2000) A non-destructive method for estimating above-ground forest biomass in threatened woodlands. *Forest Ecology and Management* 130: 37–46.

Mukhopadhyay R, Kumar S, Aghababaei H, Kulshrestha A (2021) Estimation of aboveground biomass from PolSAR and PolInSAR using regression-based modelling techniques. *Geocarto International* 23:1–27.

Naidu MT, Kumar OA (2016) Tree diversity, stand structure, and community composition of tropical forests in Eastern Ghats of Andhra Pradesh, India. *Journal of Asia-Pacific Biodiversity* 9(3):328–34. https://doi.org/10.1016/j.japb.2016.03.019.

NASA (2018, 2022) https://climate.nasa.gov/climate_resources/24/ (accessed on 01/04/2018, 14/11/2022).

Nath AJ, Das AK (2012) Carbon pool and sequestration potential of village bamboos in the agroforestry system of northeast India. *Tropical Ecology* 53(3):287–93.

Nath AJ, Das G, Das AK (2009) Above ground standing biomass and carbon storage in village bamboos in North East India. *Biomass and Bioenergy* 33(9):1188–96.

Nath S, Nath AJ, Sileshi GW, Das AK (2017) Biomass stocks and carbon storage in Barringtonia acutangula floodplain forests in North East India. *Biomass and Bioenergy* 98:37–42.

Naveenkumar J, Arunkumar KS, Sundarapandian SM (2017) Biomass and carbon stocks of a tropical dry forest of the Javadi Hills, Eastern Ghats, India. *Carbon Management* 8(5–6):351–61.

Nirala D, Khanduri VP, Sankanur MS (2018) Biomass and carbon stock assessment in different age group plantations of teak (Tectona grandis Linn. F.) in Bhabar and Shivalik regions of Uttarakhand. *International Journal of Bio-resource and Stress Management* 9(3):379–82.

Osuri AM, Kumar VS, Sankaran M (2014) Altered stand structure and tree allometry reduce carbon storage in evergreen forest fragments in India's Western Ghats. *Forest Ecology and Management* 329:375–83. doi:10.1016/j.foreco.2014.01.039

Panda MR, Oraon PR, Tirkey P (2020) Distribution of woody biomass reserves in tropical dry Sal (Shorea robusta roth.) forests of Ranchi. *The Pharma Innovation Journal* 9(6): 477–82.

Pande PK (2005) Biomass and productivity in some disturbed tropical dry deciduous teak forests of Satpura plateau, Madhya Pradesh. *Tropical Ecology* 46(2):229–40.

Patil P, Dutta D, Biradar C, Singh M (2015) Quantification of the terrestrial phytomass and carbon in the mountainous forest ecosystem using remote sensing and in-situ observations. *The International Archives of Photogrammetry, Remote Sensing and Spatial Information Sciences* 40(7):483. https://doi.org/10.5194/isprsarchives-XL-7-W3-483-2015 a

Patil P, Singh S, Dadhwal VK (2011) Above ground forest phytomass assessment in southern Gujarat. *Journal of the Indian Society of Remote Sensing* 40(1):37–46.

Pattnayak S, Kumar M, Dhal NK, Sahu SC (2020) Estimation of carbon pools in secondary tropical deciduous forests of Odisha, India. *Journal of Forestry Research* 32(2):663–73.

Pawar GV, Singh L, Jhariya MK, Sahu KP (2014) Effect of anthropogenic disturbances on biomass and carbon storage potential of a dry tropical forest in India. *Journal of Applied and Natural Science* 6(2):383–92.

Pradhan A, Ormsby AA, Behera N. A (2019) comparative assessment of tree diversity, biomass and biomass carbon stock between a protected area and a sacred forest of Western Odisha, India. *Écoscience* 26(3):195–204.

Purohit S, Aggarwal SP, Patel NR (2021) Estimation of forest aboveground biomass using combination of Landsat 8 and Sentinel-1A data with random forest regression algorithm in Himalayan Foothills. *Tropical Ecology* 62(2):288–300.

Rai SN, Proctor J (1986) Ecological studies on four rainforests in Karnataka, India: I. Environment, structure, floristics and biomass. *The Journal of Ecology* 1:439–54. www.jstor.org/stable/2260266

Ramachandran A, Jayakumar S, Haroon RM, Bhaskaran A, Arockiasamy DI (2007) Carbon sequestration: Estimation of carbon stock in natural forests using geospatial technology in the Eastern Ghats of Tamil Nadu, India. *Current Science* 92 (3): 323–31.

Rathore AC, Kumar A, Tomar JM, Jayaprakash J, Mehta H, Kaushal R, Alam NM, Gupta AK, Raizada A, Chaturvedi OP (2018) Predictive models for biomass and carbon stock estimation in Psidium guajava on bouldery riverbed lands in North-Western Himalayas, India. *Agroforestry Systems* 92(1):171–82.

Raupach MR, Marland G, Ciais P, Le Quéré C, Canadell JG, Klepper G, Field CB (2007) Global and regional drivers of accelerating CO2 emissions. *Proceedings of the National Academy of Sciences* 104(24):10288–93.

Ravindranath NH, Ostwald M (2008) *Carbon Inventory Methods: Handbook for Greenhouse Gas Inventory, Carbon Mitigation and Roundwood Production Projects*. Berlin: Springer Science & Business Media.

Ray R, Ganguly D, Chowdhury C, Dey M, Das S, Dutta MK, Mandal SK, Majumder N, De TK, Mukhopadhyay SK, Jana TK (2011) Carbon sequestration and annual increase of carbon stock in a mangrove forest. *Atmospheric Environment* 45(28):5016–24. http://dx.doi.org/10.1016/j.atmosenv.2011.04.074

Rizvi RH, Dhyani SK, Yadav RS, Singh R (2011) Biomass production and carbon stock of poplar agroforestry systems in Yamunanagar and Saharanpur districts of northwestern India. *Current Science* 100(5) 736–42.

Rodger AS (1993) The carbon cycle and global forest ecosystem. *Water, Air and Soil Pollution* 70: 295–307.

Roy PS (1989) Spectral reflectance characteristics of vegetation and their use in estimating productive potential. *Proceedings: Plant Sciences* 99(1):59–81.

Roy PS, Kumar S (1986) Advanced Very High Resolution Radiometer (AVHRR) satellite data for vegetation monitoring. In *Proc. International Seminar on Photogrammetry and Remote Sensing for Developing Countries*.

Roy PS, Ravan SA (1996) Biomass estimation using satellite remote sensing data—an investigation on possible approaches for natural forest. *Journal of Biosciences* 21(4):535–61.

Roy S, Mudi S, Das P, Ghosh S, Shit PK, Bhunia GS, Kim J (2021) Estimating Above Ground Biomass (AGB) and Tree Density using Sentinel-1 Data. In *Spatial Modeling in Forest Resources Management* (pp. 259–280). Cham: Springer.

Sahu KP, Singh L, Alone RA, Jhariya MK, Pawar GV (2013) Biomass and carbon storage pattern in an age series of teak plantation in dry tropics. *Vegetos-An International Journal of Plant Research* 26(1):205–17. http://dx.doi.org/10.5958/j.2229-4473.26.1.030

Sahu SC, Suresh HS, Ravindranath NH (2016) Forest structure, composition and above ground biomass of tree community in tropical dry forests of Eastern Ghats, India. *Notulae Scientia Biologicae* 8(1):125–133. http://dx.doi.org/10.15835/nsb.8.1.9746

Sajad S, Haq SM, Yaqoob U, Calixto ES, Hassan M (2021) Tree composition and standing biomass in forests of the northern part of Kashmir Himalaya. *Vegetos* 24:1–0.

Salunkhe O, Khare PK, Sahu TR, Singh S (2016) Estimation of tree biomass reserves in tropical deciduous forests of Central India by non-destructive approach. *Tropical Ecology* 57(2):153–61.

Shah S, Sharma DP, Pala NA, Tripathi P, Kumar M (2014) Temporal variations in carbon stock of Pinus roxburghii Sargent forests of Himachal Pradesh, India. *Journal of Mountain Science* 11(4):959–66.10.1007/s11629-013-2725-2

Shahid M, Joshi SP (2015) Biomass and carbon stock assessment in moist deciduous forests of Doon valley, western Himalaya, India. *Taiwania* 60(2):71–6. http://dx.doi.org/10.6165/tai.2015.60.71

Sharma CM, Baduni NP, Gairola S, Ghildiyal SK, Suyal S (2010) Tree diversity and carbon stocks of some major forest types of Garhwal Himalaya, India. *Forest Ecology and Management* 260(12):2170–9.

Sharma CM, Gairola S, Baduni NP, Ghildiyal SK, Suyal S (2011) Variation in carbon stocks on different slope aspects in seven major forest types of temperate region of Garhwal Himalaya, India. *Journal of Biosciences* 36(4):701–8.

Sharma LK, Nathawat MS, Sinha S (2013) Top-down and bottom-up inventory approach for above ground forest biomass and carbon monitoring in REDD framework using multi-resolution satellite data. *Environmental Monitoring and Assessment* 185(10):8621–37. http://dx.doi.org/10.1007/s10661-013-3199-y

Sharma V, Chaudhry S (2013) An Overview of Indian Forestry Sector with REDD. *ISRN Forestry*. http://dx.doi.org/10.1155/2013/298735

Shukla G, Chakravarty S (2018) Biomass, primary nutrient and carbon stock in a Sub-Himalayan Forest of West Bengal, India. *Journal of Forest and Environmental Science* 34(1):12–23.

Singh JS, Tiwari AK, Saxena AK (1985) Himalayan forests: A net source of carbon for the atmosphere. *Environmental Conservation* 12(1):67–9.

Singh L, Singh JS (1991) Species structure, dry matter dynamics and carbon flux of a dry tropical forest in India. *Annals of Botany* 68(3):263–73.

Singh TP, Das S (2014) Predictive analysis for vegetation biomass assessment in Western Ghat region (WG) using geospatial techniques. *Journal of the Indian Society of Remote Sensing* 42(3):549–57.

Singh V, Tewari A, Kushwaha SP, Dadhwal VK (2011) Formulating allometric equations for estimating biomass and carbon stock in small diameter trees. *Forest Ecology and Management* 261(11):1945–9.

Sinha S, Jeganathan C, Sharma LK, Nathawat MS (2015) A review of radar remote sensing for biomass estimation. *International Journal of Environmental Science and Technology* 12(5):1779–92. http://dx.doi.org/10.1007/s13762-015-0750-0

Sinha S, Santra A, Das AK, Sharma LK, Mohan S, Nathawat MS, Mitra SS, Jeganathan C (2019) Accounting tropical forest carbon stock with synergistic use of space-borne ALOS PALSAR and COSMO-Skymed SAR sensors. *Tropical Ecology* 60(1):83–93.

Smith TJ III, Whelan KRT (2006) Development of allometric relations for three mangrove species in South Florida for use in the Greater Everglades Ecosystem restoration. *Wetlands Ecology and Management* 14:409–419f.

Srinivas K, Sundarapandian S (2019) Biomass and carbon stocks of trees in tropical dry forest of East Godavari region, Andhra Pradesh, India. *Geology, Ecology, and Landscapes* 3(2):114–22.

Subbanna S, Viswanath S (2021) Above-ground biomass and carbon content in select bamboo species across humid and semiarid zones of Karnataka, India. In *Forest Resources Resilience and Conflicts* (pp. 153–63). Amsterdam: Elsevier.

Suganthi K, Das KR, Selvaraj M, Kurinji S, Goel M, Govindaraju M (2017) Assessment of altitudinal mediated changes of co2 sequestration by trees at pachamalai reserve forest, Tamil Nadu, India. In *Carbon Utilization* (pp. 89–99). New York: Springer. http://dx.doi.org/10.1007/978-981-10-3352-0_7.

Sundarapandian SM, Dar JA, Gandhi DS (2013) Kantipudi S, Subashree K. Estimation of biomass and carbon stocks in tropical dry forests in Sivagangai District, Tamil Nadu, India. *International Journal of Environmental Science and Engineering Research* 4(3):66–76.

Suryawanshi MN, Patel AR, Kale TS, Patil PR (2014) Carbon sequestration potential of tree species in the environment of North Maharashtra University Campus, Jalgaon (MS) India. *Bioscience Discovery* 5(2):175–9.

Thakur M, Verma RK (2019) Biomass and Soil Organic Carbon Stocks Under Cedrus deodara Forests in Mandi District of Himachal Pradesh. *Nature Environment and Pollution Technology* 1;18(3):879–87.

Thakur T, Swamy SL (2010) Analysis of land use, diversity, biomass, C and nutrient storage of a dry tropical forest ecosystem of India using satellite remote sensing and GIS techniques. *International Forestry and Environment Symposium* 15:273–278.

Thumaty KC, Fararoda R, Middinti S, Gopalakrishnan R, Jha CS, Dadhwal VK (2015) Estimation of above ground biomass for central Indian deciduous forests using ALOS PALSAR L-band data. *Journal of the Indian Society of Remote Sensing* 44(1):31–9.

Tiwari AK, Singh JS (1984) Mapping forest biomass in India through aerial photographs and non destructive field sampling. *Applied Geography* 2:151–65.

Tucker CJ, Vanpraet C, Boerwinkle E and Gaston A (1983) Satellite remote sensing of total dry matter accumulation in the Senegalese Sahel. *Remote Sensing of Environment* 13: 461–74.

Udayakumar M, Manikandan S, Selvan BT, Sekar T (2016) Density, Species richness and aboveground biomass of trees in 10 hectare permanent study plot, Pachaimalai, Tamil Nadu. *Scholars Academic Journal of Bioscience* 4(4):342–7.

Udayakumar M, Selvam A, Sekar T (2018) Aboveground biomass stockpile and carbon sequestration potential of Albiziasaman in Chennai Metropolitan City, India. *Plant* 6(3):60–66

Upadhaya K, Thapa N, Barik SK (2015) Tree diversity and biomass of tropical forests under two management regimes in Garo hills of north-eastern India. *Tropical Ecology* 56(2):257–68.

Vashum KT, Jayakumar S (2012) Methods to estimate aboveground biomass and carbon stock in natural forests-a review. *Journal of Ecosystem & Ecography* 2(4):1–7. http://dx.doi.org/10.4172/2157-7625.1000116

Verwijst T, Telenius B (1999) Biomass estimation procedures in short rotation forestry. *Forest Ecology and Management* 121(1–2):137–46.

Vinod K, Asokan PK, Zacharia PU, Ansar CP, Vijayan G, Anasukoya A, Kunhi Koya VA, Nikhiljith M (2019) Assessment of biomass and carbon stocks in mangroves of Thalassery estuarine wetland of Kerala, south west coast of India. *Journal of Coastal Research* 86(SI):209–17.

Waikhom AC, Nath AJ, Yadava PS (2017) Aboveground biomass and carbon stock in the largest sacred grove of Manipur, Northeast India. *Journal of Forestry Research* 29(2):425–8.

Wani AA, Bhat AF, Gatoo AA, Zahoor S, Mehraj B, Mir NA, Wani N, Qasba SS, ul Islam MA, Masoodi TH (2019) Relationship of forest biomass carbon with biophysical parameters in north Kashmir region of Himalayas. *Environmental Monitoring and Assessment* 191(9):1–3. https://doi.org/10.1007/s10661-019-7669-8

Yadav BK, Nandy S (2015) Mapping aboveground woody biomass using forest inventory, remote sensing and geostatistical techniques. *Environmental Monitoring and Assessment* 187(5):1–2.

Yuen JQ, Fung T, Ziegler AD (2016) Review of allometric equations for major land covers in SE Asia: Uncertainty and implications for above-and below-ground carbon estimates. *Forest Ecology and Management* 360:323–40. https://doi.org/10.1016/j.foreco.2015.09.016

# 21 Brassinosteroid Hormones

*A Promising Strategy for Abiotic Stress Management in Plants under Changing Climate*

Sandeep Kumar

## 21.1 INTRODUCTION

It is estimated that the population of the world will continue to inflate rapidly in the next century, and the production of agricultural food around the world will also need to increase a great deal in order to accommodate this vastly increased population (FAO et al., 2017). The reality is, however, that a number of environmental factors—including biotic and abiotic stressors—have a substantial impact on the availability of food and nutrition for this rapidly growing population of the globe (Elferink & Schierhorn, 2016). In order to accommodate an additional 2.3 billion people by 2050, productivity will need to improve by 70% (Vincent et al., 2016; Ali et al., 2019; Bhadouria et al., 2019; Mohammad et al., 2022). Animals' strategies for coping with and tolerating environmental stress are different and less complex than those of plants. For plant physiologists and biotechnologists, determining how plants react to diverse environmental pressures is one of the most crucial factors. Drought, salt, and extremely high temperatures are the most frequent and significant factors limiting agricultural productivity (Kashyap et al., 2021; Dos Santos et al., 2022). The increase in global temperature is due to the unobstructed growth in greenhouse gas emissions and their consequences due to extreme weather events and seasonal shifts.

Worldwide food security is in danger due to accelerating climate change affecting global supply and creating a problem for expanding populations. Climate change poses a significant threat to food security—especially high temperatures, which lead to lower yields of commercially important food crops and the proliferation of weeds and pests. Regardless, abiotic stress is perhaps the greatest challenge presently faced by the agriculture sector (Nelson et al., 2009; Bhadouria et al., 2019; Singh et al., 2020; Godde, 2021). As an abiotic stress, the high temperature tenaciously limits the choice of crops and agricultural production over large areas. Extreme weather events can incite total crop failure because of low seed germination rates, reduction in growth, and reduced photosynthesis, and ultimately lead to the death of seedlings (Ali et al., 2019; Singh et al., 2020). Almost every physiological and biochemical process is extremely sensitive to heat. However, different plants have different strategies to develop tolerances to unfavorable conditions, particularly against high temperature stress. Further, to survive such adverse environmental conditions, plants have evolved complex strategies to recognize extreme signals and stimulate optimal responses to them. These extremes including alteration in membrane permeability, delay in seed germination, leaves formation, alterations in photosynthesis rate and respiration rate, and reduction in growth can harm plant tissues and eventually cause plant death (Guilioni et al., 1997; Wahid et al., 2007; Hasanuzzaman et al., 2013).

Additionally, brassinosteroids (BRs) promote photosynthetic activities and also help to photosystem antennae molecules to absorb more light energy; plants used this additional absorbed energy to combat harmful impacts of biotic and abiotic stresses. On the other hand, this extra energy produce lots of reactive oxygen species (ROS), including singlet oxygen ($^1O_2$), hydroxyl radicals, superoxide ($O^{2-}$), $H_2O_2$ and $OH^-$ (Das & Roychoudhury, 2014; Sharma et al., 2012; Sachdev et al., 2021). These additionally produced ROS adversely affect the biomolecules and bio-membranes in plant cells and ultimately create problem for cellular efficiency. Metabolic efficiency of plant cells suffers as a result of structural and functional disruptions of organelles such as the nucleus, mitochondria, Golgi complex, and chloroplast. Furthermore, the decrease in photosynthesis caused by high-temperature stress is associated with the inactivation of numerous chloroplast and mitochondrial catalytic enzymes, primarily leading to oxidative stress (Sachdev et al., 2021).

Oxidative stress causes lipid peroxidation, which ultimately leads to membrane damage, enzyme inactivation, and protein degradation. Hence, the ability to maintain cell membrane integrity and reduce oxidative stress have been proposed as good indicators of thermo-tolerance in plants (Bano et al., 2021; Dos Santos et al., 2022). It has been found through numerous studies that plants have developed a wide spectrum of defense mechanisms to deal with the consequences of oxidative stress on morphological, physiological, biochemical, and molecular levels as a result of environmental changes (Sairam & Saxena, 2000; Barrera, 2012; Das & Roychoudhury, 2014). At the biochemical level, plants have developed both enzymatic and non-enzymatic defensive mechanisms to scavenge the detrimental effects of the ROS and protect cells from oxidative damage (Bano et al., 2021). A very few of the key enzymes involved in the process of catalysis include superoxide dismutase (SOD, EC 1.15.1.1), peroxidase (POD, EC 1.11.1.7), catalase (CAT, EC 1.11.1.6), ascorbate peroxidase (APOX, EC 1.11.1.11) monodehydroascorbate reductase (MDHAR, EC 1.6.5.4), dehydroascorbate reductase (DHAR, EC 1.8.5.1), glutathione reductase (GR, EC 1.6.4.2), glutathione peroxidase (GPOX, EC 1.15.1.1) and glutathione S-transferase (GST, EC 2.5.1.18); key non-enzymes include proline and sugars, ascorbate (AsA), phenolics, reduced glutathione (GSH), tocopherol, flavonoids, alkaloids, carotenoids, and anthocyanin (Blokhina et al., 2003; Apel & Hirt, 2004). Due to antioxidant protection, the process of thermal tolerance may be acquired, and the mechanisms underlying the synthesis of heat-shocked protein (HSP) may be improved. However, due to the intricacy of stress tolerance related response, traditional breeding methods have had limited success and necessitate effective evolution of methodologies in order to fulfil the rising global food demands (Kovtun et al., 2000; Vahala et al., 2003). New and effective approaches should therefore be formulated in this direction. Studies also suggest that the expression of antioxidant enzymes can be enhanced through the use of plant hormones (Saini et al., 2015). There are many plant hormones that have been implicated in modulating responses to various stressors, including ethylene (ET), abscisic acid (ABA), jasmonic acid (JA), salicylic acid (SA), and BRs. The engineering of phytohormones likely can be an effective way of producing climate-resilient plants with high yields that can tolerate extreme weather conditions (Metwally et al., 2003; Özdemir et al., 2004).

A phytohormone is an organic molecule that is produced by plants at very low concentrations in order to regulate a wide range of cellular processes. Despite the fact that they are produced in plants at low concentrations, they are capable of communicating cellular activity by circulating a message throughout the body (Weyers & Paterson, 2001; Spartz & Gray, 2008). Furthermore, they play an important role in the coordination of various signaling pathways during the response to abiotic and biotic stress. It has been reported that very few phytohormones are being identified as stress hormones in recent reports, including ABA, BRs, SAs, JAs, and PA (Bhardwaj et al., 2022). In plants, these phytohormones play imperative roles in a number of processes, including seed latency, growth regulation, inhibition of germination, stomatal closure, fruit shedding, defense against both biotic and abiotic stresses, and facilitating their response to them (Sabagh et al., 2021). Using phytohormones could be an excellent alternative strategy for farmers looking to improve the nutritional and economic value of their crops. In this chapter, the focus is on a comprehensive inspection of BRs and how they influence plant life and response to abiotic stress, as well as their metabolic engineering so that crops can tolerate abiotic stress in order to achieve a balance between food abundance and quality and the prospects for the future (Sharma et al., 2017).

BRs are an impressive class of plant hormones due to their multipurpose roles in plants, followed by auxin, gibberellins, cytokinins, ABA, and ET (Saini et al., 2015; Peres et al., 2019). In terms of structural similarity, they accelerate the process of embryogenesis and are comparable to an animal hormone. They are crucial in a variety of other areas of plant science, as well, including cell division, cell elongation, photo-morphogenesis, stomatal opening or closing, vascular tissue differentiation, seed germination, and most importantly, modulating metabolism under stress conditions by regulating the defense system (Nolan et al., 2020; Tripathi et al., 2021). BRs also have a role in regulating plant oxidation radical metabolism, ET production, and root gravitropic response, as well as modulating plant responses to stressors such as frost, drought, salt, disease, heat, and nutritional inadequacy. However, the most bioactive and commercially accessible BRs are 24-epiBL and 28-homoBL, which are commonly used in farming as well as in physiological and molecular research (Sirhindi et al., 2018). As far as BRs signaling is concerned, it has a similar mechanism in plants as well as in animals, such that it up-regulates thousands of genes involved in cell division and differentiation, resulting in a controlled overall process of plant development (Anwar et al., 2018).

## 21.2 DISCOVERY AND STRUCTURE OF BRASSINOSTEROIDS

The search for evidence that animal steroid hormones were present in plants intensified throughout the second half of the twentieth century as scientists around the globe vigorously pursued finding proof that such substances were present (Gustafson, 1937). There is no doubt that all the various animal steroid hormones are present in plants in various forms or as their analogues. Despite this report, the growth of stigmas was reported to be accelerated with the use of spore extracts of *Lycopodium* (Mitchell et al., 1970). Moreover, *Zea mays* L. pollen extract was studied by Mitchell and Whitehead (1941), who found that the ether extract increased intermodal elongation and length of cells rather than the number of cells. The bioactive extracts of immature bean seeds were studied in the following studies. In comparison to plants that had not been treated, a significant increase in length was noted in the second internode when applied to the first, second, third, fourth, and fifth internodes of bean plants. This increase was most pronounced in the second internode. In the 1950s, Mitchell (1950) used the BR bioassay as a way of detecting responses to BR from the beans' second internode, and this was known as the bean second internode bioassay. Marumo et al. (1968) were credited with initiating research on BRs in Japan when they isolated the active form of these compounds from the seeds of *Brassica napus* L. for the first time. As a result of this research, Mitchell et al. (1970) purified a lipophilic chemical from the pollen of *B. napus* L. that demonstrated hormonal activity

**FIGURE 21.1** Structures of some brassinosteroids isolated from different plants are depicted.

when the second internode experiment was done. There is a compound known as *Brassin* that was named after the plant from which it was isolated for the first time (Peres et al., 2019). Several extensive studies have demonstrated the widespread presence of black locusts (BLs) (Figure 21.1), as well as their unique physiological properties that affect plant growth and development, as well as their potential application in agriculture in the future. Several independent studies were conducted in 1996, and one of them involved the isolation of BR-insensitive and BR-deficient mutant *Arabidopsis thaliana* plants, establishing that BRs are a significant endogenous growth regulator (Clouse et al., 1996).

## 21.3 ROLE OF BRASSINOSTEROIDS IN GROWTH AND DEVELOPMENT

BRs are crucial to ensure that plants grow and develop properly, and they also assist the plant cells in making connections with the external environment (Halliday, 2004). Plant hormones interact with each other at the signaling level and activate the signal transduction pathway in coordination to achieve adequate plant growth. They also perform a critical role in almost all physiological and metabolic processes of plants and are accountable for achieving adequate growth and development under normal, as well as inadequate, environmental conditions (Kolachevskaya et al., 2021). At the cellular level, BRs encouraged elongation, enzyme activation and DNA/RNA (deoxyribonucleic acid/ribonucleic acid) synthesis, alleviated photosynthetic activities, and altered the mechanical properties of the cell wall and the cell's membrane permeability (Clouse, 2008). At the morphological level, BRs help in promoting growth and reproductive organ development, shortening vegetative growth, increasing crop yield, and improving fruit quality. As a physiological process, seed germination is controlled by a variety of plant hormones at the cellular and molecular levels in order to ensure its success. In the cases of *L. esculentum*, *O. sativa*, *T. aestivum*, *B. juncea*, *E. camaldulensis*, *L. sativus*, *A. hypogea*, and *O. minor*, BRs improved seed germination rate (Sharma & Bhardwaj, 2007; Kumar et al., 2014). BRs help seed germination, but in a very dose-dependent manner. In contrast to *Arabidopsis* mutant plants, BRs have few distinctive characteristics, but they can be distinguished by their short, vigorous stems; dark green, round leaves; and abnormal de-etiolation, resulting in short hypocotyls and long cotyledons in dark-grown seedlings. Contrary to that, exogenous BRs cause the mutant BRs phenotype to replicate rather than restore it. The application of exogenous BRs is different from the application of auxins or gibberellins (Park et al., 2017).

## 21.4 BRASSINOSTEROIDS IN REPRODUCTION AND FRUIT DEVELOPMENT

Fruit ripening is a distinct stage of plant life that directly affects food productivity, nutrition, and health. BRs have recently been implicated for their role in fruit ripening. BRs not only affected fruit ripening but also reduced physiological shedding of fruit (mainly in citrus, peaches, apples, pears, or persimmons) from trees (Batista-Silva et al., 2018; Maduwanthi & Marapana, 2019). They were successfully applied to the above-ground portion of a growing tree and reduced fruit drop. In addition to regulating elongation, division, and differentiation of cells at the cellular level, BRs can also have a significant impact on the behavior of whole plants. BRs also regulate hypocotyl elongation and leaf development, shoot and root development, male fertility, and senescence, as well as the responses to a variety of stresses (Clouse, 2011). The transition of the plant from the vegetative to the reproductive phase is controlled by both internal levels of metabolites and growth hormones as well as external environmental elements (Egamberdieva et al., 2017). The stimulatory role of BR on vegetative—as well as reproductive growth parameters has been studied by different researchers on different plant species (Li, Lei et al., 2020). It was found that BRs adversely affected the reproductive structure of the plants in a dynamic and antagonistic manner, and interestingly, there was a relationship between BR and flowering time regulation (Egamberdieva et al., 2017; Li, Wu et al., 2020).

According to reports, BRs play a significant role in flower production and fruit development, and they also regulate the quality and quantity of fruits as physiological and biochemical processes are coordinated during fruit ripening, including the accumulation of metabolites and chromo-pigments and the degradation of organic acids. They enhance the softness of fruit and the development of flavor and aroma during ripening (Soliman et al., 2020; Sabagh et al., 2021). It has also been reported that BR hormones are essential for growth and development, as well as for the ripening of non-menopausal fruits such as grapes. According to this study, grape berry ripening was associated with an increased concentration of endogenous BRs rather than indole-3-acetic acid (IAA) or gibberellic acid (GA) concentrations. As a result of changes in endogenous BR levels, they demonstrated that fruit ripening was negatively affected. As a result of the treatment with 24-epiBL, corn, canola, orange, wheat, tobacco, grape, and sugarbeet yields were significantly increased (Jediyi et al., 2019). According to the results of a large-scale field study, the application of 24-epiBL to wheat resulted in increased percentages of kernel set, the number of caryopsis per ear, and a greater weight per 1,000 kernels of wheat. The reduction in the number of sterile spikelets per year and the increased rate of photosynthesis were all associated with this increase in yields, the translocation of the photosynthesis to the years, and the increase in the rate of photosynthesis (Symons et al., 2006; Kurepin et al., 2012).

## 21.5 PHOTOSYNTHESIS, WATER RELATIONS, AND NUTRIENT UPTAKE

Brassinosteroid treatments in the form of foliar spray enhance the photo-quenching activities by altering the efficiency of PS-II, photochemical extinction coefficient, and quantum efficiency. Exogenous BRs treatment has significantly assisted the plants to survive under extreme conditions—including low temperature, drought, low light, and heavy metal concentration stress—by enhancing the photosynthetic activities in various plants (Yang et al., 2018; Castañeda-Murillo et al., 2022). Moreover, the role of BR is to protect chlorophyll pigments by enhancing the stability of the thylakoidal membranes by regulating the functions of the enzymes chlorophyllase enzyme activities. It also modulates the immune system by modifying the gene expression of defense-related genes, alleviating discomfort during various stresses like extreme heat and water deficit for wheat and cucumber plants (Szafrańska et al., 2017; Li et al., 2022). Additionally, BRs treatment helped the wheat seedlings grown under stressful conditions as it activates the defense enzymes, along with other physiological and metabolic activities (Mohsin et al., 2022). Seedlings treated with BRs also showed improvements in stomatal permeability, quantum yield, PS-II activities (specifically, RuBisCo activities), and carbon dioxide absorption. It also regulates the activities of respiratory burst oxidase homolog (RBOH), which increases the activity of alternative oxidase (AOX) (Dietz et al., 2016). High levels of AOX dissipate excess energy, thus regulating the flow of electrons in chloroplasts and mitochondria, reducing the accumulation of ROS and increasing the protection of photosynthetic mechanisms. BRs hae been reported to be coupled with external stimuli to modulate stomatal opening, an important component of stress due to salinity and dehydration (Xu et al., 2011). According to Haubrick et al. (2006), it is essential that potassium modifying channels (K+) are blocked or occluded at the stoma, which inhibits epidermal desquamation, and that protecting cells' ability to absorb K+ also reduces its opening. In addition to modulating the expression of TF MAPK (MAP kinase) and YODA (YDA), BRs also contribute to dryness and salt stress in the mouth. BRs decrease toxic ion accumulation, increase absorption of essential inorganic ions, and improve cell homeostasis, especially in the epidermis, roots, and leaves of rapeseed during salt stress. An exogenous treatment of BRs increased nitrogen metabolism in tomato seedlings, as well as glutamate dehydrogenase, glutamate synthetase, and glutamine synthetase activity. BR supplementation also increases the activity of $Ca^{2+}$ ATPase and $H^+$ ATPase in the leaves and roots, which helps to modulate the electrochemical gradient and maintain ionic balance in order to overcome the disadvantages of stress conditions (Shu et al., 2016).

## 21.6 SYNERGISTIC ROLE OF BRASSINOSTEROIDS METABOLISM AND SIGNALING

It has been demonstrated that a wide range of phytohormones—including BR, GA, JA, SA, and ABA—regulate the expression of genes associated with stress in plants grown in unfavorable environments (Peres et al., 2019). These hormones and their interactions are also important for various physiological responses, including plant development, seed germination, and responses associated with defense in plants grown under unfavorable conditions, as shown in Figure 21.2 (Li, Wu et al., 2020). In recent molecular studies, it has been demonstrated that BRs have been observed to interact with other phytohormones like auxins, GA, ABA, ET, and JA in order to regulate the various physiological activities of plants and to provide resistance to a variety of stresses. It also helps to regulate the expression of several genes in the *Arabidopsis* genome by activating a signaling pathway, as studies indicate that BRs interact with important receptors like BIN2, BZR1, and BES1 to activate this vast signaling cascade in *Arabidopsis* (Li, Lei et al., 2020). There is 88% similarity at the protein level and 97% similarity at the DNA regulatory domain level in these receptors. It is believed that the majority of variations in these receptors contain a DNA binding domain (DBD), a phosphorylation domain containing 22 putative phosphorylation sites for BIN2 protein, and a PEST motif (rich in Pro, Glu, Ser, and Thr) (Jiang et al., 2015). They promote protein degradation, and a 14–3–3 binding motif that interacts with the 14–3 infinite distribution of BES1 and BZR1. In order to interact with BIN2, BZR1 and BES1 are believed to have a unique normalized C-terminal region that is unique to both proteins (Cao et al., 2020). It has been found that experiments (BZR1-1D and BS1-D), which aimed to identify the communities of BZR1, BZR2/BS1, and other proteins, have revealed inaccurate protein communities. These communities exhibit diversity in their

# Brassinosteroid Hormones

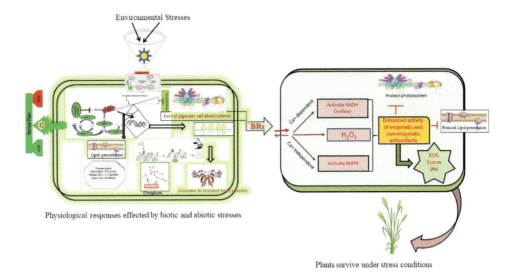

**FIGURE 21.2** An illustration of the negative effects of abiotic factors on cells and the mechanism by which BRs reduce those effects is provided.

associations with various biochemical factors. (Sun et al., 2010; Gudesblat & Russinova, 2011; Yu et al., 2011). Secure chromatin (CIP) with *Arabidopsis* tiles and component investigations have shown that both BZR1 and BS1 have unique DNA control features. Therefore, you should be looking for a quality enhancer and keeping an eye on the different qualities that are compatible with the dimerization partner, as well. A positive feedback loop occurs during the initiation of PR signaling when PES1 and PZR1 bind to their own promoter sequences in order to direct their behavior. BES1 and BZR1 are both transcription factors that act as regulators of BR-responsive functions, as they recruit or regulate transcription factors such as AtIWS, BIM, and GATA that are responsible for controlling TFs (Sun et al., 2010). BR-activated transcription factors cooperated to bind to specific E groups (These E groups include ERF1, ERF2, ERF3, ERF4, ERF5, and ERF6 and activation of these E groups by BRs regulates various physiological processes, including growth, development, and stress responses in plants) and induce high-quality BR-activated expression elongation, cargo RNA and histone modifications, and BES1 can indeed interact with each other (Yin et al., 2005; Guo et al., 2013). BES1 adds IWS1 to the target qualities to improve quality expression during registration renewal. REF6 and its homolog ELF6 (early flowering 6), which act directly on histone demethylases, regulate a variety of physiological cycles, including flowering. Space group complex 8 (SDG8), a histone lysine methyltransferase involved in the di- and trimethylation of histone H3 Lys-36, can be recruited by BES1. The control of some BR targets is also affected by epigenetic processes (Xu et al., 2014; Wang et al., 2014). A transcriptional repressor module formed by BES1, TPL, TPR, and HDAC19 was recently found to be associated with BES1.

## 21.7 ABIOTIC STRESSES: CHALLENGES IN A CHANGING WORLD

In order to increase the productivity of plants, it is essential to understand how environmental extremes disrupt plant life, as well as how these disruptions take place at the physiological, biochemical, and molecular levels. The root cause of widespread crop losses in the world is stress, so one of the main research priorities should be identifying how these disruptions occur (Arora et al., 2012; Manghwar et al., 2022). There are a number of environmental factors to which plants are exposed throughout the course of their lives, all of which have different impacts on them at different stages as well as in different ways. Despite the fact that some plants may become affected at an early stage, there is a chance that they will recover and ultimately survive the disease if they are not exposed at an early stage. There is a wide range of variation when it comes to the susceptibility of plants to stress or their ability to resist it, depending on their species or genotypes, due to these difficult environmental conditions and the difficult economic conditions in these difficult times. Environmental factors are indeed one of the most important factors that contribute to the decline of global agricultural production (Apel & Hirt, 2004) and phytohormones work on plants in more than just one way (Fahad et al., 2015). For example, it affects the way plants grow; the integrity of the membrane, pigment content, and the process of photosynthesis; and its many other effects on plants. Among the stress factors that reduce crop productivity, salinity has been identified as the second most significant factor after drought, according to Wani et al. (2020). There has been significant progress in selecting particular hormones specific to various types of stress—as well as specific stress combinations—in the past, and the plant has been shown to be protected against abiotic stress by inducing defense-related mechanisms. It has been shown that some hormones can perform more than one stress-resistance function (Noctor & Foyer, 1998).

## 21.8 BRASSINOSTEROIDS IN ABIOTIC STRESS MANAGEMENT

The ability of plant hormones to deal with both biotic and abiotic stresses makes them capable of dealing with a broad range of difficulties, since plant hormones are not only capable of regulating plant growth, but can also confer resistance against a wide range of environmental stressors (Wani et al., 2016). It is evident that BRs can tolerate a wide spectrum of stress conditions, including thermal stress, drought stress, heavy metal stress, salt stress, infections, pesticides, and even viruses, etc. Antioxidant enzymes in the plant cell are regulated by BRs in order to prevent the effects of oxidative stress (Verma & Kumar, 2016). BRs also regulate the level of lipid peroxidation to make sure that plants are resistant to oxidative stress (Kumar et al., 2014). There is no doubt that one of the major consequences of these environmental stresses is the production of excessive amounts of ROS which leads the oxidative damage inside the plant cells (Noctor & Foyer, 1998). To achieve tolerance, there should be equilibrium between the production and elimination of these ROS, as they are primarily produced in the cell organelles in plant cells such as chloroplasts, mitochondria, and peroxisomes. Further, many stress factors interrupt the equilibrium between ROS production and removal in the said organelles, leading to the accumulation of these ROS. Moreover, this elevated amount of ROS initiates lipid peroxidation and leads to protein degradation, nucleic acid degradation, and lipid peroxidation (Noctor & Foyer, 1998). Ultimately, these ROS could lead to plant cell death by enhancing the expression of ROS-dependent and cell death-related genes as given in Figure 21.2. The ROS generated are removed by an array of enzymatic and non-enzymatic antioxidant defense systems (Asif et al., 2022).

BRs also regulated the contents of ascorbate, reduced glutathione (GSH), tocopherol, proline, glycine betaine flavonoids, sugar alkaloids, and carotenoids which are components of a non-enzymatic antioxidant system Kumar et al., 2014; Sirhindi et al., 2018). Tocopherol and ascorbate are effective antioxidants due to their ability to scavenge and remove singlet oxygen. Through the removal of surplus excitation energy and quenching of singlet oxygen, carotenoids shield chlorophyll (Vanacker et al., 2000). Some of the antioxidant enzymes included in the enzymatic defense system are super oxide dismutase (SOD), catalase (CAT), ascorbate peroxidase (APOX), peroxidase (POD), glutathione reductase (GR), monodehydroascorbate reductase (MDHAR), and dehydroascorbate reductase (DHAR). It is important to understand that the first line of defense against ROS is SOD enzyme, which converts ROS into hydrogen peroxide (Yao et al., 2001). APOX and CAT in chloroplasts, and PODs in the cytoplasm, detoxify the hydrogen peroxide in water and molecular oxygen (Fariduddin et al., 2019). Additionally, separate ascorbate and glutathione cycles in chloroplasts and mitochondria also convert hydrogen peroxide into water. Hydrogen peroxide is oxidized by APOX and converted into water and molecular oxygen. The regenerating reaction can then be catalyzed by MDHAR and NADPH, which act as reducing powers. MDHAR and DHAR can spontaneously regenerate, then ascorbate regeneration is mediated by DHAR, which is served by the oxidation of reduced glutathione (GSH) to oxidized glutathione (GSSG). Finally, GR converts GSSG into GSH by using NADPH as a reducing agent. The amount of oxidative stress in a cell is determined by the amounts of $O_2$, $H_2O_2$, and $OH^-$ concentrations present. Antioxidant enzyme activity needs to be balanced in order to mitigate the deleterious effect of these alleviated ROS (Asada & Takahashi, 1987).

## 21.9 BRASSINOSTEROIDS IN BIOTIC STRESS MANAGEMENT

Krishna (2003) reviewed the role of BRs in developing plant resistance to bacterial and viral pathogens, also identifying a new foreign disease. Further, neither acidic nor basic pathogenesis-related (PR) genes were expressed by BR treatments, suggesting that resistance induced by BR differs from disease resistance induced by systemic acquired resistance (SAR) and wounds. As a crucial component of tobacco defense mechanisms, steroid hormones contributed significantly to disease resistance (Hao et al., 2018). Plant defense responses are influenced by other significant BR signaling components. Similarly, in another study conducted by Zhang et al. (2018), it was found that BRs can enhance plant defenses against a variety of pathogens. In addition, plants that were treated with BRs were more resistant to pathogens than plants that had not been treated (Yi et al., 2009). Barley and cucumber plants were also resistant to disease when treated with BR (Elsharkawy et al., 2012). Additionally, increased activity of defense enzymes including PODs and polyphenol oxidases (PPOs) are associated with BR-induced disease resistance genes in cucumber (Ali et al., 2013). However, Hayat et al. (2012) reported that BRs involved in growth development and metabolism collaborate to stimulate defense responses in tissues exposed to biotic stress, but in a dose-dependent manner. High concentrations of BRs induce programmed cell death (PCD) through BR-associated kinase-1 (BAK-1) and regulate basal defense. It also imparted the role in embryogenesis and increase the expression of proteins BAK-1 and SERK-3 (somatic embryogenesis-related kinase-3) in embryonic stem cells, as per the three independent research groups (Chinchilla et al., 2007).

## 21.10 CONCLUSIONS

This chapter indicates that brassinosteroids (BRs) contribute a significant role in controlling physiological, biochemical, and molecular activities of plants to mitigate the deleterious effects of extreme environmental conditions. Therefore, the most important assignment for plant scientists is to develop an accurate model that elucidates the

# TABLE 21.1
## Reports Showing the Diverse Roles of BRs in Various Plants under Various Abiotic Stresses

| Sr No. | Name of Plants | Mode of Application | Stress Condition | BR Concentration | Impact on Physiological Response | References |
|---|---|---|---|---|---|---|
| 1. | Brassica Juncea L. | Priming and foliar treatment | High temperature | 24-epiBL and 28-homoBL ($10^{-6}$, $10^{-8}$ and $10^{-10}$ M) | Batter germination, Chlorophyll a and b, and carotenoids content, enhanced activity of antioxidant enzymes like SOD, POD CAT, DHAR and MDHAR | Kumar et al. (2012, 2014) |
| 2. | Brassica rapa L. | Foliar treatment | Optimum temperature conditions 28–30°C | BRs (200 nM), | Improved the antioxidant enzyme activity and physiological response in plants | Lee et al. (2020) |
| 3. | Zea mays L. | Priming and foliar | 8.0 dS m$^{-1}$ NaCl | 2 μM, 24-epiBL | Reduced Na$^+$ and Cl$^-$ concentrations but increased nitrogen (N), calcium (Ca$^{2+}$), K$^+$, and phosphorus (P). | Kaya et al. (2018) |
| 4. | Oryza sativa L. | Priming | Salinity | 0.5 μM, 1 μM and 3 μM 24-epiBL and 28-homoBL | Enhanced seed germination and batter seedlings growth | Anuradha and Rao et al. (2001) |
| 5. | Triticum aestivum L. | Priming and foliar | Late sowing | 1 μM, 24-epiBL | Higher total chlorophyll content, nitrate reductase, and SOD, CAT, and POD activity | Kumari and Hemantaranjan (2019) |
| 6. | Solanum lycopersicum L. | Foliar | 100 mM NaCl | 1 μM, 24-epiBL | Enhanced growth, photosynthesis, homeostasis, ascorbic acid and reduced glutathione, and the secondary metabolites assisting the enzymatic antioxidant system | Ahanger et al. (2020) |
| 7. | Cucumis sativus L. | Priming | Cd, 2.5 mM | 5 μM, 24-epiBL | Photosynthetic rate, CAT, SOD, APOX and POD activity increased, while H$_2$O$_2$, MDA and ELP decreased | Shah et al. (2019) |
| 8. | Festuca arundinacea Schreb. | Foliar | Pb, 1000 mg kg$^{-1}$ | 0.2 mg L$^{-1}$, 24-epiBL | Total chlorophyll, carotenoids, TSP, free proline, SOD, and CAT were increased while decreased the contents of APX, H$_2$O$_2$ and MDA | Zhong et al. (2021) |
| 9. | Zea mays L. | Foliar | 4°C | 1 mg L$^{-1}$, brassinolide | Increased the germination rate and improved antioxidants, and reduced reactive | Sun et al. (2020) |
| 10. | Oryza sativa L. | Foliar | ≥35°C | 1 μmol L$^{-1}$ 24-EBL | Ameliorate pistil fertilization ability, activate the defense system, enhancing osmotic regulation, protecting photosynthesis | Yang et al. (2021) |

*(Continued)*

## TABLE 21.1
## Continued

| Sr No. | Name of Plants | Mode of Application | Stress Condition | BR Concentration | Impact on Physiological Response | References |
|---|---|---|---|---|---|---|
| 11. | Oryza sativa L. | Priming and foliar | Salinity: NaCl 200 mM | BR (0.1, and 1 µM) | Overexpression of SERK2, enhanced grain size, and salt resistance | Dong et al. (2020) |
| 12. | Raphanus sativus | Priming and foliar | Cu stress | 24-epiEBL | Modulated the expression of genes, impacting the metabolism of IAA and ABA, resulting in enhanced Cu stress tolerance | Choudhary et al. (2012) |
| 13. | Brassica juncea L | Priming and foliar | Pb stress | 24-epiBL($10^{10}$, $10^{-8}$, and $10^{-6}$ M) | The activity of the enzymes superoxide dismutase, catalase, peroxidase, and ascorbate peroxidase were enhanced in the seedlings treated with BRs | Soares et al. (2020) |
| 14 | Brassica juncea L | Priming and foliar | Salt and high temperature stress | 28-homoBL($10^{-10}$, $10^{-8}$, and $10^{-6}$ M) | Activities of SOD, CAT, APOX, GR, DHAR, and MDHAR increased in response to 28-homoBL | Kaur et al. (2018) |
| 15 | Brassica oleracea | Priming | | 24-epiBL and 28-homoBL ($10^{-6}$, $10^{-9}$, and $10^{-12}$ M) | Seedling growth, chlorophyll a and b, carotenoids, and antioxidant enzyme activity were influenced by BRs | Nirmal et al. (2016) |

precise pathway of the antioxidant defense system for abiotic stress tolerance by plants. Although several concepts and theories have indeed been put forth by many authors, it is still unclear what function BRs serves in the ability to withstand abiotic stress. However, evidences shows that BRs increase the host's resistance to diseases and reduces abiotic and biotic stresses, protecting the plants—particularly when the plant is under abiotic stress. Therefore, the use of BRs in crop production may ensure safe food production, as well as the protection of the environment. In addition, there remains a lot more to learn about the biosynthesis of BRs and the possible mechanisms by which it is transported in the plants. Future research should look for evidence that BRs are transferred out of cells and find BR transporters. Finally, we need to learn more about how BRs contribute to the overall developmental program of plants in specific situations such as shade, high temperatures, and drought. Long-term goals should be to modify the BR pathway in crops and other plants so that plants can cope with the stress caused by greater changes in various environmental conditions.

## REFERENCES

Ahanger, Mohammad Abass, Rayees Ahmad Mir, Mohammed Nasser Alyemeni, and Parvaiz Ahmad. "Combined effects of brassinosteroid and kinetin mitigates salinity stress in tomato through the modulation of antioxidant and osmolyte metabolism." *Plant Physiology and Biochemistry* 147 (2020): 31–42.

Ali, Raza, Ali Razzaq, Sundas Saher Mehmood, Xiling Zou, Xuekun Zhang, Yan Lv, and Jinsong Xu. "Impact of climate change on crops adaptation and strategies to tackle its outcome: A review." *Plants* 8, no. 2 (2019): 34.

Anuradha, S., and S. Seeta Ram Rao. "Effect of brassinosteroids on salinity stress induced inhibition of seed germination and seedling growth of rice (Oryza sativa L.)." *Plant Growth Regulation* 33, no. 2 (2001): 151–153.

Anwar, Ali, Yumei Liu, Rongrong Dong, Longqiang Bai, Xianchang Yu, and Yansu Li. "The physiological and molecular mechanism of brassinosteroid in response to stress: A review." *Biological Research* 51 (2018).

Apel, Klaus, and Heribert Hirt. "Reactive oxygen species: Metabolism, oxidative stress, and signaling transduction." *Annual Review of Plant Biology* 55 (2004): 373.

Arora, Priya, Renu Bhardwaj, and Mukesh Kumar Kanwar. "Effect of 24-epibrassinolide on growth, protein content and antioxidative defense system of *Brassica juncea* L. Subjected to cobalt ion toxicity." *Acta Physiologiae Plantarum* 34, no. 5 (2012): 2007–2017.

Asada, K., and M. Takahashi. "Production and scavenging of active oxygen in photosynthesis." In Kyle, D. J., Osmond, C. B., Arntzen, C. J., eds., *Photoinhibition* (pp. 227–287). Elsevier, Amsterdam, 1987.

Asif, Muhammad, Ayesha Aziz, Ghazala Ashraf, Tayyaba Iftikhar, Yimin Sun, Fei Xiao, and Hongfang Liu. "Unveiling microbiologically influenced corrosion engineering to transfigure damages into benefits: A textile sensor for $H_2O_2$ detection in clinical cancer tissues." *Chemical Engineering Journal* 427 (2022): 131398.

Bano, Ambreen, Anmol Gupta, Smita Rai, Touseef Fatima, Swati Sharma, and Neelam Pathak. "Mechanistic role of reactive oxygen species and its regulation via the antioxidant system under environmental stress." *Plant Stress Physiology—Perspectives in Agriculture* (2021): 1–18.

Barrera, Giuseppina. "Oxidative stress and lipid peroxidation products in cancer progression and therapy." *International Scholarly Research Notices* 2012 (2012).

Batista-Silva, Willian, Vitor L. Nascimento, David B. Medeiros, Adriano Nunes-Nesi, Dimas M. Ribeiro, Agustín Zsögön, and Wagner L. Araújo. "Modifications in organic acid profiles during fruit development and ripening: Correlation or causation?." *Frontiers in Plant Science* 9 (2018): 1689.

Bhadouria, R., R. Singh, V. K. Singh, A. Borthakur, A. Ahamad, G. Kumar, and P. Singh. "Agriculture in the era of climate change: Consequences and effects." In *Climate Change and Agricultural Ecosystems* (pp. 1–23). Woodhead Publishing, Sawston, 2019.

Bhardwaj, Savita, Dhriti Sharma, Simranjeet Singh, Praveen C. Ramamurthy, Tunisha Verma, Mamta Pujari, Joginder Singh, Dhriti Kapoor, and Ram Prasad. "Physiological and molecular insights into the role of silicon in improving plant performance under abiotic stresses." *Plant and Soil* (2022): 1–19.

Blokhina, Olga, Eija Virolainen, and Kurt V. Fagerstedt. "Antioxidants, oxidative damage and oxygen deprivation stress: A review." *Annals of Botany* 91, no. 2 (2003): 179–194.

Cao, Yunpeng, Dandan Meng, Xiaoxu Li, Lihu Wang, Yongping Cai, and Lan Jiang. "A Chinese white pear (*Pyrus bretschneideri*) BZR gene PbBZR1 act as a transcriptional repressor of lignin biosynthetic genes in fruits." *Frontiers in Plant Science* 11 (2020): 1087.

Castañeda-Murillo, Cristian Camilo, Javier Gustavo Rojas-Ortiz, Alefsi David Sánchez-Reinoso, Cristhian Camilo Chávez-Arias, and Hermann Restrepo-Díaz. "Foliar brassinosteroid analogue (DI-31) sprays increase drought tolerance by improving plant growth and photosynthetic efficiency in lulo plants." *Heliyon* 8, no. 2 (2022): e08977.

Chinchilla, Delphine, Cyril Zipfel, Silke Robatzek, Birgit Kemmerling, Thorsten Nürnberger, Jonathan DG Jones, Georg Felix, and Thomas Boller. "A flagellin-induced complex of the receptor FLS2 and BAK1 initiates plant defence." *Nature* 448, no. 7152 (2007): 497–500.

Choudhary, S. P., Oral, H. V., Bhardwaj, R., Yu, J. Q., Tran, L. S. P. "Interaction of brassinosteroids and polyamines enhances copper stress tolerance in *Raphanus sativus*." *Journal of Experimental Botany* 63, no. 15 (2012): 5659–5675.

Clouse, Steven D. "Brassinosteroids." *The Arabidopsis Book/American Society of Plant Biologists* 9 (2011).

Clouse, Steven D. "The molecular intersection of brassinosteroid-regulated growth and flowering in *Arabidopsis*." *Proceedings of the National Academy of Sciences* 105, no. 21 (2008): 7345–7346.

Clouse, Steven D., Mark Langford, and Trevor C. McMorris. "A brassinosteroid-insensitive mutant in Arabidopsis thaliana exhibits multiple defects in growth and development." *Plant Physiology* 111, no. 3 (1996): 671–678.

Das, Kaushik, and Aryadeep Roychoudhury. "Reactive oxygen species (ROS) and response of antioxidants as ROS-scavengers during environmental stress in plants." *Frontiers in Environmental Science* 2 (2014): 53.

Dietz, Karl-Josef, Ron Mittler, and Graham Noctor. "Recent progress in understanding the role of reactive oxygen species in plant cell signaling." *Plant Physiology* 171, no. 3 (2016): 1535–1539.

Dong, Nana, Wenchao Yin, Dapu Liu, Xiaoxing Zhang, Zhikun Yu, Wei Huang, Jihong Liu et al. "Regulation of brassinosteroid signaling and salt resistance by SERK2 and potential utilization for crop improvement in rice." *Frontiers in Plant Science* 11 (2020): 621859.

Dos Santos, Tiago Benedito, Alessandra Ferreira Ribas, Silvia Graciele Hülse de Souza, Ilara Gabriela Frasson Budzinski, and Douglas Silva Domingues. "Physiological responses to drought, salinity, and heat stress in plants: A review." *Stresses* 2, no. 1 (2022): 113–135.

Egamberdieva, Dilfuza, Stephan J. Wirth, Abdulaziz A. Alqarawi, Elsayed F. Abd_Allah, and Abeer Hashem. "Phytohormones and beneficial microbes: Essential components for plants to balance stress and fitness." *Frontiers in Microbiology* 8 (2017): 2104.

Elferink, Maarten, and Florian Schierhorn. "Global demand for food is rising. Can we meet it." *Harvard Business Review* 7, no. 4 (2016): 2016.

Elsharkawy, Mohsen Mohamed, Masafumi Shimizu, Hideki Takahashi, and Mitsuro Hyakumachi. "The plant growth-promoting fungus Fusarium equiseti and the arbuscular mycorrhizal fungus Glomus mosseae induce systemic resistance against Cucumber mosaic virus in cucumber plants." *Plant and Soil* 361, no. 1 (2012): 397–409.

Fahad, Shah, Lixiao Nie, Yutiao Chen, Chao Wu, Dongliang Xiong, Shah Saud, Liu Hongyan, Kehui Cui, and Jianliang Huang. "Crop plant hormones and environmental stress." *Sustainable Agriculture Reviews* (2015): 371–400.

FAO, World Health Organization, and WHO Expert Committee on Food Additives. *Evaluation of Certain Contaminants in Food: Eighty-third Report of the Joint FAO/WHO Expert Committee on Food Additives.* World Health Organization, Geneva, 2017.

Godde, C. M., Daniel Mason-D'Croz, D. E. Mayberry, Philip K. Thornton, and M. Herrero. "Impacts of climate change on the livestock food supply chain; A review of the evidence." *Global Food Security* 28 (2021): 100488.

Gudesblat, Gustavo E., and Eugenia Russinova. "Plants grow on brassinosteroids." *Current Opinion in Plant Biology* 14, no. 5 (2011): 530–537.

Guilioni, Lydie, Jacques Wery, and Francois Tardieu. "Heat stress-induced abortion of buds and flowers in pea: Is sensitivity linked to organ age or to relations between reproductive organs?" *Annals of Botany* 80, no. 2 (1997): 159–168.

Guo, Hongqing, Lei Li, Maneesha Aluru, Sriniva Aluru, and Yanhai Yin. "Mechanisms and networks for brassinosteroid regulated gene expression." *Current Opinion in Plant Biology* 16, no. 5 (2013): 545–553.

Gustafson, Felix G. "Parthenocarpy induced by pollen extracts." *American Journal of Botany* (1937): 102–107.v

Halliday, Karen J. "Plant hormones: The interplay of brassinosteroids and auxin." *Current Biology* 14, no. 23 (2004): R1008-R1010.

Hao, Yi, Wen Yuan, Chuanxin Ma, Jason C. White, Zetian Zhang, Muhammad Adeel, Tao Zhou, Yukui Rui, and Baoshan Xing. "Engineered nanomaterials suppress Turnip mosaic virus infection in tobacco (Nicotiana benthamiana)." *Environmental Science: Nano* 5, no. 7 (2018): 1685–1693.

Hasanuzzaman, Mirza, Kamrun Nahar, Md Mahabub Alam, Rajib Roychowdhury, and Masayuki Fujita. "Physiological, biochemical, and molecular mechanisms of heat stress tolerance in plants." *International Journal of Molecular Sciences* 14, no. 5 (2013): 9643–9684.

Haubrick, Laura Lillian, Gro Torsethaugen, and Sarah M. Assmann. "Effect of brassinolide, alone and in concert with abscisic acid, on control of stomatal aperture and potassium currents of *Vicia faba* guard cell protoplasts." *Physiologia Plantarum* 128, no. 1 (2006): 134–143.

Hayat, Shamsul, Mohammed Nasser Alyemeni, and Syed Aiman Hasan. "Foliar spray of brassinosteroid enhances yield and quality of *Solanum lycopersicum* under cadmium stress." *Saudi Journal of Biological Sciences* 19, no. 3 (2012): 325–335.

Jediyi, Hicham, Khalid Naamani, Abderrahim Ait Elkoch, and Naima Lemjiber. "Changes in grapes composition during ripening of five *Vitis vinifera* L varieties as related to *Tephritidae* and *Drosophilidae* infestations." *Physiology and Molecular Biology of Plants* 25, no. 6 (2019): 1407–1418.

Jiang, Jianjun, Chi Zhang, and Xuelu Wang. "A recently evolved isoform of the transcription factor BES1 promotes brassinosteroid signaling and development in Arabidopsis thaliana." *The Plant Cell* 27, no. 2 (2015): 361–374.

Kashyap, Vikrant Hari, Isha Kohli, Abhinav Singh, Aishi Bhattacharya, Prashant Kumar Singh, Ajit Varma, and Naveen Chandra Joshi. "Physiological, biochemical, and morphological approaches to mitigate the effects of abiotic stress in plants." In *Stress Tolerance in Horticultural Crops* (pp. 193–212). Woodhead Publishing, Sawston, 2021.

Kaur, Harpreet, Geetika Sirhindi, Renu Bhardwaj, M. N. Alyemeni, Kadambot HM Siddique, and Parvaiz Ahmad. "28-homobrassinolide regulates antioxidant enzyme activities and gene expression in response to salt-and temperature-induced oxidative stress in Brassica juncea." *Scientific Reports* 8, no. 1 (2018): 1–13.

Kaya, Cengiz, Salih Aydemir, Nudrat Aisha Akram, and Muhammad Ashraf. "Epibrassinolide application regulates some key physio-biochemical attributes as well as oxidative defense system in maize plants grown under saline stress." *Journal of Plant Growth Regulation* 37, no. 4 (2018): 1244–1257.

Kolachevskaya, Oksana O., Yulia A. Myakushina, Irina A. Getman, Sergey N. Lomin, Igor V. Deyneko, Svetlana V. Deigraf, and Georgy A. Romanov. "Hormonal regulation and crosstalk of auxin/cytokinin signaling pathways in potatoes in vitro and in relation to vegetation or tuberization stages." *International Journal of Molecular Sciences* 22, no. 15 (2021): 8207.

Kovtun, Yelena, Wan-Ling Chiu, Guillaume Tena, and Jen Sheen. "Functional analysis of oxidative stress-activated mitogen-activated protein kinase cascade in plants." *Proceedings of the National Academy of Sciences* 97, no. 6 (2000): 2940–2945.

Krishna, Priti. "Brassinosteroid-mediated stress responses." *Journal of Plant Growth Regulation* 22, no. 4 (2003): 289–297.

Kumar, Sandeep, Sirhindi Geetika, Bhardwaj Renu, and Kumar Manish. "Brassinosteroids denigrate the seasonal stress through antioxidant defense system in seedlings of Brassica juncea L." *Journal of Stress Physiology & Biochemistry* 10, no. 2 (2014): 74–83.

Kumar, Sandeep, Sirhindi Geetika, Bhardwaj Renu, Kumar Manish, and Arora Priya. "Role of 24-epibrassinolide in amelioration of high temperature stress through antioxidant defense system in *Brassica juncea* L." *Plant Stress* 6, no. 1 (2012): 55–58.

Kumari, Asha, and A. Hemantaranjan. "Brassinosteroids: The master regulators." *Plant Abiotic Stresses Physiological Mechanisms Tools and Regulation* (2019): 273.

Kurepin, Leonid V., Se-Hwan Joo, Seong-Ki Kim, Richard P. Pharis, and Thomas G. Back. "Interaction of brassinosteroids with light quality and plant hormones in regulating shoot growth of young sunflower and Arabidopsis seedlings." *Journal of Plant Growth Regulation* 31, no. 2 (2012): 156–164.

Lee, Hong Gil, Jin Hoon Won, Yee-Ram Choi, Kyounghee Lee, and Pil Joon Seo. "Brassinosteroids regulate circadian oscillation via the BES1/TPL-CCA1/LHY module in Arabidopsis thaliana." *Iscience* 23, no. 9 (2020): 101528.

Li, Bei-Bei, Yu-Shi Fu, Xiao-Xia Li, Hai-Ning Yin, and Zhu-Mei Xi. "Brassinosteroids alleviate cadmium phytotoxicity by minimizing oxidative stress in grape seedlings: Toward regulating the ascorbate-glutathione cycle." *Scientia Horticulturae* 299 (2022): 111002.

Li, Taotao, Wei Lei, Ruiyuan He, Xiaoya Tang, Jifu Han, Lijuan Zou, Yanhai Yin, Honghui Lin, and Dawei Zhang. "Brassinosteroids regulate root meristem development by mediating BIN2-UPB1 module in *Arabidopsis*." PLoS Genetics 16, no. 7 (2020): e1008883.

Li, Xi, Pingfan Wu, Ying Lu, Shaoying Guo, Zhuojun Zhong, Rongxin Shen, and Qingjun Xie. "Synergistic interaction of phytohormones in determining leaf angle in crops." *International Journal of Molecular Sciences* 21, no. 14 (2020): 5052.

Maduwanthi, S. D. T., and R. A. U. J. Marapana. "Induced ripening agents and their effect on fruit quality of banana." *International Journal of Food Science* 2019 (2019).

Manghwar, Hakim, Amjad Hussain, Qurban Ali, and Fen Liu. "Brassinosteroids (BRs) role in plant development and coping with different stresses." *International Journal of Molecular Sciences* 23, no. 3 (2022): 1012.

Marumo, Singo, Hiroyuki Hattori, Hiroshi Abe, and Katsura Munakata. "Isolation of 4-chloroindolyl-3-acetic acid from immature seeds of *Pisum sativum*." Nature 219, no. 5157 (1968): 959–960.

Metwally, Ashraf, Iris Finkemeier, Manfred Georgi, and Karl-Josef Dietz. "Salicylic acid alleviates the cadmium toxicity in barley seedlings." *Plant Physiology* 132, no. 1 (2003): 272–281.

Mitchell, Herschel K. "Vitamins and metabolism in Neurospora." In *Vitamins & Hormones* (vol. 8, pp. 127–150). Academic Press, Cambridge, 1950.

Mitchell, John W., N. Mandava, J. F. Worley, J. R. Plimmer, and M. V. Smith. "Brassins—A new family of plant hormones from rape pollen." *Nature* 225, no. 5237 (1970): 1065–1066.

Mitchell, John W., and Muriel R. Whitehead. "Responses of vegetative parts of plants following application of extract of pollen from Zea mays." *Botanical Gazette* 102, no. 4 (1941): 770–791.

Mohammad, Mustafa, Zita Szalai, Anna Divéky-Ertsey, Izóra Gál, and László Csambalik. "Conceptualizing Multiple Stressors and Their Consequences in Agroforestry Systems." *Stresses* 2, no. 3 (2022): 242–255.

Mohsin, Sayed Mohammad, Jannatul Fardus, Atsushi Nagata, Nobuhisa Tamano, Hirofumi Mitani, and Masayuki Fujita. "Comparative Study of Trehalose and Trehalose 6-Phosphate to Improve Antioxidant Defense Mechanisms in Wheat and Mustard Seedlings under Salt and Water Deficit Stresses." *Stresses* 2, no. 3 (2022): 336–354.

Nelson, Erik, Guillermo Mendoza, James Regetz, Stephen Polasky, Heather Tallis, DRichard Cameron, Kai MA Chan et al. "Modeling multiple ecosystem services, biodiversity conservation, commodity production, and tradeoffs at landscape scales." *Frontiers in Ecology and the Environment* 7, no. 1 (2009): 4–11.

Nirmal Spall Kaur, Sirhnidi Geetika, and Kumar Sandeep. "Comparative influence of Brassinosteroids correspondents (24-Epibl and 28-Homobl) on the morpho-physiological constraints of brassica oleracea (Cabbage, Cauliflower and Broccoli)." *Biochem Physiol* 5 (2016): 193. doi:10.4172/2168-9652.1000193

Noctor, Graham, and Christine H. Foyer. "Ascorbate and glutathione: Keeping active oxygen under control." *Annual Review of Plant Biology* 49, no. 1 (1998): 249–279.

Nolan, Trevor M., Nemanja Vukašinović, Derui Liu, Eugenia Russinova, and Yanhai Yin. "Brassinosteroids: Multidimensional regulators of plant growth, development, and stress responses." *The Plant Cell* 32, no. 2 (2020): 295–318.

Özdemir, Filiz, Melike Bor, Tijen Demiral, and İsmail Türkan. "Effects of 24-epibrassinolide on seed germination, seedling growth, lipid peroxidation, proline content and antioxidative system of rice (*Oryza sativa* L.) under salinity stress." *Plant Growth Regulation* 42, no. 3 (2004): 203–211.

Park, Jiyoung, Youngsook Lee, Enrico Martinoia, and Markus Geisler. "Plant hormone transporters: What we know and what we would like to know." *BMC Biology* 15, no. 1 (2017): 1–15.

Peres, Ana Laura G. L., José Sérgio Soares, Rafael G. Tavares, Germanna Righetto, Marco A. T. Zullo, N. Bhushan Mandava, and Marcelo Menossi. "Brassinosteroids, the sixth class of phytohormones: A molecular view from the discovery to hormonal interactions in plant development and stress adaptation." *International Journal of Molecular Sciences* 20, no. 2 (2019): 331.

Sabagh, Ayman E. L., Sonia Mbarki, Akbar Hossain, Muhammad Aamir Iqbal, Mohammad Sohidul Islam, Ali Raza, Analía Llanes et al. "Potential role of plant growth regulators in administering crucial processes against abiotic stresses." *Frontiers in Agronomy* 3 (2021): 648694.

Sachdev, Swati, Shamim Akhtar Ansari, Mohammad Israil Ansari, Masayuki Fujita, and Mirza Hasanuzzaman. "Abiotic stress and reactive oxygen species: Generation, signaling, and defense mechanisms." *Antioxidants* 10, no. 2 (2021): 277.

Saini, Shivani, Isha Sharma, and Pratap Kumar Pati. "Versatile roles of brassinosteroid in plants in the context of its homoeostasis, signaling and crosstalks." *Frontiers in Plant Science* 6 (2015): 950.

Sairam, R. K., and D. C. Saxena. "Oxidative stress and antioxidants in wheat genotypes: Possible mechanism of water stress tolerance." *Journal of Agronomy and Crop Science* 184, no. 1 (2000): 55–61.

Shah, Anis A., Shakil Ahmed, and Nasim A. Yasin. "24-epibrassinolide triggers cadmium stress mitigation in Cucumis sativus through intonation of antioxidant system." *South African Journal of Botany* 127 (2019): 349–360.

Sharma, Isha, Navdeep Kaur, and Pratap K. Pati. "Brassinosteroids: A promising option in deciphering remedial strategies for abiotic stress tolerance in rice." *Frontiers in Plant Science* 8 (2017): 2151.

Sharma, Pallavi, Ambuj Bhushan Jha, Rama Shanker Dubey, and Mohammad Pessarakli. "Reactive oxygen species, oxidative damage, and antioxidative defense mechanism in plants under stressful conditions." *Journal of Botany* 2012 (2012).

Sharma, Priyanka, and Renu Bhardwaj. "Effects of 24-epibrassinolide on growth and metal uptake in *Brassica juncea* L. under copper metal stress." *Acta Physiologiae Plantarum* 29, no. 3 (2007): 259–263.

Shu, Kai, Xiao-dong Liu, Qi Xie, and Zu-hua He. "Two faces of one seed: Hormonal regulation of dormancy and germination." *Molecular Plant* 9, no. 1 (2016): 34–45.

Singh, V. K., R. Singh, S. Tripathi, R. S. Devi, P. Srivastava, P. Singh, A. Kumar, and R. Bhadouria. "Seed priming: State of the art and new perspectives in the era of climate change." In Prasad, M. N. V., & Pietrzykowski, M. (Eds.), *Climate Change and Soil Interactions* (pp. 143–170). Elsevier, 2020.

Sirhindi, Geetika, Renu Bhardwaj, Manish Kumar, Sandeep Kumar, Neha Dogra, Harpreet Sekhon, Shruti Kaushik, and Isha Madaan. "Physiological roles of brassinosteroids in conferring temperature and salt stress tolerance in plants." In Ramakrishna, A., & Gill, S. S. (Eds.), *Metabolic Adaptations in Plants during Abiotic Stress* (pp. 339–358). CRC Press, Boca Raton, FL, 2018.

Soares, Tássia Fernanda Santos Neri, Denise Cunha Fernandes dos Santos Dias, Ariadne Morbeck Santos Oliveira, Dimas Mendes Ribeiro, and Luiz Antônio dos Santos Dias. "Exogenous brassinosteroids increase lead stress tolerance in seed germination and seedling growth of Brassica juncea L." *Ecotoxicology and Environmental Safety* 193 (2020): 110296.

Soliman, Mona, Amr Elkelish, Trabelsi Souad, Haifa Alhaithloul, and Muhammad Farooq. "Brassinosteroid seed priming with nitrogen supplementation improves salt tolerance in soybean." *Physiology and Molecular Biology of Plants* 26, no. 3 (2020): 501–511.

Spartz, Angela K., and William M. Gray. "Plant hormone receptors: New perceptions." *Genes & Development* 22, no. 16 (2008): 2139–2148.

Sun, Lin, Elena Feraru, Mugurel I. Feraru, Sascha Waidmann, Wenfei Wang, Gisele Passaia, Zhi-Yong Wang, Krzysztof Wabnik, and Jürgen Kleine-Vehn. "PIN-LIKES coordinate brassinosteroid signaling with nuclear auxin input in Arabidopsis thaliana." *Current Biology* 30, no. 9 (2020): 1579–1588.

Sun, Yu, Xi-Ying Fan, Dong-Mei Cao, Wenqiang Tang, Kun He, Jia-Ying Zhu, Jun-Xian He et al. "Integration of brassinosteroid signal transduction with the transcription network for plant growth regulation in *Arabidopsis*." *Developmental Cell* 19, no. 5 (2010): 765–777.

Symons, Gregory M., Christopher Davies, Yuri Shavrukov, Ian B. Dry, James B. Reid, and Mark R. Thomas. "Grapes on steroids. Brassinosteroids are involved in grape berry ripening." *Plant Physiology* 140, no. 1 (2006): 150–158.

Szafrańska, Katarzyna, Russel J. Reiter, and Małgorzata M. Posmyk. "Melatonin improves the photosynthetic apparatus in pea leaves stressed by paraquat via chlorophyll breakdown regulation and its accelerated de novo synthesis." *Frontiers in Plant Science* 8 (2017): 878.

Tripathi, Durgesh Kumar, Kanchan Vishwakarma, Vijay Pratap Singh, Ved Prakash, Shivesh Sharma, Sowbiya Muneer, Miroslav Nikolic, Rupesh Deshmukh, Marek Vaculík, and Francisco J. Corpas. "Silicon crosstalk with reactive oxygen species, phytohormones and other signaling molecules." *Journal of Hazardous Materials* 408 (2021): 124820.

Vahala, Jorma, Markku Keinanen, Andres Schutzendubel, Andrea Polle, and Jaakko Kangasjarvi. "Differential effects of elevated ozone on two hybrid aspen genotypes predisposed to chronic ozone fumigation. Role of ethylene and salicylic acid." *Plant Physiology* 132, no. 1 (2003): 196–205.

Vanacker, Helene, Tim LW. Carver, and Christine H. Foyer. "Early $H2O2$ accumulation in mesophyll cells leads to induction of glutathione during the hyper-sensitive response in the barley-powdery mildew interaction." *Plant Physiology* 123, no. 4 (2000): 1289–1300.

Verma, Sneh, and Vijay L. Kumar. "Attenuation of gastric mucosal damage by artesunate in rat: Modulation of oxidative stress and NFκB mediated signaling." *Chemico-Biological Interactions* 257 (2016): 46–53.

Vincent, Gitz, Alexandre Meybeck, L. Lipper, C. De Young, and S. Braatz. "Climate change and food security: Risks and responses." *Food and Agriculture Organization of the United Nations (FAO) Report* 110 (2016): 2–4.

Wahid, Abdul, Saddia Gelani, M. Ashraf, and Majid R. Foolad. "Heat tolerance in plants: An overview." *Environmental and Experimental Botany* 61, no. 3 (2007): 199–223.

Wang, Wenfei, Ming-Yi Bai, and Zhi-Yong Wang. "The brassinosteroid signaling network—A paradigm of signal integration." *Current Opinion in Plant Biology* 21 (2014): 147–153.

Wani, Shabir Hussain, Vinay Kumar, Tushar Khare, Rajasheker Guddimalli, Maheshwari Parveda, Katalin Solymosi, Penna Suprasanna, and P. B. Kavi Kishor. "Engineering salinity tolerance in plants: Progress and prospects." *Planta* 251, no. 4 (2020): 1–29.

Wani, Shabir Hussain, Vinay Kumar, Varsha Shriram, and Saroj Kumar Sah. "Phytohormones and their metabolic engineering for abiotic stress tolerance in crop plants." *The Crop Journal* 4, no. 3 (2016): 162–176.

Weyers, Jonathan DB. and Neil W. Paterson. "Plant hormones and the control of physiological processes." *New Phytologist* 152, no. 3 (2001): 375–407.

Xu, Fei, Shu Yuan, and Hong-Hui Lin. "Response of mitochondrial alternative oxidase (AOX) to light signals." *Plant Signaling & Behavior* 6, no. 1 (2011): 55–58.

Xu, Peng, Shou-Ling Xu, Zi-Jing Li, Wenqiang Tang, Alma L. Burlingame, and Zhi-Yong Wang. "A brassinosteroid-signaling kinase interacts with multiple receptor-like kinases in *Arabidopsis*." *Molecular Plant* 7, no. 2 (2014): 441.

Yang, Chao, Yamei Ma, Yong He, Zhihong Tian, and Jianxiong Li. "Os OFP 19 modulates plant architecture by integrating the cell division pattern and brassinosteroid signaling." *The Plant Journal* 93, no. 3 (2018): 489–501.

Yang, Jianchang, Wenqian Miao, and Jing Chen. "Roles of jasmonates and brassinosteroids in rice responses to high temperature stress–A review." *The Crop Journal* 9, no. 5 (2021): 977–985.

Yao, Jeffrey K., Ravinder D. Reddy, and Daniel P. Van Kammen. "Oxidative damage and schizophrenia." *CNS Drugs* 15, no. 4 (2001): 287–310.

Yi, Hwe-Su, Martin Heil, Rosa M. Adame-Alvarez, Daniel J. Ballhorn, and Choong-Min Ryu. "Airborne induction and priming of plant defenses against a bacterial pathogen." *Plant Physiology* 151, no. 4 (2009): 2152–2161.

Yin, Yanhai, Dionne Vafeados, Yi Tao, Shigeo Yoshida, Tadao Asami, and Joanne Chory. "A new class of transcription factors mediates brassinosteroid-regulated gene expression in *Arabidopsis*." *Cell* 120, no. 2 (2005): 249–259.

Yu, Xiaofei, Lei Li, Jaroslaw Zola, Maneesha Aluru, Huaxun Ye, Andrew Foudree, Hongqing Guo et al. "A brassinosteroid transcriptional network revealed by genome-wide identification of BES1 target genes in *Arabidopsis thaliana*." *The Plant Journal* 65, no. 4 (2011): 634–646.

Zhang, Lan, Golam Jalal Ahammed, Xin Li, Ji-Peng Wei, Yang Li, Peng Yan, Li-Ping Zhang, and Wen-Yan Han. "Exogenous brassinosteroid enhances plant defense against *Colletotrichum gloeosporioides* by activating phenylpropanoid pathway in *Camellia sinensis* L." *Journal of Plant Growth Regulation* 37, no. 4 (2018): 1235–1243.

Zhong, Chunmei, Barunava Patra, Yi Tang, Xukun Li, Ling Yuan, and Xiaojing Wang. "A transcriptional hub integrating gibberellin–brassinosteroid signals to promote seed germination in Arabidopsis." *Journal of Experimental Botany* 72, no. 13 (2021): 4708–4720.

# 22 Phytohormones
## *Role in Ecophysiological Responses of Tropical Plants to Varying Resource Availability*

*Pallavi B. Dhal, Sadhna, Rajkumari Sanayaima Devi, Rahul Bhadouria, and Sachchidanand Tripathi*

## 22.1 INTRODUCTION

The tropical region lies at 23°N–23°S latitude, while the subtropical region is 25°–40°N and 25°–40°S (https://content.meteoblue.com/). The tropics are located between the northern Tropic of Cancer and the southern Tropic of Capricorn. The equator and some regions of North America, South America, Africa, Asia, and Australia comprise the tropics. The tropical rainforests of the world thrive in the warm, wet conditions found around the equator. Temperature fluctuations in the tropics are often much smaller than those outside the tropics. Annual temperature variations are minor in the tropics, but significant shifts in cloud cover and precipitation define the seasons. Pressure patterns in the tropics tend to be more stable, making for significantly less variation in wind direction.

Precipitation, elevation, and soil type define tropical forests. Precipitation in the tropics averages at least 100 millimeters each month throughout the year. The other two forms are "tropical dry forests" and "semi-evergreen rainforests," with the former experiencing a severe dry season and a period of leaf fall and the latter experiencing a less severe drought period. There are several distinct types of tropical forests found all over the world, including lowland evergreen rainforests, dry deciduous forests, low montane forests, upper montane forests, heath forests, mangrove forests, freshwater swamp forests, and peat swamp forests. The climate of the tropics is seasonal, with warm and wet conditions favoring a stable environment for a greater number of species. The lowland evergreen rainforest, semi-evergreen rainforest, freshwater swamp forest, and peat swamp forest are typical habitats for emergent trees (Thomas & Baltzer, 2002). Lianas can be found in great abundance in lowland evergreen, dry deciduous, freshwater swamp, and semi-evergreen forests. Whereas non-vascular epiphytes are plentiful in peat swamp forests, vascular epiphytes are found in all other types of tropical forests (Thomas & Baltzer, 2002).

As a result of the heavy precipitation and the rapid uptake of nutrients by vegetation, the soils in the tropical rainforest are typically nutrient-poor and non-fertile (Thomas & Baltzer, 2002). Soils rich in iron and aluminum oxides but lacking in natural fertility, such as those found in tropical rainforests, are primarily oxisols and ultisols. The majority of tropical soils have low cation exchange capacities and are acidic, leached, and well-weathered. Due to drought stress, plants may be more open to nutrient inputs when water is scarce or after the drought has ended, as they will have allocated more biomass underground to conserve resources.

In moist tropical and subtropical forests, light is one of the most influential environmental factors on plant development and survival (Tripathi et al., 2020; Bhadouria et al., 2020). In tropical and subtropical forests, irradiance varies greatly across space and time. Plants here have adapted in a wide variety of ways to the shifting light conditions that they encounter regularly. Plants in the understory of a tropical rainforest receive as little as 1% of the light that reaches the canopy, making light the scarcest resource (Chazdon & Fetcher, 1984). Under the dense canopy of tall trees, light is often a limiting factor for the growing seedlings that compose the understory of tropical forests. In low-light environments, the seedlings and shade-tolerant species direct nutrient allocation primarily toward leaf development and longevity, resulting in a smaller root mass and a slower growth rate (Cai et al., 2008). Other limiting factors for plant development in terrestrial ecosystems are typically nitrogen (N) or phosphorus (P) (Ågren et al., 2012). The productivity of most tropical forests is thought to be constrained by the availability of P. For montane forests, nitrogen or both N and P serve as limiting nutrients (Huang et al., 2013), but in lowland tropical forests, P is the limiting nutrient (Fay et al., 2015). In general, species that prefer sunlight respond to fertilization by expanding their growth, while many that can grow in shade boost their foliar concentrations of N and P (Tripathi & Raghubanshi, 2014). As atmospheric carbon dioxide ($CO_2$) rises as a result of climate change, their ability to enhance rates of photosynthetic carbon uptake is hampered by deficits of a few essential nutrients like N and P in the tropical system. Fertilization experiments with various concentrations of the limiting nutrient are commonly used to conclude the nutrient's role in limiting growth or other physiological processes. If there is no change after adding nutrients, then it is assumed that the supplementary nutrient was not limiting.

Nitrogen applied to heavily weathered tropical forest soils cause cations to be released and reach further into the soil profile, as shown by Cusack et al. (2016). Plant P concentrations and soil labile P have been shown to decrease in tropical forests, but N addition considerably increases both in grasslands, wetlands, and temperate forests (Deng et al., 2017).

Root biomass can account for more than half of a tree's total biomass in certain settings. Some tropical savanna trees experience a high physiological and metabolic rate when subjected to drought for 4–5 months (Bucci et al., 2006). To survive dry seasons, deeper roots are required to access water in deeper soil layers. Thus, keeping up with the metabolic demands of maintaining optimal root activity is costly but necessary for preserving the tree's overall carbon, nutrient, and water balance. Moist tropical forests with not much variation in precipitation pattern assign a major fraction of their total biomass to their roots.

To respond to shifts in nutrient availability and other constraints, plants have developed sophisticated mechanisms for detecting and communicating the external and internal concentrations of each of these nutrients. Phytohormones, also known as plant growth hormones, play a crucial role in regulating plant growth and development in response to environmental stimuli. In this chapter, we will further characterize the role of hormones in the response of plants (Table 22.1) to limited nutrients in the soil and other stresses, we will discuss (Figure 22.1) evidence and the possible hormonal and nutritional responses.

## 22.2 ETHYLENE

Ethylene promotes reactive oxidative species (ROS) build-up, which induces plant defense mechanisms. Ethylene assists GSH de novo synthesis (Yoshida et al., 2009). The effect on the production of ethylene due to metal stress has been reported to be metal- and concentration-specific (Thao

**TABLE 22.1**
**Roles of Varying Hormones in the Ecophysiology of Tropical Plants**

| S. No. | Target Species | Phytohormones | Impact on Ecophysiology/Stress Condition | Reference |
|---|---|---|---|---|
| 1. | Neolamarckia cadamba | Auxin | Altered root morphology in K+ deficit soil conditions | Liu et al. (2021) |
| 2. | Arabidopsis thaliana | Auxin | Lateral root (LR) branching in response to low nitrogen settings | Meier et al. (2020) |
| 3. | Arabidopsis thaliana | Auxin | Exogenous auxin positively regulates phosphate starvation response 1 (PHR1) | Huang et al. (2018) |
| 4. | Solanum lycopersicum | Auxin | A key regulator in fleshy fruit production and fruit size | Fenn & Giovannoni (2021) |
| 5. | Arabidopsis thaliana | Auxin | Interaction along with COP1 and PIF in shade avoidance | Pacín et al. (2016) |
| 6. | Arabidopsis thaliana | Auxin | Petiole elongation in response to light deprivation | Sasidharan et al. (2015) |
| 7. | Arabidopsis thaliana | Auxin | Unidirectional petiole development during shade avoidance involving cMTs | Kakar et al. (2013) |
| 8. | Petunia hybrida | Auxin | Involved in dark-stimulated adventitious root formation in nitrogen deficit conditions | Yang et al. (2019) |
| 9. | Arabidopsis thaliana | Auxin | Promote primary root elongation in sulfur deficit conditions | Zhao et al. (2022) |

| S. No. | Target Species | Phytohormones | Impact on Ecophysiology/Stress Condition | Reference |
|---|---|---|---|---|
| 10. | *Pisum sativum* | Auxin | The key regulator in vein formation and maximum gas exchange, indirectly influencing the photosynthetic rate | McAdam et al. (2017) |
| 11. | *Arabidopsis thaliana* | Auxin | Branching of root hairs in iron and sulfur branching | Leitner et al. (2012) |
| 12. | *Corchus capsularis* | Phosphorus + gibberellin | Synergistic effects of P and GA3 reduced Cu toxicity by phytoextraction, decreasing the release of reactive oxygen species and organic acids, and improving the efficiency of antioxidants | Alatawi et al. (2022) |
| 13. | *Jatropha curcas* | Gibberellin | Positive regulator of shoot branching | Ni et al. (2015) |
| 14. | *Betula platyphylla* | Gibberellin | Xylem development and induction of secondary wall biosynthesis–related genes | Guo et al. (2015) |
| 15. | *Portulaca grandiflora* | Gibberellin | Improves photosynthetic performance under salinity stress. | Shaikha et al. (2018) |
| 16. | *Zea mays.*, *Pisum sativum Var. abyssinicum*, and *Lathyrus sativus* | Gibberellin | Seed priming with GA3 improves the percentage of germination and shoot and root growth, along with reduced mean germination time. | Tsegay et al. (2018) |
| 17. | *Brassica juncea* | Gibberellin | Increases sulfur assimilation under cadmium toxicity | Masood et al. (2016) |
| 18. | *Olea europaea* | Gibberellin | Foliar GA3 application mitigates the adverse effects of salinity stress | Moula et al. (2020) |
| 19. | *Zea mays* | Gibberellin | External application of increased corn grain yield in severe drought stress | Maleki et al. (2020) |
| 20. | *Nicotiana tabacum* | Gibberellin | Improves light capture efficiency and triggers biomass accumulation under low-irradiance conditions. | Falcioni et al. (2018) |
| 21. | *Coriandrum sativum* | Gibberellin + NPK application | Decreases the oxidative stress due to boron toxicity and enhances nutritional status | Saleem et al. (2021) |
| 22. | *Corchorus capsularis* | Gibberellin | Improved biomass and growth, photosynthetic pigments content under copper toxicity | Saleem et al. (2020) |

*(Continued)*

**TABLE 22.1**
**Continued**

| S. No. | Target Species | Phytohormones | Impact on Ecophysiology/Stress Condition | Reference |
|---|---|---|---|---|
| 23. | *Daucus carota* | Gibberellin | Exogenous application improved growth and chlorophyll content in leaves, increased concentration of phenolic compounds under lead stress | Ghani et al. (2021) |
| 24. | *Pisum sativum* | Gibberellin | Foliar application promoted tolerance to copper stress by enhancing growth, antioxidants under copper induced stress | Javed et al. (2021) |
| 25. | *Arabidopsis thaliana* | Gibberellin | Exogenous GA decreases cadmium accumulation in tissues and improves root growth under cadmium stress | Zhu et al. (2012) |
| 26. | *Mangifera indica* | Gibberellin | MiFT plays a vital function in flower induction in mango; GA3 limits flowering by inhibiting MiFT expression, suggesting that fruit load influences GA metabolism | Nakagawa et al. (2012) |
| 27. | *Petunia × hybrida* | Gibberellin | The influence of blue light on the elongation of the main stem; red (R) light that inhibits this elongation is reliant on the regulation of gibberellin content | Fukuda et al. (2016) |
| 28. | *Arabidopsis thaliana* | Gibberellin | The concentration of gibberellic acid (GA) observed to rise following shade treatment, rendering elongation | Bou-Torrent et al. (2014) |
| 29. | *Sorghum sp.* | Exogenous GA3 and nitrogen | *Sorghum* seedling emergence, growth, and antioxidant enzymes and here, nitrogen and GA3 improved seedling emergence | Ali et al. (2021) |
| 30. | *Rubus sp.* | Gibberellin | Exogenous GA3 reduction fruit ripening time and breaking bud dormancy | Lin and Agehara (2020) |
| 31. | *Jatropha* | Gibberellin | Flowering suppressed by exogenous GA whereas flowering stimulated in application of GA biosynthesis inhibitors | Li et al. (2018) |
| 32. | *Arabidopsis thaliana* | Gibberellin | The role in transitioning from the juvenile to the generative state | Yamaguchi et al. (2014) |

| S. No. | Target Species | Phytohormones | Impact on Ecophysiology/Stress Condition | Reference |
|---|---|---|---|---|
| 33. | Solanum lycopersicum | Gibberellin | Mitigation of injury due to chill of the fruit during cold storage | Ding et al. (2015) |
| 34. | Triticum aestivum | Gibberellin | Determine the physiochemical changes responsible for inducing salt tolerance in wheat through GA3 priming under salinity stress | Iqbal et al. (2013a) |
| 35. | Nicotiana tabacum | Gibberellin | C19-GAs restore cambial proliferation, xylem fiber differentiation in leaves, and stem elongation in the absence of developing leaves, demonstrating the importance of GAs in regulating secondary growth and establishing the presence of leaves as a prerequisite for GA signaling in stems | Dayan et al. (2018) |
| 36. | Cucurbita maxima | Cytokinin | Exogenous application improved photosynthetic parameters, starch content, and growth under drought stress. Soluble sugars decreased post-application of cytokinin | Niakan and Habibi (2016) |
| 37. | Annona leptopetala | ABA (abscisic acid) | ABA levels higher in the dry period, rendering a small stomatal aperture and decreasing ethylene precursor (ACC) | Figueiredo-Lima et al. (2018) |
| 38. | Annona leptopetala | Jasmonic acid | Higher in the rainy season and positive influence on stomatal conductance | Figueiredo-Lima et al. (2018) |
| 39. | Arabidopsis thaliana | Ethylene | Induce cell expansion via reorientation of cortical microtubules along the petiole | Polko et al. (2012) |
| 40. | Coffea spp. | Ethylene | Re-watering under drought stress conditions restored ethylene production in shoots critical for anthesis | Lima et al. (2021) |
| 41. | Helianthus annuus | Abscisic acid | Had higher amounts of ABA when grown in shade | Cagnola et al. (2012) |
| 42. | Arabidopsis thaliana | Abscisic acid | The build-up of ABA in buds prevents branching from occurring | Reddy et al. (2013) |
| 43. | Tropical Forest trees | Abscisic acid | ABA controls the stomatal opening and stress signaling to regulate water and carbon fluxes in response to high temperature and drought stress | Sampaio et al. (2018) |

*(Continued)*

### TABLE 22.1
### Continued

| S. No. | Target Species | Phytohormones | Impact on Ecophysiology/Stress Condition | Reference |
| --- | --- | --- | --- | --- |
| 44. | *Arabidopsis thaliana* | Abscisic acid | Roots exposed to NO$_3$. increased the ABA signal threefold, mostly in the endodermis and stele of the growing tip | Ondzighi-Assoume et al. (2016) |
| 45. | *Arabidopsis thaliana* | Abscisic acid | The possibility exists that ABA generated in response to drought stress can limit NO$_3^-$ sensing, leading to diminished NO$_3^-$ absorption | Léran et al. (2015) |
| 46. | *Arabidopsis thaliana* | Abscisic acid | Roots cultivated under low NO$_3^-$ conditions induce LR branching via ABA signaling | Forde (2014) |
| 47. | *Arabidopsis thaliana* | Brassinosteroid (BR) | Increases LR elongation under mild N deficit | Jia et al. (2019) |

et al., 2015). Cadmium (Cd) is known to promote the synthesis of ethylene in young leaves (Arteca & Arteca, 2007). Masood et al. (2012) demonstrated that ethylene is essential for metal stress tolerance in mustard under Cd stress.

Additionally, it has been found that copper (Cu) causes the up-regulation of ACO genes and aids in ethylene synthesis in *Nicotiana glutinosa* (Keunen et al., 2016). Under the influence of various abiotic stresses, AP2/ERF (the ethylene-responsive transcription factor) is the link between hormonal control and ROS signaling. Ethylene is known to be a flood tolerance regulator in a rice variety that grows in flooding conditions (Bailey-Serres et al., 2012). In paddies, water lodging increases ethylene, which opposes abscisic acid (ABA) (Yang et al., 2004).

The sudden thrust into a terrestrial setting after evacuating from a flood results in the increased production of ROS and harmful metabolites that are closely linked to re-oxygenation. Plants employ several chemical mechanisms and signaling processes; central among them is ethylene's function in preventing these adverse effects. Re-oxygenation was found to promote an increase in ethylene production (Tsai et al., 2014). The maximum ethylene response was seen in the flooding escape plant, and this allowed for continuous shoot elongation even though the plant's leaf tip was submerged in water. Ethylene reduces submergence in paddies, and during re-oxygenation, it regulates other hormonal signals, ROS production, and metabolic activities (Tsai et al., 2014). Molecular evidence links ethylene with hypoxia, although here the signaling route is unclear (Sasidharan et al., 2018).

*Arabidopsis* responds to re-oxygenation by increasing the expression of the gene for ethylene biosynthesis (Tsai et al., 2014). *Arabidopsis* expresses ethylene-responsive genes containing ERF/AP2 during oxygen deficit owing to water overload (Bailey-Serres et al., 2012); this improves plant stress in hypoxia by activating genes associated with ethylene biosynthesis and sugar metabolism (Hinz et al., 2010).

Stomata in wild *Arabidopsis* close faster than ethylene flash flood mutant eto1 under drought stress (Tanaka et al., 2005). By inhibiting ROS generation, ethylene modulates stomatal closure in an EIN2-mediated mechanism (Watkins et al., 2014). Similarly, under drought stress, *Vicia faba*, orange, French bean, etc., produce more ethylene. In maize, ethylene inhibits leaf development and ACC, a transduction molecule during drought (Sairam et al., 2008). In the absence of oxygen, ethylene increases the adventitious root, aerenchyma, the height of the petiole or stem, and hyponasty response (Voesenek & Sasidharan, 2013).

The apical hook shields the meristem as the hypocotyl emerges above the soil. The apical hook undergoes a gravity-driven reorientation which is critical for successful germination and seedling emergence. In order to safeguard the meristem, ethylene acts as a developmental and thigmomorphogenic signal that strengthens the apical hook (Harpham et al., 1991). In the absence of light, auxins play a supporting role in apical hook development following ethylene signaling (Abbas et al., 2013). The hormone ethylene stimulates gravitropic reorientation in seedlings grown in light (Guo et al., 2004). There is a dose-dependent dual effect of ethylene on negative gravitropism in light in *Arabidopsis*. Long-term (>12 h) exposure increases gravibending, while shorter exposures (12 h) decrease it (Li, 2008).

Cysteine is a key intermediate in the biosynthesis of ethylene and the non-protein thiol GSH, a significant

# Phytohormones

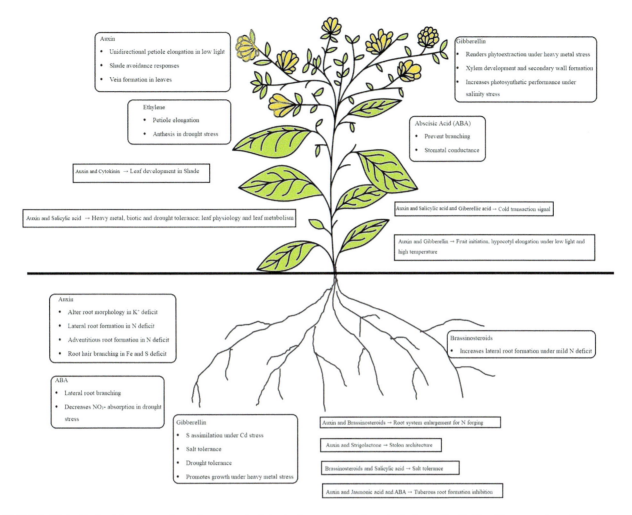

**FIGURE 22.1** Role of hormones in various ecophysiological responses of tropical plants. S—Sulfur; K—potassium; N—nitrogen; Fe—iron; $NO_3^-$—nitrate; Cd—cadmium. "And" has been used for various hormones cross-talk and synergistic effects for a particular ecophysiological response.

component of stored S in plants. Sulfur (S) is a constituent of cysteine (Cys) and S-adenosyl methionine (SAM), critical for reduced glutathione (GSH) production and also the precursor to ethylene biosynthesis. Iqbal et al. (2013b) discussed that sulfur (S) and ethylene control physiological activities in a mutually dependent fashion under favorable environmental conditions, but under stressful conditions, S absorption capability and GSH synthesis increase. To help plants adjust to these circumstances, ethylene signaling also controls GSH production. The growth of plants in both ideal and stressful conditions is thus regulated by a relationship between S and ethylene.

## 22.3 CYTOKININ

### 22.3.1 Cytokinin in Triggering the Development of Lateral Roots in Response to Changes in Nitrogen Levels

The regulation of root architecture development is another important function of cytokinin (CKs). Exogenous CKs have been shown to impede lateral root (LR) development in several studies: reduced CK levels improved root development and branching (Laplaze et al., 2007). CKs limit lateral root primordium (LRP) production by interfering with the auxin gradient, likely during LRP initiation and organization (Laplaze et al., 2007). Overall, a high nitrogen supply prevents roots from spreading outward, while low nitrogen speeds up both root development and branching of roots (Tranbarger et al., 2003). Since nitrogen status and CK levels are strongly correlated, CKs may affect root architecture in varying nitrogen availability.

### 22.3.2 Cytokinin in Response to Nitrogen Availability

Among the various roles that CK plays in plants, nitrogen signaling is a notable one (Sakakibara, 2006). Growth-limiting effects caused by inadequate nitrogen availability can be somewhat mitigated with exogenous CK treatment (Kuiper et al., 1989). Previous studies have shown that CKs may either serve as a local signal or as a long-distance signal traveling from the shoot to the roots (Matsumoto-Kitano et al., 2008). Nitrogen supplementation increases CK concentration in maize xylem sap, roots, and shoots, suggesting that CKs signal nitrogen supplementation from roots to leaves (Sakakibara et al., 2006).

CKs may detect the presence of sufficient nitrogen, possibly suppressing nitrogen uptake gene expression. The suppression of AtNRT genes by exogenous cytokinin in the root under both high-nitrogen (HN) and low-nitrogen (LN) conditions demonstrates the CK action is nitrogen-independent (Kiba et al., 2011). No significant suppression was detected with other phytohormones, suggesting that the action is CK-specific.

However, an increase in nitrate supply causes a rise in the concentration of root cap synthesized CK (Aloni et al., 2006), promoting bud and leaf development. Root epidermal or cortical cells express cytokinin-repressive nutrient acquisition genes. CKs regulate epidermal-expressed iron starvation-inducible genes (Séguéla et al., 2008).

### 22.3.3 Role of Cytokinin in Shade Avoidance Response

Plant reactions to changing light levels at different heights in the canopies led scientists to the discovery of CKs participation in shade avoidance responses (Pons et al., 2001). Reduced photosynthetic ability and eventual senescence result from a slow delivery rate of CKs in darkened leaves, where transpiration rate and stomatal conductance are already low (Boonman & Pons, 2007).

Inhibiting leaf growth when exposed to shade was identified as yet another function of CKs. Furthermore, it has been revealed that the CK receptor AHK3 mediates the root-to-hypocotyl ratio response in the presence of shadow (Novák et al., 2015). Reduced photosynthetic capacity and a temporary halt in leaf development are triggered by the absence of bioactive CKs, guaranteeing that energy is diverted into extension growth under cover.

### 22.3.4 Cytokinin in Regulating the Cambial Activity

In temperate deciduous hardwood trees, CKs from the root tips stimulate activity and sensitivity of cambium, allowing differentiation required to produce ring-porous wood. The function of the environment in vascular adaptation and evolution is examined in light of these mechanisms. It has been shown that cytokinins released by the root tips (Aloni et al., 2005, 2006) make the cambium more responsive to the auxin signal from developing leaves. The cambium becomes more responsive to the dilute concentration of IAA streams from expanding buds during increased concentration of CK (Coenen & Lomax, 1997). Earlywood-wide vessel differentiation is only possible during this short window of time in the early spring, due to the special physiological conditions mentioned previously (Aloni, 2001).

### 22.3.5 Cytokinin in Tuber Development

CKs have been speculated to have a role in tuberization by encouraging cell division at the outset of tuberization, leading to the formation of a local sink (Guivarc'h et al., 2002). The overexpression of the CK oxidase (CKX1) enzyme in *Arabidopsis* reduces tuber formation (Guivarc'h et al., 2002). Recent research has demonstrated that ectopic expression of the tomato LONELY GUY 1 gene (TLOG1), which encodes a CK-activating enzyme, confers on tomato basal axillary meristems the potential to produce aerial mini tubers resembling potato tubers (Abelenda & Prat, 2013).

## 22.4 GIBBERELLIN

### 22.4.1 Gibberellin in Regulating the Conducting Elements of the Vascular System

Gibberellin (GA), in combination with auxin, stimulates long tracheids and promotes cambial activity, the particular signal triggering fiber development (Aloni, 2013; Aloni et al., 2013). Gibberellin is produced in response to an increase in mature leaf biomass, which in turn promotes cambial activity, tracheid development, and fiber production, all of which contribute to an increase in woodiness. Differentiation of tracheary components, fibers, and xylogenesis are all induced by GA. Fusiform cambial initials elongate over time in woodier species, indicating increased GA stimulation from older leaves (Carlquist, 2013).

### 22.4.2 Gibberellin in Shade Avoidance Response

Internode length and light absorption optimization may boost plant performance and biomass accumulation. In the low-irradiance environment, GA is especially important for promoting growth and biomass accumulation. Also, gibberellin promotes etiolated, thinner leaves, along with larger leaf areas, thinner leaves, and lower pigment content, while PAC (a gibberellin biosynthesis inhibitor) causes smaller, dark-green plants with more pigment and two layers of palisade parenchyma cells (Falcioni et al., 2018).

## 22.5 AUXIN

### 22.5.1 Auxin in Vascular Development

Several lines of evidence imply that the plant hormone auxin (often indole-3-acetic acid) plays a function in vessel development (Kaneda et al., 2011), in addition to its well-established involvement in governing vascular development in plants (Scarpella et al., 2010). It is known that auxin operates in concert with other plant hormones (namely cytokinins and brassinosteroids) during xylem formation. A crucial regulator connecting foliar and xylem development is polar auxin transfer through the vascular cambium. It is believed that hormones such as auxin play a crucial role in controlling the diameter of xylem vessels. Transportation is likely enhanced in certain plant families (including chaparral species and lianas) by the combination of broader vessels with narrow conduits (vessels and/or tracheids).

One of the most crucial metrics for assessing plant water relations is the degree to which the diameter of xylem vessels varies; in the tropical rainforest, the mean hydraulic vessel diameter (Dv) measured in young branches is widely variable (Hacke et al., 2017).

The signaling steps that lead to the chemical alteration of pectin, which allows for slippage between adjacent cells, may be triggered by auxin, which could then facilitate vessel expansion in the cambial zone (Braybrook & Peaucelle, 2013).

According to the vascular adaptation hypothesis (Aloni, 1987), environmental factors regulate plant growth, height, and shape, and so exert influence over the vascular system. The length of a stem and the diameter of a stem both positively affect the size of a vessel. Vessels in dry-land plants are often narrower than those in tall, woody plants and lianas found in forests (Hacke et al., 2017). According to this theory, secondary wall deposition is regulated by a decreasing auxin gradient from leaves to roots along the plant axis.

## 22.5.2 Auxin in Shade Avoidance

Shade-intolerant plants detect a decline in red (660 nm) to far-red (730 nm) light when light intensity decreases. These light signals cause shadow avoidance syndrome (SAS), which results in hypocotyl and petiole rather than leaves, fruits, and seeds. SAS causes leaf hyponasty, extended petioles, hypocotyl, and internode elongation. Long-term exposure to shade causes plants to bloom early, have fewer branches, be more susceptible to insects, and produce fewer seeds.

The stem and petiole elongation as the manifestation of SA response is partially regulated by auxin. Expansins and xyloglucan endotransglucosylase/hydrolases (XTHs) (Sasidharan et al., 2010) are two such protein families thought to be involved in shade avoidance. Also, shade-induced hypocotyl or petiole elongation growth requires a variety of hormones in addition to auxin, including gibberellins, brassinosteroids, and ethylene.

## 22.5.3 Auxin in the Regulation of Root Architecture

Increased levels of carbon dioxide ($CO_2$) have a few positive physiological impacts on tropical plants (Bhadouria et al., 2016). Increases in net photosynthesis, for instance, are largely responsible for the greater biomass accumulation of plants in environments of elevated $CO_2$ (Bhadouria et al., 2017). Growing more root hairs is a practical method for overcoming inadequate soil nutrients. Studies show that higher levels of carbon dioxide ($CO_2$) not only prompted the development of longer root hairs but also increased their number. Microscopical analysis reveals that higher $CO_2$ led to a greater number of trichoblast files than baseline $CO_2$. As a result, under high $CO_2$ levels, new trichoblast files may be initiated alongside the development of ectopic root hairs. This further confirmed the importance of auxin in increasing root hair density and length in response to high $CO_2$.

Phosphate (P) deficit in Arabidopsis triggers the onset of LR by altering the plant's sensitivity to auxin. Low P promotes this first stage of LR development by increasing auxin sensing and activating the auxin response factor 19 (ARF19) in pericycle cells (Bustos et al., 2010). Exogenous auxin positively regulates PHR1, a phosphate starvation response 1 (PHR1), the central transcription factor coordinating P deficiency responses (Huang et al., 2018).

Plants respond to low levels of nitrogen (N) by expanding the number and length of their LRs (Gruber et al., 2013) so that they can better search for and absorb N from their environment, although the prior research demonstrates that auxin mediates nutritional status-dependent signal transduction from shoots to roots and communication between root cells (Liu & von Wirén, 2022).

LR elongation is encouraged by local nitrate (Remans et al., 2006), while LR branching is encouraged by local ammonium (Lima et al., 2010). Since auxin accumulates in LR primordia and tips to promote LR emergence and elongation (Mounier et al., 2014), local nitrate input activates LR elongation via nitrate transporters (Remans et al., 2006). Auxin levels dramatically increase in the stolon following tuber start and remain reasonably high during subsequent tuber growth (Kloosterman et al., 2007).

## 22.5.4 Auxin in Secondary Root Growth

During early secondary root growth, auxin-related transcripts and secondary growth factor transcriptional activity increased, but gibberellin-related transcripts decreased. During early xylem proliferation in the root vasculature, cell wall biosynthesis increased. After the vascular cambium forms, the metabolic pathway responsible for storing starch is switched on. Research by Rüscher et al. (2021) demonstrated that storage roots developed laterally by laying down non-lignified parenchyma cells and formed a single vascular cambium.

## 22.5.5 Auxin in Response to Sulfur Deficit Conditions

Sulfur (S) is essential for cellular metabolism, plant development, and stress responses—both abiotic and biotic. Lateral root growth is inhibited in S-deficiency circumstances. Dan et al. (2007) explored that S deficit suppresses auxin accumulation in *Arabidopsis thaliana*. Sulfate level, according to Zhao et al. (2014), impacts primary root extension by controlling endogenous auxin levels and fostering the survival of root stem cells. The cysteine, glutathione, and IAA levels in roots are favorably associated with external sulfate concentrations within the physiological range, which ultimately influence root system design. Falkenberg et al. (2008) found a link between auxin-related transcriptional regulators (IAA13, IAA28, and ARF-2) and fundamental plant metabolism. As a result, it was shown that transcription factors related to auxin are a component of a comprehensive response pattern to nutrient deprivation that coordinates metabolic changes that are responsible for sulfur homeostasis.

### 22.5.6 Auxin in Response to Potassium Deficit Conditions

Primary root elongation is inhibited in response to K+ deficient stress because auxin accumulates after ethylene and PIN3 and PIN7 transcription are upregulated (Muday et al., 2012). Ethylene, meanwhile, boosts auxin transport by boosting PIN2 and AUX1 transcription (Muday et al., 2012).

### 22.5.7 Auxin in Response to Phosphate Deficit

Foraging for nutrients in the rhizosphere necessitates a shift from primary to lateral root development when phosphate (Pi) is scarce. Pi-deficient seedlings developed normally, even though their roots had normal auxin concentration and transport (Pérez-Torres et al., 2008). Pi-deficient seedlings had auxin-hypersensitive main root pericycle cells. Thus, root responses to low Pi are auxin-dependent. Root elongation in response to iron deprivation is also mediated by auxin transport via the auxin transporter AUX1 (Giehl et al., 2012).

## 22.6 ABSCISIC ACID

### 22.6.1 Abscisic Acid in Response to Shade Avoidance

In the immature buds, shade encourages the build-up of ABA, which represses branching and prevents the plant from growing more branches. Sunflower (*Helianthus annuus*) and tomato leaves (Cagnola et al., 2012) had higher amounts of abscisic acid (ABA) when grown in shade (Kurepin et al., 2007). Following the findings of Reddy et al. (2013), the build-up of ABA in buds prevents branching from occurring. In responsive buds, ABA synthesis and signal transduction genes expressed differently as R/FR rises. ABA biosynthetic mutants branch more under low R/FR. ABA did not affect shade-induced petiole elongation (Pierik et al., 2011), suggesting organ-specific roles for ABA in the SAS.

### 22.6.2 Abscisic Acid in Response to Water Stress

High temperatures and droughts cause unsustainable water loss and ROS generation, damaging photosynthetic membranes. ABA signaling pathways are intertwined with ROS: closing stomata, decreasing transpiration and net photosynthesis, modifying hydraulic conductivities, and activating antioxidant systems in defense. ABA controls the relationship between carbon flux, water, and ROS in response to high temperature and drought stress (Sampaio et al., 2018).

### 22.6.3 Abscisic Acid in Response to Nitrate Deficit Conditions

Low $NO_3^-$ conditions cause LR branching, especially from roots in direct touch with the supply (Forde, 2014). High $NO_3^-$ inhibits *Arabidopsis* root branching via ABA signaling. Drought stress releases biologically active ABA from glucosidase (Xu et al., 2012). Additionally, $NO_3^-$ sensing and signaling of the $NO_3^-$ transporters may be regulated by stress transduction pathways. The possibility exists that ABA generated in response to drought stress can limit $NO_3^-$ sensing, leading to diminished $NO_3^-$ absorption (Léran et al., 2015).

Nitrogenase activity is inhibited in response to drought through a pathway involving ureide and ABA. Exogenous ABA supply to peas has been demonstrated to limit modulation and regulate nodule development by inhibiting Nod factors, cytokinin induction, and signal transduction of the nodule primordia, providing more evidence for the interaction between allantoin and ABA in legumes (Ding et al., 2008). Also, ABA plays a role in lowering N fixing rates by as much as 80% (González et al., 2001). An enzymatic connection is present between drought tolerance and decreased N stress in plants. A plant's ability to break down allantoin is inhibited while it is experiencing drought stress. Accumulation of allantoin results in both the induction of de novo ABA synthesis and the activation of ABA from the inactive glycosylated state.

### 22.6.4 Abscisic Acid in Response to Low Phosphorus Conditions

Low phosphorus (LP) induces ABA accumulation and promotes Pi uptake in *Arabidopsis thaliana* (Zhang et al., 2022). In response to Pi availability, LP strongly stimulates ABA biosynthesis, metabolism, and stress responses in plants. The mutant lacking Beta-glucosidase1 (BG1), the enzyme responsible for converting conjugated ABA to active ABA, exhibited shorter roots and lower Pi content compared to the wild type when grown in LP conditions. LP also increased ABA build-up and phosphate transporter 1;1/1;4 (PHT1;1/1;4) expression. ABA signaling transcription factor ABI5 mutants showed a similar pattern. Thus, ABA is involved in response to phosphate deprivation in *Arabidopsis*, favorably controlling the phosphate uptake.

## 22.7 JASMONIC ACID IN PLANT DEFENSE MECHANISMS

Insects and diseases are often able to easily infest plants that are either partially shaded or grown in an FR-rich environment (Ballaré, 2014). Insect herbivory is greater on FR-exposed plants compared to wild-type plants (Moreno et al., 2009), demonstrating that shade can down-regulate the jasmonic acid (JA) pathway to decrease plant immunity.

Pratelli et al. (2014) reviewed the expression of genes encoding enzymes involved in the production and degradation of various amino acids increased in *Arabidopsis* in response to the administration of exogenous JA. This shows that JA is involved in shaping the transition between primary and secondary metabolic pathways. In general, the down-regulation of NiR, GS1;2, and other transporters caused by JA signaling–suppressed nitrogen absorption in roots and decreased translocation from roots to leaves. In plants, both NiR and GS1;2 enzymes are necessary for the assimilation of inorganic nitrogen, although GS1;2 is especially important for the main assimilation of ammonium in rice roots (Funayama et al., 2013). Wu et al. (2019) concluded that JA signaling affects N uptake and allocation in rice. (+)-12-oxo-phytodienoic acid (OPDA), a precursor of the jasmonate family, regulates stress-responsive cellular redox homeostasis. A hetero-oligomeric cysteine (Cys) synthase complex (CSC) boosts cellular reduction potentials (Xiang et al., 2018) by promoting Cys (sulfur assimilation) and thiol metabolites production. Here, it may imply that sulfur assimilation and JA are related (Haas et al., 2008).

## 22.8 STRIGOLACTONE IN RESPONSE TO VARIOUS NUTRIENT DEFICIT CONDITIONS

Root-produced strigolactone (SL) suppresses lateral bud growth to avoid shoot branching (Shinohara et al., 2013). The root synthesizes SL, which travels up the stem via the xylem to restrict lateral bud formation (Kohlen et al., 2012). Interactions between SL signaling and the auxin signaling system increase xylem differentiation and are adequate for cambium stimulation (Agusti et al., 2011). Cambial activity and vascular differentiation are both positively regulated by SL (Agusti et al., 2011). SL responds to phosphate stress by encouraging the plant to grow a longer main stem at the expense of branching out in low-light environment. Heavy rainfall, with an average of 2,100 millimeters per year (Prior & Bowman, 2014), may be a contributing factor to a lack of soil minerals and, consequently, an insufficient supply of inorganic phosphate, which could explain why Eucalyptus trees in southeast Australia's moist forests grow so tall so quickly. SL increases nutrient delivery to the actively growing apical bud (Agusti et al., 2011). Unlike SL, which promotes solely main stem elongation by limiting lateral bud formation, the cytokinin produced in root caps (Aloni et al., 2006) signals bud and leaf development, and cytokinin concentration in root cap cells increases with nitrate ($NO_3^-$) availability (Ruffel et al., 2011).

## 22.9 CROSS-TALK RELATED TO VARIOUS PHYTOHORMONES IN RESPONSE TO THE ECOPHYSIOLOGY OF TROPICAL PLANTS

Here, we will discuss at how the antagonistic and synergistic effects of hormones through cross-talk between them can produce a wide range of ecophysiological responses (Table 22.2). For example, the cytokinin–auxin cross-talk involves in delaying leaf senescence, auxin–strigolactone interaction governs the stolon architecture, salicylic acid–auxin cross-talk is involved in plant defenses, etc.

### 22.9.1 CYTOKININ–AUXIN CROSS-TALK

Cytokinin from root tips increases cambium sensitivity to auxin from immature leaves (Nieminen et al., 2008). In their experiments, Barbez et al. (2012) show that the hormones auxin and cytokinin alter the synthesis rate of each other, thereby changing the hormone's relative concentrations. Accumulated auxin in leaves triggers the expression of cytokinin oxidase, which degrades cytokinin and slows leaf development during SAS.

### 22.9.2 AUXIN–STRIGOLACTONE CROSS-TALK GOVERNING STOLON ARCHITECTURE

The strigolactones (SLs) also play a crucial role in regulating shoot branching (Hayward et al., 2009). Auxin, in conjunction with strigolactones, exert control over stolon architecture in a manner analogous to that overshoot branching (Pasare et al., 2013). Specifically, they inhibit axillary bud expansion. Inhibiting shoot branching and maintaining tuber bud dormancy are two of SL's most prominent functions, so silencing the SL biosynthetic StCCD8 gene in potatoes has a dramatic effect on both functions (Pasare et al., 2013). In the in vitro tuberization trials, Roumeliotis et al. (2012) used a synthetic SL analog (GR24) and observed a significant reduction in bud outgrowth from the axillary stolons, which ultimately led to a reduction in tuber development in the potato. However, SLs were quantified in potato roots for the first time, and their existence in root extracts was confirmed. This evidence suggests that auxin plays a crucial role in tuber development. Auxin and SL might also play a function in controlling the development of the stolon axillary bud, according to this in vitro tuberization research.

### 22.9.3 BRASSINOSTEROIDS, AUXIN, AND ETHYLENE CROSS-TALK

To compensate for the lack of light they encounter after germinating beneath the surface, seedlings grow upward. The negative gravitropic capability helps them control their development upward. It is critical for a seedling's survival to emerge from the ground after germinating underground,

## TABLE 22.2
### Roles of Various Hormonal Cross-Talks in Ecophysiology of Tropical Plants

| S. No. | Target Species | Phytohormones | Impact on Ecophysiology/Stress Condition | References |
|---|---|---|---|---|
| 1. | Arabidopsis thaliana | Brassinosteroid-auxin module | Root system enlargement for N foraging is influenced by natural variation at many loci in the BR-auxin signaling module | Jia et al. (2019); Jia et al. (2020); Jia et al. (2021) |
| 2. | Zoysia matrella | Auxin, gibberellins, ethylene, and calcium | Involved in cold signal transduction | Long et al. (2020) |
| 3. | Oryza sativa | Auxin–salicylic acid cross-talk | Involved in heavy metal, drought, and fungal stress | Tiwari et al. (2020) |
| 4. | Ipomoea batatas | Auxin, cytokinin, strigolactone, and ethylene | Induction of adventitious roots from the cuttings | Steffens & Rasmussen (2016) |
| 5. | Ipomoea batatas | Auxin + ethylene | High number of adventitious roots around petiole–stem junction | Ma et al. (2014) |
| 6. | Zea mays | Auxin–salicylic acid cross-talk | Regulate leaf physiology, leaf metabolism, and resource allocation patterns affecting root growth | Agtuca et al. (2014) |
| 7. | Arabidopsis | Brassinosteroids and salicylic acid | Salt tolerance | Ashraf et al. (2010) |
| 8. | Arabidopsis | Cytokinin–auxin cross-talk | Slows the leaf development during shade avoidance | Barbez et al. (2012) |
| 9. | Solanum tuberosum | Auxin–strigolactone cross-talk | Control over stolon architecture | Pasare et al. (2013) |
| 10. | Arabidopsis | Brassinosteriods, light, and ethylene | BR affects apical hook formation after germination by moderating the light/dark and ethylene pathways | Herrera-Ubaldo, H. (2022) |
| 11. | Glycine max | Gibberellins and auxin | Hypocotyl elongation under low light and high temperature | Bawa et al. (2020) |
| 12. | Arabidopsis | ABA–gibberellin–auxin cross-talk with melatonin | Seed germination inhibited | Lv et al. (2021) |
| 13. | Solanum lycopersicum | Gibberellins and Auxin | Controls fruit initiation | Hu et al. (2018) |
| 14. | Solanum lycopersicum | GRAS transcription factors, gibberellin–auxin cross-talk | Tolerance to multiple abiotic stresses | Habib et al. (2019) |
| 15. | Zea mays | Ethylene-modulated auxin and gibberellins signaling | Cell wall modification to control elongation | Zhang et al. (2020) |
| 16. | Manihot esculenta | JA, ABA, and auxin | Tuberous root formation inhibited | Utsumi et al. (2020) |
| 17. | Solanum nigrum | NO with auxins | Cadmium stress | Xu et al. (2011) |
| 18. | Arabidopsis thaliana | Auxin + brassinosteroid | BR–auxin hormonal module involved in root foraging for nitrogen | Jia et al. (2021) |

and this requires both enough elongation and negative gravitropism. Multiple hormones exert strict control over negative gravitropism. When it comes to controlling gravitropism in shoots, ethylene and brassinosteroids (BR) have opposing effects. Two-day-old seedlings are more likely to undergo gravitropic reorientation when BR is absent than when ethylene is present (Vandenbussche et al., 2013). Increased sensitivity to BR can be shown when ethylene signaling is inhibited. Ethylene production in wild-type tobacco can be stimulated by a low R/FR (Pierik et al., 2004). The absence of shade-induced petiole elongation in the ethylene-insensitive mutants *ein2–1* and *ein3–1eil1–3* of the plant *Arabidopsis thaliana* suggests that ethylene is a positive regulator of this process (Pierik et al., 2009). Recent studies have shown that the ethylene signaling pathway's main transcription factor, EIN3, is quickly degraded in response to light stimulation of the photoreceptor phyB (Shi et al., 2016). Root development under $K^+$ deficient circumstances has been linked to ethylene and auxin (Jung et al., 2009).

Some AUX/IAA genes are regulated by both ethylene and BRs. Unlike ethylene, BRs control many components of the auxin signaling pathway. Brassinosteroids (BR) were recently shown to be antagonistic regulators of gravitropism in shoots. It has been hypothesized (Vandenbussche et al., 2013) that auxin signaling is necessary for the effect of BRs to take place. Therefore, it is probable that a distinctive equilibrium governing vertical development is maintained by the opposing signals of ethylene and BR in elongation and negative gravitropism. Light-induced loss of endodermal amyloplasts, however, greatly diminish graviperception, and hence, counteract negative gravitropism (Kim et al., 2011).

### 22.9.4 Gibberellin and Auxin Cross-Talk

In the presence of auxin, gibberellin acts as a particular signal causing fiber differentiation by promoting cambial activity and encouraging long tracheids (Aloni, 2013; Aloni et al., 2013). There is a lot of evidence to suggest that GA is the primary hormone responsible for initiating tubers. The active GA content rapidly decreases during the early stages of tuber formation due to an increased profile of genes involved in GA breakdown, which in turn allows for the expansion of the stolon tip. It is well established that GA plays a role in the elongation of shoots and stolons, but the formation of the new tuber organ demands a transition in meristem identity and a rotation in the cell division plane. Auxin is also crucial in the processes of embryo patterning, flower development, and the commencement of lateral roots. Furthermore, Xu et al. (1998) concluded that GA is a dominant regulator, acting as a stimulant during stolon elongation and an inhibitor during tuber initiation and growth in response to environmental cues in potatoes.

Auxin and GAs pathways have also been found to be necessary for hypocotyl elongation and the control of plant growth in response to both shade and high temperature (Kurepin et al., 2007). Increased amounts of IAA and GA, as well as potential cell expansion (Yang et al., 2018), caused plants growing in the shade and high ambient temperature to exhibit hypocotyl elongation features (Yang et al., 2018). Soybean hypocotyl elongation was shown to be regulated by GAs and auxin under low-light and high-temperature conditions, as observed by Bawa et al. (2020).

### 22.9.5 Salicylic Acid–Auxin Cross-Talk

Reductions in salicylic acid (SA) production and responsiveness have been linked to phyB inactivation (De Wit et al., 2013). Shade decreases SA-dependent disease resistance in part because of a low R/FR ratio, which decreases the phosphorylation level of the SA-signaling component nonexpressor of pathogenesis-related gene 1 (NPR1). As a result of SA's inhibition of auxin receptors, Aux/IAA proteins are stabilized and the plant's resilience to disease is enhanced (Wang et al., 2007). To maintain a healthy equilibrium between growth and defense, plants respond to shade by suppressing jasmonate- and salicylic acid–induced defense responses.

## 22.10 RESEARCH GAPS AND FUTURE PERSPECTIVES

Hormones involved in ecophysiological responses to stress and variable nutrient availability are primarily studied in commercially significant crop plants rather than in forest trees. The research on tropical forest vegetation, especially those plants found in different tropical settings, has a significant knowledge deficit. This disjuncture may make it more difficult to comprehend ecophysiological responses to the changing climate in the future. There has to be more extensive research to discover the role of hormones in triggering certain ecophysiological reactions with respect to environmental shifts in the tropics.

## 22.11 CONCLUSION

This chapter concludes with current studies on role of hormones in ecophysiological responses to heavy metal stress, drought stress, salinity stress, and nitrogen, phosphorus, and sulfate deficits. Included in this context is the function of hormones both independently and in concert with other hormones, either in a synergistic or antagonistic manner, to control plant growth and development in response to varying environmental conditions.

## REFERENCES

Abbas, Mohamad, David Alabadí, and Miguel A. Blázquez. "Differential growth at the apical hook: All roads lead to auxin." *Frontiers in Plant Science* 4 (2013): 441.

Abelenda, J. A., and Salomé Prat. "Cytokinins: Determinants of sink storage ability." *Current Biology* 23, no. 13 (2013): R561–R563.

Ågren, Göran I., JÅ Martin Wetterstedt, and Magnus FK Billberger. "Nutrient limitation on terrestrial plant growth–modeling the interaction between nitrogen and phosphorus." *New Phytologist* 194, no. 4 (2012): 953–960.

Agtuca, Beverly, Elisabeth Rieger, Katharina Hilger, Lihui Song, Christelle AM Robert, Matthias Erb, Abhijit Karve, and Richard A. Ferrieri. "Carbon-11 reveals opposing roles of auxin and salicylic acid in regulating leaf physiology, leaf metabolism, and resource allocation patterns that impact root growth in *Zea mays*." *Journal of Plant Growth Regulation* 33, no. 2 (2014): 328–339.

Agusti, J., S. Herold, M. Schwarz, P. Sanchez, K. Ljung, E. A. Dun, P. B. Brewer, C. A. Beveridge, T. Sieberer, E. M. Sehr, and T. Greb. "Strigolactone signaling is required for auxin-dependent stimulation of secondary growth in plants." *Proceedings of the National Academy of Sciences* 108, no. 50 (2011): 20242–20247.

Alatawi, Aishah, Xiukang Wang, Muhammad Hamzah Saleem, Muhammad Mohsin, Muzammal Rehman, Kamal Usman, Shah Fahad, Manar Fawzi Bani Mfarrej, Daniel Ingo Hefft, and Shafaqat Ali. "Individual and synergic effects of phosphorus and gibberellic acid on organic acids exudation pattern, ultra-structure of chloroplast and stress response gene expression in cu-stressed jute (*Corchorus Capsularis* L.)." *Journal of Plant Growth Regulation* (2022): 1–26.

Ali, Adam Yousif Adam, Muhi Eldeen Hussien Ibrahim, Guisheng Zhou, Nimir Eltyb Ahmed Nimir, Aboagla Mohammed Ibrahim Elsiddig, Xiurong Jiao, Guanglong Zhu, Ebtehal Gabralla Ibrahim Salih, Mohamed Suliman Eltyeb Suliman Suliman, and Safiya Babiker Mustafa Elradi. "Gibberellic acid and nitrogen efficiently protect early seedlings growth stage from salt stress damage in Sorghum." *Scientific Reports* 11, no. 1 (2021): 1–11.

Aloni, Roni. "Differentiation of vascular tissues." *Annual review of plant physiology* 38. no. 1 (1987): 179–204.

Aloni, Roni. "Foliar and axial aspects of vascular differentiation: Hypotheses and evidence." *Journal of Plant Growth Regulation* 20, no. 1 (2001).

Aloni, Roni. "Role of hormones in controlling vascular differentiation and the mechanism of lateral root initiation." *Planta* 238, no. 5 (2013): 819–830.

Aloni, Roni, Adam Foster, and Jim Mattsson. "Transfusion tracheids in the conifer leaves of *Thuja plicata* (Cupressaceae) are derived from parenchyma and their differentiation is induced by auxin." *American Journal of Botany* 100, no. 10 (2013): 1949–1956.

Aloni, Roni, Erez Aloni, Markus Langhans, and Cornelia I. Ullrich. "Role of auxin in regulating *Arabidopsis* flower development." *Planta* 223, no. 2 (2006): 315–328.

Aloni, Roni, Markus Langhans, Erez Aloni, Ellen Dreieicher, and Cornelia I. Ullrich. "Root-synthesized cytokinin in *Arabidopsis* is distributed in the shoot by the transpiration stream." *Journal of Experimental Botany* 56, no. 416 (2005): 1535–1544.

Arteca, Richard N., and Jeannette M. Arteca. "Heavy-metal-induced ethylene production in *Arabidopsis thaliana*." *Journal of Plant Physiology* 164, no. 11 (2007): 1480–1488.

Ashraf, M., N. A. Akram, R. N. Arteca, and Majid R. Foolad. "The physiological, biochemical and molecular roles of brassinosteroids and salicylic acid in plant processes and salt tolerance." *Critical Reviews in Plant Sciences* 29, no. 3 (2010): 162–190.

Bailey-Serres, Julia, Takeshi Fukao, Daniel J. Gibbs, Michael J. Holdsworth, Seung Cho Lee, Francesco Licausi, Pierdomenico Perata, Laurentius ACJ Voesenek, and Joost T. van Dongen. "Making sense of low oxygen sensing." *Trends in Plant Science* 17, no. 3 (2012): 129–138.

Ballaré, Carlos L. "Light regulation of plant defense." *Annual Review of Plant Biology* 65 (2014): 335–363.

Barbez, Elke, Martin Kubeš, Jakub Rolčík, Chloé Béziat, Aleš Pěnčík, Bangjun Wang, Michel Ruiz Rosquete et al. "A novel putative auxin carrier family regulates intracellular auxin homeostasis in plants." *Nature* 485, no. 7396 (2012): 119–122.

Bawa, George, Lingyang Feng, Guopeng Chen, Hong Chen, Yun Hu, Tian Pu, Yajiao Cheng et al. "Gibberellins and auxin regulate soybean hypocotyl elongation under low light and high-temperature interaction." *Physiologia Plantarum* 170, no. 3 (2020): 345–356.

Bhadouria, Rahul, Pratap Srivastava, Rishikesh Singh, Sachchidanand Tripathi, Hema Singh, and Akhilesh Singh Raghubanshi. "Tree seedling establishment in dry tropics: An urgent need of interaction studies." *Environment Systems and Decisions* 37, no. 1 (2017): 88–100.

Bhadouria, Rahul, Pratap Srivastava, Rishikesh Singh, Sachchidanand Tripathi, Pramit Verma, and Akhilesh Singh Raghubanshi. "Effects of grass competition on tree seedlings growth under different light and nutrient availability conditions in tropical dry forests in India." *Ecological Research* 35, no. 5 (2020): 807–818.

Bhadouria, Rahul, Rishikesh Singh, Pratap Srivastava, and Akhilesh Singh Raghubanshi. "Understanding the ecology of tree-seedling growth in dry tropical environment: A management perspective." *Energy, Ecology and Environment* 1, no. 5 (2016): 296–309.

Boonman, Alex, and Thijs L. Pons. "Canopy light gradient perception by cytokinin." *Plant Signaling & Behavior* 2, no. 6 (2007): 489–491.

Bou-Torrent, Jordi, Anahit Galstyan, Marçal Gallemí, Nicolás Cifuentes-Esquivel, Maria José Molina-Contreras, Mercè Salla-Martret, Yusuke Jikumaru, Shinjiro Yamaguchi, Yuji Kamiya, and Jaime F. Martínez-García. "Plant proximity perception dynamically modulates hormone levels and sensitivity in Arabidopsis." *Journal of Experimental Botany* 65, no. 11 (2014): 2937–2947.

Braybrook, Siobhan A., and Alexis Peaucelle. "Mechano-chemical aspects of organ formation in *Arabidopsis thaliana*: The relationship between auxin and pectin." *PLoS One* 8, no. 3 (2013): e57813.

Bucci, Sandra J., Fabian G. Scholz, Guillermo Goldstein, Frederick C. Meinzer, Augusto C. Franco, Paula I. Campanello, R. A. N. D. O. L. Villalobos-Vega, Mercedes Bustamante, and F. E. R. N. A. N. D. O. Miralles-WILHELM. "Nutrient availability constrains the hydraulic architecture and water relations of savannah trees." *Plant, Cell & Environment* 29, no. 12 (2006): 2153–2167.

Bustos, Regla, Gabriel Castrillo, Francisco Linhares, María Isabel Puga, Vicente Rubio, Julian Perez-Perez, Roberto Solano, Antonio Leyva, and Javier Paz-Ares. "A central regulatory system largely controls transcriptional activation and repression responses to phosphate starvation in *Arabidopsis*." *PLoS Genetics* 6, no. 9 (2010): e1001102.

Cagnola, Juan Ignacio, Edmundo Ploschuk, Tomás Benech-Arnold, Scott A. Finlayson, and Jorge José Casal. "Stem transcriptome reveals mechanisms to reduce the energetic cost of shade-avoidance responses in tomato." *Plant Physiology* 160, no. 2 (2012): 1110–1119.

Cai, Z. Q., L. Poorter, Q. Han, and F. Bongers. "Effects of light and nutrients on seedlings of tropical *Bauhinia* lianas and trees." *Tree Physiology* 28 (2008): 1277–1285.

Carlquist, Sherwin. "More woodiness/less woodiness: Evolutionary avenues, ontogenetic mechanisms." *International Journal of Plant Sciences* 174, no. 7 (2013): 964–991.

Chazdon, R. L., and N. Fetcher. "Light environments of tropical forests." In *Physiological Ecology of Plants of the Wet Tropics* (pp. 27–36). Springer, Dordrecht, 1984.

Coenen, Catharina, and Terri L. Lomax. "Auxin—cytokinin interactions in higher plants: Old problems and new tools." *Trends in Plant Science* 2, no. 9 (1997): 351–356.

Cusack, Daniela F., Jordan Macy, and William H. McDowell. "Nitrogen additions mobilize soil base cations in two tropical forests." *Biogeochemistry* 128, no. 1 (2016): 67–88.

Dan, Hanbin, Guohua Yang, and Zhi-Liang Zheng. "A negative regulatory role for auxin in sulphate deficiency response in *Arabidopsis thaliana*." *Plant Molecular Biology* 63 (2007): 221–235.

Dayan, Franck E., Abigail Barker, and Patrick J. Tranel. "Origins and structure of chloroplastic and mitochondrial plant protoporphyrinogen oxidases: Implications for the evolution of herbicide resistance." *Pest Management Science* 74, no. 10 (2018): 2226–2234.

De Wit, Mieke, Steven H. Spoel, Gabino F. Sanchez-Perez, Charlotte MM Gommers, Corné MJ Pieterse, Laurentius ACJ Voesenek, and Ronald Pierik. "Perception of low red:

Far-red ratio compromises both salicylic acid-and jasmonic acid-dependent pathogen defences in *Arabidopsis*." *The Plant Journal* 75, no. 1 (2013): 90–103.

Deng, Qi, Dafeng Hui, Sam Dennis, and K. Chandra Reddy. "Responses of terrestrial ecosystem phosphorus cycling to nitrogen addition: A meta-analysis." *Global Ecology and Biogeography* 26, no. 6 (2017): 713–728.

Ding, Yang, Jiping Sheng, Shuying Li, Ying Nie, Jinhong Zhao, Zhen Zhu, Zhidong Wang, and Xuanming Tang. "The role of gibberellins in the mitigation of chilling injury in cherry tomato (*Solanum lycopersicum* L.) fruit." *Postharvest Biology and Technology* 101 (2015): 88–95.

Ding, Yiliang, Peter Kalo, Craig Yendrek, Jongho Sun, Yan Liang, John F. Marsh, Jeanne M. Harris, and Giles ED Oldroyd. "Abscisic acid coordinates nod factor and cytokinin signaling during the regulation of nodulation in *Medicago truncatula*." *The Plant Cell* 20, no. 10 (2008): 2681–2695.

Falcioni, Renan, Thaise Moriwaki, Evanilde Benedito, Carlos Moacir Bonato, Luiz Antônio de Souza, and Werner Camargos Antunes. "Increased gibberellin levels enhance light capture efficiency in tobacco plants and promote dry matter accumulation." *Theoretical and Experimental Plant Physiology* 30, no. 3 (2018): 235–250.

Falkenberg, Bettina, Isabell Witt, Maria Inés Zanor, Dirk Steinhauser, Bernd Mueller-Roeber, Holger Hesse, and Rainer Hoefgen. "Transcription factors relevant to auxin signalling coordinate broad-spectrum metabolic shifts including sulphur metabolism." *Journal of Experimental Botany* 59, no. 10 (2008): 2831–2846.

Fay, Philip A., Suzanne M. Prober, W. Stanley Harpole, Johannes MH Knops, Jonathan D. Bakker, Elizabeth T. Borer, Eric M. Lind et al. "Grassland productivity limited by multiple nutrients." *Nature Plants* 1, no. 7 (2015): 1–5.

Fenn, Matthew A., and James J. Giovannoni. "Phytohormones in fruit development and maturation." *The Plant Journal* 105, no. 2 (2021): 446–458.

Figueiredo-Lima, Karla V., Silvia Pereira, Hiram M. Falcão, Emilia CP Arruda, Alfonso Albacete, André LA Lima, and Mauro G. Santos. "Stomatal conductance and foliar phytohormones under water status changes in *Annona leptopetala*, a woody deciduous species in tropical dry forest." *Flora* 242 (2018): 1–7.

Forde, Brian G. "Nitrogen signalling pathways shaping root system architecture: An update." *Current Opinion in Plant Biology* 21 (2014): 30–36.

Fukuda, Naoya, Chiho Ajima, Tomohisa Yukawa, and Jorunn E. Olsen. "Antagonistic action of blue and red light on shoot elongation in petunia depends on gibberellin, but the effects on flowering are not generally linked to gibberellin." *Environmental and Experimental Botany* 121 (2016): 102–111.

Funayama, Kazuhiro, Soichi Kojima, Mayumi Tabuchi-Kobayashi, Yuki Sawa, Yosuke Nakayama, Toshihiko Hayakawa, and Tomoyuki Yamaya. "Cytosolic glutamine synthetase1; 2 is responsible for the primary assimilation of ammonium in rice roots." *Plant and Cell Physiology* 54, no. 6 (2013): 934–943.

Ghani, Muhammad Awais, Muhammad Mehran Abbas, Basharat Ali, Rukhsanda Aziz, Rashad Waseem Khan Qadri, Anam Noor, Muhammad Azam et al. "Alleviating role of gibberellic acid in enhancing plant growth and stimulating phenolic compounds in carrot (*Daucus carota* L.) under lead stress." *Sustainability* 13, no. 21 (2021): 12329.

Giehl, Ricardo FH, Joni E. Lima, and Nicolaus von Wirén. "Localized iron supply triggers lateral root elongation in *Arabidopsis* by altering the AUX1-mediated auxin distribution." *The Plant Cell* 24, no. 1 (2012): 33–49.

González, Esther M., Loli Gálvez, and Cesar Arrese-Igor. "Abscisic acid induces a decline in nitrogen fixation that involves leghaemoglobin, but is independent of sucrose synthase activity." *Journal of Experimental Botany* 52, no. 355 (2001): 285–293.

Gruber, Benjamin D., Ricardo FH Giehl, Swetlana Friedel, and Nicolaus von Wirén. "Plasticity of the *Arabidopsis* root system under nutrient deficiencies." *Plant Physiology* 163, no. 1 (2013): 161–179.

Guivarc'h, Anne, Jacques Rembur, Marc Goetz, Thomas Roitsch, Michèle Noin, Thomas Schmülling, and Dominique Chriqui. "Local expression of the ipt gene in transgenic tobacco (*Nicotiana tabacum* L. cv. SR1) axillary buds establishes a role for cytokinins in tuberization and sink formation." *Journal of Experimental Botany* 53, no. 369 (2002): 621–629.

Guo, Hongwei, and Joseph R. Ecker. "The ethylene signaling pathway: New insights." *Current Opinion in Plant Biology* 7, no. 1 (2004): 40–49.

Guo, Huiyan, Yucheng Wang, Huizi Liu, Ping Hu, Yuanyuan Jia, Chunrui Zhang, Yanmin Wang, Shan Gu, Chuanping Yang, and Chao Wang. "Exogenous GA3 application enhances xylem development and induces the expression of secondary wall biosynthesis related genes in *Betula platyphylla*." *International Journal of Molecular Sciences* 16, no. 9 (2015): 22960–22975.

Haas, Florian H., Corinna Heeg, Rafael Queiroz, Andrea Bauer, Markus Wirtz, and Rudiger Hell. "Mitochondrial serine acetyltransferase functions as a pacemaker of cysteine synthesis in plant cells." *Plant Physiology* 148, no. 2 (2008): 1055–1067.

Habib, Sidra, Muhammad Waseem, Ning Li, Lu Yang, and Zhengguo Li. "Overexpression of SlGRAS7 affects multiple behaviors leading to confer abiotic stresses tolerance and impacts gibberellin and auxin signaling in tomato." *International Journal of Genomics* 2019 (2019).

Hacke, U. G., R. Spicer, S. G. Schreiber, and L. Plavcová. "An ecophysiological and developmental perspective on variation in vessel diameter." *Plant, Cell & Environment* 40, no. 6 (2017): 831–845.

Harpham, N. V. J., A. W. Berry, E. M. Knee, G. Roveda-Hoyos, Ilya Raskin, I. O. Sanders, A. R. Smith, C. K. Wood, and M. A. Hall. "The effect of ethylene on the growth and development of wild-type and mutant *Arabidopsis thaliana* (L.) Heynh." *Annals of Botany* 68, no. 1 (1991): 55–61.

Hayward, Alice, Petra Stirnberg, Christine Beveridge, and Ottoline Leyser. "Interactions between auxin and strigolactone in shoot branching control." *Plant Physiology* 151, no. 1 (2009): 400–412.

Herrera-Ubaldo, Humberto. "Crosstalk between ethylene, light, and brassinosteroid signaling in the control of apical hook formation." *The Plant Cell* (2022).

Hinz, Manuela, Iain W. Wilson, Jun Yang, Katharina Buerstenbinder, Danny Llewellyn, Elizabeth S. Dennis, Margret Sauter, and Rudy Dolferus. "*Arabidopsis* RAP2.2: An ethylene response transcription factor that is important for hypoxia survival." *Plant Physiology* 153, no. 2 (2010): 757–772.

Hu, Jianhong, Alon Israeli, Naomi Ori, and Tai-ping Sun. "The interaction between DELLA and ARF/IAA mediates crosstalk between gibberellin and auxin signaling to control fruit initiation in tomato." *The Plant Cell* 30, no. 8 (2018): 1710–1728.

Huang, Ke-Lin, Guang-Jing Ma, Mei-Li Zhang, Huan Xiong, Huan Wu, Cai-Zhi Zhao, Chun-Sen Liu et al. "The ARF7 and ARF19 transcription factors positively regulate phosphate starvation response1 in *Arabidopsis* roots." *Plant Physiology* 178, no. 1 (2018): 413–427.

Huang, Zhiqun, Xiaohua Wan, Zongming He, Zaipeng Yu, Minghuang Wang, Zhenhong Hu, and Yusheng Yang. "Soil microbial biomass, community composition and soil nitrogen cycling in relation to tree species in subtropical China." *Soil Biology and Biochemistry* 62 (2013): 68–75.

Iqbal, Muhammad, and Muhammad Ashraf. "Gibberellic acid mediated induction of salt tolerance in wheat plants: Growth, ionic partitioning, photosynthesis, yield and hormonal homeostasis." *Environmental and Experimental Botany* 86 (2013a): 76–85.

Iqbal, Noushina, Asim Masood, M. Iqbal R. Khan, Mohd Asgher, Mehar Fatma, and Nafees A. Khan. "Cross-talk between sulfur assimilation and ethylene signaling in plants." *Plant Signaling & Behavior* 8, no. 1 (2013b): e22478.

Javed, Talha, Muhammad Moaaz Ali, Rubab Shabbir, Raheel Anwar, Irfan Afzal, and Rosario Paolo Mauro. "Alleviation of copper-induced stress in pea (*Pisum sativum* L.) through foliar application of gibberellic acid." *Biology* 10, no. 2 (2021): 120.

Jia, Zhongtao, Ricardo FH. Giehl, and Nicolaus von Wirén. "Local auxin biosynthesis acts downstream of brassinosteroids to trigger root foraging for nitrogen." *Nature Communications* 12, no. 1 (2021): 1–12.

Jia, Zhongtao, Ricardo FH. Giehl, and Nicolaus von Wirén. "The root foraging response under low nitrogen depends on DWARF1-mediated brassinosteroid biosynthesis." *Plant Physiology* 183, no. 3 (2020): 998–1010.

Jia, Zhongtao, Ricardo FH. Giehl, Rhonda C. Meyer, Thomas Altmann, and Nicolaus von Wirén. "Natural variation of BSK3 tunes brassinosteroid signaling to regulate root foraging under low nitrogen." *Nature Communications* 10, no. 1 (2019): 2378.

Jung, Ji-Yul, Ryoung Shin, and Daniel P. Schachtman. "Ethylene mediates response and tolerance to potassium deprivation in Arabidopsis." *The Plant Cell* 21, no. 2 (2009): 607–621.

Kakar, Klementina, Hongtao Zhang, Ben Scheres, and Pankaj Dhonukshe. "CLASP-mediated cortical microtubule organization guides PIN polarization axis." *Nature* 495, no. 7442 (2013): 529–533.

Kaneda, Minako, Mathias Schuetz, Billy SP Lin, Carolina Chanis, B. Hamberger, T. L. Western, J. Ehlting, and A. L. Samuels. "ABC transporters coordinately expressed during lignification of *Arabidopsis* stems include a set of ABCBs associated with auxin transport." *Journal of Experimental Botany* 62, no. 6 (2011): 2063–2077.

Keunen, Els, Kerim Schellingen, Jaco Vangronsveld, and Ann Cuypers. "Ethylene and metal stress: Small molecule, big impact." *Frontiers in Plant Science* 7 (2016): 23.

Kiba, Takatoshi, Toru Kudo, Mikiko Kojima, and Hitoshi Sakakibara. "Hormonal control of nitrogen acquisition: Roles of auxin, abscisic acid, and cytokinin." *Journal of Experimental Botany* 62, no. 4 (2011): 1399–1409.

Kim, Keunhwa, Jieun Shin, Sang-Hee Lee, Hee-Seok Kweon, Julin N. Maloof, and Giltsu Choi. "Phytochromes inhibit hypocotyl negative gravitropism by regulating the development of endodermal amyloplasts through phytochrome-interacting factors." *Proceedings of the National Academy of Sciences* 108, no. 4 (2011): 1729–1734.

Kloosterman, Bjorn, Christina Navarro, Gerard Bijsterbosch, Theo Lange, Salomé Prat, Richard GF Visser, and Christian WB Bachem. "StGA2ox1 is induced prior to stolon swelling and controls GA levels during potato tuber development." *The Plant Journal* 52, no. 2 (2007): 362–373.

Kohlen, Wouter, Tatsiana Charnikhova, Michiel Lammers, Tobia Pollina, Peter Tóth, Imran Haider, María J. Pozo et al. "The tomato CAROTENOID CLEAVAGE DIOXYGENASE 8 (S l CCD 8) regulates rhizosphere signaling, plant architecture and affects reproductive development through strigolactone biosynthesis." *New Phytologist* 196, no. 2 (2012): 535–547.

Kuiper, Daan, Pieter JC Kuiper, Hans Lambers, Jacqueline Schuit, and Marten Staal. "Cytokinin concentration in relation to mineral nutrition and benzyladenine treatment in Plantago major ssp. pleiosperma." *Physiologia Plantarum* 75, no. 4 (1989): 511–517.

Kurepin, Leonid V., RJ Neil Emery, Richard P. Pharis, and David M. Reid. "Uncoupling light quality from light irradiance effects in *Helianthus annuus* shoots: Putative roles for plant hormones in leaf and internode growth." *Journal of Experimental Botany* 58, no. 8 (2007): 2145–2157.

Laplaze, Laurent, Eva Benkova, Ilda Casimiro, Lies Maes, Steffen Vanneste, Ranjan Swarup, Dolf Weijers et al. "Cytokinins act directly on lateral root founder cells to inhibit root initiation." *The Plant Cell* 19, no. 12 (2007): 3889–3900.

Leitner, Johannes, Katarzina Retzer, Barbara Korbei, and Christian Luschnig. "Dynamics in PIN2 auxin carrier ubiquitylation in gravity-responding *Arabidopsis* roots." *Plant Signaling & Behavior* 7, no. 10 (2012): 1271–1273.

Léran, Sophie, Kai H. Edel, Marjorie Pervent, Kenji Hashimoto, Claire Corratgé-Faillie, Jan Niklas Offenborn, Pascal Tillard, Alain Gojon, Jörg Kudla, and Benoît Lacombe. "Nitrate sensing and uptake in *Arabidopsis* are enhanced by ABI2, a phosphatase inactivated by the stress hormone abscisic acid." *Science Signaling* 8, no. 375 (2015): ra43.

Li, Jialong, Bang-Zhen Pan, Longjian Niu, Mao-Sheng Chen, Mingyong Tang, and Zeng-Fu Xu. "Gibberellin inhibits floral initiation in the perennial woody plant *Jatropha curcas*." *Journal of Plant Growth Regulation* 37, no. 3 (2018): 999–1006.

Li, Ning. "The dual-and-opposing-effect of ethylene on the negative gravitropism of *Arabidopsis* inflorescence stem and light-grown hypocotyls." *Plant Science* 175, no. 1–2 (2008): 71–86.

Lima, Andre Almeida, Iasminy Silva Santos, Marlon Enrique López Torres, Carlos Henrique Cardon, Cecílio Frois Caldeira, Renato Ribeiro Lima, William John Davies, Ian Charles Dodd, and Antonio Chalfun-Junior. "Drought and re-watering modify ethylene production and sensitivity, and are associated with coffee anthesis." *Environmental and Experimental Botany* 181 (2021): 104289.

Lima, Joni E., Soichi Kojima, Hideki Takahashi, and Nicolaus von Wirén. "Ammonium triggers lateral root branching in *Arabidopsis* in an Ammonium transporter1; 3-dependent manner." *The Plant Cell* 22, no. 11 (2010): 3621–3633.

Lin, Syuan-You, and Shinsuke Agehara. "Exogenous gibberellic acid advances reproductive phenology and increases early-season yield in subtropical blackberry production." *Agronomy* 10, no. 9 (2020): 1317.

Liu, Yi, Qi Yin, Baojia Dai, Kai-li Wang, Lu Lu, Mirza Faisal Qaseem, Jinxiang Wang, Huiling Li, and Ai-Min Wu. "The key physiology and molecular responses to potassium deficiency in *Neolamarckia cadamba*." *Industrial Crops and Products* 162 (2021): 113260.

Liu, Ying, and Nicolaus von Wirén. "Integration of nutrient and water availabilities via auxin into the root developmental program." *Current Opinion in Plant Biology* 65 (2022): 102117.

Long, Sixin, Fengying Yan, Lin Yang, Zhenyuan Sun, and Shanjun Wei. "Responses of Manila Grass (*Zoysia matrella*) to chilling stress: From transcriptomics to physiology." *PloS One* 15, no. 7 (2020): e0235972.

Lv, Yan, Jinjing Pan, Houping Wang, Russel J. Reiter, Xia Li, Zongmin Mou, Jiemei Zhang, Zhengping Yao, Dake Zhao, and Diqiu Yu. "Melatonin inhibits seed germination by crosstalk with abscisic acid, gibberellin, and auxin in *Arabidopsis*." *Journal of Pineal Research* 70, no. 4 (2021): e12736.

Ma, Wenying, Jingjuan Li, Baoyuan Qu, Xue He, Xueqiang Zhao, Bin Li, Xiangdong Fu, and Yiping Tong. "Auxin biosynthetic gene TAR 2 is involved in low nitrogen-mediated reprogramming of root architecture in *Arabidopsis*." *The Plant Journal* 78, no. 1 (2014): 70–79.

Maleki, Abbas, Amin Fathi, and Sadegh Bahamin. "The effect of gibberellin hormone on yield, growth indices, and biochemical traits of corn (*Zea Mays* L.) under drought stress." *Journal of Iranian Plant Ecophysiological Research* 15, no. 59 (2020): 1–16.

Masood, Asim, M. Iqbal R. Khan, Mehar Fatma, Mohd Asgher, Tasir S. Per, and Nafees A. Khan. "Involvement of ethylene in gibberellic acid-induced sulfur assimilation, photosynthetic responses, and alleviation of cadmium stress in mustard." *Plant Physiology and Biochemistry* 104 (2016): 1–10.

Masood, Asim, Noushina Iqbal, and Nafees A. Khan. "Role of ethylene in alleviation of cadmium-induced photosynthetic capacity inhibition by sulphur in mustard." *Plant, Cell & Environment* 35, no. 3 (2012): 524–533.

Matsumoto-Kitano, Miho, Takami Kusumoto, Petr Tarkowski, Kaori Kinoshita-Tsujimura, Kateřina Václavíková, Kaori Miyawaki, and Tatsuo Kakimoto. "Cytokinins are central regulators of cambial activity." *Proceedings of the National Academy of Sciences* 105, no. 50 (2008): 20027–20031.

McAdam, Scott AM, Morgane P. Eléouët, Melanie Best, Timothy J. Brodribb, Madeline Carins Murphy, Sam D. Cook, Marion Dalmais et al. "Linking auxin with photosynthetic rate via leaf venation." *Plant Physiology* 175, no. 1 (2017): 351–360.

Meier, Markus, Ying Liu, Katerina S. Lay-Pruitt, Hideki Takahashi, and Nicolaus von Wirén. "Auxin-mediated root branching is determined by the form of available nitrogen." *Nature Plants* 6, no. 9 (2020): 1136–1145.

Moreno, Javier E., Yi Tao, Joanne Chory, and Carlos L. Ballaré. "Ecological modulation of plant defense via phytochrome control of jasmonate sensitivity." *Proceedings of the National Academy of Sciences* 106, no. 12 (2009): 4935–4940.

Moula, Ines, Olfa Boussadia, Georgios Koubouris, Mortadha Ben Hassine, Wafa Boussetta, Marie Christine Van Labeke, and Mohamed Braham. "Ecophysiological and biochemical aspects of olive tree (*Olea europaea* L.) in response to salt stress and gibberellic acid-induced alleviation." *South African Journal of Botany* 132 (2020): 38–44.

Mounier, Emmanuelle, Marjorie Pervent, Karin Ljung, Alain Gojon, and Philippe Nacry. "Auxin-mediated nitrate signalling by NRT 1.1 participates in the adaptive response of *Arabidopsis* root architecture to the spatial heterogeneity of nitrate availability." *Plant, Cell & Environment* 37, no. 1 (2014): 162–174.

Muday, Gloria K., Abidur Rahman, and Brad M. Binder. "Auxin and ethylene: Collaborators or competitors?." *Trends in Plant Science* 17, no. 4 (2012): 181–195.

Nakagawa, Masahiro, Chitose Honsho, Shinya Kanzaki, Kousuke Shimizu, and Naoki Utsunomiya. "Isolation and expression analysis of flowering locus t-like and gibberellin metabolism genes in biennial-bearing mango trees." *Scientia Horticulturae* 139 (2012): 108–117.

Ni, Jun, Congcong Gao, Mao-Sheng Chen, Bang-Zhen Pan, Kaiqin Ye, and Zeng-Fu Xu. "Gibberellin promotes shoot branching in the perennial woody plant *Jatropha curcas*." *Plant and Cell Physiology* 56, no. 8 (2015): 1655–1666.

Niakan, M., and M. Habibi. "The effect of cytokinin on growth indicators and photosynthesis of *Cucurbita maxima* L. under different levels of drought." *Journal of Iranian Plant Ecophysiological Research* 11, no. 42 (2016): 56–65.

Nieminen, Kaisa, Juha Immanen, Marjukka Laxell, Leila Kauppinen, Petr Tarkowski, Karel Dolezal, Sari Tähtiharju et al. "Cytokinin signaling regulates cambial development in poplar." *Proceedings of the National Academy of Sciences* 105, no. 50 (2008): 20032–20037.

Novák, Jan, Martin Černý, Jaroslav Pavlů, Jana Zemánková, Jan Skalák, Lenka Plačková, and Břetislav Brzobohatý. "Roles of proteome dynamics and cytokinin signaling in root to hypocotyl ratio changes induced by shading roots of *Arabidopsis* seedlings." *Plant and Cell Physiology* 56, no. 5 (2015): 1006–1018.

Ondzighi-Assoume, Christine A., Sanhita Chakraborty, and Jeanne M. Harris. "Environmental nitrate stimulates abscisic acid accumulation in *Arabidopsis* root tips by releasing it from inactive stores." *The Plant Cell* 28, no. 3 (2016): 729–745.

Pacín, Manuel, Mariana Semmoloni, Martina Legris, Scott A. Finlayson, and Jorge J. Casal. "Convergence of constitutive photomorphogenesis 1 and phytochrome interacting factor signalling during shade avoidance." *New Phytologist* 211, no. 3 (2016): 967–979.

Pasare, Stefania A., Laurence JM Ducreux, Wayne L. Morris, Raymond Campbell, Sanjeev K. Sharma, Efstathios Roumeliotis, Wouter Kohlen et al. "The role of the potato (*Solanum tuberosum*) CCD8 gene in stolon and tuber development." *New Phytologist* 198, no. 4 (2013): 1108–1120.

Pérez-Torres, Claudia-Anahí, José López-Bucio, Alfredo Cruz-Ramírez, Enrique Ibarra-Laclette, Sunethra Dharmasiri, Mark Estelle, and Luis Herrera-Estrella. "Phosphate availability alters lateral root development in *Arabidopsis* by modulating auxin sensitivity via a mechanism involving the TIR1 auxin receptor." *The Plant Cell* 20, no. 12 (2008): 3258–3272.

Pierik, Ronald, Mieke De Wit, and Laurentius ACJ. Voesenek. "Growth-mediated stress escape: Convergence of signal transduction pathways activated upon exposure to two different environmental stresses." *New Phytologist* 189, no. 1 (2011): 122–134.

Pierik, Ronald, Mieke LC. Cuppens, Laurentius ACJ. Voesenek, and Eric JW. Visser. "Interactions between ethylene and gibberellins in phytochrome-mediated shade avoidance responses in tobacco." *Plant Physiology* 136, no. 2 (2004): 2928–2936.

Pierik, Ronald, Tanja Djakovic-Petrovic, Diederik H. Keuskamp, Mieke de Wit, and Laurentius ACJ. Voesenek. "Auxin and ethylene regulate elongation responses to neighbor proximity signals independent of gibberellin and DELLA proteins in *Arabidopsis*." *Plant Physiology* 149, no. 4 (2009): 1701–1712.

Polko, Joanna K., Martijn van Zanten, Jop A. van Rooij, Athanasius FM Marée, Laurentius ACJ Voesenek, Anton JM Peeters, and Ronald Pierik. "Ethylene-induced differential petiole growth in *Arabidopsis thaliana* involves local microtubule reorientation and cell expansion." *New Phytologist* 193, no. 2 (2012): 339–348.

Pons, Thijs L., Wilco Jordi, and Daan Kuiper. "Acclimation of plants to light gradients in leaf canopies: Evidence for a possible role for cytokinins transported in the transpiration stream." *Journal of Experimental Botany* 52, no. 360 (2001): 1563–1574.

Pratelli, Réjane, and Guillaume Pilot. "Regulation of amino acid metabolic enzymes and transporters in plants." *Journal of Experimental Botany* 65, no. 19 (2014): 5535–5556.

Prior, Lynda D., and David MJS Bowman. "Across a macroecological gradient forest competition is strongest at the most productive sites." *Frontiers in Plant Science* 5 (2014): 260.

Reddy, Srirama Krishna, Srinidhi V. Holalu, Jorge J. Casal, and Scott A. Finlayson. "Abscisic acid regulates axillary bud outgrowth responses to the ratio of red to far-red light." *Plant Physiology* 163, no. 2 (2013): 1047–1058.

Remans, Tony, Philippe Nacry, Marjorie Pervent, Sophie Filleur, Eugene Diatloff, Emmanuelle Mounier, Pascal Tillard, Brian G. Forde, and Alain Gojon. "The *Arabidopsis* NRT1.1 transporter participates in the signaling pathway triggering root colonization of nitrate-rich patches." *Proceedings of the National Academy of Sciences* 103, no. 50 (2006): 19206–19211.

Roumeliotis, Efstathios, Bjorn Kloosterman, Marian Oortwijn, Wouter Kohlen, Harro J. Bouwmeester, Richard GF Visser, and Christian WB Bachem. "The effects of auxin and strigolactones on tuber initiation and stolon architecture in potato." *Journal of Experimental Botany* 63, no. 12 (2012): 4539–4547.

Ruffel, Sandrine, Gabriel Krouk, Daniela Ristova, Dennis Shasha, Kenneth D. Birnbaum, and Gloria M. Coruzzi. "Nitrogen economics of root foraging: Transitive closure of the nitrate–cytokinin relay and distinct systemic signaling for N supply vs. demand." *Proceedings of the National Academy of Sciences* 108, no. 45 (2011): 18524–18529.

Rüscher, David, José María Corral, Anna Vittoria Carluccio, Patrick AW Klemens, Andreas Gisel, Livia Stavolone, H. Ekkehard Neuhaus, Frank Ludewig, Uwe Sonnewald, and Wolfgang Zierer. "Auxin signaling and vascular cambium formation enable storage metabolism in cassava tuberous roots." *Journal of Experimental Botany* 72, no. 10 (2021): 3688–3703.

Sairam, R. K., D. Kumutha, K. Ezhilmathi, P. S. Deshmukh, and G. C. Srivastava. "Physiology and biochemistry of waterlogging tolerance in plants." *Biologia Plantarum* 52, no. 3 (2008): 401–412.

Sakakibara, H., Takei, K., & Hirose, N. (2006). Interactions between nitrogen and cytokinin in the regulation of metabolism and development. *Trends in Plant Science*, 11(9), 440–448.

Saleem, Muhammad Hamzah, Shafaqat Ali, Muhammad Kamran, Naeem Iqbal, Muhammad Azeem, Muhammad Tariq Javed, Qasim Ali et al. "Ethylenediaminetetraacetic acid (EDTA) mitigates the toxic effect of excessive copper concentrations on growth, gaseous exchange and chloroplast ultrastructure of *Corchorus capsularis* L. and improves copper accumulation capabilities." *Plants* 9, no. 6 (2020): 756.

Saleem, Muhammad Hamzah, Shah Fahad, Muhammad Adnan, Mohsin Ali, Muhammad Shoaib Rana, Muhammad Kamran, Qurban Ali et al. "Foliar application of gibberellic acid endorsed phytoextraction of copper and alleviates oxidative stress in jute (*Corchorus capsularis* L.) plant grown in highly copper-contaminated soil of China." *Environmental Science and Pollution Research* 27, no. 29 (2020): 37121–37133.

Saleem, Muhammad Hamzah, Xiukang Wang, Sajjad Ali, Sadia Zafar, Muhammad Nawaz, Muhammad Adnan, Shah Fahad et al. "Interactive effects of gibberellic acid and NPK on morpho-physio-biochemical traits and organic acid exudation pattern in coriander (*Coriandrum sativum* L.) grown in soil artificially spiked with boron." *Plant Physiology and Biochemistry* 167 (2021): 884–900.

Sampaio Filho, Israel De Jesus, Kolby Jeremiah Jardine, Rosilena Conceição Azevedo De Oliveira, Bruno Oliva Gimenez, Leticia Oliveira Cobello, Luani Rosa de Oliveira Piva, Luiz Antonio Candido, Niro Higuchi, and Jeffrey Quintin Chambers. 2018. "Below versus above Ground Plant Sources of Abscisic Acid (ABA) at the Heart of Tropical Forest Response to Warming" *International Journal of Molecular Sciences* 19, no. 7: 2023.

Sasidharan, Rashmi, and Laurentius ACJ. Voesenek. "Ethylene-mediated acclimations to flooding stress." *Plant Physiology* 169, no. 1 (2015): 3–12.

Sasidharan, Rashmi, C. C. Chinnappa, Marten Staal, J. Theo M. Elzenga, Ryusuke Yokoyama, Kazuhiko Nishitani, Laurentius ACJ. Voesenek, and Ronald Pierik. "Light quality-mediated petiole elongation in *Arabidopsis* during shade avoidance involves cell wall modification by xyloglucanendotransglucosylase/hydrolases." *Plant Physiology* 154, no. 2 (2010): 978–990.

Sasidharan, Rashmi, Sjon Hartman, Zeguang Liu, Shanice Martopawiro, Nikita Sajeev, Hans van Veen, Elaine Yeung, and Laurentius ACJ. Voesenek. "Signal dynamics and interactions during flooding stress." *Plant Physiology* 176, no. 2 (2018): 1106–1117.

Scarpella, Enrico, Michalis Barkoulas, and Miltos Tsiantis. "Control of leaf and vein development by auxin." *Cold Spring Harbor Perspectives in Biology* 2, no. 1 (2010): a001511.

Séguéla, Mathilde, Jean-François Briat, Grégory Vert, and Catherine Curie. "Cytokinins negatively regulate the root iron uptake machinery in *Arabidopsis* through a growth-dependent pathway." *The Plant Journal* 55, no. 2 (2008): 289–300.

Shaikha, A. S. A. S., S. S. A. S. Shamsa, A. R. Gabriel, S. S. Kurup, and A. J. Cheruth. "Exogenous gibberellic acid ameliorates salinity-induced morphological and biochemical alterations in *Portulaca grandiflora*." *Planta Daninha* 35 (2018).

Shi, Hui, Xing Shen, Renlu Liu, Chang Xue, Ning Wei, Xing Wang Deng, and Shangwei Zhong. "The red-light receptor phytochrome B directly enhances substrate-E3 ligase interactions to attenuate ethylene responses." *Developmental Cell* 39, no. 5 (2016): 597–610.

Shinohara, Naoki, Catherine Taylor, and Ottoline Leyser. "Strigolactone can promote or inhibit shoot branching by triggering rapid depletion of the auxin efflux protein PIN1 from the plasma membrane." *PLoS Biology* 11, no. 1 (2013): e1001474.

Steffens, Bianka, and Amanda Rasmussen. "The physiology of adventitious roots." *Plant Physiology* 170, no. 2 (2016): 603–617.

Tanaka, Yoko, Toshio Sano, Masanori Tamaoki, Nobuyoshi Nakajima, Noriaki Kondo, and Seiichiro Hasezawa. "Ethylene inhibits abscisic acid-induced stomatal closure in *Arabidopsis*." *Plant Physiology* 138, no. 4 (2005): 2337–2343.

Thao, Nguyen Phuong, M. Iqbal R. Khan, Nguyen Binh Anh Thu, Xuan Lan Thi Hoang, Mohd Asgher, Nafees A. Khan, and Lam-Son Phan Tran. "Role of ethylene and its cross talk with other signaling molecules in plant responses to heavy metal stress." *Plant Physiology* 169, no. 1 (2015): 73–84.

Thomas, S. C., & Baltzer, J. L. (2002). *Tropical Forests. Encyclopaedia of Life Sciences*. Nature Publishing Group. www.els.net. DOI: https://doi.org/10.1038/npg.els.0003179. Accessed on 11/8/2002.

Tiwari, Poonam, Yuvraj Indoliya, Abhishek Singh Chauhan, Puja Singh, Pradyumna Kumar Singh, Poonam C. Singh, Suchi Srivastava, Veena Pande, and Debasis Chakrabarty.

"Auxin-salicylic acid cross-talk ameliorates OsMYB–R1 mediated defense towards heavy metal, drought and fungal stress." *Journal of Hazardous Materials* 399 (2020): 122811.

Tranbarger, T. J., Y. Al-Ghazi, Bertrand Muller, B. Teyssendier De La Serve, Patrick Doumas, and B. Touraine. "Transcription factor genes with expression correlated to nitrate-related root plasticity of *Arabidopsis* thaliana." *Plant, Cell & Environment* 26, no. 3 (2003): 459–469.

Tripathi, S. N., and A. S. Raghubanshi." Seedling growth of five tropical dry forest species in relation to light and nitrogen gradients." *Journal of Plant Ecology* 7 (2014): 250–263.

Tripathi, Sachchidanand, Rahul Bhadouria, Pratap Srivastava, Rajkumari S. Devi, Ravikant Chaturvedi, and A. S. Raghubanshi. "Effects of light availability on leaf attributes and seedling growth of four tree species in tropical dry forest." *Ecological Processes* 9, no. 1 (2020): 1–16.

Tsai, Kuen-jin, Shu-jen Chou, and Ming-che Shih. "Ethylene plays an essential role in the recovery of *Arabidopsis* during post-anaerobiosis reoxygenation." *Plant, Cell & Environment* 37, no. 10 (2014): 2391–2405.

Tsegay, Berhanu Abraha, and Melkamu Andargie. "Seed priming with gibberellic acid (GA3) alleviates salinity induced inhibition of germination and seedling growth of *Zea mays* L., *Pisum sativum* var. *abyssinicum* A. Braun and *Lathyrus sativus* L." *Journal of Crop Science and Biotechnology* 21, no. 3 (2018): 261–267.

Utsumi, Yoshinori, Maho Tanaka, Chikako Utsumi, Satoshi Takahashi, Akihiro Matsui, Atsushi Fukushima, Makoto Kobayashi et al. "Integrative omics approaches revealed a crosstalk among phytohormones during tuberous root development in cassava." *Plant Molecular Biology* (2020): 1–21.

Vandenbussche, Filip, Pieter Callebert, Petra Zadnikova, Eva Benkova, and Dominique Van Der Straeten. "Brassinosteroid control of shoot gravitropism interacts with ethylene and depends on auxin signaling components." *American Journal of Botany* 100, no. 1 (2013): 215–225.

Voesenek, L. A. C. J., and R. Sasidharan. "Ethylene–and oxygen signalling–drive plant survival during flooding." *Plant Biology* 15, no. 3 (2013): 426–435.

Wang, Dong, Karolina Pajerowska-Mukhtar, Angela Hendrickson Culler, and Xinnian Dong. "Salicylic acid inhibits pathogen growth in plants through repression of the auxin signaling pathway." *Current Biology* 17, no. 20 (2007): 1784–1790.

Watkins, Justin M., Paul J. Hechler, and Gloria K. Muday. "Ethylene-induced flavonol accumulation in guard cells suppresses reactive oxygen species and moderates stomatal aperture." *Plant Physiology* 164, no. 4 (2014): 1707–1717.

Wu, Xiaoying, Chaohui Ding, Scott R. Baerson, Fazhuo Lian, Xianhui Lin, Liqin Zhang, Choufei Wu, Shaw-Yhi Hwang, Rensen Zeng, and Yuanyuan Song. "The roles of jasmonate signalling in nitrogen uptake and allocation in rice (Oryza sativa L.)." *Plant, Cell & Environment* 42, no. 2 (2019): 659–672.

Xiang, Xiaoli, Yongrui Wu, José Planta, Joachim Messing, and Thomas Leustek. "Overexpression of serine acetyltransferase in maize leaves increases seed-specific methionine-rich zeins." *Plant Biotechnology Journal* 16, no. 5 (2018): 1057–1067.

Xu, Jin, Wenying Wang, Jianhang Sun, Yuan Zhang, Qing Ge, Liguo Du, Hengxia Yin, and Xiaojing Liu. "Involvement of auxin and nitric oxide in plant Cd-stress responses." *Plant and Soil* 346, no. 1 (2011): 107–119.

Xu, Xin, André AM van Lammeren, Evert Vermeer, and Dick Vreugdenhil. "The role of gibberellin, abscisic acid, and sucrose in the regulation of potato tuber formation in vitro." *Plant Physiology* 117, no. 2 (1998): 575–584.

Xu, Zheng-Yi, Kwang Hee Lee, Ting Dong, Jae Cheol Jeong, Jing Bo Jin, Yuri Kanno, Dae Heon Kim et al. "A vacuolar β-glucosidase homolog that possesses glucose-conjugated abscisic acid hydrolyzing activity plays an important role in osmotic stress responses in *Arabidopsis*." *The Plant Cell* 24, no. 5 (2012): 2184–2199.

Yamaguchi, Nobutoshi, Cara M. Winter, Miin-Feng Wu, Yuri Kanno, Ayako Yamaguchi, Mitsunori Seo, and Doris Wagner. "Gibberellin acts positively then negatively to control onset of flower formation in *Arabidopsis*." *Science* 344, no. 6184 (2014): 638–641.

Yang, Feng, Yuanfang Fan, Xiaoling Wu, Yajiao Cheng, Qinlin Liu, Lingyang Feng, Junxu Chen et al. "Auxin-to-gibberellin ratio as a signal for light intensity and quality in regulating soybean growth and matter partitioning." *Frontiers in Plant Science* 9 (2018): 56.

Yang, Huaiyu, Yvonne Klopotek, Mohammad R. Hajirezaei, Siegfried Zerche, Philipp Franken, and Uwe Druege. "Role of auxin homeostasis and response in nitrogen limitation and dark stimulation of adventitious root formation in petunia cuttings." *Annals of Botany* 124, no. 6 (2019): 1053–1066.

Yang, J. C., Zhang, J. H., Ye, Y. X., Wang, Z. Q., Zhu, Q. S., & Liu, L. J. (2004). Involvement of abscisic acid and ethylene in the responses of rice grains to water stress during filling. *Plant, Cell & Environment*, 27(8), 1055–1064.

Yoshida, Saiko, Kuninori Iwamoto, Taku Demura, and Hiroo Fukuda. "Comprehensive analysis of the regulatory roles of auxin in early transdifferentiation into xylem cells." *Plant Molecular Biology* 70, no. 4 (2009): 457–469.

Zhang, Yu, Ting-Ting Li, Lin-Feng Wang, Jia-Xing Guo, Kai-Kai Lu, Ru-Feng Song, Jia-Xin Zuo, Hui-Hui Chen, and Wen-Cheng Liu. "Abscisic acid facilitates phosphate acquisition through the transcription factor ABA insensitive5 in *ARABidopsis*." *The Plant Journal* 111, no. 1 (2022): 269–281.

Zhang, Yushi, Yubin Wang, Delian Ye, Jiapeng Xing, Liusheng Duan, Zhaohu Li, and Mingcai Zhang. "Ethephon-regulated maize internode elongation associated with modulating auxin and gibberellin signal to alter cell wall biosynthesis and modification." *Plant Science* 290 (2020): 110196.

Zhao, Qing, Ping-Xia Zhao, Yu Wu, Chang-Quan Zhong, Hong Liao, Chuan-You Li, Xiang-Dong Fu, Ping Fang, Ping Xu, and Cheng-Bin Xiang. "SUE4, a novel PIN1-interacting membrane protein, regulates acropetal auxin transport in response to sulfur deficiency." *New Phytologist* (2022).

Zhao, Qing, Yu Wu, Lei Gao, Jun Ma, Chuan-You Li, and Cheng-Bin Xiang. "Sulfur nutrient availability regulates root elongation by affecting root indole-3-acetic acid levels and the stem cell niche." *Journal of Integrative Plant Biology* 56, no. 12 (2014): 1151–1163.

Zhu, Xiao Fang, Tao Jiang, Zhi Wei Wang, Gui Jie Lei, Yuan Zhi Shi, Gui Xin Li, and Shao Jian Zheng. "Gibberellic acid alleviates cadmium toxicity by reducing nitric oxide accumulation and expression of IRT1 in *Arabidopsis thaliana*." *Journal of Hazardous Materials* 239 (2012): 302–307.

# 23 Next-Generation Techniques in Ecophysiology
## Metabolomics, Proteomics, SAR/QSAR

*Priyanka Rathore and Rashmi Shakya*

## 23.1 INTRODUCTION

Global climate change greatly affects the productivity and physiological aspects of plants. These different stress conditions can disturb homeostasis and imbalance the normal developmental pathways. Being sessile organisms, plants can survive and respond to adverse climatic conditions by controlling different physiological processes. Abiotic factors severely affect the physiology of the plant and thus lead to crop yield loss (Zörb et al. 2019). In response to abiotic stresses such as drought stress, temperature stress, metal stress and oxidative stress, plants activate various biochemical, physiological and metabolic pathways that help to combat stress. Next-generation "omics" technologies such as genomics, proteomics, metabolomics and transcriptomics provides insight to the plant stress mechanisms by analyzing the complex biological networks (Shu et al., 2022). It provides novel information that helps in design of strategies for plant stress response pathways. All "omics" approaches are largely dependent on various advanced bioinformatic tools for the analysis and interpretation of large datasets. Raw data generated is further processed for noise filtering, peak detection and alignment to generate homogenous information and comparison of samples by statistical methods (Kaur et al., 2021). Multivariate analysis can then be performed to reduce the dimensionality of data, clusters of samples, simplification, and uniformity of variables, such as principal component analysis (PCA), hierarchical cluster analysis (HCA) and analysis of variance (ANOVA) (Worley & Powers, 2013). PCA is a multivariate method used for the interpretation of large datasets and the reduction of dimensions by creating principal components and observing the variance in data. Omics technologies have advanced analytical tools with large-scale computational facilities and high-throughput data analysis pipelines by compiling information from expression, protein profiling with metabolic pathways of the genes under various biotic and abiotic conditions in ecophysiological aspects of the plants. Transcriptome does not represent exact transcriptional activations of the of cellular messenger ribonucleic acid (mRNA). Proteomics display the expressed genes and the functional proteins which play critical role in stress pathways. However, metabolic pathways account for the important metabolites providing adaptions and combating stress mechanism at the physiological level in adverse environmental conditions. All these omics approaches—including transcriptomics, proteomics and metabolomics—have contributed in deciphering the mechanisms involved in abiotic stresses to mitigate them. However, a key challenge in such studies is to integrate large complex datasets from different studies and evaluation of "omics" datasets for the identification of genes, transcripts, proteins and metabolites.

## 23.2 PROTEOMICS IN PLANT STRESS

Being sessile, plants are more prone to stress conditions in changing environmental conditions such as drought, temperature shift, salinity and heavy metals that alter many physiological processes by altering the molecular mechanisms to tackle adverse stress conditions (He et al., 2018). Plants develop different strategies to cope up the stresses at the physiological, biochemical, cellular and molecular levels (Hasanuzzaman et al., 2013). Proteins are biomolecules that perform a vital role in cellular functions and maintaining homeostasis. Proteomics is a branch of molecular biology that deals with the characterization of the entire protein complement and quantitative analysis of protein profiles of the organs, tissues and cells of living organisms under defined conditions. It includes an integrative view of different perspectives of molecular biology disciplines of protein studies with analyses such as genomics, transcriptomics and yeast two-hybrid analysis. Currently, much of the knowledge is available mainly from genomic or transcriptomic approaches in providing an essential understanding of cellular function. However, information from the study of genes alone does not provide enough information about the mechanisms of various diseases/stress and their environments (Manzoni et al., 2018).

The proteome is dynamic and it represents the environment of the cell in which it is analyzed. Any change in intrinsic and extrinsic factors leads to a change in the abundance of proteins in a cell. Additionally, proteomics is a methodological approach to the study of the proteome of the living organism, as the whole set of proteins in a spatially and temporally studied biological entity: organelles, cell, tissue, organ, organisms, population (Chandramouli & Qian, 2009). Therefore, the study of proteomics requires very sensitive techniques and different aspects of proteomics studies such as protein expression profiling, protein-protein interaction, proteome-mining, structural proteomics, functional proteomics and post-translational modifications (Graves & Haystead, 2002). All these studies deal with annotation, functional characterization and localization of proteins, which can be studied through both in vitro techniques and in situ approaches by isolating proteins from biological sources or inside the tissue or particularly cell level.

Plants are constantly affected by different stress conditions in their natural environments, which affects their growth and productivity. In changing environmental conditions, plants respond quickly by perceiving and activating the molecular programs by expressing different proteins (Ding et al., 2020). Cross-talk between multiple signals perceived by plants results in complex and diverse sets of stress-related protein expressions which play multiple roles in the stress response (Vo et al., 2021). Therefore, it is important to understand proteins' functions under stress by examining the proteome as a whole. Furthermore, protein function depends on the molecular structure, post-translational modifications (PTMs) and subcellular localization (Duan & Walther, 2015; Liu et al., 2019). Different subcellular localizations, and different physiological and biochemical environments, are critical for the current folding of proteins to function properly. PTMs are a critical regulatory process that provides functional diversity to a protein to function spatially and temporally, such as signal transduction responses maintained by different modifications of protein in providing cell defense homeostasis (Han et al., 2022).

Abiotic stress alters gene expression of proteins and thus leads to plant responses toward stress. Most of the proteomics works have been done in model systems and crop species such as *Arabidopsis thaliana* and rice (Goff et al., 2002; Yu et al., 2002; Jorrin-Novo, 2014). Proteomics has become an important field in plant sciences for the identification of organelle-specific proteins and cells, and at different stress conditions. Different accumulations of proteins and activation of pathways in organelles is observed under stress to maintain homeostasis such as increased accumulation of mitochondrial proteins ATP citrate lyase enzyme and NAD dehydrogenase, or activating a set of proteins involved in tricarboxylic acid and production of secondary metabolites and in elevated chloroplast expression of the pyruvate dehydrogenase, acetyl-CoA synthetase and acetaldehyde dehydrogenase (Hossain et al., 2012; Liu et al., 2019). Increased production of fatty acids, phytohormones and glucoronate cycle are also elevated in response to stress (Lin et al., 2022).

Proteomic technology has many advancements and many dimensions of research based on its application. Overall impression of proteomics can be subdivided into different research areas, such as descriptive proteomics, post-translational modifications, interactomics, differential expression proteomics, and proteomics (Jorrin-Novo, 2014). Additionally, proteomic techniques are also categorized into gel-based and gel-free proteomics.

## 23.3 PROTEOMICS TECHNIQUES

Systematic efforts are being made to map the whole localized plant proteome, which depends entirely on emerging advanced technologies and methods. Many traditional and recent techniques have been developed to identify the protein and its functionality. Most of the techniques involve the common processing steps for subcellular separation and purification, such as the enrichment of proteins by selective fractionation and centrifugal technologies, chromatography or electrophoresis (Liu et al., 2019; Christopher et al., 2021). Spatial proteomics is an important field of research that includes common techniques such as centrifugation-based separation and affinity purification (Aslam et al., 2017). However, these methods are time-consuming, as it is difficult to get pure fractions and depend on the availability of specificity of antibodies. New methods have been developed in recent years based on different separation principles such as vacuum infiltration/centrifugation, flow cytometric sorting, free-flow electrophoresis, etc., in plants (Parsons et al., 2014; Galbraith, 2014; Liu et al., 2019). All these proteomic methods provide a deeper understanding of the specialized proteomic behavior of tissues spatially much information at the subcellular proteome level. Flow field-flow fractionation (FlFFF) is a liquid separation method based on elution for separating biomolecules and particles according to their sizes. Ribosome profiling was done in *Nicotiana benthamiana* by using FlFFF, whereby intact ribosomes and free molecules from ribosomal subunits were separated (Pitkänen et al., 2014).

In recent years, numerous advanced techniques have been developed by various methodologies such as sample pre-treatment and multiplexing the reactions. Qualitative proteomic assessment provides high coverage and consistent quality for accurate and reliable quantitative analysis. two types of approaches are used based on the method of quantitation, i.e., labeling-based and label-free quantitation (Rozanova et al., 2021). Labeling-based approaches include isobaric tags for relative and absolute quantification (iTRAQ), isotope-coded affinity tag (ICAT), tandem mass tag (TMT), stable isotope labelling with amino acids in cell culture (SILAC), and uniform 15N/18O labelling and peptides, while label-free techniques include liquid chromatography–mass spectrometry (LC–MS) and spectral counting sequential windowed acquisition of all theoretical fragment ion (SWATH) mass spectrometry (Škultéty et al., 2010; Vo et al., 2021).

iTRAQ and TMT are the most prevalent isobaric tag technologies for relative and absolute quantitation of proteins (Rozanova et al., 2021). The isobar tag used in these techniques is composed of a reporter group, a mass balancing group and an amine-reactive region. The reporter group of isobaric tags generates reporter ions in the MS2 which can be used to obtain a quantification of the labelled peptides among different samples (Pappireddi et al., 2019). The multiplexing channels in iTRAQ can go up to 2- and 8-plex, while TMT reagents can go up to 2–16-plex in a single run. SILAC is an in vivo metabolic labelling-based approach in which isotopic labels are incorporated into amino acids in proteins during cellular growth through the growth medium containing isotopically distinct amino acids (Wang et al., 2018), then MS analysis is performed and quantification of proteins is done by comparing the heavy or non-labeled proteins. The techniques for quantitative proteomics are listed in Table 23.1.

### 23.3.1  2-DE-MS and LC–MS

Two-dimensional mass spectrometry (MS) is the standard method for quantitative analysis of protein separation by first isoelectric focusing and second by SDS-PAGE with MS for the identification of proteins (Aslam et al., 2017). High sensitivity, resolution and throughput of proteome analysis is provided by 2-DE with protein MS. Multidimensional separation of proteins provides higher levels of resolution which are based on two or more independent physical properties of the proteins. However, the 2-DE-MS approach becomes challenging for the global detection of proteins expressed by the cell as it is not able to detect low-abundance proteins. LC–MS can measure complex protein mixtures for the identification of the localization of proteins and PTMs (Tuli & Ressom, 2009). LC–MS/MS can identify and sequence proteins directly from complex mixtures. It allows quantification of proteome, and it can therefore be employed in unraveling cellular signaling networks and understanding the molecular mechanisms involved in plant abiotic stress response (Aebersold & Mann, 2016).

MS-based protein identification is a standard platform in proteomics studies that allows the direct identification of the protein constituents from protein complexes (Aslam et al., 2017). However, in the case of a highly complex mixture of peptides, separation of proteins is needed before MS analysis. In MS, analyte molecules are converted into an ionized state that generates multiples of fragmented ions, followed by analysis of fragmented ions formed during the ionization process based on their mass-to-charge ratio (m:z) (Glish & Vachet, 2003).

The protein identified from MS or tandem mass spectrometry (MS/MS) from the experimental data (m:z ratio) is then deduced from the protein/peptide sequence from custom-built protein databases or identification from orthologs available at different protein databases, such as Uniprot, NCBI and TAIR (Table 23.2). All these databases tell us how confident the identification of protein should be, and based on the parameters such as bit score, e-value, number of hits and percentage of sequence covered. Furthermore, algorithms and bioinformatics packages are now developed for the analysis and identification of the quantification of proteins (Nesvizhskii, 2007). These algorithm-based programs identify proteins by comparing the mass of parental

**TABLE 23.1**
**Methods for Quantitative Proteomics**

| S. No. | Technique | Level | | Multiplexing | Type of Labeling | Application | Drawback |
|---|---|---|---|---|---|---|---|
| 1. | ICAT | peptide (cysteine) | | 2-plex | Chemical | In vitro, low sample complexity applicable to any sample | Limited multiplexing capability |
| 2. | iTRAQ | peptide (Lysine) | In vitro | 8-plex | Amine-specific with isobaric tags | In vitro, efficient labeling, high multiplexing capability, applicable to any sample | Expensive |
| 3. | TMT | Peptide | In vitro | 10-plex | Amine-specific with isobaric tags | In vitro, efficient labeling, simple, applicable to any sample | Expensive |
| 4. | SILAC | Protein | | 5-plex | Metabolic | In vitro and in vivo, efficient labeling, applicable to cells | Expensive and tedious |
| 5. | 15N labelling | Protein | | 2-plex | Metabolic | In vitro and ex vivo, efficient labeling, applicable to cells and model organisms | Complex data analysis, limited multiplexing capability, expensive |
| 6. | 18O labelling | Protein | | 2-plex | Enzymatic | In vitro, simple handling, applicable to any sample | Incomplete labeling, limited multiplexing capability |
| 7. | ICPL | Peptide | | 4-plex | Chemical | In vitro, improved fragmentation efficiency during MS/MS | Requires complex computational analysis |

## TABLE 23.2
### Protein Databases Used in Proteomics Studies

| S. No. | Database | URL |
| --- | --- | --- |
| 1. | PeptideSearch | www.narrador.embl-heidelberg.de/GroupPages/Homepage.html |
| 2. | CD server protein domain servers | www.ncbi.nlm.nih.gov/Structure/cdd/cdd.shtml |
| 3. | EBI Protein structure modeling server | http://biotech.ebi.ac.uk:8400/ |
| 4. | Gibbs Motif based alignment server | http://bayesweb.wadsworth.org/gibbs/gi |
| 5. | MASCOT | www.matrixscience.com/search form select.html |
| 6. | MEME motif-based alignment server | http://meme.sdsc.edu/meme/website/meme.html |
| 7. | MS-FIT | www.prospector.ucsf.edu/ |
| 8. | PDB protein structure databases | www.rcsb.org/pdb/ |
| 9. | PredictProtein protein structure prediction sever | www.embl-heidelberg.de/predictprotein |
| 10. | Profound | www.prowl.rockefeller.edu/ |
| 11. | SwissModel protein structure databases | http://swissmodel.expasy.org/repository/ |
| 12. | SwissModel protein structure modeling server | www.expasy.org/swissmod |

peptide peak with the in silico predicted (peptide mass fingerprinting) (Tuli & Ressom, 2009). Another algorithm determines peptide sequences directly from fragment ion spectra generated experimentally.

## 23.4 APPLICATION OF PROTEOMICS IN PLANT STRESS RESPONSES

Crop plants also survive and thrive with other living organisms in biotic and abiotic stress conditions. Current rapid changes in climate change can affect plant growth and yield through numerous stresses, such as cold stress and water stress—both drought and waterlogging. Winter wheat was studied under variable climatic conditions during late vegetative and reproductive development using physiological, proteomic and transcriptional approaches (Li et al., 2014). Table 23.3 summarizes region-wise case studies of proteomics application in response to plant stress.

## 23.5 METABOLOMICS IN PLANT STRESS RESPONSE

Plants respond to environmental stress very rapidly by triggering biochemical pathways that include signal transduction. The complex interactions between plants and the environment have developed a number of advanced techniques to enhance understanding of the physiology of these interactions (Ghatak et al., 2018). Most of the plant environmental interactions in plants result in the activation of different metabolic pathways. However, plants also produce different non-protein molecules by activating different biochemical pathways. Plants synthesize countless numbers of metabolites to combat stress, metabolites act as mediators and signal molecules and play an important role in plant defense response to regulate various cellular functions (Allwood et al., 2008). Metabolome refers to the complete and functional set of metabolites produced in a plant.

Environmental stress alters the plant's metabolic homeostasis, which allows plants to withstand constant exposure to diverse abiotic and biotic stresses. Any change in environmental condition activates a network of signal transduction pathways by the plants which triggers the production of different compounds that provide homeostasis (Shulaev et al., 2008). A diverse group of metabolites is produced by plants that are involved in defense responses to different processes such as antioxidants, osmoprotectants, phenolics, reactive oxygen species (ROS) and signal transduction molecules produced by signal transduction in response to stresses (Apel & Hirt, 2004; Alcázar et al., 2006). These metabolites can be byproducts of metabolic pathways induced by stress signaling.

Metabolites are mainly divided into two groups, primary and secondary metabolites. Primary metabolites include carbohydrates, lipids and proteins which are involved in basic biological processes, while secondary metabolites are not exclusively important for plant survival; instead, they help plants in extreme environmental conditions, such as flavonoids, antioxidant molecules, etc. (Razzaq et al., 2019). In response to biotic stress such as herbivory and pathogen infestation, plants activate well-coordinated defense pathways which recognize the herbivore at the biochemical level and rapidly activate the defense pathways by producing defense metabolites.

Metabolomics is an analytical approach to studying complex biochemical interactions between plants and stress response. It includes very recently developed approaches to metabolic fingerprinting and metabolite profiling (Wang et al., 2010; Ghatak et al., 2018). The commonly used analytical techniques in metabolomics are liquid and gas chromatography (LC and GC, respectively) coupled with mass spectrometry (MS) and nuclear magnetic resonance (NMR) spectroscopy (Ghatak et al., 2018). Metabolomics can be studied at various organismal levels from a single cell, tissue, an organ or the entire organism. The metabolome of

## TABLE 23.3
### Overview of Proteomics Techniques in Plant Environmental Stress

| S. No. | Study Location | Type of Stress | Proteomics Techniques Used | Species | Comments/Concluding Remark | Reference |
|---|---|---|---|---|---|---|
| 1. | China | Drought stress | iTRAQ, LC MS/MS | Zea mays | Photosynthesis, synthesis of redundant proteins and stress defense proteins | Jiang et al. (2019) |
| 2. | Sweden | Agrobacterium | SDS-PAGE, LC–MS/MS | Solanum tuberosum | LRR-like receptor protein kinase, photosynthetic chlorophyll a/b binding proteins, cytochrome bf-6 complex | Burra et al. (2018) |
| 3. | Japan | Biotic stress | 2-DE, MALDI-TOF MS | Oryza sativa | Thaumatin-like protein and probenazole | Mahmood et al. (2006) |
| 4. | Germany | Biotic stress | LC-MALDI-MS/MS | Oryza sativa | Virulence-associated factors | González et al. (2012) |
| 5. | Czech Republic | Biotic stress (Fusarium culmorum) | 2D-DIGE, LC–MS/MS | Horduem vulgare | USP A-like protein, z-serpin, inhibitors of lytic enzymes and chaperones | Kosová et al. (2017) |
| 6. | China | Cold stress | SDS-PAGE, iTRAQ, SCX-LCMS/MS | Simmondsia chinensis | Ferredoxin 3, NADP-malic enzyme, glyceraldehyde-3-phosphate dehydrogenase, RuBisCo activase | Gao et al. (2019) |
| 7. | China | Drought stress | LC–MS/MS | Tibetan hull-less Horduem vulgare | ABA-induced and ethylene synthesis pathways | Wang et al. (2020) |
| 8. | India | Drought stress | 2DE, PMF, MALDI-TOF/TOF MS | Arachis hypogaea | LEA protein, calcium ion binding protein and sucrose synthase isoform | Thangella et al. (2018) |
| 9. | Slovenia | Drought stress | 2DE, LC–MS/MS | Phaseolus vulgaris | Ferredoxin-NADP reductase, photosystem II assembly factor, thylakoid protein, plastidial membrane proteins | Zadražnik et al. (2019) |
| 10. | Japan | Drought stress | IPG, SDS-PAGE, nanoLC-MS/MS | Glycine max | Ascorbate peroxidase, HSP, methionine | Mohammadi et al. (2012) |
| 11. | USA | Drought Stress | 2-DE, nanoLC–MS/MS | Zea mays | Reactive oxygen species (ROS), metabolism, hydrolases, carbohydrate metabolism | Zhu et al. (2007) |
| 12. | Australia | Environmental stress | 2-DE, nanoLC–MS/MS | Pisum sativum | HSPs, non-phosphorylating respiratory pathways | Taylor et al. (2005) |
| 13. | Japan | Flooding stress | 2-DE SDS-PAGE; LC–MS | Glycine max | Ribosomal Proteins, Voltage-Dependent Anion Channel Protein, Isocitrate Dehydrogenase | Mustafa & Komatsu (2016) |
| 14. | China | Heat stress | EI-tandem MS/MS, Q Exactive, iTRAQ, SCX, LC–MS/MS | Zea mays | HSPs, ethylene responsive proteins | Zhao et al. (2016) |
| 15. | China | Heat stress | 2DE, iTRAQ, LC–MS | Spinacia oleracea | ROS scavenging pathways, protein synthesis and turnover, carbohydrate and amino acid metabolism | Li et al. (2019) |

| S. No. | Study Location | Type of Stress | Proteomics Techniques Used | Species | Comments/Concluding Remark | Reference |
|---|---|---|---|---|---|---|
| 16. | China | Heat stress | 2-DE, SDS-PAGE, MALDITOF-MS | *Zea mays* | ATP synthase subunit b, chlorophyll a–b binding protein, oxygen-evolving enhancer protein, photosystem I reaction center subunit IV | Hu et al. (2015) |
| 17. | Israel | Heat stress | LC–MS/MS Orbitrap LTQ XLMS | *Solanum lycopersicum* | Glutathione-disulfide reductase, glutaredoxin, protein disulfide isomerase | Jegadeesan et al. (2018) |
| 18. | Japan | Osmotic stress | 2DE-LC MS/MS, nanoLC–MS/MS | *Glycine max* | Plasma membrane $H^+$ atpase, calnexin | Nouri & Komatsu (2010) |
| 19. | China | Salt stress | iTRAQ, LC–MS/MS | *Gossypium herbaceum* | Phosphoethanolamine N-methyltransferase, 14–3-3 protein | Gong et al. (2017) |
| 20. | South Africa | Salt stress | 2-DE-MALDI-TOF-TOF-MS | *Sorghum* | Malate dehydrogenase, APX | Ngara et al. (2012) |
| 21. | Italy | Salt stress | 2-DE-MALDI-TOF-MS | *Triticum aestivum* | Glutamine synthase, glycine dehydrogenase | Caruso et al. (2008) |

living plants is dynamic and shows variation in metabolite presence and concentrations over time. It is also very difficult to detect unstable metabolites which are present in trace concentrations. Therefore, this requires metabolomics techniques to be highly sensitive to detect all key metabolites. Metabolic fingerprinting involves the identification of characteristics or patterns of metabolites associated with different stress responses and tells relevant differences of the sample based on metabolites (Wang et al., 2010). It aims to compare the patterns in different plant stress states, and the data obtained is used to identify features specific to a fingerprint. (Kopka et al., 2004). However, the steps in the metabolic profiling analysis mainly involve sample preparation followed by extraction, separation and detection—the complex dataset thus obtained is then analyzed using different software packages (Piasecka et al., 2019). Before detection, the termination step or quenching of selected biochemical processes in the system must be performed. The recent advancement of omics technologies in studying plant defense mechanisms—such as genomics, transcriptomics, proteomics, metabolomics, and phenomics—allows us to collectively examine the molecular mechanisms involved. Therefore, omics techniques will help to decipher the biosynthesis pathways and candidate genes involved that are actively regulated by abiotic stresses (Kaur et al., 2021). The major problem with integrated omics approaches is data integration from a wide variety of data (Piasecka et al., 2019). It requires further development of data integration and data analysis software to easily integrate data from different sources. Additionally, omics data should be combined with mathematical modeling to analyze the dynamics of the complex networks involved in stress responses (Shulaev, 2008).

## 23.6 METABOLOMIC TECHNIQUES

### 23.6.1 Chromatography Coupled with MS

Metabolomics can be generally studied using a variety of platforms that can detect the total metabolite content of a plant. Plant metabolomes are dynamic and vary in both composition and abundance according to different developmental or stress conditions. Metabolite profiling includes analyses of all or a subset of metabolites simultaneously in a sample (Kopka et al., 2004). Multiple platforms with a diversity of techniques have been employed in metabolite profiling such as different chromatography techniques coupled with MS or NMR analysis (Ghatak et al., 2018). GC–MS is a chromatographic technique with high sensitivity to detect molecules that can be volatile, while LC–MS is largely used for non-volatile and can be used for structurally different compounds. On the other hand, NMR spectroscopy can be used to estimate the chemical profile of the metabolites with qualitative and quantitative analysis. Capillary electrophoresis–MS (CE–MS) is another alternative for metabolite profiling because of its low sample requirements and high resolving power (Ramautar et al., 2019). CE–MS is used for the separation of polar and charged molecules based on their electrophoretic mobility (Ramautar et al., 2009). However, detection is challenging in CE because of the small dimension of the capillary tube and the difficult to handle large volumes of samples.

### 23.6.2 Nuclear Magnetic Resonance (NMR)

NMR is a spectroscopic technique for the study of proteins and metabolites that detects compounds based on the energy absorbed by changes in the nuclear spin state (Emwas et al., 2019). NMR spectroscopy can be applied in non-targeted metabolomic analysis, as it is non-selective and delivers structural information at the atomic level. NMR is non-destructive and can also be used with in vivo samples without any sample preparation. NMR can be applied to both metabolite fingerprinting and metabolic profiling. The limitation of NMR is its low sensitivity for detecting metabolites present in low amounts, and also the size of the compound under study (Pan & Raftery, 2007). NMR/MS requires minimal sample preparation as compared to other techniques discussed in this chapter. NMR is mainly used for the analysis of metabolites present in higher concentrations in the sample and the less stable compounds (Emwas et al., 2019). However, NMR spectroscopy is less sensitive to detecting low-abundance metabolites, and signal overlapping could also affect analyte identification (Razzaq et al., 2019). Therefore, MS is mostly preferred over NMR for high sensitivity and resolution, enabling it to detect a larger set of metabolites.

### 23.6.3 Application of Metabolomics Approaches for the Study of Plant Response to Stress

Metabolomics is arising as a new-era strategy to examine the plant biochemistry at the cellular level in stress mechanisms in plants by evaluating the primary and secondary metabolites produced by the plants which regulate different molecular mechanisms (Table 23.4).

## 23.7 STRUCTURE–ACTIVITY RELATIONSHIP (SAR)/QUANTITATIVE STRUCTURE–ACTIVITY RELATIONSHIP (QSAR)

### 23.7.1 Introduction to SAR and QSAR

Structure–activity relationship (SAR) is based on the fact that more or less similar activities are exhibited by similar molecules. It represents the relationship between the structure of a molecule and its biological activity. It allows the determination of the functional group of the molecule which is responsible for eliciting a specific response on a target in the organism. This response in an organism could be modified by changing the chemical structure of the compound. This technique is used by pharmacologists to synthesize new chemicals by inserting new functional groups into biochemical compounds and to test the modifications carried out for their biological effects (Mohapatra, 2020). The underlying problem with this principle is an inability

**TABLE 23.4**
**Overview of Metabolomics Techniques in Plant Environmental Stress**

| S. No. | Study Location | Type of Stress | Proteomics Techniques Used | Species | Comments/Concluding Remark | Reference |
|---|---|---|---|---|---|---|
| 1. | Germany | Cold stress | GC–MS | Arabidopsis thaliana | Fumaric acid, succinic acid, fructose, galactose, glucose, raffinose, sucrose, galactinol, glycine, proline | Korn et al. (2010) |
| 2. | China | Drought | LC–MS/MS | Nicotiana tabacum | Flavanoids | Hu et al. (2019) |
| 3. | China | Drought | H-NMR | Z. mays | Alanine, triacylglyceride, malate, glutamate | Sun et al. (2015) |
| 4. | Argentina | Drought | GC–MS | Arachis hypogea | Trehalose, Proline, GABA | Furlan et al. (2017) |
| 5. | Portugal | Drought | GC-TOF | Vigna unguiculata | Galactinol, proline, quercetin | Goufo et al. (2017) |
| 6. | Portugal | Drought and Heat Stress | GC–MS | Eucaliptus globule | Mannose, galactose, sorbitol, inositol, ctrate, glutamate, proline, GABA | Correia et al. (2018) |
| 7. | China | Heavy metal response | GC–MS | Raphanus sativus | Gluconate, fructose, galactose and glucose, citrate | Wang et al. (2015) |
| 8. | Belgium | Low oxygen stress | GC–MS | Lycopersicum esculentum | Lactate and sugar alcohols | Ampofo-Asiama et al. (2013) |
| 9. | Bulgaria | Low temperature stress | GC–MS | Haberlea rhodopensis, Thellungiella halophyla, Arabidopsis thaliana | Sucrose, fructose and glucose, raffinose, sucrose, proline, 4-hydroxyproline | Benina et al. (2013) |
| 10. | USA | Oxidative stress | GC–MS | Cucumis sativus | Tricarboxylic acid cycle intermediates, lactulose, raffinose | Zhang et al. (2018) |
| 11. | Australia | Salt Stress | GC–MS, HPLC–MS | Hordeum spp. | 4-hydroxy-proline, alanine, asparagine, glutamine, proline, ABA, SA | Cao et al. (2017) |
| 12. | Australia | Salt stress | GC–MS | Oryza sativa | Sugars, fatty acids, IAA | Liu et al. (2013) |

to relate minor differences at the level of the molecule, as different biological activities—such as absorption, solubility, reactivity, biotransformation, target activity and toxicity—might be dependent on another difference. This is referred to as the SAR paradox, which implies that not necessarily all similar molecules have similar activities. As a result, refinements in the existing method were brought about to develop mathematical equations which can precisely relate the chemical structure of the compound and its corresponding biological activity, known as quantitative structure–activity relationship (QSAR). The quantitative structure–property relationship (QSPR) technique was developed to correlate the chemical and physical properties of a molecule to its structure (Tandon et al., 2019). SAR and QSAR are theoretical models which are crucial for the prediction of physicochemical, biological and environmental attributes of substances. SAR is based on the relationship between the qualitative attributes of a compound containing a chemical structure (or functional group) and its ability to perform a certain biological activity. Contrary to this, QSAR is a quantitative method that relies on mathematical modeling methods to explore the relationships that exist between a compound's chemical structure and biological activities. The underlying principle of QSAR states that the change in the structures causes a change in biological activity (Muhammad et al., 2018). Both these related approaches of computational modeling attempt to predict the function or toxicity of an understudied or unknown chemical compound based on the similarities with chemical structures for which the requisite data exists (Basant et al., 2015). Through these tools, the predictions can be made in a cost-effective and timely manner without engaging in labor-intensive and time-consuming wet laboratory experiments. For convenience, the term QSAR will be used to represent both types of approaches as being done in research. QSAR models are a widely accepted, predictive and scientifically reliable tool that is significant for drug discovery, pharmacy and environmental toxicology. QSAR studies are widely used in computer-aided drug designing (CADD) processes and are indispensable for the global pharmaceutical industry, which starts with the discovery of a lead and goes on through its development. The algorithms used in QSAR software during the initial stages of CADD are crucial in the elimination of those compounds which lack drug-like properties and those compounds also which might elicit a toxic response. It is being used as a tool in natural sciences for learning the behavior of biological systems as physiological function is related to the chemical nature of compounds.

### 23.7.2 History of QSAR

The history of QSAR dates back to 1863, when Cros correlated the water solubility of aliphatic alcohols with their toxicity (Cros, 1863). Later, based on the investigation of different alkaloids, a mathematical equation was published which was considered a milestone for the first generation of QSAR (Crum-Brown, 1868). Assessment of the narcotic activity of various drugs paved the way for systematic QSAR studies (Cantor, 2001; Pohorille et al., 1998). An important method that accounts for the effect of substituent on the mechanism of reaction was introduced by Hammet (1935). Based on Hammet's model, an approach for separation of various effects of a chemical substituents—such as steric, polar and resonance—was proposed by Taft in 1956. The pioneering work of Hansch et al. (1963) through a classical approach to QSAR resulted in the development of the linear Hansch equation. Several advancements in QSAR methodologies have been proposed since then, such as 2D QSAR (Hansch et al., 1990); 3D QSAR, which corresponds to CoMSIA, i.e., comparative molecular similarity indices analysis (Klebe & Mietzner, 1994; Klebe & Abraham, 1999) and CoMFA, i.e., comparative molecular field analysis (Marshall & Cramer, 1988); 4D QSAR (Hopfinger et al., 1997); and HQ SAR, i.e., hologram quantitative structure–activity relationship (Tong et al., 1998). QSAR is the end result of a computational method that starts with an appropriate description of the chemical structure of the compound and provides hypotheses, inferences and predictions of the behavior of molecules in the biological, environmental and physicochemical system under study (Eriksson et al., 2003).

## 23.8 QSAR METHODOLOGY

The classification of QSAR is broadly on the basis of the dimensions of the descriptors involved in the model and the type of biological activity predicted as a dependent variable (Qiao et al., 2014). The latter includes, Quantitative structure-toxicity relationship (QSTR), Quantitative structure reactivity relationship (QSRR), Quantitative structure-metabolism relationship (QSMR), Quantitative structure-pharmacokinetics relationship (QSPR), and Quantitative structure-bioavailability relationship (QSBR) (Can, 2014; Goryński et al., 2013). QSAR models can also be classified on the basis of the binding nature of the molecule, receptor-receptor dependent, and receptor-independent as well as based on analysis of correlation—linear and nonlinear (Peter et al., 2019). Obtaining a good quality model that can predict the activity of a chemical compound that is not present in the training set is dependent on various aspects and the execution of three main steps which are involved in QSAR modeling. These three steps are:

Step 1: Quality of data
Step 2: Selection of descriptor
Step 3: Statistical methods

### 23.8.1 Quality of Data

The choice of algorithm for selecting the training sets and test sets of chemicals is important for the development of the QSAR model. The model is constructed based on the training sets, whereas external validation of the model is accomplished with the help of test sets. The stability of the statistical methods is ensured if the number of chemicals which are used for building the QSAR model is large. The chemical compounds chosen for the training set are expected to be of sufficient diversity in structure to study the biological activity in its entirety.

Leonard and Roy (2006) demonstrated that predictive QSAR models are not obtained when the generation of test and training sets was done by random division or by the activity-range algorithm. There is no correlation between external and internal validation statistics when test and training sets were generated by random division of the data. However, when training and test sets were selected on the basis of K-means clusters of factor scores of the descriptor space, along with/without the biological activity values, good external validation statistics resulted (Roy et al., 2008). Therefore, selecting the test and training sets based on the closeness of the representative points of the test set to the corresponding points of the training set in the multidimensional descriptor space is important for the validation of the QSAR model.

### 23.8.2 Selection of Molecular Descriptors

Molecular descriptors provide the mathematical values that describe the structure or shape of molecules, which help in predicting the properties and activity of compounds in complex experiments. Descriptors are essentially the chemical characteristic of a molecule in numerical form. A better understanding of the biological action mechanism is essential for the choice of a molecular descriptor (Yang et al., 2018). Additionally, in the algorithm used in computing descriptors, the dimensionality of the molecule is important for the determination of details by the descriptor (Khan, 2016). Quantities which encode structural attributes and physicochemical properties are called 1D descriptors. The problem with 1D descriptors is their insensitivity to molecular topology and that they eventually may be identical for different chemical compounds. Therefore, they are preferably used in combination with other molecular descriptors. The 2D descriptors are employed for the description of chemical space. These are molecular conformation–independent and graph-invariant descriptors. A rigorous description of molecular structures is offered by 3D descriptors, which involve conformational probing and can classify isomers via a computationally exhaustive process. Like 3D descriptors, 4D descriptors simultaneously assess several structural conformations (Tandon et al., 2019).

### 23.8.3 Statistical Methods

Several statistical methods for QSAR study are available based on the size of datasets and the descriptors. These criteria can be presented as follows.

1. Simple linear or multiple linear regression: this is a good choice if the number of chemicals and descriptors is small.
2. Partial least squares (PLS) or principal components regression (PCR): either of these methods can be used with a larger number of chemicals and a larger pool of descriptors.

To develop more robust QSAR equations, some of the methods can be used in combination with various variable selection methods. For example, the combination of genetic algorithm (GA) with multiple linear regression (MLR), PLS, and artificial neural networks (ANNs) are reported to be effective approaches for QSAR (Roy et al., 2015).

## 23.9 APPLICATION OF QSAR APPROACHES FOR THE STUDY OF PLANT RESPONSE TO STRESS

### 23.9.1 Designing and Development of Coumarin Derivatives as New Plant Protection Products (PPP)

Plant protection products (PPPs) are defined as biological or chemical products which are used for protecting plants against diseases and pests, such as insecticides, herbicides and fungicides. The synthetic active ingredients of PPPs are often associated with environmental risks, as these are harmful to pests but may also adversely affect other non-target organisms. This situation called for the development of highly specific and ecofriendly PPPs by focusing on their active component. In this quest, the QSAR technique has been used as a rational approach for the screening of novel potential candidates and their optimization. QSAR and molecular docking are critical in designing new agents which can act as plant protectants by improving their activity. To combat rice blasts caused by the fungus, *Magnaporthe griesa*, QSAR models were developed for predicting fungicidal activities of thazoline derivatives (Du et al., 2008). Recently, using QSAR modeling, Rastija et al. (2021) reported that coumarin derivatives possess significant antifungal activity against *Macrophomina phaseolina* and *Sclerotinia sclerotiourum*. It was shown that the presence of several electron-withdrawal groups at the C-3 position in coumarin derivatives increased antifungal activity against *M. phaseolina*. However, benzoyl group at C-3 in coumarin derivatives strongly inhibited *S. sclerotiourum*. The tested compounds were not found to be harmful to nematodes and bacteria inhabiting soil.

### 23.9.2 QSAR Analysis for Inhibition of Soybean Lipoxygenase (LOX) Activity by Coumarin Derivatives

Compounds produced during the LOX pathway in plants—such as jasmonates, alcohols and divinyl ethers—play an important role in plants' pathogenic interactions and abiotic stress, but some metabolites may have negative effects on human health. Therefore, compounds that can inhibit LOX activity are researched. The best QSAR model to find coumarin derivatives as an inhibitor of soybean LOX activity was obtained using a GA. LOX activity was inhibited in the 7.1–96.6% range. This study has also revealed compounds with benzoyl substitution at the C-3 position as potential inhibitors of LOX activity (Lončarić et al., 2020).

## 23.10 SUMMARY

Combining the techniques of transcriptomics, proteomics, metabolomics and QSAR approaches greatly enhances the understanding of the plant stress response and cellular physiology. The defense mechanisms of plants under different stresses can be unfolded by analyzing the gene expression and protein abundance, and metabolic response candidate genes can be identified by correlating available transcriptomic data, proteomics and expressed metabolites. Future studies in proteomics and metabolomics can be aimed toward the construction of mega-libraries of metabolites, the construction of databases, and development of bioinformatic software for rapid and high throughput identification of metabolites in higher plant species to improve adaptability to adverse environmental conditions. Proteomics provides more understanding of targets genes and regulators of plant immunity, both at the cellular and subcellular levels. 2-DE and LC–MS are the two commonly used techniques for the identification of a large number of proteins at a time due to their high resolution and high efficiency. Coupling these techniques with MS in tandem with advanced data analysis and statistical methods gives more efficient data analysis and has greatly improved the knowledge of gene expression. This recent development of labeling-based and label-free approaches allows more quantification of proteomics and much reduced periods of time.

Metabolomics helps to understand the compounds secreted by plants and further identification of the candidate genes and pathways involved in the production of metabolites in plant defense response to abiotic and biotic stresses. QSAR approach is instrumental in providing insight into the correlation between the structure of chemicals and their biological activity, and in paving an alternative for the development and designing of novel molecules with improved biological activity. The screening, optimization and designing of antifungal compounds and PPPs which are target specific and ecofriendly are instrumental in combating pathogen attacks and development of agents used for plant protection.

## REFERENCES

Aebersold, R., and Mann, M. 2016. Mass-spectrometric exploration of proteome structure and function. *Nature* 537(7620): 347–55. https://doi.org/10.1038/nature19949

Alcázar, R., Cuevas, J.C., Patron, M. et al. 2006. Abscisic acid modulates polyamine metabolism under water stress in Arabidopsis thaliana. *Physiol Plant.* 128: 448–455. https://doi.org/10.1111/j.1399-3054.2006.00780.x

Allwood, J.W., Ellis, D.I., and Goodacre, R. 2008. Metabolomic technologies and their application to the study of plants and plant—host interactions. *Physiol Plant.* 132: 117–135. https://doi.org/10.1111/j.1399-3054.2007.01001.x

Ampofo-Asiama, J., Baiye, V.M.M., Hertog, M.L.A.T.M., Waelkens, E., Geeraerd, A., and Nicolai, B.M. 2013. The metabolic response of cultured tomato cells to low oxygen stress. *Plant Biol.* 16: 594–606. https://doi.org/10.1111/plb.12094

Apel, K., and Hirt, H. 2004Reactive oxygen species: Metabolism, oxidative stress, and signal transduction. *Annu Rev Plant Biol.* 55: 373–399. https://doi.org/10.1146/annurev.arplant.55.031903.141701

Aslam, B., Basit, M., Nisar, M.A., Khurshid, M., and Rasool, M.H., 2017. Proteomics: Technologies and their applications. *Journal of Chromatographic Science* 55(2):182–196. https://doi.org/10.1093/chromsci/bmw167

Basant, N., Gupta, S., and Singh, K. P. 2015. Predicting toxicities of diverse chemical pesticides in multiple avian species using tree-based QSAR approaches for regulatory purposes. *J Chem Inf Model.* 55(7): 1337–1348. https://doi.org/10.1021/acs.jcim.5b00139

Benina, M., Obata, T., Mehterov, N., Ivanov, I., Petrov, V., Toneva, V., Fernie, A.R., and Gechev, T.S. 2013. Comparative metabolic profiling of Haberlea rhodopensis, Thellungiella halophyla, and Arabidopsis thaliana exposed to low temperature. *Front Plant Sci.* 4: 499. https://doi.org/10.3389/fpls.2013.00499

Burra, D. D., Lenman, M., Levander, F., Resjö, S., & Andreasson, E. 2018. Comparative membrane-associated proteomics of three different immune reactions in potato. *Int J Mo Sci.* 19(2): 538. https://doi.org/10.3390/ijms19020538

Can, A. 2014. Quantitative structure–toxicity relationship (QSTR) studies on the organophosphate insecticides. *Toxicol Lett.* 230: 434–443. https://doi.org/10.1016/j.toxlet.2014.08.016

Cantor, R.S. 2001. Breaking the Meyer–Overton rule: Predicted effects of varying stiffness and interfacial activity on the intrinsic potency of anesthetics. *Biophys J.* 80(5): 2284–2297. https://doi.org/10.1016/S0006-3495(01)76200-5

Cao, D., Lutz, A., Hill, C.B., Callahan, D.L., and Roessner, U. 2017. A Quantitative Profiling Method of Phytohormones and Other Metabolites Applied to Barley Roots Subjected to Salinity Stress. *Front Plant Sci.* 7: 207. https://doi.org/10.3389/fpls.2016.02070

Caruso, G., Cavaliere, C., Guarino, C., Gubbiotti, R., Foglia, P., and Lagana, A. 2008. Identification of changes in Triticum durum L. leaf proteome in response to salt stress by two-dimensional electrophoresis and MALDI-TOF mass spectrometry. *Anal Bioanal Chem.* 391: 381–390. https://doi.org/10.1007/s00216–008–2008-x

Chandramouli, K., and Qian, P.Y. 2009. Proteomics: Challenges, techniques and possibilities to overcome biological sample complexity. *Human genomics and proteomics: HGP.* 2009: 239204. https://doi.org/10.4061/2009/239204

Christopher, J.A., Stadler, C., Martin, C.E. et al. 2021. Subcellular proteomics. *Nat Rev Methods Primers.* 1: 32. https://doi.org/10.1038/s43586-021-00029-y

Correia, B., Hancock, R.D., Amaral, J., Gomez-Cadenas, A., Valledor, L., and Pinto, G. 2018. Combined drought and heat activates protective responses in eucalyptus globulus that are not activated when subjected to drought or heat stress alone. *Front Plant Sci.* 9: 819. https://doi.org/10.3389/fpls.2018.00819

Cros, A. F. 1863. *Action de l'alcool amylique sur l'organisme* (Doctoral dissertation), Faculté de médecine de Strasbourg.

Crum–Brown, A.F.T. 1868. On the connection between chemical constitution and physiological action. Pt 1. On the physiological action of the salts of the ammonium bases, derived from Strychnia, Brucia. Thebia, Codeia, Morphia, and Nicotia. *T Roy Soc Edin.* 25:151–203.

Ding, Y., Shi, Y. and Yang, S., 2020. Molecular regulation of plant responses to environmental temperatures. *Mol. Plant.* 13(4): 544–564. https://doi.org/10.1016/j.molp.2020.02.004

Du, H., Wang, J., Hu, Z., Yao, X., and Zhang, X. 2008. Prediction of fungicidal activities of rice blast disease based on least-squares support vector machines and project pursuit regression. *J Agric Food Chem.* 56:10785–10792. https://doi.org/10.1021/jf8022194

Duan, G., and Walther, D. 2015. The roles of post-translational modifications in the context of protein interaction networks. *PLoS Computational Biology* 11(2): e1004049. https://doi.org/10.1371/journal.pcbi.1004049

Emwas, A. H., Roy, R., McKay, R. T., Tenori, L., Saccenti, E., Gowda, G., Raftery, D., Alahmari, F., Jaremko, L., Jaremko, M., and Wishart, D. S. 2019. NMR spectroscopy for metabolomics research. *Metabolites* 9(7): 123. https://doi.org/10.3390/metabo9070123

Eriksson, L., Jaworska, J., Worth, A. P., et al. 2003. Methods for reliability and uncertainty assessment and for applicability evaluations of classification–and regression–based QSARs. *Environ Health Perspect.* 111(10): 1361–1375. https://doi.org/10.1289/ehp.5758

Furlan, A.L., Bianucci, E., Castro, S., and Dietz, K.J. 2017. Metabolic features involved in drought stress tolerance mechanisms in peanut nodules and their contribution to biological nitrogen fixation. *Plant Sci.* 263: 12–22. https://doi.org/10.1016/j.plantsci.2017.06.009

Galbraith, D. W. 2014. Flow cytometry and sorting in Arabidopsis. *Methods in Molecular Biology (Clifton, N.J.).* 1062: 509–537. https://doi.org/10.1007/978-1-62703-580-4_27

Gao, F., Ma, P., Wu, Y., Zhou, Y., and Zhang, G. (2019). Quantitative proteomic analysis of the response to cold stress in jojoba, a tropical woody crop. *Int J Mol Sci.* 20, 243. https://doi.org/10.3390/ijms20020243

Ghatak, A., Chaturvedi, P., and Weckwerth, W. 2018. Metabolomics in Plant Stress Physiology. *Advances in Biochemical Engineering/Biotechnology* 164: 187–236. https://doi.org/10.1007/10_2017_55

Glish, G., and Vachet, R. 2003. The basics of mass spectrometry in the twenty-first century. *Nat Rev Drug Discov.* 2: 140–150. https://doi.org/10.1038/nrd1011

Goff, S.A., Ricke, D., Lan, T.H., Presting, G., Wang, R., Dunn, M., Glazebrook, J., Sessions, A., Oeller, P., and Varma, H. 2002. A Draft Sequence of the Rice Genome (Oryza sativa L. ssp. japonica). *Science* 296: 92–100. https://doi.org/10.1126/science.1068275

Gong, W., Xu, F., Sun, J., Peng, Z., He, S., Pan, Z., and Du, X. 2017. iTRAQ-Based Comparative Proteomic Analysis of Seedling Leaves of Two Upland Cotton Genotypes Differing in Salt Tolerance. *Front Plant Sci.* 8: 2113. https://doi.org/10.3389/fpls.2017.02113

González, J.F., Degrassi, G., Devescovi, G., et al. 2012. A proteomic study of Xanthomonas oryzae pv. Oryzae in rice xylem sap. *J Proteomics.* 75: 5911–5919. https://doi.org/10.1016/j.jprot.2012.07.019

Goryński, K., Bojko, B., Nowaczyk, A., Buciński, A., Pawliszyn, J. and Kaliszan, R. 2013. Quantitative structure–retention relationships models for prediction of high performance liquid chromatography retention time of small molecules: Endogenous metabolites and banned compounds. *Analytica Chimica Acta.* 797: 13–19. https://doi.org/10.1016/j.aca.2013.08.025

Goufo, P., Moutinho-Pereira, J.M., Jorge, T.F., Correia, C.M., Oliveira, M.R., Rosa, E.A.S., António, C., and Trindade, H. 2017. Cowpea (*Vigna unguiculata* L. Walp.) Metabolomics: Osmoprotection as a Physiological Strategy for Drought Stress Resistance and Improved Yield. *Front Plant Sci.* 8:586. https://doi.org/10.3389/fpls.2017.00586

Graves, P.R., and Haystead, T.A. 2002. Molecular biologist's guide to proteomics. *Microbiology and Molecular Biology Reviews: MMBR* 66(1): 39–63. https://doi.org/10.1128/MMBR.66.1.39-63.2002

Hammett, L.P. 1935. Some Relations between reaction rates and equilibrium constants. *Chemical Reviews* 17(1):125–136. https://doi.org/10.1021/cr60056a010

Han, D., Yu, Z., Lai, J., et al. 2022. Post-translational modification: A strategic response to high temperature in plants. *aBIOTECH* 3: 49–64 https://doi.org/10.1007/s42994-021-00067-w

Hansch, C., Muir, R.M., Fujita, T., Maloney, P.P., Geiger, F., et al. 1963. The correlation of biological activity of plant growth regulators and chloromycetin derivatives with Hammett constants and partition coefficients. *J Am Chem Soc.* 85 (18): 2817–2824. https://doi.org/10.1021/ja00901a033

Hansch, C., Sinclair, J.F., Sinclair, P.R. 1990. Induction of Cytochrome P450 by Barbiturates in Chick Embryo Hepatocytes: A Quantitative Structure- Activity Analysis. *Quantitative Structure-Activity Relationships* 9(3): 223–226. https://doi.org/10.1002/qsar.19900090306

Hasanuzzaman, M., Nahar, K., Alam, M.M., Roychowdhury, R., and Fujita, M. 2013. Physiological, Biochemical, and Molecular Mechanisms of Heat Stress Tolerance in Plants. *Int J Mol Sci.* 14(5): 9643–9684. https://doi.org/10.3390/ijms14059643

He, M., He, C.Q., and Ding, N.Z. 2018. Abiotic Stresses: General Defenses of Land Plants and Chances for Engineering Multistress Tolerance. *Front Plant Sci.* 9: 1771. https://doi.org/10.3389/fpls.2018.01771

Hopfinger, A.J., Wang, S., Tokarski, J.S., Jin, B., Albuquerque, M., et al. 1997. Construction of 3D-QSAR models using the 4D-QSAR analysis formalism. *J Am Chem Soc.* 119 (43): 10509–10524. https://doi.org/10.1021/ja9718937

Hossain, Z., Nouri, M. Z., and Komatsu, S. 2012. Plant cell organelle proteomics in response to abiotic stress. *Journal of Proteome Research* 11(1): 37–48. https://doi.org/10.1021/pr200863r

Hu, B., Yao, H., Peng, X., Wang, R., Li, F., Wang, Z., Zhao, M.; Jin, L. (2019). Overexpression of Chalcone synthase improves flavonoid accumulation and drought tolerance in Tobacco. Preprint: 2019060103. https://doi.org/10.20944/preprints201906.0103.v1

Hu, X., Yang Y., Gong F., Zhang D., Zhang L., Wu L., Li C., and Wang W. 2015. Protein sHSP26 improves chloroplast performance under heat stress by interacting with specific chloroplast proteins in maize (*Zea mays*). *J. Proteomics.* 115: 81–92. https://doi.org/10.1016/j.jprot.2014.12.009.

Jegadeesan, S., Chaturvedi, P., Ghatak, A., Pressman, E., Meir, S., Faigenboim, A., Rutley, N., Beery, A., Harel, A., Weckwerth, W., and Firon, N. 2018. Proteomics of Heat-Stress and Ethylene-Mediated Thermotolerance Mechanisms in Tomato Pollen Grains. *Front Plant Sci.* 9: 1558. https://doi.org/10.3389/fpls.2018.01558

Jiang, Z., Jin, F., Shan, X., and Li, Y. 2019. iTRAQ-based proteomic analysis reveals several strategies to cope with drought stress in maize seedlings. *Int J Mol Sci.* 20(23): 5956. https://doi.org/10.3390/ijms20235956

Jorrin-Novo, J.V. 2014. Plant proteomics methods and protocols. *Methods Mol Biol* (Clifton, N.J.). 1072: 3–13. https://doi.org/10.1007/978-1-62703-631-3_1

Kaur, B., Sandhu, K. S., Kamal, R., Kaur, K., Singh, J., Röder, M. S., and Muqaddasi, Q. H. 2021. Omics for the improvement of abiotic, biotic, and agronomic traits in major cereal crops: Applications, challenges, and prospects. *Plants (Basel, Switzerland)* 10(10): 1989. https://doi.org/10.3390/plants10101989

Khan, A.U. 2016. Descriptors and their selection methods in QSAR analysis: Paradigm for drug design. *Drug Discov.* 21(8): 1291–1302. https://doi.org/10.1016/j.drudis.2016.06.013

Klebe, G., and Abraham, U. 1999. Comparative molecular similarity index analysis (CoMSIA) to study hydrogen-bonding properties and to score combinatorial libraries. *J Comput Aided Mol Des.* 13(1): 1–10. https://doi.org/10.1023/a:1008047919606

Klebe, G., and Mietzner, T. 1994. A fast and efficient method to generate biologically relevant conformations. *J Comput Aided Mol Des.* 8(5): 583–606. https://doi.org/10.1007/BF00123667

Kopka, J., Fernie, A., Weckwerth, W., Gibon, Y., & Stitt, M. 2004. Metabolite profiling in plant biology: Platforms and destinations. *Genome Biology* 5(6): 109. https://doi.org/10.1186/gb-2004-5-6-109

Korn, M., Gärtner, T., Erban, A., Kopka, J., Selbig, J., Hincha, D.K. 2010. Predicting arabidopsis freezing tolerance and heterosis in freezing tolerance from metabolite composition. *Mol Plant.* 3: 224–235. https://doi.org/10.1093/mp/ssp105

Kosová, K., Chrpová, J., Šantrůček, J., Hynek, R., Štěrbová, L., Vítámvás, P., Bradová, J., & Prášil, I. T. (2017). The effect of Fusarium culmorum infection and deoxynivalenol (DON) application on proteome response in barley cultivars Chevron and Pedant. *J Proteomics.* 169: 112–124. https://doi.org/10.1016/j.jprot.2017.07.005

Leonard, J.T., and Roy, K. 2006. On selection of training and test sets for the development of predictive QSAR models. *QSAR & Combinatorial Science.* 25(3): 235–251. https://doi.org/10.1002/qsar.200510161

Li, S., Yu, J., Li, Y., Zhang, H., Bao, X., Bian, J., Xu, C., Wang, X., Cai, X., Wang, Q., Wang, P., Guo, S., Miao, Y., Chen, S., Qin, Z., and Dai, S. 2019. Heat-responsive proteomics of a heat-sensitive spinach variety. *Int J Mol Sci.* 20(16): 3872. https://doi.org/10.3390/ijms20163872

Li, X., Cai, J., Liu, F., Dai, T., Cao, W., and Jiang, D. 2014. Physiological, proteomic and transcriptional responses of wheat to combination of drought or waterlogging with late spring low temperature. *Funct Plant Biol.* 41: 690–703. https://doi.org/10.1071/FP13306

Lin, Y., Dai, Y., Xu, W., Wu, X., Li, Y., Zhu, H., and Zhou, H. 2022. The Growth, Lipid Accumulation and Fatty Acid Profile Analysis by Abscisic Acid and Indol-3-Acetic Acid Induced in Chlorella sp. FACHB-8. *Int J Mol Sci.* 23: 4064. https://doi.org/10.3390/ijms23074064

Liu, D., Ford, K.L., Roessner, U., Natera, S., Cassin, A.M., Patterson, J.H., and Bacic, A. 2013. Rice suspension cultured cells are evaluated as a model system to study salt responsive networks in plants using a combined proteomic and metabolomic profiling approach. *Proteomics* 13: 2046–2062. https://doi.org/10.1002/pmic.201200425

Liu, Y., Lu, S., Liu, K. et al. 2019. Proteomics: A powerful tool to study plant responses to biotic stress. *Plant Methods* 15: 135. https://doi.org/10.1186/s13007-019-0515-8

Lončarić, M., Strelec, I., Pavić, V., Šubarić, D., Rastija, V., and Molnar, M. 2020. Lipoxygenase inhibition activity of coumarin derivatives—qsar and molecular docking study. *Pharmaceuticals* 13(7): 154. https://doi.org/10.3390/ph13070154

Mahmood, T., Jan, A., Kakishima, M., et al. 2006. Proteomic analysis of bacterial blight defense-responsive proteins in rice leaf blades. *Proteomics* 6: 6053–6065. https://doi.org/10.1002/pmic.200600470

Manzoni, C., Kia, D.A., Vandrovcova, J., Hardy, J., Wood, N.W., Lewis, P.A., and Ferrari, R. 2018. Genome, transcriptome and proteome: The rise of omics data and their integration in biomedical sciences. *Briefings in Bioinformatics* 19(2): 286–302. https://doi.org/10.1093/bib/bbw114

Marshall, G.R., and Cramer III, R. D. 1988. Three-dimensional structure-activity relationships. *Trends Pharmacol Sci.* 9 (8): 285–289. https://doi.org/10.1016/0165-6147(88)90012-0

Mohammadi, P.P., Moieni, A., Hiraga, S., and Komatsu, S. 2012. Organ-specific proteomic analysis of drought-stressed soybean seedlings. *J Proteomics.* 75: 1906–1923. https://doi.org/10.1016/j.jprot.2011.12.041

Mohapatra, A. 2020. Software tools for toxicology and risk assessment. In *Information Resources in Toxicology*, pp. 791–812. Academic Press. https://doi.org/10.1016/B978-0-12-813724-6.00072-4

Muhammad, U., Uzairu, A., and Ebuka Arthur, D. 2018. Review on: Quantitative structure activity relationship (QSAR) modeling. *J Anal Pharm Res.* 7(2): 240–242. https://doi.org/10.15406/japlr.2018.07.00232

Mustafa, G., and Komatsu, S. 2016. Insights into the response of soybean mitochondrial proteins to various sizes of aluminum oxide nanoparticles under flooding stress. *J Proteome Res.* 15(12): 4464–4475. https://doi.org/10.1021/acs.jproteome.6b00572

Nesvizhskii, A.I. 2007. Protein identification by tandem mass spectrometry and sequence database searching. In Matthiesen, R. (eds), *Mass Spectrometry Data Analysis in Proteomics. Methods in Molecular Biology*, vol. 367. Humana Press. https://doi.org/10.1385/1-59745-275-0:87

Ngara, R., Ndimba, R., Borch-Jensen, J., Jensen, O.N., and Ndimba, B. 2012. Identification and profiling of salinity stress-responsive proteins in Sorghum bicolor seedlings. *J Proteomics.* 75: 4139–4150. https://doi.org/10.1016/j.jprot.2012.05.038

Nouri, M.Z., and Komatsu, S. 2010. Comparative analysis of soybean plasma membrane proteins under osmotic stress using gel-based and LC MS/MS-based proteomics approaches. *Proteomics* 10: 1930–1945. https://doi.org/10.1002/pmic.200900632

Pan, Z., and Raftery, D. 2007. Comparing and combining NMR spectroscopy and mass spectrometry in metabolomics. *Anal Bioanal Chem.* 387: 525–527. https://doi.org/10.1007/s00216-006-0687-8

Pappireddi, N., Martin, L., and Wühr, M. 2019. A review on quantitative multiplexed proteomics. *Chembiochem: A European Journal of Chemical Biology* 20(10): 1210–1224. https://doi.org/10.1002/cbic.201800650

Parsons, H.T., Fernández-Niño, S.M., and Heazlewood, J.L. (2014). Separation of the plant Golgi apparatus and endoplasmic reticulum by free-flow electrophoresis. *Methods in Molecular Biology* (Clifton, N.J.) 1072: 527–539. https://doi.org/10.1007/978-1-62703-631-3_35

Peter, S.C., Dhanjal, J.K., Malik, V., Radhakrishnan, N., Jayakanthan, M., and Sundar, D. 2019. Quantitative structure-activity relationship (QSAR): Modeling approaches to biological applications. In Ranganathan, S., Gribskov, M., Nakai, K., and Schönbach, C. (eds), *Encyclopedia of Bioinformatics and Computational Biology*, pp. 661–676. Academic Press, Oxford. https://doi.org/10.1016/B978-0-12-809633-8.20197-0

Piasecka, A., Kachlicki, P., and Stobiecki, M. 2019. Analytical methods for detection of plant metabolomes changes in response to biotic and abiotic stresses. *Int J Mol Sci.* 20(2): 379. https://doi.org/10.3390/ijms20020379

Pitkänen, L., Tuomainen, P., and Eskelin, K. 2014. Analysis of plant ribosomes with asymmetric flow field-flow fractionation. *Anal Bioanal Chem.* 406(6): 1629–1637. https://doi.org/10.1007/s00216-013-7454-4

Pohorille A, Wilson MA, New MH, et al. 1998. Concentrations of anesthetics across the water–membrane interface, the meyer–overton hypothesis revisited. *Toxicol Lett.* 100: 421–430. https://doi.org/10.1016/S0378-4274(98)00216-1

Qiao, L.S., Cai, Y.L., He, Y.S., Jiang, L.D., Huo, X.Q., and Zhang, Y.L. 2014. Trend of multi-scale QSAR in drug design. *Asian J Chem.* 26: 5917.

Ramautar, R., Somsen, G.W., and de Jong, G.J. 2009. CE-MS in metabolomics. *Electrophoresis* 30(1): 276–291. https://doi.org/10.1002/elps.200800512

Ramautar, R., Somsen, G.W., and de Jong, G.J. 2019. CE-MS for metabolomics: Developments and applications in the period 2016–2018. *Electrophoresis* 40(1): 165–179. https://doi.org/10.1002/elps.201800323

Rastija, V., Vrandečić, K., Ćosić, J., Majić, I., Šarić, G. K., Agić, D., and Molnar, M. 2021. Biological activities related to plant protection and environmental effects of coumarin derivatives: QSAR and molecular docking studies. *Int J Mol Sci.* 22(14):7283. https://doi.org/10.3390/ijms22147283

Razzaq, A., Sadia, B., Raza, A. et al. 2019. Metabolomics: A way forward for crop improvement. *Metabolites* 9(12): 3030. https://doi.org/10.3390/metabo9120303

Roy, K., Kar, S., and Das, R.N. 2015. Statistical methods in QSAR/QSPR. In *A Primer on QSAR/QSPR Modeling. SpringerBriefs in Molecular Science*, pp. 37–59. Springer, Cham. https://doi.org/10.1007/978-3-319-17281-1_2

Roy, K., Roy, P., and Leonard, J. 2008. On some aspects of validation of predictive QSAR models. *Chemistry Central Journal* 2: 9. https://doi.org/10.1186/1752-153X-2-S1-P9

Rozanova, S., Barkovits, K., Nikolov, M., Schmidt, C., Urlaub, H., and Marcus, K. 2021. Quantitative mass spectrometry-based proteomics: An overview. In Marcus, K., Eisenacher, M., and Sitek, B. (eds), *Quantitative Methods in Proteomics. Methods in Molecular Biology*, vol. 2228, pp. 85–116. Humana, New York, NY. https://doi.org/10.1007/978-1-0716-1024-4_8

Shu, J., Ma, X., Ma, H., Huang, Q., Zhang, Y., Guan, M., and Guan, C. 2022. Transcriptomic, proteomic, metabolomic, and functional genomic approaches of Brassica napus L. during salt stress. *PLoS One* 17(3): e0262587. https://doi.org/10.1371/journal.pone.0262587

Shulaev, V., Cortes, D., Miller, G., and Mittler, R. 2008. Metabolomics for plant stress response. *Physiol Plant.* 132(2): 199–208. https://doi.org/10.1111/j.1399-3054.2007.01025.x

Škultéty, L., Danchenko, M., Preťová, A., and Hajduch, M. 2010. Techniques in plant proteomics. In Jain, S., and Brar, D. (eds), *Molecular Techniques in Crop Improvement*, pp. 469–491. Springer, Dordrecht. https://doi.org/10.1007/978-90-481-2967-6_19

Sun, C., Gao, X., Fu, J., Zhou, J., and Wu, X. 2015. Metabolic response of maize (Zea mays L.) plants to combined drought and salt stress. *Plant Soil* 388: 99–117. https://doi.org/10.1007/s11104-014-2309-0

Tandon, H., Chakraborty, T., and Suhag, V. 2019. A concise review on the significance of QSAR in drug design. *Chem Biomol Eng.* 4(4): 45–51. https://doi.org/10.11648/j.cbe.20190404.11

Taylor, N.L., Heazlewood, J.L., Day, D.A., and Millar, A.H. 2005. Differential impact of environmental stresses on the pea mitochondrial proteome. *Mol Cell Proteomics.* 4(8): 1122–1133. https://doi.org/10.1074/mcp.M400210-MCP200

Thangella, P.A.V., Pasumarti, S.N.B.S., Pullakhandam, R., Geereddy, B.R., and Daggu, M.R. 2018. Differential expression of leaf proteins in four cultivars of peanut (Arachis hypogaea L.) under water stress. *3 Biotech* 8(3): 157. https://doi.org/10.1007/s13205-018-1180-8

Tong, W., Lowis, D.R., Perkins, R., Chen, Y., Welsh, et al. 1998. Evaluation of quantitative structure–activity relationship methods for large-scale prediction of chemicals binding to the estrogen receptor. *J Chem Inf Comput Sci.* 38 (4): 669–677. https://doi.org/10.1021/ci980008g

Tuli, L., and Ressom, H.W. 2009. LC-MS Based Detection of Differential Protein Expression. *J Proteomics Bioinformat.* 2: 416–438. https://doi.org/10.4172/jpb.1000102

Vo, K.T.X., Rahman, M.M., Rahman, M.M., et al. 2021. Proteomics and metabolomics studies on the biotic stress responses of rice: An update. *Rice* 14: 30. https://doi.org/10.1186/s12284-021-00461-4

Wang, J.H., Byun, J., and Pennathur, S. 2010. Analytical approaches to metabolomics and applications to systems biology. *Semin Nephrol.* 30(5): 500–511. https://doi.org/10.1016/j.semnephrol.2010.07.007

Wang, X., He, Y., Ye, Y., et al. 2018. SILAC–based quantitative MS approach for real-time recording protein-mediated cell-cell interactions. *Sci Rep.* 8: 8441. https://doi.org/10.1038/s41598-018-26262-2

Wang, Y., Sang, Z., Xu, S., Xu, Q., Zeng, X., Jabu, D., and Yuan, H. 2020. Comparative proteomics analysis of Tibetan hull-less barley under osmotic stress via data-independent acquisition mass spectrometry. *GigaScience* 9(3): giaa019. https://doi.org/10.1093/gigascience/giaa019

Wang, Y., Xu, L., Shen, H., Wang, J., Liu, W., Zhu, X., Wang, R., Sun, X., Liu, L. 2015. Metabolomic analysis with GC-MS to reveal potential metabolites and biological pathways involved in Pb & Cd stress response of radish roots. *Sci Rep.* 5: 18296. https://doi.org/10.1038/srep18296

Worley, B., and Powers, R. 2013. Multivariate Analysis in Metabolomics. *Curr Metabolomics.* 1(1): 92–107. https://doi.org/10.2174/2213235X11301010092

Yang, H., Sun, L., Li, W., Liu, G., Tang, Y. 2018. In silico prediction of chemical toxicity for drug design using machine learning methods and structural alerts. *Front Chem.* 6: 30. https://doi.org/10.3389/fchem.2018.00030

Yu, J., Hu, S., Wang, J., Wong, G.K.S., Li, S., Liu, B., Deng, Y., Dai, L., Zhou, Y., and Zhang, X. 2002. A Draft Sequence of the Rice Genome (Oryza sativa L. ssp. indica). *Science* 296: 79–92. https://doi.org/10.1126/science.1068037

Zadražnik, T., Moen, A., and Šuštar-Vozlič, J. 2019. Chloroplast proteins involved in drought stress response in selected cultivars of common bean (Phaseolus vulgaris L.). *3 Biotech* 9(9): 331. https://doi.org/10.1007/s13205-019-1862-x

Zhang, H., Du, W., Peralta-Videa, J.R., Gardea-Torresdey, J.L., White, J.C., Keller, A.A., Guo, H., Ji, R., and Zhao, L. 2018. Metabolomics reveals how cucumber (Cucumis sativus) reprograms metabolites to cope with silver ions and silver nanoparticle-induced oxidative stress. *Environ Sci Technol*. 52: 8016–8026. https://doi.org/10.1021/acs.est.8b02440

Zhao, F., Zhang, D., Zhao, Y., Wang, W., Yang, H., Tai, F., Li, C., and Hu, X. 2016. The Difference of Physiological and Proteomic Changes in Maize Leaves Adaptation to Drought, Heat, and Combined Both Stresses. *Front Plant Sci*. 7: 1471. https://doi.org/10.3389/fpls.2016.01471

Zhu, J., Alvarez, S., Marsh, E.L., LeNoble, M.E., Cho, I.J., Sivaguru, M., Chen, S., Nguyen, H.T., Wu, Y., Schachtman, D.P., et al. 2007. Cell wall proteome in the maize primary root elongation zone. II. Region-specific changes in water soluble and lightly ionically bound proteins under water deficit. *Plant. Physiol*. 145: 1533–1548. https://doi.org/10.1104/pp.107.107250.

Zörb, C., Geilfus, C. M., Dietz, K. J. 2019. Salinity and crop yield. *Plant Biol*. 21: 31–38. https://doi.org/10.1111/plb.12884

# 24 Markers of Oxidative Stress in Plants

*Rashmi Shakya and Deepali*

## 24.1 INTRODUCTION

Oxidative stress is a condition that results due to overproduction and accumulation of reactive oxygen species (ROS) in plants due to extreme abiotic conditions. These extreme abiotic conditions or stresses—which include salinity, high temperature, drought, floods, heavy metals, mechanical injury, pest infestation, and ultraviolet radiation—are responsible for major losses in terms of yield and quality of crops (Qamer et al., 2021). At optimum levels, ROS are very important because they play an important role in various physiological processes in plants, including their interactions with abiotic and biotic surroundings (Kreslavski et al., 2012; Mhamdi & Van Breusegem, 2018). They act as signaling molecules and participate in the regulation of various cellular processes. In plants, ROS are inevitably produced in mitochondria, chloroplasts, peroxisomes, cytoplasm, and apoplastic regions by the action of various enzymes (Berwal & Ram, 2018), but mitochondria are the major producer of ROS in non-photosynthetic tissues and under dark conditions (Das & Roychoudhury, 2014). Under optimal conditions, as well, the generation of ROS happens during various metabolic processes taking place in these compartments, especially via electron transport systems (ETSs). ROS includes compounds containing activated oxygen atoms, not necessarily radicals e.g., hydrogen peroxide [$H_2O_2$]). Due to the stability and high penetrability through biological membranes, $H_2O_2$ is the most widely studied (Wu et al., 2011). Other common ROS include superoxide anion ($O_2^{\cdot-}$), hydroxyl radical (HO$^\cdot$), peroxynitrite anion (ONOO$^-$), singlet oxygen ($^1O_2$), nitric oxide (NO), etc. Free radicals, however, are chemical compounds that contain unpaired electrons and exist independently. Oxygen atoms are not present in some free radicals, e.g., transition metals. Oxidative stress in a cell is promoted by elevated ROS via the oxidation of compounds present in the cell. It is a physiological state when the rate of oxidation (loss of electrons) is higher than the reduction (gain of electrons), resulting in lipid peroxidation, DNA damage, protein oxidation, and PCD (Singh et al., 2022). In simple terms, the long-term redox imbalance in a cell leads to oxidative stress and it is among one of the stress factors which is responsible for causing damage to the cell and triggering defense reactions via signaling.

The electron leakage in ETSs taking place in chloroplasts, mitochondria, and peroxisomes during the major metabolic processes like photosynthesis, oxidative phosphorylation, photorespiration, tricarboxylic acid cycle, and β-oxidation, is the major contributor to oxidative stress in the plants. Some of the electrons lost during ETSs activate $O_2$ leading to the formation of $O_2^{\cdot-}$. ROS are generally also produced under non-stressed conditions but antioxidants present in the cell detoxify them, thereby limiting the oxidative damage (Figure 24.1). However, during unfavorable conditions e.g., abiotic stress, the ROS homeostasis is disturbed due to the overproduction of ROS, and antioxidants present in the cell are exhausted during the early oxidative stress stage. Furthermore, the ROS, specially $O_2^{\cdot-}$ generated due to enhanced activities of NADPH oxidases and peroxidases, leads at later stages to oxidative stress which causes damage to lipids by lipid peroxidation, DNA damage by affecting nucleic acids, protein oxidation leading to enzyme inhibition, and degeneration of antioxidants (Demidchik, 2014). These events lead to programmed cell death (PCD) of plant cells. ROS production commonly occurs during almost all types of abiotic stresses, such as drought, salinity, and heavy metals. It is important that ROS homeostasis is tightly regulated so that plants can adapt to stress and survive (Choudhury et al., 2013). However, the mechanisms involved during in vivo spatiotemporal regulation of ROS are yet to be explored. The measurement of ROS in biological systems is a challenging task due to their highly reactive and short-lived natures. Furthermore, ROS production and their subsequent detoxification in specific subcellular compartments (mitochondria, chloroplasts, and peroxisomes) require detection methods specific to these organelles. The quantification of ROS can be done through direct or indirect methods following the formation of oxidative stress byproducts of proteins, lipids, and nucleic acids. In this chapter, we present a detailed account of various oxidative stress markers, both enzymatic and non-enzymatic. The description of each oxidative stress marker is followed by its specific quantification method and identification/localization method.

## 24.2 MARKERS OF OXIDATIVE STRESS IN PLANTS

The accumulation of ROS in response to various abiotic stresses overwhelming the antioxidant machinery of the cell is deleterious to plants. To combat these elevated levels

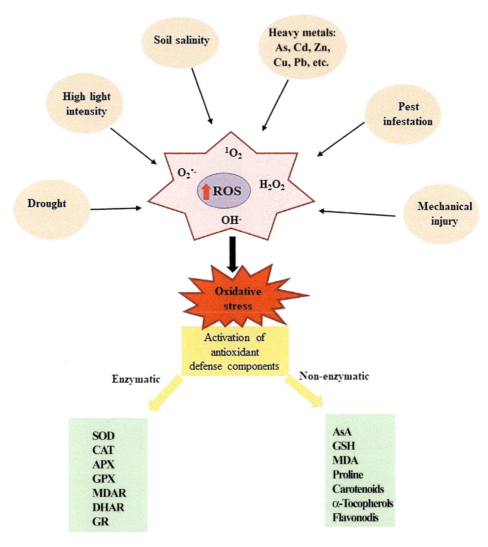

**FIGURE 24.1** Production of reactive oxygen species (ROS) due to different types of stress conditions and activation of different components of antioxidants components following oxidative stress.

of ROS, plants have strategically evolved a wide spectrum of detoxification mechanisms—both enzymatic and non-enzymatic. The byproducts of these mechanisms constitute a pool of compounds that are referred to as 'markers of oxidative stress'. The enzymatic antioxidant markers include catalase (CAT), superoxide dismutase (SOD), glutathione peroxidase (GPX), ascorbate peroxidases (APX), glutathione S-transferase (GST), dehydroascorbate reductase (DHAR), mono-hydro ascorbate reductase (MDAR) and glutathione reductase (GR) (Table 24.1). Among the noteworthy non-enzymatic markers are $H_2O_2$, malondialdehyde (MDA), glutathione (GSH), proline, α-tocopherols, carotenoids, and flavonoids (Table 24.2). Recently, ophthalmic acid (OPH), a tripeptide analog of glutathione (GSH), has also been reported as a marker of oxidative stress in plants (Servillo et al., 2018). Both types of mechanisms are important for ROS homeostasis in plants, and it has been reported that high antioxidative activity leads to better stress tolerance (Nadarajah, 2020; Mahmood et al., 2020). The relative balance between the production of ROS and the activating of enzymes involved in the scavenging of ROS is decisive in determining the signaling or oxidative stress due to ROS (Qamer et al., 2021).

Detection, quantification, and visualization of short-lived ROS can be achieved by electron paramagnetic resonance (EPR) spectroscopy technique. Nevertheless, it is not sensitive enough for in vivo quantification of the highly reactive radicals in biological systems due to their low concentration, specific compartmentalization, and enzymatic defense mechanisms. This issue is overcome by using spin trapping or spin probing technique, by which ROS are reacted with trap molecules, facilitating the formation of stable and less reactive compounds that can be easily detected with EPR (Steffen-Heins & Steffens, 2015). In the following section, a comprehensive account of different markers of oxidative stress in plants is presented. The most commonly used methods for their quantification in plants have also been briefly discussed.

## TABLE 24.1
### Enzymatic Markers of Oxidative Stress in Plants

| S. No. | Enzymatic Markers | EC Number | Molecular Weight (KDa)[#] | Subcellular Location[*] | Function |
|---|---|---|---|---|---|
| 1. | Superoxide dismutase (SOD) | 1.15.1.1 | 22–34 | M, P, Ch, Cy | Catalyzes the dismutation of $O_2^{\cdot-}$ to $O_2$ and $H_2O_2$ |
| 2. | Catalase (CAT) | 1.11.1.6 | ~57 | M, P | Converts $H_2O_2$ into $H_2O$ |
| 3. | Glutathione peroxidase (GPX) | 1.11.1.9 | ~19 | M, Ch, Cy, ER | Converts GSH into GSSG and $H_2O$ using $H_2O_2$ |
| 4. | Ascorbate peroxidase (APX) | 1.1.11.1 | 40.41 | M, P, Ch, Cy | Converts $H_2O_2$ into $H_2O$ and DHA using AsA |
| 5. | Monodehydro ascorbate reductase (MDAR) | 1.6.5.4 | 53.53 | M, P, Ch, Cy | Recycles ascorbate by converting MDHA to AsA |
| 6. | Dehydroascorbate reductase (DHAR) | 1.8.5.1 | 23–28 | M, Ch, Cy | Recycles ascorbate by converting DHA to AsA |
| 7. | Glutathione reductase (GR) | 1.6.4.2 | 53.87 | M, Ch, Cy | Converts GSSG to GSH |

*Source:* [#] Molecular weight of enzymes as reported for *Arabidopsis thaliana* mentioned in uniport.org database
*Notes:* M = mitochondria, P = peroxisomes, Ch = chloroplasts, Cy = cytosol, ER = endoplasmic reticulum

## TABLE 24.2
### Structure, Subcellular Location and Functions of Non-Enzymatic Markers of Oxidative Stress in Plants

| S. No. | Non-enzymatic Markers | Subcellular Location | Structure | Function |
|---|---|---|---|---|
| 1. | Hydrogen peroxide ($H_2O_2$) | BM, M, P, Ch, Cy | HYDROGEN PEROXIDE | Signal molecule (at low conc.), oxidative damage (high conc.) |
| 2. | Ascorbic acid (AsA) | M, P, Ch, Cy, V, Ap | | Scavenges $H_2O_2$ through the action of APX |
| 3. | Glutathione (GSH) | M, P, Ch, Cy, V, Ap | | Controls ROS levels via glutathione/ascorbate cycle |
| 4. | Malondialdehyde (MDA) | BM, Cy, Nu | | One of the final products of LPO |
| 5. | Proline | M, Ch, Cy | | Osmoprotectant, helps plants to adapt salt stress |
| 6. | Carotenoids | Ch and other plastids | B-carotene | Activate cellular defense via oxidation |

| S. No. | Non-enzymatic Markers | Subcellular Location | Structure | Function |
|---|---|---|---|---|
| 7. | α-Tocopherols | BM | | Scavenges lipid peroxyl radicals in membranes and halts LPO |
| 8. | Flavonoids | V | Isoflavan | Reduces $H_2O_2$ under excess light |

*Notes:* * BM = biological membranes, M = mitochondria, P = peroxisomes, Ch = chloroplasts, Cy = cytosol, Ap = apoplast, V = vacuole

## 24.3 ENZYMATIC MARKERS OF OXIDATIVE STRESS

### 24.3.1 Superoxide Dismutase (SOD)

SOD (E.C.1.15.1.1) are metalloenzymes that catalyze the dismutation of $O_2^{•-}$ to $O_2$ and $H_2O_2$ and form the first line of defense against the oxidative stress caused by ROS. This scavenging of $O_2^{•-}$ prevents the formation of OH- radicals via Haber-Weiss reaction.

$$O_2^{•-} + O_2^{•-} + 2 \times H^+ \rightarrow 2 \times H_2O_2 + O_2 \quad (24.1)$$

There are three types of SOD isozymes, depending on the metal ion bound to them. Of these, Fe-SOD and Mn-SOD are localized in chloroplasts and mitochondria, respectively, whereas Cu/Zn-SOD is localized in chloroplasts, peroxisomes, and cytosol (Nishiyama et al., 2017). Those SODs which accommodate both $Mn^{2+}$ and $Fe^{2+}$ as a cofactor for catalysis are referred to as cambialistic Fe/Mn-SOD. However, they show different ion preferences which are substantiated by the studies based on their crystal structure that have revealed their different binding forms with these ions (Dong et al., 2021). A recent study based on the selection of SOD as a selection criterion for grain yield under drought stress has reported SOD to be the most important enzymatic antioxidant to be considered as a selection criterion for transgenic purposes. Mostly, the Mn-SOD isozyme is highly expressed, whereas Cu/Zn-SOD is more likely to be expressed in tolerant genotypes (Saed-Moucheshi et al., 2021). The SOD activity can be determined spectrophotometrically by the inhibition of the photochemical oxidation of nitroblue tetrazolium (NBT) by SOD (Sarker & Oba, 2018). Another assay known as xanthine/xanthine oxidase (X-XOD) involves the reaction of xanthine with XOD to generate $O_2^{•-}$ which reacts with 2-(4-iodophenyl)-3-(4-nitrophenol)-5-phenyltetrazolium chloride (INT) to form a red formazan dye (Gunes et al., 2015).

### 24.3.2 Catalase (CAT)

CAT (E.C.1.11.1.6) is a 250 kDa (MW), the tetrameric heme-containing antioxidant enzyme which causes the dismutation of $H_2O_2$ into $H_2O$ and $O_2$. CAT is abundantly found in peroxisomes where $H_2O_2$ is primarily generated. It has also been reported from other organelles, such as mitochondria, chloroplast, and cytosol (Das & Roychoudhury, 2014).

$$H_2O_2 \rightarrow H_2O + \tfrac{1}{2} \times O_2 \quad (24.2)$$

Inside peroxisomes, $H_2O_2$ is produced due to the enzymatic activity of different oxidases including glycolate oxidase, acyl-CoA oxidase, and urate oxidase. During the β-oxidation and glyoxylate cycle, fatty acids are converted into carbohydrates in specialized peroxisomes known as glyoxysomes. The enzyme acyl-CoA oxidase is a part of the β-oxidation pathway. Glycolate produced in the chloroplasts during photorespiration is converted into glyoxylate by glycolate oxidase. Urate oxidase is present in peroxisomes and is involved in the function of ureides. Despite the pivotal role played by catalases in regulating $H_2O_2$ levels within the plant cell, their role during abiotic stress conditions is little known, owing to their peroxisomal location. The presence of multiple isoforms suggests the versatile roles played by them in plant systems. Based on their structure and sequence, they can be recognized into three classes, namely: (1) typical or monofunctional catalase; (2) catalase-peroxidase; and (3) pseudo catalase or Mn-catalase (Liu & Kokare, 2017).

CAT acts as a major $H_2O_2$ sink, and plants deficient in CAT activity have been reported to be susceptible to stress; thus, they are indispensable for defense in plants against various stresses (Zafar et al., 2020; Sarker & Oba, 2018). The quantitative estimation of CAT is done most commonly through the UV spectrophotometric method. This method monitors the change in absorbance at 240 nm at high $H_2O_2$ levels in solution (~30 mM). Other CAT estimation methods include iodometry, chemiluminescence, polarimetry, and titration of the unreacted excess of hydrogen peroxide (Poli et al., 2018; Hadwan, 2018).

### 24.3.3 Glutathione Peroxidases (GPX)

GPX (EC 1.11.1.9) contain selenium, belong to non–heme-containing peroxidase family, and are important for scavenging of $H_2O_2$ and other organic peroxides from the cytosol. The reduction of organic hydroperoxides and lipid peroxides is more efficient as compared to $H_2O_2$.

$$H_2O_2 + 2 \times GSH \rightarrow H_2O + GSSG \quad (24.3)$$

GPX contains multiple isoenzymes which have been reported to be present in different subcellular compartments (mitochondria, chloroplast, endoplasmic reticulum, and cytosol) and show tissue-specific expression patterns under abiotic stress. In plants, GPX contains cysteine residue in their active sites instead of selenocysteine. Besides the detoxification of $H_2O_2$ and other organic hydroperoxides, they regulate the redox homeostasis in the cell via the maintenance of NADPH/NADP+ or thiol/disulfide balance (Bela et al., 2015). GPX in plants is generally regarded as thioredoxin peroxidases because they use thioredoxin (Trx) as a reducing agent instead of glutathione. In *Lycopersicon esculentum* and *Helianthus annuus*, they show both thioredoxin peroxidase and glutathione peroxidase functions, however, Trx regenerating system is more efficient (Herbette et al., 2002). The quantification of GPX activity is based on colorimetric, spectrophotometric, fluorometric, and polarographic methods. These methods are based on either measurement of ROOH or GSH consumption periodically or GSSG production. The colorimetry-based method uses Ellman's reagent (DTNB) for the evaluation of GSH consumption. The reagent is unstable and this method is insensitive compared to other methods. The estimation of the GPXs can be done by employing the NADPH-coupled spectrophotometric method (Navrot et al., 2006). The fluorometric method using the GPX activity assay kit uses a proprietary NADP sensor which generates a fluorescent probe after reacting with GSH. The signal is measured at Ex/Em = 420/480 nm. This method is sensitive enough to detect activity as low as 1.25 mU/mL in solution (Mett & Müller, 2021).

### 24.3.4 Ascorbate Peroxidase (APX)

APX (E.C.1.1.11.1) is a member of the class I heme-peroxidases (known as peroxidase-catalase superfamily) and participates in $H_2O_2$ scavenging through ascorbate-glutathione (ASC-GSH) pathway. Contrary to major $H_2O_2$ scavenging by CAT in the peroxisomes, the same function is carried out by APX in the chloroplast and the cytosol. The activity of the APX is influenced by the availability of the reduced AsA and reduced GSH (Caverzan et al., 2012). The availability of these antioxidants in their reduced state is controlled by the activity of enzymes such as mono-hydro ascorbate reductase (MDAR), dehydroascorbate reductase (DHAR), and glutathione reductase (GR), which utilize NAD(P)H for regeneration of oxidized GSH and AsA. $H_2O_2$ is reduced to $H_2O$ and DHA (dehydroascorbate) by APX using reduced AsA.

$$H_2O_2 + AsA \rightarrow 2 \times H_2O + DHA \quad (24.4)$$

The different isoforms of APX are classified based on their location in the subcellular compartment. The soluble isoforms, namely cAPX, mitAPX, and sAPX, have been localized in the cytosol, mitochondria, and stroma, respectively. On the other hand, membrane-bound isoforms, namely mAPX and tAPX, are present in microbodies (peroxisomes and glyoxisomes) and thylakoids, respectively. In chloroplasts, the reduction of $H_2O_2$ by tAPX is considered the first line of antioxidant defense, whereas $H_2O_2$ removal by sAPX in the stroma is regarded as the second layer (Jardim-Messeder et al., 2022). All these isoforms are significant and play protective functions under different abiotic stress conditions like salinity, high light, drought, and heavy metals. The detection of APX activity in the gel can be accomplished by inhibition of ascorbate-dependent reduction of nitroblue tetrazolium (Mittler & Zilinskas, 1993). This method is specific as well as sensitive, i.e., it can detect less than 0.01 units of APX activity. The spectrophotometric estimation of different APX isozymes can also be done as described by Elavarthi & Martin (2010).

### 24.3.5 Monodehydroascorbate Reductase (MDAR)

MDAR (E.C.1.6.5.4) is involved ascorbate recycling in plants. It converts in the conversion of oxidized AsA (monodehydroascorbate, MDHA) back to reduced AsA. Both NADH and NADPH can be used by MDARs as reducing agents, but they show a high affinity for NADH, even in chloroplast where NADPH is abundant relatively. MDHA is also reduced by ferredoxin in an illuminated chloroplast. It is co-localized with the APX in the mitochondria and peroxisomes, where APX scavenges $H_2O_2$ and oxidizes AsA in the process.

$$MDHA + NADPH \rightarrow AsA + DHA + NADP^+ \quad (24.5)$$

Five isoforms of MDAR have been reported from *Arabidopsis*. Among these, two isoforms are cytosol specific, one is found in cytosol and peroxisome, another one is attached to the peroxisomal membrane, and one is chloroplastic/mitochondrial form (Tanaka et al., 2021). In tomatoes, however, only three isoforms have been reported which excludes cytosol-specific isoforms. The activity of MDAR can be estimated spectrophotometrically (Park et al., 2016).

### 24.3.6 Dehydroascorbate Reductase (DHAR)

DHAR (E.C.1.8.5.1) catalyzes the reduction of DHA to ascorbate using reduced GSH as an electron donor, thus regenerating the cellular AsA pool similar to MDAR (Terai et al., 2020). It is instrumental in maintaining the redox state in plant cells via regulation of AsA pool size in apoplast, as well as in symplast (Das & Roychoudhury, 2014).

$$DHA + 2 \times GSH \rightarrow AsA + GSSG \quad (24.6)$$

In *Arabidopsis*, three different isoforms of DHAR— namely DHAR1, DHAR2, and DHAR3—have been reported. The subcellular location of DHAR2 and DHAR3 is cytosol and stroma, respectively. However, the DHAR1 subcellular location is still not clear as one study reported it to be peroxisomal (Reumann et al., 2009) whereas another reported it to be cytosolic (Rahantaniaina et al., 2017). The DHAR activity can be estimated spectrophotometrically according to Trümper et al. (1994).

### 24.3.7 GLUTATHIONE REDUCTASE (GR)

GR (E.C.1.6.4.2) is a flavoprotein oxidoreductase that reduces GSSG to GSH in the presence of NADPH and flavine adenine dinucleotide (FAD). Reduced GSH is used up to regenerate AsA from MDHA and DHA, and as a result, it is converted to its oxidized form i.e. GSSG. It is an important enzyme of the ascorbate-glutathione pathway and catalyzes the formation of a disulfide bond in GSSG to maintain a high cellular GSH/GSSG ratio. It is abundantly present in chloroplasts but also present in small amounts in cytosol and mitochondria as well.

$$GSSG + NADPH \rightarrow 2 \times GSH + NADP^+ \quad (24.7)$$

The GR activity can be estimated fluorometrically and spectrophotometrically (Mannervik, 1999). Piggott & Karuso, 2007 described a highly sensitive fluorometric assay for free sulfhydryl groups based on FRET. This method is based on the increase in fluorescence intensity that occurs at 520 nm when a probe containing two molecules of fluorescein linked via a disulfide group is cleaved by glutathione. Table 24.3 highlights some of the recent reports on enzymatic markers of oxidative stress and their outcomes, which can be exploited for future research.

## 24.4 NON-ENZYMATIC MARKERS OF OXIDATIVE STRESS

### 24.4.1 HYDROGEN PEROXIDE ($H_2O_2$)

$H_2O_2$ is a major non-radical redox metabolite produced in plants. It has an intermediate oxidation number (−1), suggesting both oxidizing and reducing properties. At nanomolar concentration, it acts as a signaling molecule and also as a phytohormone, but at high concentrations, it causes oxidative damage to biomolecules (Černý et al., 2018). $H_2O_2$ is moderately long-lived in vivo with half-life ranging between ms and s—in contrast to its counterparts, superoxide and hydroxyl radicals, which have half-lives of 1 μs and 1 ns, respectively. Additionally, it is smaller in size (Petrov & Van Breusegem, 2012) and present in lower mm concentration, i.e., relatively abundant as its concentration is generally reported to be present ~ 1μmolg$^{-1}$ of fresh weight under optimum conditions (Cheeseman, 2006). It is also able to cross cell membranes, and thus can easily move to different cell compartments (Bienert & Chaumont, 2014). The synthesis of $H_2O_2$ takes place during aerobic metabolism through many enzymatic and non-enzymatic routes in plants. Enzymes involved in $H_2O_2$ synthesis include cell wall peroxidases, glycolate oxidases, glucose oxidases, flavin-containing enzymes, amine oxidases, sulfite oxidases, and oxalate (Chang & Tang, 2014; Francoz et al., 2015; Niu & Liao, 2016). NADPH oxidases under heavy metal stress, also leading to the accumulation of $H_2O_2$, which causes a significant rise in proline (Ben Rejeb et al., 2015). Non-enzymatic reactions taking place during photosynthesis, respiration, and photorespiration also generate $H_2O_2$. During the photorespiration process in peroxisomes, $H_2O_2$ is produced during the conversion of lycollate to glyoxylate. In chloroplast, $O_2$ reduction leads to the formation of $H_2O_2$ during light reaction. The ETSs in chloroplast and mitochondria generate superoxide radical which is reduced to $H_2O_2$ by the action of SOD. The regulation of excess $H_2O_2$ is achieved through the scavenging action of CAT (peroxisomes), POX, APX (cytosol, chloroplasts, mitochondria), and GR, whereas ascorbate (AsA), a key antioxidant, eliminates $H_2O_2$ by reacting with it directly. Glutathione (GSH) oxidizes excess $H_2O_2$ via the generation of AsA (Kapoor et al., 2015; Niu & Liao, 2016).

$H_2O_2$ participates in an array of physiological and developmental processes, such as seed germination, flowering, stomatal closure, root development, PCD, and senescence (Hernández-Barrera et al., 2015; Niu & Liao, 2016). The signaling role of $H_2O_2$ has been well documented in both animal and plant systems (Černý et al., 2018; Sies, 2017; Kim et al., 2015). Despite its importance in cellular processes and participation as a signaling molecule in the defense mechanism of the plant, it is very challenging to determine the accurate levels of $H_2O_2$ in plant tissues. The most commonly used means used for in vitro estimation of $H_2O_2$ are spectrophotometric (Zhou et al., 2006) and luminescence-based (Lu et al., 2009) methods. The methods based on the detection of ROS using dichlorofluorescein-based fluorescent dye have been considered more robust for both relative and absolute estimation of ROS (Swanson et al., 2011). In vivo imaging of $H_2O_2$ in cellular compartments has been achieved using boronate-based dyes like Peroxy Green and Peroxy Crimson (Miller et al., 2005). Chakraborty et al. (2016) optimized an Amplex Red (10-acetyl-3,7-dihydroxyphenoxazine)-based method for fluorimetric quantification of $H_2O_2$ in three model systems, namely *Arabidopsis thaliana*, *Oryza sativa*, and *Pinus nigra*. For quantitative estimation of $H_2O_2$, the method employed for its extraction from plant tissues and the composition of extraction buffers should be given due importance due to the presence of interfering substances and soluble proteins. Compounds such as polyvinylpolypyrrolidone (PVPP) and trichloroacetic acid (TCA) are used for the removal of interfering substances. It is not advisable to use activated charcoal as an additive because it degrades $H_2O_2$ upon contact (Zhou et al., 2006; Lu et al., 2009).

### 24.4.2 ASCORBIC ACID (AsA)

Ascorbic acid (AsA), i.e., vitamin C is one of the most powerful antioxidants due to its regenerative nature and is important for detoxifying ROS (Anjum et al., 2014). The

## TABLE 24.3
### Recent Reports About Various Enzymatic Markers Implicated in Oxidative Stress in Plants along with Their Outcomes

| S. No. | Enzymatic Marker (s) | Type of Stress | Plant Species | Country | Research Outcome | Reference |
|---|---|---|---|---|---|---|
| 1. | SOD | Salinity | *Brassica napus* L. (Rapeseed) | Shenzhen, China | Tolerance to salinity is improved by using $CeO_2$ nanoparticles via modulation Cu/Zn-SOD and LOX-IV isozyme activities | Li et al. (2022b) |
| 2. | SOD | As-induced stress | *Brassica oleracea* | Faisalabad, Pakistan | As-induced toxicity and tolerance vary greatly with As levels, plant physiological response, and plant organ type | Shahid et al. (2022) |
| 3. | SOD and CAT | Chilling stress | *Ziziphus jujube* Mill. (Jujube) | Ningxia, China | Exogenous NO inhibits the chilling injury by maintaining cellular redox homeostasis via S-nitrosylation of SOD and CAT | Zhang et al. (2023) |
| 4. | CAT | Water stress | *Zea mays* (maize) | Tabriz, Iran | DREB2 gene of SC706 genotype has better tolerance for water deficiency and can be used in molecular breeding programs | Mafakheri et al. (2022) |
| 5. | GPX | Cold stress | *Chrysanthemum morifolium* Ramat. | | Lysine decrotonylation of DgGPX1 enhances cold resistance via incresing the GPX activity | Yang et al. (2022) |
| 6. | GR | Heat, drought, salt, As-induced stress | *Triticum aestivum* L. (bread wheat) | Chandigarh, India | Increased GR activity under abiotic stress can be used as a foundation for validation of precise GR genes in future research for crop improvement programs | Kaur et al. (2022) |
| 7. | GR | Cu-induced stress | *Brassica rapa* ssp. *Chinensis* cv. Suzhouqing (Chinese cabbage) | Nanjing, China | BcGR1.1 gene can be used as a reference by breeders for the selection of Cu-tolerant crops, and to explore possibility to reuse soils with excessive copper content | Li et al. (2022a) |
| 8. | APX | Salt, drought, heat, freezing | *Cryptomeria fortunei* | Nanjing, China | CfAPX expression was mostly upregulated under abiotic stresses | Zhang et al. (2022) |
| 9. | APX | Salt stress | *Actinidia chinensis* (kiwifruit) | Hefei, China | Enhanced expression of cytosolic ascorbate peroxidases, AcAPX1/2 during salt stress can be exploited in molecular breeding for the crop improvement of kiwifruit | Guo et al. (2022) |
| 10. | MDAR | UV-B Radiation | *Lycopersicon esculentum* (tomato) | Seville, Spain | LCOs improve the physiological status of plants, lowering oxidative stress as indicated by MDA, and could be used as elicitors in agricultural systems subjected to UV-B stress. | Lucas et al. (2022) |
| 11. | SOD, CAT, APX, GR, MDAR, DHAR | Cr-stress | *Oryza sativa* L. Damini (paddy rice) | Prayagraj, India | ZnO nanoparticles regulate ascorbate-glutathione cycle and protect rice from Cr(VI) toxicity via enhancement of antioxidant enzymes | Prakash et al. (2022) |

physiologically active form of AsA is ascorbate, which is formed after deprotonation of OH group at the C3 position. It is abundant in the cytoplasm of the cell and approximately 5% is transported to the apoplast region, where it participates in oxidative stress signaling. MAPK signal transduction, redox homeostasis hormonal balance, and activities of antioxidant enzymes under normal conditions, as well as during abiotic stress conditions (Akram et al., 2017; Zechmann, 2011; Pignocchi et al., 2006). Ascorbate-glutathione pair is involved in the manipulation of oxidative metabolism and regulation of plant development processes. The ascorbate peroxidase-glutathione reductase (APX-GR) is more efficient in $H_2O_2$ detoxification as compared to CAT and POD. Ascorbate oxidase (AO) mediates oxidation of ascorbate to dehydroascorbate (DHA) in the apoplast and then transported into the cytosol, where DHA reductase (DHAR) reduces DHA to AsA using glutathione as an electron donor. The oxidized forms of AsA, namely MDAR and DHAR, are very significant in physiological processes (Pandey et al., 2015; Anjum et al., 2014; Gallie, 2013). These enzymes have already been discussed in detail in Section 24.3.5 and Section 24.3.6.

Ascorbate is one of the main redox metabolites, along with glutathione and pyridine nucleotides NAD(P)H, involved in non-enzymatic pathways for ROS scavenging that maintain ROS at basal non-toxic levels (Álvarez-Robles et al., 2022; Noctor et al., 2018). In addition to its role in redox homeostasis, AsA helps in the regulation of growth and development of plants, and in ameliorating arsenic- and cadmium-induced oxidative stress in *Solanum melongena* and *Triticum aestivum* L. (Zhou et al., 2022; Alamri et al., 2021). The qualitative and quantitative estimation of AsA can be accomplished through chromatography (Han et al., 2004), spectrophotometry (Dürüst et al., 1997), titration (Verdini & Lagier, 2000), and electroanalytical methods (Terry et al., 2005). However, these methods have issues, such as degradation of AsA, high over-potential of AsA oxidation, and selectivity. Consequently, high-pressure liquid chromatography (HPLC) technique is employed as the standard method for AsA determination. Recently, Abera et al. (2020) proposed a modified spectrophotometric method through titration with hexavalent chromium in the presence of Mn(II) as a catalyst.

### 24.4.3 Glutathione (GSH)

Glutathione (γ-glutamyl-cysteinyl-glycine) is the most abundant non-protein thiol found in plants in a free or bound state with concentrations ranging at low mM. In free form, it exists in two reversible states, i.e., GSH (reduced form, sulfhydryl) and GSSG (oxidized form, disulfide dimer); however, GSH is the abundant form (Dorion et al., 2021). The biosynthesis of GSH involves the sequential activity of GCS and GS enzymes. The rate limiting γ-glutamylcysteine synthetase (GCS) catalyzes the formation of γ-Glu-Cys from the amino acids and ATP, whereas glutathione synthetase (GS) forms GSH from γ-Glu-Cys, glycine, and ATP. During oxidative stress, GSH levels decrease. It has been reported to be vital for the survival of plants. Studies on *Arabidopsis* mutants having reduced GSH levels showed their survival with increased sensitivity to metal toxicity and defense responses during stress (Shanmugam et al., 2012). In association with ascorbate, GSH is instrumental in controlling elevated ROS levels (via glutathione/ascorbate cycle) or GSH peroxidases (GPX) during stress in plants, as has been very well documented (Ding et al., 2020; Zechmann, 2020).

Meyer et al. (2001) used monochlorobimane (MCB) as a reliable probe for the measurement of *in situ* cytoplasmic GSH concentration of a population of *Arabidopsis* cells using fluorometry and in single cells using confocal laser scanning microscopy (CLSM) and a two-photon laser scanning microscope (TPLSM). The quantitative estimation of GSH can be done using fluorescence, spectrophotometric, and HPLC methods (Sahoo et al., 2017).

### 24.4.4 Malondialdehyde (MDA)

Oxidative damage mediated by accumulation of ROS in plant cells leads to damage of lipids present in cellular membranes via lipid peroxidation (LPO). ROS mainly targets the ester linkage between glycerol and fatty acids and the double bond between C-atoms in phospholipids present in biological membranes. The polyunsaturated fatty acids (PUFA) present in membranes are mainly targeted by ROS. Malondialdehyde (MDA) is among one of the final products of lipid peroxidation. Lipid peroxidation comprises three stages, namely initiation, propagation, and termination. During the initial step, oxidation of PUFA present in membranes to produce fatty acid radicals occurs. The ROS species, namely $O_2^-$ and $OH^-$ radicals, react with the methylene group present in PUFA to generate lipid hydroperoxides, peroxy radicals, and conjugated dienes. The decomposition of lipid hydroperoxides leads to the production of aldehydes, namely MDA, crotonaldehyde, lipid epoxides, lipid alkoxyl radicals alkane, and alcohols. MDA is one of the most widely accepted markers of oxidative stress in plants. The most popular method for quantitative estimation is the thiobarbituric acid reactive substances (TBARS) assay, in which MDA quantity can be estimated spectrophotometrically or fluorometrically (Hodges et al., 1999). However, other methods—such as gas chromatography, liquid chromatography, and mass spectrometry—involve derivatization, followed by these separation techniques (Davey et al., 2005).

### 24.4.5 Proline

Salt stress is one of the major abiotic factors worldwide that impacts crop productivity. To cope with salinity stress, various regulatory processes involving osmotic adjustments in cells via biosynthesis of osmoprotectants such as proline come into play, which helps plants to adapt. Apart from being an antioxidant, it also acts as a signaling molecule and stabilizes proteins, DNA, membranes, and enzymes, thereby helping the cell conserve energy for further growth and development. It can scavenge free radicals and is capable of lipid peroxidation, thus inhibiting PCD (Singh et al., 2022). Endogenous accumulation of proline under various abiotic stress conditions takes place most commonly. The exogenous application of proline has been reported to improve salt tolerance (El Moukhtari et al., 2020). Accumulation of proline causes a drop in the osmotic potential of cells facilitating osmotic adjustments and stability to membranes, thereby preventing them from osmotic stress. Overproduction of proline facilitates the maintenance of water uptake, osmotic adjustment, and redox homeostasis, thus helping in restoring the cell structures and mitigation of oxidative damage (Ghosh et al., 2022).

### 24.4.6 CAROTENOIDS

These are a diverse group of isoprenoid compounds that carry out important functions in plants as accessory light-harvesting pigments, part of the antioxidants system, and pollinator attractants. Under photooxidative stress conditions, ROS mediates the non-enzymatic oxidation of carotenoids like β-carotene into bioactive compounds like β–cyclocitral and dihydroactinidiolide. These act as stress signals and activate cellular defense in the plant. The antioxidant activities of carotenoids depend on their interaction with other antioxidants like vitamins C and E (Swapnil et al., 2021).

### 24.4.7 α-TOCOPHEROLS

These are major vitamin E compounds. They are lipophilic antioxidants that efficiently scavenge ROS and lipid radicals. In response to various ecophysiological stresses, the levels of α-tocopherols are differentially expressed, depending on the magnitude of stress and the ability of a particular species to withstand the stress. The accumulation of α-tocopherols is an indicator of plant stress tolerance, whereas decreased levels lead to oxidative damage (Sadiq et al., 2019). Among the four isomers of tocopherol—namely α-, β-, γ- and δ-—the α-tocopherol possesses the highest antioxidant capacity. They are located in photosynthetic tissue, like an envelope of chloroplast, the membrane of thylakoid, and plastoglobuli. They are an important component of cell biological membranes, and thus are indispensable for cellular defense. They protect the PS II both structurally and functionally via quenching of excess energy from $O_2$. They also play an important role in the lipid peroxidation cycle, as they effectively scavenge lipid peroxyl radicals in thylakoid membranes and halt the propagation step, thereby modulating the formation of oxylipins like phytohormone jasmonic acid (Munné-Bosch, 2007). The α-tocopherols reduce the radicals at the interface of water and membrane. Consequently, the so-formed TOH⁻ radicals interact with GSH and AsA to get recycled into their reduced form.

### 24.4.8 FLAVANOIDS

Flavanoids are diverse secondary metabolites that are ubiquitous in plants. Based on their structure, they are mainly classified into four categories: flavanols, flavones, isoflavones, and anthocyanins (Agati et al., 2012). The role of flavonoids is controversial due to their exclusive presence in the vacuoles of epidermal cells where ROS is not present and the fact that plants have very efficient antioxidant machinery to cope with environmental stresses. It is a known fact that under severe stress conditions, the activities of certain antioxidant enzymes like APX decline (Hatier & Gould, 2008). Thus, it has been proposed that conditions that are responsible for the inactivation of antioxidant enzymes may induce the biosynthesis of flavonoids like anthocyanins. Fini et al. (2011) showed that accumulation of flavonoids takes place in the vacuoles of mesophyll cells of leaves under excessive light and anthocyanins reduce the $H_2O_2$ which is generated after mechanical injury.

### 24.4.9 OPHTHALMIC ACID (OPH): A NOVEL OXIDATIVE STRESS MARKER IN PLANTS

It is a tripeptide analog of GSH. Chemically, it is γ-glutamyl-L-2-aminobutyryl-glycine, in which 2-aminobutyric acid (2-ABA) substitutes cysteine residue. It is similar to GSH in some of its chemical properties and biosynthesis, but lacks reducing properties due to the absence of cysteine residue. The oxidative stress detoxification work on cyanobacterium *Synechocystis* reported OPH accumulation under glucose stress (Narainsamy et al., 2016). An inverse correlation between intracellular GSH and OPH levels has been reported during oxidative stress. Servillo et al. (2018) reported the use of OPH as a possible marker of oxidative damage in plant cells like it is in animals. They reported increased OPH levels by employing HPLC-ESI-MS/MS analysis under different types of oxidative stress conditions.

## 24.5 SUMMARY

In plants, ROS are inevitable byproducts of aerobic plant metabolism. In recent decades, climate change has led to an increase in the frequency and intensity of biotic and abiotic stress events. The production of ROS is increased under stressful environmental conditions, which leads to extensive damage to DNA, lipids, and proteins, and damage to the integrity of the cell, and is eventually responsible for PCD. On one hand, the accumulation of ROS leads to oxidative damage to cells, but on the other hand, it triggers a cascade of stress-signaling pathways via antioxidant components for preventing further damage. The components of antioxidative systems in plants include enzymatic markers (SOD, CAT, GPX, APX, MDAR, DHAR and GR) and non-enzymatic markers ($H_2O_2$, AsA, GSH, proline, MDA, carotenoids, α-tocopherol, and flavonoids). OPH has been recently reported to be another marker of oxidative stress. The activity of some antioxidant enzymes like

APX decreases under severe stress conditions, but these conditions activate antioxidant flavonoids to take charge of the defense system. The markers of oxidative stress can be qualitatively and qualitatively estimated through techniques like spectrophotometry, fluorometry, CLSM, zymography, etc. Maintenance of ROS homeostasis is crucial for various physiological processes. Better knowledge of ROS homeostasis and its regulation is the key to understanding the ecophysiology of the plants at the molecular level due to climate change and will help in addressing global food security and biodiversity conservation challenges. Considering the fact that plants are exposed to different environmental stresses simultaneously, it is imperative that future studies should be targeted not only for understanding the expression and regulation of the antioxidant enzymes and other markers, but also their interaction networks to further decipher their roles in tolerance of these stresses. Subsequently, the developed knowledge base can be exploited for the production of economically important crops with better tolerance for oxidative stress by genetic manipulation.

## REFERENCES

Abera, H., Abdisa, M., & Washe, A. P. (2020). Spectrophotometric method to the determination of ascorbic acid in M. stenopetala leaves through catalytic titration with hexavalent chromium and its validation. *International Journal of Food Properties*, 23(1), 999–1015. https://doi.org/10.1080/10942912.2020.1775249

Agati, G., Azzarello, E., Pollastri, S., & Tattini, M. (2012). Flavonoids as antioxidants in plants: Location and functional significance. *Plant Sciences*, 196, 67–76. https://doi.org/10.1016/j.plantsci.2012.07.014

Akram, N. A., Shafiq, F., & Ashraf, M. (2017). Ascorbic acid-a potential oxidant scavenger and its role in plant development and abiotic stress tolerance. *Frontiers in Plant Science*, 8, 613. https://doi.org/10.3389/fpls.2017.00613

Alamri, S., Kushwaha, B. K., Singh, V. P., Siddiqui, M. H., Al-Amri, A. A., Alsubaie, Q. D., & Ali, H. M. (2021). Ascorbate and glutathione independently alleviate arsenate toxicity in brinjal but both require endogenous nitric oxide. *Physiologia Plantarum*, 173(1), 276–286. https://doi.org/10.1111/ppl.13411

Álvarez-Robles, M. J., Clemente, R., Ferrer, M. A., Calderón, A., & Bernal, M. P. (2022). Effects of ascorbic acid addition on the oxidative stress response of *Oryza sativa* L. plants to As (V) exposure. *Plant Physiology and Biochemistry*, 186, 232–241. https://doi.org/10.1016/j.plaphy.2022.07.013

Anjum N. A., Gill S., Gill R., Hasanuzzaman M., Duarte A. C., Tuteja N., et al. (2014). Metal/metalloid stress tolerance in plants: Role of ascorbate, its redox couple, and associated enzymes. *Protoplasma* 251, 1265–1283. https://doi.org/10.1007/s00709-014-0636-x

Bela, K., Horváth, E., Gallé, Á., Szabados, L., Tari, I., & Csiszár, J. (2015). Plant glutathione peroxidases: Emerging role of the antioxidant enzymes in plant development and stress responses. *Journal of Plant Physiology*, 176, 192–201. https://doi.org/10.1016/j.jplph.2014.12.014

Ben Rejeb, K., Vos, L. D., Le Disquet, I., Leprince, A. S., Bordenave, M., Maldiney, R., et al. (2015). Hydrogen peroxide produced by NADPH oxidases increases proline accumulation during salt or mannitol stress in Arabidopsis thaliana. *New Phytologist*, 208, 1138–1148. https://doi.org/10.1111/nph.13550

Berwal, M. K., & Ram, C. (2018). Superoxide dismutase: A stable biochemical marker for abiotic stress tolerance in higher plants. In *Abiotic and Biotic Stress in Plants*. IntechOpen. https://doi.org/10.5772/intechopen.82079

Bienert, G. P., & Chaumont, F. (2014). Aquaporin-facilitated transmembrane diffusion of hydrogen peroxide. *Biochimica Biophysica Acta* 1840, 1596–1604. https://doi.org/10.1016/j.bbagen.2013.09.017

Caverzan, A., Passaia, G., Rosa, S. B., Ribeiro, C. W., Lazzarotto, F., & Margis-Pinheiro, M. (2012). Plant responses to stresses: Role of ascorbate peroxidase in the antioxidant protection. *Genetics and Molecular Biology*, 35, 1011–1019. https://doi.org/10.1590/S1415-47572012000600016

Černý, M., Habánová, H., Berka, M., Luklová, M., & Brzobohatý, B. (2018). Hydrogen peroxide: Its role in plant biology and crosstalk with signalling networks. *International Journal of Molecular Sciences*, 19(9), 2812. https://doi.org/10.3390/ijms19092812

Chakraborty, S., Hill, A. L., Shirsekar, G., Afzal, A. J., Wang, G. L., Mackey, D., & Bonello, P. (2016). Quantification of hydrogen peroxide in plant tissues using Amplex Red. *Methods*, 109, 105–113. https://doi.org/10.1016/j.ymeth.2016.07.016

Chang, Q., and Tang, H. (2014). Optical determination of glucose and hydrogen peroxide using a nanocomposite prepared from glucose oxidase and magnetite nanoparticles immobilized on graphene oxide. *Microchimica Acta*, 181, 527–534. https://doi.org/10.1007/s00604-013-1145-x

Cheeseman, J. M. (2006). Hydrogen peroxide concentrations in leaves under natural conditions. *Journal of Experimental Botany*, 57(10), 2435–2444. https://doi.org/10.1093/jxb/erl004

Choudhury, S., Panda, P., Sahoo, L., & Panda, S. K. (2013). Reactive oxygen species signaling in plants under abiotic stress. *Plant Signaling & Behavior*, 8(4), e23681. https://doi.org/10.4161/psb.23681

Das, K., & Roychoudhury, A. (2014). Reactive oxygen species (ROS) and response of antioxidants as ROS-scavengers during environmental stress in plants. *Frontiers in Environmental Science*, 2, 53. https://doi.org/10.3389/fenvs.2014.00053

Davey, M. W., Stals, E., Panis, B., Keulemans, J., & Swennen, R. L. (2005). High-throughput determination of malondialdehyde in plant tissues. *Analytical Biochemistry*, 347(2), 201–207. https://doi.org/10.1016/j.ab.2005.09.041

Demidchik, V. (2015). Mechanisms of oxidative stress in plants: From classical chemistry to cell biology. *Environmental and Experimental Botany*, 109, 212–228. https://doi.org/10.1016/j.envexpbot.2014.06.021

Ding, H., Wang, B., Han, Y., & Li, S. (2020). The pivotal function of dehydroascorbate reductase in glutathione homeostasis in plants. *Journal of Experimental Botany*, 71(12), 3405–3416. https://doi.org/10.1093/jxb/eraa107

Dong, X., Wang, W., Li, S., Han, H., Lv, P., & Yang, C. (2021). Thermoacidophilic Alicyclobacillus superoxide dismutase: Good candidate as additives in food and medicine. *Frontiers in Microbiology*, 12, 577001. https://doi.org/10.3389/fmicb.2021.577001

Dorion, S., Ouellet, J. C., & Rivoal, J. (2021). Glutathione metabolism in plants under stress: Beyond reactive oxygen species detoxification. *Metabolites*, 11(9), 641. https://doi.org/10.3390/metabo11090641

Dürüst, N., Sümengen, D., & Dürüst, Y. (1997). Ascorbic acid and element contents of foods of Trabzon (Turkey). *Journal of Agricultural and Food Chemistry*, 45(6), 2085–2087. https://doi.org/10.1021/jf9606159

El Moukhtari, A., Cabassa-Hourton, C., Farissi, M., & Savouré, A. (2020). How does proline treatment promote salt stress tolerance during crop plant development? *Frontiers in Plant Science*, 11, 1127. https://doi.org/10.3389/fpls.2020.01127

Elavarthi, S., & Martin, B. (2010). Spectrophotometric assays for antioxidant enzymes in plants. In Sunkar, R. (eds), *Plant Stress Tolerance. Methods in Molecular Biology*, vol. 639. Humana Press. https://doi.org/10.1007/978-1-60761-702-0_16

Fini, A., Brunetti, C., Di Ferdinando, M., Ferrini, F., & Tattini, M. (2011). Stress-induced flavonoid biosynthesis and the antioxidant machinery of plants. *Plant Signaling & Behavior*, 6(5), 709–711. https://doi.org/10.4161/psb.6.5.15069

Francoz, E., Ranocha, P., Nguyen-Kim, H., Jamet, E., Burlat, V., & Dunand, C. (2015). Roles of cell wall peroxidases in plant development. *Phytochemistry*, 112, 15–21. https://doi.org/10.1016/j.phytochem.2014.07.020

Gallie D. R. (2013). Increasing vitamin C content in plant foods to improve their nutritional value—successes and challenges. *Nutrients*, 5, 3424–3446. https://doi.org/10.3390/nu5093424

Ghosh, U. K., Islam, M. N., Siddiqui, M. N., Cao, X., & Khan, M. A. R. (2022). Proline, a multifaceted signalling molecule in plant responses to abiotic stress: Understanding the physiological mechanisms. *Plant Biology*, 24(2), 227–239. https://doi.org/10.1111/plb.13363

Gunes, S., Tamburaci, S., Imamoglu, E., & Dalay, M. C. (2015). Determination of superoxide dismutase activities in different cyanobacteria for scavenging of reactive oxygen species. *Journal of Biologically Active Products from Nature*, 5(1), 25–32. https://doi.org/10.1080/22311866.2014.983973

Guo, X. H., Yan, H. E., Zhang, Y., Yi, W. A. N. G., Huang, S. X., Liu, Y. S., & Wei, L. I. (2022). Kiwifruit (Actinidia chinensis 'Hongyang') cytosolic ascorbate peroxidases (AcAPX1 and AcAPX2) enhance salinity tolerance in *Arabidopsis thaliana*. *Journal of Integrative Agriculture*, 21(4), 1058–1070. https://doi.org/10.1016/S2095-3119(21)63652-3

Hadwan, M. H. (2018). Simple spectrophotometric assay for measuring catalase activity in biological tissues. *BMC Biochemistry*, 19(1), 1–8. https://doi.org/10.1186/s12858-018-0097-5

Han, J. S., Kozukue, N., Young, K. S., Lee, K. R., & Friedman, M. (2004). Distribution of ascorbic acid in potato tubers and in home-processed and commercial potato foods. *Journal of Agricultural and Food Chemistry*, 52(21), 6516–6521. https://doi.org/10.1021/jf0493270

Hatier, J. H. B., & Gould, K. S. (2008). Foliar anthocyanins as modulators of stress signals. *Journal of Theoretical Biology*, 253(3), 625–627. https://doi.org/10.1016/j.jtbi.2008.04.018

Herbette, S., Lenne, C., Leblanc, N., Julien, J. L., Drevet, J. R., & Roeckel-Drevet, P. (2002). Two GPX-like proteins from Lycopersicon esculentum and Helianthus annuus are antioxidant enzymes with phospholipid hydroperoxide glutathione peroxidase and thioredoxin peroxidase activities. *European Journal of Biochemistry*, 269(9), 2414–2420. https://doi.org/10.1046/j.1432-1033.2002.02905.x

Hernández-Barrera, A., Velarde-Buendía, A., Zepeda, I., Sanchez, F., Quinto, C., Sánchez-Lopez, R., et al. (2015). Hyper, a hydrogen peroxide sensor, indicates the sensitivity of the Arabidopsis root elongation zone to aluminum treatment. *Sensors*, 15, 855–867. https://do.org/10.3390/s150100855

Hodges, D. M., DeLong, J. M., Forney, C. F., & Prange, R. K. (1999). Improving the thiobarbituric acid-reactive-substances assay for estimating lipid peroxidation in plant tissues containing anthocyanin and other interfering compounds. *Planta*, 207(4), 604–611. https://doi.org/10.1007/s004250050524

Jardim-Messeder, D., Zamocky, M., Sachetto-Martins, G., & Margis-Pinheiro, M. (2022). Chloroplastic ascorbate peroxidases targeted to stroma or thylakoid membrane: The chicken or egg dilemma. *FEBS Letters*. https://doi.org/10.1002/1873-3468.14438

Kapoor, D., Sharma, R., Handa, N., Kaur, H., Rattan, A., Yadav, P., et al. (2015). Redox homeostasis in plants under abiotic stress: Role of electron carriers, energy metabolism mediators and proteinaceous thiols. *Frontiers in Environmental Sciences*, 3, 13. https://doi.org/10.3389/fenvs.2015.00013

Kaur, A., Tyagi, S., Singh, K., & Upadhyay, S. K. (2022). Exploration of glutathione reductase for abiotic stress response in bread wheat (*Triticum aestivum* L.). *Plant Cell Reports*, 41(3), 639–654. https://doi.org/10.1007/s00299-021-02717-1

Kim, Y. J., Lee, Y. H., Lee, H. J., Jung, H., and Hong, J. K. (2015). $H_2O_2$ production and gene expression of antioxidant enzymes in kimchi cabbage (*Brassica rapa* var. glabra Regel) seedlings regulated by plant development and nitrosative stress-triggered cell death. *Plant Biotechnology Reports*, 9, 67–78. https://doi.org/10.1007/s11816-015-0343-x

Kreslavski, V. D., Los, D. A., Allakhverdiev, S. I., & Kuznetsov, V. V. (2012). Signaling role of reactive oxygen species in plants under stress. *Russian Journal of Plant Physiology*, 59(2), pp. 141–154. https://doi.org/10.1134/S1021443712020057

Li, Y., Huang, F., Tao, Y., Zhou, Y., Bai, A., Yu, Z., . . . & Li, Y. (2022a). BcGR1.1, a cytoplasmic localized glutathione reductase, enhanced tolerance to copper stress in *Arabidopsis thaliana*. *Antioxidants*, 11(2), 389. https://doi.org/10.3390/antiox11020389

Li, Y., Liu, J., Fu, C., Khan, M. N., Hu, J., Zhao, F., . . . & Li, Z. (2022b). $CeO_2$ nanoparticles modulate Cu–Zn superoxide dismutase and lipoxygenase-IV isozyme activities to alleviate membrane oxidative damage to improve rapeseed salt tolerance. *Environmental Science: Nano*, 9(3), 1116–1132. https://doi.org/10.1039/D1EN00845E

Liu, X., & Kokare, C. (2017). Microbial enzymes of use in industry. In *Biotechnology of Microbial Enzymes* (pp. 267–298). Academic Press. https://doi.org/10.1016/B978-0-12-803725-6.00011-X

Lu, S., Song, J., & Campbell-Palmer, L. (2009). A modified chemiluminescence method for hydrogen peroxide determination in apple fruit tissues. *Scientia Horticulturae*, 120(3), 336–341. https://doi.org/10.1016/j.scienta.2008.11.003

Lucas, J. A., García-Villaraco, A., Ramos-Solano, B., Akdi, K., & Gutierrez-Mañero, F. J. (2022). Lipo-Chitooligosaccharides (LCOs) as elicitors of the enzymatic activities related to ros scavenging to alleviate oxidative stress generated in tomato plants under stress by UV-B radiation. *Plants*, 11(9), 1246. https://doi.org/10.3390/plants11091246

Mafakheri, K., Valizadeh, M., & Mohammadi, S. A. (2022). Evaluation of catalase and DREB-2 gene expression in Maize (*Zea mays* L.) genotypes under water deficit stress condition. *Crop Biotechnology*, 11(36), 75–93. https://doi.org/10.30473/cb.2022.62784.1869

Mahmood, T., Khalid, S., Abdullah, M., et al. (2020). Insights into drought stress signaling in plants and the molecular genetic basis of cotton drought tolerance. *Cells*, 9(1), 105. https://doi.org/10.3390/cells9010105

Mannervik, B. (1999). Measurement of glutathione reductase activity. *Current Protocols in Toxicology*, 1, 7–2. https://doi.org/10.1002/0471140856.tx0702s00

Mett, J., & Müller, U. (2021). The medium-chain fatty acid decanoic acid reduces oxidative stress levels in neuroblastoma cells. *Scientific Reports*, 11, 6135.

Meyer, A. J., May, M. J., & Fricker, M. (2001). Quantitative in vivo measurement of glutathione in Arabidopsis cells. *The Plant Journal*, 27(1), 67–78. https://doi.org/10.1046/j.1365-313x.2001.01071.x

Mhamdi, A., & Van Breusegem, F. (2018). Reactive oxygen species in plant development. *Development*, 145(15), dev164376. https://doi.org/10.1242/dev.164376

Miller, E. W., Albers, A. E., Pralle, A., Isacoff, E. Y., & Chang, C. J. (2005). Boronate-based fluorescent probes for imaging cellular hydrogen peroxide. *Journal of the American Chemical Society*, 127(47), 16652–16659. https://doi.org/10.1021/ja054474f

Mittler, R., & Zilinskas, B. A. (1993). Detection of ascorbate peroxidase activity in native gels by inhibition of the ascorbate-dependent reduction of nitroblue tetrazolium. *Analytical Biochemistry*, 212(2), 540–546. https://doi.org/10.1006/abio.1993.1366

Munné-Bosch, S. (2007). α-Tocopherol: A multifaceted molecule in plants. *Vitamins & Hormones*, 76, 375–392. https://doi.org/10.1016/S0083-6729(07)76014-4

Nadarajah, K. K. (2020). ROS homeostasis in abiotic stress tolerance in plants. *International Journal of Molecular Sciences*, 21(15), 5208. https://doi.org/10.3390/ijms21155208

Narainsamy, K., Farci, S., Braun, E., Junot, C., Cassier-Chauvat, C., & Chauvat, F. (2016). Oxidative-stress detoxification and signalling in cyanobacteria: The crucial glutathione synthesis pathway supports the production of ergothioneine and ophthalmate. *Molecular Microbiology*, 100(1), 15–24. https://doi.org/10.1111/mmi.13296

Navrot, N., Collin, V., Gualberto, J., Gelhaye, E., Hirasawa, M., Rey, P., . . . & Rouhier, N. (2006). Plant glutathione peroxidases are functional peroxiredoxins distributed in several subcellular compartments and regulated during biotic and abiotic stresses. *Plant Physiology*, 142(4), 1364–1379. https://doi.org/10.1104/pp.106.089458

Nishiyama, Y., Fukamizo, T., Yoneda, K., & Araki, T. (2017). Complete amino acid sequence of a copper/zinc-superoxide dismutase from ginger rhizome. *The Protein Journal*, 36(2), 98–107. https://doi.org/10.1007/s10930-017-9700-7

Niu, L., & Liao, W. (2016). Hydrogen peroxide signaling in plant development and abiotic responses: Crosstalk with nitric oxide and calcium. *Frontiers in Plant Science*, 7, 230. https://doi.org/10.3389/fpls.2016.00230

Noctor, G., Reichheld, J. P., & Foyer, C. H. (2018). ROS-related redox regulation and signaling in plants. In *Seminars in Cell & Developmental Biology*, vol. 80, pp. 3–12. Academic Press. https://doi.org/10.1016/j.semcdb.2017.07.013

Pandey, P., Singh, J., Achary, V., & Reddy, M. K. (2015). Redox homeostasis via gene families of ascorbate-glutathione pathway. *Frontiers in Environmental Sciences*, 3, 25 https://doi.org/10.3389/fenvs.2015.00025

Park, A. K., Kim, I. S., Do, H., Jeon, B. W., Lee, C. W., Roh, S. J., Shin, S. C., Park, H., Kim, Y. S., Kim, Y. H., & Yoon, H. S. (2016). Structure and catalytic mechanism of monodehydroascorbate reductase, MDHAR, from *Oryza sativa* L. japonica. *Scientific Reports*, 6(1), 1–10. https://doi.org/10.1038/srep33903

Petrov, V. D., & Van Breusegem, F. (2012). Hydrogen peroxide—a central hub for information flow in plant cells. *AoB Plants*, pls014. https://doi.org/10.1093/aobpla/pls014

Piggott, A. M., & Karuso, P. (2007). Fluorometric assay for the determination of glutathione reductase activity. *Analytical Chemistry*, 79(22), 8769–8773. https://doi.org/10.1021/ac071518p

Pignocchi, C., Kiddle, G., Hernández, I., Foster, S. J., Asensi, A., Taybi, T., et al. (2006). Ascorbate oxidase-dependent changes in the redox state ofthe apoplast modulate gene transcript accumulation leading to modified hormone signaling and orchestration of defense processes in tobacco. *Plant Physiolog*, 141, 423–435. https://doi.org/10.1104/pp.106.078469

Poli, Y., Nallamothu, V., Balakrishnan, D., Ramesh, P., Desiraju, S., Mangrauthia, S. K., . . . & Neelamraju, S. (2018). Increased catalase activity and maintenance of photosystem II distinguishes high-yield mutants from low-yield mutants of rice var. Nagina22 under low-phosphorus stress. *Frontiers in Plant Science*, 9, 1543. https://doi.org/10.3389/fpls.2018.01543

Prakash, V., Rai, P., Sharma, N. C., Singh, V. P., Tripathi, D. K., Sharma, S., & Sahi, S. (2022). Application of zinc oxide nanoparticles as fertilizer boosts growth in rice plant and alleviates chromium stress by regulating genes involved in regulating oxidative stress. *Chemosphere*, 134554. https://doi.org/10.1016/j.chemosphere.2022.134554

Qamer Z., Chaudhary, M. T., Du, X., Hinze, L., & Azhar, M. T. (2021). Review of oxidative stress and antioxidative defense mechanisms in *Gossypium hirsutum* L. in response to extreme abiotic conditions. *Journal of Cotton Research*, 4(1), 1–9. https://doi.org/10.1186/s42397-021-00086-4

Rahantaniaina, M. S., Li, S., Chatel-Innocenti, G., Tuzet, A., Issakidis-Bourguet, E., Mhamdi, A., & Noctor, G. (2017). Cytosolic and chloroplastic DHARs cooperate in oxidative stress-driven activation of the salicylic acid pathway. *Plant Physiology*, 174(2), 956–971. https://doi.org/10.1104/pp.17.00317

Reumann, S., Quan, S., Aung, K., Yang, P., Manandhar-Shrestha, K., Holbrook, D., . . . & Hu, J. (2009). In-depth proteome analysis of Arabidopsis leaf peroxisomes combined with in vivo subcellular targeting verification indicates novel metabolic and regulatory functions of peroxisomes. *Plant Physiology*, 150(1), 125–143. https://doi.org/10.1104/pp.109.137703

Sadiq, M., Akram, N. A., Ashraf, M., Al-Qurainy, F., & Ahmad, P. (2019). Alpha-tocopherol-induced regulation of growth and metabolism in plants under non-stress and stress conditions. *Journal of Plant Growth Regulation*, 38(4), 1325–1340. https://doi.org/10.1007/s00344-019-09936-7

Saed-Moucheshi, A., Sohrabi, F., Fasihfar, E., Baniasadi, F., Riasat, M., & Mozafari, A. A. (2021). Superoxide dismutase (SOD) as a selection criterion for triticale grain yield under drought stress: A comprehensive study on genomics and expression profiling, bioinformatics, heritability, and phenotypic variability. *BMC Plant Biology*, 21(1), 1–19. https://doi.org/10.1186/s12870-021-02919-5

Sahoo, S., Awasthi, J. P., Sunkar, R., & Panda, S. K. (2017). Determining glutathione levels in plants. In *Plant Stress Tolerance* (pp. 273–277). Humana Press, New York, NY. https://doi.org/10.1007/978-1-4939-7136-7_16

Sarker, U., & Oba, S. (2018). Catalase, superoxide dismutase and ascorbate-glutathione cycle enzymes confer drought tolerance of Amaranthus tricolor. *Scientific Reports*, 8(1), 1–12. https://doi.org/10.1038/s41598-018-34944-0

Servillo, L., Castaldo, D., Giovane, A., Casale, R., D'Onofrio, N., Cautela, D., & Balestrieri, M. L. (2018). Ophthalmic acid is a marker of oxidative stress in plants as in animals. *Biochimica et Biophysica Acta (BBA)-General Subjects*, 1862(4), 991–998. https://doi.org/10.1016/j.bbagen.2018.01.015

Shahid, M., Khalid, S., Bibi, I., Khalid, S., Masood, N., Qaisrani, S. A., . . . & Dumat, C. (2022). Arsenic-induced oxidative stress in Brassica oleracea: Multivariate and literature data analyses of physiological parameters, applied levels and plant organ type. *Environmental Geochemistry and Health*, 44(6), 1827–1839. https://doi.org/10.1007/s10653-021-01093-9

Shanmugam, V., Tsednee, M., & Yeh, K. C. (2012). Zinc tolerance induced by iron 1 reveals the importance of glutathione in the cross-homeostasis between zinc and iron in Arabidopsis thaliana. *The Plant Journal*, 69(6), 1006–1017. https://doi.org/10.1111/j.1365-313X.2011.04850.x

Sies, H. (2017). Hydrogen peroxide as a central redox signaling molecule in physiological oxidative stress: Oxidative eustress. *Redox Biology*, 11, 613–619. https://doi.org/10.1016/j.redox.2016.12.035

Singh, A., Mehta, S., Yadav, S., Nagar, G., Ghosh, R., Roy, A., . . . & Singh, I. K. (2022). How to cope with the challenges of environmental stresses in the era of global climate change: An update on ros stave off in plants. *International Journal of Molecular Sciences*, 23(4), 1995. https://doi.org/10.3390/ijms23041995

Steffen-Heins, A., & Steffens, B. (2015). EPR spectroscopy and its use in planta—a promising technique to disentangle the origin of specific ROS. *Frontiers in Environmental Science*, 3, 15. https://doi.org/10.3389/fenvs.2015.00015

Swanson, S. J., Choi, W. G., Chanoca, A., & Gilroy, S. (2011). In vivo imaging of $Ca^{2+}$, pH, and reactive oxygen species using fluorescent probes in plants. *Annual Review of Plant Biology*, 62, 273–297. https://doi.org/10.1146/annurev-arplant-042110-103832

Swapnil, P., Meena, M., Singh, S. K., Dhuldhaj, U. P., & Marwal, A. (2021). Vital roles of carotenoids in plants and humans to deteriorate stress with its structure, biosynthesis, metabolic engineering and functional aspects. *Current Plant Biology*, 26, 100203. https://doi.org/10.1016/j.cpb.2021.100203

Tanaka, M., Takahashi, R., Hamada, A., Terai, Y., Ogawa, T., Sawa, Y., . . . & Maruta, T. (2021). Distribution and functions of monodehydroascorbate reductases in plants: Comprehensive reverse genetic analysis of *Arabidopsis thaliana* enzymes. *Antioxidants*, 10(11), 1726. https://doi.org/10.3390/antiox10111726

Terai, Y., Ueno, H., Ogawa, T., Sawa, Y., Miyagi, A., Kawai-Yamada, M., Ishikawa, T., & Maruta, T., 2020. Dehydroascorbate reductases and glutathione set a threshold for high-light–induced ascorbate accumulation. *Plant Physiology*, 183(1), 112–122. https://doi.org/10.1104/pp.19.01556

Terry, L. A., White, S. F., & Tigwell, L. J. (2005). The application of biosensors to fresh produce and the wider food industry. *Journal of Agricultural and Food Chemistry*, 53(5), 1309–1316. https://doi.org/10.1021/jf040319t

Trümper, S., Follmann, H., & Häberlein, I. (1994). A novel dehydroascorbate reductase from spinach chloroplasts homologous to plant trypsin inhibitor. *FEBS Letters*, 352(2), 159–162. https://doi.org/10.1016/0014-5793(94)00947-3

Verdini, R. A., & Lagier, C. M. (2000). Voltammetric iodometric titration of ascorbic acid with dead-stop end-point detection in fresh vegetables and fruit samples. *Journal of Agricultural and Food Chemistry*, 48(7), 2812–2817. https://doi.org/10.1021/jf990987s

Wu, P., Cai, Z., Gao, Y., Zhang, H., and Cai, C. (2011). Enhancing the electrochemical reduction of hydrogen peroxide based on nitrogen-doped graphene for measurement of its releasing process from living cells. *Chemical Communications*, 47(40), 11327–11329. https://doi.org/10.1039/C1CC14419G

Yang, X., Lin, P., Luo, Y., Bai, H., Liao, X., Li, X., . . . & Liu, Q. (2022). Lysine decrotonylation of glutathione peroxidase at lysine 220 site increases glutathione peroxidase activity to resist cold stress in chrysanthemum. *Ecotoxicology and Environmental Safety*, 232, 113295. https://doi.org/10.1016/j.ecoenv.2022.113295

Zafar, S. A., Hameed, A., Ashraf, M., Khan, A. S., Li, X., & Siddique, K. H. (2020). Agronomic, physiological and molecular characterisation of rice mutants revealed the key role of reactive oxygen species and catalase in high-temperature stress tolerance. *Functional Plant Biology*, 47(5), 440–453. https://doi.org/10.1071/FP19246

Zechmann, B. (2011). Subcellular distribution of ascorbate in plants. *Plant Signaling and Behavior*, 6, 360–363. https://doi.org/10.4161/psb.6.3.14342

Zechmann, B. (2020). Subcellular roles of glutathione in mediating plant defense during biotic stress. *Plants*, 9(9), 1067. https://doi.org/10.3390/plants9091067

Zhang, S., Liu, L., Wu, Z., Wang, L., & Ban, Z. (2023). S-nitrosylation of superoxide dismutase and catalase involved in promotion of fruit resistance to chilling stress: A case study on Ziziphus jujube Mill. *Postharvest Biology and Technology*, 197, 112210. https://doi.org/10.1016/j.postharvbio.2022.112210

Zhang, Y., Yang, L., Zhang, M., Yang, J., Cui, J., Hu, H., & Xu, J. (2022). CfAPX, a cytosolic ascorbate peroxidase gene from Cryptomeria fortunei, confers tolerance to abiotic stress in transgenic Arabidopsis. *Plant Physiology and Biochemistry*, 172, 167–179. https://doi.org/10.1016/j.plaphy.2022.01.011

Zhou, B., Wang, J., Guo, Z., Tan, H., & Zhu, X. (2006). A simple colorimetric method for determination of hydrogen peroxide in plant tissues. *Plant Growth Regulation*, 49(2), 113–118. https://doi.org/10.1007/s10725-006-9000-2

Zhou, Z., Wei, C., Liu, H., Jiao, Q., Li, G., Zhang, J., . . . & Yang, S. (2022). Exogenous ascorbic acid application alleviates cadmium toxicity in seedlings of two wheat (*Triticum aestivum* L.) varieties by reducing cadmium uptake and enhancing antioxidative capacity. *Environmental Science and Pollution Research*, 29(15), 21739–21750. https://doi.org/10.1007/s11356-021-17371-z

# Index

Note: Page numbers in *italics* indicate a figure and page numbers in **bold** indicate a table on the corresponding page.

## A

abiotic stress, 15, 16, 92, 102, 106, 128, 253, 259, 284, 298, 302
above ground biomass, 41, 47, 101, 167, 233, 239
abscisic acid, 86, 92, 110, 152, 220, 249, 274
aceto-carmine, 121, 122
adaptation, 3, 5, 8, 16, 77, 122, 132, 165, 181, 196, 200, 209
African baobab (*Adansonia digitata* L.), 8
Air Pollutants and Ecophysiology of Tropical Plants, **38**
air pollution, 61, 66, 77, 221
allometric equations, 234, 235, 237, 240
altitude and structure of the plant community, *210*
altitudinal gradient, 209, 221
altitudinal shift, 209, 211, 212, 214
ameliorative role of phytohormones under heat stress, *129*
anatomy, 167, 211
anthocyanin, 198, 200, 201, 205
anthropogenic factors, 26, 68, 163, 176
antioxidant, 29, 31, 45, 59, 92, 111, 128, 131, 133, 306
antioxidant enzyme, 111, 131, 133, 258, 306
ascorbate peroxidase, 302
atmospheric pressure, 195, 213
auxin, 273, 275

## B

below ground biomass, 179, 183, 233
bifacial chlorosis, 61
biochemical, 66, 75, 90
biodiversity, 3, 100, 102, 146, 161, 193
biodiversity hotspots, 24, 209
bioindicators, 74, 77
biomass, 46, 233, 234, 235, 244, 266
biome, 3, 8, 10, 176
biotic interactions, 213
biotic stress, 113, 287, 288
borneo, 25, 138
brassinosteroid, 132, 253, 255, 258

## C

$C_3$ plants, 150
carbon allocation, 47
carbon balance, 86
carbon dioxide, 85, 128, 146, 234
carbon dynamics, 89
carbon sequestration, 8, 99
carbon sink, 26, 102, 233
carbon stock, 90, 100, 233
carotenoids, 45, 133, 214, 306
catalase, 301
categories of fire-affected ecosystems, *169*
cerrado, 7, 177
chlorophyll, 44, 61, 68, 74, 111, 132, 260
classification of plant functional traitS, 74
climate, 3–10, 15–20, 24–32, 98–102, 146–154

climate change and the earth, *27*
climatic change, 10, 15, 24, 26, 73, 91, 146, 153, 163, 175, 197, 211
Colombia, 193, 195, 197, 203
conservation, 125, 162
crops, 41, 48, 101, 138, 224, 260, 287
cytokinin, 265, 272, 275

## D

deciduous, 4, 20, 91, 161, 196, 204
deciduous plant, 9
deficit conditions, 274
deforestation, 13, 162
dehydroascorbate reductase, 133, 305
destructive method, 234, 239
diameter at breast height, 239
dipterocarp(s)/dipterocarpaceae, 138, 142
distribution, 6, 44, 148, 211, 219
disturbance, 175
dormancy, 20, 166, 213
dose-response relationship, 46, 48
drought, 38, 18, 91, 104, 108, 110, 219
drought tolerance and water-use efficiency, *6*
dry forest, 3, 5, 162, 244
dust pollution, 73

## E

ecological hotspots, **25**
ecology, 3, 163
ecophysiological response of tropical plants, **220**
ecophysiology, 37, 66, 85, 163, 211, 275
ecosystem functioning, 100, 161, 165, 221
ecosystem services, 101, 146, 180, 219
ectomycorrhiza, 138
effect of NOX on some tropical plants, **58**
effects of elevated $CO_2$ on tropical plants, *31*
effects of fire on plant ecophysiology, *164*
effects of high temperature on tropical plants, *29*
elevated $CO_2$, 19, 31, 146, 152, 214
elevated temperature, 11, 28, 119, 150
elevation, 20, 193, 200
elevation ranges in commercial fruit grown in Colombia, **197**
endemic plants, 92
energy production, 163
environmental conditions, 11, 104, 151, 180, 193, 271
environmental disturbances, 166
environmental stress, 74, 89, 92, 107, 220, 288, 306
enzymatic markers, 298, 303
ethylene, 92, 130, 266, 270, 276
evapotranspiration, 10, 41, 74, 93, 201
evolutionary, 15, 19, 164, 219

## F

FACE technique, 30
features of tropical ecosystems, *147*

feijoa, 193, 201
fertilization, 19, 150, 203, 213, 265
fire, 161, 169, 175, 181
fire-adaptive traits, 169, 170
fire frequency, 176
fire intensity, 166, 177
fire regimes, 167, 176, 179
fire-return interval and tree density, *167*
fire-sensitive species, 177
flavonoids, 301
flora, 25
flow cytometry, 228
flowering, 9, 268
flower production, 119, 125
foliar factors affecting the absorption of PM, *67*
foliar injury, 41
food safety, 101
food security, 101, 253, 307
forest, 3, 85, 91, 162, 234
forest biomass, 233
forest cover (percentage) in India, 236
forest fires, 162
fruits, 25, 59, 120, 178, 194
fungi, 7, 20, 70, 138, 140, 143, 221
fungicide, 21, 138, 144, 221, 292
$F_v/F_m$, 43, 131, 200

## G

gas chromatography, 226, 305
gaseous pollutants, 39
gene expression, 109, 132, 272
genetic engineering, 110, 134
genomics, 106, 284
geographical, 20, 25, 148, 163
germination, 85, 20, 92, 103, 104, 168, 212, 272, 276
gibberellin, 272, 276
global climate change, 15, 41, 161, 193
global warming, 26, 98, 119, 161, 185
glutathione, 69, 133, 301, 303, 305
glutathione peroxidase, 301
glutathione reductase, 254, 299, 303, 305
glutathione S-transferase, 68, 254, 299
goldenberry, 194
grasses, 3, 5, 7, 37, 180, 184
green technologies, 70
green wall, 70, 71
growth, 4, 60, 255, 273

## H

heat, 7, 17, 57, 99, 108, 119, 124, 128
heat shock proteins, 105, 129
heat stress, 17, 108, 130, 223, 289
heavy metals, 68, 221, 223, 279, 290
highlands, 196, 211
high-throughput phenotyping, 104
hopea nervosa, 138, 141
hormonal stress, 92
hotspots, 24, 193, 210
hybridization method, 223
hydrogen peroxide, 220, 259, 303

## I

impact of temperature in the tropical forest, *88*
increased temperature, 10, 28, 149, 209
infrared thermometry, 225
interaction, 15, 20, 213
invasiveness, 228
isotope, 226
iTRAQ, 107, 286, 288

## J

Jammu and Kashmir, 235
jasmonic acid, 223, 275, 306

## K

Karnataka, 244

## L

labeling-based quantitation, 285, 293
Leaf Area Index, 67, 70, 90, 183
leaf functional traits, **74**
leaf necrosis, 58
leaf traits, 39, 75
life cycle, 5, 149, 214, 233
light exposure, 85, 93
light quality, 273
lignotubers, 170
lipid peroxidation, 45, 58, 258, 285
liquid chromatography, 225, 305
litter or fuel load, 177
local extinctions, 148, 209

## M

Malaysia, 24, 47, 149
malondialdehyde, 305
Mancozeb, 138
mass spectrometry, 225, 285
Megadiverse Countries, 25
Megadiverse Nations, **25**
metabolites, 44, 213
metabolomics, 225, 284, 289
methods for quantitative proteomics, **286**
methods of estimation of above-ground biomass, **239**
microbe-centric approach, 89
microbe/plant-based technology for phytoremediatio, *70*
microbial diversity, 99
microbiomes, 100, 101
microscopy, 121, 228
mineralization, 99, 152
Miombo, 8
miRNA, 112
molecular response of a tropical plant, *17*
monodehydroascorbate reductase, 302
morphological responses, 90
mountain species, 209
multivariate analysis, 284
*Muntingia calabura*, 119
mycorrhizae, 100

## N

negative impact of NO X on tropical plants, *60*
net photosynthetic rate, 10, 128, 150
nitrate assimilation, 58
nitrate metabolization, 58
nitric oxide, 130
nitrogenous pollution, 61
nitrogen oxide, 57, 62
NO-mediated signaling mechanisms in abiotic stress, *59*
non-destructive method, 240
non-enzymatic markers, 303
North Zonal Council, 235
nuclear magnetic resonance, 289
nursery, 139, 142
nutrient, 90, 98, 120, 140, 256, 275
nutrient availability, 90, 98, 266
nutrient translocation, 90
nutrient uptake, 138, 256

## O

omics, 104, 106, 284
ophthalmic acid, 306
oskar syndrome, 167
oxidative stress, 298, 306
ozone, 37, 39, 48, 57, 223

## P

Panama, 120
*Parashorea tomentella*, 138
particulate matter, 41, 74, 221
permafrost, 99
pests, 152
ph, 76
phenology, 20, 125, 149, 213, 214
phenols, 200, 214
phosphorus, 274, 278
photochemical reactions, 37
photosynthesis, 6, 59, 65, 107
photosynthetic acclimation, 150
photosynthetic pigments, 9, 61, 201
photosynthetic rate, 10, 61, 77, 128, 212
phyllosphere, 70
physiological aberrations, 57
physiological and biochemical effects of pm on plants, *68*
physiological mechanisms, 108, 165
physiological process, 9, 11, 57, 61, 106, 284
physiology, 37, 58, 61, 66, 73, 211, 275, 287
phytohormone, 128, 129, 257
phytoremediation, 66, 70, 74
phytotoxin, 57
pigments, 9, 45, 61, 74, 167, 200, 267
plant adaptation in elevated $CO_2$ and temperature, *19*
plant community, 3, 16, 166, 171, 210
plant ecophysiological responses to fire, *166*
plant functional traits, 73, 77
plant growth regulators, 77
plant–herbivore interaction, *18*
plant–herbivore interaction in drought, **19**
plant hormones, 92, 254, 273
plant metabolism, 61, 71, 221, 274, 306
plant physiology, 58, 219, 223
plant responses, 9, 15, 37, 48, 148, 182, 285
plant responses to elevated carbon dioxide, *148*
plant sexual reproduction, 24, 46, 47, 90, 119, 149, 153, 179, 193, 244
plant species, 133, 270, 284
plant species preferred for greenbelt plantings, **78**
plant stress, 133, 166, 284
plant yield, 17, 32
pollen, 119, 123, 125
pollen abortion, 119, 125, 131
pollen tube growth, 125
polyphenol oxidases, 258
precipitation, 5, 148, 153, 165, 194, 204, 213, 265
precipitation patterns, 98, 209
primary productivity, 48, 213
programmed cell death, 228, 298
proline, 75, 76, 111, 128, 130, 132, 292
proline biosynthesis, 181
proteomics, 225, 284, 293
proteomics techniques in plant environmental stress, **288**

## R

rainforest loss, 91
reactive oxygen species, 29, 60, 92, 105
REDD+, 233
regeneration, 69, 138, 178, 180, 258
regression equation, 234, 237, 240
remote sensing, 24, 239, 246
reproductive phenology, 20, 179
resistance, 168, 210
resource availability, 86, 265
resource availability in tropical rainforest, *86*
respiration, 10, 30, 41, 87, 119, 253, 301
resprouters, 169
rhizosphere, 68, 99, 274
ribulose 3, 5-bisphosphate carboxylase, 128
roadside, 66, 75, 77
root architecture, 273

## S

salicylic acid, 130
salt stress, 105, 107, 111, 258, 300
salt tolerance, 111, 269, 306
SAR/QSAR, 284
satellite method, 234
savanna, 3, 175, 181, 233
seasonal variation, 90, 211, 213
secondary metabolites, 18, 167, 213, 285, 287
seedling, 20, 43, 131, 138
seed set, 133
serotiny, 170
Shorea, 43, 142, 238, 244
shrubs, 4, 181, 183
SILAC, 285
soil biodiversity, 100
soil carbon pool, 99
soil degradation, 101
soil health, 88, 102
soil organic matter, 98, 181, 233
soil salinity, 108, 219
soil warming, 98
soluble protein, 44, 58, 106, 130, 214
sources of air pollutants and effects, *39*
sources of gaseous and particulate pollutants, *38*
sources of heavy metals and impacts on plants, *69*
South America, 25, 27, 89, 163, 176, 197, 265
SPAD value, 128
species abundance, 16
species extinction, 27, 92
species richness, 180, 209, 211
statistical methods, 284, 292
sterility, 119
stomata, 7, 9, 41, 43, 48, 61, 74, 92, 111, 198, 224, 269
stomatal behavior, 224
stomatal conductance, 43, 53, 58, 61, 128, 210, 224

# Index

stress tolerance, 106, 111, 270, 306
strigolactone, 275
subtropics, 37, 196
sugarcane, 104, 114
superoxide dismutase, 68, 254, 299, 301
sustainability, 91
systemic acquired resistance, 258

## T

temperature stress, 128, 253, 260
thiobarbituric acid reactive substances, 128
TMT, 285
transcription factors, 32, 108, 130, 276
transportation and metabolism of $SO_2$ and $NO_2$ in plants, *40*
tree height, 175, 224, 244

tree mortality, 87, 91, 166
trichomes, 66, 68
tropical Andes, 193, 204
tropical ecosystems, 7, 62, 89, 146, 161, 171, 193, 221
tropical forest, 86, 89, 91, 125, 162, 210, 265
tropical rainforests, 90
tropospheric ozone, 40

## U

urban areas, 73, 77
urban green space, 77
urbanization, 73, 78
urbanization and air pollution, 73
UV radiation, 194, 201

## V

vehicular pollution, 66, 75

## W

water loss, 4, 7, 26, 92, 105, 131, 198
water sourcing, 7
water use, 6, 58, 131, 167

## Y

Yamunanagar, 235

## Z

zonal councils, 234, 244